Synthesis and Applications of Carbohydrates, Lipids, and Steroids

This definitive text provides in-depth information on chemical, photochemical, and microbiological synthesis of carbohydrates and their derivatives. Recent developments in the synthesis and applications of carbohydrates involved in food industries, biosurfactants, nano-catalysts, and nucleosides are thoroughly discussed in this innovative work. The text also provides information about synthesis and applications of artificial lipids and steroids. The approach of this book is to fulfill the requirements of graduate, postgraduate students, and scientists belonging to various fields.

Key features

- Provides in-depth knowledge of synthesis and applications of carbohydrates, lipids, and steroids.
- Elaborates strategies for stereocontrolled glycosylation and their progress in synthetic carbohydrate chemistry.
- Discusses examples derived from drug development and regulatory applications and recent research results.
- Enlightens from basics to recent advances in stereoselective glycosylation.
- Includes problems with answers based on each chapter.

Synthesis and Applications of Carbohydrates, Lipids, and Steroids

Edited by
Ahindra Nag

CRC Press is an imprint of the
Taylor & Francis Group, an **informa** business

First edition published 2025
by CRC Press

2385 NW Executive Center Drive, Suite 320, Boca Raton FL 33431

and by CRC Press
4 Park Square, Milton Park, Abingdon, Oxon, OX14 4RN

CRC Press is an imprint of Taylor & Francis Group, LLC

ISBN: 9781032568607 (hbk)
ISBN: 9781032568614 (pbk)
ISBN: 9781003437413 (ebk)

DOI: 10.1201/9781003437413

Typeset in Times LT Std
by Apex CoVantage, LLC

Contents

Preface

A recent development in carbohydrates research explores the diverse applications of carbohydrate materials due to their ecofriendly nature, cost-effectiveness, flexibility in modifying physical and chemical properties, and renewability. Moreover, environmental restrictions in terms of biodegradability and biocompatibility on the use of a number of chemical products stimulate research in this field. Both simple and derived carbohydrate products are used in food industries as food components, biosurfactants, food packaging, and bioplastics. Carbohydrates due to their high solubility, organ specificity, broad efficacy, protected mode of action, and safety profile hold potential therapeutic development. Carbohydrates are also useful raw materials for producing bulk chemicals at low prices; they can also yield high-value fine chemicals after undergoing chemical conversions because of their enantiomeric purity.

Lipid research is challenging and limited in scope, not only because of their structural diversity but also because of the difficulty in reconstructing their natural synthetic pathways. To better understand the structure and function of lipids, there is an enormous interest in developing simple methodologies to generate artificial lipids in a controlled manner that contributors discuss in this book.

Recently, there is growing demand for research on steroids as they constitute an extensive and important class of biologically active polycyclic compounds that are widely used for therapeutic purposes, for example as drugs for prevention and therapy for disorders like anti-inflammatory; diuretic; anabolic; contraceptive; antiandrogenic; and progestational as well as anticancer agents for breast, prostate, colon, and many other cancers. Steroids constitute the largest group of pharmaceuticals next to antibiotics.

There are specialized books available on carbohydrates, lipids, and steroids, but the subject material in most [textbooks] is presented in a diffused or highly specialized form, lacking detailed discussion of theoretical and experimental processes. As a result, students and researchers must search textbooks, journals, and pharmacopoeias for the information. The major objective of writing this book is to present to students and researchers the current state of carbohydrate, lipid, and steroid research for greener synthesis and newer applications in a lucid, condensed and cohesive form.

I express my indebtedness and gratitude to all the professors who have contributed chapters in the book. I also acknowledge my indebtedness to my wife, Jayita Datta, to Subrata Bairy, and to my sons Aritra and Anindya for their encouragement and sustained cooperation. I am also thankful to my research scholars, Dr. Bipasa Halder and Dr. Dipak Patil, for their assistance in the completion of the book. The cooperation of publishers Mrs. Renu Upadhya and Mrs. Jytosna Jangra and of Taylor and Francis, UK, in bringing out the book is very much appreciated.

Ahindra Nag

About the Editor

Professor **Ahindra Nag** B.Tech, M.Sc, Ph.D, Member and Fellow of the Institute of Chemistry, DEM, has over 34 years of teaching and research experience in the Chemistry Department, Indian Institute of Technology, Kharagpur, India. He has published 100 research papers and 14 textbooks and holds 3 patents. He has guided 18 research scholars and served as a visiting professor to University of Rome (Italy), Academia Sinica (Taiwan), and Tennessee (USA). He is an editorial member of different international journals and was secretary of the International Green Chemistry Society. He has been the keynote speaker at multiple international conferences.

List of Contributors

Luigi A. Agrofoglio
Institute of Organic and Analytical Chemistry-Centre National de la Recherche Scientifique (CNRS)
University of Orléans
Orléans, France

Angelo Albini
Department of Chemistry
University of Pavia
Pavia, Italy

Athar Ata
Department of Chemistry
Michigan Technological University
Houghton, Michigan, USA

Zanjila Azeem
Medicinal and Process Chemistry Division
Central Drug Research Institute
Lucknow, India

Nabamita Basu
Department of Chemistry
Nabagram Hiralal Paul College
West Bengal, India

Farahnaz K. Behbahani
Department of Chemistry
Karaj Branch, Islamic Azad University
Karaj, Iran

Roberto J. Brea
BioinspiredNanochemistry Group-Centro Interdisciplinar de Química e Bioloxía
Universidade da Coruña
Coruña, Spain

Mathias Brouillard
Institute of Organic and Analytical Chemistry CNRS, University of Orléans
Orléans, France

Rosa E. del Río
Instituto de Investigaciones Químico Biológicas
Universidad Michoacana de San Nicolás de Hidalgo
Michoacán, México

Ikechukwu P. Ejidike
Department of Biology
The University of Winnipeg
Winnipeg, Canada

Himanshu Gangwar
Medicinal and Process Chemistry Division
Central Drug Research Institute
Lucknow, India

Juan-Pablo García-Merinos
Instituto de Investigaciones Químico Biológicas
Universidad Michoacana de San Nicolás de Hidalgo
Michoacán, México

Rina Ghosh
Department of Chemistry
Jadavpur University
West Bengal, India

Bipasa Halder
Chemistry Department
Indian Institute of Technology
Khargpur, India

Renata Kołodziejska
Faculty of Pharmacy, Department of Organic Chemistry
Nicolaus Copernicus University in Toruń
Bydgoszcz, Poland

Daria Kupczyk
Faculty of Medicine, Department of Medical Biology and Biochemistry
Nicolaus Copernicus University in Toruń
Bydgoszcz, Poland

Adam Kurčina
Department of Organic Chemistry
Faculty of Science
Charles University
Czech Republic

Christophe Len
Chimie ParisTech, PSL Research University CNRS
Institute of Chemistry for Life and Health Sciences
Paris, France

Yliana López
Instituto de Investigaciones Químico Biológicas
Universidad Michoacana de San Nicolás de Hidalgo
Michoacán, México

Pintu Kumar Mandal
Medicinal and Process Chemistry Division
Central Drug Research Institute
Lucknow, India

Sunny Maurya
Medicinal and Process Chemistry Division
Central Drug Research Institute
Lucknow, India

Mana Mohan Mukherjee
National Institute of Diabetes and Digestive and Kidney Diseases
National Institutes of Health
Bethesda, Maryland, USA

Ahindra Nag
Chemistry Department
Indian Institute of Technology
Khargpur, India

Samina Naz
Department of Biology, Department of Chemistry
The University of Winnipeg
Winnipeg, Canada

Hanna Pawluk
Faculty of Medicine, Department of Medical Biology and Biochemistry
Nicolaus Copernicus University in Toruń
Bydgoszcz, Poland

Susmita Poddar
Department of Chemistry
Jadavpur University
West Bengal, India

Brenda Portasany
BioinspiredNanochemistry Group
Centro Interdisciplinar de Química e Bioloxía
Universidade da Coruña
Coruña, Spain

Vincent Roy
Institute of Organic and Analytical Chemistry CNRS, University of Orléans
Orléans, France

Evan Saillard
Institute of Organic and Analytical Chemistry CNRS, University of Orléans
Orléans, France

Simran Sandhu
Department of Biology
The University of Winnipeg
Winnipeg, Canada

Rosa Santillan
Departamento de Química
Centro de Investigación y de Estudios Avanzados del IPN
México, D.F., México

Federica A. Souto-Trinei
BioinspiredNanochemistry Group
Centro Interdisciplinar de Química e Bioloxía
Universidade da Coruña
Coruña, Spain

Renata Studzińska
Faculty of Pharmacy, Department of Organic Chemistry
Nicolaus Copernicus University in Toruń
Bydgoszcz, Poland

Jan Veselý
Department of Organic Chemistry
Faculty of Science
Charles University
Czech Republic

1 Concepts of Carbohydrates, Lipids, and Steroids

Ahindra Nag and Bipasa Halder

1.1 INTRODUCTION OF CARBOHYDRATES

Carbohydrates are widespread in nature and produced in huge amounts. Carbohydrates have multiple primary characteristics: i) high density of functional groups, ii) diversity of structures based on different configurations, iii) ideal biocompatibility as they are ubiquitous in the body, and iv) they are inexpensive and possess high chemical as well as enantiomeric purity.

In response to the demand for environmentally friendly processes and products, sugars have been revealed to be competent raw materials. In the last decade, the use of carbohydrates as industrial chemical feedstock has increased noticeably; in particularly, scientists have reported on using chitin, chitosan, alginate, and agar, among the main advances published over the last 10–15 years [1–7].

Carbohydrates such as sucrose and glucose (e.g., derived from starch) [2] are sufficiently available and are produced in bulk amounts for nonchemical uses. Starch is used in adhesives, thickeners, and paper additives and as a source for maltodextrins, glucose, and high-fructose corn syrups. A recent development [4] in this field is the synthesis of fully biodegradable bioplastics. The diversity of sucrose and glucose for utilization in chemistry and nonfood purposes is lower than 5%. The prices of these carbohydrates are comparable with those of commonly used organic solvents, which makes them inexpensive compared with other homo-chiral compounds [5]. In addition to being inexpensive, the enantiomeric purity of carbohydrates makes these compounds particularly interesting as starting materials for the preparation of homo-chiral pharmaceutical and agrochemical products [5].

Carbohydrates [6–12] are composed of carbon, hydrogen, and oxygen. They correspond to the formula $C_x(H_2O)_y$. For example glucose has the formula $C_6H_{12}O_6$ or $C_6(H_2O)_6$. It was later found that certain compounds that are not carbohydrates correspond to this formula such as formaldehyde (CH_2O,) and lactic acid ($C_3H_6O_3$); even some of the carbohydrates do not follow the general formula, such as rhamnose ($C_6H_{12}O_5$) and digitoxose ($C_6H_{12}O_4$).

1.2 CLASSIFICATION OF CARBOHYDRATES

Carbohydrates [10, 11] are classified as sugars and nonsugars. Sugars are sweet, crystalline, and soluble in water. Their molecular weights are fixed for a particular compound. They are subdivided into monosaccharides or oligosaccharides. Monosaccharides are simple sugars that cannot be hydrolyzed further. The most abundant monosaccharide in nature is D-glucose, a hexose sugar; other common hexose monosaccharides include galactose and fructose. Monosaccharides can be further subclassified in accordance with the nature of the carbonyl: Aldoses have an aldehydic group, and ketoses have a ketonic group. Aldoses in which the –OH group of the asymmetric farthest from the aldehyde group is written on the right side are known as ***D***-sugars, and those with the –OH group on left side are known as ***L***-sugars, as shown in Figure 1.1.

Monosaccharides also vary by number of carbon atoms [6, 7]. Triose (D-glyceraldehyde) is a three-carbon monosaccharide; tetrose is a monosaccharide with four carbons; pentose has five carbons; and hexose has six (Figure 1.2(a)). Thus D-aldoses series [9] are shown in Figure 1.2(b).

DOI: 10.1201/9781003437413-1

D and L Designation

Highest number asymmetric carbon

D-Glucose	L-Glucose
1 CHO	1 CHO
H–2C–OH	HO–2C–H
HO–3C–H	H–3C–OH
H–4C–OH	HO–4C–H
H–5C–OH	HO–5C–H
6 CH_2OH	6 CH_2OH

FIGURE 1.1 D and L configurations of glucose.

Glyceraldehyde (triose)	Ribose (pentose)	Fructose (hexose)	Glucose (hexose)
CHO	CHO	CH_2OH	CHO
H–OH	H–OH	H=O	H–OH
CH_2OH	H–OH	HO–H	HO–H
	H–OH	H–OH	H–OH
	CH_2OH	H–OH	H–OH
		CH_2OH	CH_2OH

FIGURE 1.2(A) Linear structures of glyceraldehyde, ribose, fructose, and glucose.

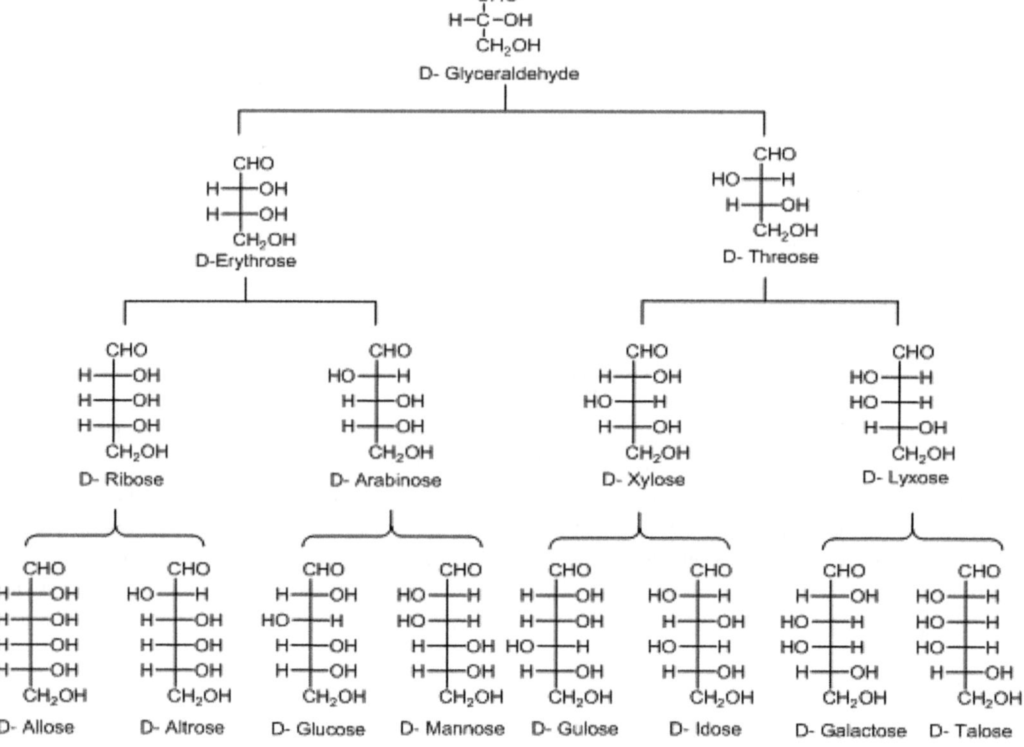

FIGURE 1.2(B) D-family of aldose.

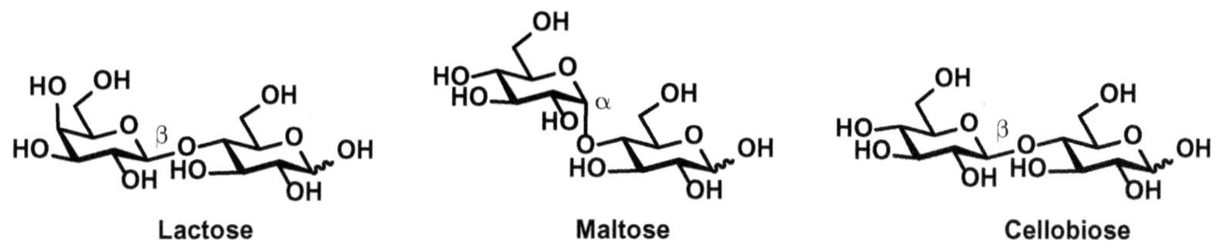

FIGURE 1.3 Glycosidic linkages between lactose, maltose, and cellobiose.

The second category of sugars, oligosaccharides, are compounds that yield two to ten monosaccharide molecules on hydrolysis and are again classified by number of monosaccharide units. Disaccharides give two sugars on hydrolysis (the same two or different) such as sucrose, lactose, maltose, and cellobiose, but there are differences among the sugars [10–13]. For instance, maltose, lactose, and cellobiose give two glucoses on hydrolysis, but sucrose yields a glucose (an aldohexose) and a fructose (a ketohexose) joined by an $\alpha \rightarrow 1$ and $\beta \rightarrow 2$ glycosidic link. And lactose comprises the condensation of galactose and glucose and forms a β-1$\rightarrow$4 glycosidic link. Maltose and cellobiose comprise the condensation of two glucose units and form an α-1$\rightarrow$4 glycosidic link and a β-1$\rightarrow$4 glycosidic link, respectively (Figure 1.3).

Trisaccharides and polysaccharides are also oligosaccharides. Maltotriose is a trisaccharide that yields three glucose units on hydrolysis, and raffinose yields three different monosaccharides: glucose, galactose, and fructose. Similarly, there are tetra-, penta- and hexa-saccharides. Polysaccharides are carbohydrates [14] that yield large monosaccharide units on hydrolysis, for example starch and glycogen. The general formula of polysaccharides is $(C_6H_{10}O_5)_n$. They are usually tasteless, amorphous solids, and they are miscible in water or they form colloidal suspension. Their molecular weights vary considerably.

1.3 ASCENDING AND DESCENDING INTERCONVERSION OF MONOSACCHARIDE

1.3.1 Conversion from Lower to Higher Carbohydrates (Ascending Series)

Kilani–Fischer synthesis [6, 9, 10] is the process of lengthening monosaccharide chains by treating aldose with hydrogen cyanide. Adding cyanide to the aldehydic group lengthens the chain by one carbon and generates a new stereocenter around which two configurations are possible. The resulting cyanohydrin compound is hydrolyzed into higher aldonic acid, which is dehydrated by heating to form lactone followed by reduction with Na-Hg into two aldoses (monosaccharides) with one carbon more than the parent aldose; these are epimers as they differ in the configuration of their second carbon atom C2 (Scheme 1.1). This process can be used to prepare aldoses up to ten carbon atoms, but the process is limited by its low yield and use of toxic reagents. In addition, the process requires having a supply of the previous sugar in the series, which itself may require substantial synthetic work if it is not readily available.

1.3.2 Conversion from Higher to Lower Carbohydrates (Descending Series)

Ruff's degradation is a process to easily convert an aldose to another aldose with one carbon less than the parent aldose (shortening the monosaccharide chain). In this process, the terminal aldehydic group is selectively oxidized with bromine water to the corresponding acid. Then, it is

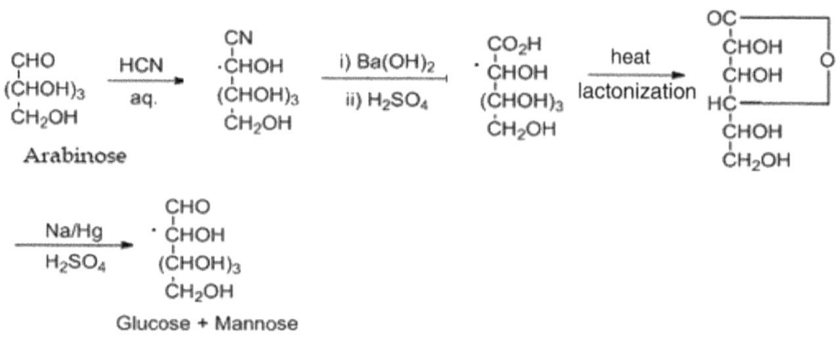

SCHEME 1.1 Kilani–Fischer synthesis.

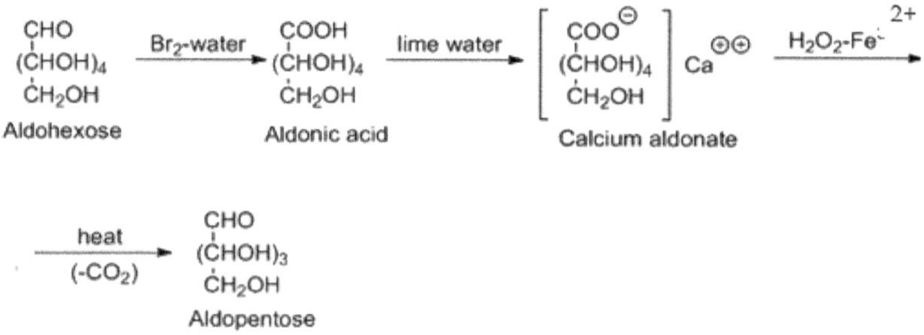

SCHEME 1.2 Ruff's degradation.

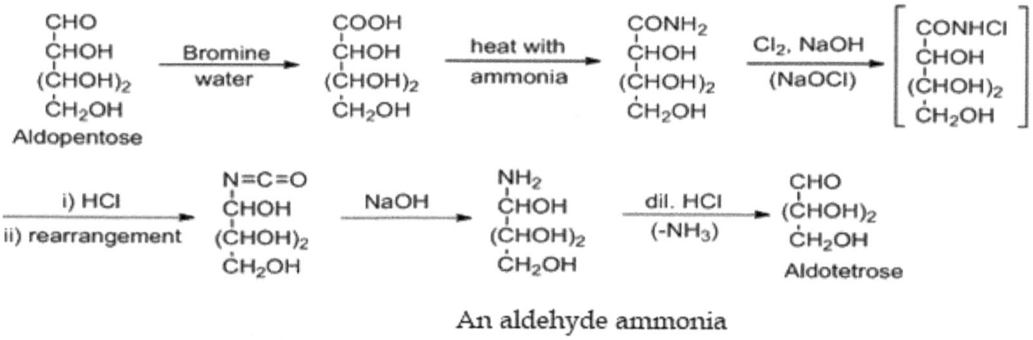

SCHEME 1.3 Weerman's degradation.

treated with lime water to obtain a calcium salt of acid that is oxidized by hydrogen peroxide in the presence of ferrous salt as a catalyst called Fenton's reagent $\left(H_2O_2^{2+}, Fe^{2+}\right)$ to form α-keto acid. Ferric ion catalyzes the oxidation reaction, which cleaves the bond between $C-1$ and $C-2$, forming an aldehyde. This keto acid is readily decarboxylated to the corresponding descended aldose (Scheme 1.2).

Weerman's degradation is named for Rudolf Adrian Weerman, [10, 11] who discovered the reaction in 1910. In this reaction, amides are degraded by sodium hypochlorite, forming an aldehyde with one less carbon atom in carbohydrate chemistry, but the process is slow yielding and rarely used (Scheme 1.3).

In Weygand's method, aldohexose is treated with hydroxyl amine, and the product, aldose oxime, is treated with 1-fluro2,4-dinitrobenzene. The desired product of aldopentose is obtained after warming with aqueous sodium bicarbonate solution (Scheme 1.4).

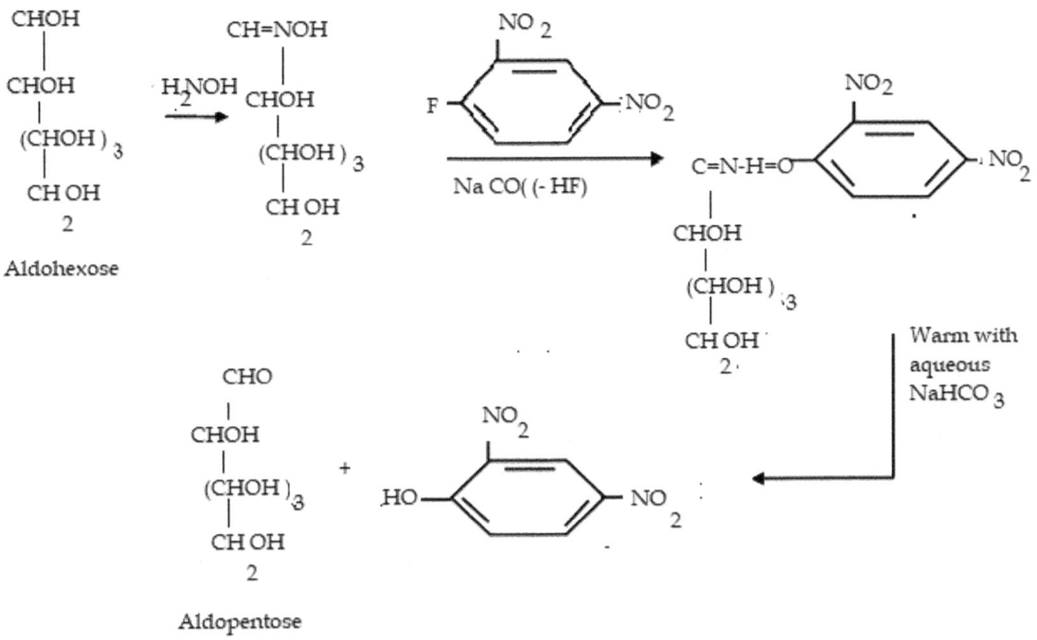

SCHEME 1.4 Weygand's method.

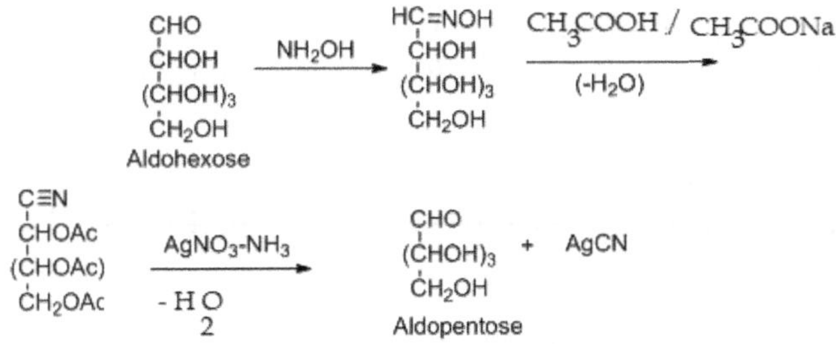

SCHEME 1.5 Whol's method.

Whol's method (1893) consists of heating aldoxime (Scheme 1.5) with acetic anhydride in the presence of sodium acetate, when the oxime group is dehydrated to a nitrile group with simultaneous acetylation of hydroxyl groups. After that, the acetylated nitrile compound is warmed with ammoniacal silver nitrate solution, which removes the nitrile group as silver cyanide and the acetyl groups as acetamide to form a new aldose. However, the yield is only 50% to 60%. This method was later modified by Zemplen [10] in 1917 by using a solution of sodium methoxide in chloroform instead of aqueous solution of ammoniacal silver nitrate to remove cyanide and acetyl groups. The yield was 70%.

1.4 PRODUCTION OF GLUCOSE AND FRUCTOSE

1.4.1 Glucose

In the laboratory [6, 9], glucose is prepared by boiling hydrochloric acid (HCl) in an alcohol solution to hydrolyze sucrose; glucose and fructose are obtained in equal amounts, and on cooling, glucose separates out from fructose as it is much less soluble than fructose. Commercially, glucose is obtained by the hydrolysis of starch by heating with dilute HCl or sulfuric acid at 120 °C under pressure of 2–3 atmospheres. The resulting solution is neutralized with calcium carbonate, and the precipitated calcium salts are filtered off. The filtrate is decolorized by animal charcoal and then concentrated under reduced pressure. Currently, glucose [2] is mainly produced by adding the enzyme α-amylase to a mixture of starch and water. A-amylase is secreted by various species of the bacterium *Bacillus*, and it breaks the starch into dilute glucose syrup that is finally evaporated under vacuum to raise the solid concentration.

1.4.2 Fructose

We have described the laboratory-scale separation of fructose formation from sucrose by acid hydrolysis. The well-known conventional approach to the commercial production of fructose from starch [14] is the enzyme hydrolysis of starch using amylolytic enzymes such as glucoamylase and subsequent isomerization of dextrose to fructose by isomerase under controlled temperature, pressure, and reaction time. The product yield by this method is only 42%, and the product mixture also contains 50% dextrose and 8% other organic impurities (e.g., proteins, fatty acids long-chain polysaccharides). These impurities are removed using activated charcoal and diatomaceous earth and by filtering the solution using a filter press and a rotary vacuum filter. Recently, ion exchange chromatography has been used for purification. The refined fructose solution is concentrated by the evaporator and held for storage.

Fructose can also be produced from inulin by single-step hydrolysis [14] using inulinase enzyme. Inulin is a β-(2, 1) polysaccharide of fructose that is found in artichoke and dahlias. In the single-step enzymatic method, inulinase acts sequentially on β-(2, 1) inulin linkages to release the fructose units. This method gives yields of approximately 95%. Acid hydrolysis of inulin/starch is not recommended because it imparts color to the product and introduces undesired constituents like difructose anhydride.

1.4.3 Conversion of an Aldose into the Corresponding Ketose (Glucose to Fructose)

Glucose on reaction with three moles of phenyl hydrazine gives glucosazone [6, 9]. The reaction is based on Amadori rearrangement. The rearrangement usually proceeds via initial formation of phenyl hydrazone followed by its Amadori rearrangement [6] with generation of 2-ketoimine by simultaneous elimination of aniline. The ketoimine then condenses with two molecules of phenyl hydrazine affording osazone. Glucosazone on hydrolysis results in the formation of osone. Reduction of osone with sodium amalgam causes selective reduction of aldehyde group, resulting in the formation of D-fructose (Scheme 1.6).

1.4.3.1 Conversion of Aldose to the Next Higher Ketone (Arabinose to Fructose)

This method involves the use of the Arndt–Eistert reaction (Scheme 1.7). The various steps and reagent required are shown below in the conversion of D-arabinose (an aldopentose) to D-fructose (a keto hexose).

1.4.4 Conversion of a Ketone to the Corresponding Aldose (Fructose into Glucose)

In this process, fructose is catalytically reduced by H_2-Ni, and the corresponding alcohol is carefully oxidized to a monocarboxylic acid. The carboxylic acid on heating is converted to lactone, which on reduction with sodium amalgam yields glucose (Scheme 1.8).

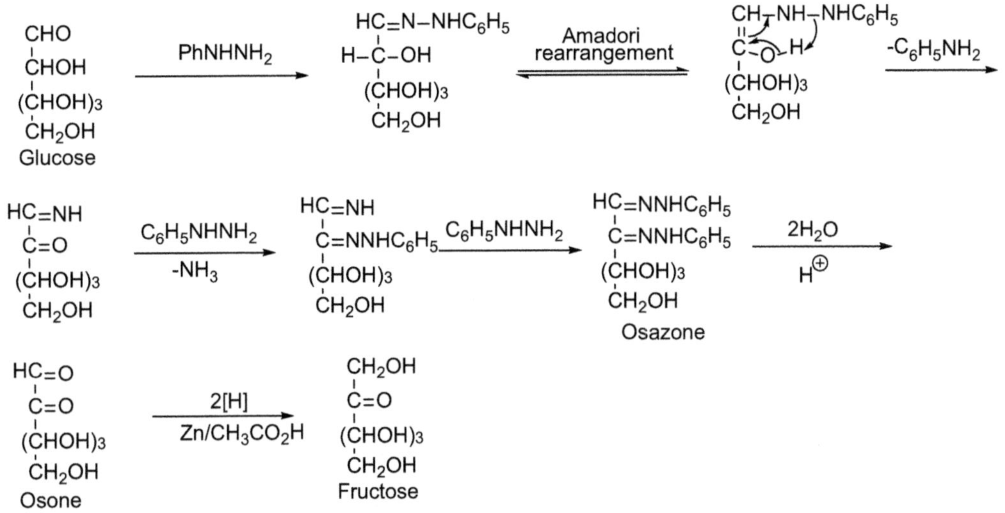

CHO–CHOH–$(CHOH)_3$–CH_2OH (Glucose) $\xrightarrow{PhNHNH_2}$ $HC{=}N{-}NHC_6H_5$ / H–C–OH / $(CHOH)_3$ / CH_2OH $\xrightleftharpoons{\text{Amadori rearrangement}}$ $CH_2{-}NH{-}NHC_6H_5$ / C=O–H / $(CHOH)_3$ / CH_2OH $\xrightarrow{-C_6H_5NH_2}$

HC=NH / C=O / $(CHOH)_3$ / CH_2OH $\xrightarrow[-NH_3]{C_6H_5NHNH_2}$ HC=NH / $C{=}NNHC_6H_5$ / $(CHOH)_3$ / CH_2OH $\xrightarrow{C_6H_5NHNH_2}$ $HC{=}NNHC_6H_5$ / $C{=}NNHC_6H_5$ / $(CHOH)_3$ / CH_2OH (Osazone) $\xrightarrow[H^{\oplus}]{2H_2O}$

HC=O / C=O / $(CHOH)_3$ / CH_2OH (Osone) $\xrightarrow[Zn/CH_3CO_2H]{2[H]}$ CH_2OH / C=O / $(CHOH)_3$ / CH_2OH (Fructose)

SCHEME 1.6 Glucose to fructose.

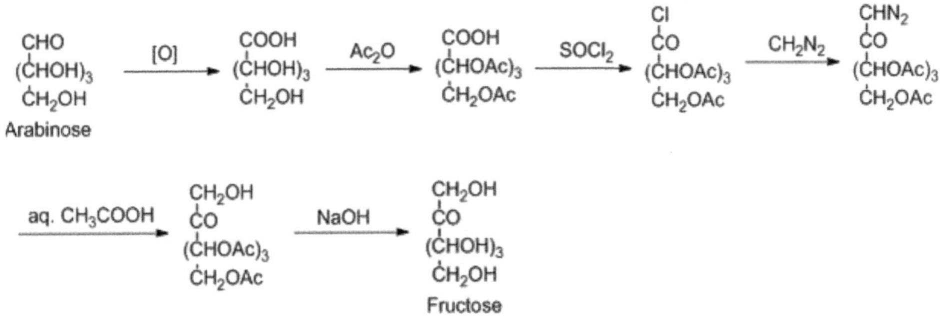

CHO / $(CHOH)_3$ / CH_2OH (Arabinose) $\xrightarrow{[O]}$ COOH / $(CHOH)_3$ / CH_2OH $\xrightarrow{Ac_2O}$ COOH / $(CHOAc)_3$ / CH_2OAc $\xrightarrow{SOCl_2}$ Cl / CO / $(CHOAc)_3$ / CH_2OAc $\xrightarrow{CH_2N_2}$ CHN_2 / CO / $(CHOAc)_3$ / CH_2OAc

$\xrightarrow{\text{aq. } CH_3COOH}$ CH_2OH / CO / $(CHOAc)_3$ / CH_2OAc $\xrightarrow{NaOH}$ CH_2OH / CO / $(CHOH)_3$ / CH_2OH (Fructose)

SCHEME 1.7 Arabinose to fructose.

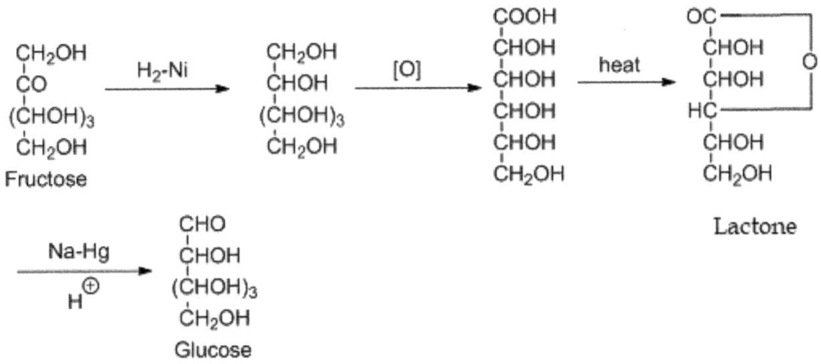

CH_2OH / CO / $(CHOH)_3$ / CH_2OH (Fructose) $\xrightarrow{H_2\text{-}Ni}$ CH_2OH / CHOH / $(CHOH)_3$ / CH_2OH $\xrightarrow{[O]}$ COOH / CHOH / CHOH / CHOH / CHOH / CH_2OH $\xrightarrow{heat}$ OC / CHOH / CHOH / HC (–O– ring) / CHOH / CH_2OH (Lactone)

$\xrightarrow[H^{\oplus}]{Na\text{-}Hg}$ CHO / CHOH / $(CHOH)_3$ / CH_2OH (Glucose)

SCHEME 1.8 Fructose into glucose.

1.5 COMPARISON OF GLUCOSE AND FRUCTOSE

1.5.1 Similarities between Glucose and Fructose

Glucose	Fructose
1) Glucose is a monosaccharide with molecular formula $C_6H_{12}O_6$.	1) Fructose is a monosaccharide with molecular formula $C_6H_{12}O_6$.
2) Glucose is a colorless sweet crystalline compound with a melting point of 146 °C. It effortlessly dissolves in water but is not easy to dissolve in alcohol or ether.	2) Fructose is a white crystalline compound with a melting point of 102 °C. It is much more soluble than glucose but is insoluble in alcohols.
3) Glucose shows mutarotation.	3) Fructose shows mutarotation.
4) Glucose on reduction gives sorbitol. On complete reduction with HI and red P, sorbitol gives n-hexane, which indicates that it contains an unbranched chain of six carbon atoms. In nature, however, the linear chain structures exist in equilibrium with their cyclized forms.	4) Fructose on reduction gives sorbitol and on complete reduction with HI and red P gives n-hexane, which indicates that it contains an unbranched chain of six carbon atoms.
5) Glucose reacts with acetic anhydride to form pentaacetate, which shows the presence of the hydroxy groups; each is attached to a separate carbon atom.	5) Fructose reacts with acetic anhydride to form pentaacetate, which shows the presence of hydroxy groups; each is attached to a separate carbon atom.
6) Glucose reacts with hydroxyl amine to form monoxime. This result shows the presence of carbonyl group in the structure.	6) Fructose reacts with hydroxyl amine to form monoxime. This result shows the presence of carbonyl group.
7) Glucose reacts with phenyl hydrazine at boiling temperature to form osazone.	7) Fructose reacts with phenyl hydrazine at boiling temperature to form osazone.
8) When glucose is treated with methyl alcohol in presence of HCl (gas), acetal is formed that is referred to as glucoside.	8) When fructose is treated with methyl alcohol in presence of HCl (gas), a ketal is formed that is referred to as fructoside.
9) Glucose does not form addition products with sodium bisulfate due to hemiacetal formation.	9) Fructose does not form addition products with sodium bisulfate.
10) Glucose reacts with Tollens' and Fehling's solution due to the presence of aldehydic group.	10) Fructose is a carbohydrate that does not contain the aldehyde group, but it forms keto-enol tautomerism in the presence of alkali and thus isomerizes to glucose. Because of this, it gives positive results on Tollens' reagent test and Fehling's test.

Tollens' reagent

$$\mathrm{CHO{-}(CHOH)_4{-}CH_2OH} + 2\mathrm{Ag(NH_3)_2OH} \longrightarrow \mathrm{COONH_4{-}(CHOH)_4{-}CH_2OH} + \underset{\text{Silver mirror}}{2\mathrm{Ag}} + 3\mathrm{NH_3} + \mathrm{H_2O}$$

Fehling's solution

$$\underset{\text{D-Glucose}}{\mathrm{CHO{-}(CHOH)_4{-}CH_2OH}} + \underset{\text{Deep blue}}{2\mathrm{Cu(OH)_2}} + \mathrm{NaOH} \longrightarrow \mathrm{COONa{-}(CHOH)_4{-}CH_2OH} + \underset{\text{Red ppt}}{\mathrm{Cu_2O}} + 3\mathrm{H_2O}$$

D-Fructose (H–C(H)–OH, C=O, HO–C–H, H–C–OH, H–C–OH, CH_2OH) ⇌ (tautomerism) enediol intermediate (H–C–OH, =C–OH, HO–C–H, H–C–OH, H–C–OH, CH_2OH) ⇌ (tautomerism) D-Glucose (H–C=O, H–C–OH, HO–C–H, H–C–OH, H–C–OH, CH_2OH)

1.5.2 Differences between Glucose and Fructose

Glucose	Fructose
1) Glucose is known as grape sugar and is the preferred energy source for the body.	1) Fructose is known as fruit sugar and is not the preferred energy source for muscle and brain. While fructose and glucose have the same calorific value, the two sugars are metabolized differently in the body. Fructose has a lower glycemic index than glucose but has a much higher glycemic load. Fructose causes seven times as much cell damage as does glucose because it binds to cellular proteins seven times faster, and it releases 100 times the number of oxygen radicals (such as hydrogen peroxide, which kills everything in sight).
2) Glucose is less lipogenic and less fat producing; it relies on hexokinase/glucokinase for initiating metabolism.	2) Fructose is more lipogenic and more fat producing and relies on fructokinase for initiating metabolism.
3) Glucose is an aldohexose and forms a pyranose ring structure. When it is cyclized by the formation of hemiacetal, glucose becomes a sugar of six-member ring.	3) Fructose is a ketohexose and forms a furan ring structure. When fructose is cyclized by the formation of hemiketal, it becomes a five-member ring sugar.
4) In glucose, the anomeric carbon is the first carbon.	4) In fructose, the anomeric carbon is in the second carbon.
5) Glucose on mild oxidation with bromine water gives gluconic acid, a monocarboxylic acid, which indicates the presence of aldehyde group. $CHO-(CHOH)_4-CH_2OH$ (D-Glucose) $\xrightarrow{Br_2\text{-water}}$ $COOH-(CHOH)_4-CH_2OH$ (D-Gluconic acid)	5) Fructose does not react with bromine water. Thus the carbonyl is a keto carbonyl.
6) When glucose reacts with nitric acid, it oxidizes to glucaric acid. 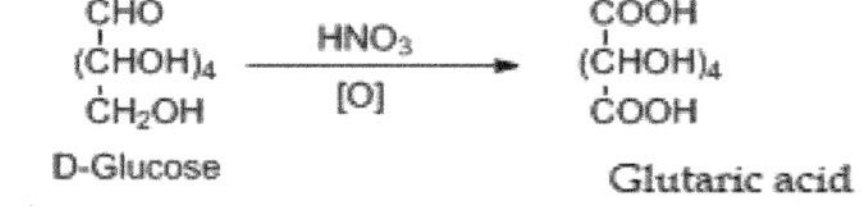	6) Fructose on reaction with nitric acid yields a mixture of glycolic acid and tartaric acid. 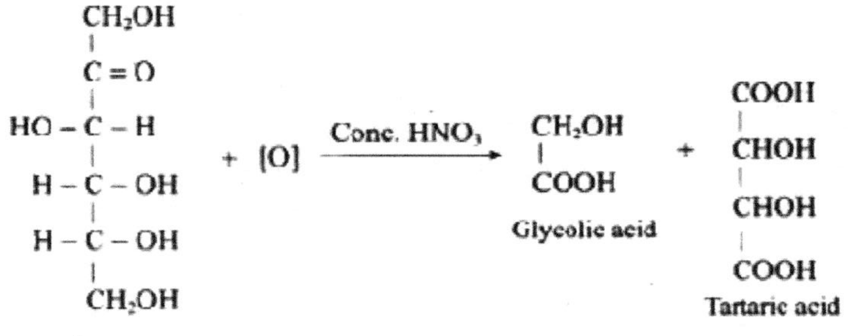
7) Seliwanoff's test: Add 1 gm of sample in 2c,c conc. HCl and a pinch of resorcinol in a dry test tube. Place the test tube in boiling water for 3 minutes deep red color usually followed by brownish precipitate. Dissolve precipitate in alcohol yielding a deep red solution.	7) Fructose responds this test but glucose does not respond this test.

1.6 SOME BASIC CARBOHYDRATE CHEMICAL REACTIONS

On Fischer projection [11–13], it appears that the glucose molecule is flat, but it is in fact a three-dimensional projection of carbohydrate (Figure 1. 4). This is important for the chirality of a molecule and also for distinguishing between a pair of enantiomers, but this projection is not suitable for noncarbohydrates.

Haworth [12, 13] proposed a six-member ring formula based on the pyran ring applicable to all sugars in 1926 and gave the evidence of X-ray analysis of sugars. Figure 1.5 displays α-D (+) glucopyranose and β-D (+) glucopyranose.

Arrangements of a molecule that can be obtained by rotation around C-C single bonds are called conformations. Reeves [6, 12] showed the conformation of α-D (+)glucopyranose and β-D (+) glucopyranose in 1950 (Figure 1.6). Both have chair form, but in the former case, the glycosidic hydroxyl group is axial, and the latter is equatorial, which is more predominate in the equilibrium mixture.

Hemiacetal [12] is an intermediate with one –OH and one –OR group at the central carbon atom bonded with four groups. The general formula for hemiacetal is $R_1R_2C(OH)OR$. However, acetal with two –OR groups at the central carbon atom has the general formula as $R_1R_2C(OR)_2$. An acetal forms when the hydroxyl group of a hemiacetal becomes protonated and loses water. Regarding stability, acetals are more stable than hemiacetals (Figure 1.7).

Generally, the hemiacetal form of glucose can react with alcohol in acidic solution to produce an acetal. When glucose is treated with dry methyl alcohol in the presence of HCl, the two hemiacetal

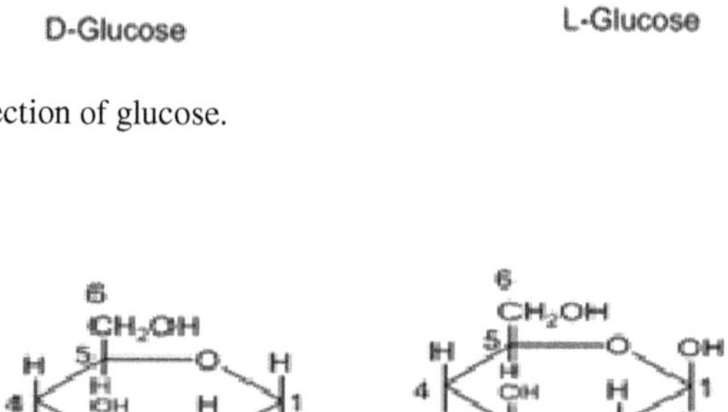

FIGURE 1.4 Fischer projection of glucose.

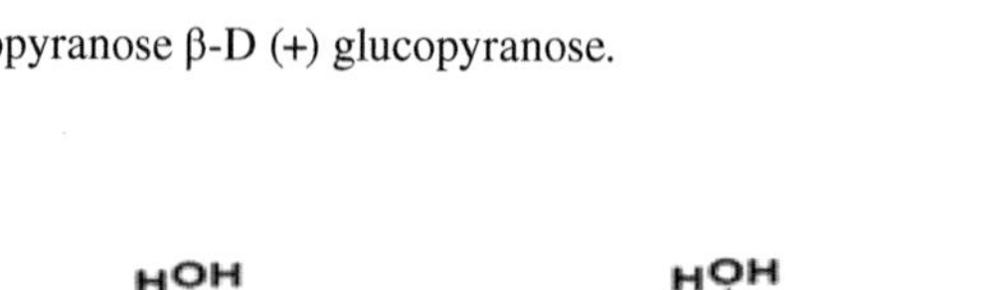

FIGURE 1.5 α-D (+)glucopyranose β-D (+) glucopyranose.

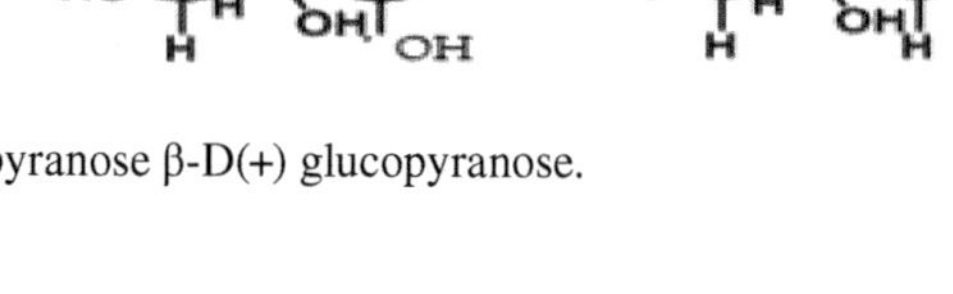

FIGURE 1.6 α-D(+) glucopyranose β-D(+) glucopyranose.

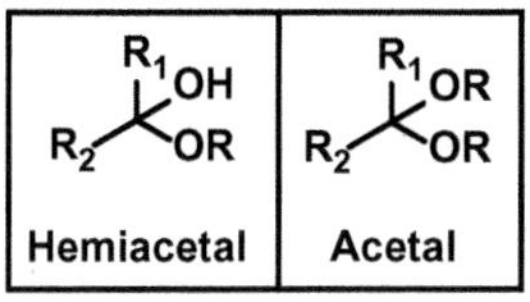

FIGURE 1.7 Hemiacetal and acetal.

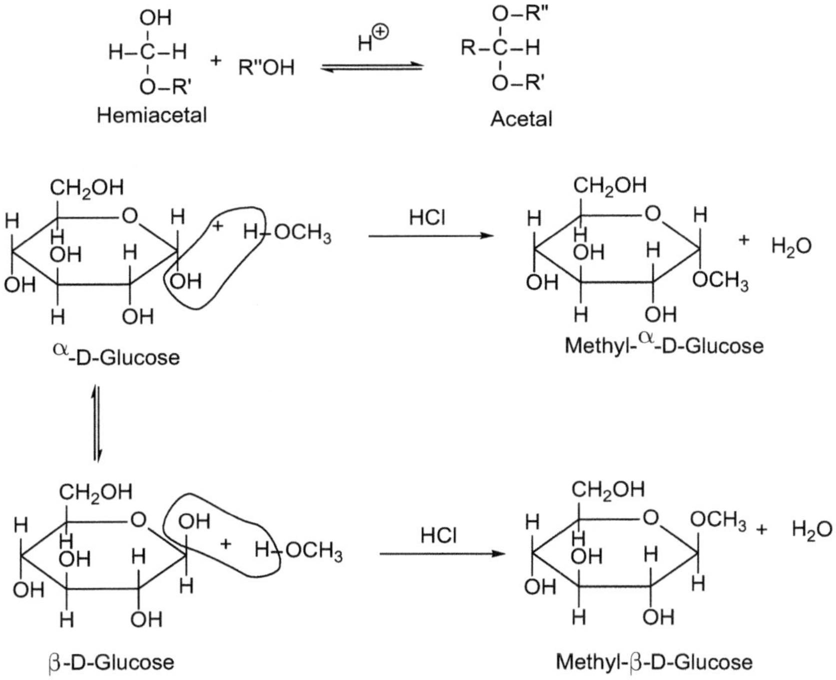

FIGURE 1.8 α- and β-anomers hemiacetal forms of glucose.

forms (α- and β-anomers) of glucose, which are in equilibrium, react separately to yield different compounds, resulting in an anomeric mixture (Figure 1.8).

Carbohydrates that differ in stereochemistry at only one of the many asymmetric centers are known as epimers. That is, α-D glucose and β-D glucose are stereoisomers that differ at only one chiral center, namely C1, and are known as C1 epimers, but D-galactose and D-mannose are not epimers as they differ in at two asymmetric centers, C2 and C4. In carbohydrate chemistry C1 epimers (cyclic hemiacetal form) are known as anomers, and the asymmetric carbon at C1 is called anomeric carbon.

Epimerization [12, 13] is the process under mutarotation that occurs at a position other than an anomeric position and involves the interconversion of one epimer to another. For example, the β-D-glucopyranose and β-D-manopyranose are epimers because they differ only in the stereochemistry at the C-2 position. In β-D-mannopyranose, the C-2 hydroxy group is axial (up from the "plane" of the ring), while in β-D-glucopyranose is equatorial (within the plane of the ring). These two molecules are epimers but because they are not mirror images of each other, they are not enantiomers (Figure 1.9).

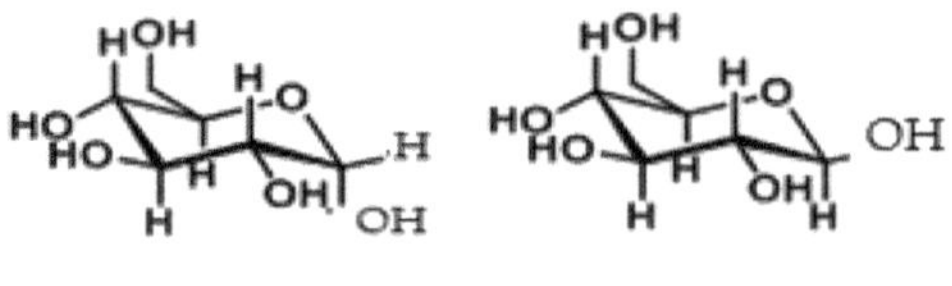

FIGURE 1.9 α-D-glucopyranose β-D-monopyranose.

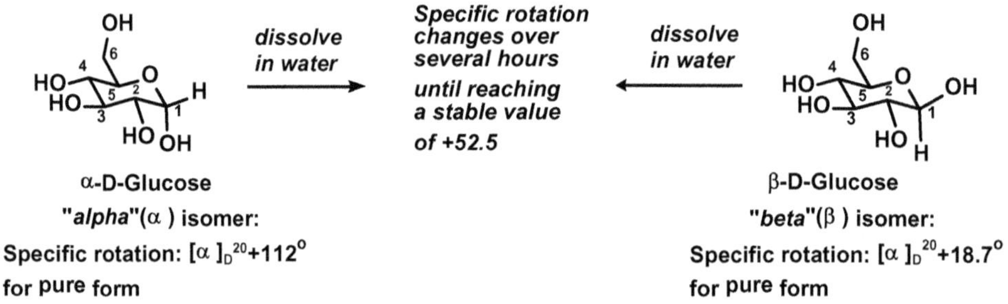

FIGURE 1.10 Mutarotation of α-D-glucose and β-D-glucose.

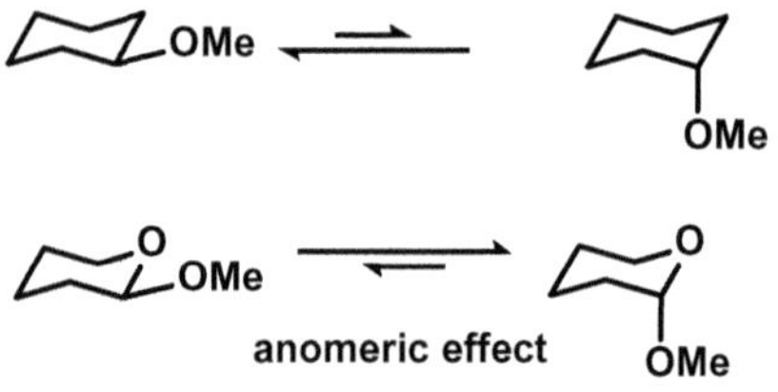

FIGURE 1.11 Anomeric effect.

The specific rotation of α- and β- forms of sugars gradually change until the constant specific rotation is attained in a solution discovered in 1844 by Dubrunfaut [12, 13], a French chemist. This change in specific rotation is known as mutarotation. The α and β anomers are diastereomers of each other and show different specific rotations. The process involves a solution or liquid sample of a pure α anomer that will rotate the plane of polarized light by a different amount and/or in the opposite direction than the pure β anomer of that compound. The optical rotation of the solution depends on the optical rotation of each anomer and its ratio in the solution. For example, if a solution of β-D-glucopyranose is dissolved in water, its specific optical rotation will be +18.7°. Over time, some of the β-D-glucopyranose will undergo mutarotation to become α-D-glucopyranose, which has an optical rotation of +112.2°. The rotation of the solution will increase from +18.7° to an equilibrium value of +52.7° (Figure 1.10).

The anomeric effect is a stereoelectronic [12, 13] effect wherein the anomeric substituents prefer the axial orientation instead of the less hindered equatorial orientation under the steric considerations. The anomeric effect was originally called the Edward–Lemieux effect but was named the anomeric effect by Lemieux [12] and Chu in 1958. For example, the anomeric effect in glucose is that α-methyl glucoside is more stable than ß-methyl glucoside (Figure 1.11) due to hyperconjugation. In β-methyl glucoside, the methoxy group is at the equatorial position and cannot be involved in hyperconjugation.

Glycosylation [15] is the reaction in which a carbohydrate, a glycosyl donor, couples with a glycosyl acceptor to construct a glycoconjugate. The glycosylation reaction requires a suitable

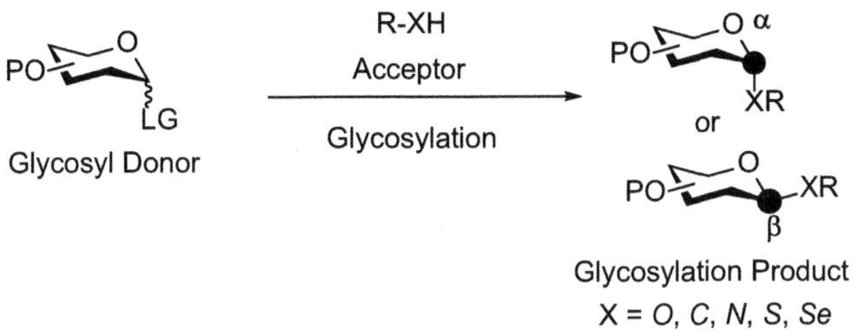

FIGURE 1.12 Glycosylation.

activating condition to form glycoside. In glycosylation, the glycosyl donor generally has a leaving group at the anomeric position, and the glycosyl acceptor has a nucleophilic acceptor that is coupled under optimum conditions to undergo glycosylation (Figure 1.12).

There are four types of glycosides [15] based on the nature of the atom the sugar attaches to. In **o-glycosides**, the compound sugar portion attaches to the oxygen atom (e.g., amygdalin, indicant, arbutin). In **c-glycosides**, the sugar portion attaches to the carbon atom (e.g., clinically approved C-glycoside drugs for the treatment of type II diabetics such as dapagliflozin, canagliflozin, and empagliflozin). In **s-glycosides**, the sugar portion attaches to the sulfur atom (e.g., sinigrin); thioglycosides are efficient metabolic decoys of glycosylation that reduce selection-dependent leukocyte adhesion. In **n-glycosides**, a sugar component attaches to an aglycone through a nitrogen atom, resulting in a C-N-C linkage (e.g., deoxyadenosine). Glycosylation is discussed in detail in Chapter 3.

Nucleoside analogues, with more than 40 derivatives marketed to treat viral DNA and RNA infections [16], represent some of the most promising and leading drugs, with Sofosbuvir (hepatitis C virus, HCV) or Remdesivir (Epstein-Barr virus, EBV) being recent examples. Nucleoside analogues are the most promising broad-spectrum antiviral against emerging RNA viruses. The basic concept of a carbohydrate nucleoside is a glycosylamine that consists of a nitrogenous base covalently attached to C1 of a sugar (ribose or deoxyribose) but without the phosphate group. Here the nitrogen-containing compound is either a pyrimidine or a purine. A nucleotide consists of a nitrogenous base, ribose or deoxyribose, and one to three phosphate groups. They take part in the molecular mechanisms of conservation, replication, and transcription of the genetic information. Carbohydrate nucleosides are discussed in detail in Chapter 6.

1.7 DISACCHARIDES

A disaccharide is a sugar that contains two molecules of monosaccharide. When two monosaccharide molecules combine, they undergo a dehydration reaction, forming a glycosidic bond. Common examples are sucrose, maltose, lactose, etc.

1.7.1 Sucrose

Sucrose is a common disaccharide [6, 11] $(C_{12}H_{22}O_{11})$ and is commonly known as table sugar. Sucrose tastes less sweet than fructose but sweeter than glucose. Sucrose is generally broken down by *sucrase* enzyme (majority into the mouth and small intestine) into simple sugars (glucose and fructose) before their absorption into the bloodstream of the intestine.

Lemieux et al. [10] synthesized sucrose in 1953 by treating penta-O-acetyl β-D-glucose with phosphorous pentachloride and treating the resulting chloride derivative with ammonia in ether and again with ammonia in benzene medium. The product was 1,2 –anhydro –α-D-glucopyranose3,4,6-triacetate.

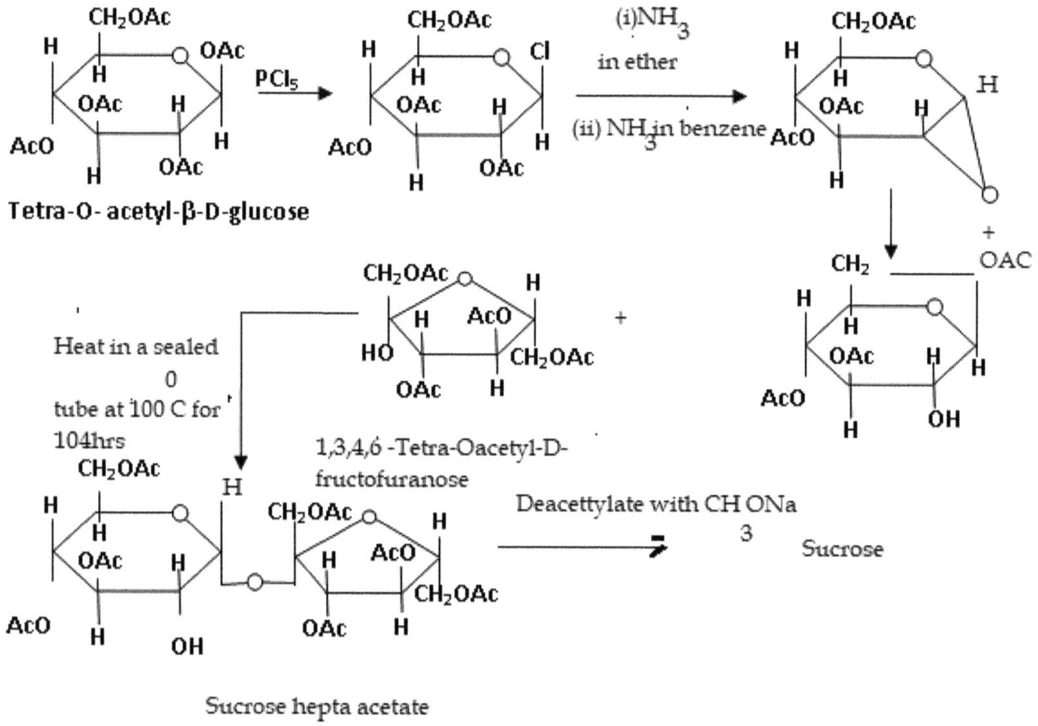

SCHEME 1.8B Sucrose synthesis.

The product was heated in a sealed tube at 100 °C for 24 hours. The mechanism the researchers proposed for product formation was that fructofuranose, a hindered secondary alcohol, had reacted with anhydroglucopyranose to form an α-glucose link. The solvent benzene evaporated off, and the residue deacetylated with sodium methoxide to give sucrose (Scheme 1.8b). However, the synthesized product yield was low, which is not commercially applicable.

Commercially [11], sucrose can be prepared from sugar cane by the following steps: i) juice extraction and purification, ii) decolorization, iii) concentration and crystallization, iv) centrifugation, and v) drying. First, sugar cane is crushed by a crusher to make small pieces that are then passed through a roller mill to squeeze out the juice into a tank. The maximum extraction of juice containing sucrose is achieved by washing with hot water and diluting the juice on the counter-current principle. The waste cellulosic material discharged from the diffuser is called bagasse and is used as fuel under boilers. Next, the juice containing different impurities such as organic acids, inorganic salts, proteins, and coloring matters is purified by two methods: high-pressure treatment and carbonation. In the first method, the juice is heated with high-pressure steam in the presence of 2%–3% lime in a steel tank, which forms insoluble calcium salts, coagulated protein, and coagulated matter; that precipitate is removed by filtration, and then carbon dioxide is passed through the juice. This second step is carbonation, and it removes the excess of lime as calcium carbonate, coloring matter, and some inorganic salts. The mud that is separated from the juice is removed by filtration.

In the second extraction step, the juice is decolorized by passing sulfur dioxide gas. The process is called sulphitation, and it entails bleaching the brown juice and completing the neutralization of lime. The insoluble calcium sulfite is removed by filtration. In the third step, the clear solution is then concentrated by boiling under reduced pressure in multiple effect evaporators. The concentrated juice is then passed through a vacuum pan where further evaporation reduces the water content to 6%–8%. The mixture of syrup and partially separated crystals known as massecuite is then discharged into a larger tank fitted with cooling pipes, called the crystallizing tank; crystals grow and form a thick crop.

Fourth, the massecuite is then sent to a centrifuge where sugar crystals are separated from the syrup; here, the crystals are sprinkled with a small amount water to wash away any syrup sticking to the surfaces of the sugar crystals. Finally, the wet sugar is passed down a rotating drum with a steam by counter-current process to make completely dry sugar. The residual mother liquor separated from the crystals is known as molasses, a valuable product for alcohol production by fermentation. Different structures of sucrose are shown in Figure 1.13.

Sucrose reacts with acetic anhydride in the presence of sodium acetate to form sucrose octa acetate, which indicates the presence of eight hydroxy groups in the sucrose molecule. On hydrolysis with dilute acids, sucrose gives equimolecular amounts of glucose and fructose, the two monosaccharides. Sucrose does not reduce in Tollens' and Fehling's reagents and does not undergo mutarotation. It reacts with dimethyl sulfate in alkaline medium to form octa methyl sucrose, which on hydrolysis in acidic medium gives a mixture of 2,3,4,6-tetramethyl-D-glucopyronose and 1,3,4,6-tetramethyl-D-fructofuranose. These results indicate that glucose unit in sucrose has pyranose form (six-membered ring) and furanose form (five-membered ring).

1.7.2 Lactose

Lactose is a disaccharide that is found in milk and milk products, in both human milk (5% to 8%) and cow milk (4% to 6%). Lactose is produced from whey, a byproduct of cheese making and casein production, by crystallizing an oversaturated solution of whey concentrate; lactose can also be isolated by dilution of whey with ethanol. Crystallization of lactose from saturated solutions below 93.5 °C yields crystals of α-lactose monohydrate, while crystallization at temperatures higher than 93.5 °C yields anhydrous crystals of β-lactose. The structure of lactose [11] is the condensation of galactose and glucose, which forms a β-1→4 glycosidic link of 4-O-β-D-galactopyranosyl-D-glucopyranose. The different structures of lactose are shown in Figure 1.14.

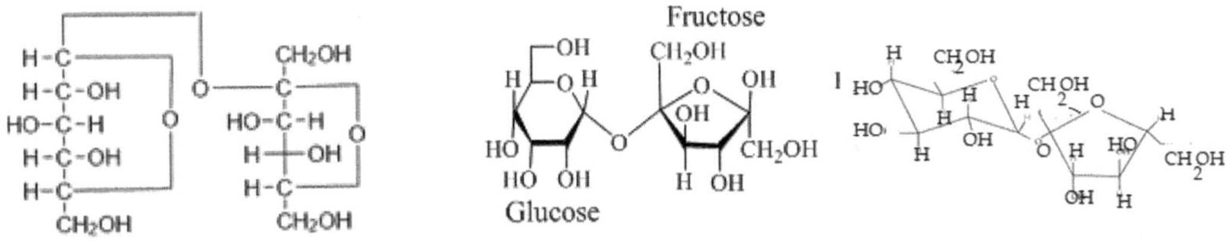

FIGURE 1.13 Sucrose structures: (left) Fischer structure, (middle) Haworth structure, (right) conformation structure.

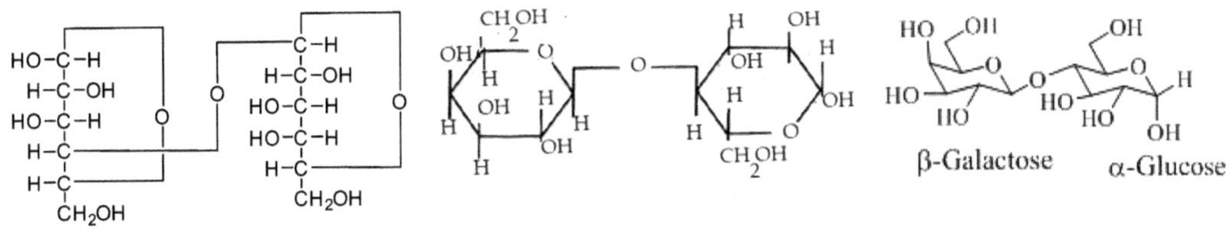

FIGURE 1.14 Lactose structures: (left) Fischer structure, (middle) Haworth structure, (right) conformation structure.

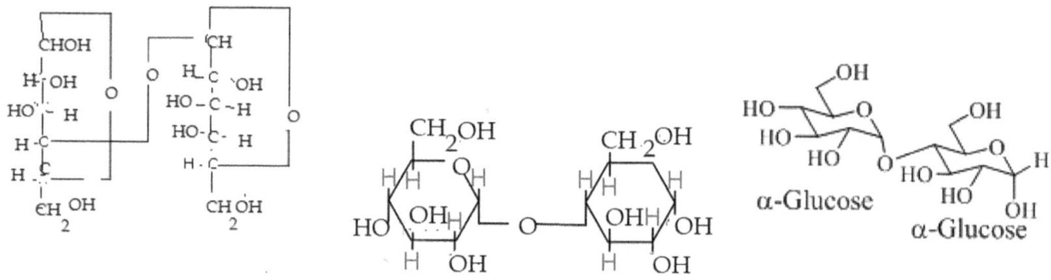

FIGURE 1.15 Maltose structures: (left) Fischer structure, (middle) Haworth structure, (right) conformation structure.

Lactose is a white crystalline solid with melting point of $(C_5H_8)n$, which conforms to the characteristics of reducing sugar. Lactose formed cotton-ball-shaped osazone and oxime and also shows mutarotation. On acidic or enzymatic (*lactase*) hydrolysis, lactose gives equimolecular amount of D(+) glucose and D (+) galactose. Lactose on oxidation with bromine water yields lactobionic acid.

1.7.3 Maltose

Maltose [12] is formed from two units of glucose joined with an α(1→4) bond. It is also present in highly variable quantities in partially hydrolyzed starch products like maltodextrin, corn syrup, and acid-thinned starch. Maltose is also known as malt sugar and is obtained by the action of *diastase* upon starch. Rafael et al. [9] produced maltose using starch from cassava bagasse catalyzed by crosslinked β-amylase aggregates. An isomer of maltose is isomaltose, which is similar to maltose but instead of a bond in the α(1→4) position, it is in the α(1→6) position (Figure 1.15).

Maltose is a reducing sugar as it reduces Fehling's solution, Benedict's solution, and Tollens' reagent. It exhibits mutarotation and forms osazone with phenyl hydrazine; it is also oxidized by bromine water to maltobionic acid. Maltose on acidic hydrolysis gives only D-glucose, indicating that glucose is the only monosaccharide unit present.

1.8 POLYSACCHARIDES

Polysaccharides are complex carbohydrates composed of multiple monosaccharide units bonded by glycosidic bonds. They are large, linear, or branched polymers with hundreds of repeating monosaccharide units (e.g., starch, cellulose, and glycogen).

1.8.1 Starch

Starch [11, 14] is a polysaccharide found in large amounts in staple foods such as wheat, potatoes, maize (corn), rice, and cassava (manioc). It is produced by most green plants for energy storage.

Commercially starch is produced from potato or other cellulosic materials in the following steps. Potatoes are cleaned during transport to the scrubber by channels to separate straw, stones, light waste, and sludge. Potato cell walls are ruptured by a rotary grater to make potato pulp; the pulp rapidly turns dark because tyrosine in the potato is oxidized by *polyphenol oxidase,* which is located in the cellular juice. Therefore, cellular juice must be separated as soon as possible.

Next, raw starch milk is produced by adding water, and then it is refined; this involves removing small fibers from the starch milk. Then the juice water is removed, and the starch milk is condensed; screens and hydro cyclones are commonly used for this purpose. To prevent the enzymatic darkening of potato juice, the starch is chemically refined using sulfurous acid. In the final stage, dewatering the refined starch milk is carried out in two stages. In the first stage, the excess water is removed using a rotary vacuum filter, and in the second, moist starch is dried without starch pasting using a pneumatic dryer. In this device, moist starch (water content 36%–40%) floats in strong, hot (160 °C) air flow and then is dried for 2–3 seconds. Then, the starch is separated from hot air in a cyclone. Because of the short period of high-temperature drying and the intensive water evaporation from the starch granules, its surface is heated only to 40 °C.

A side product of potato pulp washing is the extracted mash, which contains all the nonstarchy substances that are insoluble in water, such as cell wall fragments (i.e., fibers) and bound starch, which cannot be mechanically separated from the blended portions of potato. The dry pulp contains 30% starch, which makes it a good source for animal feed.

Starch is a mixture of two polysaccharides, amylose and amylopectin, that are made of D-glucose units. When starch is heated with hot water, it can be separated into these components. The part that is soluble in water is amylose, and the part which is insoluble in water is amylopectin (Figure 1.16). In amylose, the C1 of one glucopyranoside ring is attached to C4 of another ring. In this manner, thousands of glucopyranosides join linearly to form a compact molecule that is soluble in water. In amylopectin, the branching of glucopyranoside units takes place between C1 and C6 of different glucose units. In general, branching occurs at interval, of 20 glucose units.

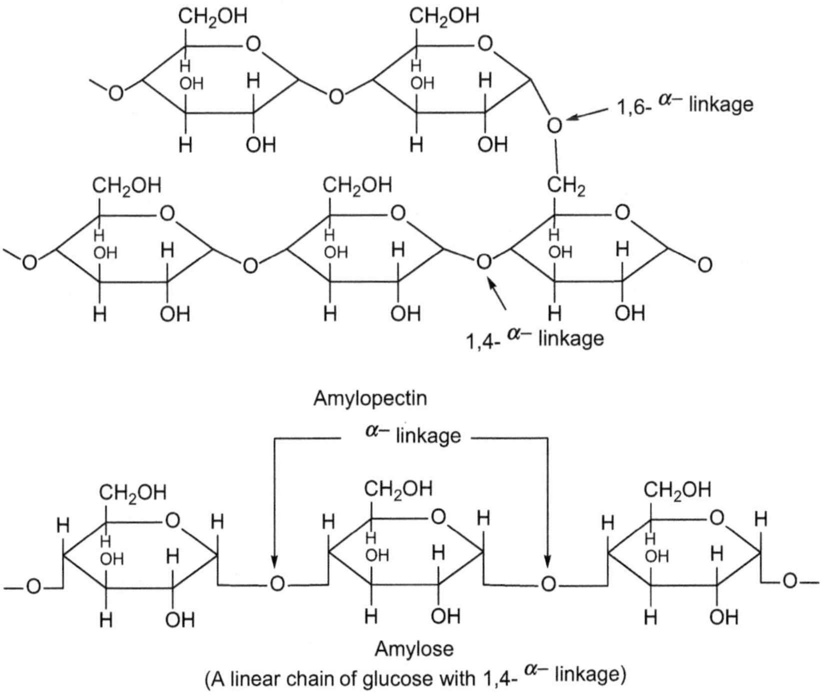

Amylopectin

Amylose
(A linear chain of glucose with 1,4- α- linkage)

FIGURE 1.16 Amylose and amylopectin structures.

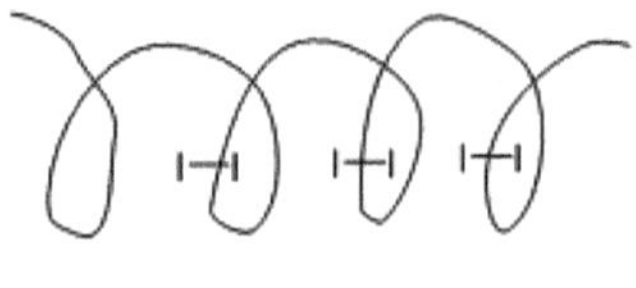

FIGURE 1.17 Starch–iodine complex.

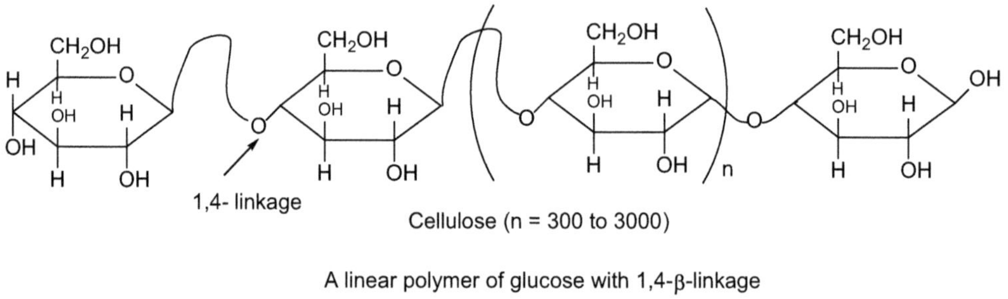

1,4- linkage

Cellulose (n = 300 to 3000)

A linear polymer of glucose with 1,4-β-linkage

FIGURE 1.18 Cellulose structure.

Pure starch [14] is a white, tasteless, odorless powder that is insoluble in cold water or alcohol, but on heating with water, it forms a gelatinous mass due to the swelling of amylopectin. Starch is easily hydrolyzed by dilute solutions of strong acids to form glucose. It shows blue with iodine, indicating that amylose molecules are present, which form helical coils. The iodine molecules insert within the coil to form a blue starch–iodine complex (inclusion complex) (Figure 1.17).

Starch does not reduce Tollens' reagent or Fehling's solution, and it does not form an osazone with phenylhydrazone. They are used in making cheap adhesives (postage stamps and envelopes) because of their sticky nature. The principal nonfood industrial use of starch is as an adhesive in paper making.

1.8.2 Cellulose

Cellulose [14] is an important polysaccharide that forms the main constituent of plant cell walls. Other sources of cellulose are cotton, flax, jute, and hemp. Cellulose is generally extracted from either (a) wood pulp or (b) cotton linters.

Cellulose from wood pulp: This process is used in the paper industry where fatty and waxy impurities are removed by organic solvents. Lignin and hemicelluloses are removed by treating with dilute acids or with calcium and sodium sulfite under high temperature and pressure. The fibers are then bleached with sodium hypochlorite and finally purified with 15%–20% sodium hydroxide.

Cotton linters: For this process, the first step is removing all impurities such as sands, small stones, and coloring matters from the cotton linters by mechanical cleaning. Then, the linters are boiled with dilute aqueous sodium hydroxide in an inert atmosphere to remove pectin, oil, and wax. The linters are then bleached with sodium hypochlorite, and the product is obtained 99.7% pure.

Cellulose is a colorless amorphous solid which are insoluble in water and organic solvents. On acidic hydrolysis it gives quantitative yield (about 96%) of crystalline D-glucose indicating that cellulose is composed of only D-glucose units. Cellulose does not reduce Tollens' reagent or gives color with Fehling's solution. It does not form osazone with phenyl hydrazine. Measurements of physical criteria such as viscosity, osmotic pressure, and ultracentrifuge sedimentation indicate that unreacted native cellulose has molecular weights on the order of 300,000–500,000 and consists of a chain length of nearly 2000–3000 glucose units. Its straight chain is composed of D-glucose units where units are joined by β-glycosidic between C-1 of one glucose unit and C-4 of the next glucose unit (Figure 1.18). X-ray analysis of cellulose indicates that glucose units in the primary chains

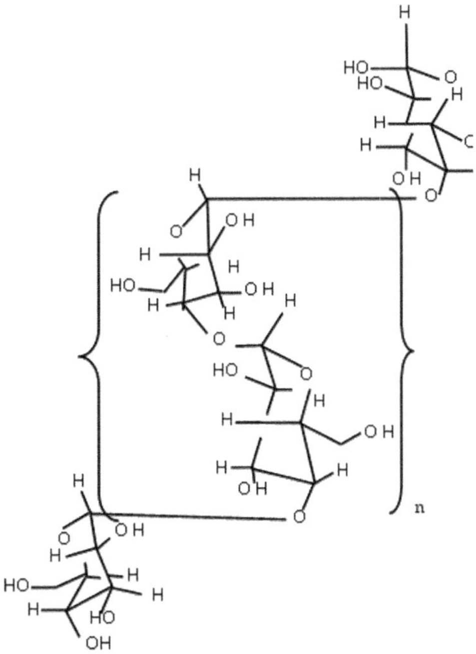

FIGURE 1.19 Glycogen structure.

are held together by primary covalent bonds, while the adjacent chains are linked by hydrogen bonding between hydroxyl groups. The presence of the covalent and hydrogen bonds in the cellulose structure explains the strength and general nonreactivity of native cellulose fibers.

Cellulose is used extensively in manufacturing paper, rayon (fiber with silklike structure), celluloid, cellophane, lacquers, adhesives, explosives (gun cotton), and cellulose acetate (motion picture film).

1.8.3 Glycogen

Glycogen is the reserve polysaccharide of animals; its stores are present in liver muscle and in smaller amounts in other tissues including red blood cells, white blood cells, kidney cells, and some glial cells [6,17,10]. The liver helps to regulate blood glucose (sugar) levels primarily with the hormones glucagon and insulin. Glycogen has a similar structure to that of amylopectin except it has more crosslinking. In the linear chains, the glucose residues are connected by α-1,4-glycosidic linkages, while α-1,6-glycosidic bonds create the branch points (Figure 1.19). The difference between glycogen and starch is that glycogen is the reserve polysaccharide of animals, whereas starch is reserve for plants.

1.8.4 Cyclodextrins

Cyclodextrins are oligosaccharides of 6–8 glucopyranose units formed by partial hydrolysis of starch [6, 9, 10]. These units form a doughnut-shaped ring and like crown ethers, they can act as a host for guest molecules. However, unlike crown ethers, cyclodextrins have a nonpolar lyophobic inside and a polar lyophilic outside as the capturing guest are molecules and not ions. Cyclodextrins catalyze organic reactions, often with regiospecificity and some steroselectivity in the entire space. The entire structure is water soluble because of OHs on the outside (Figure 1.20). They also serve as models for enzyme action.

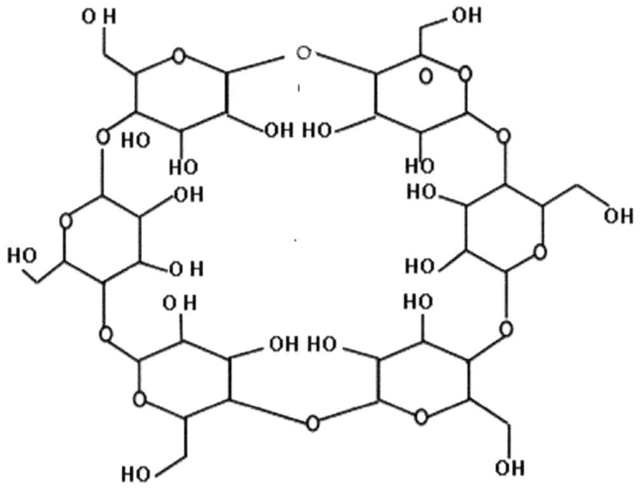

FIGURE 1.20 Cyclodextrin structure.

1.9 LIPIDS

Lipids are a group of organic compounds that are widely distributed in living systems [17, 19–23]. They are defined as naturally occurring compounds that are insoluble in water but soluble in nonpolar organic solvents such as ether, chloroform or carbon tetrachloride. Lipids have the following properties: i) They are oily or greasy nonpolar molecules stored in the adipose tissue of the body. ii) They are heterogeneous, mainly composed of hydrocarbon chains. iii) They are energy-rich organic molecules that provide energy for different life processes. iv) They are a characterized by their solubility in nonpolar solvents such as acetone, ether, chloroform, or benzene but insoluble in water. v) They are significant in biological systems as they form a mechanical barrier dividing a cell from the external environment known as the cell membrane.

Lipids help with the absorption of fat-soluble vitamins, and they transport lipoproteins as well. Lipids serve surfactants by reducing surface tension, and they function as energy sources. Lipids improve taste and palatability, and they also have a role in endocrine and nervous system function. Lipids also act as electronic insulators in neurons.

1.9.1 Classification of Lipids

Lipids [6, 19] are mainly divided into three types depending on their compositions: simple, compound, and derived. The simple lipids are esters of fatty acids and alcohols that on hydrolysis give fatty acids and alcohols. These are two types of simple lipid, fats and oils and waxes.

1.9.1.1 Fats and Oils

Fats and oils [20] make up over 90% of the lipid of adipose tissue in mammals. Fats and oils, obtained from plants as well as animals, are tremendous in nutrition (in calories value), and they rank highest among food. Fats play several major roles in our body and in fact are necessary in the correct amounts for the proper functioning of our bodies. Many vitamins are fat soluble and require fats for effective absorption by the body; fats also provide insulation to the body. They are also an efficient way apart from glycogen to store energy for longer periods.

Fats and oils are triesters of glycerol and acids that are known as triglycerides. When three OH groups of glycerol are esterified with the same acid, the triester is known as simple glyceride but mixed glyceride with two or three different acids. The acids that comprise triester are called fatty acids, and they come in long chains that are saturated or unsaturated. The natural fatty acids include all the saturated acids with even numbers of carbon atoms (4 to 40 atoms), whereas unsaturated fatty acids have 18 and 20 carbon atoms. In certain fats, the straight-chain fatty acids are linked with a cyclic or aromatic group. There are various hydroxy acids such as ricinoleic acid from castor oil and a few cyclic fatty acids with double bonds in the ring such as chaulmoogeric and hydnocapric acids from chaumoogra oil (Figure 1.21). The glycerides are of considerable therapeutic value; they have antimetabolite properties and are effectively used to combat leprosy and other external infections.

Chemically, fats contain large proportions of saturated fatty acids and melt at higher temperatures, while oils contain a larger portion of unsaturated fatty acids and melt at lower temperatures [20]. All natural fats and oils invariably consist of a mixture of glycerides; for instance, lard is a mixture of oleo-palmito-stearin, palmito-distearin, stearo-dipalmitin, and palmito-diolein. Detailed images of saturated and unsaturated fatty acids are shown in Figure 1.22.

Saturated fatty acids are long-chain fatty acids (12 to 26 carbon atoms) that are found in meats and fish, whereas medium- (6 to 10 atoms) and short-chain (fewer than 6 atoms) fatty acids occur primarily in dairy products. Saturated fatty acid chains contain only carbon-carbon single bonds. Their general formula is $C_nH_{2n+1}COOH$ as n-butyric (C_4), n-caprice (C_6), n-caprylic (C_8), n-capric (C_{10}), n-lauric (C_{12}), n-myristic (C_{14}), n-palmitic (C_{16}), n-stearic (C_{18}), and archidic (C_{20}) acids among many others. However, among the saturated fatty acids, palmitic and stearic acids occur most widely. Caprylic acid is solid at 25 °C, and the lower fatty acids up to capric acid are steam volatile. Butyric acid is soluble in water, while others are insoluble.

In terms of industrial applications, higher fatty acids [20]: are incorporated in the manufacture of shaving creams and candles; as emulsifying agents in polishes, cosmetics, and pharmaceuticals; in preparing metal stearates such as calcium and zinc stearates used in plastics and in producing greases; magnesium stearate is used in paints and talc powders, and aluminum stearate is used to thicken lubricants; in the compounding of rubber materials and the production of esters such as butyl stearate, and in manufacturing alkyd resins and plasticizers.

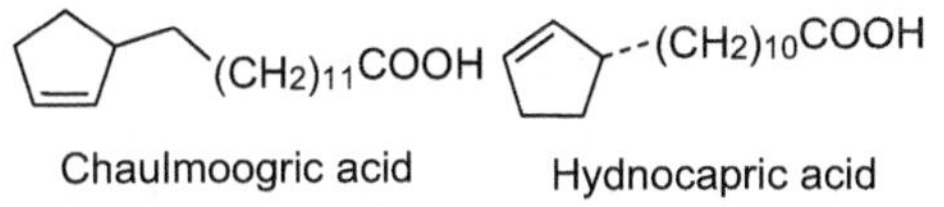

$(CH_2)_{11}COOH$ $(CH_2)_{10}COOH$

Chaulmoogric acid Hydnocapric acid

FIGURE 1.21 Hydnocapric and chulmooric acids.

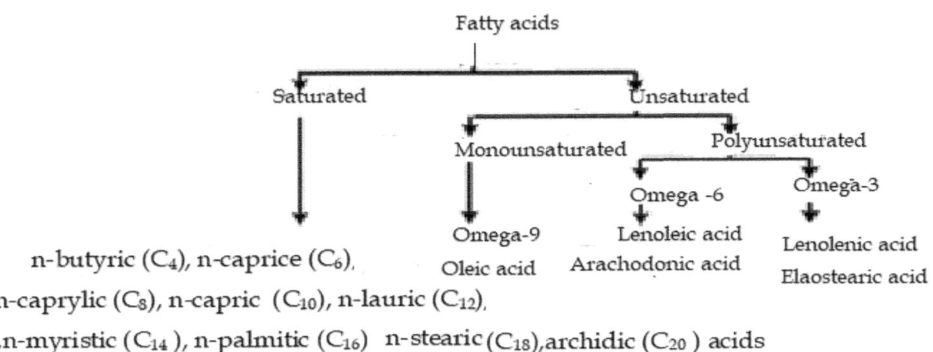

FIGURE 1.22 Classification of fatty acids.

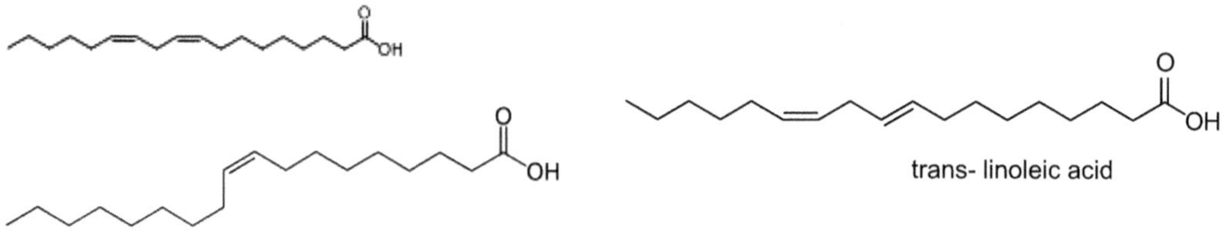

trans- linoleic acid

FIGURE 1.23 Linoleic acid structures: (left top) *cis*; (left bottom) *trans*; (right) transoleic acid.

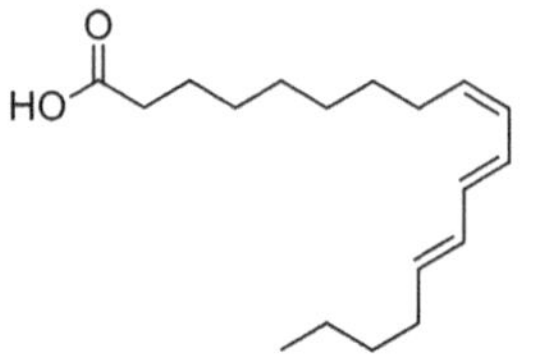

FIGURE 1.24 Tung oil ($C_{18}H_{23}O_2$).

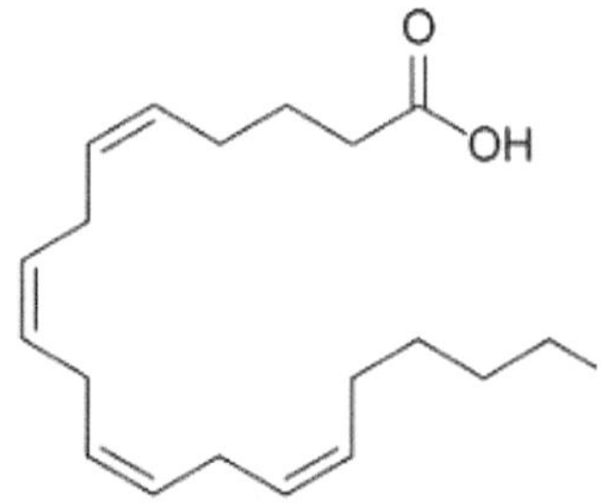

FIGURE 1.25 Arachidonic acid ($C_{20}H_{32}O_2$).

Unsaturated fatty acids have the general formula $C_nH_{2n-1}COOH$. They contain at least one double bond, which are either *cis* or *trans* form, depending on the geometry of the bond. *Cis* and *trans* fatty acids are geometrical isomers, and *cis* fat is considered to be better for the human body in increasing good cholesterol within the body. *Cis* fat is found naturally in liquid form, which limits its ability and potential to clog human arteries. More *cis* configurations along the carbon chain make the chain curved in its conformations, such as in *cis*-linoleic acid (Figure 1.23). In trans fatty acid containing long aliphatic carbon chains, two hydrogen bonds are on the opposite sides of the carbon chain in double bonds; the carbon chains do not bend but hold a straight saturated fatty acid conformation, such as transoleic acid (Figure 1.23).

Elaostearic acid, a conjugated fatty acid, is an isomer of linolenic acid. It is present in Chinese wood oil, also called Tung oil, and takes the following structure: 9,11,13-octadecatrienioic acid which is present in Chinese wood oil (Figure 1.24).

Because of the presence of conjugated double bonds, fat containing this acid as its glyceride polymerizes very rapidly to gelatinous form. Glycerides of very highly unsaturated acids containing 20 and 22 carbon atoms and four to six double bonds are present in marine animal oils. Arachidonic acid has 20 carbon chain with four double bonds (Figure 1.25). The acid has important biological functions such as growth development and a precursor of numerous lipid mediators.

Only one hydroxyl unsaturated acid, ricinoleic acid (12-hydroxy oleic acid), has been found in natural fat. It is obtained from castor oil, which is widely used for medicinal purposes (purgative) and for the production of surfactant. It can be converted to nondrying oil.

FIGURE 1.26 A-linolenic acid.

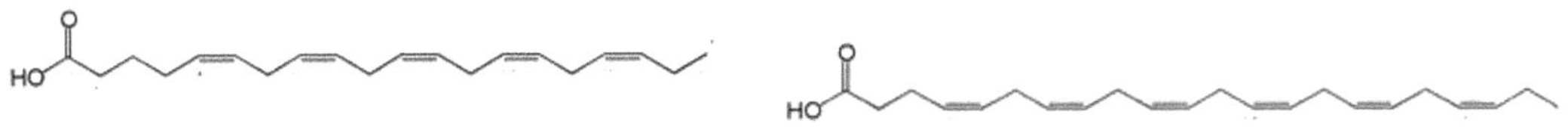

FIGURE 1.27 EPA and DHA structures.

Essential fatty acids are substances that humans and other animals must ingest because the body requires them for good health but cannot synthesize them [17, 22]. Normally important unsaturated fatty acids include linoleic acid (ω-6 fatty acid), α-linolenic acid (ω-3 fatty acid), eicosapentaenoic acid (20:5, n-3; EPA), and docosahexaenoic acid (22:6, n-3; DHA). Fatty acids have many uses [17]. For instance, they lower total cholesterol and triglycerides in people with high cholesterol, and they lower blood pressure in people with hypertension. So-called good fats lower the risk of heart attack and abnormal heart rhythms, and they prevent and treat atherosclerosis (hardening of arteries) by slowing the development of plaque and blood clots. Fatty acids are also useful in several diseases like diabetes, depression, and arthritis, and they help to reduce colorectal cancer.

Linoleic acid is dienoic acid ($C_{18}H_{32}O_2$), and the double bonds are present in 9,10 and 12,13. It is called omega-6 fatty and is available from vegetable oils such as soybean, safflower, and corn oils as well as from nuts and seeds. Figure 1.26 shows the structure of α-linolenic acid, an essential 18-carbon polyunsaturated fatty acid with three double bonds at 9,10, 12,13, and 15,16 positions. This acid is called omega -3 fatty acid, and its intake can decrease the risk of cardiovascular disease and also prevent arrhythmias and serum triglyceride levels. The acid is found in flaxseed, canola, soy, perilla, and walnut oils.

Figure 1.27 shows eicosapentaenoic acid (20:5, *n*-3; EPA) and docosahexaenoic acid (22:6, *n*-3; DHA) structures that are polyunsaturated fatty acids. Both are found in animals, plants, and oils including flax seed, hemp seed, olive, soya, canola, pumpkin seed, and sunflower seed; they are also found in milk, eggs, meat, algae, fungi, and bacteria. Fish oil is widely used as a source of ω-3 fatty acid, which is also produced from fish and poultry processing wastes. Nag et al. [22] extracted these fatty acids from the marine fish hilisa and vetki and from chicken viscera. Fatty acids lower total cholesterol and triglycerides.

Beta oxidation [24] is a catabolic process that occurs in mitochondria and some portions of proxisomes. In that process, the bond is broken between the second carbon/beta carbon and the third carbon/gamma carbon chain of fatty acid. Two carbon atoms are removed in the form of acetyl-Co A. Mitochondrial fatty acid beta oxidation disorders show either neonatal onset with hyperammonemia, transient hypoglycemia, metabolic acidosis, cardiomyopathy, and sudden death or late onset with neuropathy, myopathy, and retinopathy.

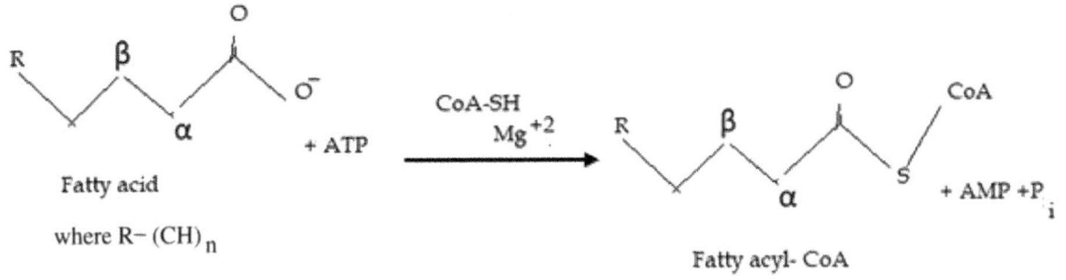

R β O O^- α + ATP CoA-SH Mg^{+2} R β O CoA α S + AMP + P_i

Fatty acid

where R− $(CH)_n$

Fatty acyl- CoA

After the conversion of fatty acids into fatty acyl-CoA, there are four steps as follows. In dehydrogenation, fatty acid is catalyzed by *acyl-CoA dehydrogenase*, which removes two hydrogen between carbons 2 and 3 from fatty acid chain.

Fattyacyl-CoA → Acyl-CoA-dehydrogenase (FAD → $FADH_2$) → trans-Δ^2-enoyl-CoA

Hydration adds water across the double bond catalyzed by *enoyl-CoA hydratase.*

Trans-Δ^2-Enoyl=CoA → Enoyl-CoA-hydrata: (H_2O) → β-Hydroxyacyl-CoA

3-Hydroxyacyl-CoA dehydrogenase catalyzes the dehydrogenation of 3-hydroxyacyl-CoA, which generates NADH.

β-Hydroxyacyl-CoA → β-Hydroxyacyl-CoA-dehydrogenase (NAD^+ → NADH + H^+) → β-Keto acyl CoA

Thiolytic cleavage is catalyzed by *thiolase*, which cleaves the terminal acetyl-CoA group and forms a new acyl-CoA which is two carbons shorter than the previous one.

β-Keto acyl CoA → Thiolase (CoA-SH) → Fatty acyl- CoA +CH-Co-S-CoA

Repeating the same cycle·

Rancidity is the term used to represent the deterioration of fats and oils resulting in an unpleasant taste. Rancid oils and fats are unsuitable for human consumption [17, 20]. Fats containing unsaturated fatty acids are more susceptible to rancidity, which is caused by one of two reactions: oxidative hydrolysis or hydrolytic rancidification. Oxidative hydrolysis occurs when fats and oils are exposed to air, moisture, light, bacteria, etc. The oxidation of unsaturated fatty acids results in the formation of unpleasant products such as dicarboxylic acids, aldehydes, and ketones. Hydrolytic rancidity occurs due to the partial hydrolysis of triacylglycerols by bacterial enzymes; bacteria from the air furnish enzymes that promote reactions. Antioxidants are added in many edible fats and oils to retard rancidification.

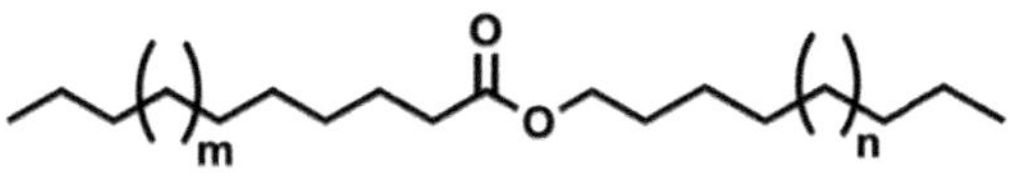

FIGURE 1.28 Schematic representation of waxes.

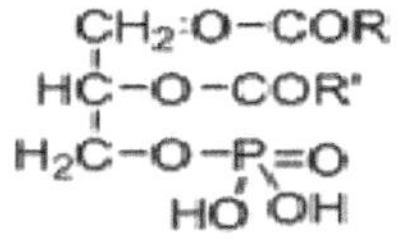

CH_2-O-COR
HC-O-COR'
H_2C-O-P=O
HO OH

FIGURE 1.29 Phospholipids.

1.9.1.2 Waxes

Waxes are esters of fatty acids with higher molecular weight and long chain of monohydric alcohols or sterols [6, 20]. Examples are lanolin, bees wax, and whale sperm oil. Waxes are insoluble in water but soluble in nonpolar organic solvents such as benzene, hexane, and chloroform; they are more difficult to saponify than fats, although it is possible by treating a solution of wax in petroleum ether with sodium ethoxide. On the saponification of waxes, fatty acids appear as water-soluble soaps, while long-chain alcohols, being insoluble in water, appear in the unsaponifiable matter (31% to 55% in respect to 1% to 2% of fats or oils.). Waxes serve a number of functions. Wax acts as a protective agent on plant surfaces against excessive loss of moisture and against fungi and bacteria. There are different industrial applications of waxes such as candles, complex coatings for wood products, and formulation of colorants for plastics. Polyethylene waxes are incorporated into inks in the form of dispersions to decrease friction. See Figure 1.28 for a schematic representation.

Compound lipids are esters of fatty acid with alcohol containing additional prosthetic or chemical groups [6, 7, 24]. Based on the additional chemical group, compound lipids are classified as phospholipids or galactolipids.

1.9.1.3 Phospholipids

Phospholipids are a key component of all cell membranes and can form lipid bilayers because of their amphiphilic characteristic. The lipid molecule has a hydrophilic "head" containing a phosphate group and two hydrophobic "tails" derived from fatty acids, joined by an alcohol residue. Thus, on hydrolysis, phospholipids yield an alcohol, a fatty acid, and phosphoric acid. This group is most abundant among the complex lipids and makes up as much as 70% of the complex liquid contents of the tissue. Sometimes phospholipids are named as a derivative of the parent compound, such as phosphatidic acid (Figure 1.29).

In aqueous solutions, phospholipids are driven by hydrophobic interactions that result in the fatty acid tails aggregating to minimize interactions with the water molecules. The result is often a phospholipid's bilayer: a membrane that consists of two layers of oppositely oriented phospholipid molecules, with their heads exposed to the liquid on both sides and with the tails directed into the membrane. The functions of phospholipids are as follows:

- In association with proteins, phospholipids form the structural components of membranes and regulate membrane permeability
- They absorb fat from the intestine and prevent accumulation of fat in the liver (fatty liver); they are lipotropic.
- They participate in the transport of lipids and are essential for the synthesis of different lipoproteins

- Arachidonic acid, an unsaturated fatty acid liberated from phospholipids, serves as a precursor for the synthesis of eicosanoids (prostaglandins, prostacyclins, thromboxanes, etc.).
- Phospholipids participate in the reverse cholesterol transport and thus help remove cholesterol from the body.
- Phospholipids act as surfactants (agent that lowers surface tension).
- Phospholipids participate in blood clotting.
- Phospholipids (phosphatidylinositols) are involved in signal transmission across membranes.

Phospholipids are classified into four compounds: i) lecithin, ii) cephalins, iii) plasmalogens, and (iv) sphingomyelins.

(i) Lecithin [24] is a poorly crystalline substance that occurs in nearly all animals (specially brain, nerve tissues, sperm, and egg yolk) and plants in seeds and sprouts. It has low solubility in water; it usually swells in water and forms colloidal solution. It is an excellent emulsifier and in aqueous solution, its phospholipids can form either liposome, bilayer sheets, micelles, or lamellae depending on hydration and temperature. Lecithin is dextro-rotatory and is readily oxidized standing in air. Therefore, lecithin should be preserved in vacuum and dark.

Lecithin can be easily extracted chemically using solvents such as hexane, ethanol acetone, petroleum ether, or benzene. Commercially, lecithin can be obtained by water degumming from the extracted oil of seeds. Lecithin can be modified for specific intended uses via enzymatic hydrolysis; in hydrolyzed lecithin, a portion of the phospholipids has one fatty acid removed by *phospholipase*. When lecithin is heated with dilute acids or alkalis, it produces choline, active glycerol -α- phosphoric acid, and fatty acids. Thus, lecithins are esters of glycerol in whose molecule two hydroxy groups have been esterified by fatty acids, while the third has been esterified by phosphoric acid, which in turn formed an ester and choline (Figure 1.30). The nature of fatty acids depends on the source of lecithin. If the phosphoric acid is tied to the end of glycerol, the compound is called α-lecithin, and if it is tied to the central carbon atom, it is called β-lecithin.

Among its common applications, lecithin is added to chocolate candies to prevent waxy white spots on the candy surfaces. When added to oleomargarine, it gives the product a consistency similar to that of butter, and when added to animal feed, it enriches fat and protein and improves palletization. Industrially, lecithin is a release agent for plastics and acts as an anti-sludge additive in motor lubricants and an anti-gumming agent in gasoline. Lecithin is an emulsifier, a spreading agent, and an antioxidant in textile, rubber, and other industries; in the paint industry, it forms protective coatings for surfaces with painting and printing ink, and it also helps as a rust inhibitor. In the pharmaceutical industry, lecithin acts as a wetting agent, a stabilizing agent, and a choline enrichment carrier. Lecithin is commonly consumed in foods, but it is likely safe when taken as a supplement in doses up to 30 grams daily for up to 6 weeks; it can, however, cause some side effects including diarrhea, nausea, stomach pain or fullness. Lecithin might cause allergic reactions in people with egg or soy allergies.

(ii) Cephalins are found in all human cells, particularly in nervous tissue such as the white matter of brain and nerve tissue in spinal cord, where they make up 45% of all phospholipids

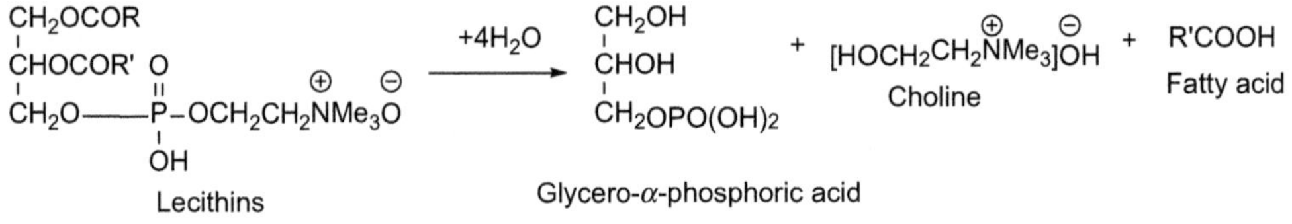

CH_2OCOR
$CHOCOR'$ O
CH_2O—$P-OCH_2CH_2\overset{\oplus}{N}Me_3\overset{\ominus}{O}$
OH
Lecithins

$\xrightarrow{+4H_2O}$

CH_2OH
$CHOH$
$CH_2OPO(OH)_2$
Glycero-α-phosphoric acid

+ $[HOCH_2CH_2\overset{\oplus}{N}Me_3]\overset{\ominus}{O}H$
Choline

+ R'COOH
Fatty acid

FIGURE 1.30 Lecithin.

(Figure 1.31). Chemically, the difference between cephalins and lecithin is that lecithin contains choline residue whereas cephalin contains coalmine (ethyl amine) residue or sometimes serine or myoinositol.

Cephalins play a role in membrane fusion and in disassembly of the contractile ring during cytokinesis in cell division. Cephalins also works with phosphatidylserine to increase the rate of thrombin formation, and they play an important role in heart blood flow. When blood flow to the heart is restricted, the asymmetrical distribution of phosphatidylethanolamine between membrane leaflets is disrupted, and as a result the membrane is disrupted. Additionally, cephalins play a role in the secretion of lipoproteins in the liver.

(iii) Plasmalogens are commonly found in cell membranes in the nervous system and also in the myelin of brain, heart, and skeletal muscle. In human heart tissue, nearly 30%–40% of choline glycerophospholipids are plasmalogens, as are 20% in brain and up to 70% of myelin sheath ethanolamine glycerophospholipids (Figure 1.32). Plasmalogens are characterized by the fact that in treatment with acids they form long chain fatty–aldehyde. The general structure of plasmalogen is similar to lecithin or cephalin in one case one of the fatty acids is replaced by aldehydic group.

Plasmalogens can protect mammalian cells against the damaging effects of reactive oxygen species. In addition, they have been implicated as being signaling molecules and modulators of membrane dynamics. The complete lack of plasmalogens, or even their reduced concentration, is associated with several neurological disorders including Alzheimer's disease, Zellweger syndrome, and Niemann–Pick type C disease. Therefore, it is of vital importance to understand how diminished plasmalogen concentrations can affect the biophysical function of brain cell membranes and whether it additionally plays a primary or secondary role in the development of those diseases.

Like all other phospholipids, **(iv) sphingomyelins** (Figure 1.33) also occur in all animal tissues, particularly in brain and spinal cord. In mammals, they ranges from 2% to 15% in most tissues, with

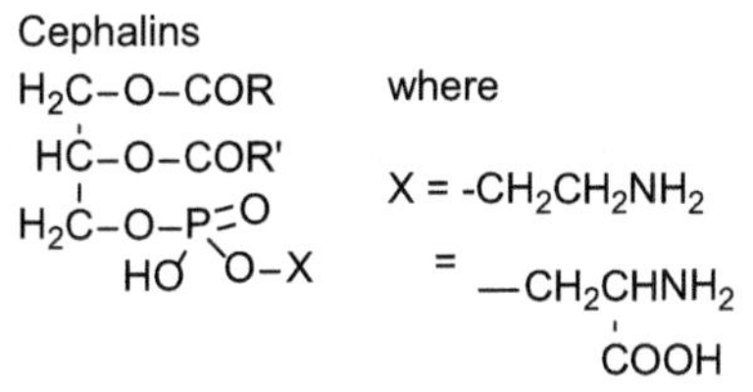

Cephalins
$H_2C{-}O{-}COR$
$HC{-}O{-}COR'$
$H_2C{-}O{-}P(=O)(OH){-}O{-}X$

where
$X = -CH_2CH_2NH_2$
$= -CH_2CH(COOH)NH_2$

FIGURE 1.31 Cephalins.

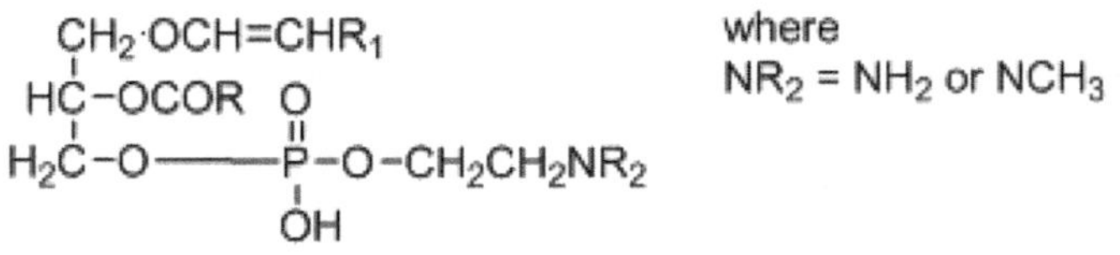

$CH_2{\cdot}OCH{=}CHR_1$
$HC{-}OCOR$
$H_2C{-}O{-}P(=O)(OH){-}O{-}CH_2CH_2NR_2$

where
$NR_2 = NH_2$ or NCH_3

FIGURE 1.32 Plasmalogen.

$CHOHCH{=}CH(CH_2)_{12}CH_3$
$CHNHCOC_{23}H_{47}$
$H_2C{-}O{-}P(=O)(OH){-}OCH_2\overset{\oplus}{N}Me_3\overset{\ominus}{O}H$

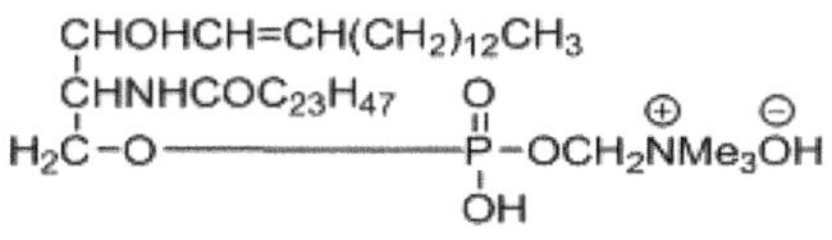

FIGURE 1.33 Sphingomyelin.

higher concentrations found in nerve tissues, red blood cells, and the ocular lenses. Sphingomyelin is also found in plant seeds; it is very sparingly soluble in ether and thus can be easily separated from lecithin and cephalin. It consists of a phosphocholine head group, a sphingosine, and a fatty acid. It is one of the few membrane phospholipids not synthesized from glycerol. Sphingosine and fatty acid can collectively be categorized as a ceramide. It differs chemically from lecithin and cephalin in two ways: In place of glycerol, its nucleus is sphingosinol, a dihydric amino alcohol with double bonds, and it contains one molecule of fatty acid, usually lignoceric ($C_{23}H_{47}$ COOH) acid, linked to an amino group rather than an alcoholic group of sphingosinol; some sphingomyelins contain stearic acid and nervonic acid in place of lignoceric acid. Thus, like lecithin and cephalins, sphingomyelins have differing fatty acid compositions. Sphingomyelins are also more prone to intermolecular hydrogen bonding than other phospholipids.

The function of sphingomyelin remained unclear until it was found to have a role in signal transduction. Additionally, the membranous myelin sheath that surrounds and electrically insulates many nerve cell axons is particularly rich in sphingomyelin; sphingomyelins have a role in cell apoptosis; and they are associated with lipid microdomains in the plasma membrane known as lipid rafts. These are characterized by the lipid molecules being in the lipid ordered phase, offering more structure and rigidity than the rest of the plasma membrane.

1.9.1.4 Galactolipids

Galactolipids are sometimes referred to as galycolipids or cerebrosides. They are found in considerable amounts in the white matter of the brain and of all nervous tissues. They are also found on the surface of all eukaryotic cell membranes, where they extend from the phospholipid bilayer into the extracellular environment. They usually occur in the amorphous state and are insoluble in ether but soluble in alcohol. On hydrolysis, galactolipids give a fatty acid, sphingosinol, and a sugar, usually galactose. Four individual groups of glycolipids differ in their fatty acids: cerasin (lignoceric fatty acid), phrenosin (α-hydroxy lignoceric fatty acid), nervone (nervonic fatty acid), and hydroxy nervone (hydroxy nervonic acid) (Figure 1.34). They act as receptors on the surface of red blood cells. They also assist the immune system by destroying and eliminating pathogens from the body.

Another class of lipids is derived lipids, which are obtained from derivatives of simple and compound lipids; terpenes and steroids are examples of derived lipids. These lipids do not contain ester linkages but can be considered to be derived from naturally occurring esterified materials; hence, they are derivatives of lipids or lipid-like substances. They are utilized in other biochemical reactions to produce other essential substances such as sterols or bile juices. Artificial membranes [23] have several advantages, including improving the stability, solubility, bioavailability and toxicity profiles of active pharmaceutical ingredients. Artificial lipids are discussed in detail in Chapter 10.

Terpenes are a class of natural hydrocarbon compounds with the formula $(C_5H_8)_n$ for $n \geq 2$; the term terpenoids refers to both hydrocarbons and oxygenated derivatives, that is, all the compounds having $(C_5)_n$. That is, all the terpenes are terpenoids but not vice versa. Certain volatile oils

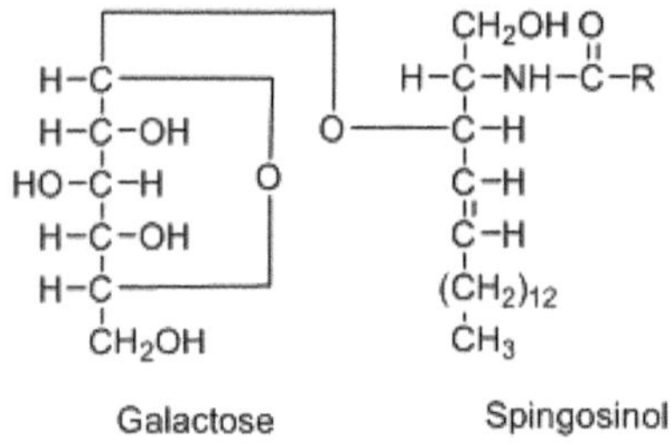

FIGURE 1.34 Galactolipids.

known as essential oils are derived from fruits, flowers, leaves, stems, bark, and roots and give pleasant smells from the presence of terpenoids in the oils and their oxygenated derivatives of such as alcohols, aldehydes, and ketones. Terpenoids are classified according to isoprene units (C_5H_8) present in the structure: i) hemiterpenes, ii) monoterpenes, iii) sesquiterpenes, iv) diterpenes, v) triterpenes, vi) tetraterpenes or carotenoids, viii) polyterpene or rubber $(C_5H_8)n$, etc. Terpenes are colorless, pleasant-smelling liquids or solids that are insoluble in water and readily volatile in steam. Since all the terpenes have one or more double bonds, they include add all the olefinic reagents such as halogens, hydrogen halide, nitroyl chloride, and bromide. Some terpenoids undergo Diels–Alder reaction or readily isomerize in the presence of acids.

Steroids are a group of structurally related compounds that are widely distributed in plants, animals, and fungi [6, 25, 26]. They are biologically active organic compounds with four rings arranged in a specific molecular configurations such as 1,2 –cyclopenteno phenanthrene nucleus in their structure. Steroids are lipid soluble and not stored in cells, so they are freely permeable to membranes. Steroid hormones are not water soluble, so they must be carried in the blood complexes to specific binding globulins. The core structure of steroid is typically 17 carbon atoms bonded in four fused rings, three six-member cyclohexane rings (rings A, B, and C) and one five-member cyclopentane ring (D) (Figure 1.35). Steroids vary by the functional groups attached to this four-ring core.

Steroids on distillation with selenium at 360 °C yield Diels hydrocarbon (3 ′-methyl-1, 2-cyclopentenophenanthrene) and some chrysene, but at 420 °C, chrysene is the main product with some picene (Figure 1.36). Diels hydrocarbon is a phenanthrene derivative obtained by the dehydrogenation of various steroids; it is synthesized as shown in Scheme 1.8c.

Sterols are forms of steroids with a hydroxy group at position three and a skeleton derived from cholestane. They are crystalline compounds and hydroxy group is secondary alcohol. Sterols can be divided into three groups: zoosterols, phytosterols, and mycosterols.

Animal steroids include cholesterol, sex hormones (which govern sex differences and reproduction), corticosteroids (regulate many metabolism functions), and anabolic steroids (interact with androgen receptors to increase muscle and bone synthesis). The main sources of cholesterol are the brain, spinal cord, gall stone, and liver cells. The human body not only can synthesis cholesterol but also can take it from food through the intestine into the blood stream.

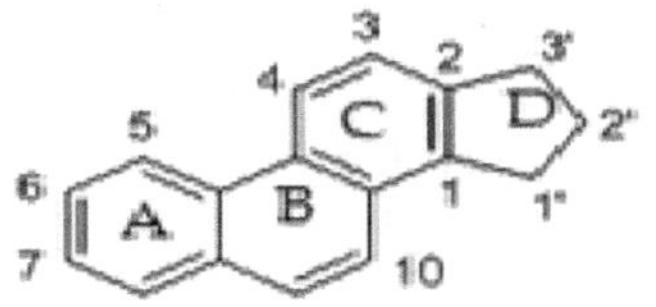

FIGURE 1.35 Steroid core structure.

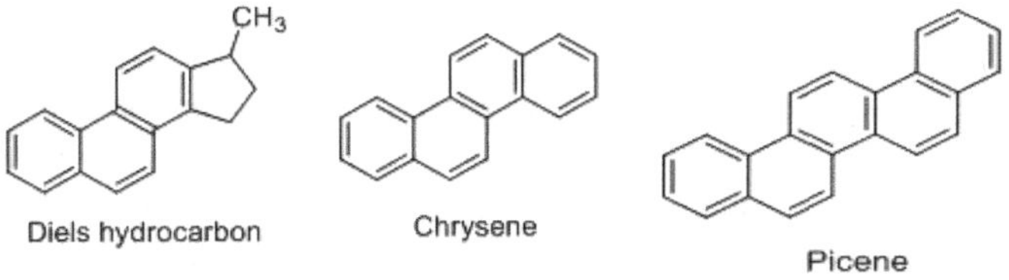

FIGURE 1.36 Steroid heating products.

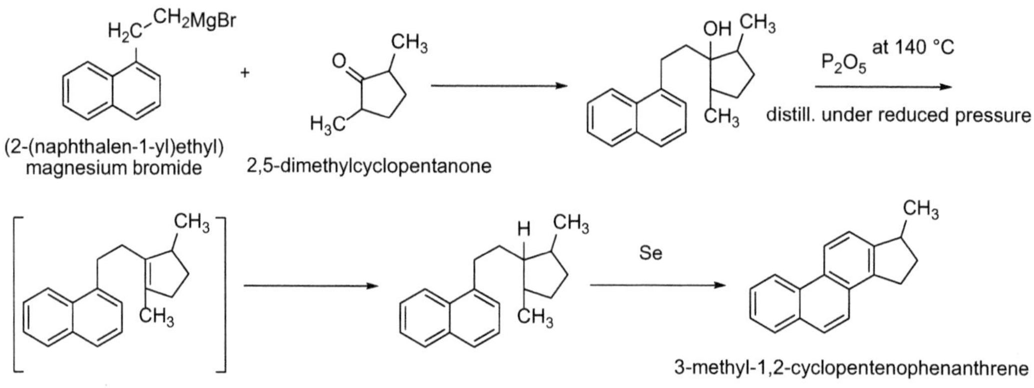

(2-(naphthalen-1-yl)ethyl) magnesium bromide + 2,5-dimethylcyclopentanone → (P_2O_5 at 140 °C, distill. under reduced pressure) → → (Se) → 3-methyl-1,2-cyclopentenophenanthrene

SCHEME 1.8C Synthesis of Diels hydrocarbon.

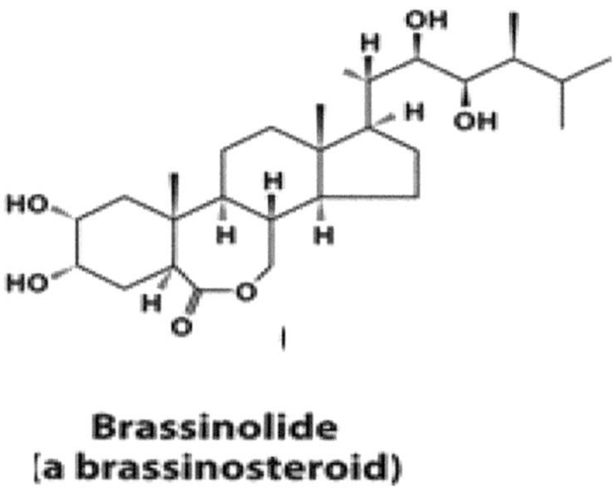

Brassinolide
[a brassinosteroid]

FIGURE 1.37 Plant growth steroid.

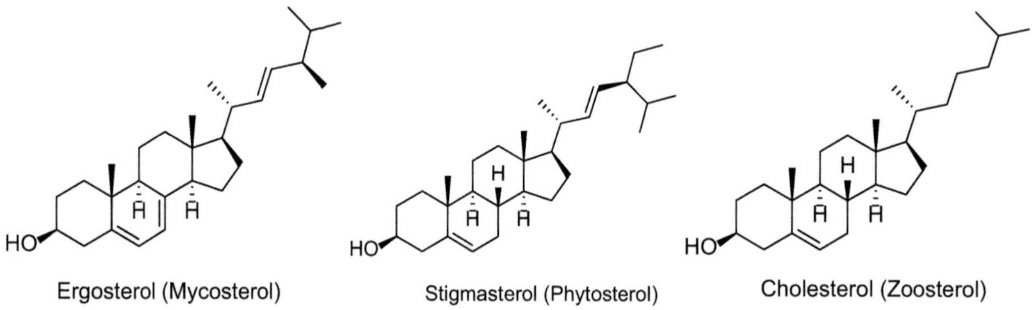

Ergosterol (Mycosterol) Stigmasterol (Phytosterol) Cholesterol (Zoosterol)

FIGURE 1.38 Different sterols.

Phytosterols are obtained from plants; examples include stigmasterol, campesterol, sitosterol, etc. They are similar to cholesterol in serving as structural components of the biological membranes of plants. Phytosterols are widely distributed in plants as a free compound or in the form of glycoside. The important sources are soybean and Calabar beans. The plant steroid brassinolide (Figure 1.37) is a potent growth regulator that increases the rate of stem elongation and affects the orientation of cellulose microfibrils in the cell wall during growth.

Mycosterols is the collective term for the fungal and yeast sterols (Figure 1.38). Fungal extracts rich in sterols of various forms are valuable and promising ingredients. One of the best-known

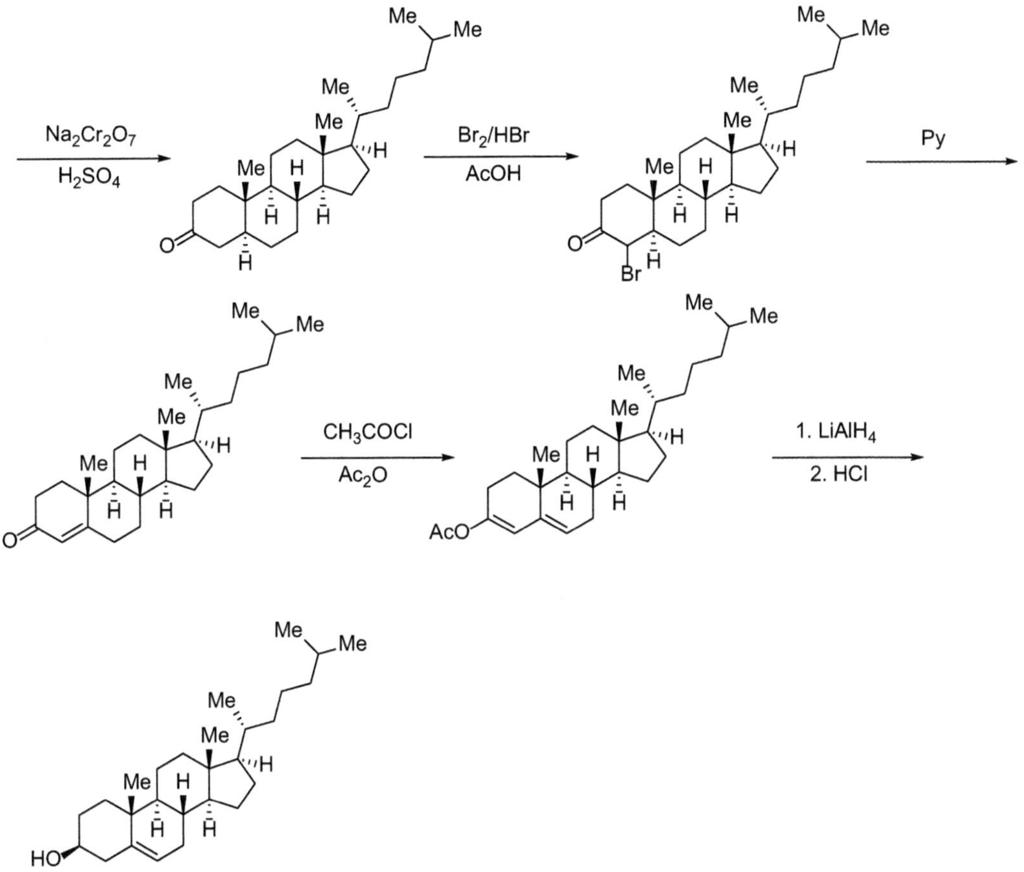

$Na_2Cr_2O_7$ / H_2SO_4 → Br_2/HBr / AcOH → Py → CH_3COCl / Ac_2O → 1. $LiAlH_4$ 2. HCl →

Cholesterol

SCHEME 1.9 Woodward cholesterol synthesis.

benefits of mycosterols is their inhibitory actions on cholesterol absorption and biosynthesis. Ergosterol (mycosterol) is a provitamin D_2 and is found in plants as well as in animals and yeast. Ergosterol is involved in maintaining the integrity of the fungal cellular membrane.

1.10 SYNTHESIS OF CHOLESTEROL

The main problems of synthesis of cholesterol [6, 23, 25] that there are 256 optically active isomers that are possible for cholesterol. During synthesis of cholesterol, each step yields products that must be separated further; indeed, the total synthesis of cholesterol is a great scientific achievement. Woodward synthesis (1951) is summarized in Scheme 1.9. Other cholesterol schemes include W.S. Johnson's 1966 synthesis of racemic cholesterol [27] and the enantiomer of natural cholesterol reported in 1996 by Rychnovsky and Mickus [28]. Other recent developments in synthesizing steroids are discussed in Chapters 4 and 8.

REFERENCES

[1] P.N. Sudha; A. Sundarajan; R. Nithya and V. Kumar; Industrial applications of marine carbohydrates; *Advances of Food and Nutrition Research*, 73, 145, 2014.

[2] S.M. Canter; *Use of Sugars and Other Carbohydrates in the Food Industry*; ACS Publication, Vol. 12, June 1955.

[3] S. Kokkinidou; D. Peterson; T. Bloch and A. Bronton; The importance role of carbohydrates in the flavor, function and formulation of oral nutritional supplements; *Nutrients*, 10(6), 742, 2018.

[4] S. Dhawan and J.N. Srivastava; Polyesters, carbohydrates and proteins based bio-plastics, their scope and applications: a review; *International Journal of Innovative Research in Science, Engineering and Technology*, 8(3), 3537, 2019.

[5] M. Kilcoyne and L. Joshi; Carbohydrates in therapeutics; *Cardiovascular & Hematological Agents in Medicinal Chemistry*, 5(3), 186, 2007.

[6] I.L. Finer; *Organic Chemistry*; Addison Wesley Longman Ltd., Vol. 2, 1973.

[7] O.P. Agrawal; *Organic Chemistry Natural Products*; Goel Publishing House, New Delhi, Vol. 1–II, 1980.

[8] Ahindra Nag, ed.; *Greener Synthesis of Organic Compounds, Drugs and Natural Products*; Taylor and Francis, 2022.

[9] Bhupinder Mehata and Manju Mehata; *Organic Chemistry*; Prentice Hall of India, New Delhi, 2005.

[10] Roy D. Guthrie and J. Honeyman; *An Introduction to Carbohydrate Chemistry*; Oxford University Press, 1974.

[11] Robert Joseph McIlroy; *Introduction to Carbohydrate Chemistry*; Butterworth, 1967.

[12] S.G. Sunderwirth and G. Olson; Conformation analysis of the Pyranoside ring; *Journal of Chemical Education*, 39, 410, 1962.

[13] Brain Capon; Mechanism in carbohydrate chemistry; *Chemical Reviews*, 69, 407, 1969.

[14] H. Honeyman, ed.; *Recent Advances of Cellulose and Starch*; Heywood & Company Ltd., London, 1959.

[15] R. Das and B. Mukhopadh; Chemical O-glycosylations: an overview; *Chemistry Open*, 8, 207, 2016.

[16] A.L. Lehninger, *Principles of Biochemistry*; Palgrave Macmillan, 8th Edition, 2021.

[17] F.D. Gunstone; *Fatty Acid and Lipid Chemistry*; Springer, 2020.

[18] R.U. Lemieux and G. Huber; A chemical synthesis of sucrose; *Journal of the American Chemical Society*, 75(16), 4116, 1953.

[19] Casimir C. Akoh; *Food Lipids*; Taylor and Francis, 2017.

[20] A. Dieffnbacher and W.D. Pocklington; *Standard Methods for the Analysis of Fats, Oils and Derivatives*; Blackwell Scientific Publications, London, 7th Edition, 1991.

[21] U.N. Das; E.J. Ramos and M.M. Meguid; Metabolic alterations during inflammation and its modulation by central actions of omega-3 fatty acids; *Current Opinion in Clinical Nutrition and Metabolic Care*, 6, 413–419, 2003.

[22] A. Nag and D. Patil; Production of PUFA concentrates from poultry and fish processing waste. *Journal of the American Oil Chemists' Society*, 88, 589–593, 2011.

[23] Christina G. Siotorou; Georgia-Parskevi Nikoleli; Dimitrios P. Nikolelis and Stefanos K. Karapetis; Artificial lipid membranes: past, present and future; *Membrane (Basal)*, 7(3), 38, September 2017.

[24] V. Roy and L.A. Agrofoglio; Nucleosides and emerging viruses: a new story; *Drug Discovery Today*, 27, 1945–1953, 2022.

[25] Robert Robison, ed.; *The Chemistry of Steroids by Klyne, Methune10: Introduction to Steroid Chemistry*; Elsevier, 1968.

[26] W. Klyne, *The Chemistry of Steroids*; Methuen, MA; Monographs on Biochemical Subjects, 1960.

[27] W.S. Johnson; J.A. Marshall; J.F.W. Keana; R.W. Franck; D.G. Martin and J.V. Bauer; Steroid total synthesis—hydrochrysene approach—*XVI:* Racemic conessine, progesterone, cholesterol, and some related natural products; *Tetrahedron*, 22, 541–601, 1966.

[28] Scott D. Rychnovsky; Daniel E. Micku; Synthesis of ENT-cholesterol, the unnatural enantiomer; *Journal of Organic Chemistry*; 57(9), 2732–2736, 1992.

2 Inositols and Their Derivatives
Synthesis and Application

Adam Kurčina and Jan Veselý

2.1 INTRODUCTION

Inositol (cyclohexane hexols) derivatives also known as cyclitols and are represented by nine stereoisomers, four of which (*myo*-, *neo*-, *chiro*- and *scyllo*-inositol) are naturally occurring have perform various biological functions. While Scherer described the first example of isolated inositol derivatives in 1850 [1], the first synthetic approaches to inositol derivatives were reported in the 1960s [2]. At that time, organic chemists started to investigate various routes to prepare inositol derivatives to understand their biological functions.

Today, we have knowledge about the role of inositol derivatives, especially phosphorylated *myo*-inositols, in cellular signal transduction and calcium mobilization, thanks to advantages in inositol synthesis. However, we still face the problem of obtaining enantiopure inositol derivatives with a high atom economy. With the growing utility of and interest in inositol derivatives, the number of reviews published has grown rapidly [3]. In this chapter, we focus on progress in the synthesis of inositol derivatives in the last ten years, but selected previous achievements are also discussed to illustrate the evolution of the field; we highlight recent chemical progress in the area of inositol chemistry. We pay special attention to enantioselective synthetic approaches toward chiral inositol derivatives, namely asymmetric phosphorylation, chiral resolution, and de novo synthesis. Later, we introduce the preparation of bioactive inositol derivatives focusing on phosphorylated forms and utilizing stereoselective approaches. Finally, we introduce natural product synthesis containing a natural scaffold.

2.2 STEREOSELECTIVE APPROACHES TO INOSITOL DERIVATIVES

2.2.1 Asymmetric Phosphorylation Used in Inositol Synthesis

Myo-inositol phosphates and phosphoinositides are the curiosities of the biochemical world. Their ubiquity and biological importance, ranging from occurrence in cell membranes to signaling molecules, made them interesting even to synthetic chemists. One way of their synthesis is to obtain enantiomerically pure phosphate intermediates via catalytic enantioselective transfer of P(V) reagents. In 2001, Miller et al. introduced a peptide catalyst that could promote highly enantioselective phosphorylation of partially protected *meso*-inositol derivative **1** (Scheme 2) [4–6]. The catalyst inspired by histidine-dependent kinases contains histidine residue responsible for the activation of phosphorylation reagent. Although catalytic enantioselective P(V) chemistry on inositols was further developed and applied [7–10], the substrate scope of this method is quite limited.

Alternative methods for improving its effectiveness have been explored. In 2010, the same group developed a phosphoramidite (P(III)) transfer methodology [11]. The cornerstone of their methodology was the formation of a peptide terminated with a suitable phosphoramidite activator. This combination ensured both enantioselectivity through the peptide and enhanced reactivity of the phosphoramidite reagent. The selected phosphoramidite activator moiety, Boc-L-tetrazolylalanine (Atz), participated in the catalytic cycle depicted in Scheme 2.1. A critical aspect of the cycle involved the regeneration of acidic tetrazole, which would normally undergo self-quenching due to

DOI: 10.1201/9781003437413-2

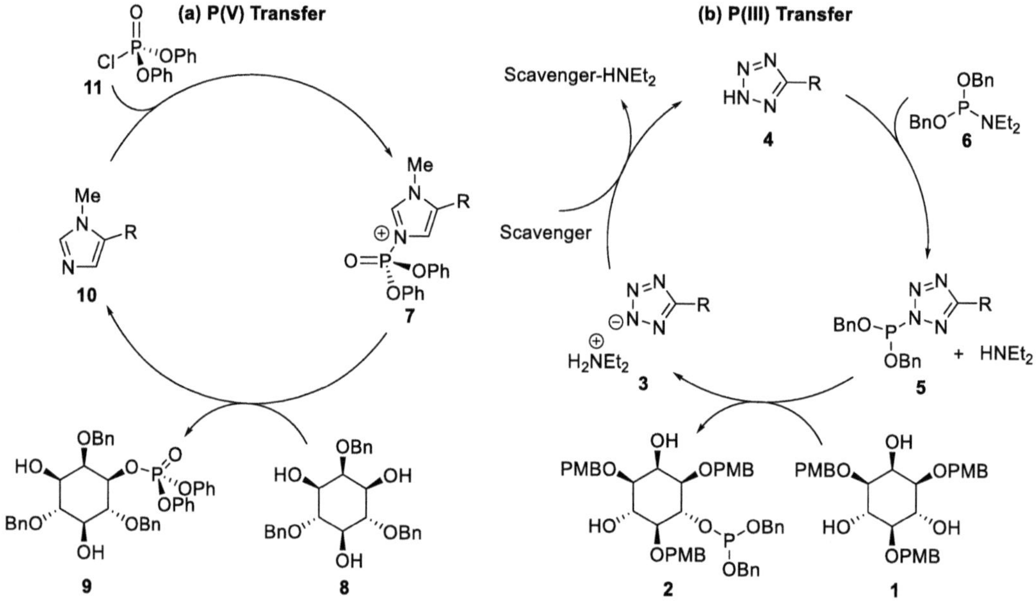

SCHEME 2.1 Cycle of P(V) and P(III) introduction depicting reagent activation and transfer.

the formation of diethylamine from phosphoramidite. This problem was mitigated by the introduction of 10 Å molecular sieves as a diethylamine scavenger.

Based on these methods, Miller et al. performed the desymmetrization of 1,3,5-tri-*O*-PMB-*myo*-inositol derivative (**1**) using peptide catalyst (**12**) and yielding nearly 70% phosphate (**13**). Subsequently, the minor isomer of (**13**) was selectively converted to bis(phosphate) under kinetic resolution under the same conditions giving highly enantiomerically enriched (**13**) (96% enantiomeric excess, 52% yield from (**1**). Final D-*myo*-inositol-6-phosphate (**14**) was obtained from (**13**) in nearly quantitative yield after deprotection using sodium in liquid ammonia (Scheme 2.2). The utility of this strategy is further exhibited as the key step in the synthesis of PI3P hapten (Section 2.3.1).

Another desymmetrization protocol was described by Jessen et al. [12] in 2013. They introduced C_2-symmetric phosphoramidite (**14**) serving as a phosphitylating reagent and chiral auxiliary for resolution. Its advantage is easy accessibility (from mandelic acid), easy deprotection (base-labile β-cyanoethyl protecting groups), and C_2 symmetry, which avoids chirality at phosphorus. Phosphoramidite (**14**) was used for desymmetrization of inositol units (**16**, **19**) bearing free hydroxyl groups at positions 1,3 and 4,6, resulting in high yields of a separable mixture of diastereomers in ratios 1:1, and 1:0.8, respectively (Scheme 2.3).

Additionally, the same group demonstrated that rare-earth metal catalysis, along with perfluorophenol (PFP) instead of 1*H*-tetrazole as a phosphoramidite activator, can alter the diastereomeric ratio to favor a specific diastereomer (Scheme 2.3) [13]. In both cases, the stereoisomers were separable by flash chromatography and/or crystallization with >99:1 diastereometric ratio. These reactions prove good step economy because desymmetrization and phosphitylation are carried out in one ste and subsequent oxidization to the phosphate can be done *in situ*.

Motivated by Jessen´s works, in 2016, Castagner et al. developed an alternative C2-symmetric diphenyl seven-membered ring phosphoramidite reagent (**25**) (Scheme 2.4), that was employed in chiral resolution of various inositol derivatives by bisphosphorylation of (**26**), (**29**), and (**30**) [14].

In the same year, Fukase and Fujimoto developed regioselective phosphorylation of *myo*-inositol derivative (**31**) with BINOL-derived phosphoramidites (**32–34**) as a desymmetrizing reagents (Scheme 2.5) [15]. The utilization of the known BINOL-derived phosphoramidite (**32**) [16],

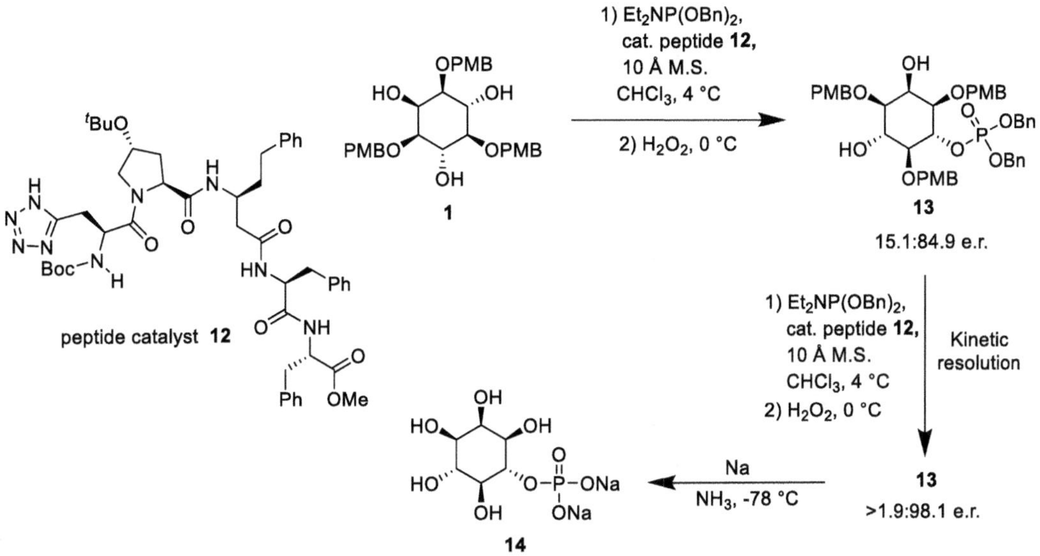

SCHEME 2.2 Synthesis of the target inositol phosphate via key peptide-catalyzed desymmetrization involving kinetic resolution.

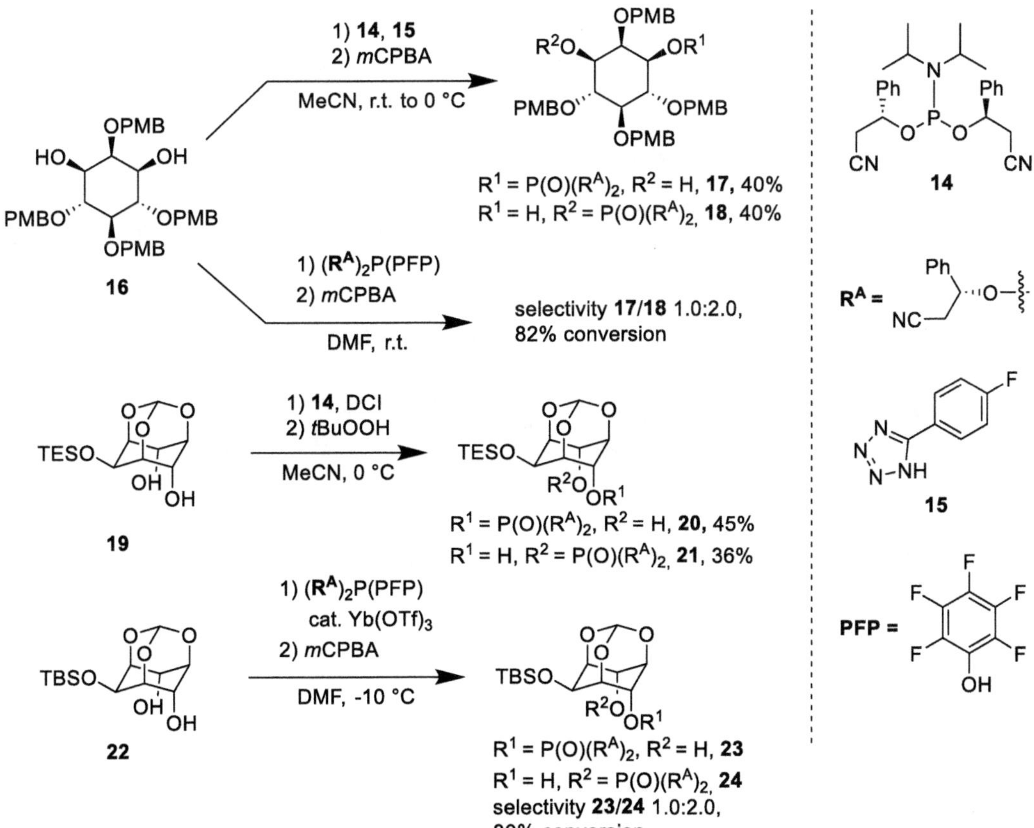

SCHEME 2.3 Chiral resolution with C2-symmetric phosphoramidite, diastereoselectivity enhanced with PFP activator and rare-earth metal catalysis.

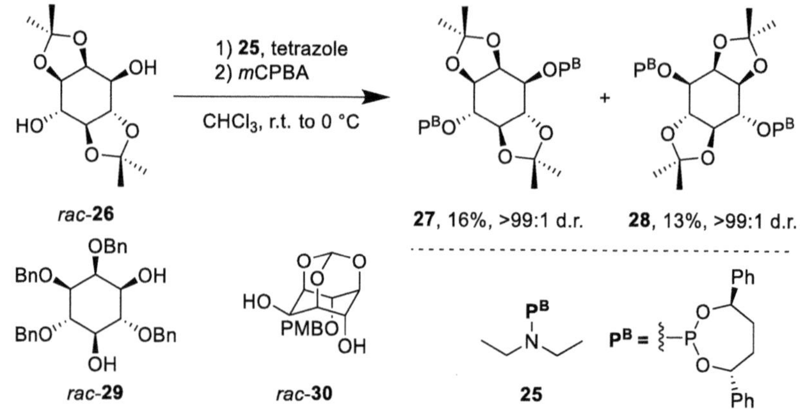

SCHEME 2.4 Chiral resolution of inositol derivatives with alternative C2-symmetric phosphoramidite. Rac-29 and rac-30 resolved with the same conditions.

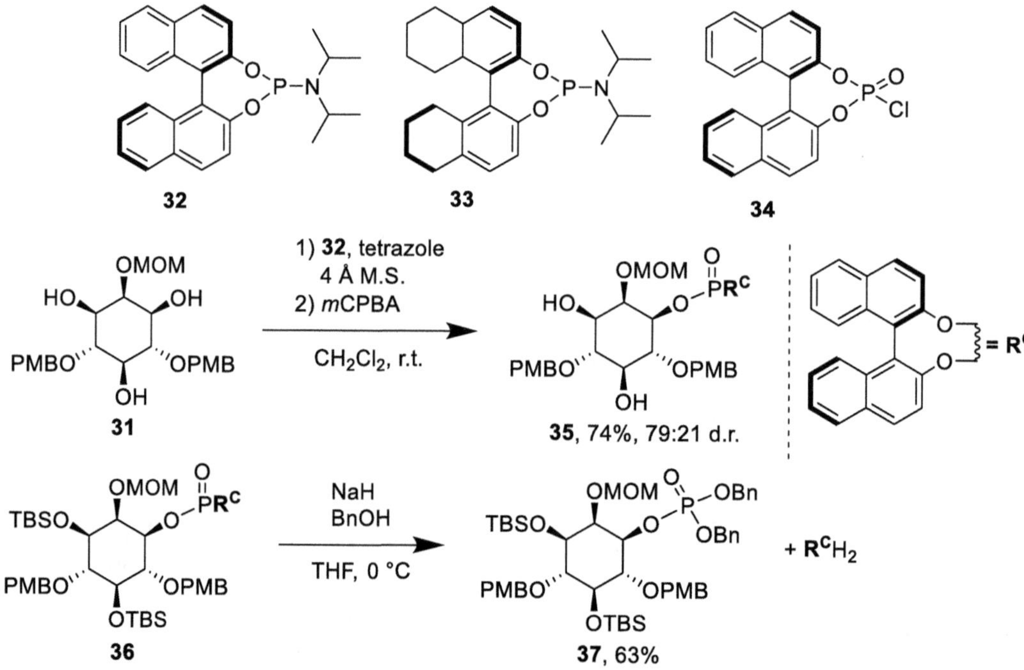

SCHEME 2.5 Desymmetrization with BINOL-derived phosphoramidite; regeneration of the chiral BINOL auxiliary in subsequent synthetic steps.

in the presence of tetrazole as an activator, resulted in the formation of inositol derivative (**35**) in high yield with a diastereomeric ratio up to 4:1. The obtained diastereomeric mixture was separable by standard chromatography, and the chiral auxiliary, BINOL, could be easily recycled. Based on the developed method, the synthesis of protozoan lysophosphatidylinositol and glycosylinositol-phospholipid from *Entamoeba histolytica* was synthesized (Section 2.3.3) [17].

2.2.2 Chiral Resolution and Desymmetrization

Several inositol phosphates and all phosphatidyl inositols possess chirality, presenting challenges in obtaining them in optically pure forms during preparation. As we showed in Section 2.2.1, various phosphorylation methods to desymmetrize inositol derivatives have been developed. Synthesis from optically active starting materials appears plausible, but it necessitates several protective steps, as detailed in Section 2.2.3. Another approach toward enantiopure inositol derivatives is based on chiral resolution/desymmetrization methods with non-phosphorous reagents, such as acyl halides and anhydrides.

Traditionally, ester and acetal groups were used to protect diol moieties in *myo*-inositol. Unsurprisingly, the introduction of chiral L-menthoxyacetyl group and D- and L-camphor acetals to perform optical resolution of *myo*-inositol to prepare naturally occurring phosphorylated phosphatidylinositols [18] Another synthetic strategy involved the formation of *myo*-inositol 1,3,5-orthoester derivative, allowing selective functionalization of free hydroxyl at position 4 and selective reduction of the orthoester group by DIBAL-H liberating the hydroxyl group at position 5 [19].

In 2010, Hung et al. developed a highly regioselective acylation of *myo*-inositol 1,3,5-orthoformate (**38**) using (1*S*)-camphanic anhydride (**39**) as the desymmetrization agent. The reaction was effectively catalyzed with $Yb(OTf)_3$, providing 6-*O*-acylated product (**40**), and 4-*O*-acylated product (**41**) at 78% yield at a ratio of 2.3:1. The pure compound (**40**) was obtained after crystallization from ethanol [20]. Later, the same group reported an extension of the developed methodology using various chiral acylation agents under rare-earth metal catalysis, enabling direct access to a wide range of chiral D-myo-inositol derivatives [21]. The proposed catalytic cycle is depicted in Scheme 2.6. First, the $M(OTf)_3$ catalyst (**42**) coordinates to the axial hydroxyl groups while simultaneously activating the acid anhydride (**45**). Considering chiral acid anhydride, this molecular arrangement allows discrimination of the 4 and 6 free hydroxyl groups. The best result in terms of yield (62%) and product ratio was observed with the acid anhydride (**39**).

Of note, in 2011, Hung et al. described another approach, regioselective desymmetrization of 2,4,5,6-tetrabenzylated-*myo*-inositol (**49**) with (1*S*)-ketopinyl chloride (**50**), giving the desired stereoisomer of acyl derivative (**51**) at ratio 3:1. The authors successfully applied the developed protocol to synthesize mycothiol (Scheme 2.7) [22].

A new, highly efficient method of obtaining enantiopure inositol derivatives was described by J. Paradies et al. in 2013 [23]. They used readily available phosphinit catalyst (**58**) that had been previously applied to desymmetrize *meso* 1,2-diols with a benzoic acid chloride [24]. By the careful optimization of reaction conditions, Paradies et al. were able to obtain orthogonally protected inositol (**55**) in high yield and enantiopurity by desymmetrization of the intermediate (**52**). The reliability of the method was further demonstrated with desymmetrization of 4,6-di-*O*-allyl and 4,6-di-*O*-(4-methoxybenzyl) substituted *myo*-inositol derivatives (**53, 54**) (Scheme 2.8).

2.2.3 Inositol Derivatives from Chiral Precursors (De Novo Synthesis)

An alternate strategy for preparing inositol derivatives is using other chiral materials that can be rearranged into an inositol scaffold. The early examples of the synthesis of inositol derivatives involved the use of chiral precursors, such as dehydroshikimic acid [25] and quebrachitol [26], but the approach developed by Prestwich et al. and starting from D-glucose became one of the most popular [27–30]. W Key intermediates in the latter approach are inososes, inositol derivatives with an oxidized hydroxyl group that serve as adaptable precursors for obtaining various inositol derivatives. Their reactive carbonyl is amenable to stereo divergent reduction by proper choice of reducing agent. An elegant way to synthesize inososes from carbohydrates is Ferrier carbocyclization [31], a robust metal-mediated molecular rearrangement of enol ether pyrans to cyclohexanones (Scheme 2.9).

In 2020, Swarts et al. [32] further extended the robustness of Ferrier carbocyclization in the synthesis of azidoinositol derivatives. The syntheses relied on the readily accessible azido methyl

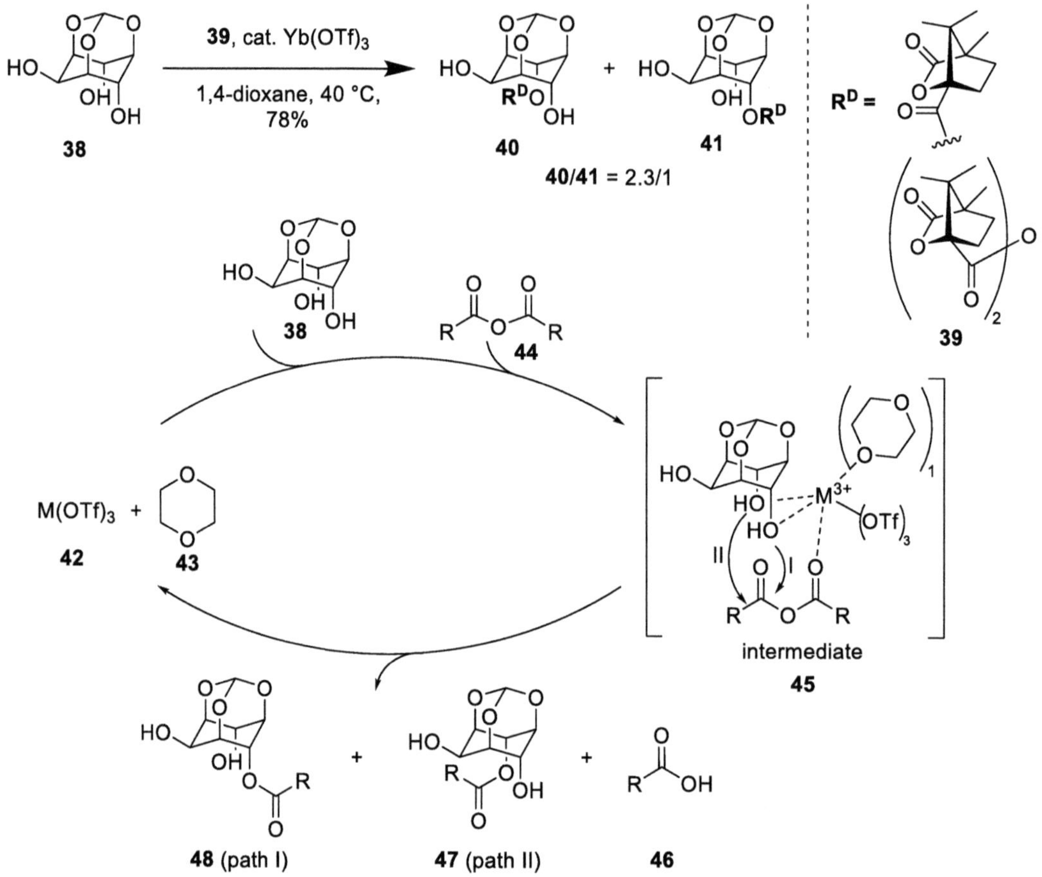

SCHEME 2.6 Selective acylation of inositol orthoformate and proposed catalytic cycle involving rare-earth metal catalysis.

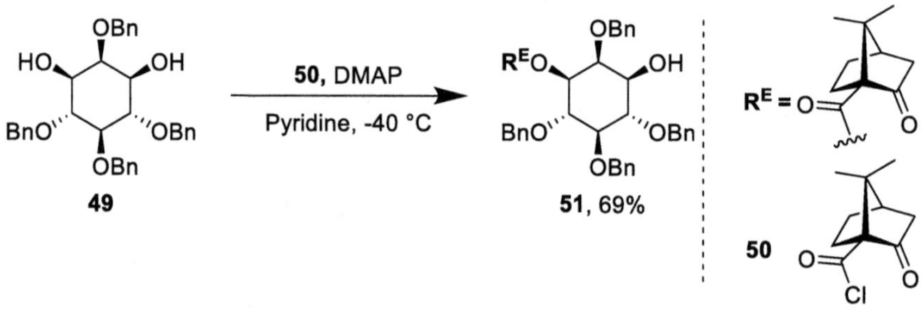

SCHEME 2.7 Utilization of (1S)-ketopinyl chloride in the key desymmetrization reaction.

glucosides **(63)**, **(70)** as starting materials. It was noteworthy that the inherent enantiopurity of the synthesized intermediates was achieved without the requirement of desymmetrization or chiral resolution. Synthesis of 3-azido-inositol(3-InoAz) **(69)** was achieved in nine steps (Scheme 2.10). Initially, starting 2-azido-2-deoxy-D-glucopyranoside **(63)** was selectively protected with *para*-methoxytrityl (PMTrt) group at the primary alcohol.

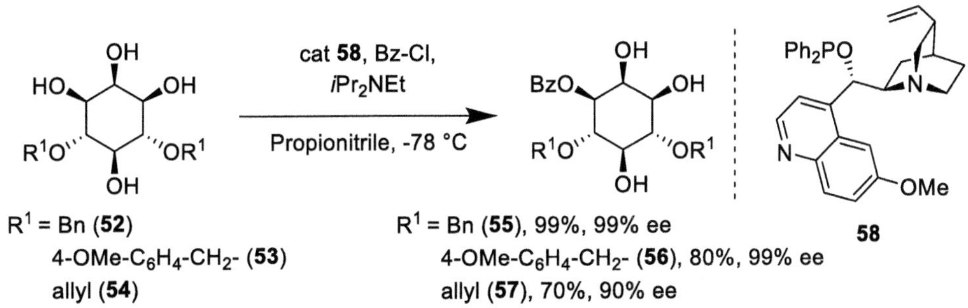

SCHEME 2.8 Efficient desymmetrization of variously substituted inositol derivatives with phosphinit catalyst.

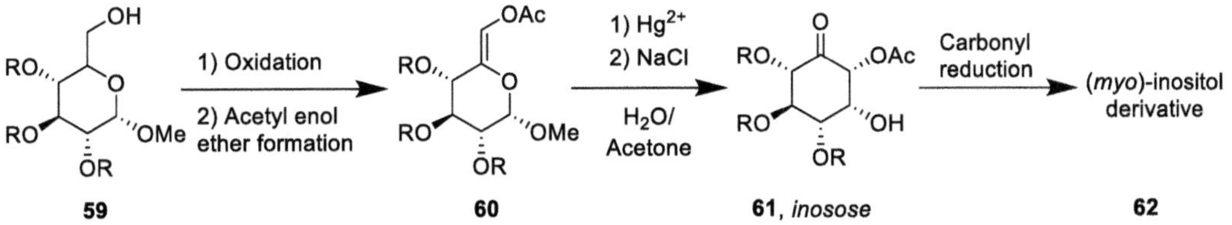

SCHEME 2.9 Ferrier carbocyclization leading to inositol derivatives; derivatives of D-glucopyranose lead to myo-inositols.

After the protection of secondary free hydroxyl groups, the PMTrt group was selectively removed, giving primary alcohol (**65**). Subsequent Swern oxidation followed by treatment with base and acetic anhydride formed acetyl enol ether (**66**). The critical point of synthesis, $Hg(OTf)_2$-catalyzed Ferrier carbocyclization, provided the desired *myo*-inosose stereoisomer (**67**) at 54% yield. Stereoselective reduction of the carbonyl moiety with sodium triacetoxyborohydride and further deprotections ultimately gave the desired enantiopure 3-InoAz (**69**). By applying the same reaction sequence together with an identical protective group on an anomeric mixture of methyl 3-azido-3-deoxy-D-glucopyranoside (**70**), (4-InoAz) (**71**) was successfully synthesized.

Unfortunately, 5-InoAz was not prepared from the corresponding azido D-glucoside (**72**) based on the developed method. The formation of acetyl enol ether [33] from the corresponding aldehyde turned out to be a problem, giving enal (**75**) as the major product (Scheme 2.11).

Although the Ferrier carbocyclization of D-glucopyranose core typically forms *myo*-inosose, which upon stereoselective reduction gives *myo*-inositol derivative, A. Walczak et al. [33] introduced post-transformations leading to *scyllo*-inositol derivatives. *Scyllo*-inositols are found in the human brain and are implicated in various neurological disorders. Current research explores *scyllo*-inositol derivatives as drugs for targeting Alzheimer's and Parkinson's diseases [34, 35]. Thus, obtaining enantiopure *scyllo*-inositol derivatives would prove interesting not only from a synthetic point of view. Starting from methyl D-glucopyranose derivative (**76**) to obtain *myo*-inositol (**79**) via familiar route, Walczak was able to selectively protect the equatorial hydroxyl at position 5, leaving the axial hydroxyl at position 1 to participate in oxidization. Subsequent reduction with sodium borohydride favored the formation of *scyllo*-inositol derivative. Although with modest yields, the synthesis to obtain *scyllo*-inositol derivatives was straightforward (Scheme 2.12).

Another method for preparing inositol derivatives and involving carbohydrates is the benzoin-type cyclization of dialdoses. Thiazolium and triazolium-type NHC catalysts can catalyze the benzoin condensation, while the use of chiral NHC catalysts offers the potential for stereoselective

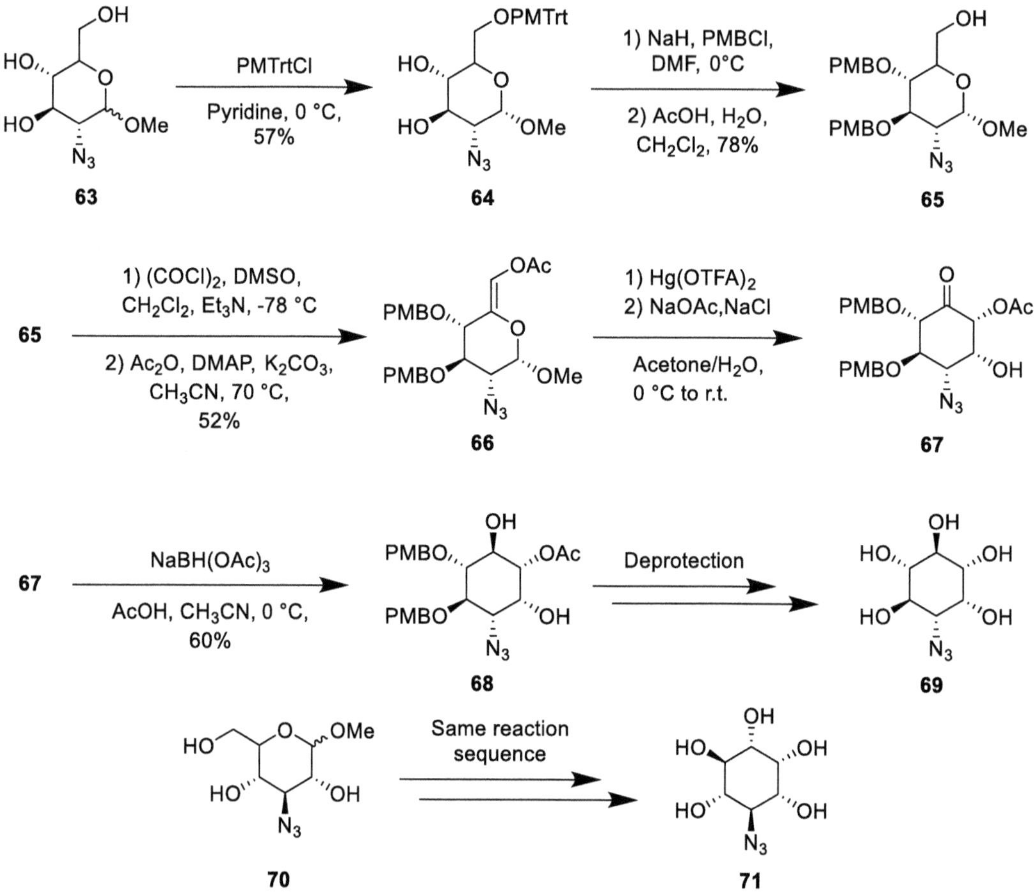

SCHEME 2.10 Synthesis of 3-InoAz and 4-InoAz with identical reaction sequences via corresponding azido-deoxy-D-glucopyranoside starting material. The Ferrier carbocyclization reaction provided desired inososes that under reduction with proper reagent provided myo-inositol derivative.

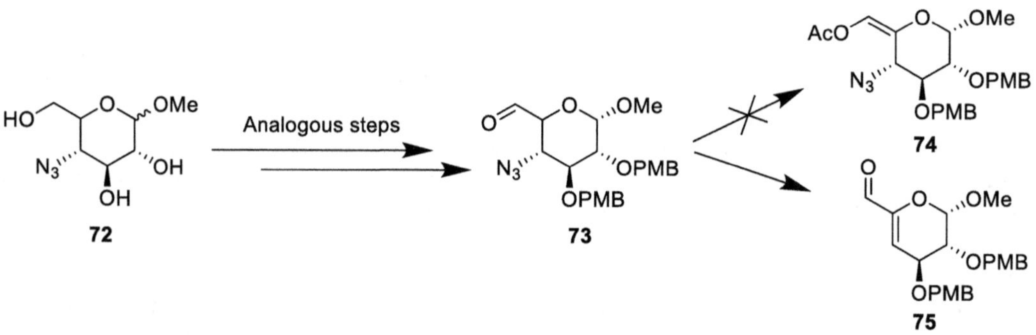

SCHEME 2.11 Utilizing analogous procedure as in Scheme 2.10, the preparation of 5-InoAz was unsuccessful; instead of acetyl enol ether (74), enal (75) was formed.

transformations [36]. One such example is benzoin condensation of 2,3,4,5-tetra-*O*-alkyl-dialdoses **(82), (85), (88)**, reported by Greatrex et al. (Scheme 2.13) [37]. The cyclization of **(82), (88)** via benzoin reaction was promoted by a triazolium carbene **(90)**, giving in the case of *manno*-configured dialdehyde predominantly single inosose stereoisomers **(83)** at up to 75% yield.

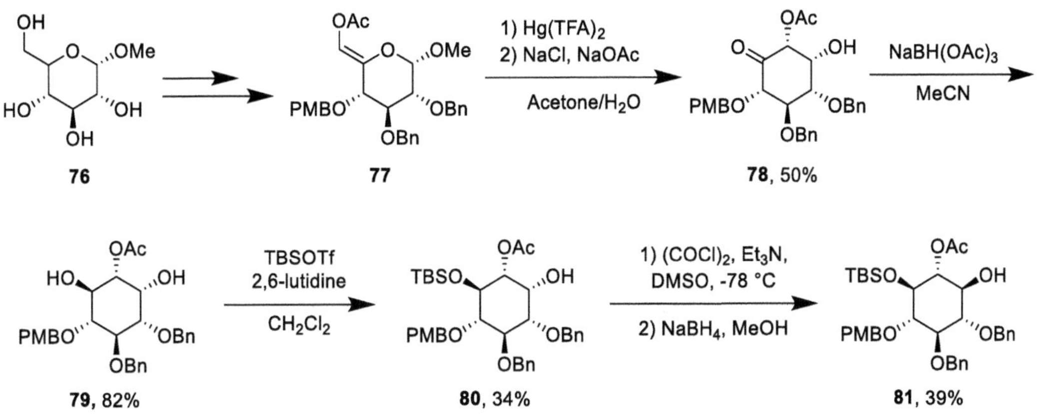

SCHEME 2.12 D-glucopyranose derivative to scyllo-inositol via Ferrier carbocyclization and post-transformations.

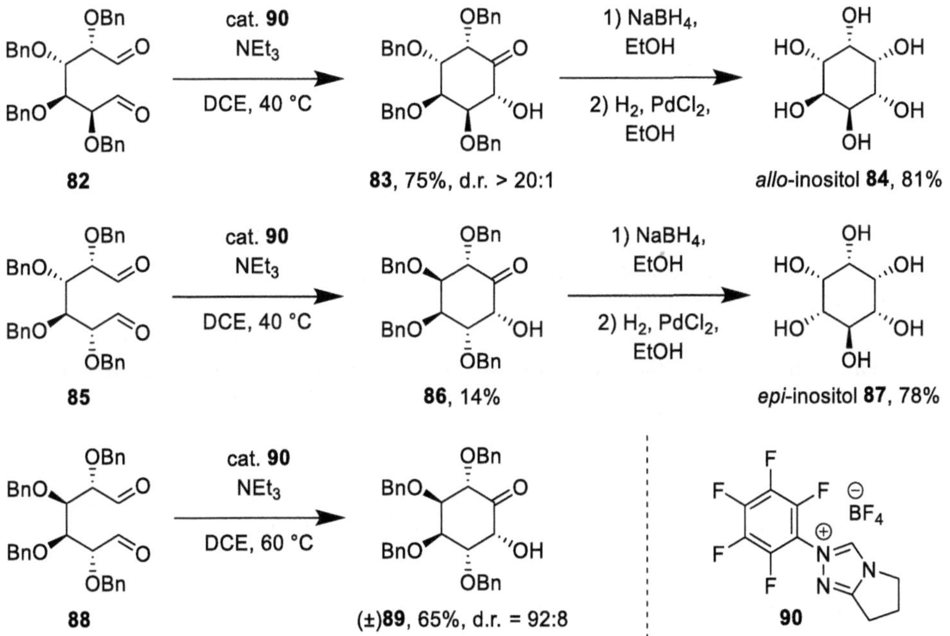

SCHEME 2.13 NHC-mediated benzoin-type cyclization of dialdoses leading to inososes.

When glucodialdehyde (**85**) was used, a mixture of three stereoisomers was observed. Subsequent stereoselective reduction using sodium borohydride followed by removal of protective groups gave *allo-* and *epi*-inositol in good yields.

Soon after, Yamada et al. disclosed a detailed study of NHC-catalyzed benzoin condensation of C^2-symmetrical dialdoses (**82**), (**91**), (**92**), and following stereoselective reduction of formed inososes (Scheme 2.14) [38]. The reaction proceeded smoothly with chiral NHC precatalyst **95** in toluene as a solvent. Interestingly, the use of less acidic triazolium salts resulted in a complex mixture. Similarly, the opposite enantiomer of salt (**95**) (*ent*-**95**) gave no product, suggesting a mismatch in the stereochemistry.

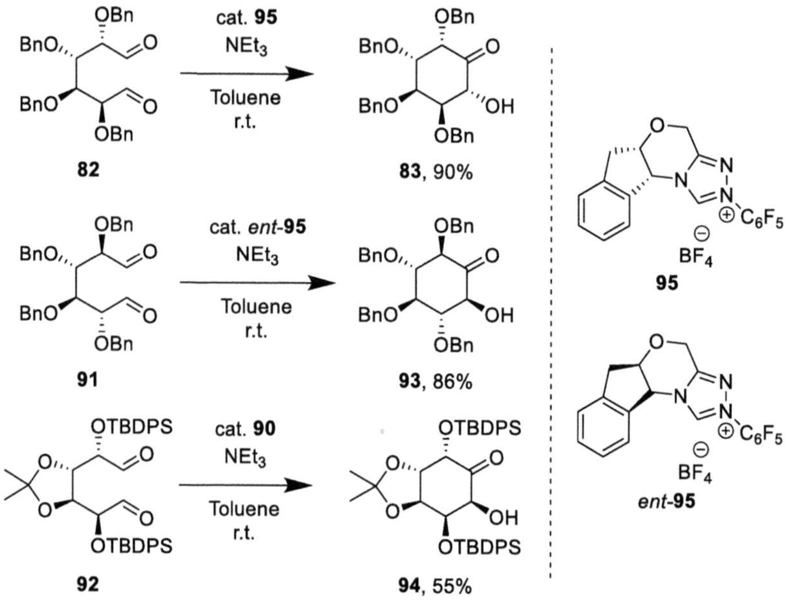

SCHEME 2.14 Extension of the NHC-mediated benzoin-type cyclization of dialdoses.

The catalyst (**95**) was also employed in reaction with L-iditol-derived dialdose (**91**) with inverse stereochemistry at α-position. Whereas the reaction catalyzed with (**95**) produced (**93**) with a moderate yield, the reaction with the opposite enantiomer of precatalyst *ent*-**10** afforded (**93**) as a single diastereomer in 86% yield. Yamada et al. explained the stereoselectivity by proposing the Breslow intermediate (Figure 2.1). The cyclization likely proceeds via chair conformation, where the benzyloxy group is equatorial because the axial position would increase allylic strain. The possibility of hydrogen bonding dictates the orientation of carbonyl, and thus, an attack at the *si*-face would occur. The *Z*-geometry in the transition state could further explain the stereochemistry and catalyst match-mismatching.

Additionally, the authors showed that acetonide protection at the 3,4-positions inverts the selectivity in the cyclization of *manno*-derived dialdose. The resulting inososes can be readily transformed into amino-, deoxy-, *O*-methyl, and *C*-methyl-inositol derivatives.

2.3 SYNTHESIS OF BIOLOGICALLY ACTIVE INOSITOL DERIVATIVES

The search for efficient methods for preparing enantiomerically pure inositol derivatives has led to the preparation of several new inositol derivatives. These new substances were constructed to study biological processes and primarily to understand the role of inositol derivatives at the cellular level. Next, we will focus on preparing and using selected examples of inositol derivatives prepared in the last decade.

2.3.1 Derivatives Bridged with Linker Moiety

Phosphoinositides, whose importance precedes their abundance, play pivotal roles within cellular biochemistry, ranging from participation in diverse membrane trafficking and signaling to yet unknown biochemical events. Phosphoinositides are a lipid family made of a phosphate group, two fatty acid chains, and one *myo*-inositol molecule. These molecules are synthesized in cells through the phosphorylation of phosphatidylinositol (PI) at positions 3, 4, and 5 (Figure 2.2), leading to a

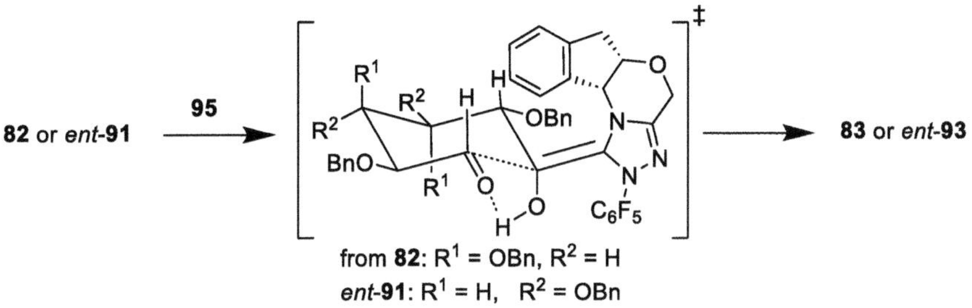

FIGURE 2.1 Proposed Breslow intermediate elucidating the observed stereochemistry.

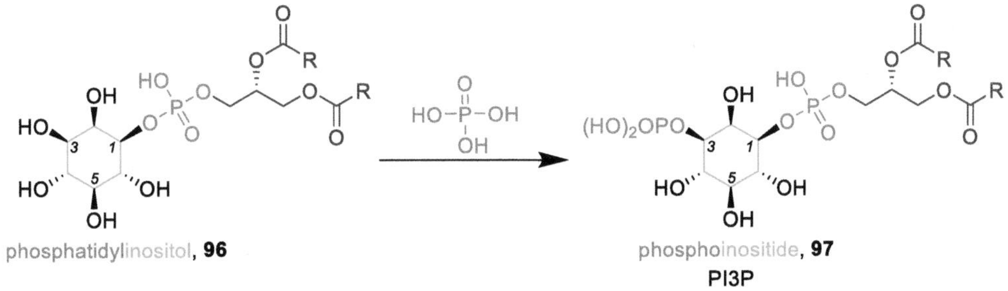

FIGURE 2.2 Phosphatidylinositols are lipids that consist of a bis-acylated glycerol backbone (blue) linked with inositol ring (green) through phosphate ester (red). Phosphorylated forms of phosphatidylinositols are called phosphoinositides.

wide variety of phosphoinositide species with unique functionalities. Studies have identified distinct localization patterns: phosphatidylinositol-3-phosphate (PI3P) primarily resides in early endosomes, PI4P at Golgi membranes, PI3,5P$_2$ in late endosomes, and PI4,5P$_2$ at the plasma membrane [39]. The phosphoinositide research has been growing since its discovery. However, the tools to probe spatial and temporal characteristics of membrane components are scarce. Utilizing antibodies presents an elegant probing method since specific antibodies can selectively bind to distinct types of endogenous phosphoinositides. Developing such antibodies has proven to be challenging, as lipids typically lack immunogenicity. Still, several methods to augment immunogenicity have been developed, for example, phospholipids carried by liposomes [40].

In 2014, a new strategy for accessing phosphoinositide antibodies was demonstrated by Miller et al.: synthesizing PI3P immunogen (**98**) capable of producing selective antibodies for PI3P in immunized rabbits (Figure 2.3) [41]. The key steps of the convergent synthesis of PI3P immunogen (**98**) were catalytic desymmetrization of inositol derivative (**8**) and the late-stage incorporation of a cysteine residue through native chemical ligation.

The initial synthetic steps consisted of desymmetrization of readily available inositol derivative (**8**) via peptide-catalyzed asymmetric phosphorylation discussed in detail previously (Section 2.2.1) [4–10] followed by the conversion of *myo*-inositol-derived diphenylphosphate (**99**) to its corresponding dibenzyl-protected derivative (**100**), necessary to enable later protecting group manipulations (Scheme 2.15).

Subsequently, intermediate (**100**) was treated with phosphoramidite (**101**), followed by *in-situ* oxidation and hydrogenolysis, providing key intermediate (**102**) in good yield. Next, coupling cysteine with a fully deprotected phosphoinositide free acid (**103**) activated with EDCI, HBTU, and CDI

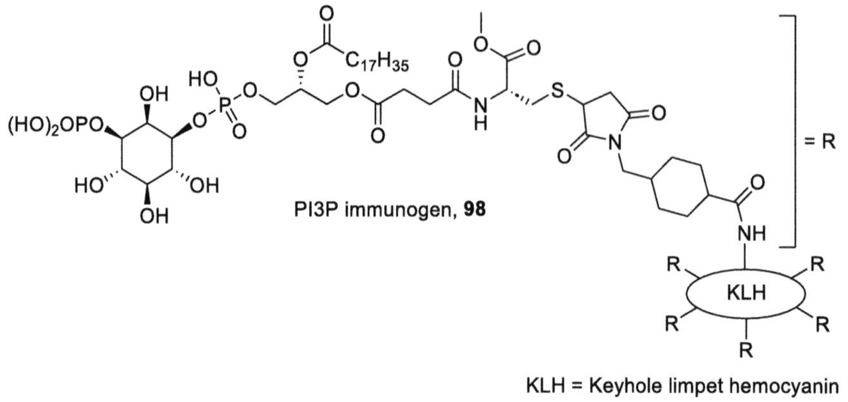

FIGURE 2.3 Target PI3P immunogen.

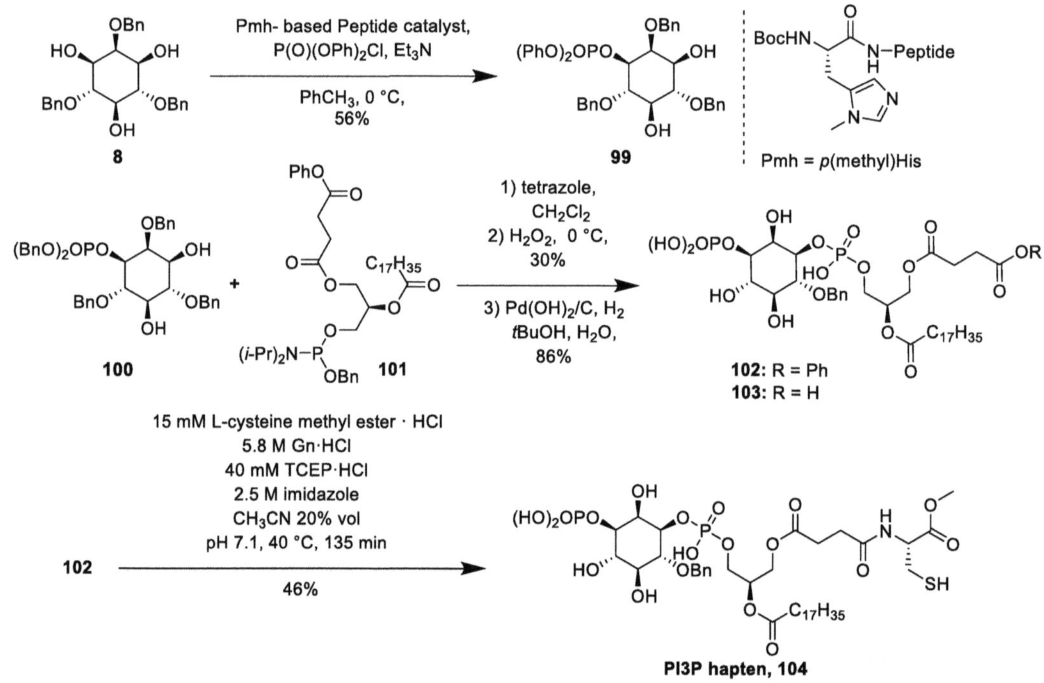

SCHEME 2.15 Synthesis of PI3P hapten (104) encompassing key desymmetrization of inositol intermediate (8) with peptide catalyst.

was not successful due to a lack of reactivity of the free acid moiety. Interestingly, modified native chemical ligation [42] successfully overcame the cysteine coupling challenge, producing the desired PI3P hapten (**104**) in moderate yield by treatment of (**102**) with cysteine methyl ester, guanidine (Gn), triscarboxyethyl phosphine (TCEP), and imidazole in a buffered solution of water and acetonitrile (4:1). The final step was the generation of the hapten–carrier protein complex, immunogene (**98**), via conjugation of hapten (**104**) to the commercially available maleimide-activated keyhole limpet hemocyanin.

Another example of a biologically active inositol derivative containing *myo*-inositol 1,2,3,4,5,6-hexakisphosphate (IP_6) was disclosed in 2015 by Hanakahi et al. [43] IP_6 is the most abundant inositol phosphate in eukaryotic cells, and it has been associated with key nuclear functions ranging from RNA editing [44] to transport of mRNA from the nucleus to cytoplasm [45] to modulation of ATP-dependent chromatin remodeling [46]. To further prove the significance of the mRNA export, Wente showed [47] that depleting yeast of IP_5-kinase (the enzyme that converts *myo*-inositol 1,3,4,5,6-pentakisphosphate, IP_5, into IP_6) resulted in defects in cell morphology, cell separation, and endocytosis. More drastic effects of depletion were observed in Zebrafish, resulting in the randomization of the placement of visceral organs [48].

Additionally, IP_6 has been identified to participate in DNA double-strand break repair *via* non-homogenous end joining (NHEJ) [49]. To further investigate the role of IP_6, Hanakahi et al. elaborated on the synthesis of biotin-labeled IP_6 ((**105**), Scheme 2.16), a versatile IP_6 analogue that can be used as a chemical probe [43]. The authors observed that (**105**) functions like IP_6 in NHEJ in vitro, and it can also be used to isolate and identify proteins with affinity for IP_6. From the synthetic point of view, the derivatization of position 2 of the *myo*-inositol unit was chosen because no plane of symmetry would be broken; thus, no further chiral resolution or desymmetrization reactions were needed.

First, the phosphitylating agent (**107**) was prepared from commercially available *N*-Boc-6-aminohexanol (**106**). Then, the penta-*O*-benzyl *myo*-inositol (**108**) (synthesized in four steps from *myo*-inositol) was phosphorylated by the reaction with (**107**) in the presence of tetrazole and subsequently oxidized to phosphate (**109**). Then, benzyl groups were removed by hydrogenolysis, and the resulting pentol (**110**) was phosphorylated by the widely known reaction with *O*-xylylene-*N,N*-diisopropylphosphoramidite followed by oxidation with *m*CPBA to give derivative (**111**). Next, the Boc-protecting group at the linker was cleaved under acidic conditions, and the free amino group was reacted with biotin *N*-hydroxysuccinimide under basic conditions to yield (**113**). The final two-step deprotection involving hydrogenolysis followed by demethylation by aqueous ammonium hydroxide and yielded the final *myo*-inositol probe (**105**).

In the same period, Holmes et al. reported the synthesis of another inositol-derived probe for biological studies, a tethered *myo*-inositol 1,3,4,5,6-pentakisphosphate derivative (**115**) (Scheme 2.17) [50]. *Myo*-inositol-1,3,4,5,6-pentakisphosphate (IP_5) is a stable constituent of mammalian cells, and it plays a major role in the regulation of transcription in response to environmental and nutritional cues. Furthermore, IP_5 has been shown to mediate Wnt/β-catenin signaling, which is a crucial molecular pathway in biological systems that regulates various cellular processes, including embryonic development, tissue regeneration, and cell proliferation [51].

To further understand the biological role and physiological function of IP_5 in cells, Holmes et al. designed and synthesized an IP_5 derivative with a linker terminated with amine group (**115**), allowing biological investigation via immobilization onto surfaces or beads [50]. After testing various alternatives, probe (**115**) was successfully immobilized using magnetic Dynabeads, superparamagnetic microspheres coated with a thin layer of hydrophilic polymer commonly utilized in various biological and biomedical applications. The preparation of an amino-modified IP_5 analogue (**115**) started with regioselective protection of both axial hydroxy groups of orthoester (**38**), readily available from *myo*-inositol [52], and followed by benzylation to give intermediate (**117**). Subsequent partial reduction with DIBAL-H, allylation, and treatment with HCl to remove acid-labile acetal and PMB groups resulted in the formation of tetraol (**119**). Phosphorylation of (**119**) with phosphoramidite (**126**) using standard phosphoramidite chemistry conditions followed by oxidation gave the expected intermediate (**120**).

It should be noted that phosphorylation with bulkier dibenzyl *N,N*-diisopropylphosphoramidite (**125**) resulted in an unreactive system in the later stage of the synthesis, in coupling with (**12**). Subsequent deallylation using $PdCl_2$ in methanol gave alcohol (**121**) for the introduction of phosphate ester with the linker. When choosing the most suitable linker for inositol-derived biological probes, several factors like length, polarity, and stability were considered. Holmes et al. utilized a moderately sized ethylene glycol-derived structure, modified with a terminal amine for future

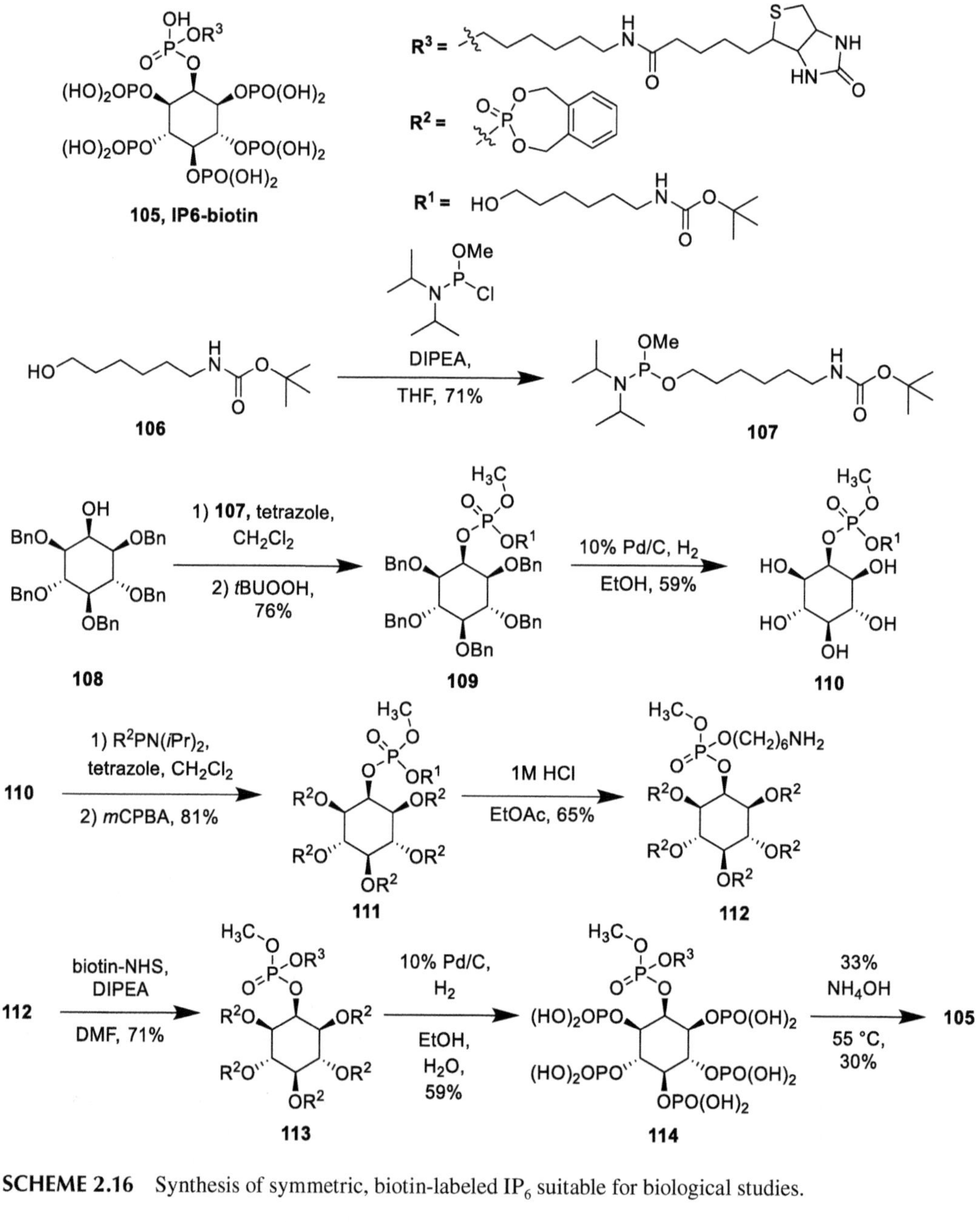

SCHEME 2.16 Synthesis of symmetric, biotin-labeled IP_6 suitable for biological studies.

conjugation, owing to its ability to maintain structural rigidity and enhanced water solubility. The corresponding phosphoramidite (**124**) was synthesized from commercially available 2-aminoethoxyethanol in two steps, protection of the free amine with benzyloxycarbonyl group and coupling with a diphosphoramidite. The fully protected IP_5 precursor (**122**) was obtained by the key reaction, coupling of the phosphoramidite (**124**) with alcohol (**121**). Finally, global deprotection *via* hydrogenolysis gave the desired IP_5 probe (**115**) with a linker group at the C-5 position bearing a free amine available for attachment to affinity beads and surfaces.

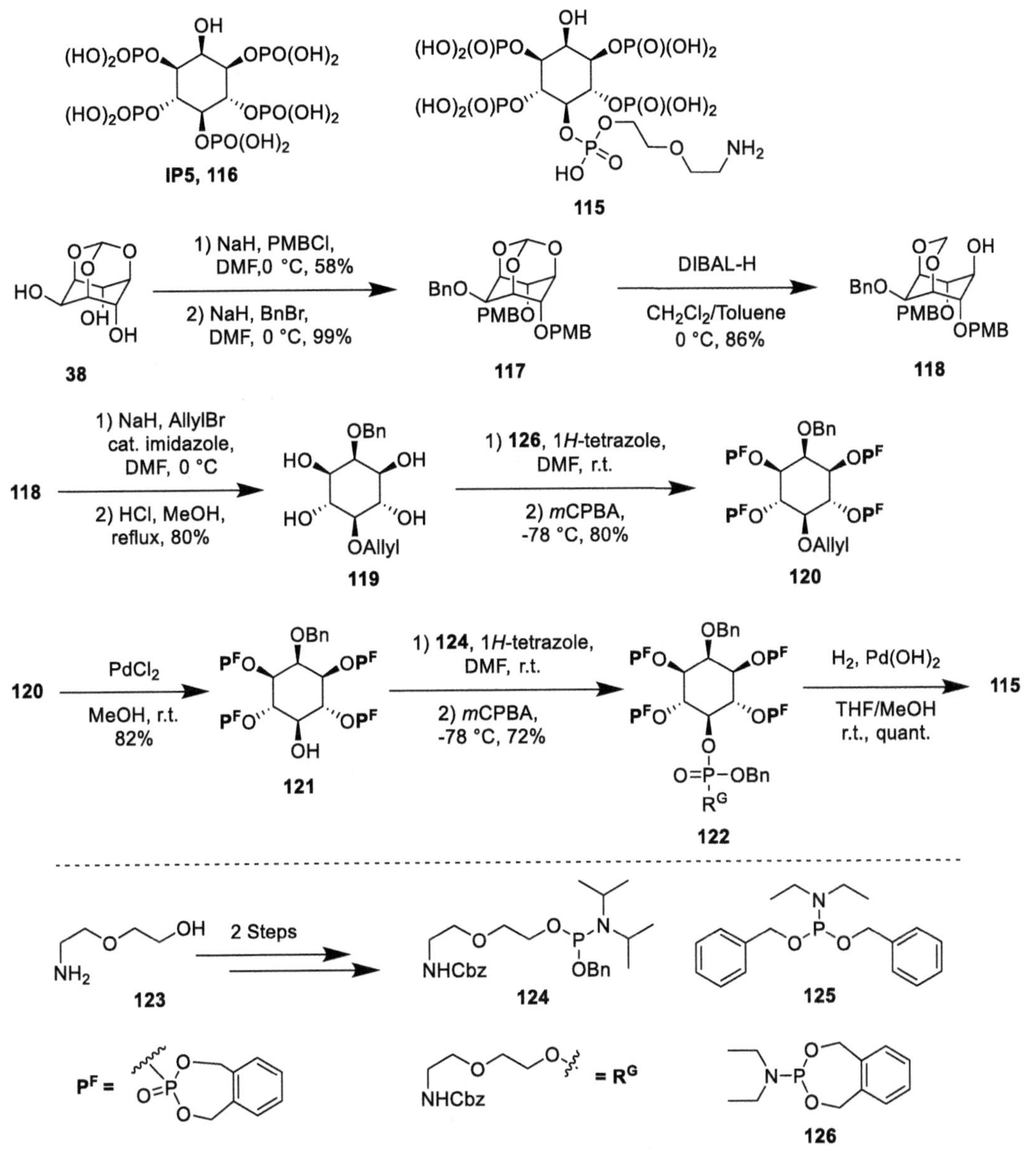

SCHEME 2.17 Synthesis of the symmetric IP5 probe with amine linker, suitable for immobilization onto Dynabeads.

2.3.2 Pyrophosphates

Inositol phosphates with diphosphate (PP) groups, in addition to monophosphates (P), are known as inositol pyrophosphates (PP-IPs). They are a member of the inositol phosphate signaling family found in all eukaryotic cells; however, due to diphosphate moiety, they are considerably energy rich. Their participation ranges from protein pyrophosphorylation and vesicle trafficking to hemostasis [53].

They are formed by kinase enzymes, namely PP-IP_5 kinases and IP_6 kinases. The inositol pyrophosphates are cleaved into phosphates via the action of diphosphoinositol polyphosphate phosphohydrolases reversibly [54]. This reversible interconversion can be thought of as an ADP/ATP analogue; therefore, a molecular probe allowing for tracking and distinguishing between most abundant inositol pyrophosphate in mammals, 5-diphosphoinositol pentakisphosphate (5-PP-IP_5), and inositol hexakisphosphate (IP_6) would be a valuable tool. A molecular probe capable of inducing luminescence upon binding to a specific ligand offers a reliable technique for quantifying target species in aqueous biological environments and living cells [55]. Recently, Potter and Butler et al. introduced a new synthetic approach to obtain 5-PP-IP_5 and subsequently discovered that the macrocyclic Eu(III)-based complex can act as a molecular probe for 5-PP-IP_5 [56].

The synthesis of 5-PP-IP5 began with selective benzylation of axial hydroxyl in position 2 of diol (**127**), directly available from *myo*-inositol in multi-gram quantities [57], allowing the free hydroxyl at position 5 to be phosphorylated by phosphoramidite (**134**) protected by methylsulfonylethyl-protecting groups (Scheme 2.18). Next, the diacetal-protecting groups in monophosphate (**129**) were removed using trifluoroacetic acid (TFA), and subsequently, the benzyl group attached to the 2-position was cleaved using the standard hydrogenation procedure, giving pentaol (**130**) in high yield.

Phosphorylation of (**130**) with an excess of dibenzyl phosphoramidite and 5-phenyl-1*H*-tetrazole activator afforded hexakisphosphate (**131**). The key stage of the synthesis was to install the pyrophosphate moiety. The first step of this stage was the deprotection of the methylsulfonylethyl group with DBU and the silylation of the exposed phosphate with *N,O*-bis(trimethylsilyl)trifluoroacetamide (BSTFA). The next steps included methanolysis using methanol and TFA and the formation of the mixed P(V)-P(III) anhydride accomplished with benzyl-protected phosphoramidite subsequently oxidized with *m*CPBA to obtain the protected pyrophosphate (**132**). The final benzyl deprotection to yield salt of 5-PP-IP_5 (**133**) was performed using Pearlman's catalyst and triethylammonium-bicarbonate buffer.

Following the successful synthesis of pyrophosphate **133**, Butler et al. identified a coordinately unsaturated cationic complex as a luminescent probe for selective binding to 5-PP-IP_5. The probe is based on the europium complex **135** and was originally developed for ADP/ATP; however, the interference is not anticipated to be an issue. Otherwise, it shows minimal interference from other molecules containing multiple phosphate and pyrophosphate.

Continuing with inositol pyrophosphates, as previously discussed, their synthesis presents a considerable challenge. Particularly, synthesizing IP_8, less studied than IP_7, has proven to be difficult, primarily due to the significant molecular crowding caused by phosphates and purification issues. In mammals, the structure of the natural IP_8 has been identified as the 1,5-$(PP)_2$-IP_4isomer, nevertheless, its enantioselective synthesis was not realized, only the preparation of unnatural and symmetric 2,5-$(PP)_2$-IP_4 isomer [58]. Thus, synthesizing enantiomeric1,5- and 3,5-$(PP)_2$-IP_4isomers would be useful for comparing their biological activities and specificity. Noteworthy, the general approach for the synthesis of non-symmetrical *myo*-inositol units involves using chiral resolution or desymmetrization along the synthetic path. With reference to previous work [12], Jessen et al. [2] developed a new strategy to synthesize 1,5- and 3,5-$(PP)_2$-IP_4 isomers based on desymmetrization by phospitylation using C^2-symmetricphosphoramidite (Scheme 2.19).

The synthesis commenced with *myo*-inositol, which in three steps was transformed into *meso*-symmetric inositol derivative (**139**) [3]. The axial hydroxyl in position 5 was phosphitylated with a bis-fluorenyl-protectedcphosphoramidite (**149**) and oxidized with *tert*-butyl hydroperoxide, and the benzylidene-protecting group was removed under acidic conditions giving diol (**140**). Next, phosphitylation of (**140**) with C^2-symmetricphosphoramidite (**14**) followed by oxidation yielded a separable mixture of the two diastereomeric inositol derivatives (**141**) and (**142**) in a ratio of 1:1. Further synthesis steps were the same for each diastereomer; PMB groups were cleaved with TFA and free hydroxyls were phosphorylated by the usual way with *o*-xylylene-derived phosphoramidite (**150**). After that, the inositol derivatives (**145**) and (**146**) were subjected to a reaction sequence of replacing the chiral auxiliary (originating from C_2-symmetric phosphoramidite) with TMS group (using BSTFA reagent). Subsequently, the TMS groups were cleaved with TFA. The monobasic phosphates were then subjected to bis-P-anhydride formation *via* phosphitylation with bis-benzyl P-amidite and subsequent oxidation. The final step of the synthesis was the removal of the benzyl protective groups with hydrogenation. The absolute configuration was determined using X-ray analysis of crystal complexes with the kinase domain of human diphosphoinositolpentakisphosphate kinase 2.

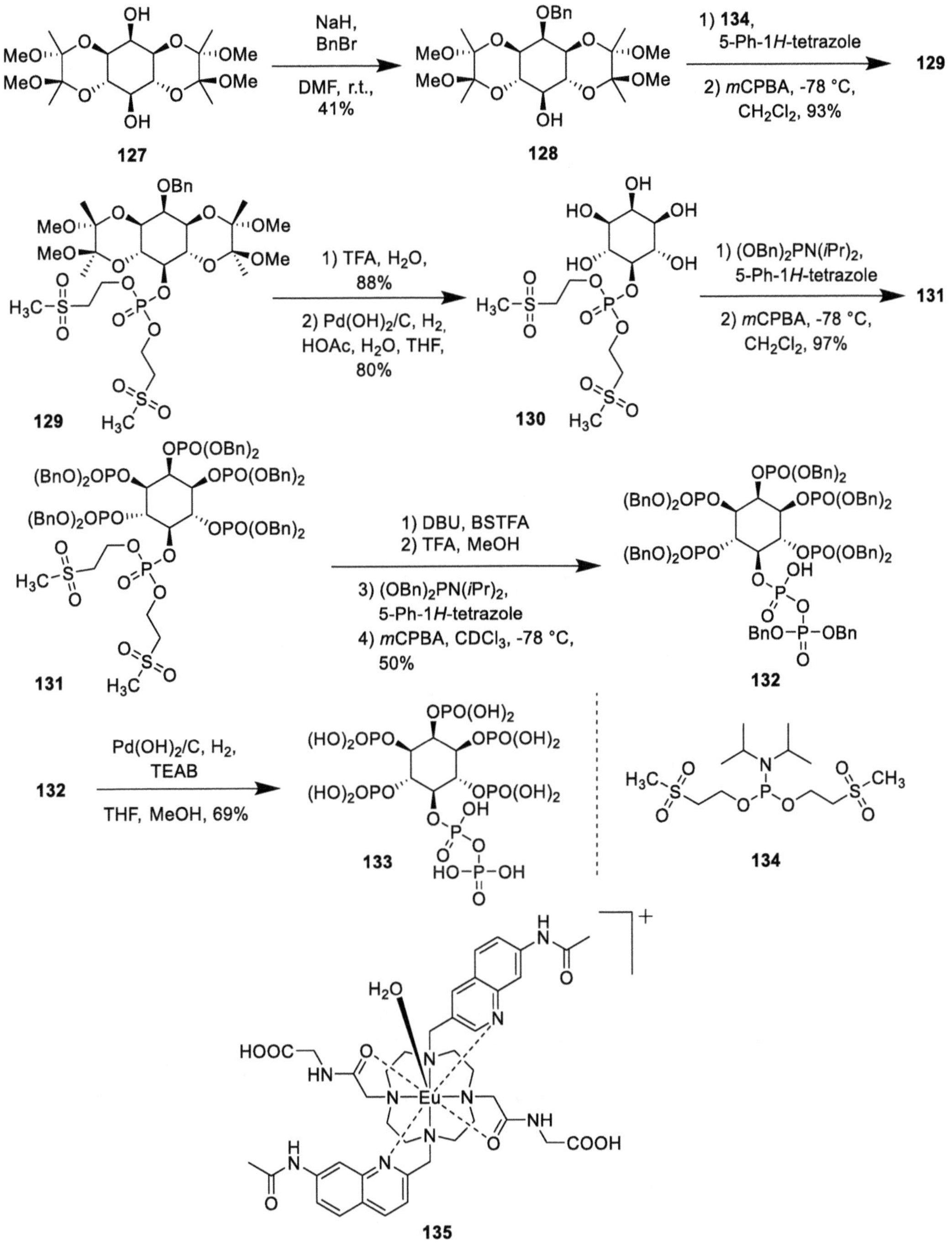

SCHEME 2.18 Synthesis of symmetric 5-PP-$IP_5$133starting from protected intermediate 127. The macrocyclic Eu(III)-based complex 135 can act as a molecular probe for 5-PP-IP_5.

As already mentioned, diphosphoinositol polyphosphates, such as 5-diphosphoinositol pentakisphosphate (5-PP-IP_5), are a family of second messengers with important roles in eukaryotic cells. Unsurprisingly, a logical extension of the research of Jessen's group was the design and step-economical synthesis of other PP-IPs, including photocaged 5-PP-IP_5 [61] and 2-PP-IP_5 (Figure 2.5) [62].

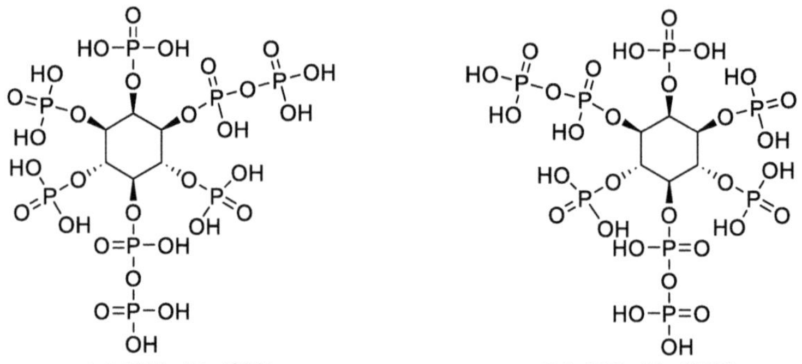

1,5-$(PP)_2$-IP_4 (**136**) 3,5-$(PP)_2$-IP_4 (**137**)

FIGURE 2.4 Natural IP_8 (**136**) and its enantiomer (**137**).

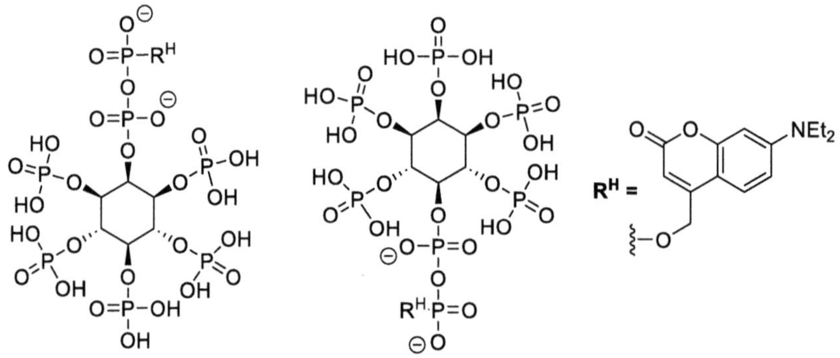

2-$(PP)_2$-IP_5 (**151**) 5-$(PP)_2$-IP_5 (**152**)

FIGURE 2.5 Structure of photocaged 2-PP-IP_5 (**151**) and 5-PP-IP_5 (**152**).

2.3.3 Phosphatidylated and Glycosylated Derivatives

Glycosylphosphatidylinositols (GPIs) are a class of complex molecules found predominantly on the surface of eukaryotic cells, functioning as protein anchors [63]. GPIs are composed of a hydrophobic phosphatidylinositol group bound with a carbohydrate linker via a glycosidic bond. The carbohydrate linker is further bound to the ethanolamine phosphate bridge to the C-terminus of the amino acid of a protein (Figure 2.6). The hydrophobic part of phosphatidylinositol (lipid tail) thus anchors the protein to the cell membrane.

GPIs are cleaved by specific phospholipases C (PLC). This process often results in the release of important second messengers, diacylglycerol (DAG) and inositol-1,4,5-triphosphate (IP_3), which then activate numerous cellular processes such as DNA synthesis. Synthesis of a non-hydrolyzable PLC-resistant GPI anchor would further aid the study of such biological processes [64]. Several methods have been employed to synthesize a range of nonhydrolyzable phosphatidylinositol (PI) substrates that resist hydrolysis by PLC. These approaches include modifying the reactive 2-OH of inositol [65], substituting the cleavable P–O bond with a C–P bond [66], and creating conformationally constrained analogues [67]. The first synthesis of PLC-resistant GPI analogue was published in 2014 by Vishwakarma et al. [68]. The synthesis encompassed three key stages: i) synthesize selectively protected pseudo-disaccharide consisting of D-*myo*-inositol and D-glucosamine unit, ii) synthesize optically pure phosphonic acid donors, and iii) couple the pseudo-disaccharide with phosphonic acid donors.

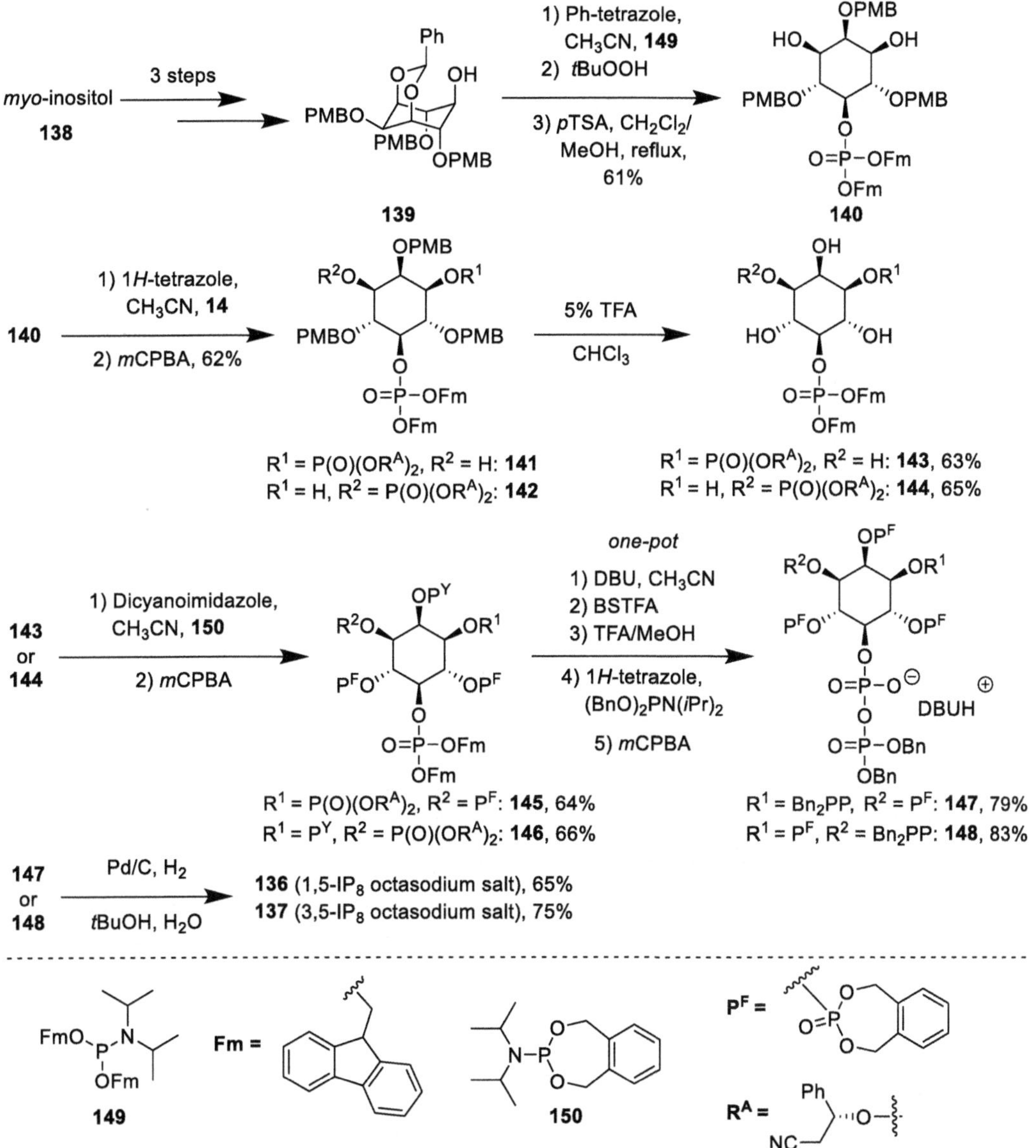

SCHEME 2.19 Synthesis of the natural pyrophosphate (138) and its enantiomer (139) with key desymmetrization utilizing $C2$-symmetric phosphoramidite.

The first stage starts with the synthesis of the selectively protected inositol unit (**154**) previously prepared by the same group [69], including chiral resolution with camphor dimethyl acetal and selective introduction of the PMB protective group by dibutyltin chemistry (Scheme 2.20). The 2-azido glycosyl donor (**155**) was synthesized from D-glucosamine through diazotransfer using triflicazide, followed by per-*O*-acetylation, selective deprotection of anomeric acetyl group using hydrazine acetate, and installation of the trichloroacetimidate group. After that, the inositol acceptor (**154**) was coupled with trichloroacetimidate (**155**) in the presence of trimethylsilyl triflate, yielding the pseudodisaccharide (**156**) as α-anomer. The acetyl groups on glycosyl moiety were exchanged with the benzyl groups by basic hydrolysis and subsequent benzylation. Finally, the PMB deprotection gave the pseudo disaccharide unit (**157**) ready for coupling with phosphonic acid donors.

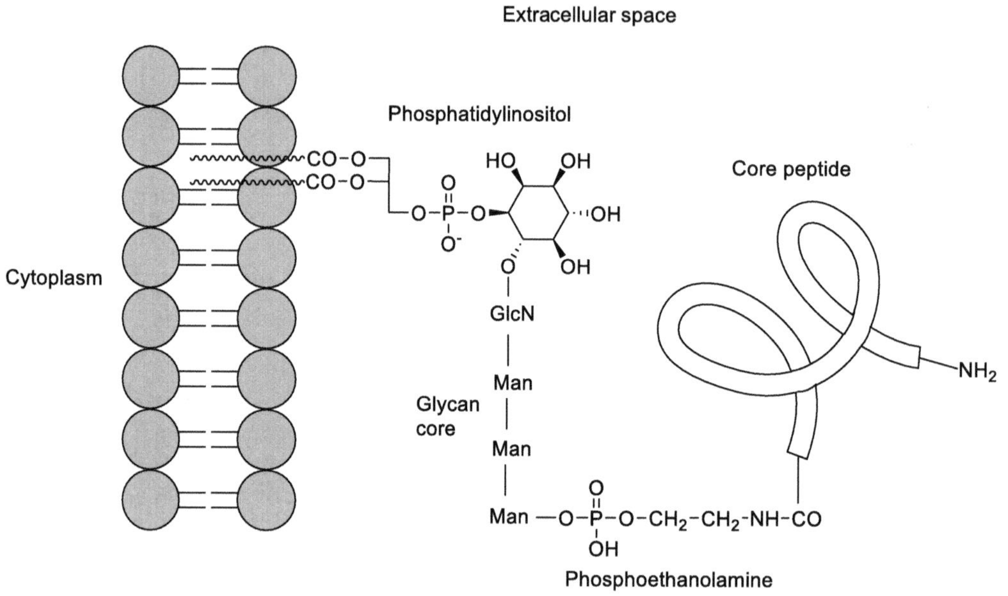

FIGURE 2.6 General structure of glycosylphosphatidylinositol.

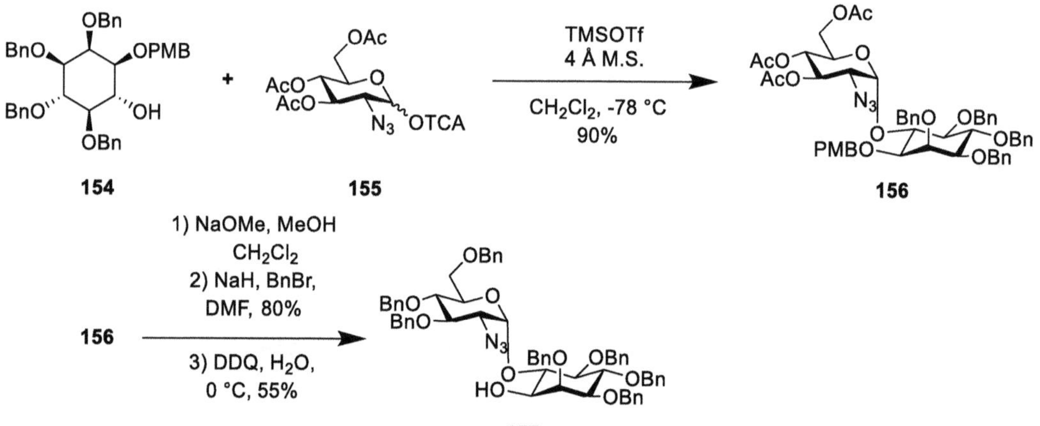

SCHEME 2.20 Synthesis of the pseudo disaccharide unit (157) suitable for coupling with phosphonic acid donors.

Moving to synthesis stage 2, the first steps in obtaining the desired phosphonic acid donor (**161**) were the formation of methyl ester of (*S*)-malic acid followed by reduction with $LiBH_4$ to obtain the (*S*)-erythritol (**159**) (Scheme 2.21). Then, regioselective protection of the vicinal diol with isopropylidene acetal group, conversion into iododerivative, and subsequent Michaelis-Arbuzov reaction using triethylphosphite yielded isopropylidene-protected phosphonate that was hydrolyzed into (**160**). Subsequent formation of palmitoyl ester and cleavage of the bis-ethyl phosphonate ester with TMSBr yielded the free phosphonate (**161**). Similarly, phosphonate (**162**) was prepared starting from 2,3-isopropylidene-protected glycerol (**165**).

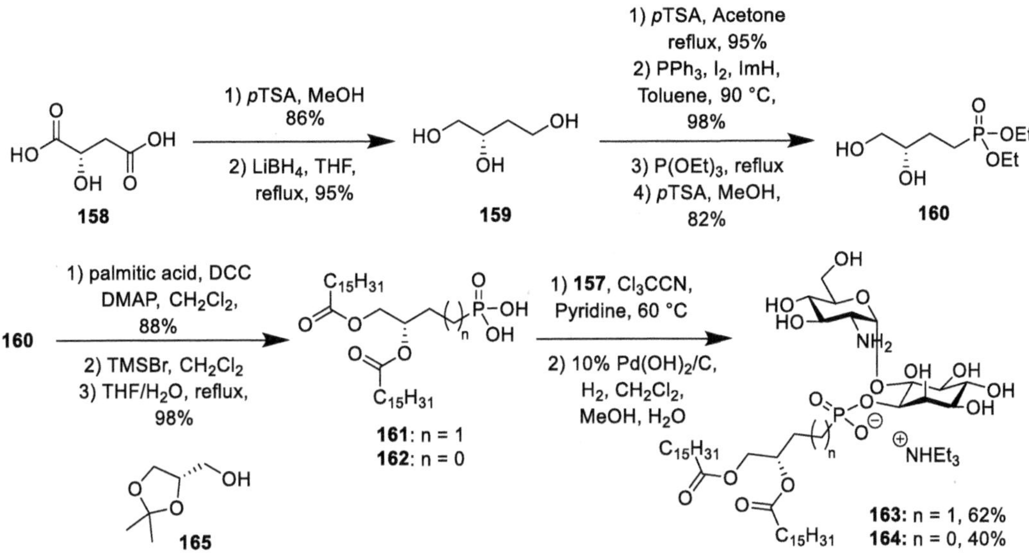

SCHEME 2.21 Synthesis of the target GPI analogues (163) and (164).

Having obtained phosphonic acids **(161, 162)** and pseudo disaccharide **(157)**, the coupling reaction was carried out, and the final step encompassed the extensive deprotection of benzyl and azide groups. The resulting GPI analogues **(163)** and **(164)** were then subjected to an enzymatic reaction with PI-PLC (phosphatidylinositol-specific phospholipase C). The TLC monitoring and *mass analysis showed that the GPI analogues were not hydrolyzed. In addition, they showed inhibitory effects, making them even more suitable for biological studies.*

Soon after, Tanaka et al. reported a synthesis of phosphatidyl inositol oligomannosides via chemoselective *O*-glycosylation [70]. Noteworthy, phosphatidylinositol mannosides (PIMs) are a class of lipids found in certain bacteria, including *Mycobacterium tuberculosis*, the causative agent of tuberculosis (TB). PIMs play important roles in the mycobacterial cell wall structure and are known to be involved in their virulence and interaction with the host immune system. They consist of a 2,6-dimannosyl *myo*-inositol core combined with a di- or monoacylated phosphatidyl lipid or *fatty acid (Figure 2.7). To perform structure activity relationship studies, Tanaka et al. synthesized the oligomannosylated phosphatidyl inositols based on a late-stage glycosylation strategy. Unfortunately, the final deprotection of the glycoconjugates failed, affording the complex mixture.*

Another example of synthesis of phosphatidyl inositols was described by Fujimoto et al. in 2016 [14] to further elucidate biological activity related to the glycerol configuration in selected natural lysophosphatidylinositols. Those motifs are present in the cell membrane of *Entamoeba histolytica*, a parasitic protozoan that causes amoebiasis in humans. It is primarily known for causing amoebic dysentery and amoebic liver abscess. *E. histolytica* is commonly found in regions with poor sanitation and limited access to clean water, particularly in tropical and subtropical areas. Even more intriguing is the cell surface of *E. histolytica*, which contains a sophisticated glycoconjugate known as lipopeptidophosphoglycan (EhLPPG) comprising a GPI anchor and the inositol phospholipid components, EhPIa and EhPIb. These components have characteristic lysophosphatidylinositol moiety (Figure 2.8). Furthermore, these inositol phospholipid components have been shown to have immunomodulatory effects, namely augmentation of natural killer T cells and production of interferon-gamma (IFN-gamma) [71].

Fujimoto et al. synthesized the phosphatidylinositol moiety EhPIa and its epimer (Figure 2.9) using regioselective phosphorylation with BINOL-derived phosphoramidites (Section 2.2.1) and Ni-catalyzed alkyl-alkyl cross-coupling reaction for long fatty acids.

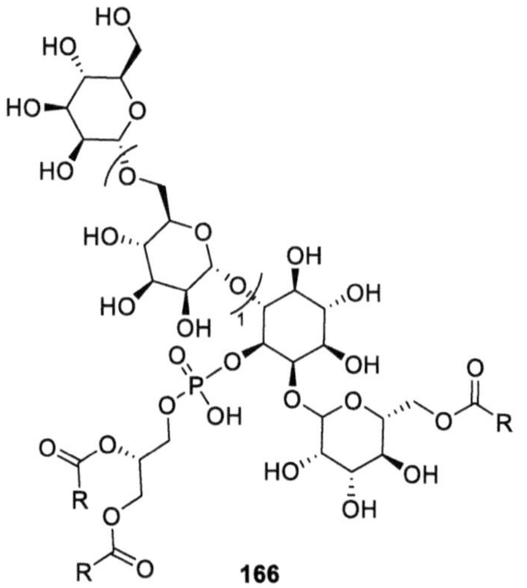

166
R = fatty acid chain

FIGURE 2.7 General structure of phosphatidyl inositol oligomannosides.

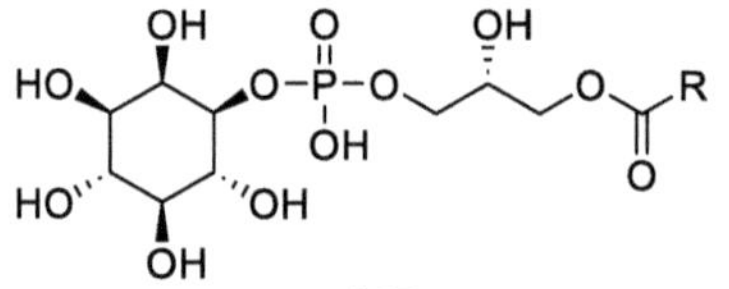

167
R = fatty acid chain

FIGURE 2.8 Characteristic lysophosphatidylinositol structure, with the characteristic absence of second O-acyl chain.

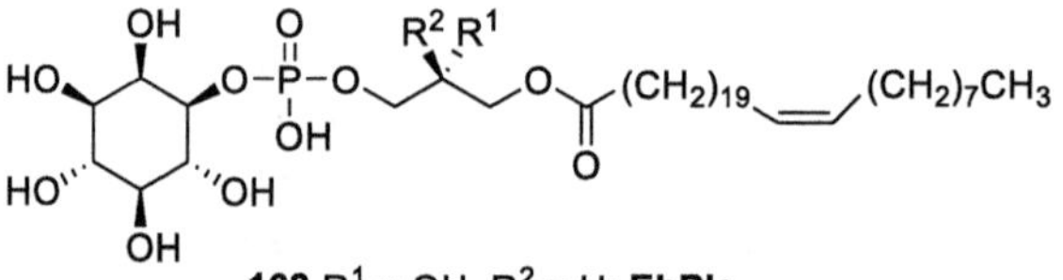

168 R^1 = OH, R^2 = H: **EhPIa**
169 R^1 = H, R^2 = OH: ***epi*-EhPIa**

FIGURE 2.9 EhPIa phosphatidylinositol moiety.

To prepare the characteristic fatty acids (**174, 175**) (Scheme 2.22), the key long-chain fatty acid derivative (**172**) was prepared by an application of nickel-catalyzed alkyl–alkyl crosscoupling reaction of organohalide (**170**) with Grignard reagent (**171**) [72]. Then, *tert*-butyl ester was cleaved with TFA® and formed free acid was subjected to subsequent esterification/deprotection procedure with either alcohol (**176**) or (**177**), yielding two enantiomeric acyl glycerols, (**174**) and (**175**).

The regioselective phosphorylation of the readily available *myo*-inositol derivative (**31**) was carried out with ®-BINOL-derived phosphoramidite (**32**) with tetrazole additive, giving (**35**) in 74% yield with acceptable selectivity (4:1 d.r., Scheme 2.23). To separate diastereomers (**35**) and (**178**)

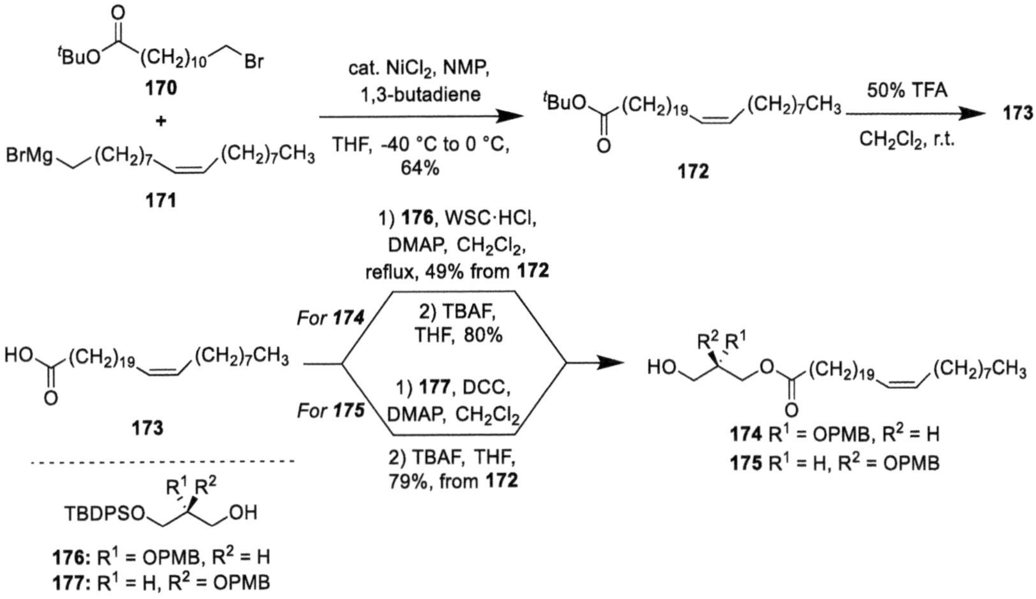

SCHEME 2.22 Synthesis of the enantiometric acyl glycerols (174) and (175) on the way to synthesizing the EhPIaphosphytidyl inositol moiety.

using silica gel column chromatography, the free hydroxyls were protected with Alloc groups. Subsequently, Alloc protective groups were exchanged with TBS groups, and BINOL group was exchanged with benzyl groups to obtain derivative (**37**) in good yield. Next, one of the benzyl group of benzylphosphate was cleaved with LiBr in acetone, andacyl glycerols (**174**), and (**175**) were introduced using the Mitsunobu conditions, yielding the fully protected intermediates (**180**) and (**181**). Final deprotection of all protecting groups successfully gave the desired inositol phospholipid (**168**) and its respective epimer (**169**).

One year later, Y. Fujimoto et al. published a synthesis of the glycosyl inositol phospholipid from *Entamoeba histolytica* (Figure 2.10), further extending and improving the utilization of BINOL-derived phosphoramidites in regioselective phosphorylation of *myo*-inositol units [73].

The overall diastereoselectivity and yield of phosphorylation of *myo*-inositol unit with BINOL derivatives improved by using a phosphatizing reagent with a selenium atom on phosphorus instead of oxygen ((**188**), Scheme 2.24). When *myo*-inositol derivative (**31**) was taken, the corresponding product (**35**) was obtained in 79% yield with increased diastereoselectivity (86:14d.r.). The authors found that atomic size influences diastereocontrol, as changing the atom on phosphorus to sulfur also increased the diastereomeric ratio of (**35**), although the yields were lower. Noteworthy, the change of the selenium atom back to oxygen can be easily performed with the *m*CPBA oxidant. After separation, *O*-allylation of pure diol (**35**) followed by selective cleavage of the 4-*O*-PMB protective group and transesterification with allyl alcohol under basic conditions gave intermediate (**189**) regenerating the BINOL. The unusual regioselectivity during the removal of the PMB group using DDQ could be explained by BINOL moiety causing steric hindrance of 6-*O*-PMB.

Subsequent protective group manipulations together with the introduction of palmitoyl ester gave the desired inositol acceptor (**191**). Next, *myo*-inositol (**191**) was glycosylated with the azido-glucose donor (**192**), providing intermediate (**193**) in 81% yield ($\alpha/\beta = 4:1$). The corresponding pseudodisaccharides (**194**) and (**195**) were obtained by esterification of the phosphate with mono-acylglycerol (**196**) and (**197**), respectively, using Mitsunobu conditions and removal of the 2-naphthylmethyl group. Deallylation and azide reduction gave glycosyl inositol phospholipids (**184**) and (**185**) that were glycosylated with the trisaccharide donor (**198**). Cleavage of the levulinate ester (Lev) group

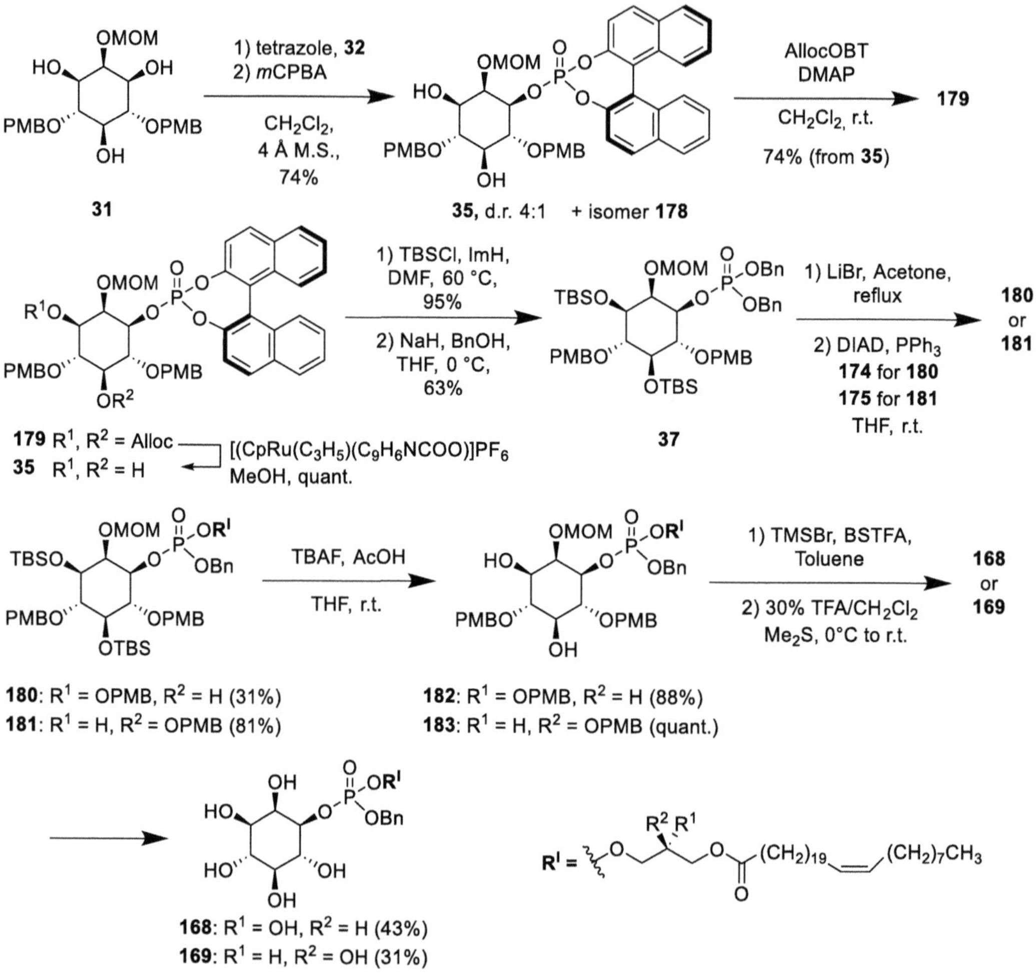

SCHEME 2.23 Synthesis of the EhPIaphosphytidyl inositol moiety utilizing desymmetrization of inositol intermediate (31).

and global deprotection produced glycosyl inositol phospholipids (**186**) and (**187**). Interestingly, based on the preliminary biological studies, the authors showed that glycosyl inositol phospholipid (**185**) was able to increase the levels of IFN-gamma more significantly than (**187**), and the major component of the cell membrane of *E. histolytica*, lipopeptidophosphoglycan was EhLPPG.

Candida albicans is a yeast that is a part of the normal microbial flora in the human body, commonly found in the gastrointestinal tract and oral cavity. In healthy individuals, *C. albicans* coexists harmlessly with other microorganisms. However, under certain conditions, it can cause infections known as candidiasis. A crucial aspect of understanding its pathogenic nature involves the investigation of factors that facilitate the shift from a harmless state to a pathogenic one. The infectivity and disease-causing properties of *C. albicans* are attributed to the cell surface glycosphingolipid known as phospholipomannan (PLM) [74]. The PLM structure consists of a mannose-inositolphosphoceramide lipid anchor that on primary hydroxyl contains phosphate ester linked to an unusual oligomeric (1 → 2)-β-mannan via anomeric oxygen (Figure 2.11). The (1 → 2)-β-mannan structural motif is the origin of the high immunogenicity. Isolation of phospholipomannans from *C. albicans* is a difficult task due to limited availability and heterogeneity. The chemical synthesis would thus mitigate both of these insufficiencies

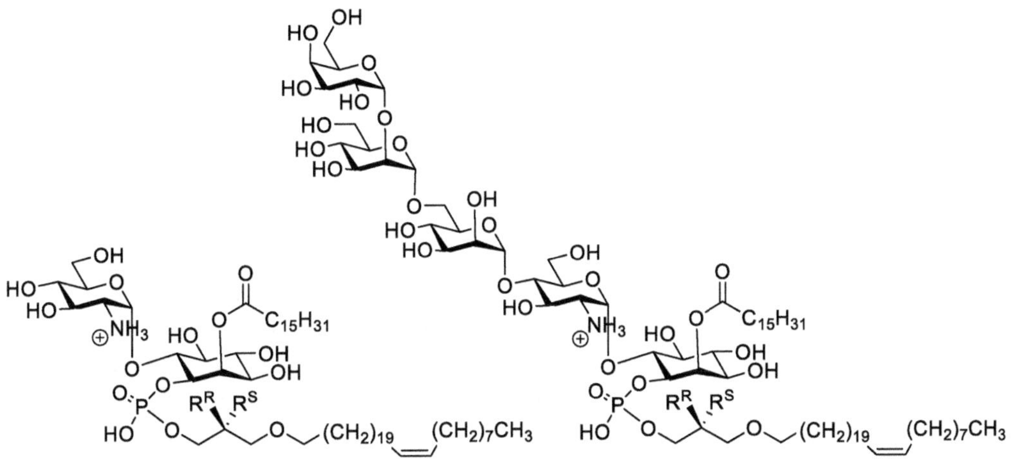

184: $R^S = OH, R^R = H$
185: $R^S = H, R^R = OH$

186: $R^S = OH, R^R = H$
187: $R^S = H, R^R = OH$

FIGURE 2.10 Glycosyl inositol phospholipids from *Entamoeba histolytica.*

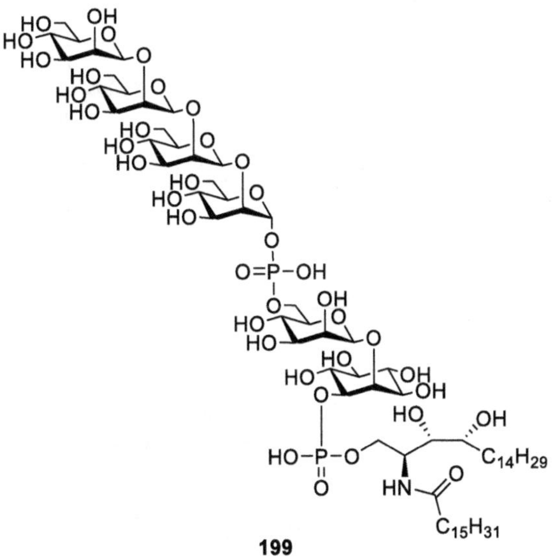

199

FIGURE 2.11 Phospholipomannan anchor of *Candida albicans.*

and in 2020, Vishwakarma et al. introduced the synthesis of a full-length PLM anchor of *C. albicans* [75].

The synthesis of tetramannoside fragment (**206**) started with the preparation of suitably protected mannose building blocks (**201**), (**202**), (**204**) by known methods, including a stannylene acetal-derived regioselective 3-*O*-benzylation (for (**201**)) and orthoester cleavage (for (**204**), Scheme 2.25). Next, optimized glycosylation conditions for (1 → 2)-β mannosylation between glycosyl thioglycoside (**202**) and glycosyl acceptor (**204**) were found (AgOTf, PhSCl, -78 °C) and applied for one-pot synthesis of tetramannoside (**205**) from building blocks (**202**) and (**201**). After that, the benzylidene acetals were cleaved with *p*TSA, and hydroxyls were subsequently protected with benzyl groups. The key tetramannosyl-*H*-phosphonate (**206**) was achieved using oxidative deprotection

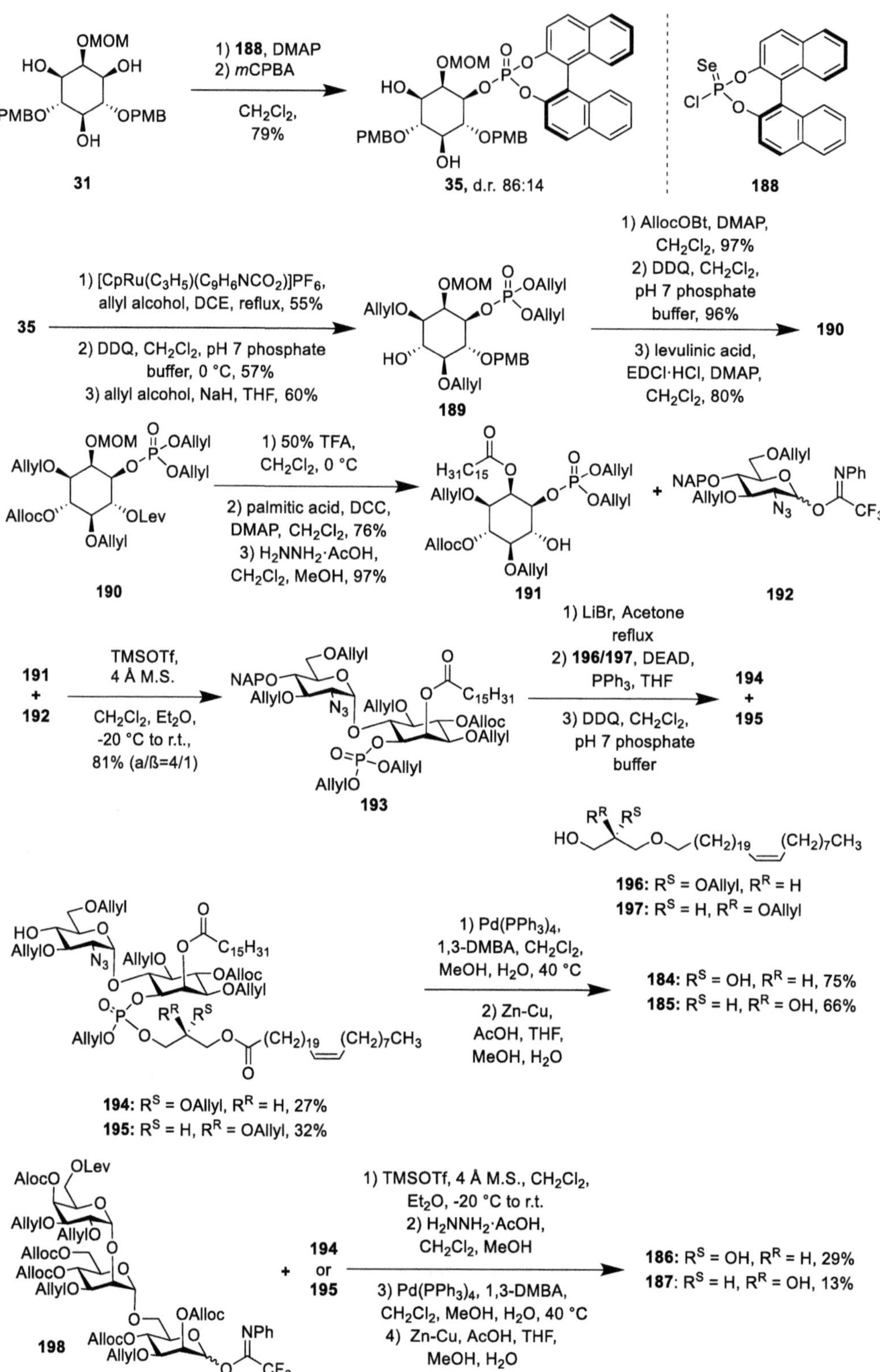

SCHEME 2.24 Synthesis of the glycosyl inositol phospholipids from *Entamoeba histolytica*.

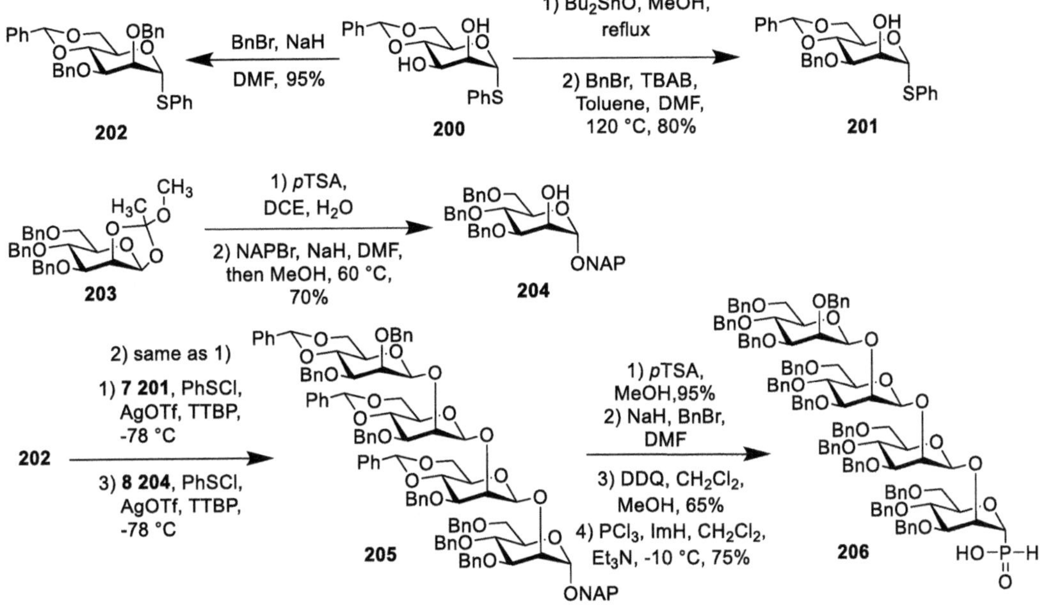

SCHEME 2.25 Synthesis of the tetramannosyl-H-phosphonate (206).

of the 2-naphthyl group and the introduction of the phosphate group at anomeric position using PCl_3 and imidazole.

The next step encompassed the synthesis of pseudodisaccharide fragments (**207**) and (**208**) from enantiometrically pure *myo*-inositol derivative (**211**) (Scheme 2.26). This intermediate was obtained by the resolution of diol (**209**) via camphanylideneacetal formation, followed by hydrolysis, and regioselective PMB group introduction. Then, the intermediate (**211**) was coupled with mannopyranosylimidat e donor (**212**), and obtained α-anomer (**213**) was deacetylated and subjected to PCl_3 and imidazole to get phosphonate (**208**). The pseudodisaccharide acceptor (**207**) was synthesized by deprotecting the PMB group of intermediate (**213**).

The last building block, phytoceramide-1-*H*-phosphonate (**215**) was prepared in five steps starting from (**214**) obtained by enantioselective A^3 coupling reaction with CuBr and acetyl glucoBox ligand. Key steps for preparing (**215**) included partial hydrogenation using Lindlar´s catalyst, Sharpless asymmetric dihydroxylation, and amidation and the introduction of the phosphate group (Scheme 2.27). Then, PLM was synthesized. The pseudodisaccharide acceptor (**216**) was coupled with the *H*-phosphonate donor of phytosphingosine (**215**), and the crude material was subsequently deacetylated to yield (**217**). The protected PLM was achieved by coupling of (**217**) with *H*-phosphonate donor (**208**). The final step was debenzylation with Pearlman's catalyst yielding the desired PLM (**199**).

As mentioned earlier, PLC hydrolyzes the phosphatidylinositol phosphates resulting in the release of water-soluble messenger IP_3. This then binds to the IP_3-binding core (IBC) of its receptor, resulting in the release of cytosolic Ca^{2+}. The natural glyconucleotidesadenophostin A and adenophostin B ((**219, 220**), Figure 2.12), extracted from certain fungal cultures, were approximately ten times more potent than IP_3 in triggering the release of calcium ions. Notably, attempts to mimicadenophostin structure resulted in compounds with less potency than IP_3 in the release of Ca^{2+}, except ribophostin, which has similar potency to that of IP_3 [76].

In 2020, Potter et al. disclosed the synthesis of D-*chiro*-inositol ribophostin (**221**), which exhibited a stronger affinity to IBC than IP_3 [77]. The first step of the synthesis was the preparation of

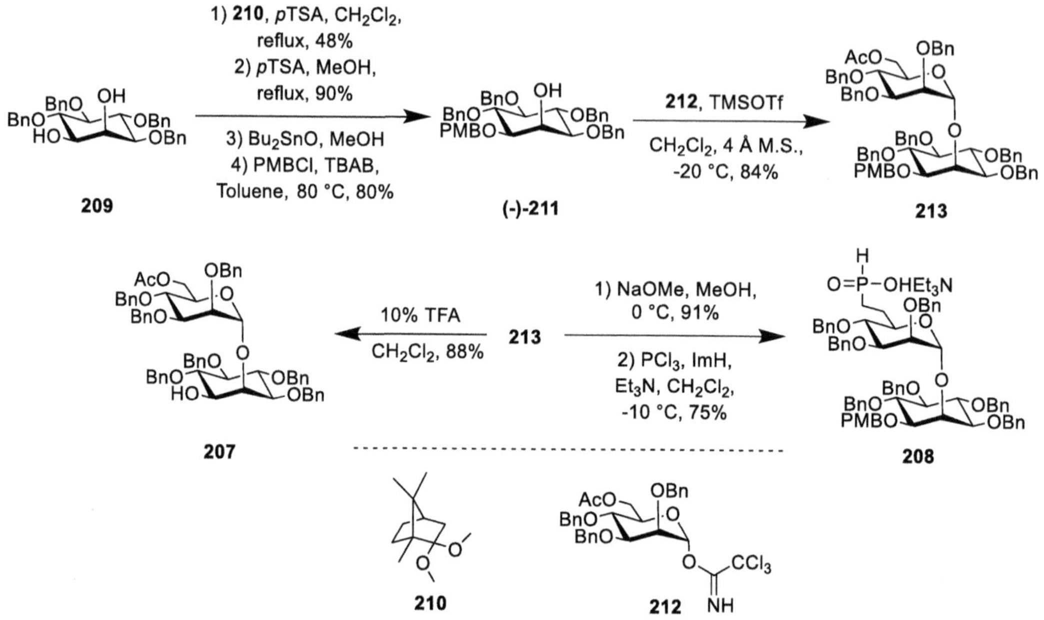

SCHEME 2.26 Synthesis of the key pseudodisaccharide units.

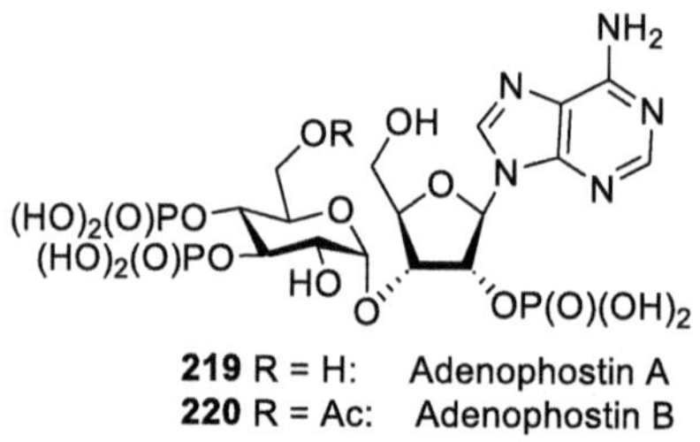

FIGURE 2.12 Structures of adenophostin A and B.

a suitably protected *myo*-inositol unit (**223**) bearing a good leaving group for S_N2 reaction with nucleophilic ribose derivative (**224**) (Scheme 2.28). The synthesis of (**223**) starting from racemic 1,2:4,5-di-*O*-isopropylidene-*myo*-inositol was described by the same group earlier and involved selective installation of PMB group and resolution using (1*S*)-camphanic chloride as key steps [78]. The coupling reaction between (**223**) and (**224**) provided intermediate (**225**), bearing two isopropylidene-protecting acetals. *Trans*-isopropylidene acetal was selectively hydrolyzed yielding diol (**226**), and using dibutyltin chemistry, monobenzylation was performed. However, no regioselectivity was observed; therefore, after separation, only intermediate (**227**) was used for selective removal of the PMB group by DDQ. Subsequent phosphitylation of (**228**) with commonly used dibenzyl diisopropylphosphoramidite followed by oxidation with *m*CPBA and removal of benzyl protecting group using hydrogenation provided target triphosphate (**221**).

2.4 INOSITOL DERIVATIVES AS KEY INTERMEDIATES IN TOTAL SYNTHESIS

The total synthesis of complex molecules continues to be an important and topical discipline even after more than a century since it was founded primarily to confirm the structure of natural substances. Today, one of its rationales can be seen in the preparation of recently discovered

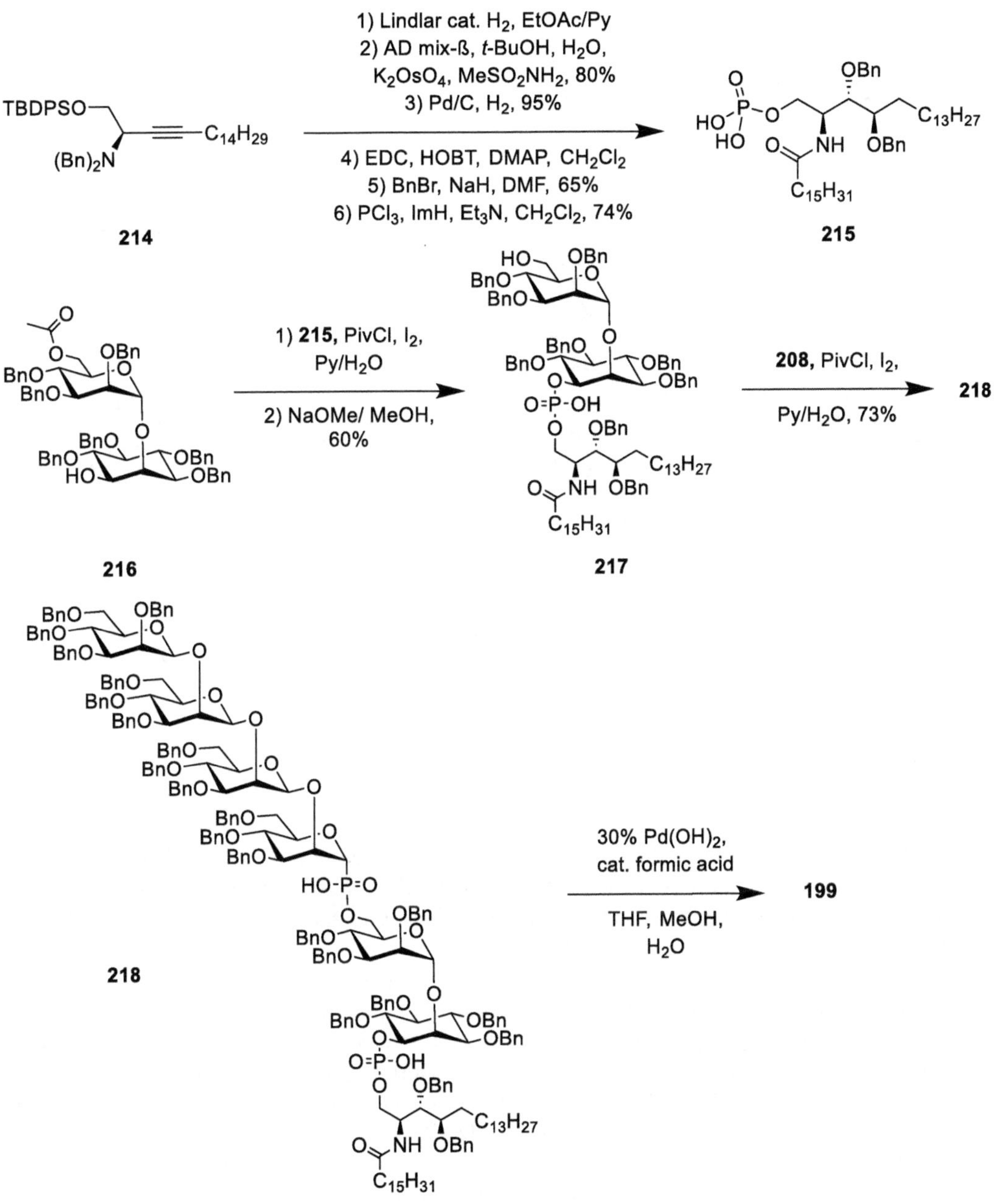

SCHEME 2.27 Synthesis of the target phospholipomannan anchor.

biologically active compounds obtained by in situ isolation techniques in insufficient quantities to fully understand their function and potential therapeutic efficiency. Another important impact of total synthesis is providing excellent training for the next generation of synthetic chemists and evaluating newly developed synthetic methodologies. Since many biologically active compounds contain stereogenic centers in their structure, the use of optically active substances readily available from natural sources makes sense. Carbohydrates have been extensively used as starting materials in enantioselective synthesis because some of them are so readily available. Similarly, inositols have been used as valuable building blocks in the last decades. Next, we discuss some selected examples of their application in total synthesis.

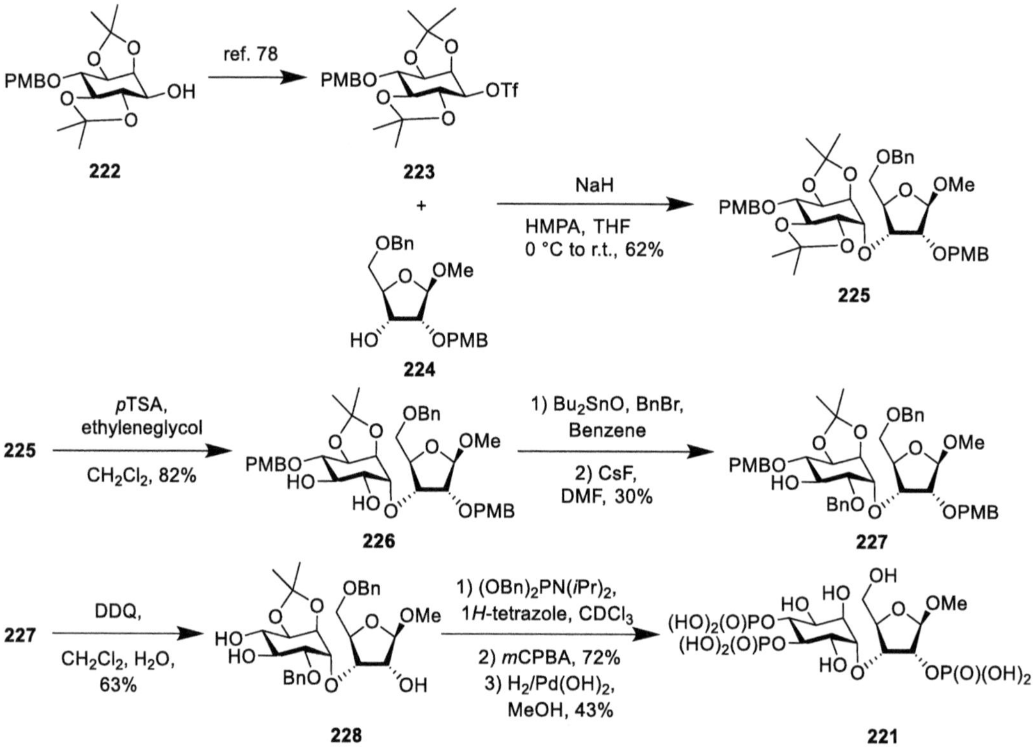

SCHEME 2.28 Synthesis of the target ribophostin (221).

In 2010, Sato et al. disclosed the total synthesis of enantiopure tetrodotoxin (**230**) (Figure 2.13), a potent neurotoxin originally isolated from puffer fish. The crucial step of the reported synthesis involved Ferrier carbocyclization of the D-glucose derivative to construct the key inositol skeleton (**229**) (Figure 2.13). Notably, this inositol derivative was synthesized before by the same group, albeit in a racemic form [80]. A year later, in 2011, Hung et al. described the synthesis of mycothiol, a natural low-molecular-weight thiol that acts as a key component of the antioxidant defense system in actinobacteria, helping to protect the bacterial cells from oxidative stress caused by reactive oxygen species (ROS) and other toxic oxidants [22]. The reported synthesis started with regioselective desymmetrization of 2,4,5,6-tetrabenzylated-*myo*-inositol as the key step, and the whole procedure was accomplished in eight steps with an overall yield 40%.

In the same year, Shashidhar et al. [81] presented the formal synthesis of racemic valiolamine (**231**) starting from *myo*-inositol (Figure 2.13). Valiolamine, a member of aminocyclitol family first isolated from *Streptomyces hygroscopicus,* was found to be an unusually potent inhibitor of α-glucosidase, maltase, and sucrase. Chemical structure modification led to voglibose (**232**), an orally active antidiabetic agent. Key reactions used in the developed synthetic approach were lithium hydride-mediated selective *O*-benzylation, a novel deoxygenation reaction of the tosylate, and stereospecific addition of dichloromethyllithium to inosose derivative. The synthesis of the key motifs of other natural aminocyclitols, hygromycin A (**233**), and methoxyhygromycin (**234**), using *myo*-inositol as a starting material, was also reported by Shashidhar et al. in 2012 [82]. It is noteworthy that the preparation of hygromycin A consists of 11 steps with only five-column chromatographic purifications, and the latter aminocyclitol required only one-column chromatography.

Recently, Hou et al. developed an asymmetric synthesis of another aminocyclitol, nabscessin A ((**235**), Figure 2.13), starting from readily available myo-inositol and d-camphor units [83].

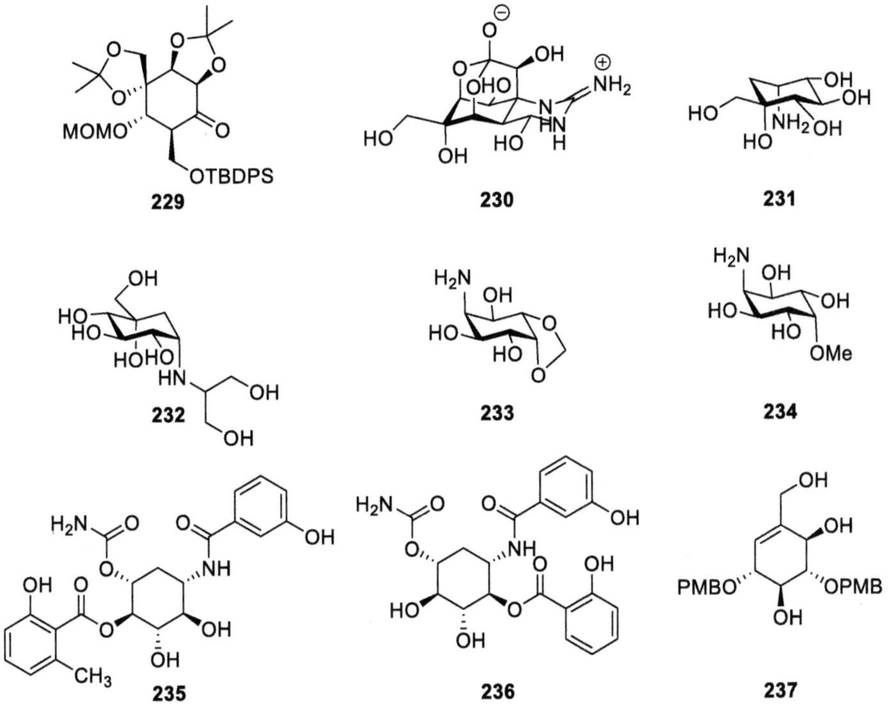

FIGURE 2.13 Compounds with inositol scaffold.

Nabscessin A belongs to the family of novel aminocyclitol amides isolated from the culture broth of a pathogenic actinomycete species, *Nocardia abscessus,* exhibiting interesting antifungal activity against *Cryptococcus neoformans* [84]. The synthesis of another nabscessin was elaborated by Banwell et al. starting from L-tartaric acid and obtaining the target nabscessin B ((**236**), Figure 2.13) in twenty steps [85]. The synthesis of nabscessin A reported by Hou et al. involved 14 steps and allowed the preparation of both enantiomers.

Carbasugars are a category of carbohydrate derivatives in which the ring oxygen atom is substituted by a methylene group, resulting in a molecule with a very similar structure but with chemically different properties. Most carbasugars are more stable than carbohydrates due to the absence of the hemiacetal moiety, and they exhibit interesting biological properties. In 2014, Sureshan et al. reported the first racemic syntheses and structural confirmation of five novel natural carbasugars: lincitol A, lincitol B, uvacalol I, uvacalol J, and uvacalol K [86]. As the key intermediate for their synthesis, *myo*-inositol derivative (**237**) was used that had been prepared by the vinylogous ring-opening of the corresponding *myo*-inositol orthoester. All the carbasugars were prepared from *myo*-inositol in 11 to 13 steps.

A few years later, in 2018, the first total synthesis of a natural compound characterized by a highly oxidized chromene core, oxirapentyl D, was reported by Ohmori et al. [87]. The authors came up with the idea of using *myo*-inositol as a starting material, and the key transformation included chelatation-directed bridgehead lithiation of hydrazon derivative. One of the last examples of using *myo*-inositol to prepare natural compounds was reported in 2021, when Okuda et al. introduced the first total synthesis of antiausterity agents uvaridacol L [88]. The synthesis comprised seven steps with 50% of the overall yield of racemic uvaridacol L, exhibiting similar cytotoxicity as the isolated natural product.

2.5 CONCLUSION

This chapter gives a thorough overview of the current research on preparing inositol derivatives, focusing on innovative methods for obtaining enantiopure inositol derivatives, phosphates, phosphoinositides, and related inositol-based intricate compounds. Synthetic chemists have addressed the challenge of desymmetrization for enantiopure inositol derivatives, which led to introduction of novel methodologies, namely peptide-catalysedphosphoramidite transfer methodology or phosphinit-catalyzed acylation. Further methodologies relied on chiral reagents itself, for example BINOL-based phosphitylating or chiral acylating reagents.

Other strategies involved developing a C_2-symmetric phosphoramidite, yielding separable diastereomeric mixtures. An alternative strategy involved de novo synthesis of an inositol scaffold from chiral compounds, specifically expanding utility of Ferrier carbocyclization, demonstrated by synthesizing azido inositol derivatives or further post-transformations leading to *scyllo*-inositol derivatives. Elegant de novo method involved benzoin-type cyclization of sugar dialdoses catalyzed by triazolium-type NHC catalysts.

In this chapter, we detailed comprehensive syntheses of diverse biologically active inositol units. These syntheses encompassed a spectrum of compounds, ranging from inositol-based immunogen and various biological probes to toribophostin, a biochemical analogue of the second messenger IP_3. Nevertheless, there is still need for catalytic approaches and methodologies for obtaining enantiopure and orthogonally protected inositol derivatives because of the ubiquitous occurrence of (masked) inositol scaffold in natural compounds. Thus, in the dynamic field of synthetic organic chemistry, the inositol moiety continues to play a significant role.

ACKNOWLEDGEMENTS

The authors gratefully acknowledge the Czech Science Foundation for financial support.

REFERENCES

[1] Scherer, J. *Justus Liebig Ann. Chem.* **1850**, *73*, 322–328.

[2] Angyal, S. J., Tate, M. E. Cyclitols: Part X. Myoinositol Phosphates. *J. Chem. Soc.* **1961**, 4122–4128. https://doi.org/10.1039/JR9610004122

[3] a) Potter, B. V. L.; Lampe, D. Chemistry of Inositol Lipid Mediated Cellular Signaling. *Angew. Chem. Int. Ed. Engl.* **1995**, *34* (18), 1933–1972. https://doi.org/10.1002/anie.199519331; b) Billington, D. C. Recent Developments in the Synthesis of Myo-Inositol Phosphates. *Chem. Soc. Rev.* **1989**, *18*, 83. https://doi.org/10.1039/cs9891800083; c) Irvine, R. F.; Schell, M. J. Back in the Water: The Return of the Inositol Phosphates. *Nat. Rev. Mol. Cell Biol.* **2001**, *2* (5), 327–338. https://doi.org/10.1038/35073015; d) Best, M. D.; Zhang, H.; Prestwich, G. D. Inositol Polyphosphates, Diphosphoinositol Polyphosphates and Phosphatidylinositol Polyphosphate Lipids: Structure, Synthesis, and Development of Probes for Studying Biological Activity. *Nat. Prod. Rep.* **2010**, *27* (10), 1403. https://doi.org/10.1039/b923844c; e) Hatch, A. J.; York, J. D. SnapShot: Inositol Phosphates. *Cell* **2010**, *143* (6), 1030–1030.e1. https://doi.org/10.1016/j.cell.2010.11.045; F) Chakraborty, A.; Kim, S.; Snyder, S. H. Inositol Pyrophosphates as Mammalian Cell Signals. *Sci. Signal.* **2011**, *4* (188). https://doi.org/10.1126/scisignal.2001958; g) Gillaspy, G. E. The Cellular Language of *Myo*-Inositol Signaling. *New Phytol.* **2011**, *192* (4), 823–839. https://doi.org/10.1111/j.1469-8137.2011.03939.x; H) Michell, R. H. Inositol and Its Derivatives: Their Evolution and Functions. *Adv. Enzyme Regul.* **2011**, *51* (1), 84–90. https://doi.org/10.1016/j.advenzreg.2010.10.002; i) Lee, J.-Y.; Kim, Y.-R.; Park, J.; Kim, S. Inositol Polyphosphate Multikinase Signaling in the Regulation of Metabolism. *Ann. N. Y. Acad. Sci.* **2012**, *1271* (1), 68–74. https://doi.org/10.1111/j.1749-6632.2012.06725.x; j) Croze, M. L.; Soulage, C. O. Potential Role and Therapeutic Interests of Myo-Inositol in Metabolic Diseases. *Biochimie* **2013**, *95* (10), 1811–1827. https://doi.org/10.1016/j.biochi.2013.05.011; k) Thomas, M. P.; Mills, S. J.; Potter, B. V. L. The “Other” Inositols and Their Phosphates: Synthesis, Biology, and Medicine (with Recent Advances in Myo-Inositol Chemistry). *Angew. Chem. Int. Ed Engl.* **2016**, *55* (5), 1614–1650. https://doi.org/10.1002/anie.201502227

[4] Sculimbrene, B. R.; Miller, S. J. Discovery of a Catalytic Asymmetric Phosphorylation Through Selection of a Minimal Kinase Mimic: A Concise Total Synthesis of D—*Myo*-Inositol-1-Phosphate. *J. Am. Chem. Soc.* **2001**, *123* (41), 10125–10126. https://doi.org/10.1021/ja016779+

[5] Sculimbrene, B. R.; Morgan, A. J.; Miller, S. J. Enantiodivergence in Small-Molecule Catalysis of Asymmetric Phosphorylation: Concise Total Syntheses of the Enantiomeric D—*mYo*-Inositol-1-Phosphate and D—*mYo*-Inositol-3-Phosphate. *J. Am. Chem. Soc.* **2002**, *124* (39), 11653–11656. https://doi.org/10.1021/ja027402m

[6] Sculimbrene, B. R.; Morgan, A. J.; Miller, S. J. Nonenzymatic Peptide-Based Catalytic Asymmetric Phosphorylation of Inositol Derivatives. *Chem. Commun.* **2003**, *15*, 1781. https://doi.org/10.1039/b304015c

[7] Sculimbrene, B. R.; Xu, Y.; Miller, S. J. Asymmetric Syntheses of Phosphatidylinositol-3-Phosphates with Saturated and Unsaturated Side Chains Through Catalytic Asymmetric Phosphorylation. *J. Am. Chem. Soc.* **2004**, *126* (41), 13182–13183. https://doi.org/10.1021/ja0466098

[8] Xu, Y.; Sculimbrene, B. R.; Miller, S. J. Streamlined Synthesis of Phosphatidylinositol (PI), PI3P, PI3,5P 2, and Deoxygenated Analogues as Potential Biological Probes. *J. Org. Chem.* **2006**, *71* (13), 4919–4928. https://doi.org/10.1021/jo060702s

[9] Wang, Y. K.; Chen, W.; Blair, D.; Pu, M.; Xu, Y.; Miller, S. J.; Redfield, A. G.; Chiles, T. C.; Roberts, M. F. Insights into the Structural Specificity of the Cytotoxicity of 3-Deoxyphosphatidylinositols. *J. Am. Chem. Soc.* **2008**, *130* (24), 7746–7755. https://doi.org/10.1021/ja710348r

[10] Kayser-Bricker, K. J.; Jordan, P. A.; Miller, S. J. Catalyst-Dependent Syntheses of Phosphatidylinositol-5-Phosphate-DiC8 and Its Enantiomer. *Tetrahedron* **2008**, *64* (29), 7015–7020. https://doi.org/10.1016/j.tet.2008.03.037

[11] Jordan, P. A.; Kayser-Bricker, K. J.; Miller, S. J. Asymmetric Phosphorylation Through Catalytic P(III) Phosphoramidite Transfer: Enantioselective Synthesis of D—*Myo*-Inositol-6-Phosphate. *Proc. Natl. Acad. Sci. U.S.A.* **2010**, *107* (48), 20620–20624. https://doi.org/10.1073/pnas.1001111107

[12] Capolicchio, S.; Thakor, D. T.; Linden, A.; Jessen, H. J. Synthesis of Unsymmetric Diphospho-Inositol Polyphosphates. *Angew. Chem. Int. Ed.* **2013**, *52* (27), 6912–6916. https://doi.org/10.1002/anie.201301092

[13] Duss, M.; Capolicchio, S.; Linden, A.; Ahmed, N.; Jessen, H. J. Desymmetrization of Myo-Inositol Derivatives by Lanthanide Catalyzed Phosphitylation with C2-Symmetric Phosphites. *Bioorg. Med. Chem.* **2015**, *23* (12), 2854–2861. https://doi.org/10.1016/j.bmc.2015.03.023

[14] Durantie, E.; Huwiler, S.; Leroux, J.-C.; Castagner, B. A Chiral Phosphoramidite Reagent for the Synthesis of Inositol Phosphates. *Org. Lett.***2016**, *18* (13), 3162–3165. https://doi.org/10.1021/acs.orglett.6b01374

[15] Aiba, T.; Sato, M.; Umegaki, D.; Iwasaki, T.; Kambe, N.; Fukase, K.; Fujimoto, Y. Regioselective Phosphorylation of Myo-Inositol with BINOL-Derived Phosphoramidites and Its Application for Protozoan Lysophosphatidylinositol. *Org. Biomol. Chem.* **2016**, *14* (28), 6672–6675. https://doi.org/10.1039/C6OB01062H

[16] De Vries, A. H. M.; Meetsma, A.; Feringa, B. L. Enantioselective Conjugate Addition of Dialkylzinc Reagents to Cyclic and Acyclic Enones Catalyzed by Chiral Copper Complexes of New Phosphorus Amidites. *Angew. Chem. Int. Ed. Engl.* **1996**, *35* (20), 2374–2376. https://doi.org/10.1002/anie.199623741

[17] Aiba, T.; Suehara, S.; Choy, S.; Maekawa, Y.; Lotter, H.; Murai, T.; Inuki, S.; Fukase, K.; Fujimoto, Y. Employing BINOL-Phosphoroselenoyl Chloride for Selective Inositol Phosphorylation and Synthesis of Glycosyl Inositol Phospholipid from *Entamoeba Histolytica*. *Chem. A European J.* **2017**, *23* (34), 8304–8308. https://doi.org/10.1002/chem.201701298

[18] For seminal work in this area, please see: Ozaki, S.; Watanabe, Y.; Ogasawara, T.; Kondo, Y.; Shiotani, N.; Nishii, H.; Matsuki, T. *Tetrahedron Lett.* **1986**, *27*, 3157–3160; b) Bruzik, K. S.; Salamonczyk, G. M. *Carbohydr. Res.* **1989**, *195*, 67–73; c) Bruzik, K. S; Tsai, M. D. *J. Am. Chem. Soc.* **1992**, *114*, 6361–6374; d) Kubiak, R. J.; Bruzik, K. S. *J. Org. Chem.* **2003**, *68*, 960–968.

[19] Lee, H. W.; Kishi, Y. *J. Org. Chem.* **1985**, 50, 4402–4404.

[20] Padiyar, L. T.; Wen, Y.-S.; Hung, S.-C. Metal Trifluoromethanesulfonate-Catalyzed Regioselective Acylation of Myo-Inositol 1,3,5-Orthoformate. *Chem. Commun.* **2010**, *46* (30), 5524. https://doi.org/10.1039/c0cc00236d

[21] Padiyar, L. T.; Zulueta, M. M. L.; Sabbavarapu, N. M.; Hung, S.-C. $Yb(OTf)_3$-Catalyzed Desymmetrization of *Myo*-Inositol 1,3,5-Orthoformate and Its Application in the Synthesis of Chiral Inositol Phosphates. *J. Org. Chem.* **2017**, *82* (21), 11418–11430. https://doi.org/10.1021/acs.joc.7b01919

[22] Chung, C.-C.; Zulueta, M. M. L.; Padiyar, L. T.; Hung, S.-C. Desymmetrization of 2,4,5,6-Tetra- *O* -Benzyl- D—*Myo*-Inositol for the Synthesis of Mycothiol. *Org. Lett.* **2011**, *13* (20), 5496–5499. https://doi.org/10.1021/ol202218n

[23] Lauber, M. B.; Daniliuc, C.-G.; Paradies, J. Desymmetrization of 4,6-Diprotected Myo-Inositol. *Chem. Commun.* **2013**, *49* (67), 7409. https://doi.org/10.1039/c3cc43663b

[24] Mizuta, S.; Sadamori, M.; Fujimoto, T.; Yamamoto, I. Asymmetric Desymmetrization of *Meso* -1,2-Diols by Phosphinite Derivatives of Cinchona Alkaloids. *Angew. Chem. Int. Ed.* **2003**, *42* (29), 3383–3385. https://doi.org/10.1002/anie.200250719

[25] Reddy, K. K.; Saady, M.; Falck, J. R.; Whited, G. Intracellular Mediators: Synthesis of L-Alpha.-Phosphatidyl-D-Myo-Inositol 3,4,5-Trisphosphate and Glyceryl Ether Analogs. *J. Org. Chem.* **1995**, *60* (11), 3385–3390. https://doi.org/10.1021/jo00116a023

[26] Qiao, L.; Hu, Y.; Nan, F.; Powis, G.; Kozikowski, A. P. A Versatile Approach to PI(3,4)P 2, PI(4,5)P and PI(3,4,5)P3 from L-(–)-Quebrachitol. *Org. Lett.* **2000**, *2* (2), 115–117. https://doi.org/10.1021/ol991188y

[27] Estevez, V. A.; Prestwich, G. D. Synthesis of Enantiomerically Pure, P-1-Tethered Inositol Tetrakis (Phosphate) Affinity Labels via a Ferrier Rearrangement. *J. Am. Chem. Soc.* **1991**, *113* (26), 9885–9887. https://doi.org/10.1021/ja00026a043

[28] Chen, J.; Feng, L.; Prestwich, G. D. Asymmetric Total Synthesis of Phosphatidylinositol 3-Phosphate and 4-Phosphate Derivatives. *J. Org. Chem.* **1998**, *63* (19), 6511–6522. https://doi.org/10.1021/jo980501r

[29] Peng, J.; Prestwich, G. D. Synthesis of L-α-Phosphatidyl-d-Myo-Inositol 5-Phosphate and l-α-Phosphatidyl-d-Myo-Inositol 3,5-Bisphosphate. *Tetrahedron Lett.* **1998**, *39* (23), 3965–3968. https://doi.org/10.1016/S0040-4039(98)00737-0

[30] Prestwich, G. D. Touching All the Bases: Synthesis of Inositol Polyphosphate and Phosphoinositide Affinity Probes from Glucose. *Acc. Chem. Res.* **1996**, *29* (10), 503–513. https://doi.org/10.1021/ar960136v

[31] Chida, N. Ferrier Carbocyclization Reaction. In *Molecular Rearrangements in Organic Synthesis*; Rojas, C. M., Ed.; Wiley, 2015; pp. 363–400. https://doi.org/10.1002/9781118939901.ch12

[32] Ausmus, A. P.; Hogue, M.; Snyder, J. L.; Rundell, S. R.; Bednarz, K. M.; Banahene, N.; Swarts, B. M. Ferrier Carbocyclization-Mediated Synthesis of Enantiopure Azido Inositol Analogues. *J. Org. Chem.* **2020**, *85* (5), 3182–3191. https://doi.org/10.1021/acs.joc.9b03064

[33] Rodriguez, J.; Walczak, M. A. Synthesis of Asymmetrically Substituted Scyllo-Inositol. *Tetrahedron Lett.* **2016**, *57* (30), 3281–3283. https://doi.org/10.1016/j.tetlet.2016.06.038

[34] Ma, K.; Thomason, L. A. M.; McLaurin, J. Scyllo-Inositol, Preclinical, and Clinical Data for Alzheimer's Disease. In *Advances in Pharmacology*; Elsevier, 2012; Vol. 64, pp. 177–212. https://doi.org/10.1016/B978-0-12-394816-8.00006-4

[35] Vekrellis, K.; Xilouri, M.; Emmanouilidou, E.; Stefanis, L. Inducible Over-Expression of Wild Type A-Synuclein in Human Neuronal Cells Leads to Caspase-Dependent Non-Apoptotic Death. *J. Neurochem.* **2009**, *109* (5), 1348–1362. https://doi.org/10.1111/j.1471-4159.2009.06054.x

[36] Enders, D.; Niemeier, O.; Henseler, A. Organocatalysis by N-Heterocyclic Carbenes. *Chem. Rev.* **2007**, *107* (12), 5606–5655. https://doi.org/10.1021/cr068372z

[37] Stockton, K. P.; Greatrex, B. W.; Taylor, D. K. Synthesis of *Allo*—and *Epi*-Inositol via the NHC-Catalyzed Carbocyclization of Carbohydrate-Derived Dialdehydes. *J. Org. Chem.* **2014**, *79* (11), 5088–5096. https://doi.org/10.1021/jo500645z

[38] Kang, B.; Sutou, T.; Wang, Y.; Kuwano, S.; Yamaoka, Y.; Takasu, K.; Yamada, K. N-Heterocyclic Carbene-Catalyzed Benzoin Strategy for Divergent Synthesis of Cyclitol Derivatives from Alditols. *Adv. Synth. Catal.* **2015**, *357* (1), 131–147. https://doi.org/10.1002/adsc.201400712

[39] Mayinger, P. Phosphoinositides and Vesicular Membrane Traffic. *Biochim. Biophys. Acta, Mol. Cell Biol. Lipids.* **2012**, *1821* (8), 1104–1113. https://doi.org/10.1016/j.bbalip.2012.01.002

[40] Atsuo, M.; Masato, U.; Tetsuro, H.; Keiji, Y.; Tohoru, Y.; Keizo, I. Production and Characterization of Monoclonal Antibodies That Bind to Phosphatidylinositol 4,5-Bisphosphate. *Mol. Immunol.* **1988**, *25* (10), 1025–1031. https://doi.org/10.1016/0161-5890(88)90010-7

[41] Chandler, B. D.; Burkhardt, A. L.; Foley, K.; Cullis, C.; Driscoll, D.; Roy D'Amore, N.; Miller, S. J. A Fully Synthetic and Biochemically Validated Phosphatidyl Inositol-3-Phosphate Hapten via Asymmetric Synthesis and Native Chemical Ligation. *J. Am. Chem. Soc.* **2014**, *136* (1), 412–418. https://doi.org/10.1021/ja410750a

[42] Kent, S. B. H. Total Chemical Synthesis of Proteins. *Chem. Soc. Rev.* **2009**, *38* (2), 338–351. https://doi.org/10.1039/B700141J

[43] Jiao, C.; Summerlin, M.; Bruzik, K. S.; Hanakahi, L. Synthesis of Biotinylated Inositol Hexakisphosphate to Study DNA Double-Strand Break Repair and Affinity Capture of IP 6 -Binding Proteins. *Biochemistry.* **2015**, *54* (41), 6312–6322. https://doi.org/10.1021/acs.biochem.5b00642

[44] Macbeth, M. R.; Schubert, H. L.; VanDemark, A. P.; Lingam, A. T.; Hill, C. P.; Bass, B. L. Inositol Hexakisphosphate Is Bound in the ADAR2 Core and Required for RNA Editing. *Science.* **2005**, *309* (5740), 1534–1539. https://doi.org/10.1126/science.1113150

[45] Cole, C. N.; Scarcelli, J. J. Transport of Messenger RNA from the Nucleus to the Cytoplasm. *Curr. Opin. Cell Biol.* **2006**, *18* (3), 299–306. https://doi.org/10.1016/j.ceb.2006.04.006

[46] Shen, X.; Xiao, H.; Ranallo, R.; Wu, W.-H.; Wu, C. Modulation of ATP-Dependent Chromatin-Remodeling Complexes by Inositol Polyphosphates. *Science.* **2003**, *299* (5603), 112–114. https://doi.org/10.1126/science.1078068

[47] Sarmah, B.; Wente, S. R. Dual Functions for the *Schizosaccharomyces Pombe* Inositol Kinase Ipk1 in Nuclear mRNA Export and Polarized Cell Growth. *Eukaryot Cell.* **2009**, *8* (2), 134–146. https://doi.org/10.1128/EC.00279-08
[48] Tsui, M. M.; York, J. D. Roles of Inositol Phosphates and Inositol Pyrophosphates in Development, Cell Signaling and Nuclear Processes. *Adv. Enzyme Reg.* **2010**, *50* (1), 324–337. https://doi.org/10.1016/j.advenzreg.2009.12.002
[49] Hanakahi, L. A.; Bartlet-Jones, M.; Chappell, C.; Pappin, D.; West, S. C. Binding of Inositol Phosphate to DNA-PK and Stimulation of Double-Strand Break Repair. *Cell.* **2000**, *102* (6), 721–729. https://doi.org/10.1016/S0092-8674(00)00061-1
[50] Gregory, M.; Catimel, B.; Yin, M.-X.; Condron, M.; Burgess, A.; Holmes, A. Synthesis of a Tethered Myo-Inositol (1,3,4,5,6) Pentakisphosphate (IP5) Derivative as a Probe for Biological Studies. *Synlett.* **2015**, *27* (1), 121–125. https://doi.org/10.1055/s-0035-1560381
[51] Gao, Y.; Wang, H. Inositol Pentakisphosphate Mediates Wnt/β-Catenin Signaling. *J. Biol. Chem.* **2007**, *282* (36), 26490–26502. https://doi.org/10.1074/jbc.M702106200
[52] H. W. Lee and Y. Kishi, *J. Org. Chem.* **1985**, *50*, 4402–4404.
[53] Shears, S. B. Inositol Pyrophosphates: Why so Many Phosphates? *Adv. Biol. Regul.* **2015**, *57*, 203–216. https://doi.org/10.1016/j.jbior.2014.09.015
[54] Thomas, M. P.; Potter, B. V. L. The Enzymes of Human Diphosphoinositol Polyphosphate Metabolism. *FEBS J.* **2014**, *281* (1), 14–33. https://doi.org/10.1111/febs.12575
[55] Ashton, T. D.; Jolliffe, K. A.; Pfeffer, F. M. Luminescent Probes for the Bioimaging of Small Anionic Species in Vitro and in Vivo. *Chem. Soc. Rev.* **2015**, *44* (14), 4547–4595. https://doi.org/10.1039/C4CS00372A
[56] Shipton, M. L.; Jamion, F. A.; Wheeler, S.; Riley, A. M.; Plasser, F.; Potter, B. V. L.; Butler, S. J. Expedient Synthesis and Luminescence Sensing of the Inositol Pyrophosphate Cellular Messenger 5-PP-InsP $_5$. *Chem. Sci.* **2023**, *14* (19), 4979–4985. https://doi.org/10.1039/D2SC06812E
[57] a) Riley, A. M.; Wang, H.; Weaver, J. D.; Shears, S. B.; Potter, B. V. L. *Chem. Commun.* **2012**, *48*, 11292–11294; b) Montchamp, J. L.; Tian, F.; Hart, M. E.; Frost, J. W. *J. Org. Chem.* **1996**, *61*, 3897–3899.
[58] Lin, H.; Fridy, P. C.; Ribeiro, A. A.; Choi, J. H.; Barma, D. K.; Vogel, G.; Falck, J. R.; Shears, S. B.; York, J. D.; Mayr, G. W. Structural Analysis and Detection of Biological Inositol Pyrophosphates Reveal That the Family of VIP/Diphosphoinositol Pentakisphosphate Kinases Are 1/3-Kinases. *J. Biol. Chem.* **2009**, *284* (3), 1863–1872. https://doi.org/10.1074/jbc.M805686200
[59] Capolicchio, S.; Wang, H.; Thakor, D. T.; Shears, S. B.; Jessen, H. J. Synthesis of Densely Phosphorylated Bis-1,5-Diphospho- *Myo* -Inositol Tetrakisphosphate and Its Enantiomer by Bidirectional P-Anhydride Formation. *Angew. Chem. Int. Ed.* **2014**, *53* (36), 9508–9511. https://doi.org/10.1002/anie.201404398
[60] Conway, S. J.; Gardiner, J.; Grove, S. J. A.; Johns, M. K.; Lim, Z. Y.; Painter, G. F.; Robinson, D. E. J. E.; Schieber, C.; Thuring, J. W.; Wong, L. S. M.; Yin, M. X.; Burgess, A. W.; Catimel, B.; Hawkins, P. T.; Ktistakis, N. T.; Stephens, L. R.; Holmes, A. B. *Org. Biomol. Chem.* **2010**, *8*, 66–76.
[61] Pavlovic, I.; Thakor, D. T.; Vargas, J. R.; McKinlay, C. J.; Hauke, S.; Anstaett, P.; Camuña, R. C.; Bigler, L.; Gasser, G.; Schultz, C.; Wender, P. A.; Jessen, H. J. Cellular Delivery and Photochemical Release of a Caged Inositol-Pyrophosphate Induces PH-Domain Translocation in Cellulo. *Nat. Commun.* **2016**, *7* (1), 10622. https://doi.org/10.1038/ncomms10622
[62] Pavlovic, I.; Thakor, D. T.; Jessen, H. J. Synthesis of 2-Diphospho-Myo-Inositol 1,3,4,5,6-Pentakisphosphate and a Photocaged Analogue. *Org. Biomol. Chem.* **2016**, *14* (24), 5559–5562. https://doi.org/10.1039/C6OB00094K
[63] Tsai, Y.; Liu, X.; Seeberger, P. H. Chemical Biology of Glycosylphosphatidylinositol Anchors. *Angew. Chem. Int. Ed.* **2012**, *51* (46), 11438–11456. https://doi.org/10.1002/anie.201203912
[64] Vishwakarma, R.; Ruhela, D. Chapter 10 Chemical Synthesis of Glycosylphosphatidylinositol (GPI) Anchors. In *The Enzymes*; Elsevier, 2009; Vol. 26, pp. 181–227. https://doi.org/10.1016/S1874-6047(09)26010-0
[65] Garigapati, V. R.; Roberts, M. F. Synthesis of Phosphatidyl-2-O-Alkylinositols as Potential Inhibitors for PI Specific PLC. *Tetrahedron Lett.* **1993**, *34* (35), 5579–5582. https://doi.org/10.1016/S0040-4039(00)73886-X
[66] Shashidhar, M. S.; Volwerk, J. J.; Keana, J. F. W.; Griffith, O. H. Inhibition of Phosphatidylinositol-Specific Phospholipase C by Phosphonate Substrate Analogues. *Biochim. Biophys. Acta, Lipids Lipid Metab.* **1990**, *1042* (3), 410–412. https://doi.org/10.1016/0005-2760(90)90172-T
[67] Mihai, C.; Yue, X.; Zhao, L.; Kravchuk, A.; Tsai, M.-D.; Bruzik, K. S. Nonhydrolyzable Analogs of Phosphatidylinositol as Ligands of Phospholipases C. *New J. Chem.* **2010**, *34* (5), 925. https://doi.org/10.1039/b9nj00629j

[68] Yadav, M.; Raghupathy, R.; Saikam, V.; Dara, S.; Singh, P. P.; Sawant, S. D.; Mayor, S.; Vishwakarma, R. A. Synthesis of Non-Hydrolysable Mimics of Glycosylphosphatidylinositol (GPI) Anchors. *Org. Biomol. Chem.* **2014**, *12* (7), 1163. https://doi.org/10.1039/c3ob42116c
[69] Saikam, V.; Raghupathy, R.; Yadav, M.; Gannedi, V.; Singh, P. P.; Qazi, N. A.; Sawant, S. D.; Vishwakarma, R. A. Synthesis of New Fluorescently Labeled Glycosylphosphatidylinositol (GPI) Anchors. *Tetrahedron Lett.* **2011**, *52* (33), 4277–4279. https://doi.org/10.1016/j.tetlet.2011.06.005
[70] Ohira, S.; Yamaguchi, Y.; Takahashi, T.; Tanaka, H. The Chemoselective O-Glycosylation of Alcohols in the Presence of a Phosphate Diester and Its Application to the Synthesis of Oligomannosylated Phosphatidyl Inositols. *Tetrahedron.* **2015**, *71* (37), 6602–6611. https://doi.org/10.1016/j.tet.2015.06.041
[71] Lotter, H.; González-Roldán, N.; Lindner, B.; Winau, F.; Isibasi, A.; Moreno-Lafont, M.; Ulmer, A. J.; Holst, O.; Tannich, E.; Jacobs, T. Natural Killer T Cells Activated by a Lipopeptidophosphoglycan from Entamoeba Histolytica Are Critically Important to Control Amebic Liver Abscess. *PLoS Pathog.* **2009**, *5* (5), e1000434. https://doi.org/10.1371/journal.ppat.1000434
[72] Iwasaki, T.; Higashikawa, K.; Reddy, V. P.; Ho, W. W. S.; Fujimoto, Y.; Fukase, K.; Terao, J.; Kuniyasu, H.; Kambe, N. *Chem. Eur. J.* **2013**, *19*, 2956–2960.
[73] https://doi.org/10.1002/chem.2017012982017_CEJ_Fujimoto__BINOL_and_SYNTHESIS
[74] Trinel, P.-A.; Maes, E.; Zanetta, J.-P.; Delplace, F.; Coddeville, B.; Jouault, T.; Strecker, G.; Poulain, D. Candida Albicans Phospholipomannan, a New Member of the Fungal Mannose Inositol Phosphoceramide Family. *J. Biol. Chem.* **2002**, *277* (40), 37260–37271. https://doi.org/10.1074/jbc.M202295200
[75] Gannedi, V.; Ali, A.; Singh, P. P.; Vishwakarma, R. A. Total Synthesis of Phospholipomannan of *Candida Albicans. J. Org. Chem.* **2020**, *85* (12), 7757–7771. https://doi.org/10.1021/acs.joc.0c00402
[76] Saleem, H.; Tovey, S. C.; Riley, A. M.; Potter, B. V. L.; Taylor, C. W. Stimulation of Inositol 1,4,5-Trisphosphate (IP3) Receptor Subtypes by Adenophostin A and Its Analogues. *PLoS One* **2013**, *8* (2), e58027. https://doi.org/10.1371/journal.pone.0058027
[77] Mills, S. J.; Rossi, A. M.; Konieczny, V.; Bakowski, D.; Taylor, C. W.; Potter, B. V. L. D *-Chiro* -Inositol Ribophostin: A Highly Potent Agonist of D *-Myo* -Inositol 1,4,5-Trisphosphate Receptors: Synthesis and Biological Activities. *J. Med. Chem.* **2020**, *63* (6), 3238–3251. https://doi.org/10.1021/acs.jmedchem.9b01986
[78] Dohle, W.; Su, X.; Mills, S. J.; Rossi, A. M.; Taylor, C. W.; Potter, B. V. L. A Synthetic Cyclitol-Nucleoside Conjugate Polyphosphate Is a Highly Potent Second Messenger Mimic. *Chem. Sci.* **2019**, *10* (20), 5382–5390. https://doi.org/10.1039/C9SC00445A.
[79] Ma, X.; Yan, Q.; Banwell, M. G.; Ward, J. S. A Total Synthesis of the Antifungal Deoxyaminocyclitol Nabscessin B from L-(+)-Tartaric Acid. *Org. Lett.* **2018**, *20* (1), 142–145. https://doi.org/10.1021/acs.orglett.7b03495.
[80] Sato, K.; Akai, S.; Sugita, N.; Ohsawa, T.; Kogure, T.; Shoji, H.; Yoshimura, J. Novel and Stereocontrolled Synthesis of (±)-Tetrodotoxin from *mYo* -Inositol. *J. Org. Chem.* **2005**, *70* (19), 7496–7504. https://doi.org/10.1021/jo050342t
[81] Jagdhane, R. C.; Shashidhar, M. S. A Formal Synthesis of Valiolamine from Myo-Inositol. *Tetrahedron.* **2011**, *67* (41), 7963–7970. https://doi.org/10.1016/j.tet.2011.08.027.
[82] Gurale, B. P.; Shashidhar, M. S.; Gonnade, R. G. Synthesis of the Aminocyclitol Units of (−)-Hygromycin A and Methoxyhygromycin from *Myo* -Inositol. *J. Org. Chem.* **2012**, *77* (13), 5801–5807. https://doi.org/10.1021/jo300444b
[83] Wu, H.-Y.; Wang, W.-Y.; Feng, C.-C.; Hou, D.-R. Asymmetric Synthesis of Nabscessin A from Inositol and D-Camphor. *J. Org. Chem.* **2020**, *85* (20), 13153–13159. https://doi.org/10.1021/acs.joc.0c01839
[84] Hara, S.; Ishikawa, N.; Hara, Y.; Nehira, T.; Sakai, K.; Gonoi, T.; Ishibashi, M. Nabscessins A and B, Aminocyclitol Derivatives from *Nocardia Abscessus* IFM 10029 T. *J. Nat. Prod.* **2017**, *80* (2), 565–568. https://doi.org/10.1021/acs.jnatprod.6b00935
[85] Ma, X.; Yan, Q.; Banwell, M. G.; Ward, J. S. A Total Synthesis of the Antifungal Deoxyaminocyclitol Nabscessin B from L -(+)-Tartaric Acid. *Org. Lett.* **2018**, *20* (1), 142–145. https://doi.org/10.1021/acs.orglett.7b03495
[86] Mondal, S.; Sureshan, K. M. Total Syntheses and Structural Validation of Lincitol A, Lincitol B, Uvacalol I, Uvacalol J, and Uvacalol K. *Org. Biomol. Chem.* **2014**, *12* (37), 7279–7289. https://doi.org/10.1039/C4OB01329H
[87] Sakai, T.; Suzuki, K.; Ohmori, K. First Total Synthesis of Oxirapentyn D, a Highly Oxidized Chromene Natural Product. *Synlett.* **2018**, *29* (10), 1351–1357. https://doi.org/10.1055/s-0036-1591563
[88] Akagi, A.; Usuguchi, K.; Takashima, I.; Okuda, K. Total Synthesis of Antiausterity Agent (±)-Uvaridacol L by Regioselective Axial Diacylation of a *Myo* -Inositol Orthoester. *Org. Lett.* **2021**, *23* (11), 4083–4087. https://doi.org/10.1021/acs.orglett.1c00079

3 Progress in Stereocontrolled 1,2-*cis* Glycosylation in Synthetic Carbohydrate Chemistry

Zanjila Azeem, Himanshu Gangwar, Sunny Maurya, and Pintu Kumar Mandal

3.1 INTRODUCTION

Carbohydrates, such as polyhydroxylated aldehydes and ketones, represent the essential category of naturally occurring synthons that are present in many biologically active molecules in nature. Carbohydrates are indispensable in eliciting a myriad of biological events in life including immune defense, signal transduction, metastasis, fertilization, cell growth, cell–cell adhesion, and neurodegenerative disease [1–2]. Carbohydrates also play a vital part in bacterial and viral infections; the pathogenesis of diabetes; and the development and growth of cancers, inflammation, septicemia, and many other diseases. For the most part, biologically beneficial sugars exist as glycoconjugates (glycoproteins or glycolipids) or glycosides, polysaccharides, lipids, proteins, peptides, etc. in nature [3].

Although these carbohydrate synthons have demonstrated supremacy, their study is challenging because it is difficult to synthesize, isolate, and characterize these molecules owing to complexity and heterogeneity in nature. This makes the chemical syntheses of the relevant carbohydrate moieties pertinent. Among various glycosidic bonds, *O*-glycosidic bonds have caught major attention and are also tough work for chemists due to their larger prosperity and complication in synthesis. Further, the challenge associated with glycoconjugates synthesis is the stereochemical outcome at the C1 or anomeric position of sugars. Generally, two types of stereoselective glycosides are observed, 1,2-*cis* and 1,2-*trans* or α- and β-glycosides. The stereoselectivity of glycosides such as 1,2-*cis* and 1,2-*trans* can be observed at the anomeric center with respect to the C2 position (Figure 3.1).

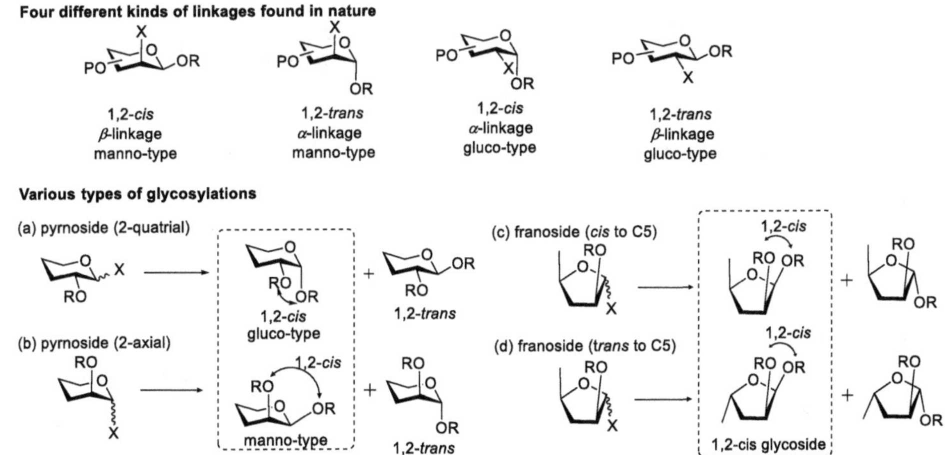

FIGURE 3.1 Structures of anomeric link, various glycosylations, and 1,2-*cis* glycosides in nature (high mannose-type *N*-glycan).

DOI: 10.1201/9781003437413-3

1,2-*cis* glycoside in nature (high mannose-type *N*-glycan is shown as the example)

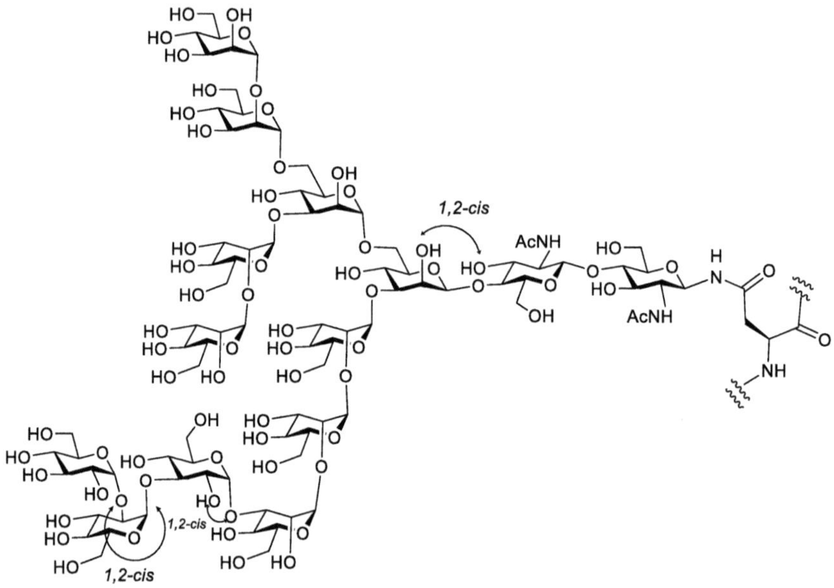

1,2-cis

1,2-cis

1,2-cis

FIGURE 3.1 (Continued)

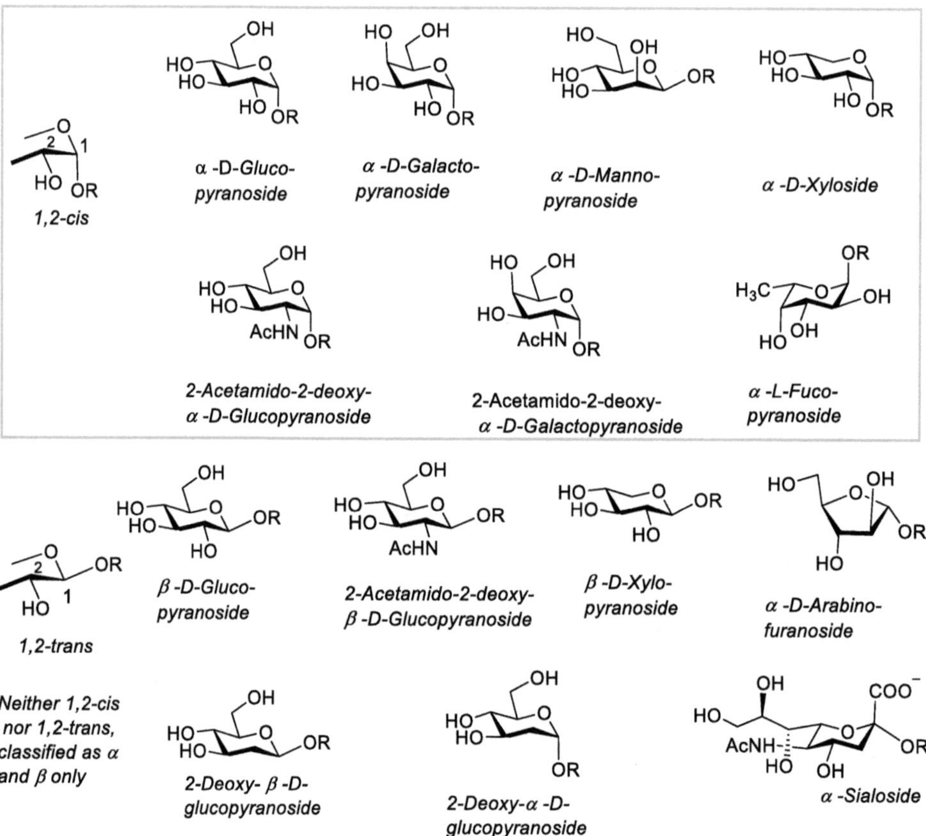

FIGURE 3.2 Most common monosaccharide residues found in the mammalian and bacterial glycome.

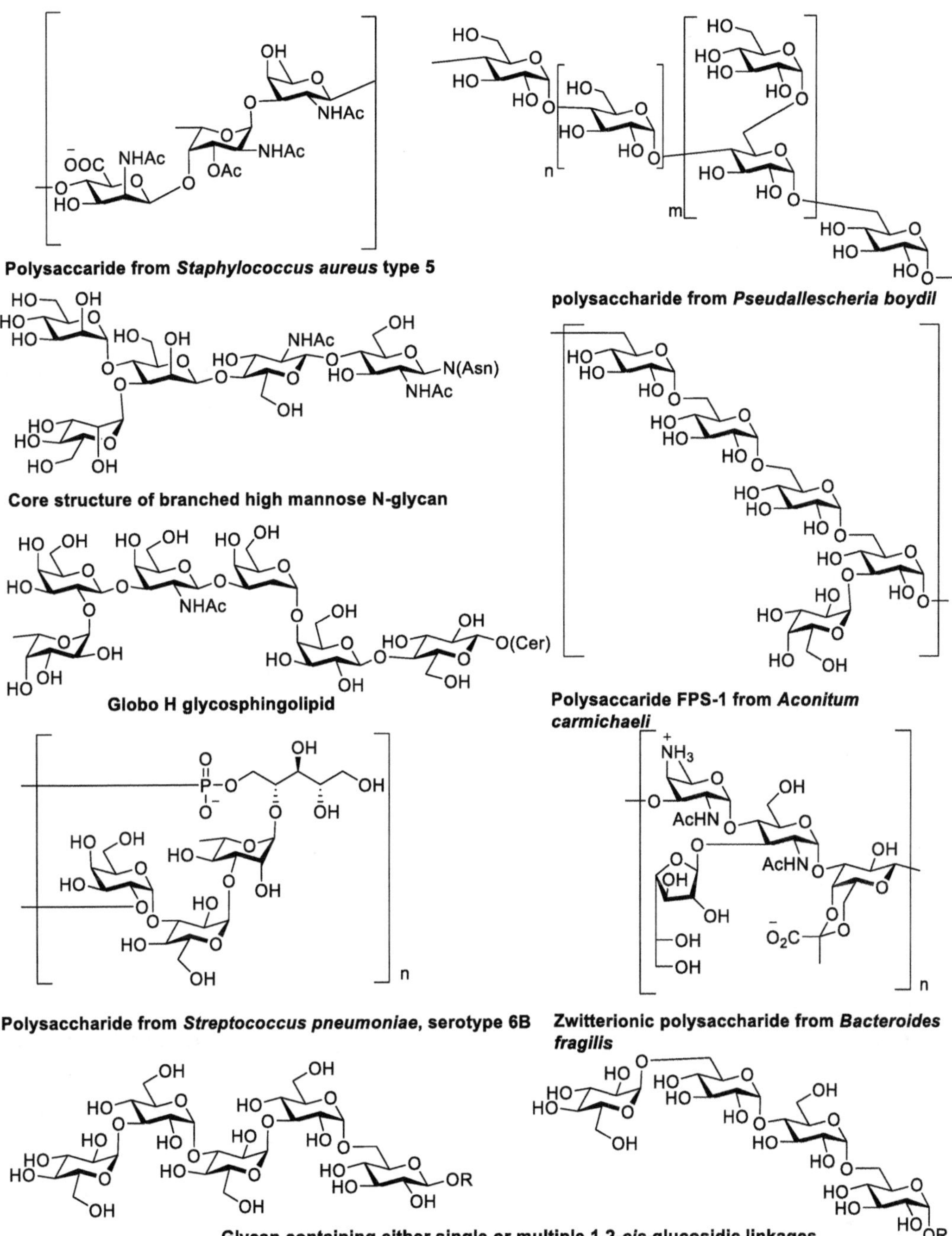

FIGURE 3.3 Naturally occurring biological importance of oligosaccharides containing 1,2-*cis* linkages.

The *cis*-configured *O*-glycosidic linkages present at the 2-positions of the nonreducing side residue of glycosides, such as α-galactopyranoside, α-glucopyranoside, β-arabinofuranoside, and β-mannopyranoside, and other common glycosides are present in natural glycans, including glycoconjugates (such as proteoglycans, glycolipids, glycoproteins, and microbial polysaccharides) as well as glycoside natural products [4].

Moreover, 1,2-*cis* glycosyl tethers such as α-D-glucoside or β-D-mannoside, α-D-galactoside, and β-D-L-rhamnoside are present in a myriad of natural compounds. Additionally, glycoside types such as 2-deoxyglycosides and sialosides do not possess adjacent substituents, and hence these compounds do not fall into the categories of 1,2-*cis* or 1,2-*trans* glycosides. Instead, they are commonly labeled α- and β-glycosides, as illustrated in Figure 3.2.

However, the difficulty in isolation led to the construction of 1,2-*cis* glycosidic linkages in the laboratory. Some 1,2-*cis* linkages containing biologically important synthons having therapeutic potential are listed in Figure 3.3 [5–10].

3.2 GENERAL PRINCIPLES AND MECHANISM OF CHEMICAL GLYCOSYLATION

In the realm of sugar chemistry, glycosylation has been found to be a challenging task owing to its structure complexity, diversity, and solubility in highly polar or aqueous solutions. In the field of glycosylation, Michael [11] and Fischer [12] reported the first aryl glycosidic bond (*p*-methoxy phenyl β-D-glucopyranoside) and alkyl glycosidic bond (methyl α-D-glucopyranoside), in the 19th century. Inspired by their work, Knoenigs and Knorr developed a more controlled and refitted glycosylation approach [13]. Generally, the construction of an *O*-glycosidic bond involves two moieties such as a glycosyl acceptor having one free nucleophilic OH group and another glycosyl donor containing a leaving group at an anomeric position (Scheme 3.1) [14].

For a more generalized mechanistic route, the donor is activated via an electrophilic promoter to generate a donor–promoter complex. Next, the formation of oxacarbenium ion proceeds in its flattened half-chair conformation. After that, the unprotected nucleophilic hydroxyl group of glycosyl acceptor may act on the carbon of the oxocarbenium ion either from the bottom or the top face of the flattened ring, forming either α or β glycosides (Scheme 3.2). The stereochemical outcome of the glycosylation depends upon the protecting group attached at the C-2 position of the glycosyl donor: A participating group such as a carboxyl group will preferentially form *trans*- or *β*-glycoside, whereas a nonparticipating group (i.e., a group other than carboxyl) will formed majorly *cis*- or *α*-glycoside.

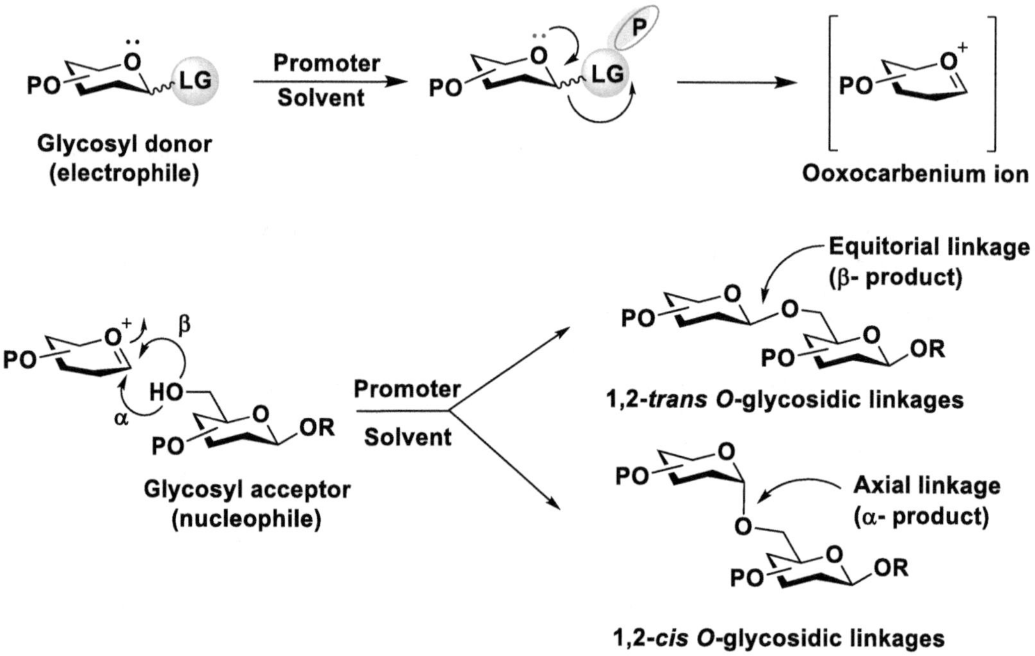

SCHEME 3.1 General process of chemical glycosylation.

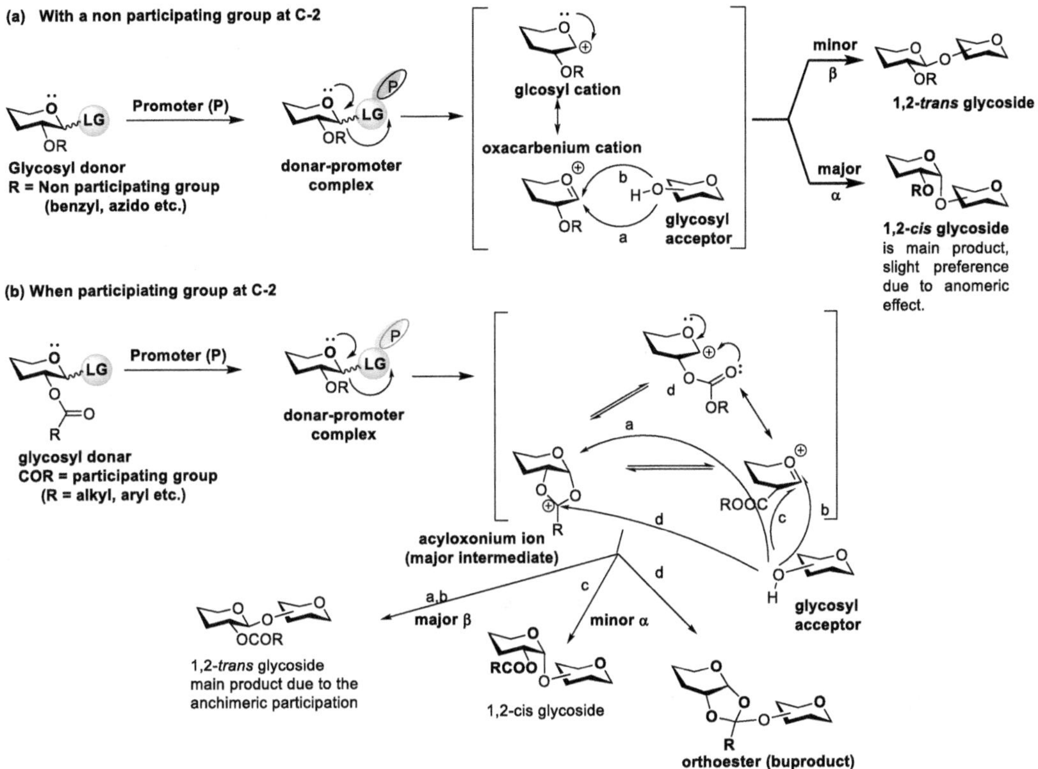

SCHEME 3.2 The general mechanism for chemical glycosylation.

Thus, the ester-protecting group at C-2 participates in glycosylation, leading to the easy formation of stereoselective 1,2-*trans*-glycosides. However, the stereoselective synthesis of 1,2 *cis*-glycosidic linkages is challenging [15], because the presence of a nonparticipating group at the C-2 position doesn't undergo neighboring group participation (NGP); therefore, a mixture of α or β stereoisomers forms that has to be separated at the end of the protocol (Scheme 3.2). Consequently, the construction of 1,2-*cis* linkages is typically more complex than forming 1,2-*trans* linkages, as the reaction often occurs via S_N^1 mechanistic pathway, and hence poor stereoselectivity control is obtained (Scheme 3.2) [16].

3.3 METHODS OF FORMING 1,2-*CIS* GLYCOSIDES

The construction of 1,2-*cis* glycosidic linkages hinges not only on the attachment of the nonparticipating group at the C-2 position but on the anomeric effect, which gave α-selective product [17] in poor selectivity, as well as on other stereocontrol factors for efficient selectivities (Scheme 3.3a). Myriad researchers have investigated the 1,2-*cis* glycosidic linkages. For instance, Barresi and Hindsgaul employed intramolecular glycosylation to build a 1,2-*cis* glycosidic bond. Further, Ito and coworkers reported naphthyl methyl ether-mediated intramolecular aglycone delivery [18] (Scheme 3.3b) via detachment of nucleophiles from the adjacent C-2 carbon and shifting to the anomeric position. This method is thoroughly employed for the stereoselective synthesis of numerous 1,2-*cis* linkages such as β-Araf, β-Manp, and α-Glcp. However, this protocol used to be technically tough for creating more than one *cis*-link.

Another strategy that follows hydrogen bond-mediated aglycan delivery involves the masked C6 position by picolinyl or picoloyl groups which drives to construction of *cis*-glycosidic link (Scheme 3.3c) [19]. Notably, this reported protocol is not worthy for *cis*-galactosidic linkages, as all of the hydroxyl groups in galactose at C3, C4, or C6 positions face upward of the rings. The

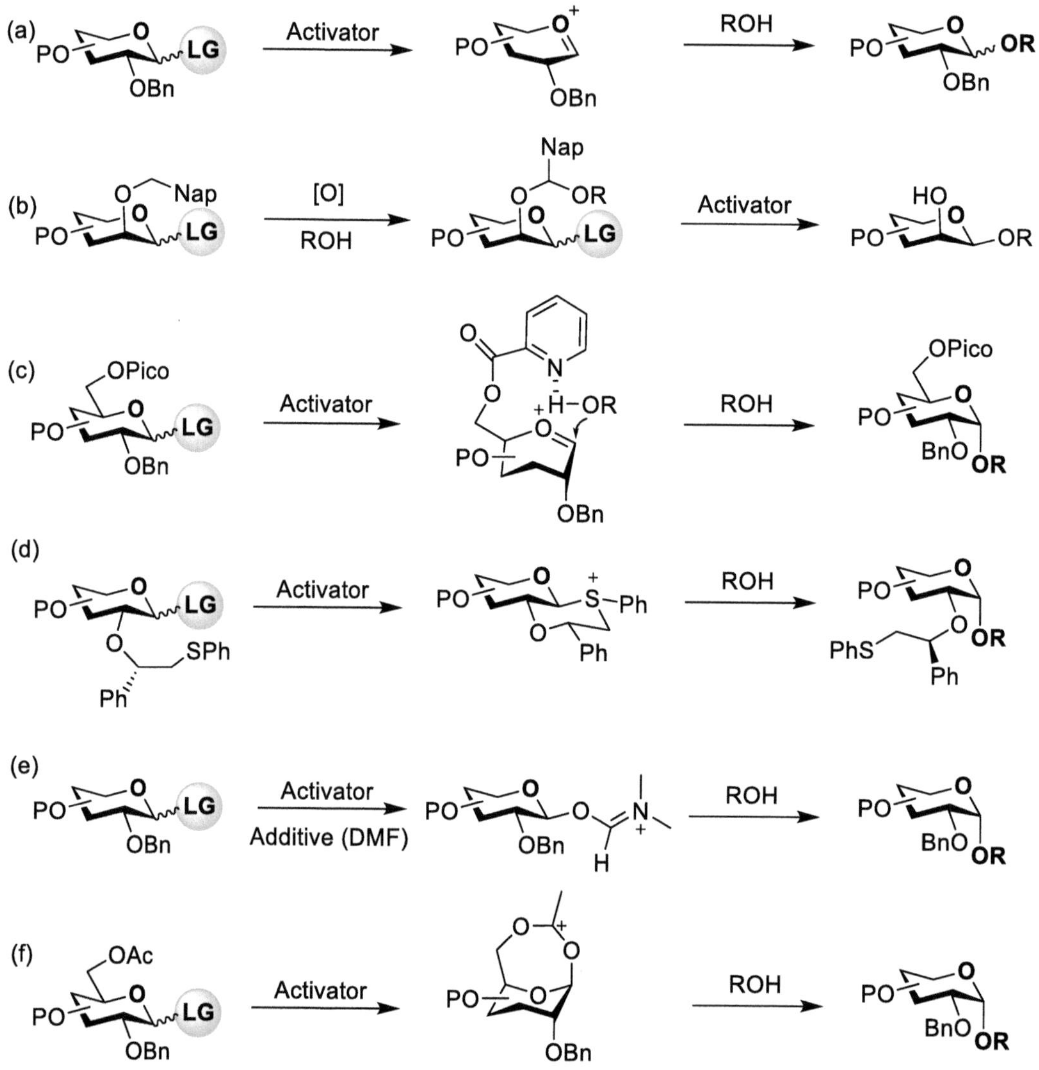

SCHEME 3.3 Well-known methods for the formation of 1,2-*cis*-glycosides: (a) A nonparticipating masking group at C-2 gives a mixture of anomers; (b) intramolecular aglycon delivery with 2-naphthylmethyl (NAP)-ether at C-2; (c) hydrogen-bond-mediated aglycan delivery; d) chiral auxiliaries assuring complete selectivity; (e) additive (or solvent) modulation; (f) remote participating groups.

idea was initially proposed by Demchenko [20]. Moreover, in 2005 Boons and colleagues introduced one of the most sophisticated and inventive methods for synthesizing 1,2-*cis*-glycosides, employing a chiral auxiliary at the O-2 position as illustrated in Scheme 3.3d [21]. Next, Koto et al. demonstrated the use of DMF as an additive to facilitate α-glycosylation; however, this approach faced challenges related to the unwanted generation of glycosyl formate as depicted in Scheme 3.3e [22]. A particularly intriguing strategy involves the remote participation of acyl groups, achieved by employing masking groups attached to the C3, C4, and/or C6 positions of glucose (Glc) and galactose (Gal) as glycosyl donors and a nonparticipating alkyl group at O-2 or an azide at C-2, which can effectively steer the stereoselectivity of glycosylations as shown in Scheme 3.3f [23, 24].

3.4 EFFECT OF THE GLYCOSYL DONOR

In addition to approaches for α- or β-glycosidic bond formation, substrate-controlled strategies show reliable and efficient stereoselective glycosylation, and shaping a glycosyl donor plays a significant role in stereoselective control glycosylation. Generally, the glycosylation reaction follows a unimolecular $S_N{}^1$ mechanistic path, so the anomeric orientation of the leaving group is of little importance. However, in the case of the $S_N{}^2$ mechanistic path, glycosylation proceeds by inverting the configuration at the anomeric center. In this context, James and coworkers presented 1,2-*cis* selectivity via β-glycosyl bromide, which formed from their α-counterpart enroute to 1,2-*cis* glycosylation (Scheme 3.4) [25].

Demchenko and group reported 1,2-*cis* linkage by α-thioglycoside in the presence of bromine, formed β-glycosyl bromide and finally 1,2-*cis* glycoside (Scheme 3.5) [26]. In the proposed methodology, α-bromide was found unreactive in the presence of bromine, and this was a particularly significant result because it confirmed that its β-bromide will be the key species en route to the glycosylation product under these reaction conditions.

Further, the same group published stereospecific 1,2-*cis* glycosylation by employing β-glycosyl thiocyanates (Scheme 3.6) [27]. This new approach employed a 1,2 *trans*-glycosyl thiocyanate with the sugar derivatives having unsubstituted hydroxyl groups in the presence of trimethylsilyl triflate and molecular sieves and follows the concerted cyclic process.

Further, Sun and Smith introduced 1,2-*cis* stereoselectivity in mannose via anomeric mannosyl triflates formed in situ from sulfoxides or thioglycosides for the synthesis of β-mannosides (Scheme 3.7) [28]. They performed the glycosylation by Tf_2O and DTMB in Et_2O; however, the reported strategy was limited to primary glycosyl acceptors, while the Smith strategy performed well with secondary acceptors and is a successful polymer-supported method for synthesizing the challenging *α*-mannopyranoside-type glycosidic bond.

3.4.1 Neighboring Group Participation of O-2 Chiral Auxiliaries

As aforementioned, the synthesis of 1,2-*trans* glycoside is feasible by virtue of NGP. The strategy generally involves a C-2 substituent such as an ester or acyl leading to exclusive 1,2-*trans* glycosides as mentioned in Scheme 3.8.

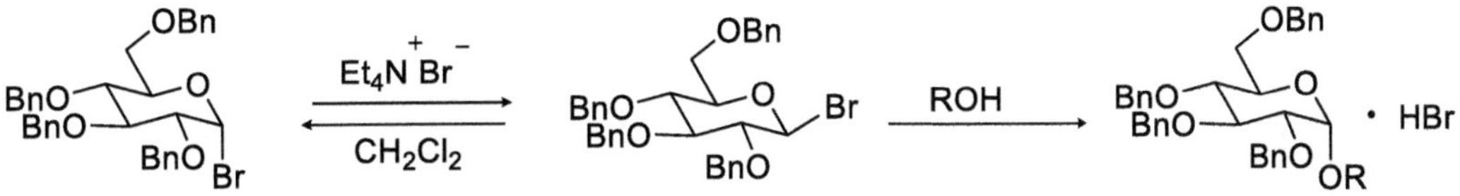

OBn BnO BnO O BnO Br $Et_4N^+Br^-$ CH_2Cl_2 OBn BnO BnO O Br BnO ROH OBn BnO BnO O BnO OR • HBr

SCHEME 3.4 Halide ion catalyzed 1,2-*cis* glycosylation.

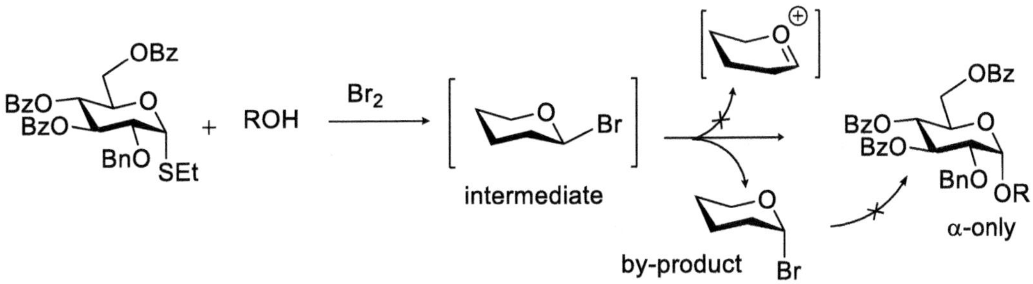

OBz BzO BzO O BnO SEt + ROH Br_2 intermediate Br by-product Br OBz BzO BzO O BnO OR α-only

SCHEME 3.5 1,2-*cis* Glycosylation of thioglycoside donor via the intermediacy of β-bromides.

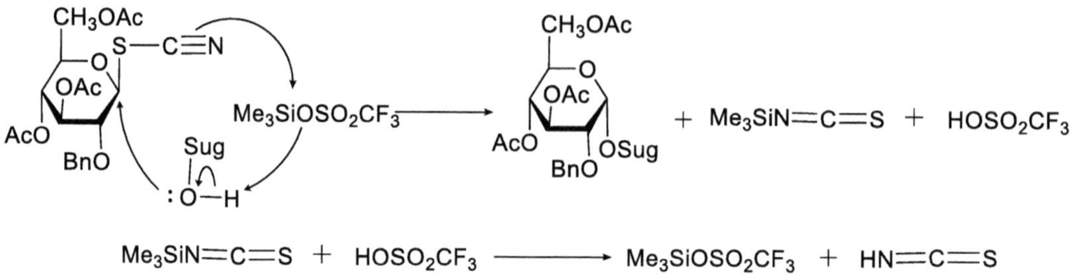

SCHEME 3.6 1,2-*cis* glycosylation by employing β-glycosyl thiocyanates.

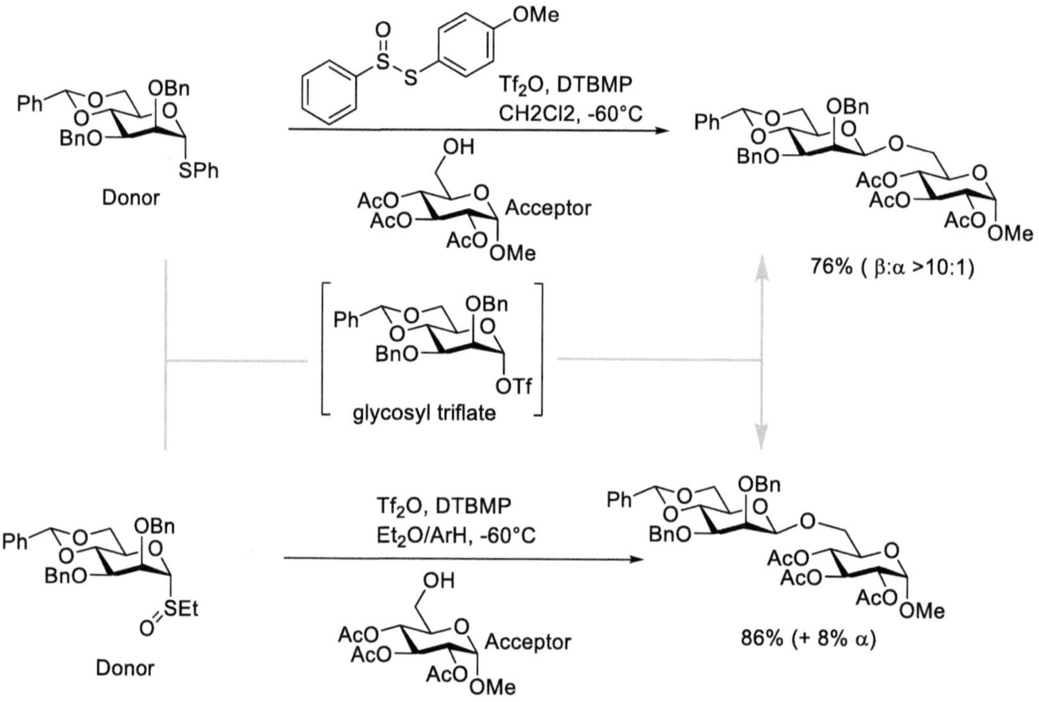

SCHEME 3.7 1,2-*cis* stereoselectivity in mannose via the in situ formation of glycosyl triflates from thioglycosides and sulfoxides.

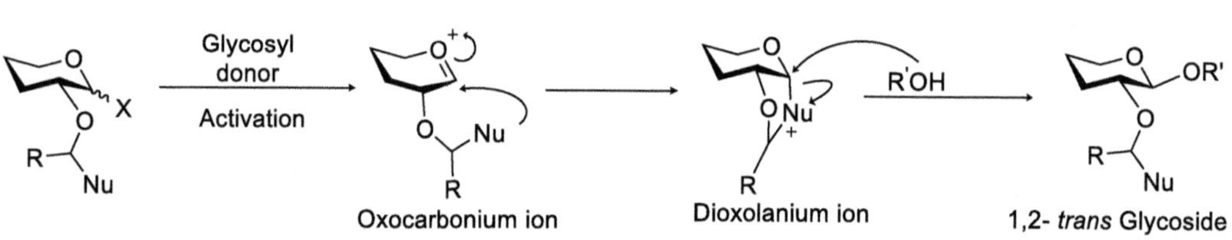

SCHEME 3.8 Neighboring group participation in the synthesis of 1,2-*trans* glycosides.

Advances with chiral auxiliaries led to remarkable achievements in the field of glycosylation, providing excellent stereocontrol. For instance, Boons and coworkers first demonstrated the solution to the anomeric stereoselectivity of glycoside via the chiral auxiliary control center. The authors proposed that anomeric selectivity hinges on the configuration of the asymmetric center of the chiral protecting group. They chose readily available ethyl mandelate in *S* and *R* form as the chiral auxiliary, and the formation of 1, 2-*cis* or 1,2-*trans* glycoside relies on *trans* and *cis* decalin intermediate, respectively. The NGP by an S auxiliary at C-2 favors the formation of a "quasi-stable" trans-decalin intermediate, thereby producing the 1,2-*cis*-glycoside. An auxiliary with (*R*)-configuration participates via a quasi-stable *cis*-decalin intermediate, leading to 1,2-*trans* glycoside (Scheme 3.9) [29].

Further, Boons developed a second-generation asymmetric chiral auxiliary to enhance the 1,2-*cis* selectivity of the glycoside. The 1,2-*cis*-selectivity depends on the fact that the (S)-asymmetric center or (S)-phenyl group occupied the equatorial position to prevent the 1,3-diaxial interaction. Therefore, the author employed 1-(S)-phenyl-2-(phenyl-sulfonyl) ethyl ether masked TCAI donor, which underwent 1,2-*cis* glycosylation exclusively with 86% yield of the product (Scheme 3.10). Further, the auxiliary was removed using BF_3-OEt_2 and acetic anhydride through acetolysis [30].

More recently, although researchers have made these achievements regarding 1,2-*cis*-selectivity, they have also faced setbacks such as the need for pure enantiomer. In this context, Turnbull employed thioglycoside converted into a bicyclic intermediate followed by sulfoxidation. Later, the sulfonium ion underwent glycosylation, leading to 1,2-*cis*-glycoside (Scheme 3.11) [31]. Boons also developed a procedure of using a bicyclic intermediate treated with a series of reagents to construct a 1,2-*cis*-glycosidic bond with 1,2-oxathiane esters [32]. The 1,2-oxathiane ethers are stable under acidic, basic, and reductive conditions, making them versatile donors that effectively couple with a wide variety of glycosyl acceptors including properly protected amino acids, primary and secondary sugar alcohols, and partially protected thioglycosides to give excellent yield and selectivity. Moreover, Turnbull also introduced oxathione spiroketal as a glycosyl donor that offered excellent 1,2-*cis*-selectivity and showed stability towards general masking group manipulations [33].

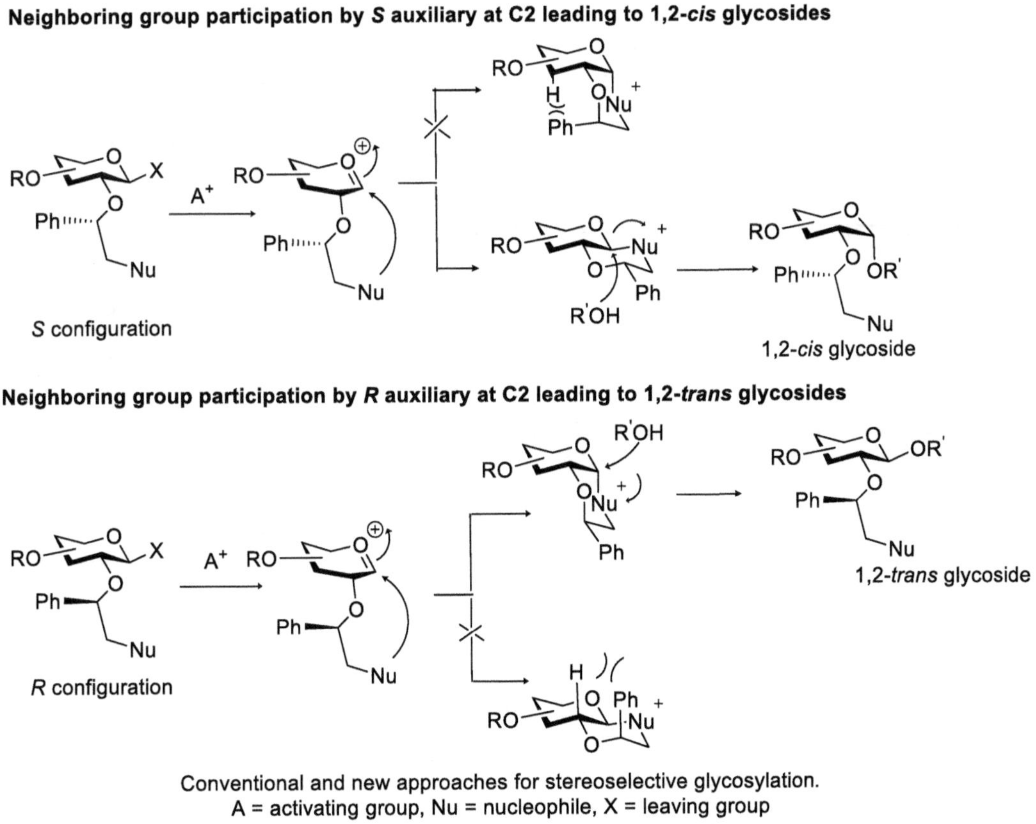

Neighboring group participation by *S* auxiliary at C2 leading to 1,2-*cis* glycosides

Neighboring group participation by *R* auxiliary at C2 leading to 1,2-*trans* glycosides

Conventional and new approaches for stereoselective glycosylation.
A = activating group, Nu = nucleophile, X = leaving group

SCHEME 3.9 Chiral auxiliaries: a) a chiral S-configured auxiliary yields 1,2-*cis* glycosides via a *trans*-configured sulfonium ion; b) a chiral R-configured auxiliary yields 1,2-*trans* glycosides via a *cis*-configured sulfonium ion.

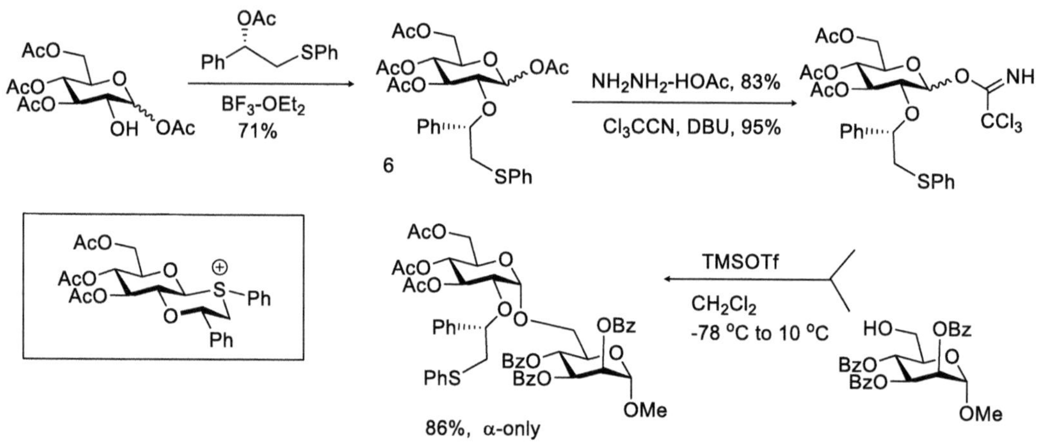
BF3-OEt2
71%
6
NH2NH2-HOAc, 83%
Cl3CCN, DBU, 95%
TMSOTf
CH2Cl2
-78 °C to 10 °C
86%, α-only

SCHEME 3.10 NGP by C-2 (1*S*)-phenyl-2-(phenylsulfanyl) ethyl moiety to 1,2-*cis* glycosides.

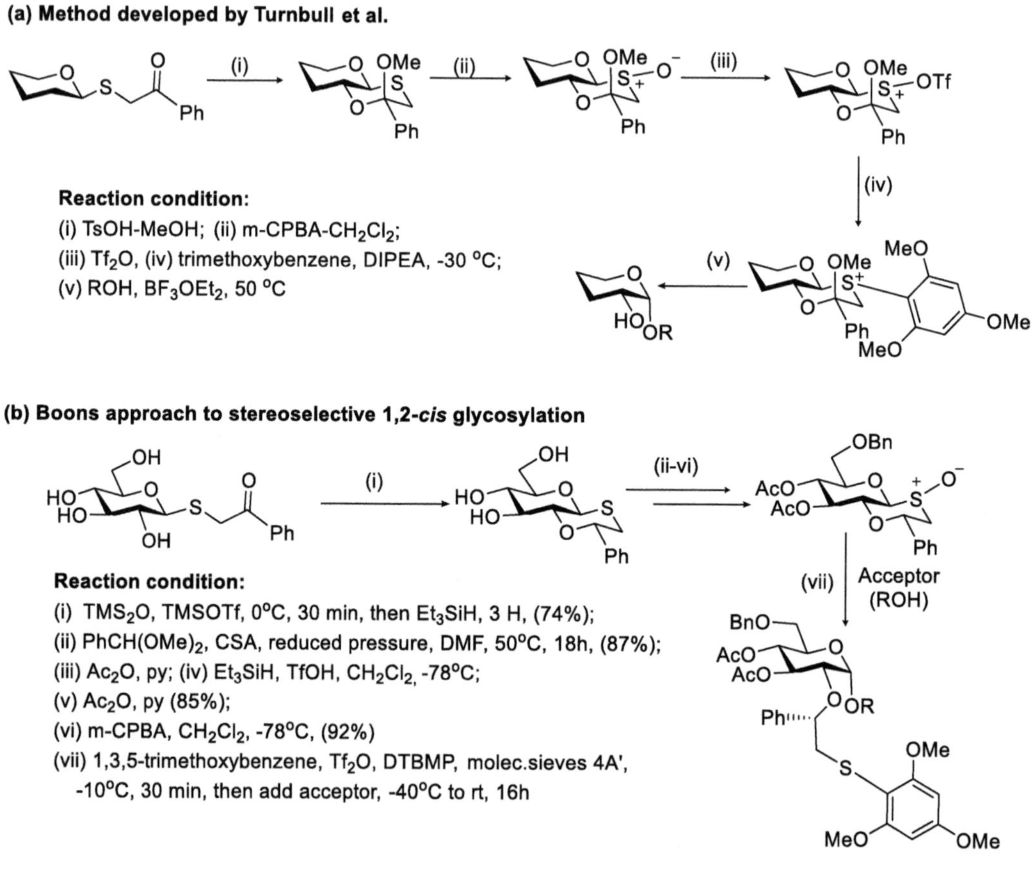
(a) Method developed by Turnbull et al.
Reaction condition:
(i) TsOH-MeOH; (ii) m-CPBA-CH2Cl2;
(iii) Tf2O, (iv) trimethoxybenzene, DIPEA, -30 °C;
(v) ROH, BF3OEt2, 50 °C
(b) Boons approach to stereoselective 1,2-cis glycosylation
Acceptor
(ROH)
Reaction condition:
(i) TMS2O, TMSOTf, 0°C, 30 min, then Et3SiH, 3 H, (74%);
(ii) PhCH(OMe)2, CSA, reduced pressure, DMF, 50°C, 18h, (87%);
(iii) Ac2O, py; (iv) Et3SiH, TfOH, CH2Cl2, -78°C;
(v) Ac2O, py (85%);
(vi) m-CPBA, CH2Cl2, -78°C, (92%)
(vii) 1,3,5-trimethoxybenzene, Tf2O, DTBMP, molec.sieves 4A',
-10°C, 30 min, then add acceptor, -40°C to rt, 16h

SCHEME 3.11 1,2-*cis* glycosylation reactions using the secondary auxiliary group.

3.4.2 Remote Participating Groups

The effects of remote participating groups are ambiguous and controversial, and they have been considered somewhat less important than NGP at the C-2 position. However, many researchers steered their attention to remote participation as the configuration at the anomeric center can be controlled by a remote protecting group, and high α-anomeric selectivity could be achieved.

3.4.2.1 At the C-4 Position

The influence of the remote protecting group at the C-4 position on stereoselectivity was reported by Ito and coworkers (Scheme 3.12) [34]. They synthesized a heptasaccharide from Gram-negative bacterium by the consecutive glycosylation of Boc-acceptor with 4-*O*-petafluoropropionyl-protected galactose donors. Herein, the role of the petafluoropropionyl group is to provide protection at the C-4 position as well as act as a remote participating group from β-face, which facilitates high α-selectivity.

3.4.2.2 At the C-6 Position

Further, the effect of the remote protecting group was studied at the C-6 position. The participation of (S)-(phenylthiomethyl) benzyl ether at C-6 led to an exclusive α-linked 2-deoxy glycoside. Also, employing (R)-(phenyl thiomethyl) benzyl ether at the C-6 position as a remote participating group provides excellent stereoselectivity, which further indicates that the anomeric outcome is not influenced by the chirality of the auxiliary group (Scheme 3.13) [35].

3.4.2.3 At the C-3 Position

In the same manner, the role of remote participating groups was investigated at the C-3 position of the sugar derivates including D-manno [36], D-gluco [37], L-rhanmno [38], and L-fuco [39]. In 2007, Xu and coworkers [40] developed a strategy to construct a 1,2-*cis*-glycosidic bond exclusively by taking advantage of 4,6-*O*-benzylidene-protected mannose donor remotely protected with *p*-trifluoromethyl benzylidene imine ether at C-3 position (Scheme 3.14). The reason for the exclusive stereoselectivity is that the comparable steric bulk of the remote participating group supported the torsional interaction of the O2-C2-C3-R3 for stereocontrolled 1,2-*cis* selectivity.

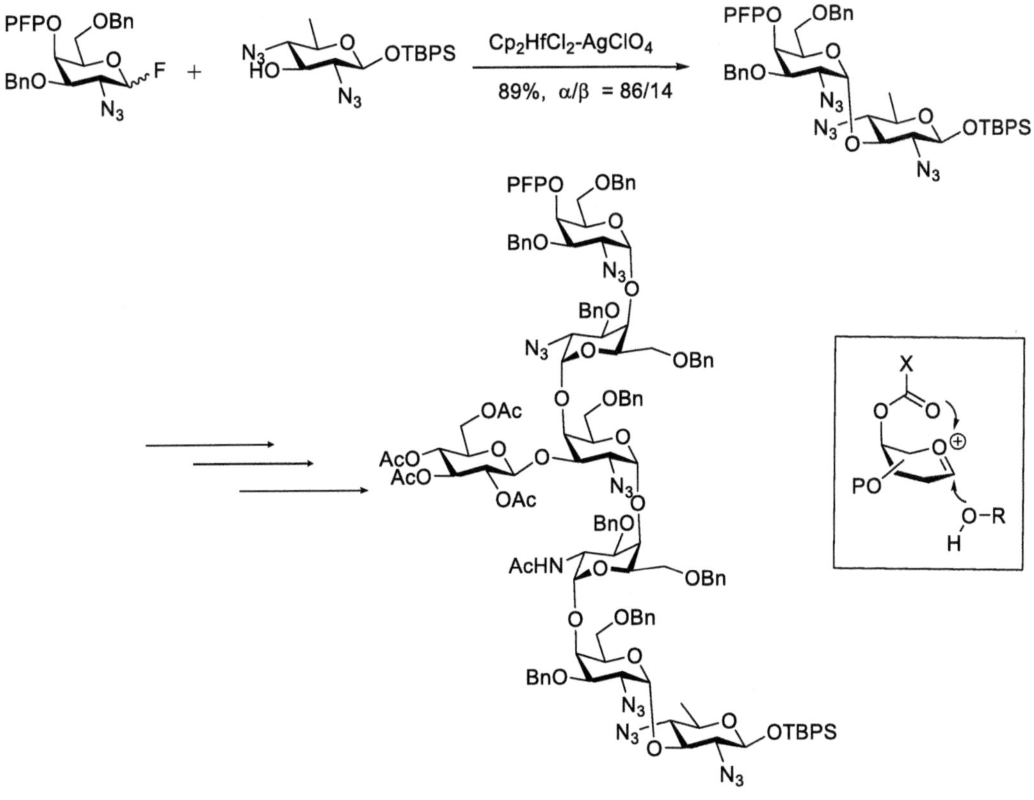

PFPO OBn BnO O F + N3 HO O OTBPS N3 — Cp_2HfCl_2-$AgClO_4$, 89%, α/β = 86/14 → PFPO OBn BnO O N3 N3 O O OTBPS N3

PFPO OBn BnO O N3 O BnO N3 O OBn O OBn OAc AcO AcO O OAc O O N3 O BnO AcHN O OBn O OBn BnO O N3 N3 O O OTBPS N3

X O O O⊕ PO O-R H

SCHEME 3.12 Synthesis of a heptasaccharide from Gram-negative bacterium by remote participation in 1,2-*cis*-galactosylation.

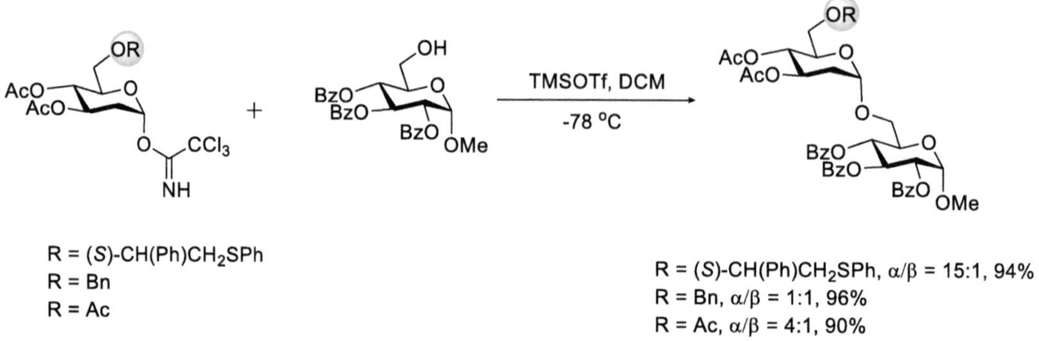

TMSOTf, DCM, -78 °C

R = (*S*)-CH(Ph)CH_2SPh
R = Bn
R = Ac

R = (*S*)-CH(Ph)CH_2SPh, α/β = 15:1, 94%
R = Bn, α/β = 1:1, 96%
R = Ac, α/β = 4:1, 90%

SCHEME 3.13 Synthesis of 2-deoxyglycosides using different protected donors at C-6 position.

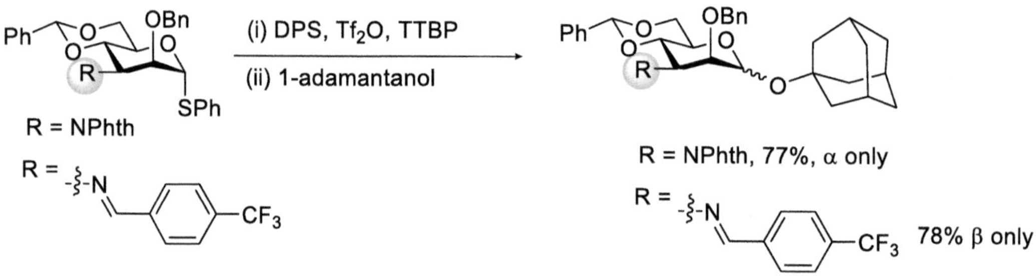

(i) DPS, Tf_2O, TTBP
(ii) 1-adamantanol

R = NPhth
R = N=CH-C_6H_4-CF_3

R = NPhth, 77%, α only
R = N=CH-C_6H_4-CF_3 78% β only

SCHEME 3.14 Influence of different groups at C3 on stereoselectivity.

In 2009, Kim reported [41] the effect of 3-*O*-acyl protection on the stereoselectivity of mannosylation (Scheme 3.15). The TCAI donors with electron-withdrawing ester groups such as acetyl or benzoyl as remote protecting at the C-3 position led to high 1,2-*cis*-selectivity. The author demonstrated an α-directing effect through the remote participation of 3-O-acyl and 6-O acetyl groups in mannopyranosylations. In Scheme 3.15, 3-*O*-acyl group protected donor and 6-*O*-acetyl group protected donor exhibit α-selectivity, whereas 4-*O*-acetyl group protected donor does not show that selectivity. To check the functionality of remote participation, experiments were designed to capture anomeric oxocarbenium ion intermediates through the intramolecular nucleophilic attack of the *tert*-butoxycarbonyl or trichloroacetimidoyl group at the O3, O4, or O6 positions of mannosyl donors. The result was a stable bicyclic product with a six-membered trichlorooxazine ring with 85% yield. The relative stability of the bicyclic product led to the stronger participation of the acyl group and hence the higher α-selectivity. That is, the remote participation by the 3-O-acyl and 6-O-acetyl groups could function, whereas no participation by the 4-O-acyl group occurred.

Further ahead, Nifantiev and coworkers [23] studied the effect of remote participation on the C-3 and C-6 positions of conformationally flexible or conformationally restricted glucosyl donors as summarized in Scheme 3.16. The reaction of N-phenyltrifluoroacetimidate donor, equipped with acetyl groups at C-3 and C-6, with glycosyl acceptor yielded a disaccharide with high selectivity α/β ratio: 11.2/1. However, in the case of the conformationally restricted 4,6-O-benzylidene-protected glucosyl donor, in contrast, excellent yield with lower stereoselectivity was achieved. The effect of steric bulkiness or strong electron-withdrawing properties of remote substituents, particularly those at C-6, have been known for a while. The beneficial effect of such substituents on 1,2-cis glucosylation and galactosylation was attributed to shielding (steric or electronic) of the top face of the ring, therefore favoring nucleophilic attack from the opposite side [42].

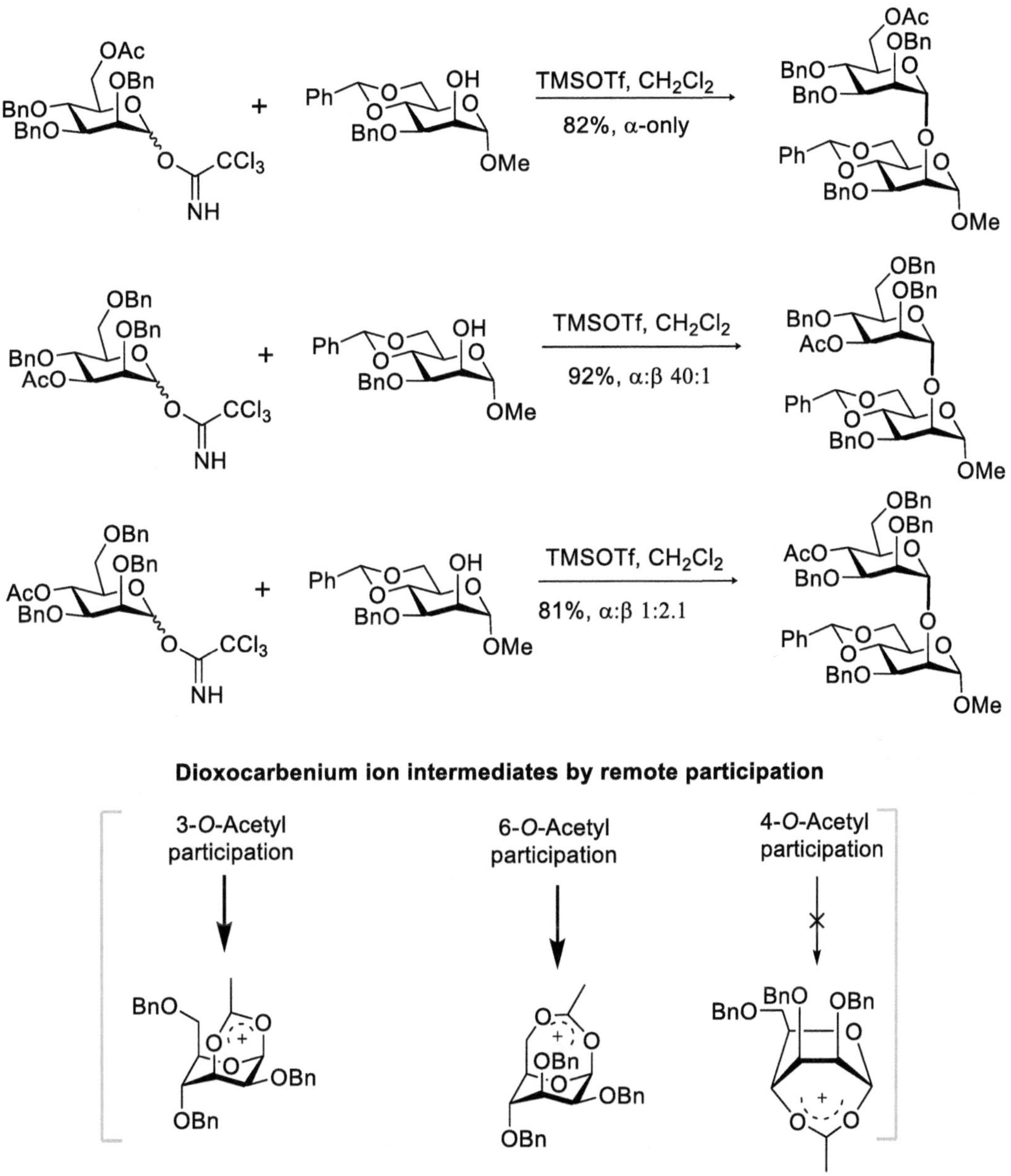

SCHEME 3.15 The effect of 3-*O*-acyl protection on the stereoselectivity of mannosylation.

3.4.3 Conformation-Constraining Protecting Groups

The torsional effects of cyclic bifunctional protecting groups, for instance, cyclic silyl, benzylidene, carbonate, and oxazolidinone, strongly influence the stereoselectivity of glycosylation. Conformation-constraining prevents the flexibility of sugar and promotes a certain conformation of the intermediate, which leads to glycosylation occurring from one side.

3.4.3.1 Benzylidene Group

The 4,6-O-benzylidene-protected mannosyl donor with a variety of leaving groups plays a crucial role in selectivity and yield. In this context, Crich et al. developed a strategy that produced high 1,2-*cis*-selectivity with the help of a conformation-constraining group. The proposed mechanism is illustrated in Scheme 3.17 [43].

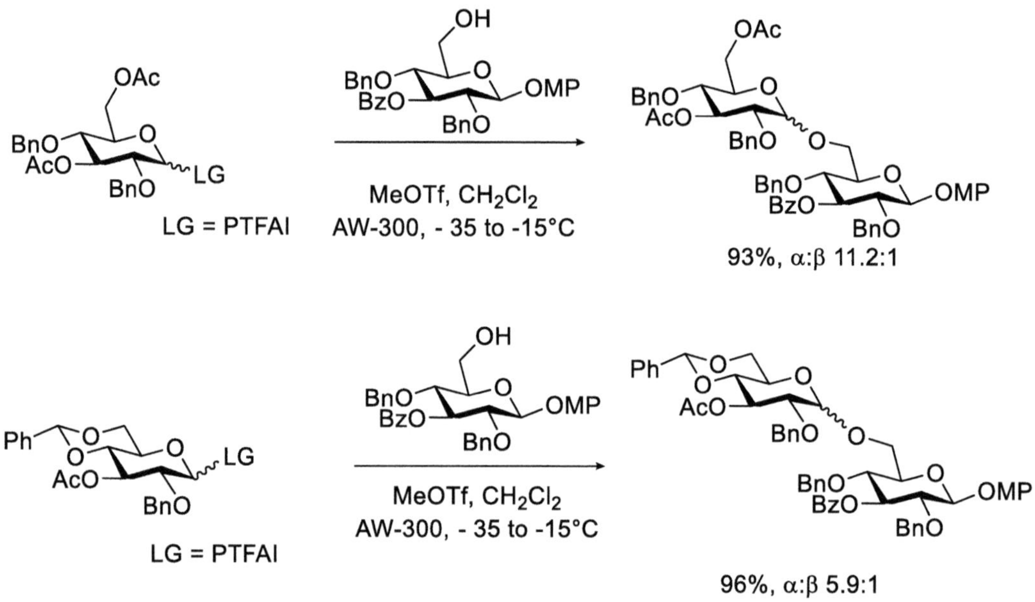

SCHEME 3.16 The α-directing effect of 3-O- and 6-O- acyl protection in mannopyranosylation.

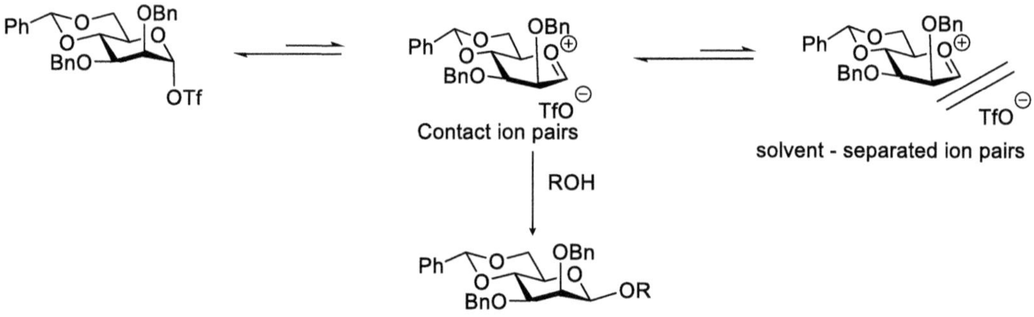

SCHEME 3.17 Proposed 1,2-*cis*-selectivity mechanism for 4,6-O-benzylidene-directed mannosylation.

The formation of the oxacarbenuim ion leads to a pathway that produces $S_N{}^1$ and gives anomeric mixtures that can be opposed by using benzylidene protection. Because of torsional strain, the α-triflate intermediate is favored and is in dynamic equilibrium with the contact ion pair. Finally, the S_N2 path leads to 1,2-*cis*-β-mannoside. Later, numerous research groups reviewed the utility of benzylidine protection on a variety of donors including mannosyl [44], glucosyl [45], and galactosyl [46].

3.4.3.2 Cyclic Silyl Groups

Di-tert-butylsilylene (DTBS) as a conformation-constraining protecting group has been extensively examined by various research groups. The steric effect of the DTBS group provides strong stabilization through space electron donation to the axially oriented oxacarbenium ion at the C4 group and opposes the attack through the up-side owing to bulkiness, which may account for the enormous α-selectivity (Scheme 3.18) [47]. In this context, Kiso reported α-selective glycosylation by employing galactosyl donors with di-tert-butylsilylene and a wide variety of leaving groups (Scheme 3.19) [48].

β-arabinofuranosides are an important component of microbial and plant polysaccharides. However, constructing 1,2-*cis*-furanosides is difficult, owing to the weak anomeric effects and

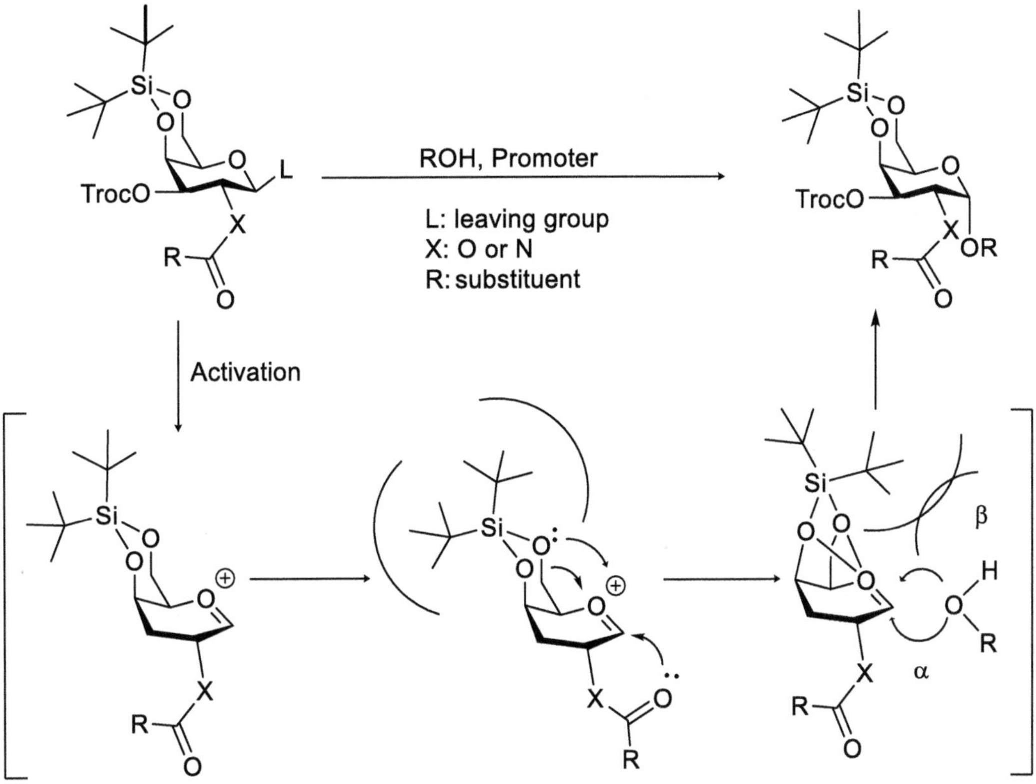

SCHEME 3.18 Proposed reaction mechanism for the DTBS-directed α-galactosylation.

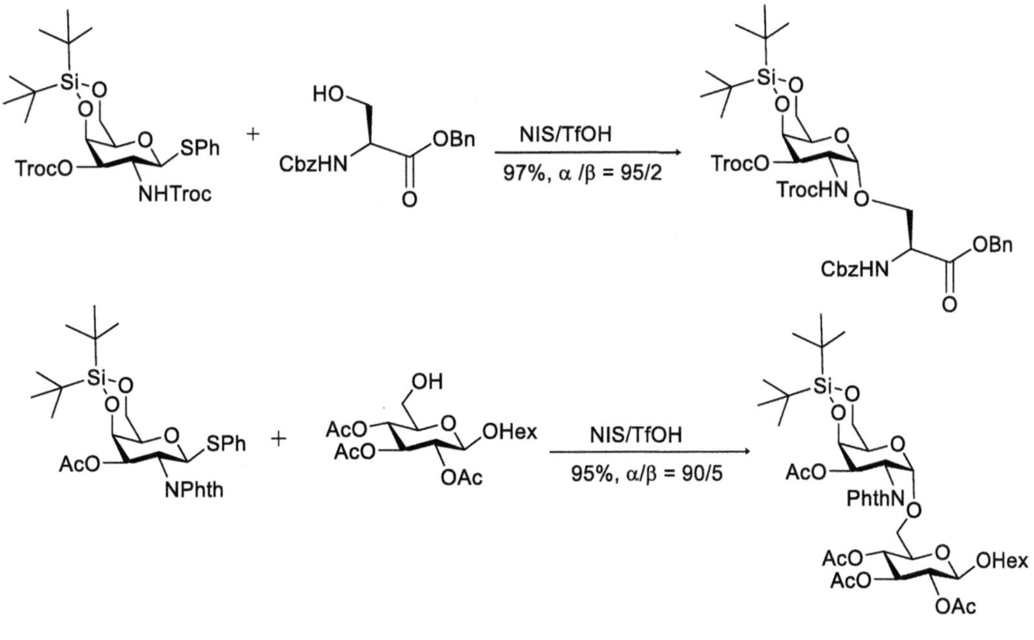

SCHEME 3.19 The glycosylation reaction with the donor having di-*tert*-butylsilylene.

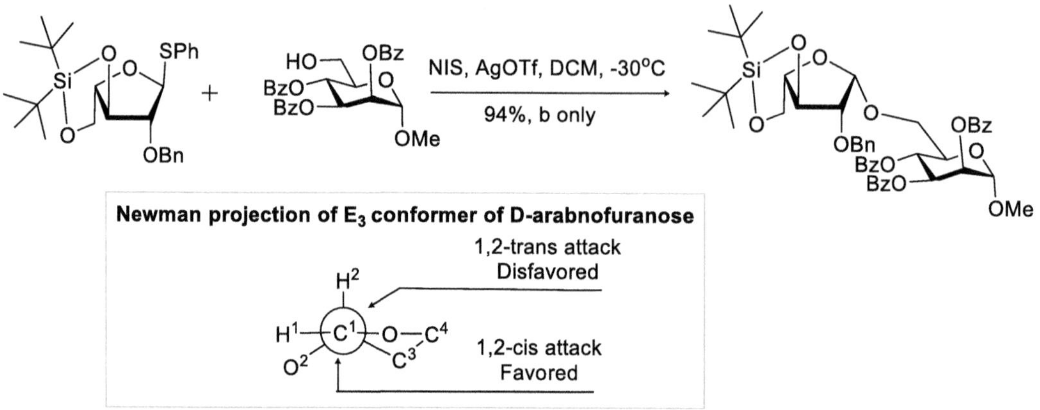

SCHEME 3.20 The glycosylation reaction with a 3,5-O-di-*tert*-butylsilylene donor.

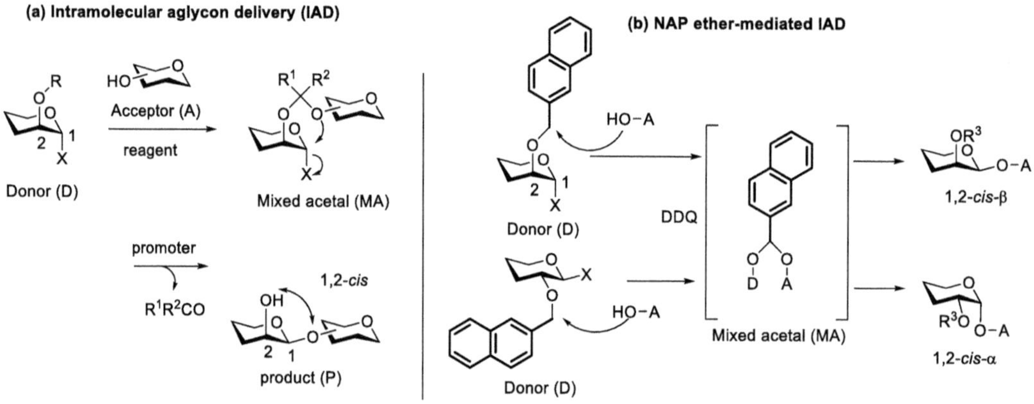

SCHEME 3.21 (a) Intramolecular aglycon delivery and (b) NAP ether-mediated IAD (R^1 = Naph, R^2 = H).

flexibilities of furan rings. β-selective furanosylation with a 3,5-O-di-tert-butylsilylene-protected donor is effectively performed using NIS and AgOTf in DCM at –30 °C under cyclic silyl group protection (Scheme 3.20) [49].

3.4.4 Intramolecular Aglycon Delivery

Achieving complete stereoselectivity in forming 1,2-*cis* glycosides is challenging, particularly for 1,2-*cis*-β (equatorial) links such as those in β-mannopyranosylation. Among various approaches, intramolecular aglycon delivery (IAD) emerges as a promising strategy for synthesizing 1,2-*cis* glycosides (Scheme 3.21). In this strategy, stereoselectivity can be independent of the acceptor's structure and allow the exclusive formation of 1,2-*cis* glycosides under kinetic control. Under IAD, tethering is an essential step wherein various mixed acetal (MA) linkages are developed to promote glycosylation in a regio- and stereoselective manner. With IAD and 2-axial-oriented substrates, such as β-mannopyranosylation, α-glycoside formation is essentially restricted. Additionally, the stereochemical outcome of forming two-equatorial-oriented 1,2-*cis* pyranosides, such as α-gluco- or α-galactopyranoside, via NAP ether-mediated IAD is less predictable.

3.4.4.1 Silicon Tethers

In 1992, Stork et al. first reported β-mannosylation using IAD with silyl acetal as a tether. The chlorodimethyl siloxy-equipped glycosyl acceptor with –SPh-containing leaving group on glycosyl donor formed silicon-tethered MA (Scheme 3.22) [50]. Then, MA was subjected to glycosylation conditions to give β-mannopyranoside. Additionally, other groups including Montgomery [51], Bols [52], and Rychnovsky [53] utilize the concept of silicon tether-dependent IAD in the synthesis of a variety of sugars.

3.4.4.2 Ketal Tethers

In 1991, Barresi and Hindsgaul used IAD to generate Ketal-tethered MA (Scheme 3.23), treating the glycosyl donor with Tebbe's reagent to prepare olephinic-containing donor. Next was donor activation followed by glycosylation under NIS with 2,6-di-*tert*-butyl-4-methylpyridine (DTBMP) to construct 1,2-*cis* glycosidic bond [54, 55].

3.4.4.3 -Iodoalkylidene Acetals as Tethers

To enhance the efficacy of the bridged intermediate, Fairbanks et al. employed an allyl ether treated with NIS or iodoniumdicollidine triflate converted into a 2-indoalkylidine mixed acetal (Scheme 3.24). Vinyl ether and propargyl ether were also used as precursors for MA. Later, treatment with DTBMP led to exclusive 1,2-*cis*-β-mannoside [56–61].

3.4.4.4 Benzylidene Acetal as Tether

An alternative strategy was introduced by Ito and Ogawa in 1994; they used *p*-methoxybenzylidene acetal as a precursor for MA [62]. This compound was obtained from the corresponding *p*-methoxybenzyl (PMB) ether (Scheme 3.25). Treatment of the donor–acceptor mixture with DDQ cleanly afforded the MA. Subsequent activation of the anomeric position of donor moiety (e.g., thioglycoside or fluoride) gave the desired β-mannopyranoside.

In this context, PMB and dimethoxybenzyl (DMB) ethers serve as donors tethered with the acceptor to form a MA. The MA construct 1,2-*cis*-glycoside gave a good yield [63,64]. For mechanistic studies, the effectiveness of the PMB-assisted β-mannosylation was optimal when

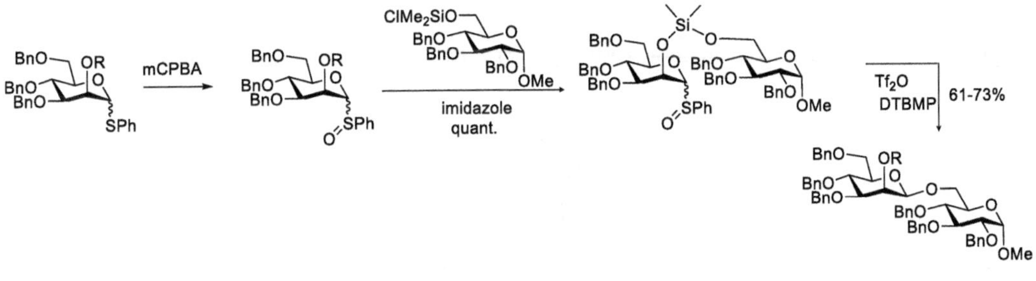

SCHEME 3.22 Silicon-tethered MA for IAD.

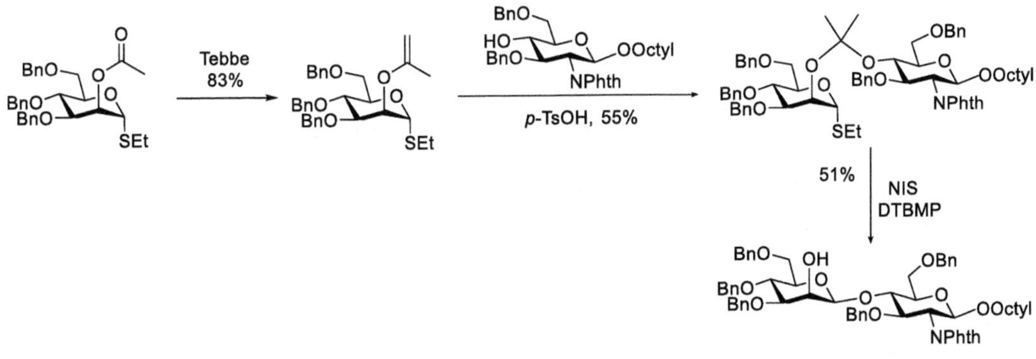

SCHEME 3.23 Ketal-tethered MA for IAD.

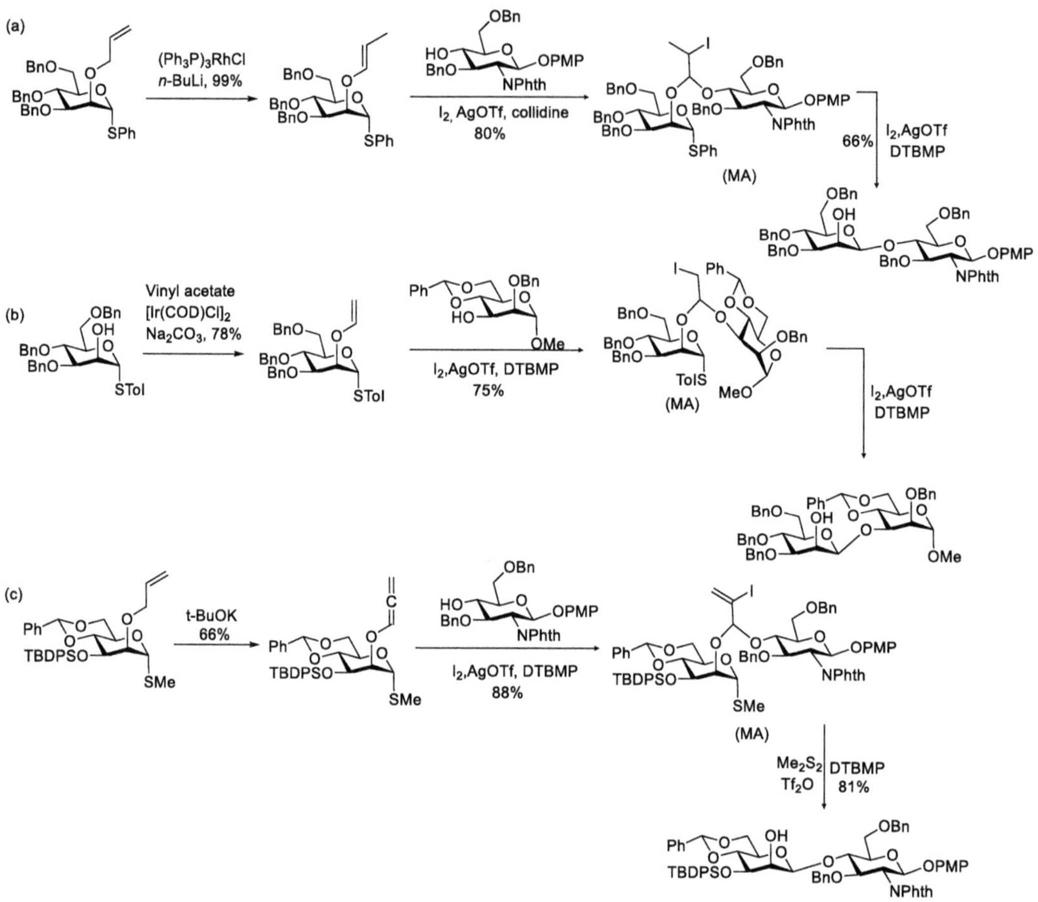

SCHEME 3.24 B-mannosylation through (a) allyl-, (b) vinyl-, and (c) propargyl-ether-mediated IADs.

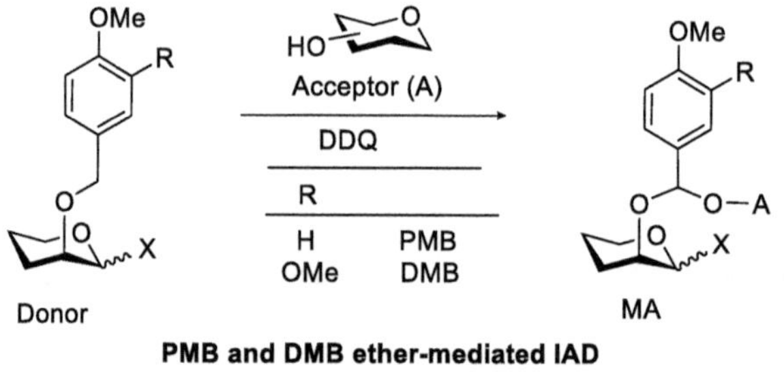

SCHEME 3.25 *p*-Methoxybenzylidene acetal (PMB) and dimethoxybenzyl (DMB) ether-mediated IAD.

a 4,6-*O*-cyclic protective group was present on the donor. Comparison of various protecting group donors that did not possess the constraining protecting group showed a diminished yield of the desired anomer (Scheme 3.26). Notably, the practicality of the PMB-IAD approach was demonstrated in that the authors were able to synthesize biologically important glycoprotein glycans such as high mannose and complex type *N*-glycans. Examples include the construction of targets such as glucosyl-, galactosyl-, and xylosyl-trehaloses through MAs obtained from the reaction of DMB-mediated IAD donor with various acceptors (Scheme 3.27). This approach permitted the rapid construction of the *Mycobacterium tuberculosis* sulfolipid-1 containing glucosyl-trehalose.

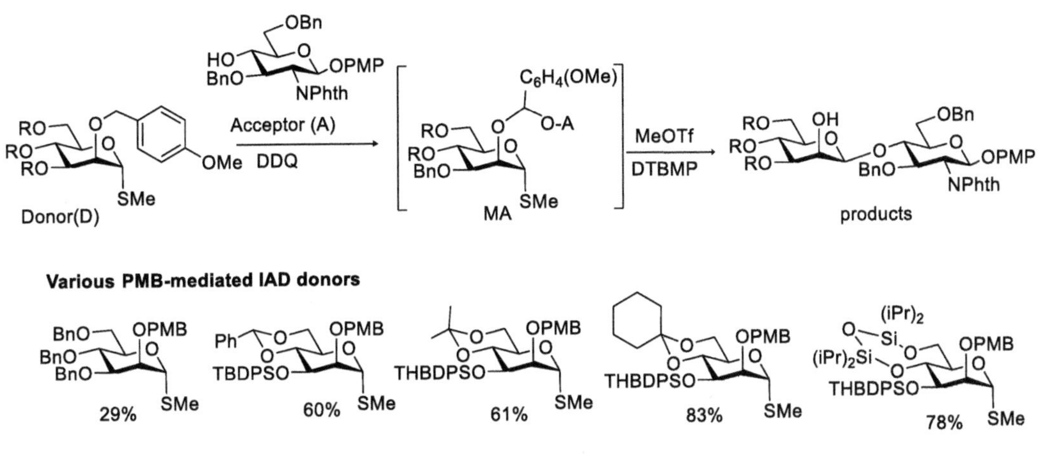

SCHEME 3.26 Protective groups on PMB-mediated IAD.

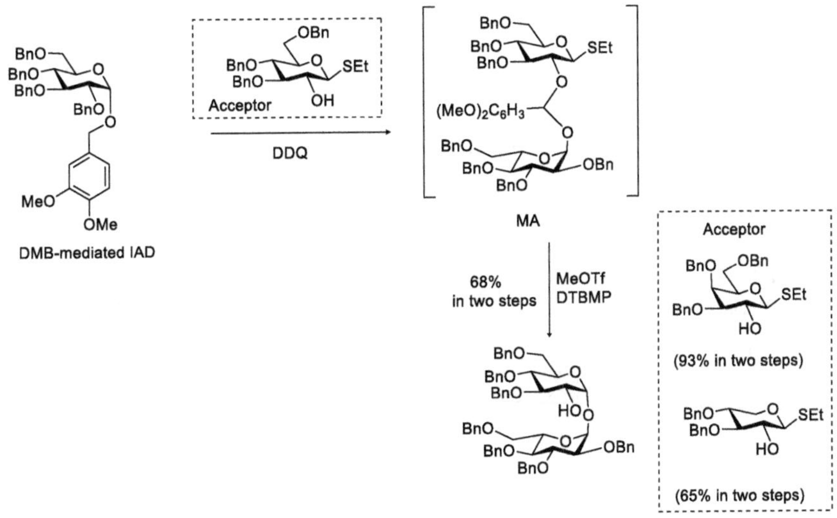

SCHEME 3.27 DMB ether-mediated IAD for β-Man*p*.

Other researchers applied 2-O-NAP with IAD and a variety of sugars including mannose, rhamnose, and glucose (Scheme 3.28) [65]. The formation of naphthylideneacetal motif with 2-O-NAP-protected thiomannoside and the acceptor proceeded quantitatively. Subsequent IAD mediated by MeOTf–DTBMP cleanly gave β-Manp, which was isolated after acetylation in 90% yield over three steps.

3.4.4.5 Hemiaminal Ethers as Tether

1,2-*cis*-α-2-deoxy-2-amino sugar is present in a myriad of glycoprotein and carbohydrate antigens, but synthesis of this analogue is hard. In this context, Knapp and group took advantage of the IAD strategy, generating aminal-tethered MA to synthesize α-linked D-glucosamine effectively (Scheme 3.29) [66].

3.5 BORONIC ESTER-MODULATED 1,2-*CIS* GLYCOSYLATION

Recently, an organoboran catalyst was applied for regio- and stereoselective synthesis of 1,2-*cis* or 1,2-*trans* glycosides. In 2015, Takahashi and Toshima demonstrated the new concept of

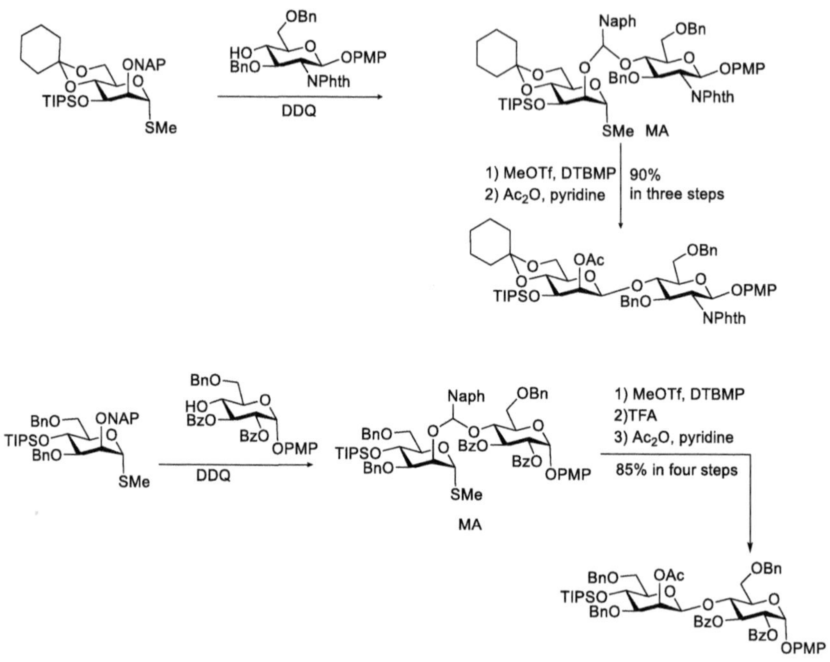

SCHEME 3.28 NAP-mediated IAD for β-Man*p*.

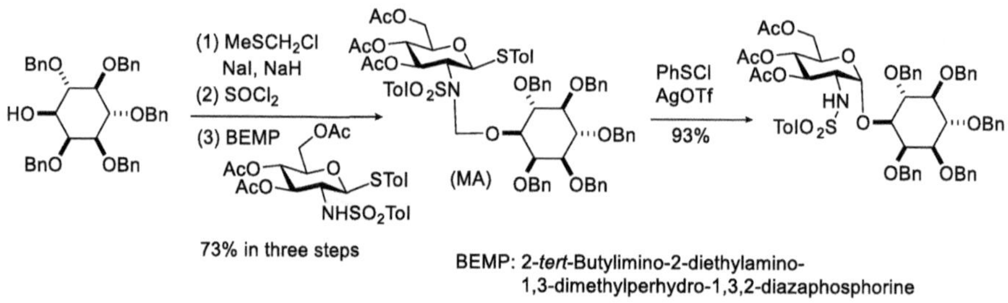

SCHEME 3.29 Aminal-tethered MA for IAD to give α-linked D-glucosamine derivative.

boronic-ester catalysis [67]. They used per-*O*-benzylated 1,2-anhydrosugar as a donor and glycosyl acceptor-derived boronic ester, which gave high yields of glycosylated product with exclusive 1,2-*cis* selectivity. This strategy directs the formation of α(1-4) glycosides using either glucose-4,6-diol acceptor or galactose 3,4-diol acceptor. Additionally, α(1-6) and α(1-3) glycosides were obtained using galactose-4,6-diol acceptor and glucose-2,3-diol acceptor, respectively (Scheme 3.30). Further, Tanaka et al. extended the concept via the boronic acid-catalyzed glycosylation of mannose sugar and the total synthesis of acremomannolipin A, GSL-1, and GSL-1' [68]. In 2017, Toshima and coworkers reported on boronic-acid-catalyzed 1,2-*cis*-β-mannosylation using a 1,2-anhydroglycosyl donor [69].

Boronate-catalyzed ring opening promoted the regio- and stereoselective formation of 1,2-*cis*-linked sugars. Then, an epoxide derived from L-rhamnose, when used as a donor with mono-ol acceptors, formed β-L-rhamnopyranoside with moderate to good yield [70]. Further, 4-nitro phenyl boronic acid catalyzed glycosylation using 4,6-diol acceptors yielded 1-4 disaccharides with complete stereoselectivity (Scheme 3.31). A published strategy was effectively applied for the construction of trisaccharide related to *Streptococcus pneumonia* serotypes 7B, 7C, and 7D.

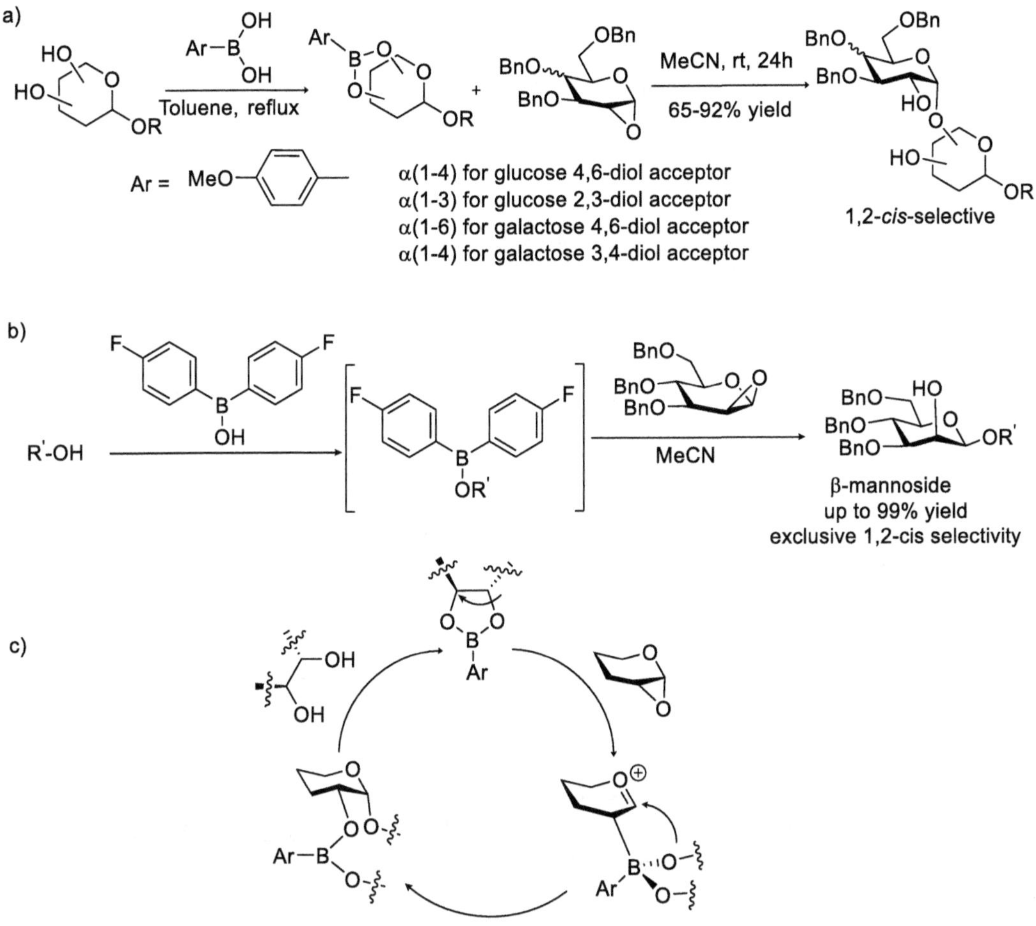

SCHEME 3.30 1,2-*cis*-selective glycosylation reaction using glycosyl acceptor-derived boronic ester catalysts.

Concomitantly, Tanaka followed the $S_N{}^i$-type mechanistic path for high region- and stereoselective 1,2-*cis* glycosylation [71]. The author introduced a *p*-nitrophenylboronic acid-catalyzed and water-mediated strategy in which 1,2-anhydro donor and unprotected sugar acceptors were utilized for stereoselective glycosylation (Scheme 3.32). In 2019, 1,2-dihydro sugar and triflate acceptor were effectively utilized for 1,2-*cis*-α-glycosylation by anomeric *O*-alkylation through borate complex with boronic acids. In 2020, Tomita et al. reported efficient regio- and 1,2-*cis*-α-stereoselective glycosylation reactions of 1,2-anhydroglucose with trans-1,2-diol sugar acceptor using diboron catalyst [72].

3.6 PHENANTHROLINE-CATALYZED α-1,2-*CIS* GLYCOSYLATION

Commercially available phenanthroline has emerged as a powerful ligand in the realm of sugar chemistry. The advantages of phenanthroline are its nucleophilicity and its fused ring, which sterically prevent the generation of unwanted intermediates. This allows sugar chemists to explore its potential in stereoselective glycosylation. A general mechanistic path is described in Scheme 3.33, wherein β-glycosyl phenonthrolinium intermediate is generated via displacement of halide by the nucleophilicity of the nitrogen atom in phenanthroline. Subsequently, the S_N2 procedure furnished the 1,2-*cis*-glycoside with predictable stereoselectivity.

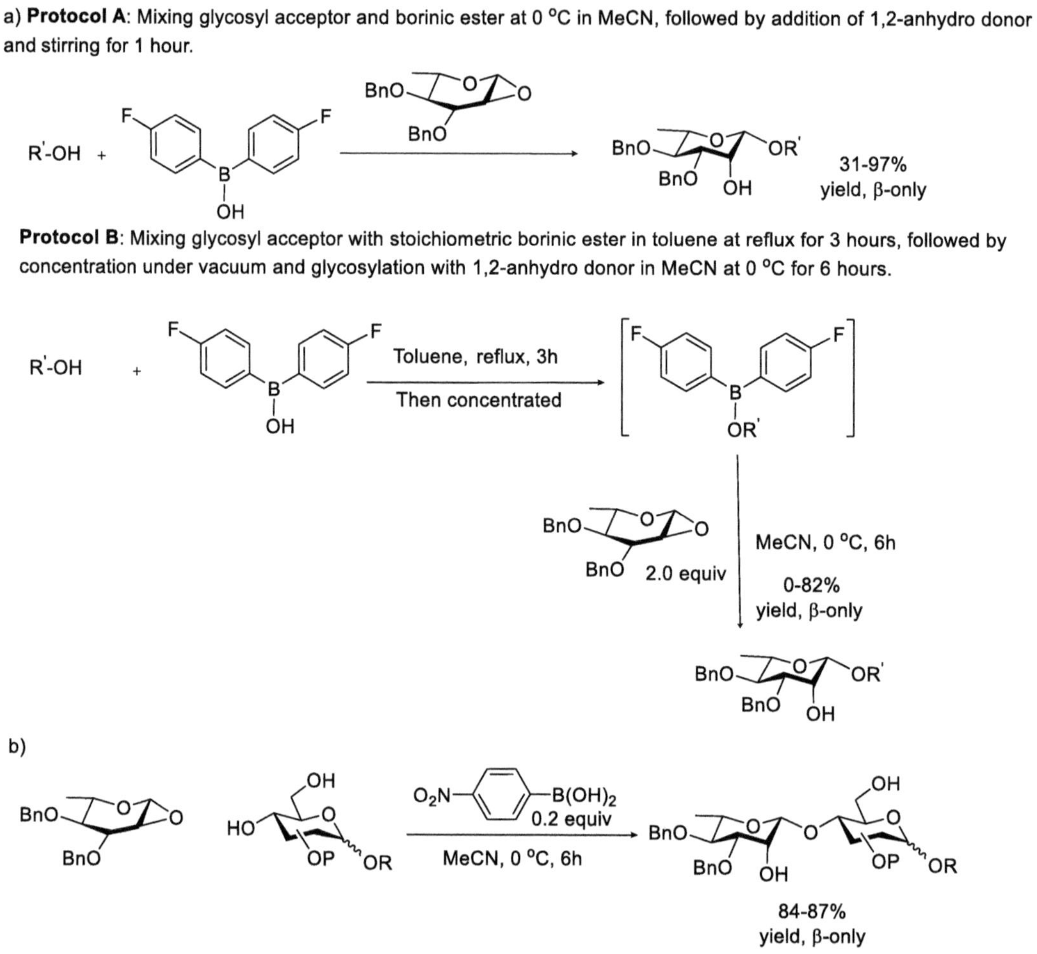

a) **Protocol A**: Mixing glycosyl acceptor and borinic ester at 0 °C in MeCN, followed by addition of 1,2-anhydro donor and stirring for 1 hour.

31-97% yield, β-only

Protocol B: Mixing glycosyl acceptor with stoichiometric borinic ester in toluene at reflux for 3 hours, followed by concentration under vacuum and glycosylation with 1,2-anhydro donor in MeCN at 0 °C for 6 hours.

Toluene, reflux, 3h

Then concentrated

2.0 equiv

MeCN, 0 °C, 6h

0-82% yield, β-only

b)

O_2N–C6H4–$B(OH)_2$ 0.2 equiv

MeCN, 0 °C, 6h

84-87% yield, β-only

SCHEME 3.31 Organoboron-catalyzed β-L-rhamnosylation.

Recently, Nguyen et al. reported the stereoselective glycosylation by wing BPen (bathophenanthroline) with isobutylene oxide as an acid scavenger in MTBE at 50 °C [73]. Later, Nguyen et al. introduced modified phenanthroline ligand to improve the glycosylation, selectively yielding the α-1,2-*cis* 2-deoxy-2-fluroglycosides [74]. Further, the utility of phenanthroline has emerged due to its pronounced effect on a variety of glycosylation processes, including chemoselective, site-selective, and orthogonal glycosylation (Scheme 3.34).

3.7 ORGANOCATALYZED SYNTHESIS OF A-SELECTIVE 2-DEOXY GLYCOSIDES

The structure of 2-deoxy sugars is present in many biologically active natural products. Researchers introduced efficient, stereoselective functionality at the 2-position of deoxy sugar via organocatalytic approaches. Thiourea as an organocatalyst has emerged as an efficient catalyst that has seen vast application in stereoselective glycosylation.

For instance, Balmond explored Schreiner's thiourea for effective stereoselective 1,2-*cis*-glycosylation (Scheme 3.35a) [75], although the reported protocol was limited to galactosides and was also detrimental to per-O-acetylated D-galactal. Further, cooperative Bronsted acid/Schreiner's thiourea-catalyzed 1,2-*cis*-α-selective glycosylation was introduced for more demanding glycosyl

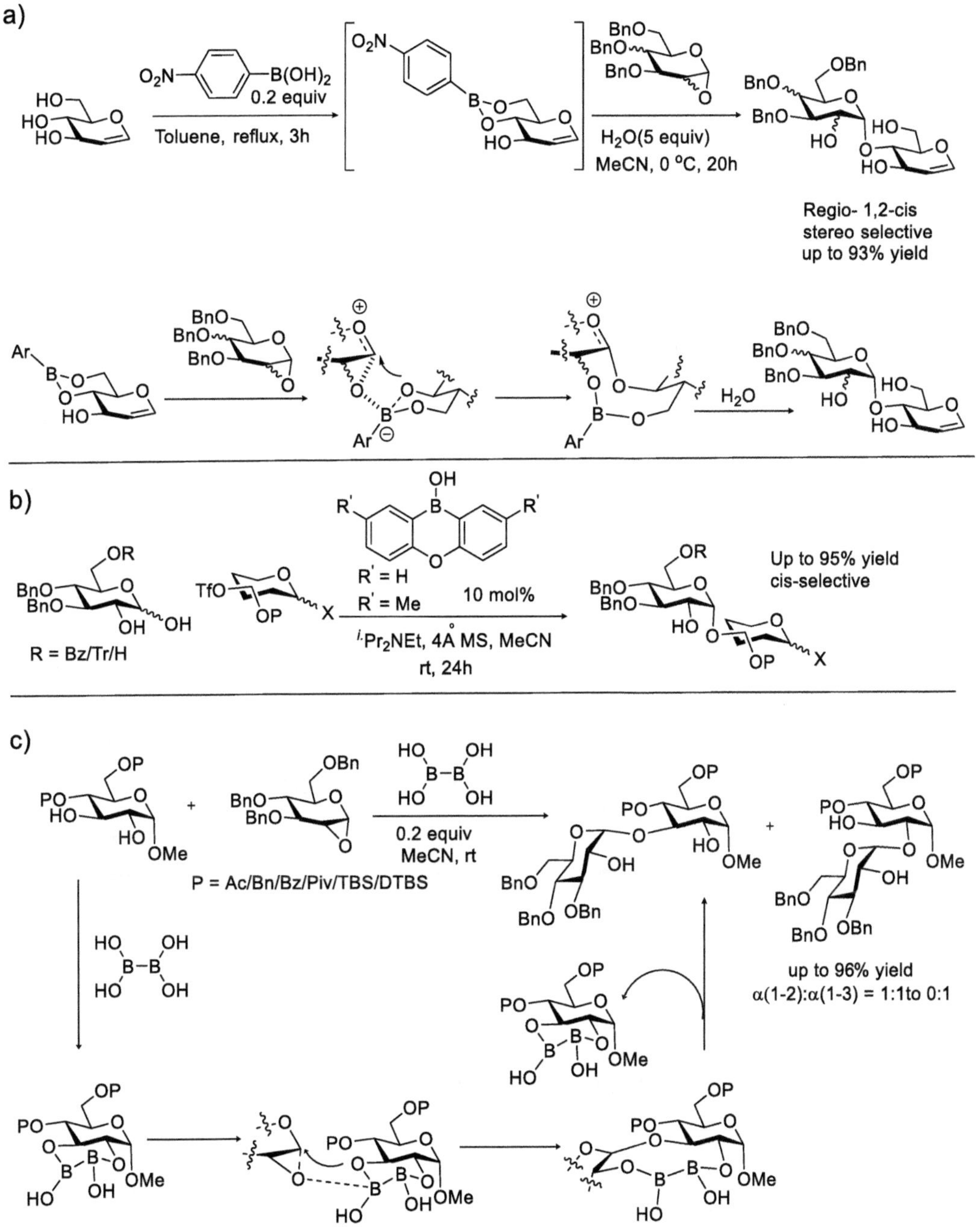

SCHEME 3.32 a) Boronic-acid-catalyzed regioselective 1,2-*cis* glycosylation reaction of unprotected sugar acceptor; b) synthesis of 1,2-*cis*-glycosides by anomeric *O*-alkylation with organoboron catalysis; and c) diboron-catalyzed regioselective and 1,2-*cis*-α-stereoselective glycosylation reaction.

donors, including glucal, galalctal, and rhamnal (Scheme 3.35b) [76], and McGarrigle et al. reported that α-stereoselective glycosylation hinges on thiouracil-based organocatalyst (Scheme 3.35c) [77]. Mandal and coworkers reported a new class of organocatalyst L-prolinethioamide that efficiently catalyzed the direct glycosylation reaction of various glycals with alcohols to give α-linkages with high stereoselectivity and good yields [78] (Scheme 3.35d).

SCHEME 3.33 Phenanthroline-catalyzed α-1,2-*cis*-glycosylation.

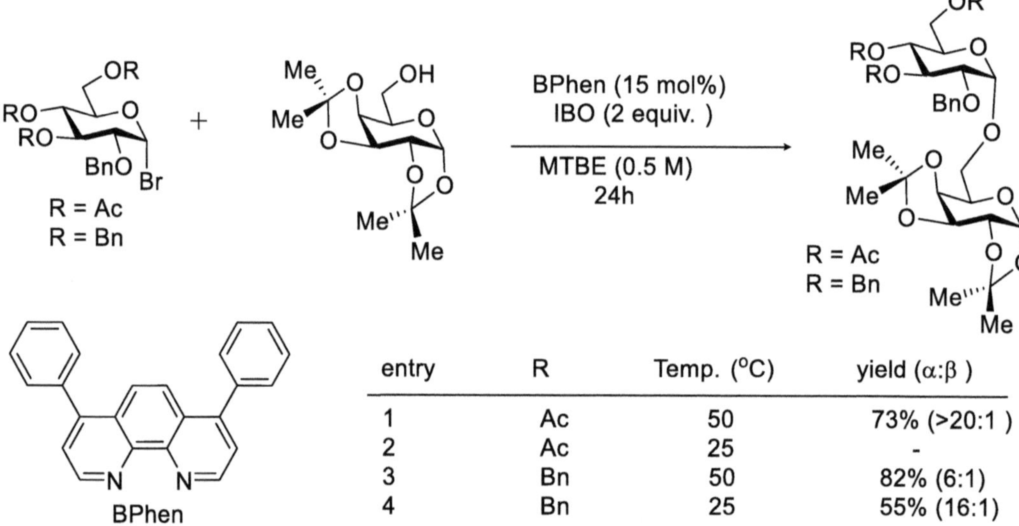

entry	R	Temp. (°C)	yield (α:β)
1	Ac	50	73% (>20:1)
2	Ac	25	-
3	Bn	50	82% (6:1)
4	Bn	25	55% (16:1)

SCHEME 3.34 Bathophenanthroline-catalyzed 1,2 -*cis* glycosylation.

3.8 EFFECTS OF REACTION CONDITIONS

3.8.1 Temperature

Temperature has a strong effect on selective glycosylation. At high temperatures, thermodynamically favored glycoside formed owing to the anomeric effect, which led to α-glycosylation [79]. Conversely, at lower temperatures, kinetically controlled glycosylation proceeded that favored β-glycosidic bond formation.

3.8.2 Solvent

Selective glycosidic construction in optimized solvent systems has been widely studied. As a rule of thumb, nitrile solvents tend to dictate the β-(equatorial) glycosylation, while ethereal solvents increase the amount of α-(axial) selective glycosylation (Scheme 3.36) [80]. However, the extent of selective glycosylation hinges on other powerful factors including temperature, additive, structure of substrates, promoters, chelators, and so on.

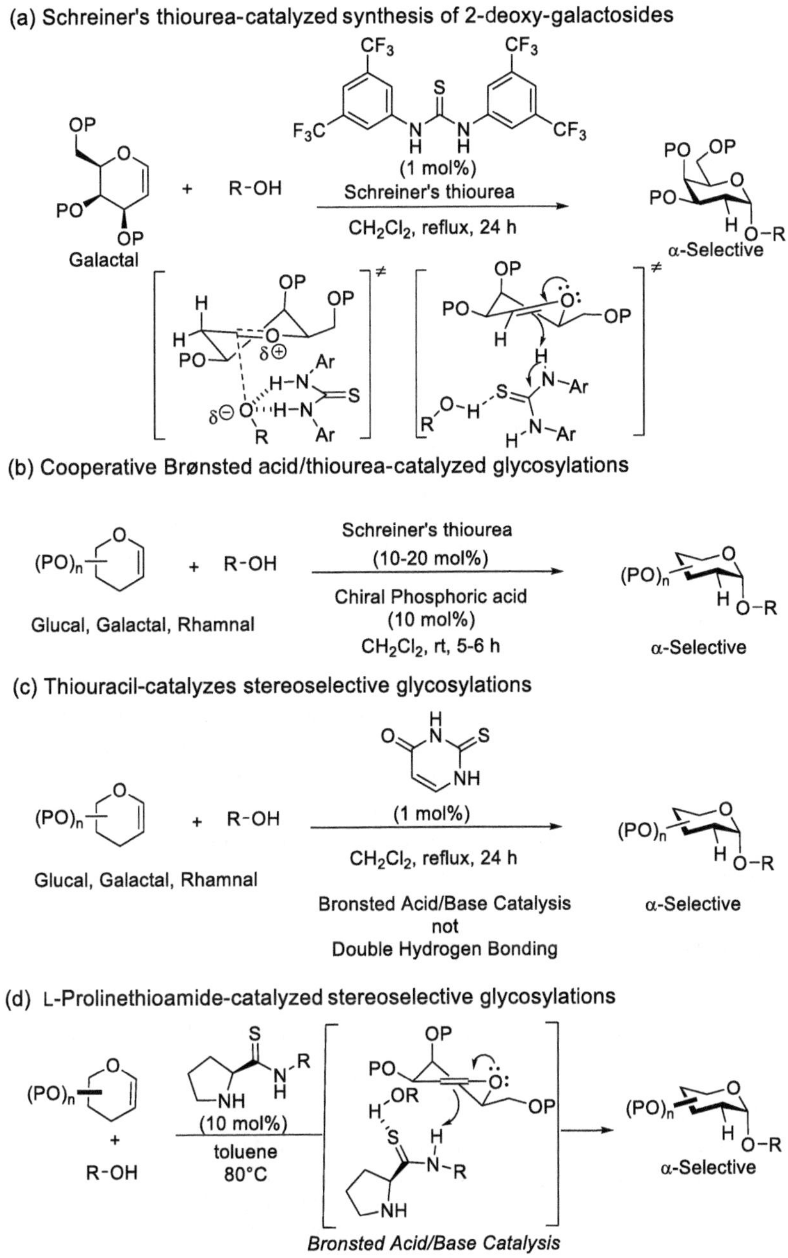

(a) Schreiner's thiourea-catalyzed synthesis of 2-deoxy-galactosides

(b) Cooperative Brønsted acid/thiourea-catalyzed glycosylations

(c) Thiouracil-catalyzes stereoselective glycosylations

(d) L-Prolinethioamide-catalyzed stereoselective glycosylations

SCHEME 3.35 α-Selective organocatalytic synthesis of 2-deoxyglycosides.

3.9 CONCLUSION

Carbohydrates such as proteoglycans, glycolipids, glycoproteins, and microbial polysaccharides exist as essential components in myriad biologically active molecules. Over the past few decades, carbohydrates have played a crucial role in many diseases, for instance, septicemia, inflammation, diabetes, and bacterial and viral infections. Although great achievements have been made to synthesize complex glycoconjugates, selective construction of glycoside is still a difficult task.

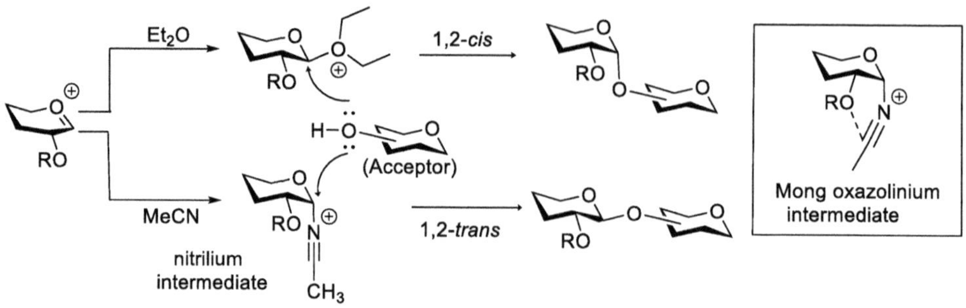

SCHEME 3.36 Effect of the reaction solvent.

The anomeric mixtures cannot be directly employed as they require separation to be introduced in the medicinal, pharmaceutical, and biological research fields; the control of stereoselectivity is highly demanded because anomeric mixtures will create complications in further steps and also in the purification of desired compounds. 1,2-*trans* selectivity is sophisticated through neighboring group participation at the C-2 position, but the construction of 1,2-*cis* glycosidic linkages is far less straightforward. The stereoselective 1,2-*cis* glycosides have already shown an excellent track record in numerous biologically important and active complex glycans. The selective construction of 1,2-*cis* glycosidic linkages is in high demand. Here, we have introduced certain concepts in detail related to achieving the targeted 1,2-*cis* glycoside: chiral auxiliaries, protecting groups, intramolecular aglycon delivery, organoboron, phenanthroline, and thiourea catalyzed glycosylation. Furthermore, we discussed key factors such as the structure of the substrate, temperature, solvent, activation system, additive, and promoter that are all promising for stereocontrolled glycosylation. We hope that this chapter guides sugar chemists to develop strategies for total stereoselectivity with high yields and large substrate scope compatibilities for 1,2-*cis* glycosylation.

ACKNOWLEDGMENTS

The authors gratefully acknowledge financial support by the Science and Engineering Research Board, DST, Govt. of India (Grant No. CRG/2022/003936), New Delhi, India CDRI communication no. 10871.

REFERENCES

[1] (a) A. Varki. Biological Roles of Oligosaccharides: All of the Theories Are Correct. *Glycobio.*, **1993**, *3* (2), 97–130; (b) A. Varki. Biological Roles of Glycans. *Glycobio.*, **2017**, *27*, 3–49. (c) A. Varki, R. D. Cummings, J. D. Esko, H. H. Freeze, C. R. Bertozzi, P. Stanley, G. W. Hart, M. E. Etzler. *Essentials of Glycobiology*, CSH Laboratory Press, New York, 2nd edn, **2009**.

[2] (a) C. R. Bertozzi, L. L. Kiessling. Chemical Glycobiology. *Science*, **2001**, *291*, 2357–2364. (b) T. J. Boltje, T. Buskas, G. J. Boons. Opportunities and Challenges in Synthetic Oligosaccharide and Glycoconjugate Research. *Nat. Chem.*, **2009**, *1*, 611–622.

[3] H. Cao, J. Hwang, X. Chen, *Opportunity, Challenge and Scope of Natural Products in Medicinal Chemistry*, Research Signpost, India, ed. V. K. Tiwari, B. B. Mishra, **2011**, pp. 411–431.

[4] X. Zhao, Y. Huang, S. Zhou, J. Ao, H. Cai, K. Tanaka, Y. Ito, A. Ishiwata, F. Ding. Recent Chemical and Chemoenzymatic Strategies to Complex-Type N-Glycans. *Front. Chem.*, **2022**, *10*.

[5] C. Zhao, M. Li, Y. Luo, W. Wu. Isolation and Structural Characterization of an Immune Stimulating Polysaccharide from Fuzi, Aconitum Carmichaeli. *Carbohydr. Res.*, **2006**, *341*, 485–491.

[6] (a) V. C. B. Bittencourt, R. T. Figueiredo, R. B. da Silva, D. S. Mour˜ao-S´a, P. L. Fernandez, G. L. Sassaki, B. Mulloy, M. T. Bozza, E. Barreto-Bergter. An α-Glucan of Pseudallescheria Boydii Is Involved in Fungal Phagocytosis and Toll-like Receptor Activation*. *J. Biol. Chem.*, **2006**, *281*, 22614–22623, (b) L. C. L. Lopes, M. I. D. da Silva, V. C. B. Bittencourt, R. T. Figueiredo, R. Rollin-Pinheiro, G. L. Sassaki, M. T. Bozza, P. A. J. Gorin, E. Barreto-Bergter. Glycoconjugates

and Polysaccharides from the Scedosporium/Pseudallescheria Boydii Complex: Structural Characterisation, Involvement in Cell Differentiation, Cell Recognition and Virulence. *Mycoses*, **2011**, *54*, 28–36.
[7] A. O. Tzianabos, A. Pantosti, H. Baumann, J. R. Brisson, H. J. Jennings, D. L. Kasper. Structural Characterization of Two Surface Polysaccharides of Bacteroides Fragilis. *J. Biol. Chem.*, **1992**, *267*, 18230–18235.
[8] J. B. Robbins, C. J. Lee, S. C. Rastogi, G. Schiffman, J. Henrichsen. Comparative Immunogenicity of Group 6 Pneumococcal Type 6A(6) and Type 6B(26) Capsular Polysaccharides. *Infect. Immun.*, **1979**, *26*, 1116–1122.
[9] C. Jones. Revised Structures for the Capsular Polysaccharides from Staphylococcus Aureus Types 5 and 8, Components of Novel Glycoconjugate Vaccines. *Carbohydr. Res.*, **2005**, *340*, 1097–1106.
[10] Z. Wang, Z. S. Chinoy, S. G. Ambre, W. Peng, R. McBride, R. P. de Vries, J. Glushka, J. C. Paulson, G. J. Boons. A General Strategy for the Chemoenzymatic Synthesis of Asymmetrically Branched N-Glycans. *Science*, **2013**, *341*, 379–383.
[11] A. Michael. On the Action of Iodine Monochloride Upon Aromatic Amines. *J. Am. Chem. Soc.*, **1879**, *1*, 305–312.
[12] E. Fischer. Ueber die Glucoside der Alkohole. *Ber.*, **1893**, *26*, 2400–2412.
[13] W. Koenigs, E. Knorr. Ueber einige Derivate des Traubenzuckers und der Galactose *Chem. Ber.,* **1901**, *34*, 957–981.
[14] F. Barresi, O. Hindsgaul. Glycosylation Methods in Oligosaccharide Synthesis. *Mod. Syn. Methods*, **1995**, *7*, 281–330.
[15] (a) A. V. Demchenko. 1,2-cis *O*-Glycosylation: Methods, Strategies, Principles *Curr.Org. Chem.*, **2003**, *7*, 35–79.
[16] I. Tvaroska, T. Bleha. Anomeric and Exo-Anomeric Effects in Carbohydrate Chemistry. *Adv. Carbohydr. Chem. Biochem.*, **1989**, *47*, 45–123.
[17] A. Ishiwata, J. Y. Lee, Y. Ito. Recent Advances in Stereoselective Glycosylation Through Intramolecular Aglycon Delivery. *Org. Biomol. Chem.*, **2010**, *8*, 3596–3608.
[18] J. P. Yasomanee, A. V. Demchenko. Effect of Remote Picolinyl and Picoloyl Substituents on the Stereoselectivity of Chemical Glycosylation. *J. Am. Chem. Soc.*, **2012**, *134*, 20097–20102.
[19] S. G. Pistorio, J. P. Yasomanee, A. V. Demchenko. Hydrogen-Bond-Mediated Aglycone Delivery: Focus on β-Mannosylation. *Org. Lett.*, **2014**, *16*, 716–719.
[20] A. V. Demchenko. Stereoselective Chemical 1,2-*cis* O-Glycosylation: From 'Sugar Ray' to Modern Techniques of the 21st Century. *Synlett.*, **2003**, 1225–1240.
[21] S. R. Lu, Y. H. Lai, J. H. Chen, C. Y. Liu, K. K. Mong. Dimethylformamide: An Unusual Glycosylation Modulator. *Angew. Chem. Int. Ed.*, **2011**, *50*, 7315–7320.
[22] Z. Li, L. Zhu, J. Kalikanda. Development of a Highly α-Selective Galactopyranosyl Donor Based on a Rational Design. *Tetrahedron Lett.*, **2011**, *52*, 5629–5632.
[23] B. S. Komarova, M. V. Orekhova, Y. E. Tsvetkov, N. E. Nifantiev. Is an Acyl Group at O-3 in Glucosyl Donors Able to Control α-Stereoselectivity of Glycosylation? The Role of Conformational Mobility and the Protecting Group at O-6. *Carbohydr. Res.*, **2014**, *384*, 70–86.
[24] D. Crich, S. Sun. Direct Formation of â-Mannopyranosides and Other Hindered Glycosides from Thioglycosides. *J. Am. Chem. Soc.*, **1998**, *120*, 435–436.
[25] R. U. Lemieux, K. B. Hendriks, R. V. Stick, K. James. Halide Ion Catalyzed Glycosidation Reactions. Syntheses of a-Linked Disaccharides. *J. Am. Chem. Soc.*, **1975**, *97* (14), 4056–4062.
[26] S. Kaeothip, J. P. Yasomanee, A. V. Demchenko. Glycosidation of Thioglycosides in the Presence of Bromine: Mechanism, Reactivity, and Stereoselectivity. *J. Org. Chem.*, **2012**, *77*, 291–299.
[27] N. K. Kochetkov, E. M. Klimov, N. N. Malysheva, A. V. Demchenko. Stereospecific 1,2-c&Glycosylation: A Modified Thiocyanate Method. *Carbohydr. Res.*, **1992**, Cl–C5.
[28] a) D. Crich, S. Sun. Formation of β-Mannopyranosides of Primary Alcohols Using the Sulfoxide Method. *J. Org. Chem.*, **1996**, 61, 4506–4507; b) D. Crich, M. Smith, S-(4-Methoxyphenyl) Benzenethiosulfinate (MPBT)/Trifluoromethanesulfonic Anhydride: A Convenient System for the Generation of Glycosyl Triflates from Thioglycosides. *Org. Lett.*, **2000**, 2, 4067–4069.
[29] J.-H. Kim, H. Yang, G.-J. Boons. Stereoselective Glycosylation Reactions with Chiral Auxiliaries. *Angew. Chem. Int. Ed.*, **2005**, *44*, 947–949.
[30] (a) J.-H. Kim, H. Yang, J. Park, G.-J. Boons. A General Strategy for Stereoselective Glycosylations. *J. Am. Chem. Soc.*, **2005**, 127, 12090–12097; (b) T. J. Boltje, J.-H. Kim, J. Park, G.-J. Boons, Stereoelectronic Effects Determine Oxacarbenium vs β-Sulfonium Ion Mediated Glycosylations. *Org. Lett.*, **2011**, 13, 284–287.

[31] M. A. Fascione, S. J. Adshead, S. A. Stalford, C. A. Kilner, A. G. Leach, W. B. Turnbull. Stereoselectiveglycosylation using oxathiane glycosyl donors. *Chem. Commun.*, **2009**, 5841–5843.
[32] T. Fang, K.-F. Mo, G.-J. Boons. Stereoselective Assembly of Complex Oligosaccharides Using Anomeric Sulfonium Ions as Glycosyl Donors. *J. Am. Chem.* Soc., **2012**, 134, 7545–7552.
[33] M. A. Fascione, N. J. Webb, C. A. Kilner, S. L. Warriner, W. B. Turnbull. Stereoselective glycosylations using oxathiane spiroketal glycosyl donors. *Carbohydr. Res.*, **2012**, 348, 6–13.
[34] M. N. Amin, A. Ishiwata, Y. Ito. Synthesis of *N*-Linked Glycan Derived from Gram-Negative Bacterium, Campylobacter Jejuni. *Tetrahedron Lett.*, **2007**, *63*, 8181–8198.
[35] J. Park, T. J. Boltje, G.-J. Boons. Direct and Stereoselective Synthesis of α-Linked 2-Deoxyglycosides. *Org. Lett.*, **2008**, *10*, 4367–4370.
[36] C. D. Meo, M. N. Kamat, A. V. Demchenko. Remote Participation-Assisted Synthesis of β-Mannosides. *Eur. J. Org. Chem.*, **2005**, 706–711.
[37] N. Ustyuzhanina, B. Komarova, N. Zlotina, V. Krylov, A. G. Gerbst, Y. Tsvetkov, N. E. Nifantiev. Stereoselective a-Glycosylation with 3-O-Acetylated D-Gluco Donors. *Synlett.*, **2006**, 921–923.
[38] Y. Takashi, N. Kazumi, T. Hiroshi, Y. Kenji, I. Toshiyuki. New Synthetic Methods and Reagents for Complex Carbohydrates. VIII. Stereoselective α- and β-Mannopyranoside Formation from Glycosyl Dimethylphosphinothioates with the C-2 Axial Benzyloxyl Group. Bull. *Chem. Soc. Jpn.*, **1994**, 67, 1359–1366.
[39] a) E. J. Corey, P. Carpino. Enantiospecific total synthesis of pseudopterosins A and E. *J. Am. Chem. Soc.*, **1989**, 111, 5472–5473; (b) A. G. Gerbst, N. E. Ustuzhanina, A. A. Grachev, D. E. Tsvetkov, E. A. Khatuntseva, N. E. Nifant'ev. Effect of the nature of protecting group at O-4 on stereoselectivity of glycosylation by 4-O-substituted 2,3-di-O-benzylfucosyl bromides. *Mendeleev Commun.*, **1999**, 114–116.
[40] D. Crich, H. Xu. Direct Stereo Controlled Synthesis of 3-Amino-3-Deoxy-β-Mannopyranosides: Importance of the Nitrogen Protecting Group on Stereoselectivity. *J. Org. Chem.*, **2007**, *72*, 5183–5192.
[41] J. Y. Baek, B.-Y. Lee, M. G. Jo, K. S. Kim. β-Directing Effect of Electron-Withdrawing Groups at O-3, O-4, and O-6 Positions and α-Directing Effect by Remote Participation of 3-O-Acyl and 6-O-Acetyl Groups of Donors in Mannopyranosylations. *J. Am. Chem. Soc.*, **2009**, *131*, 17705–17713.
[42] C. A. A. van Boeckel, T. Beetz. Substituent Effects in Carbohydrate Chemistry, Part II+. Coupling Reactions Involving Gluco- and Galacto-Pyranosyl Bromides Promoted by Insoluble Silver Salts. *Recl. Trav. Chim. Pays-Bas*, **1985**, *104*, 171–173.
[43] (a) D. Crich, S. Sun. Are Glycosyl Triflates Intermediates in the Sulfoxide Glycosylation Method? A Chemical and 1H, 13C, and 19F NMR Spectroscopic Investigation. *J. Am. Chem. Soc.*, **1997**, *119*, 11217–11223; (b) D. Crich, S. Sun. Direct Synthesis of β-Mannopyranosides by the Sulfoxide Method. *J. Org. Chem.*, **1997**, *62*, 1198–1199; (c) D. Crich, S. Sun. Formation of β-Mannopyranosides of Primary Alcohols Using the Sulfoxide Method. *J. Org. Chem.*, **1996**, *61*, 4506–4507; (d) D. Crich, N. S. Chandrasekera. Mechanism of 4,6-O-Benzylidene-Directed β- Mannosylation as Determined by α-Deuterium Kinetic Isotope Effects. *Angew. Chem. Int. Ed.*, **2004**, *43*, 5386–5389.
[44] (a) D. Crich, B. Wu. 1-Naphthylpropargyl Ether Group: A Readily Cleaved and Sterically Minimal Protecting System for Stereoselective Glycosylation. *Org. Lett.*, **2006**, *8*, 4879–4882; (b) D. Crich, P. Jayalath, T. K. Hutton. Enhanced Diastereoselectivity in β-Mannopyranosylation Through the Use of Sterically Minimal Propargyl Ether Protecting Groups. *J. Org. Chem.*, **2006**, *71*, 3064–3070; (c) D. Crich, P. Jayalath. 2-O-Propargyl Ethers: Readily Cleavable, Minimally Intrusive Protecting Groups for β-Mannosyl Donors. *Org. Lett.*, **2005**, *7*, 2277–2280; (d) D. Crich, M. S. Karatholuvhu. Application of the 4-Trifluoromethylbenzenepropargyl Ether Group as an Unhindered, Electron Deficient Protecting Group for Stereoselective Glycosylation. *J. Org. Chem.*, **2008**, *73*, 5173–5176.
[45] (a) D. Crich, W. Cai. Chemistry of 4,6-O-Benzylidene-D-Glycopyranosyl Triflates: Contrasting Behavior Between the Gluco and Manno Series. *J. Org. Chem.*, **1999**, *64*, 4926–4930. (b) D. Crich, O. Vinogradova. On the Influence of the C2-O2 and C3-O3 Bonds in 4,6-Obenzylidene-Directed β-Mannopyranosylation and α-Glucopyranosylation. *J. Org. Chem.*, **2006**, *71*, 8473–8480.
[46] L. Chen, F. Kong. Unusual α-Glycosylation with Galactosyl Donors with a C2 Ester Capable of Neighboring Group Participation. *Tetrahedron Lett.*, **2003**, *44*, 3691–3695.
[47] (a) M. Miljkovic, D. Yeagley, P. Deslongchamps, Y. L. Dory. Experimental and Theoretical Evidence of Through-Space Electrostatic Stabilization of the Incipient Oxocarbenium Ion by an Axially Oriented Electronegative Substituent During Glycopyranoside Acetolysis. *J. Org. Chem.*,**1997**, *62*, 7597–7604. (b) A. Imamura, H. Ando, H. Ishida, M. Kiso. DTBS (di-tert-butylsilylene)-Directed α Galactosylation for the Synthesis of Biologically Relevant Glycans. *Curr. Org. Chem.*, **2008**, *12*, 675–689.
[48] (a) A. Imamura, H. Ando, S. Korogi, G. Tanabe, O. Muraoka, H. Ishida, M. Kiso. Di-Tertbutylsilylene (DTBS) Group-Directed α-Selective Galactosylation Unaffected by C-2 Participating Functionalities.

Tetrahedron Lett., **2003**, *44*, 6725–6728; (b) A. Imamura, A. Kimura, H. Ando, H. Ishida, M. Kiso. Extended Applications of Di-Tertbutylsilylene-Directed α-Predominant Galactosylation Compatible with C2-Participating Groups Toward the Assembly of Various Glycosides. *Chem. Eur. J.*, **2006**, *12*, 8862–8870.

[49] X. Zhu, S. Kawatkar, Y. Rao, G.-J. Boons. Practical Approach for the Stereoselective Introduction of β-Arabinofuranosides. *J. Am. Chem. Soc.*, **2006**, *128*, 11948–11957.

[50] (a) G. Stork, G. Kim, Stereocontrolled synthesis of disaccharides via the temporary silicon connection. *J. Am. Chem. Soc.*, **1992**, 114, 1087–1088; (b) G. Stork, J. J. L. Clair. Stereoselective Synthesis of β-Mannopyranosides via the Temporary Silicon Connection Method. *J. Am. Chem. Soc.*, **1996**, 118, 247–248.

[51] (a) Z. A. Buchan, S. J. Bader, J. Montgomery. Ketone Hydrosilylation with Sugar Silanes Followed by Intramolecular Aglycone Delivery: An Orthogonal Glycosylation Strategy. Angew. *Chem. Int. Ed.*, **2009**, 48, 4840–4848; (b) K. M. Partridge, S. J. Bader, Z. A. Buchan, C. E. Taylor, J. Montgomery. A Streamlined Strategy for Aglycone Assembly and Glycosylation. Angew. *Chem. Int. Ed.*, **2013**, 52, 13647–13650.

[52] (a) M. Bols. Stereocontrolled synthesis of α-glucosides by intramolecular glycosidation. *J. Chem. Soc., Chem. Commun.*, **1992**, 913–914; (b) M. Bols. Application of intramolecular glycosidation to the stereocontrolled synthesis of disaccharides containing α-gluco and α-galacto linkages. *J Chem. Soc., Chem. Commun.*, **1993**, 791–792; (c) M. Bols. Efficient stereocontrolled glycosidation of secondary sugar hydroxyls by silicon tethered intramolecular glycosidation. *Tetrahedron*, **1993**, 49, 10049–10060; (d)M. Bols. Synthesis of Kojitriose using Silicon- tethered glycosidation. *Acta Chem. Scand.*, **1996**, 931–937; (e) M. Bols, H. C. Hansen, Stereocontrolled synthesis of glycosides was achieved by intramolecular glycosidation of an aglycon tethered to the 4,5 and 6 position of a thioglycoside donor. *Chem. Lett.*, **1994**, 1049–1052.

[53] G. K. Packard, S. D. Rychnovsky. β-Selective Glycosylations with Masked d-Mycosamine Precursors. *Org. Lett.*, **2001**, 3, 3393–3396.

[54] (a) F. Barresi, O. Hindsgaul. Synthesis of .b.-mannopyranosides by intramolecular aglycon delivery. *J. Am. Chem. Soc.*, **1991**, 113, 9376–9377; (b) F. Barresi, O. Hindsgaul. Improved Synthesis of β-Mannopyranosides by Intramolecular Aglycon Delivery. *Synlett.*, **1992**, 759–761; (c) F. Barresi, O. Hindsgaul. The synthesis of P-mannopyranosides by intramolecular aglycon delivery: scope and limitations of the existing methodology. *Can. J. Chem.*, **1994**, 72, 1447–1465.

[55] (a) S. C. Ennis, A. J. Fairbanks, R. J. Tennant-Eyles, H. S. Yeates. Stereoselective Synthesis of a-Glucosides and b-Mannosides: Tethering and Activation with N-Iodosuccinimide. *Synlett.*, **1999**, 1387–1390; (b) S. C. Ennis, A. J. Fairbanks, C. A. Slinn, R. J. Tennant-Eyles, H. S. Yeates. *N*-Iodosuccinimide-mediated intramolecular aglycon delivery. *Tetrahedron*, **2001**, 57, 4221–4230.

[56] (a) C. M. P. Seward, I. Cumpstey, M. Aloui, S. C. Ennis, A. J, Redgrave, A. J. Fairbanks. Stereoselective cis glycosylation of 2-*O*-allyl protected glycosyl donors by intramolecular aglycon delivery (IAD). *Chem. Commun.*, **2000**, 1409–1410; (b) I. Cumpstey, A. J. Fairbanks, A. J. Redgrave. Stereospecific Synthesis of 1,2-*cis* Glycosides by Allyl-Mediated Intramolecular Aglycon Delivery. 2. The Use of Glycosyl Fluorides. *Org. Lett.*, **2001**, 3, 2371–2374; (c) M. Aloui, D. J. Chambers, I. Cumpstey, A. J. Fairbanks, A. J. Redgrave, C. M. P. Seward. Stereoselective 1,2-cis Glycosylation of 2-*O*-Allyl Protected Thioglycosides. *Chem. Eur. J.*, **2002**, 8, 2608–2621; (d) I. Cumpstey, K. Chayajarus, A. J. Fairbanks, A. J. Redgrave, C. M. P. Seward. Allyl Protecting Group Mediated Intramolecular Aglycon Delivery: Optimisation of Mixed Acetal Formation and Mechanistic Investigation. *Tetrahedron Asymmetry*, **2004**, 15, 3207–3221.

[57] J. Tatai, P. Fügedi. A New, Powerful Glycosylation Method: Activation of Thioglycosides with Dimethyl Disulfide–Triflic Anhydride. *Org. Lett.*, **2007**, 9, 4647–4650.

[58] (a) I. Cumpstey, A. J. Fairbanks, A. J. Redgrave. Allyl protecting group mediated intramolecular aglycon delivery (IAD): synthesis of α-glucofuranosides and β-rhamnopyranosides. *Tetrahedron*, **2004**, 60, 9061–9074; (b) A.J. Fairbanks. Intramolecular Aglycon Delivery (IAD): The Solution to 1,2-*cis* Stereocontrol for Oligosaccharide Synthesis. *Synlett.*, **2003**, 1945–1958; (c) E. Attolino, I. Cumpstey, A.J. Fairbanks. Synthesis of the Glc3Man N-glycan tetrasaccharide by iterative allyl IAD. *Carbohydr. Res.*, **2006**, 341, 1609–1618.

[59] K. Chayajarus, D. J. Chambers, M. J. Chughtai, A. J. Fairbanks. Stereospecific Synthesis of 1,2-cis Glycosides by Vinyl-Mediated IAD. *Org. Lett.*, **2004**, 6, 3797–3800.

[60] Y. Olimoto, S. Sakaguchi, Y.Ishii. Development of a Highly Efficient Catalytic Method for Synthesis of Vinyl Ethers. *J. Am. Chem. Soc.*, **2002**, 124, 1590–1591.

[61] (a) E. Attolino, A. J. Fairbanks. β-Mannosylation of *N*-acetyl glucosamine by propargyl mediated intramolecular aglycon delivery (IAD): synthesis of the *N*-glycan core pentasaccharide. *Tetrahedron Lett.*, **2007**, 48, 3061–3064; (b) E. Attolino, T. W. D. F. Ridsing, C. D. Heidecke, A. J. Fairbanks. Propargyl mediated intramolecular aglycon delivery (IAD): applications to the synthesis of core *N*-glycan oligosaccharides. *Tetrahedron Asymmetry*, **2007**, 18, 1721–1734.

[62] Y. Ito, T. Ogawa. A Novel Approach to the Stereoselective Synthesis of β-Mannosides. Angew. *Chem. Int. Ed. Engl.*, **1994**, 33, 1765–1767.

[63] (a) M. Lergenmüller, T. Nukada, K. Kuramochi, A. Dan, T. Ogawa, Y. Ito. On the Stereochemistry of Tethered Intermediates in *p*-Methoxybenzyl-Assisted β-Mannosylation. *Eur. J. Org. Chem.*, **1999**, 1367–1376; (b) Y. Ito, H. Ando, M. Wada, T. Kawai, Y. Ohnishi, Y. Nakahara. On the mechanism of *p*-methoxybenzylidene assisted intramolecular aglycon delivery. *Tetrahedron*, **2001**, 57, 4123–4132; (c) Y. Ito, Y. Ohnishi, T. Ogawa, Y. Nakahara. Highly Optimized β-Mannosylation via *p*-Methoxybenzyl Assisted Intramolecular Aglycon Delivery. *Synlett.*, **1998**, 1102–1104.

[64] (a) M. R. Pratt, C. D. Leigh, C. R. Bertozzi. Formation of 1,1-α,α-Glycosidic Bonds By Intramolecular Aglycone Delivery. A Convergent Synthesis of Trehalose. *Org. Lett.*, **2003**, 5, 3185–3188; (b) C. D. Leigh, C. R. Bertozzi, Synthetic Studies toward *Mycobacterium tuberculosis* Sulfolipid-I. *J. Org. Chem.*, **2008**, 73, 1008–1017.

[65] A. Ishiwata, A. Sakurai, Y. Nishimiya, S. Tsuda, Y. Ito. Synthetic Study and Structural Analysis of the Antifreeze Agent Xylomannan from Upis ceramboides. *J. Am. Chem. Soc.*, **2011**, 113, 19524–19535.

[66] K. Ajayi, V. V. Thakur, R. C. Lapo, S. Knapp. Intramolecular α-Glucosaminidation: Synthesis of Mycothiol. *Org. Lett.*, **2010**, 12, 2630–2633.

[67] A. Nakagawa, M. Tanaka, S. Hanamura, D. Takahashi, K. Toshima. Regioselective and 1,2-*Cis*-α-Stereoselective Glycosylation Utilizing Glycosyl-Acceptor-Derived Boronic Ester Catalyst. *Angew. Chem. Int. Ed.*, **2015**, *54*,10935–10939.

[68] M. Tanaka, J. Nashida, D. Takahashi, K. Toshima. Glycosyl-Acceptor-Derived Borinic Ester-Promoted Direct and β-Stereoselective Mannosylation with a 1,2-Anhydromannose Donor. *Org. Lett.*, **2016**, *18*, 2288–2291; (b) M. Tanaka, D. Takahashi, K. Toshima, 1,2-cis-α-Stereoselective Glycosylation Utilizing a Glycosyl-Acceptor-Derived Borinic Ester and its Application to the Total Synthesis of Natural Glycosphingolipids. *Org. Lett.*, **2016**, *18*, 5030–5033.

[69] N. Nishi, J. Nashida, E. Kaji, D. Takahashi, K. Toshima. Boronate-Catalyzed Ring Opening Promoted the Regio- and Stereoselective Formation of 1,2-cis-Linked Sugars. *Chem. Commun.*, **2017,** *53*, 3018–3021.

[70] N. Nishi, K. Sueoka, K. Iijima, R. Sawa, D. Takahashi, K. Toshima. Stereospecific β-L-Rhamnopyranosylation Through an SNi-Type Mechanism by Using Organoboron Reagents. *Angew. Chem. Int. Ed.*, **2018**, *57*, 13858–13862.

[71] M. Tanaka, A. Nakagawa, N. Nishi, K. Iijima, R. Sawa, D. Takahashi, K. Toshima, Boronic-Acid-Catalyzed Regioselective and 1,2-Cis-Stereoselective Glycosylation of Unprotected Sugar Acceptors via SNi-Type Mechanism. *J. Am. Chem. Soc.*, **2018**, *140*, 3644–3651.

[72] S. Tomita, M. Tanaka, M. Inoue, K. Inaba, D. Takahashi, K. Toshima. Diboron-Catalyzed Regio- and 1,2-Cis-α-Stereoselective Glycosylation of Trans-1,2-Diols. *J. Org. Chem.*, **2020**, *85*, 16254–16262.

[73] F. Yu, J. Li, P. M. DeMent, Y.-J. Tu, H. B. Schlegel, H. M. Nguyen. Phenanthroline-Catalyzed Stereoretentive Glycosylations. *Angew. Chem. Int. Ed.*, **2019**, *58*, 6957–6961.

[74] P. M. DeMent, C. Liu, J. Wakpal, R. N. Schaugaard, H. B. Schlegel, H. M. Nguyen. Phenanthroline-Catalyzed Stereoselective Formation of α-1,2-cis 2-Deoxy-2-Fluoro Glycosides. *ACS Catal.*, **2021**, *11* (4), 2108–2120.

[75] E. I. Balmond, D. M. Coe, M. C. Galan, E. M. McGarrigle. α-Selective Organocatalytic Synthesis of 2-Deoxygalactosides. *Angew. Chem. Int. Ed.*, **2012**, *51*, 9152–9155.

[76] C. Palo-Nieto, A. Sau, R. Williams, M. C. Galan, Cooperative Brønsted Acid-Type Organocatalysis for the Stereoselective Synthesis of Deoxyglycosides. *J. Org. Chem.*, **2017**, *82*, 407–414.

[77] G. A. Bradshaw, A. C. Colgan, N. P. Allen, I. Pongener, M. B. Boland, Y. Ortin, E. M. McGarrigle. Stereoselective Organocatalyzed Glycosylations—Thiouracil, Thioureas and Monothiophthalimide Act as Brønsted Acid Catalysts at Low Loadings. *Chem. Sci.*, **2019**, *10*, 508–514.

[78] A. Tiwari, A. Khanam, A. Kumar, M. Lal, P. K. Mandal. L-Proline Derived Thioamide Small Organic Molecule for the α-Stereoselective Synthesis of 2-Deoxyglycosides. *Adv. Synth. Catal.*, **2023**, *365*, 2949–2958.

[79] (a) R. R. Schmidt, E. Rucker. Stereoselective Glycosidations of Uronic Acids. *Tetrahedron Lett.*, **1980**, *21*, 1421–1424; (b) H. Dohi, Y. Nishida, H. Tanaka, K. Kobayashi. O-Methoxycarbonylphenyl 1-Thio-β-d-Galactopyranoside, a Non-Malodorous Thio Glycosylation Donor for the Synthesis of Globosyl α (1-4)-Linkages. *Synlett.*, **2001**, *2001*, 1446–1448.

[80] (a) A. Ishiwata, Y. Munemura and Y. Ito. Synergistic Solvent Effect in 1,2-cis-Glycoside Formation. *Tetrahedron*, **2008**, *64*, 92–102. (b) G. Wulff, G. Rohle. Results and Problems of *O*-Glycoside Synthesis. *Angew. Chem. Int. Ed. Engl.*, **1974**, *13*, 157–170.

4 The Role of Catalysts in the Synthesis of Carbohydrates, Lipids, and Steroids

Yliana López, Juan-Pablo García-Merinos, Rosa E. del Río, and Rosa Santillan

4.1 INTRODUCTION

Carbohydrates, lipids, and steroids play important roles in the human body. Carbohydrates act as an energy source, help control blood glucose and insulin metabolism, and participate in cholesterol and triglyceride metabolism, and lipids are fatty compounds that are part of the cell membrane and that help control what goes in and out of your cells; they help with moving and storing energy, absorbing vitamins and making hormones. The steroids, meanwhile, constitute an extensive and important class of biologically active polycyclic compounds that are widely used for therapeutic purposes, for example as drugs for prevention and therapy serving functions such as antiinflammatory, diuretic, anabolic, contraceptive, antiandrogenic, and pregestational as well as anticancer agents for breast, prostate, colon, and many other cancers. Steroids in fact constitute the largest group of pharmaceuticals next to antibiotics. For these and other reasons, carbohydrates, lipids, and steroids are important scaffolds for synthesizing molecules of pharmaceutical interest.

Catalytic chemistry plays an essential role in the production and functionalization of an enormous range of chemicals, biomolecules, and materials; the resulting products have direct impacts on human health, quality of life, and the global economy. Given this scope and importance, a range of different systems, including heterogeneous solids, homogeneous small molecules, metal complexes, and enzymes have been developed as catalysts for chemical synthesis. This chapter provides an overview of the several catalytic methods applied over the last ten years for synthesizing carbohydrate, lipid, and steroid derivatives.

4.2 CARBOHYDRATE SYNTHESIS AND APPLICATIONS

4.2.1 Transforming Carbohydrates into Valuable Chemical Derivatives

Developing new strategies for converting carbohydrates into 5-hydroxymethylfurfural (HMF) (3) has attracted much attention because of fossil resource depletion and for being considered as the most important platform molecule in the utilization of biomass [1]. The POM/hypothalamic–pituitary–adrenal axis catalysts can bear one or two acid sites, acid protons, or/and metals with Lewis acidity. Both types of acidic sites can work as active sites in acidic catalysis [1a]. Thus, these compounds have provoked a huge interest for acid catalysis in the scientific community [1a]. For instance, researchers have demonstrated the effectiveness of $Cs_{2.5}H_{0.5}PW_{12}O_{40}$ [2, $Ag_3PW_{12}O_{40}$ [3], and $[MIMPS]_3PW_{12}O_{40}$ (MIMPS = 1-[3-(sulfonic acid)propyl]-3-methylimidazolium) [4] for the intramolecular dehydration of fructose or glucose to HMF (3) (Scheme 4.1). The dual acid sites guarantee the isomerization and dehydration of glucose into (3).

DOI: 10.1201/9781003437413-4

In different work, Corma and group described the utility of one-pot reactions through several representative examples of transformations carried out with mono- and multifunctional catalytic systems [5]. Corma and Villandier elegantly applied a unique catalytic system that performs the whole synthetic sequence to produce alkyl glucoside starting from cellulose (6) (Scheme 4.2) [6]. They used anionic liquid medium with sulfonic resin as catalyst and alkyl glucoside surfactants with a mass yield of 82%. They were able to achieve this methodology by properly coupling the hydrolysis of cellulose (6) with the Fischer glycosylation reaction of the monosaccharide formed during the first step (Scheme 4.2). The amount of water present in the reaction medium and its evolution during the reaction time was a key variable for the success of the global process.

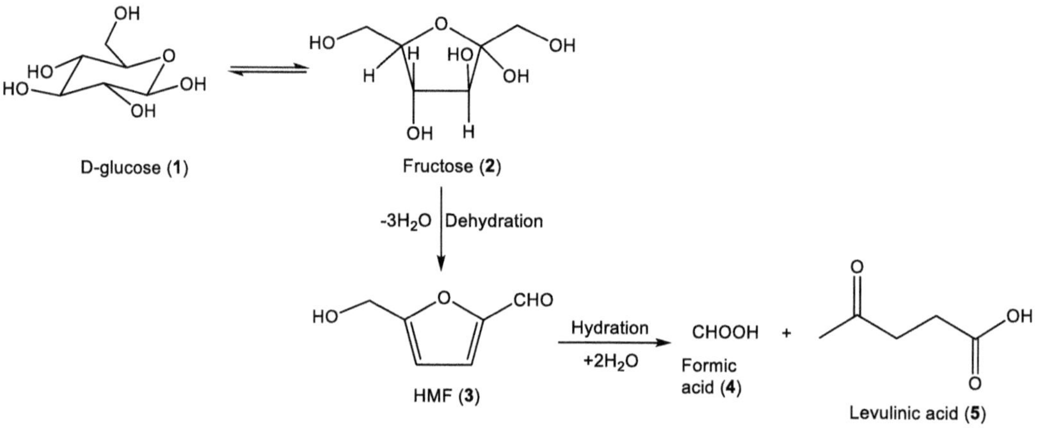

D-glucose (**1**) ⇌ Fructose (**2**)

$-3H_2O$ Dehydration

HMF (**3**) — Hydration $+2H_2O$ → CHOOH Formic acid (**4**) + Levulinic acid (**5**)

SCHEME 4.1 HMF from glucose (1) [1b].

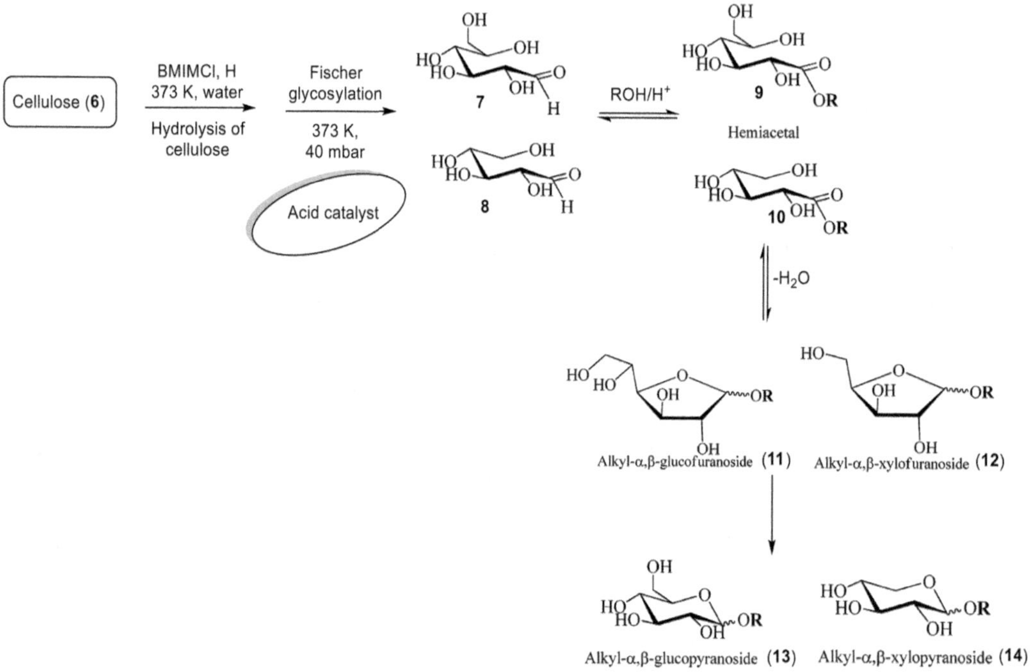

Cellulose (**6**) — BMIMCl, H, 373 K, water; Hydrolysis of cellulose → Fischer glycosylation, 373 K, 40 mbar; Acid catalyst → **7**, **8** ⇌ (ROH/H^+) **9** Hemiacetal, **10**

$-H_2O$

Alkyl-α,β-glucofuranoside (**11**) Alkyl-α,β-xylofuranoside (**12**)

Alkyl-α,β-glucopyranoside (**13**) Alkyl-α,β-xylopyranoside (**14**)

SCHEME 4.2 One-step conversion of cellulose (6) into alkyl-α,β-glycoside surfactants (13), (14) [5, 6a].

High (≥95% C) yields of *n*-hexane (20) and *n*-pentane (21) were reported by Tomishige and group *via* hydrogenolysis of aqueous sorbitol (15) and xylitol (16), respectively, using the I-ReO_x/SiO_2 catalyst combined with H-ZSM-5 as a co-catalyst (Scheme 4.3). The direct production of *n*-hexane (20) from glucose (1) or cellobiose (19) can be achieved by using the same system. The catalytic system has a good reusability. Adding H-ZSM-5 improves the dehydration of secondary alcohols, which enhances the hydrogenolysis of secondary alcohols to alkanes [6b, 7].

4.2.2 Glycosylation

Matthies *et al.* reported gold(I)-catalyzed glycosylation in a continuous flow reactor. The reaction with o-glycosyl (24) in the presence of 13% mol gold catalyst gave the desired *β*-glycosides (25). Furthermore, glycosylation in a continuous flow reactor was demonstrated. The reaction setup allowed for access to a range of glycosides in good to high yields. The glycosylation proceeded in short reaction times of only 20–30 min. The anomeric ratio could be controlled by neighboring groups (Scheme 4.4) [8, 9].

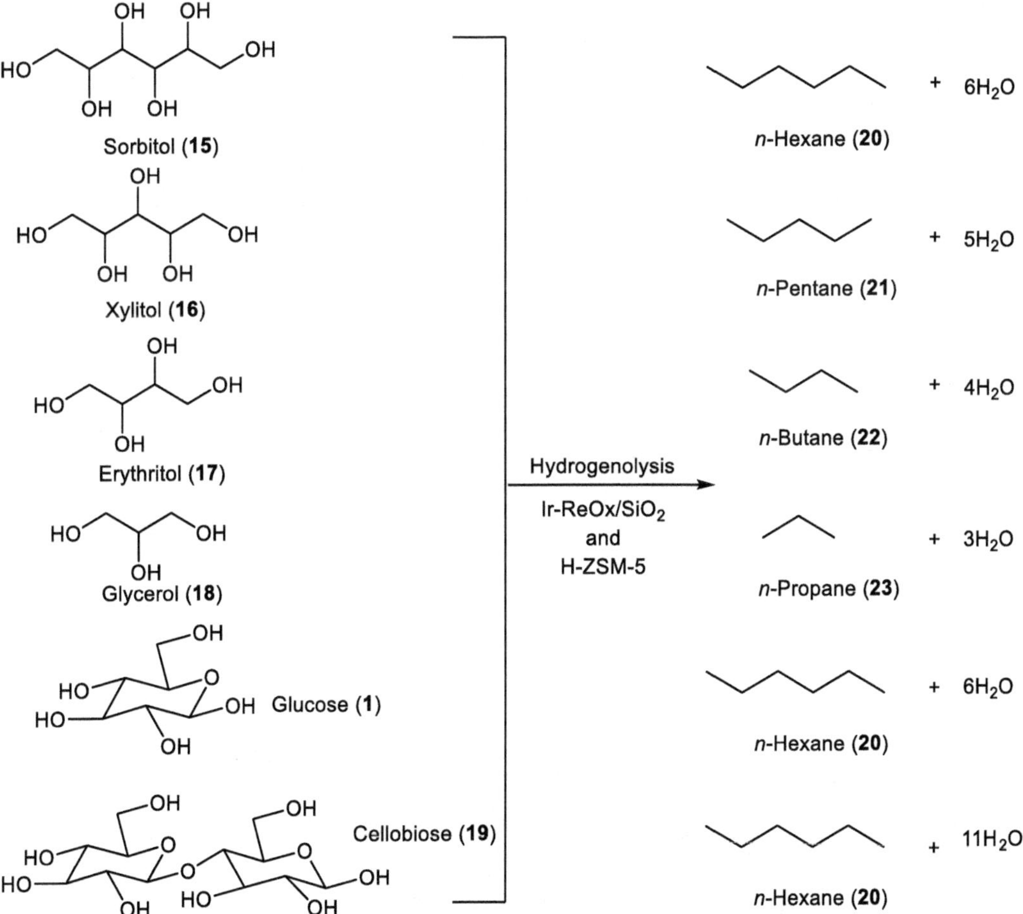

SCHEME 4.3 Complete C–O hydrogenolysis of sugars and sugar alcohols to alkanes using Ir-ReOx/SiO_2 and H-ZSM-5 [6b, 7].

Glycosylation of a sialyl donor with a thioethyl galactosyl acceptor was performed in the presence of $Bi(OTf)_3$ used in large excess compared to the sialyl donor (Scheme 4.5). In this case, the desired disaccharide (28) was obtained with nearly complete α-stereoselectivity (α/β = 20:1). Activation with other Lewis acids led to decreased stereoselectivities [10, 11].

4.2.3 Carbohydrate-Derived Catalysts and Their Applications in Organic Synthesis

Cellulose and its derivatives are a broad linear polymer with a large number of hydroxyl groups that coordinate well with metals both directly and after functional modification [12, 13]. Nanocellulose and microcrystalline are two commonly used types of cellulose to prepare carbon materials that have the advantages of better dissolution options, larger surface areas, and controllable morphology [14, 15]. Chang et al. prepared an amorphous carbon-based sulfonated catalyst by carbonizing cellulose at different temperature and then sulfonating it with concentrated sulfuric acid [16]. The as-obtained sulfonated catalyst exhibits high activity in the one-step synthesis of hydroxymethylfurfural (3) from inulin (29) in ionic liquid, and it can be easily regenerated after treatment with dilute sulfuric acid (Scheme 4.6). In short, these catalysts exhibit better catalytic activity and higher reusability than traditional solid acid catalysts [17].

Wang *et al.* reported a green and economical method for the preparation of nitrogen and phosphorus co-doped carbon-based metal-free catalysts (NPC) (30) with cellulose crystallite as carbon source and $(NH_4)_2HPO_4$ as both nitrogen and phosphorus source (Scheme 4.7). The NPC (30) presented flake-like morphology with N and P elemental doping levels of 4.3 atom% and 10.66 atom% respectively. The NPC (30) showed excellent catalytic activity for the reduction of *p*-nitrophenol (*p*-NP) (31). The TOF (mmol *p*-NP/(mg catalyst min)) of the reduced (31) is 2×10^{-5}, which is comparable with those of noble metal-based catalysts and conventional graphene-based catalysts [18].

In 2019, Elhampour *et al.* immobilized Pd nanoparticles on mesoporous triazine-based carbon (MTC) (34) to prepare a retrievable and efficient heterogeneous catalyst Pd@MTC (35) for Heck

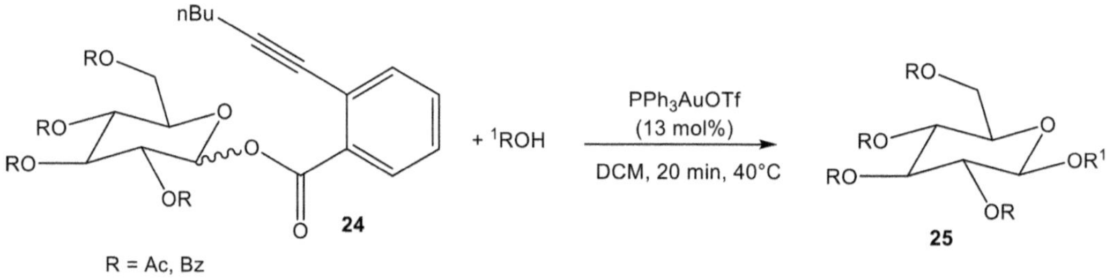

R = Ac, Bz

SCHEME 4.4 Gold-catalyzed glycosylation [8, 9].

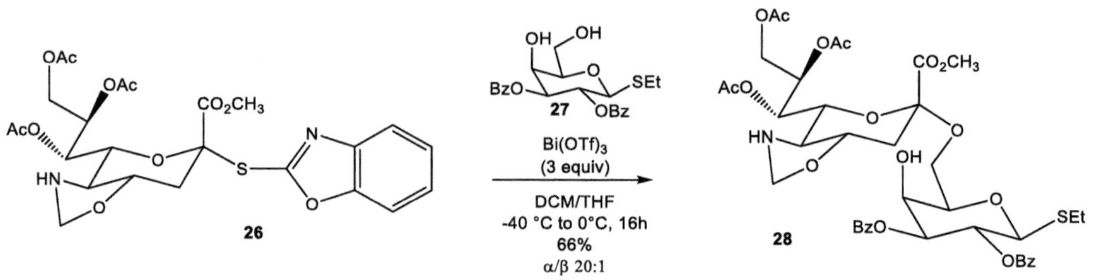

SCHEME 4.5 $Bi(OTf)_3$-catalyzed glycosylation [10, 11].

cross-coupling reaction [17, 19]. To prepare the MTC (34), cellulose modified by $ClCH_2CN$ (33) was fully mixed with $ZnCl_2$ and then carbonized at 400 °C for 40 h (Scheme 4.8). The authors observed that the presence of $ZnCl_2$ and nitrile group was extremely beneficial for forming the mesoporous structure during pyrolysis, which enhanced catalytic efficiency.

Beller *et al.* synthesized a novel sustainable Co-Co_3O_4@Chitosan-700 (38) catalyst from bio-waste chitosan (36) in combination with earth-abundant cobalt salt. This catalyst was applied successfully for hydride halogenation of alkyl and (hetero)aryl halides (>40 examples) and showed excellent chemoselectivity using molecular hydrogen (Scheme 4.9) [17, 20].

In 2018, Sels *et al.* employed sucrose (41) as sustainable carbon source to prepare meso-structured silica-carbon nanocomposites (45) with large mesopore interconnectivity using mild

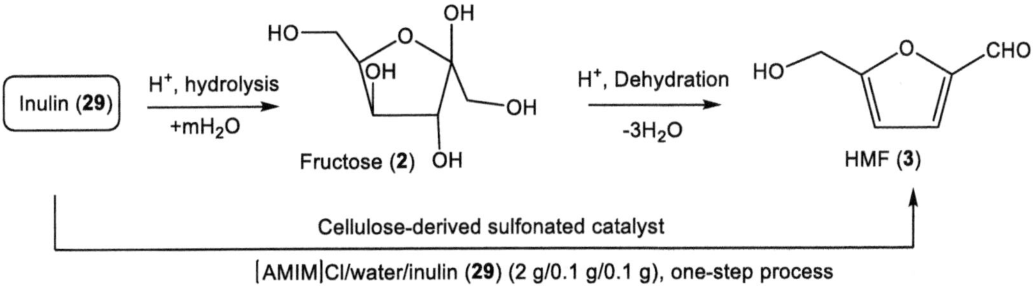

Inulin (**29**) —H⁺, hydrolysis / +mH₂O→ Fructose (**2**) —H⁺, Dehydration / -3H₂O→ HMF (**3**)

Cellulose-derived sulfonated catalyst

[AMIM]Cl/water/inulin (**29**) (2 g/0.1 g/0.1 g), one-step process

SCHEME 4.6 The conversion of inulin (29) to HMF (3) *via* one-step and two-step pathways [16, 17].

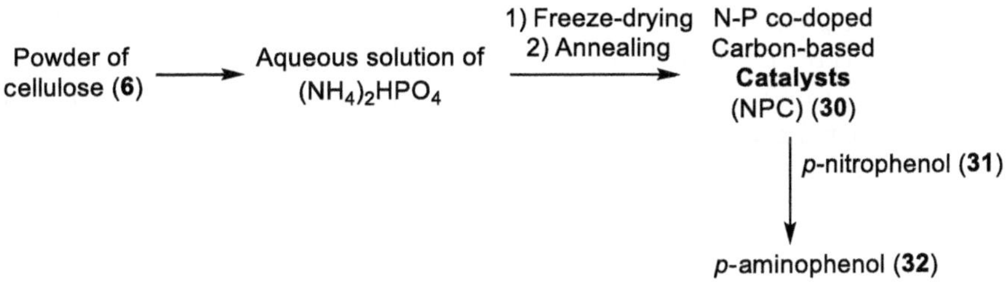

Powder of cellulose (**6**) → Aqueous solution of $(NH_4)_2HPO_4$ —1) Freeze-drying 2) Annealing→ N-P co-doped Carbon-based **Catalysts** (NPC) (**30**)

p-nitrophenol (**31**) → *p*-aminophenol (**32**)

SCHEME 4.7 Preparation of NPC (30) and its use as a catalyst in the reduction of *p*-nitrophenol (31) [18].

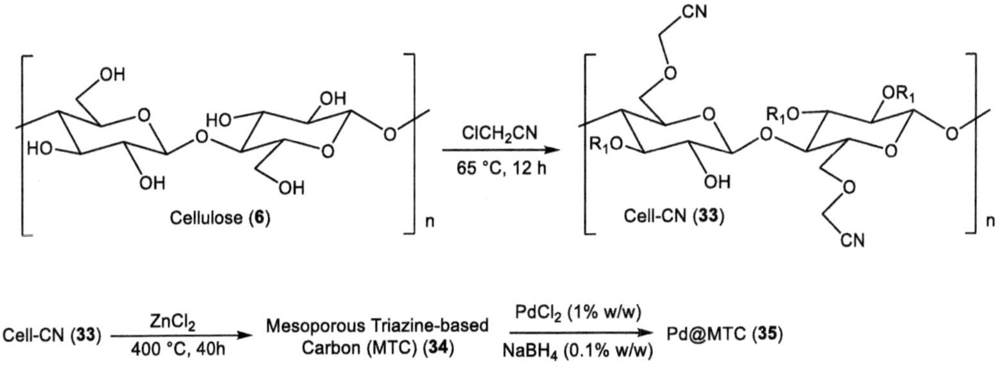

Cellulose (**6**) —$ClCH_2CN$ / 65 °C, 12 h→ Cell-CN (**33**)

Cell-CN (**33**) —$ZnCl_2$ / 400 °C, 40h→ Mesoporous Triazine-based Carbon (MTC) (**34**) —$PdCl_2$ (1% w/w) / $NaBH_4$ (0.1% w/w)→ Pd@MTC (**35**)

SCHEME 4.8 Fabrication of Pd@MTC (35) [17, 19].

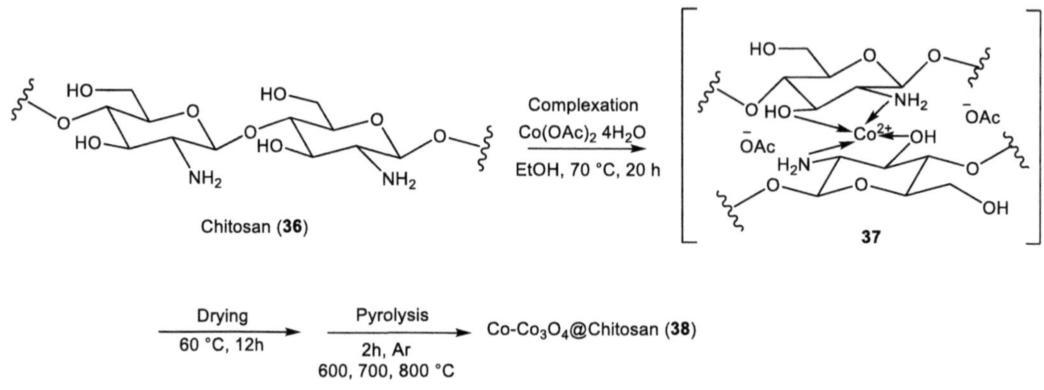

SCHEME 4.9 Synthesis of Co-Co_3O_4@chitosan materials [17, 20].

vapor-phase-assisted hydrothermal treatment. The prepared nanocomposites exhibited superior catalytic performance in the 2-methylfuran condensation with furfural to that of their counterparts from dry pyrolysis (Scheme 4.10) [21].

4.2.4 Acetylation of Carbohydrate-Derived Polyols

Acetylation of monosaccharides is usually the first step in the synthesis of complex carbohydrates [22], and Mensah *et al.* described the efficacy of the cationic Palladium (II) catalyst $Pd(PhCN)_2(OTf)_2$ in the acetylation of carbohydrate-derived polyols (55), (57), (59), and (61) (Table 4.1). They used two equivalent of acetic anhydride for every hydroxy group. In all cases, the acetylation reaction afforded the acetylated sugars (56), (58), (60), and (62) in excellent yields (Table 4.1) [23].

Mensah *et al.* also investigated the tolerance of the $Pd(PhCN)_2(OTf)_2$ catalyzing acetylation with several acid-sensitive hydroxyl-protecting groups and demonstrated that the acetonide, TBDPS, and PMP (*p*-methoxyphenyl) protecting groups were stable under the acetylation protocol, affording the corresponding acetylated products in excellent yields (Scheme 4.11 and Table 4.1); the TMS and TBS groups were unstable and subsequently hydrolyzed, and the resulting hydroxy group acetylated.

4.2.5 Deacetylation of Carbohydrates

Lin *et al.* in 2019 synthesized three dinuclear dysprosium complexes, $[Dy_2(hmb)_2(OTf)_2(H_2O)_4]$.HOTf.2THF (A.HOTf.2THF), $[Dy_2(hmb)_2(OTf)_2(H_2O)_4].(CH_3)_2CO$ (A.$(CH_3)_2CO$), and $[Dy_2(hmi)_3(H_2O)_2]$.2HOTf (B.2HOTf), and found that complex A.HOTf.2THF was an effective and chemoselective catalyst for deacetylating esters in methanol (Table 4.2). Both high efficiency and chemoselectivity were achieved for substrates bearing different aromatic and alkyl chain derivatives. This method was especially valuable for preparing peracetylated hemiacetal of monosaccharides and disaccharides with 85%–88% and 64%–84% yields, respectively. Most importantly, this approach is a green alternative to using only lanthanide salts or other described procedures, since complex A.HOTf.2THF can be effectively recycled and reused without reducing activity and selectivity (Scheme 4.12) [24].

Liu and Cao reported the effective one-step synthesis of polysubstituted guanidinoglucosides (73) from peracetylated methyl 6-deoxy-6-thioureidoglucosides (72) using HgO-MS 4Å as catalyst (Scheme 4.13) [25]. In this reaction, HgO was applied as desulfurizing agent and MS 4Å as dehydrating agent and alkaline catalyst for the one-step synthesis of a series of trisubstituted guanidino-containing sugars. When the reaction was carried out in the presence of MS 4Å, a significant increase was observed in the reaction rate, and the reaction was complete within 12 h. In the absence

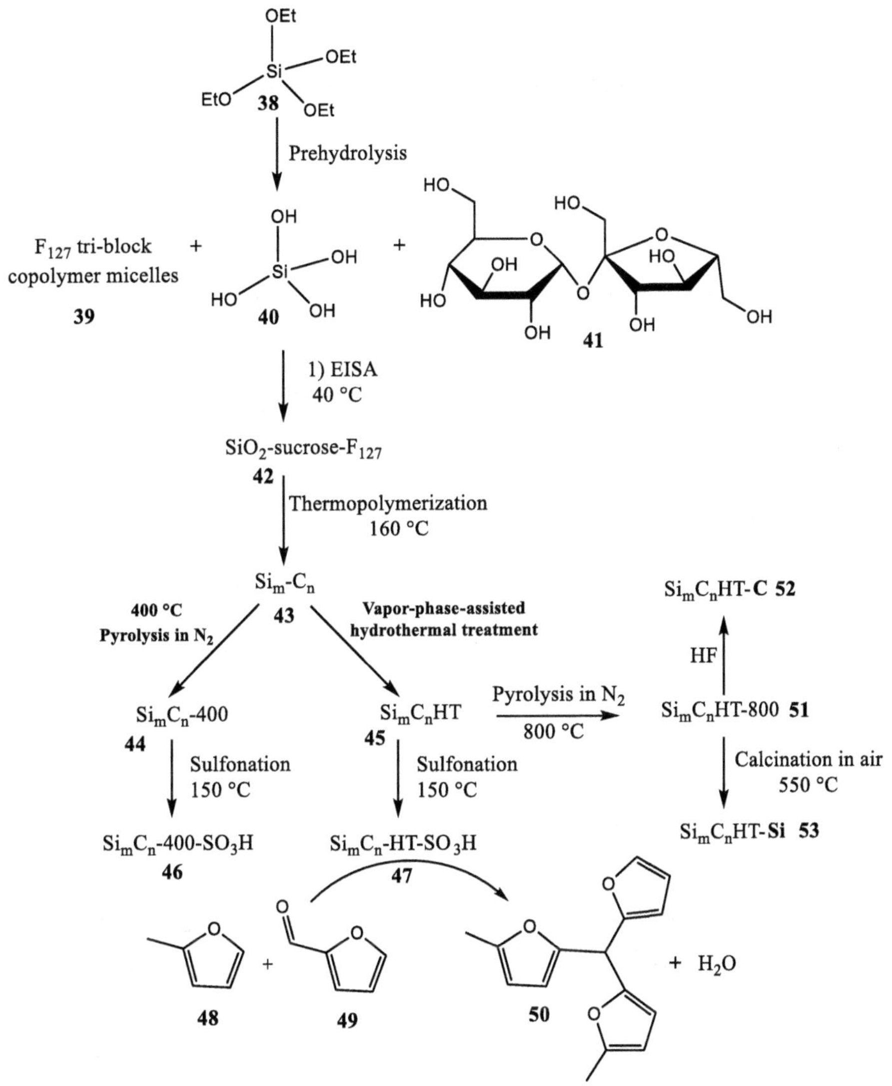

OEt
OEt
EtO Si OEt
38
Prehydrolysis
F_{127} tri-block copolymer micelles
39
+
OH
HO Si OH
OH
40
+
HO HO O OH HO O O HO OH OH OH
41
1) EISA
40 °C
SiO_2-sucrose-F_{127}
42
Thermopolymerization
160 °C
Si_m-C_n
43
400 °C
Pyrolysis in N_2
Vapor-phase-assisted hydrothermal treatment
Si_mC_n-400
44
Sulfonation
150 °C
Si_mC_n-400-SO_3H
46
Si_mC_nHT
45
Sulfonation
150 °C
Si_mC_n-HT-SO_3H
47
Pyrolysis in N_2
800 °C
Si_mC_nHT-800 51
HF
Si_mC_nHT-C 52
Calcination in air
550 °C
Si_mC_nHT-Si 53
48 + 49 → 50 + H_2O

SCHEME 4.10 The preparation of nanocomposite acid catalyst and its catalytic application [21].

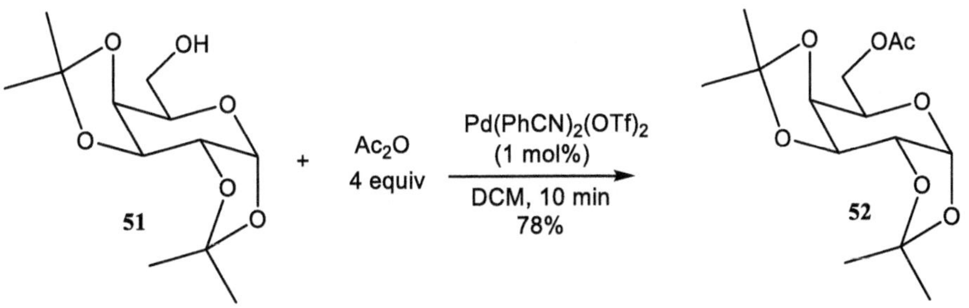

51 + Ac_2O 4 equiv
$Pd(PhCN)_2(OTf)_2$ (1 mol%)
DCM, 10 min
78%
→ 52

SCHEME 4.11 Acetylation of carbohydrate-derived polyols with acetic anhydride catalyzed by $Pd(PhCN)_2(OTf)_2$ (generated *in situ* from $Pd(PhCN)_2Cl_2$ (1 equiv.) and AgOTf (2 equiv) in CH_2Cl_2 (1.25 M) [23].

TABLE 4.1
$Pd(PhCN)_2(OTf)_2$ Catalyzed Acetylation of Carbohydrate-Derived Polyols [23].

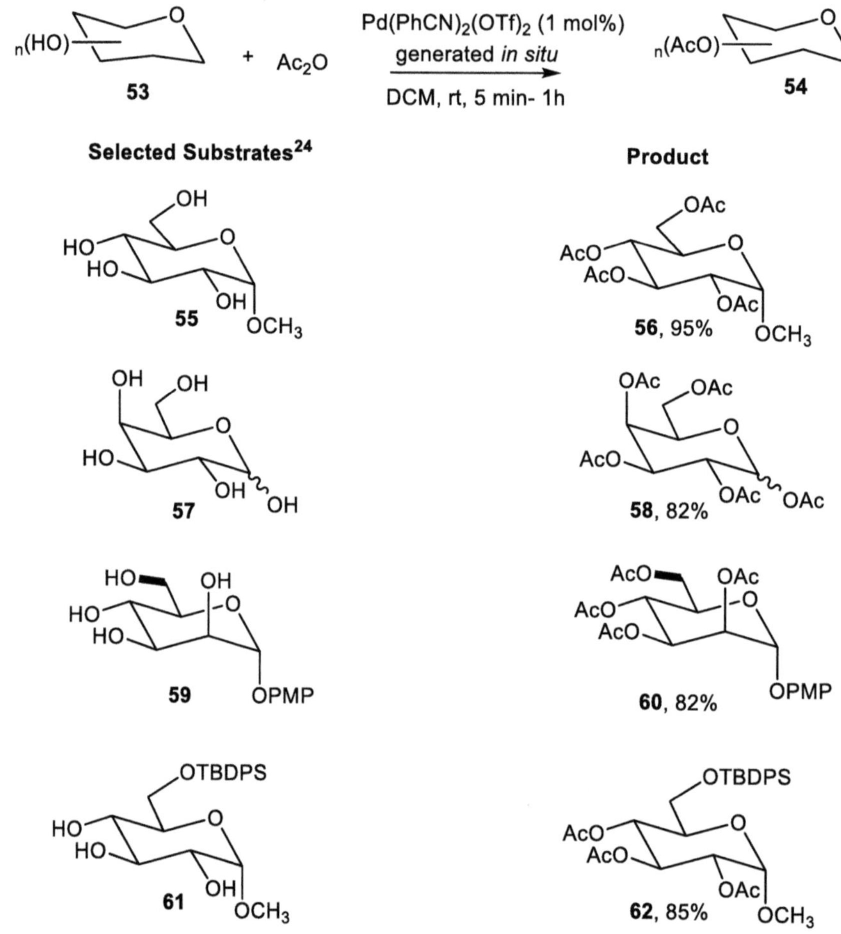

Selected Substrates[24]	Product
55	56, 95%
57	58, 82%
59	60, 82%
61	62, 85%

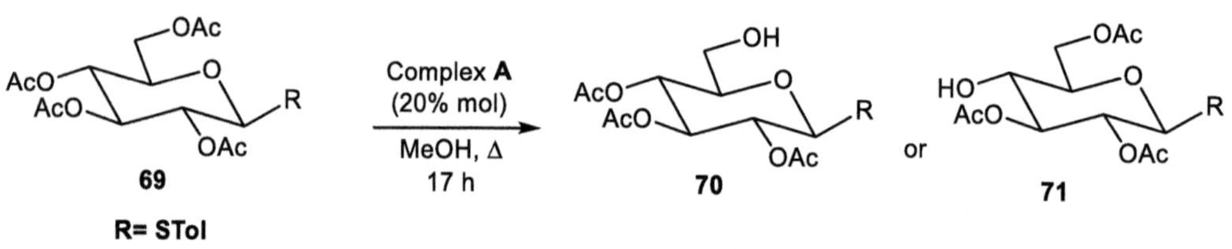

SCHEME 4.12 Regioselective deacetylation of 1-thioglucoside [24].

of molecular sieves, the reaction time was extended twofold, with low yields. Guanidinoglucosides are synthesized through a carbodiimide intermediate, where the MS 4Å facilitates the formation of the carbodiimide as a weak base and dehydrating agent.

Another reaction explored in carbohydrate chemistry was described by Bols and group in 2014: A 4,6-*O*-silylene-protected mannosyl thioglycoside (74) was β-selective in its reactions with alcohols

TABLE 4.2
Regioselective Deacetylation of Monosaccharides and Disaccharides [24].

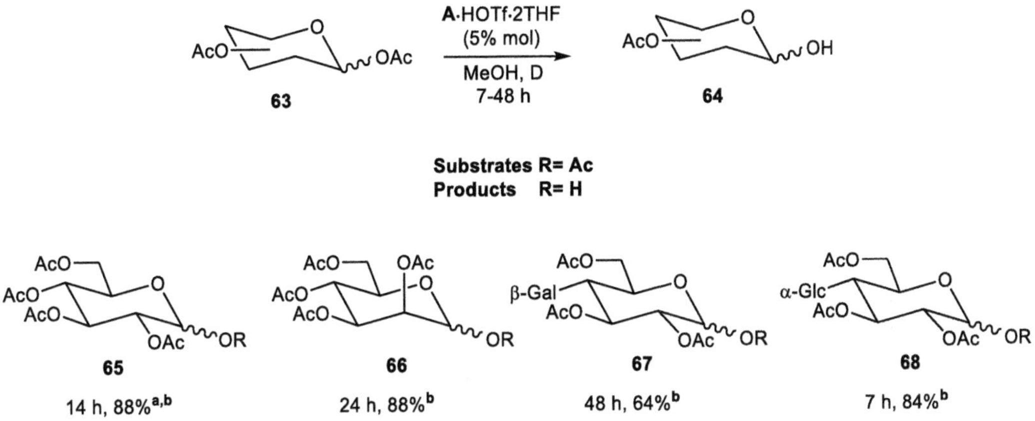

63 → 64: A·HOTf·2THF (5% mol), MeOH, D, 7-48 h

Substrates R= Ac
Products R= H

65	66	67	68
14 h, 88%[a,b]	24 h, 88%[b]	48 h, 64%[b]	7 h, 84%[b]

[a] Can also be completed after 60 h without heating.
[b] Isolated yields

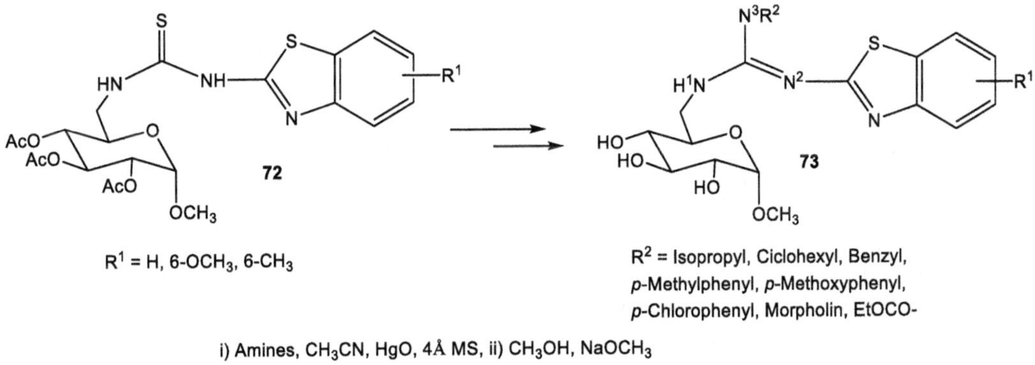

72 → 73

R^1 = H, 6-OCH_3, 6-CH_3

R^2 = Isopropyl, Ciclohexyl, Benzyl, *p*-Methylphenyl, *p*-Methoxyphenyl, *p*-Chlorophenyl, Morpholin, EtOCO-

i) Amines, CH_3CN, HgO, 4Å MS, ii) CH_3OH, $NaOCH_3$

SCHEME 4.13 Synthesis of polysubstituted guanidinoglucosides (73) using HgO-MS 4Å catalyst [25].

in reactions carried out at room temperature with activation by *N*-iodosuccinimide and only 10 mol % of triflic acid (Scheme 4.14) [26a,b]. The authors proposed that the β-mannoside was formed from β-selective glycosylation of the oxocarbenium ion in a B2,5 conformation [26b–c].

Finally, the importance of carbohydrates has also been demonstrated in areas such as organcatalysis. Recently, Marra and group provided an exhaustive review of the detailed preparations of all the sugar-based organocatalysts as well as their catalytic properties; these catalysts have been employed to perform classical organic transformations such as aldol reaction, the Mannich reaction, and Diels–Alder cycloaddition. However, despite the huge effort required for their synthesis, the actual potential of highly functionalized sugar-based organocatalysts remains largely unexplored, specifically from a green chemistry point of view [27d].

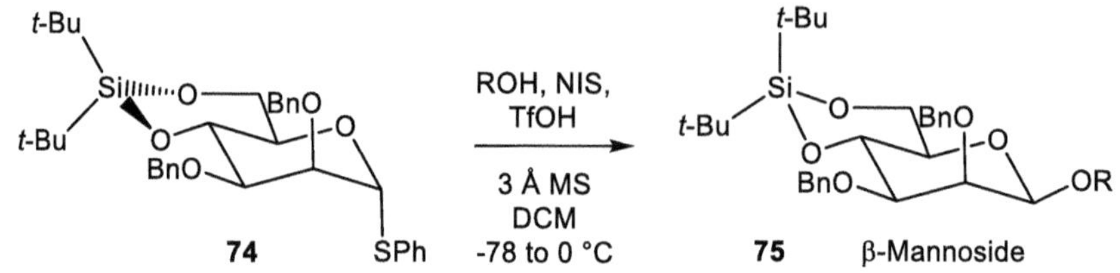

SCHEME 4.14 β-selective mannosylation with triflic acid [26a,b].

4.3 LIPID SYNTHESIS BY DIFFERENT CATALYSIS METHODS

Designer lipids are novel, health-friendly lipids with potential application in foods, nutraceuticals, and pharmaceuticals, including for treating obesity, cancer, heart disease, and inflammation [27]. These advantages arise due to chemical or enzymatic modifications of fats and oils; transforming lipids rearranges the fatty acids within a triglyceride molecule or between two different triglycerides, and the resulting lipid has superior physicochemical properties to those of the naturally occurring triglycerides.

Jadhav and Annapure in 2021 described the component fatty acids used to synthesize designer lipids along with the synthesis process, the reactors they used to increase the lipid yields, and the lipids' applications in the food and nutraceutical sectors. Among the synthesis processes the authors described were ester–ester exchange, transesterification, and acidolysis, reactions that took place in the presence of either chemical catalyst or enzymes; the synthesis routes depended on the type of substrate and triglyceride [27]. With a chemical catalyst, the reaction was rapid, but the process failed to modify a specific position on the glycerol backbone, which limited the end application of the designer lipid. Instead, enzymatic synthesis was highly recommended: The specificity of the enzymes and the mild reaction conditions allowed for producing a high-quality lipid with the desired properties for food and pharmaceutical applications [27].

Jadhav and Annapure mentioned in detail different points of comparison between chemical and enzymatic synthesis of designer lipids based on the characteristics of both. Chemically catalyzed reactions are carried out at elevated temperatures and form products in less reaction time than lipase-catalyzed reactions, but the fatty acids are randomly arranged on the glycerol backbone, thus changing the physicochemical properties of the formed lipids. Lipase-catalyzed synthesis of designer lipids has several advantages over chemically catalyzed synthesis. With the use of specific lipases, only fatty acids on specific positions are rearranged without disturbing the acid, thus forming more natural lipids with desired physicochemical properties. Because of the higher specificity of lipases, the enzymatic process does not form many byproducts, which decreases the cost of purifying the synthesized lipids; additionally, lipases are nontoxic to humans. Therefore, enzymatic synthesis is highly recommended for producing designer lipids for food and pharmaceutical applications. Figure 4.1 shows examples of some fatty acids.

Polyunsaturated fatty acids, such as docosahexanoic acid (DHA) (81), play a major role in the physiology of living organisms. Serhan research groups have established that DHA (81) is a substrate for the biosynthesis of several potent anti-inflammatory proresolving mediators, such as protectin D1 (83), among others [29]. All of these compounds have enabled new research are as related to many diseases associated with inflammation [29c]. It was reported that protectin D1 (83) is biosynthesized from DHA (81) *via* a lipoxygenase-mediated pathway that converts (81) by 15-lipoxygenase (15-LO) to a 17*S*-hydroperoxide intermediate, which is rapidly converted into the 16,17-epoxide (82), followed by enzymatic hydrolysis to the anti-inflammatory and proresolving oxygenated lipid (83) (Scheme 4.15) [29, 30]. This compound exhibited strong *in vivo* protective activity in several inflammatory as well as many other disease models.

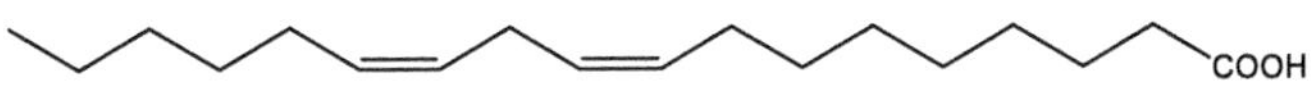

Linoleic acid (**76**)

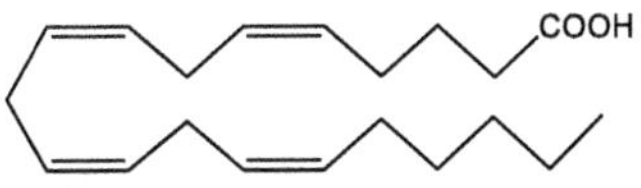

Arachidonic acid (**77**)

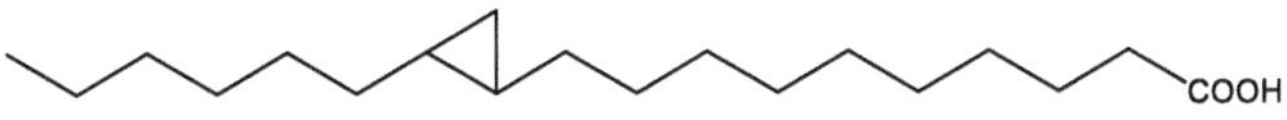

Cyclopropyl
10-(2-Hexylcyclopropanyl)decanoic acid (lactobacillic or phytomonic acid) (**78**)

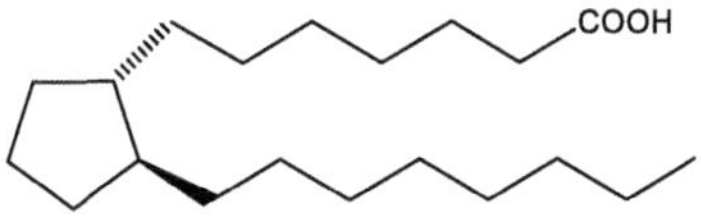

Prostanoic acid (**79**)

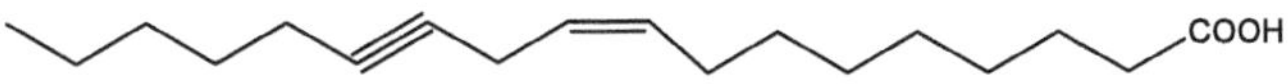

Crepenynic (*cis*-9-octadecene-11-ynoic acid) (**80**)

FIGURE 4.1 Examples of some fatty acids [28].

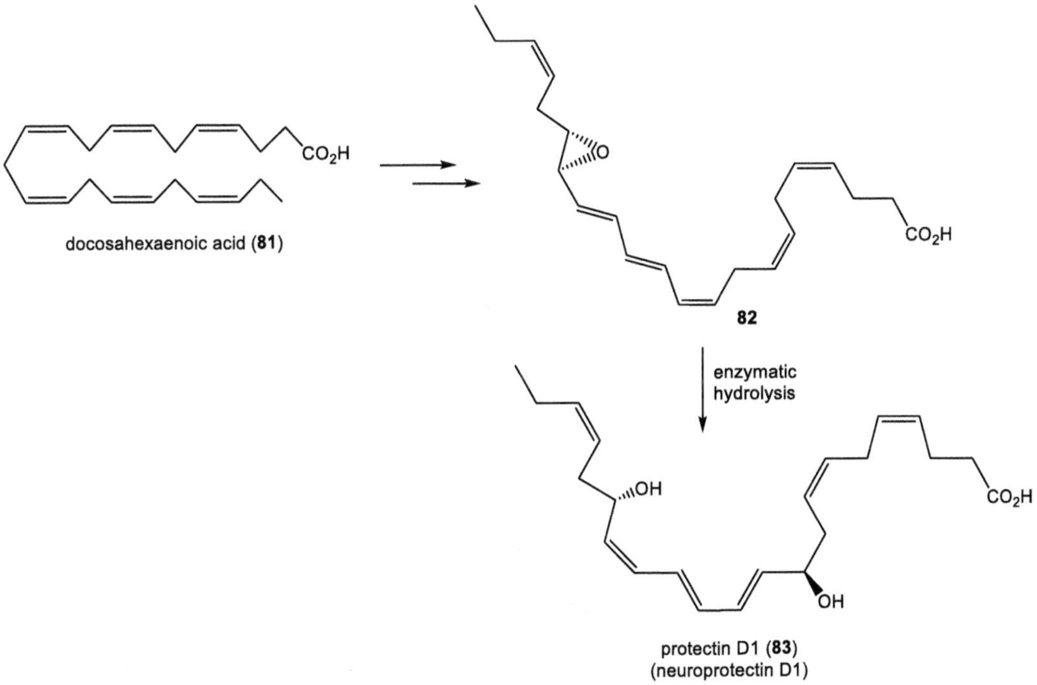

SCHEME 4.15 Biosynthesis of protectin D1 (83) [29, 30].

Hansen *et al.* reported a stereoselective synthesis of the potent anti-inflammatory, proresolving, and neuroprotective lipid mediator protectin D1 (83) (Schemes 4.16, 4.17, and 4.18). The lipid was prepared in eight steps and 15% yield from the known aldehyde (89) in a convergent manner. The key features were a stereocontrolled Evans-aldol reaction with Nagao's chiral auxiliary (90) and a highly selective reduction of internal alkyne (94) using Lindlar's catalyst, allowing the sensitive conjugated *E,E,Z*-triene to be introduced late in the preparation of (83) [30].

Galano and group reported the synthesis of alkyl-dihomo-IsoF derivatives, in which one of the main intermediates was the compound (103); the triol (101) was treated under reactions developed

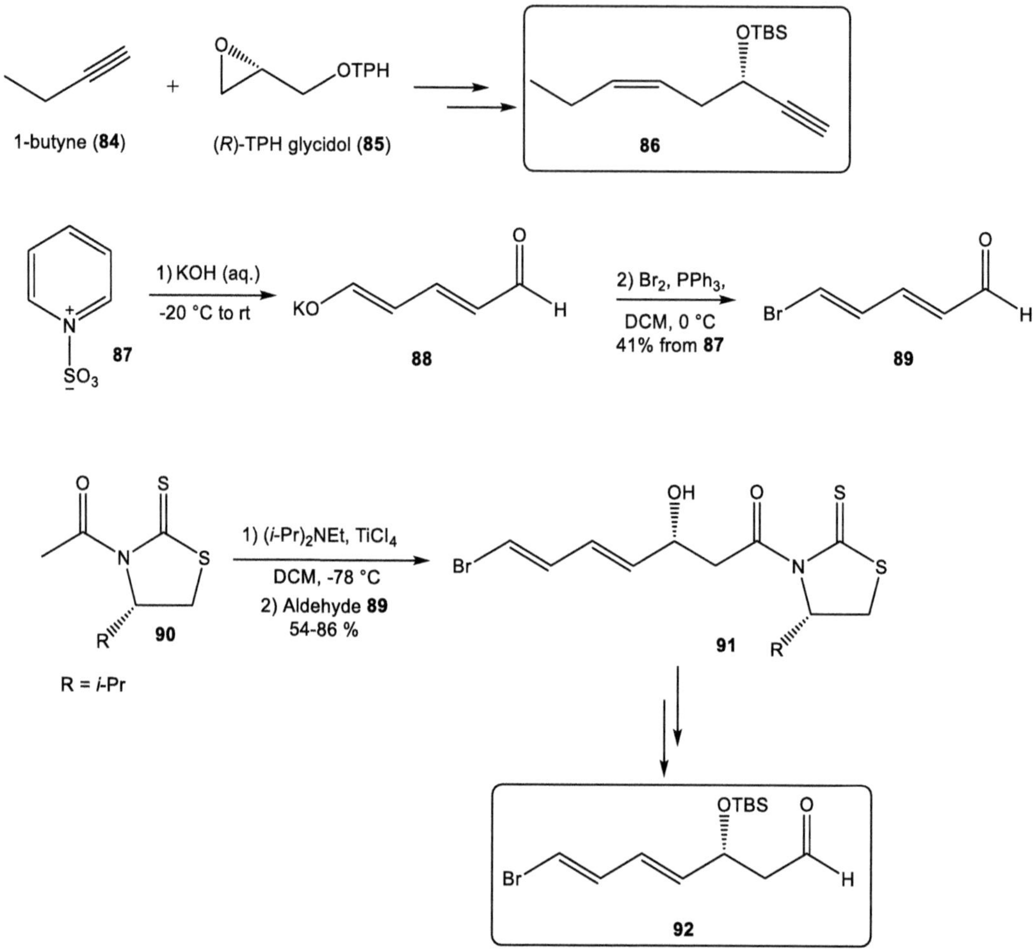

SCHEME 4.16 Synthesis of alkyne (86) and aldehyde (92) [30].

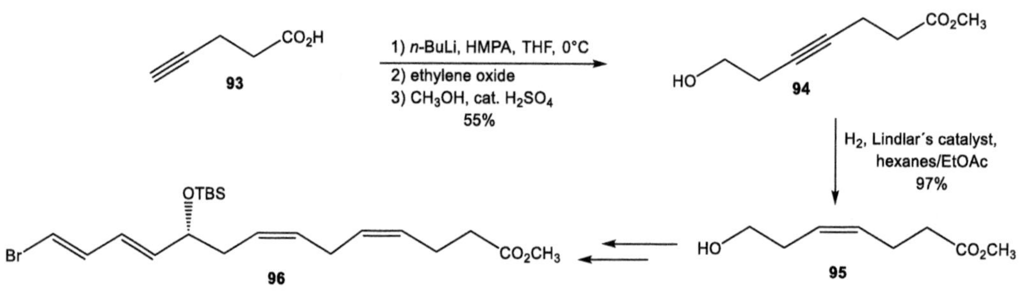

SCHEME 4.17 Synthesis of ester (96) [30].

by the group of Borhan, that is, orthoester formation with $(CH_3C(OCH_3)_3$ and PPTS, followed by *in situ* addition of catalytic $BF_3.OEt_2$ to conduct the intramolecular attack at the orthoester intermediate (Scheme 4.19). The cyclization process was extremely temperature dependent, with the best results obtained at 15 °C. Finally, after several steps, they obtained compound 10-*epi*-17(*RS*)-SC-Δ^{15}-11-dihomo-IsoF (104) at 76% yield [31].

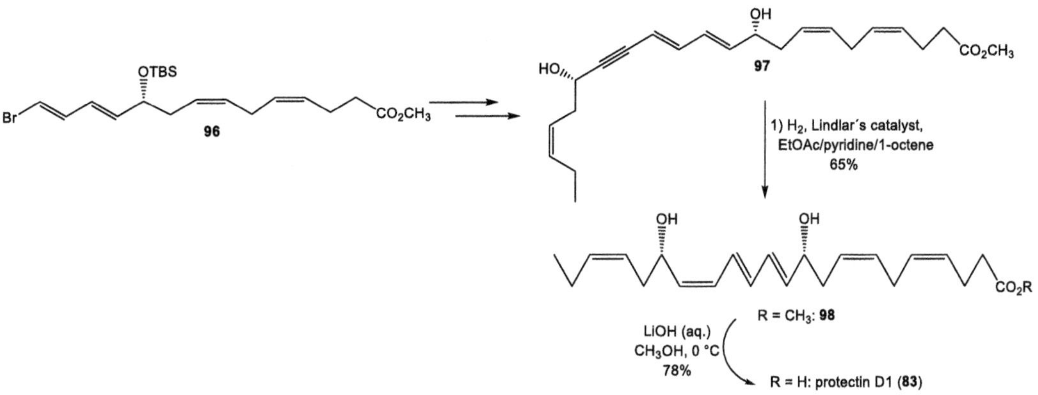

SCHEME 4.18 Synthesis of protectin D1(83) [30].

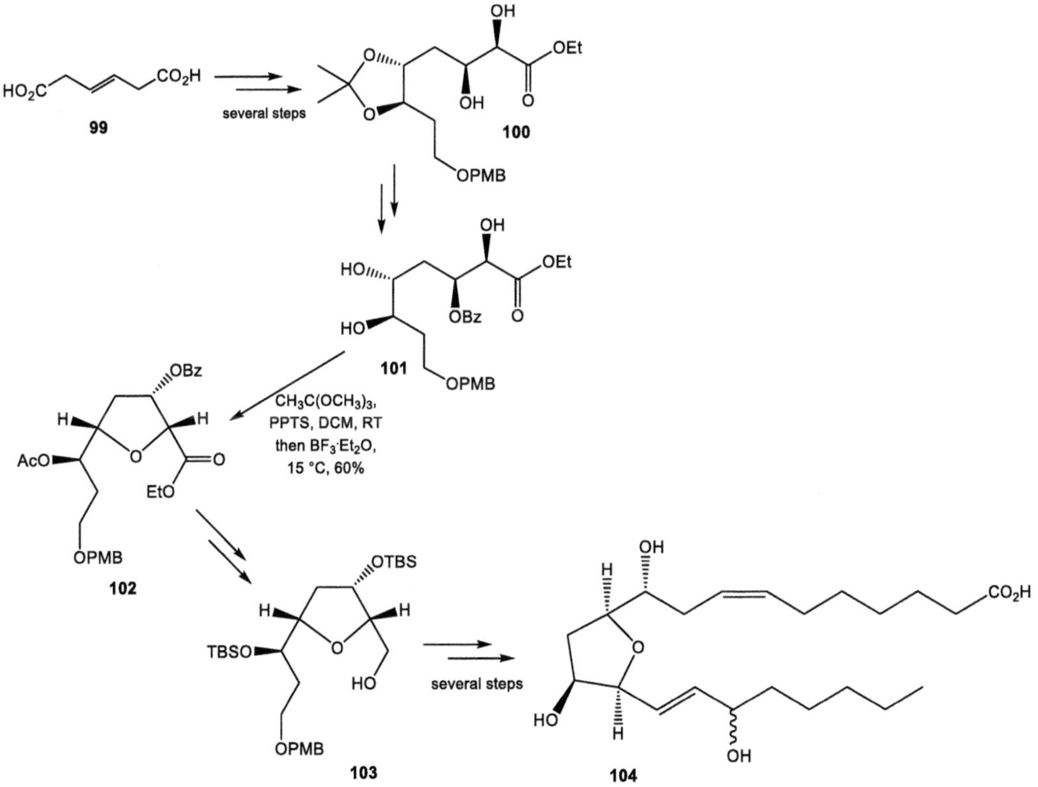

SCHEME 4.19 Synthesis of 10-*epi*-17(*RS*)-SC-Δ^{15}-11-dihomo-IsoF (104) [31].

The lipid bilayers of vertebrate cells contain a surprising diversity of glycolipids that play a number of important roles in cell biology [32]. While the bilayers of the plasma membrane and organelles are more abundant in phospholipids [33], glycolipids provide a rich source of binding epitopes and physical properties distinct from those of phospholipids and glycoproteins [34]. Similar to phospholipids in these cells, glycolipids contain a bifurcated lipid tail. However, vertebrate glycolipids are largely built on a ceramide (Cer) core, containing a sphingosine (Sph) chain and a fatty acid linked *via* an amide, as opposed to the diacyl chains attached as esters in phospholipids. These features make glycosphingolipids more akin to sphingolipids, which have received renewed attention in recent decades as their roles in signaling have been more fully alucidated [32,35].

The synthesis of the unusual human skin ceramides *via* coupling Sph with *N*-hydroxysuccinimide (NHS) fatty acid esters (106) to provide ultralong Cer analogs after several steps from 16-bromohexadecanoic acid (105) (Scheme 4.20) [36]. The NHS-ester was used to form the amide of the target and also enhanced the solubility of the fatty acid. The synthesis was carried out in 12 steps (11% yield on a small scale and 7% on a gram scale). One of the main stages of the synthesis involved the hydrogenation of the succinimidyl ester using the catalyst 10% Pd/C.

Researchers have explored cross-metathesis (CM) for the synthesis of ceramides (111) and (112) (Figure 4.2) [37].

Overkleeft *et al.* described a new synthetic route for 6-hydroxysphingosine (115) and its α-hydroxy ceramide counterpart (116), the synthesis employs a CM strategy. The crucial CM reaction on which the synthesis hinges unites two similar allylic alcohol alkenes. The researchers used

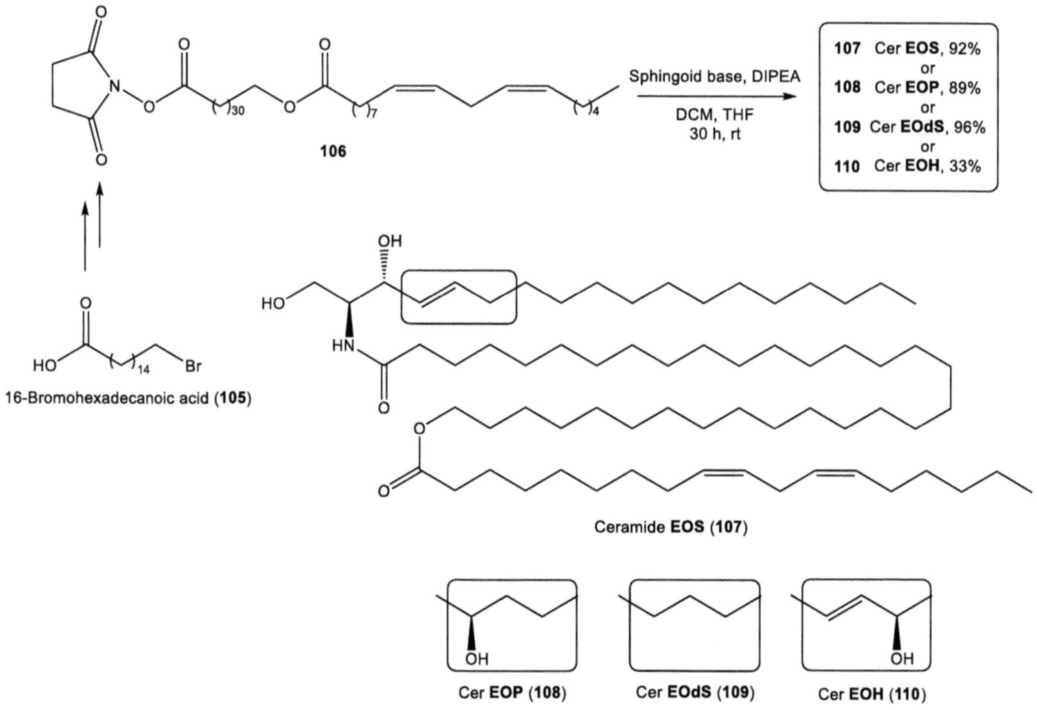

SCHEME 4.20 Synthesis of ultralong Cer (107–110) [36].

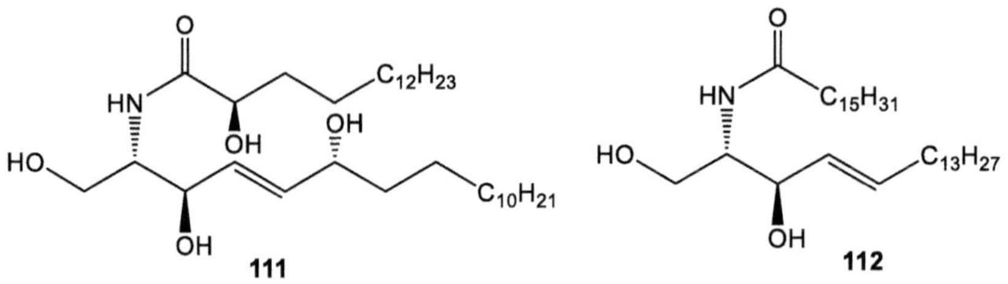

FIGURE 4.2 Ceramides prepared by cross-metathesis [32].

a cyclic carbonate group to protect the diol system present in the sphingosine head CM partner; the cyclic carbonate served two purposes: It made the double bonds electron rich, discriminating it from the long-chain allylic alcohol (its designated CM partner), and it tied back the functional groups on the olefin, making the alkene sterically most accessible for the catalyst and the CM event. This was carried out in combination with an activated catalyst system (the Grubbs II-CuI reagent pair), this led to an efficient CM connection of the sphingosine head and tail alkenes. The use of mild conditions for this connection in conjunction with the straightforward deprotection (mild base followed by mild acid) making this approach versatile and amenable to the use of a variety of CM coupling partners to generate labeled and tagged 6-hydroxysphingosine-derived probes for future biochemical studies (Scheme 4.21) [37].

One research group investigated the synthesis of deuterated analogues of archaebacteria membrane lipid tail-deuterated 1,2-di(3*RS*,7*R*,11*R*-phytanoyl)-*sn*-glycero-3-phosphocholine (DPhyPC) (117) and 1,2-di(3*RS*,7*R*,11*R*-phytanyl)-*sn*-glycero-3-phosphocholine (DPEPC) (118) from perdeuterated phytanic acid (121) (Figure 4.3) [38]. The researchers used the two lipids to build two different model membranes: Lang-muir monolayers and a tethered bilayer membrane on a solid substrate, characterized by pressure area isotherm and neutron reflectometry. Phytanic acid (120) was deuterated to give perdeuterated phytanic acid (121) in good yield using a hydrothermal H/D exchange reaction with heterogeneous platinum metal catalyst (Scheme 4.22).

Guo *et al.* recently reported the synthesis and evaluation of liposomal anti-GM3 cancer vaccine, they prepared and investigated two types of vaccine GM3-lipid/αGalCer, in which GM3 is linked with lipid anchor and coassembled with αGalCer (Figure 4.4). They demonstrated that *β*GalCer exceptionally optimized lipid anchor, which enables the noncovalent vaccine candidate,

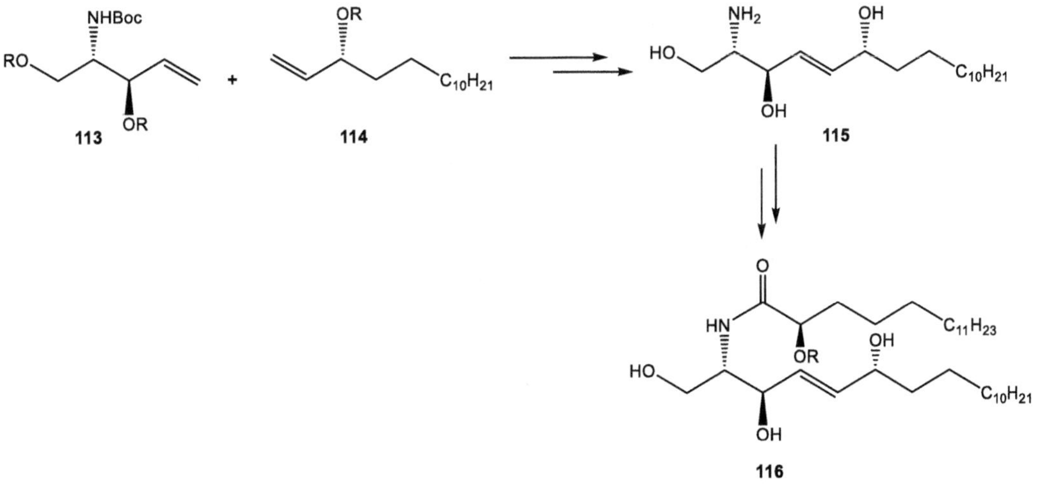

NHBoc RO OR 113 + OR $C_{10}H_{21}$ 114 → NH_2 OH HO OH $C_{10}H_{21}$ 115 → O HN OR OH $C_{11}H_{23}$ HO OH $C_{10}H_{21}$ 116

SCHEME 4.21 Synthesis of 6-hydroxysphingosine (115) and *α*-hydroxy ceramide (116) [37].

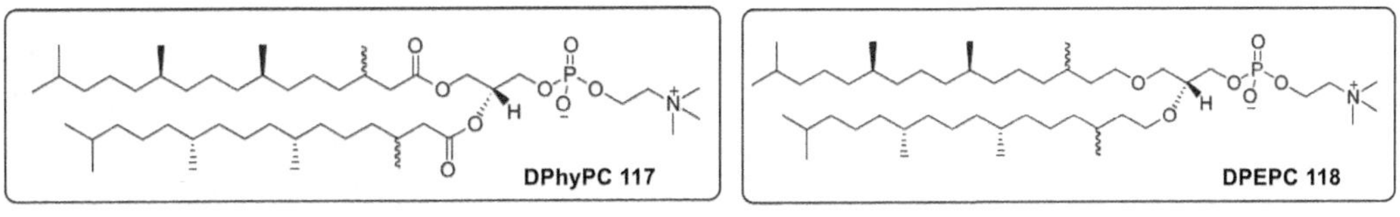

DPhyPC 117 DPEPC 118

FIGURE 4.3 Chemical structures DPhyPC (117) and DPEPC (118) [38].

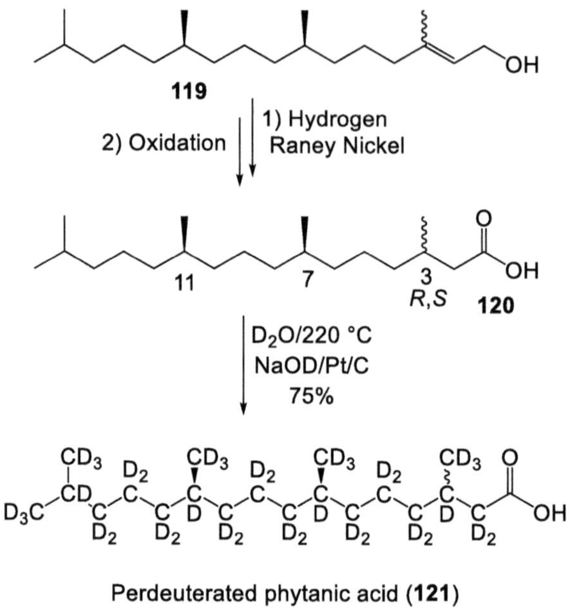

Perdeuterated phytanic acid (**121**)

SCHEME 4.22 Synthesis of perdeuterated phytanic acid (121) [38].

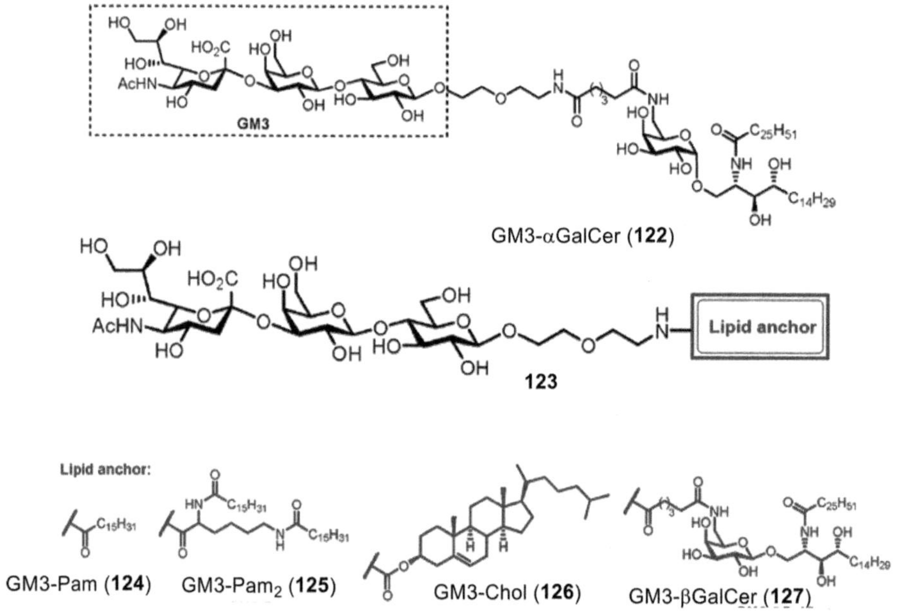

FIGURE 4.4 GM3-βGalCer/αGalCer [39].

GM3-βGalCer/αGalCer to evoke a comparable antibody level to GM3-αGalCer (122) [39]. This study highlights the importance of vaccine constructs utilizing covalent or noncovalent assembly between αGalCer with carbohydrate antigens and choosing an appropriate lipid anchor for use in noncovalent vaccine formulation. GM3-αGalCer (122) was prepared from glycolipid conjugate by hydrogenation using Pearlman's catalyst, and the same reaction conditions were used for the synthesis of GM3-Pam (124), GM3-βGalCer (127), and GM3-Chol (in the final step) (126).

Undoubtedly, catalysis has played a very important role in the synthesis of lipids. Methods have involved using metals (e.g., Pearlman's catalyst) [40], Pd/C, Lindlar's catalyst, Zn, AcOH [41], Zn/Cu/Ag amalgam [42], Brown's catalyst, $Ni(OAc)_2$ in the presence of $NaBH_4$ or Rosenmund's catalyst, or $Pd/BaSO_4$ poisoned with quinoline [43] Some retrosynthetic analyses are based on C-C bond-forming reactions (e.g., Sonogashira coupling [42], cross-metathesis reactions, the Zn-mediated addition of alkynyl nucleophiles to aldehydes in the presence of chiral aminoalcohols [44]), or functional layout (e.g., the asymmetric organo-or metallo-catalytic reduction of ynones [44]). Ultimately, enzymatic catalysis is a dependable and effective method of synthesizing lipids for future purposes [45].

4.4 STEROID SYNTHESIS THROUGH CLASSIC CATALYSIS

Steroid chemistry attracted great interest in the 1940s as a result of the transformation of natural products into progesterone, a discovery that caused a revolution in the synthesis of steroidal hormones [46]. Subsequently, thanks to the contributions related with conformation of the steroid nucleus carried out by Barton, winning him the Nobel Prize in Chemistry in 1969, a great number of studies on structural modification of steroidal molecules have been described [47], some of these molecules have potential application as drugs, while others have found application in other fields, such as materials science, as plant growth promoters, tumor markers, to mention a few. The wide number of applications and remarkable bioactivity of steroidal derivatives have led to the development of new methodologies that involve the functionalization of the steroid nucleus with other bioactive molecules [46].

In the search for more active compounds with good biological activities, different structural modifications have been reported worldwide. This chapter provides an overview of the several strategies for the synthesis of the steroid skeleton employing different catalytic methods. Classical reactions for steroid synthesis have been very important; some reactions, for example reductions, coupling reactions, multicomponent, isomerization, and condensation reactions, among others, have been reconsidered but now from the point of view applying new catalytic methods as a powerful tool. Many of these have features such as being solvent free and having short reaction times, water-mediated reactions, and fewer reaction steps and also contribute in an environmentally friendly way.

4.4.1 Metal Catalysts

Skoda-Földes and group in 2013 reported the synthesis of 13α-18-nor-16-carboxamido steroids *via* a palladium-catalyzed aminocarbonylation reaction of the corresponding iodoalkenes. The starting material was an unnatural 13α-16-keto steroid, previously prepared by a Wagner–Meerwein rearrangement of a 16α,17α-epoxide in the presence of [BMIM][BF_4]. The 13α-16-keto steroid (128) was converted to a mixture of 16-iodo-16-ene (129) and 16-iodo-15-ene (130) derivatives in two steps by Barton´s methodology. The aminocarbonylation of the steroidal alkenyl iodides was carried out using different primary and secondary amines as nucleophiles. Under this methodology, the products, 16-carboxamido-16-ene (131) and 16-carboxamido-15-ene (132) derivatives, were obtained in good yields. In addition, reduction with Pd/C in formic acid of the mixture of the isomeric unsaturated carboxamides (131) and (132) leads to the 16α-carboxamide derivative (133) as a single product (Scheme 4.23) [48].

In 2013, researchers reported a general approach based on the Cu^{I}-catalyzed azide-alkyne 1,3-dipolar cycloaddition reaction for the conjugation of two spirostanic steroids. This process provided rapid access to a small library of triazole-based-bis-spirostanic conjugates (134) with varied functionalization patterns as well as different stereochemistry of the linkage (Figure 4.5). This approach was promising for the discovery of novel bis-steroidal conjugates with potential applications in medicinal chemistry [49]. Continuing with other examples related to the use of copper, in 2014, Nenajdenko and Sokolova systematically evaluated the capacity of synthetic chiral α, β-azidoisocyanides for use as bifunctional building blocks for multicomponent Passerini and Ugi reactions and also for Cu(I)-catalyzed [3+2]-cycloaddition to form 1,2,3-triazoles. The bifunctional

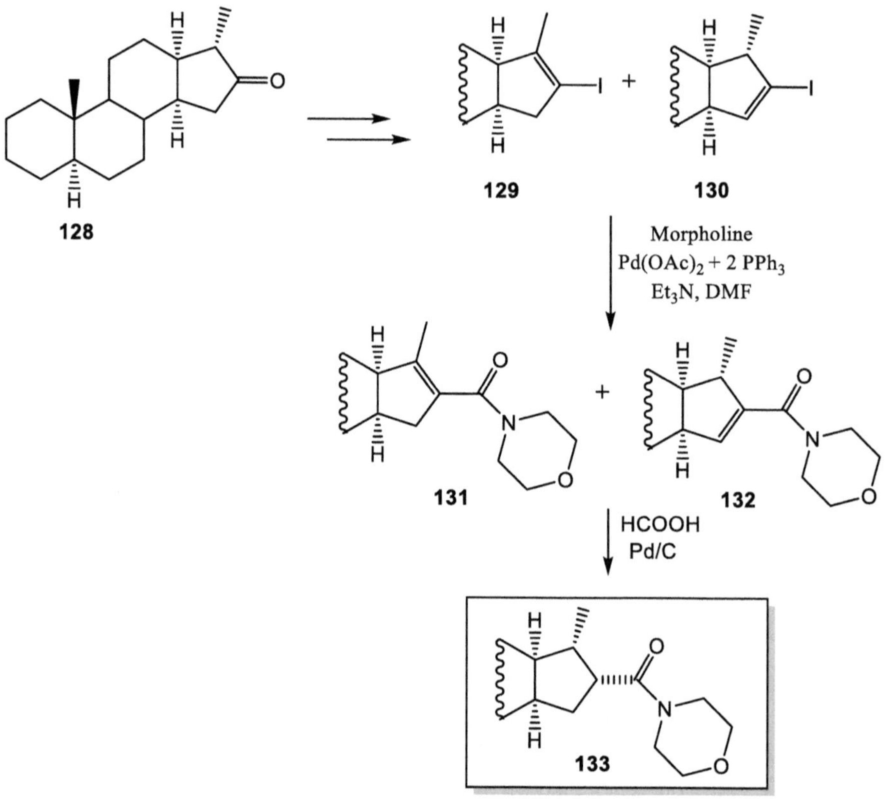

SCHEME 4.23 Synthesis of the 16α-carboxamide derivative (133) [48].

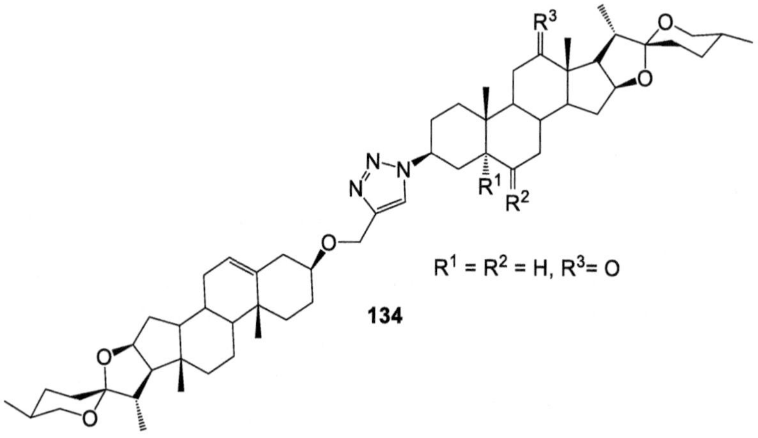

FIGURE 4.5 Synthesis of triazole-linked bis-spirostanic conjugated (134) by Cu^{I}-catalyzed 1,3-dipolar cycoaddition [49].

nature of these compounds enabled the development of a methodology for preparing peptides containing an azide group and conjugating them subsequently with various biologically active compounds containing an ethynyl group for targeted modification of biomolecules. For this synthesis, they used biologically active chelesterol and ethynyl estradiol 3-methyl ether (Scheme 4.24) [50].

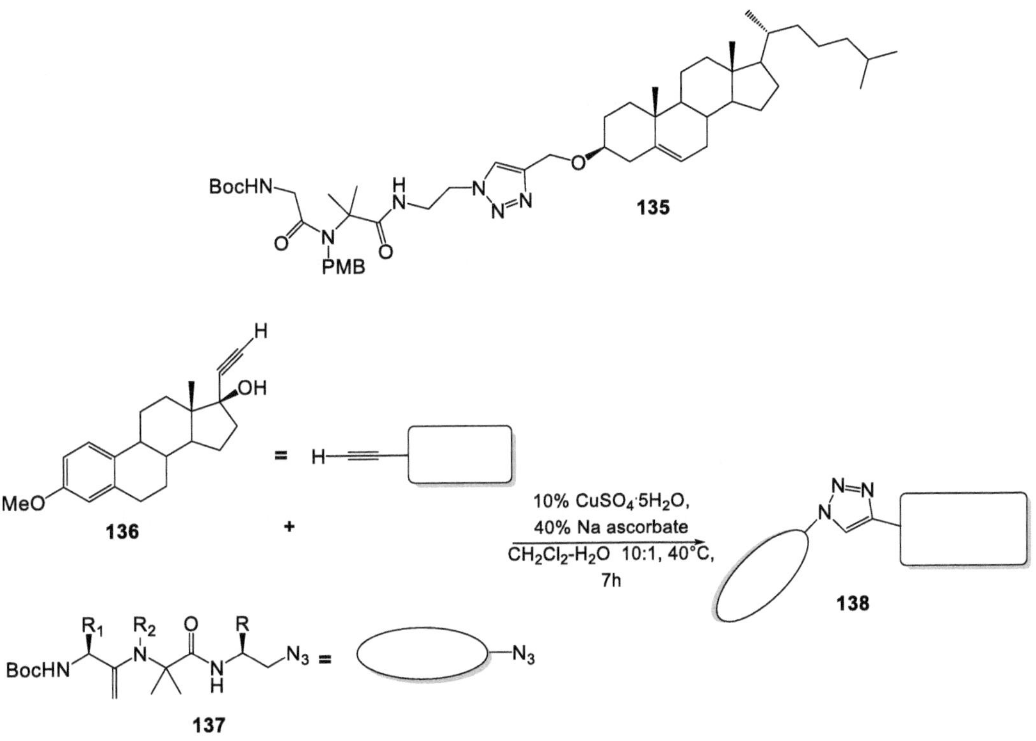

BocHN
N
PMB
H
N
N
N=N
O
135

OH
H
MeO
136
+
10% $CuSO_4 \cdot 5H_2O$,
40% Na ascorbate
CH_2Cl_2-H_2O 10:1, 40°C,
7h
N=N
N
138

R_1 R_2 O R
BocHN N N N_3 =
H
N_3
137

e.g. R = H ; R = Me; R_2 = 4-$MeOC_6H_4CH_2$

SCHEME 4.24 Peptide–steroid bioconjugates (138) *via* Cu(I)-Cataly [3+2]-cycloaddition of azidoisocyanide Cu complexes [50].

In 2019, scientists reported a Cu(II)-catalyzed diastereoselective Michael/aldol cascade approach to accomplish concise total synthesis of cardiotonic steroids with varying degrees of oxygenation including cardenolides ouabagenin (139), sarmentologenin (140), 19-hydroxy-sarmentogenin (141), and 5-*epi*-panogenin (142). These syntheses enabled the subsequent structure activity relationship studies on 37 synthetic and natural steroids to elucidate the effect of oxygenation, stereochemistry, C3-glycosylation, and C-17-heterocyclic ring. With their evaluation, they found that glycosylated steroids cannogenol-*L*-α-rhamnoside (143), strophanthidol-*L*-α-rhamnoside (144), and digitoxigenin-*L*-α-rhamnoside (145) were the most potent steroids, demonstrating broad anticancer activity and selectivity (Figure 4.6) [51].

Farfán *et al.* reported on a family of five steroid-flanked derivatives of *p*-nitroaniline; they studied the influence of steroidal framework functionalization over A-ring on solid-state estructure, finding ethynyl estradiol, ethisterone, and norethisterone led primarily to poorly crystalline solids, while mestranol and 17α-ethynyl-5α-androst-2-en-17β-ol led to crystalline solids, whose structures were solved and discussed via SXRD. The steroidal molecular compasses (146–150) were synthesized *via* double Sonogashira crosscoupling reactions between the corresponding 17-α-ethynylsteroids and 2,5-dibromo-4-nitroaniline with $Pd(PPh_3)_2Cl_2$ and CuI as catalysts, and using *i*-Pr_2NH as base (Figure 4.7) [52].

A series of keto-steroids and hydroxyl-steroids were prepared in good yield with Jones reaction at the positions 3 and 6 of the steroid ring. The keto-steroids (151a-d) were reduced by sodium borohydride with different catalysts to produce diverse products. To boost the speed of the reaction, cerium trichloride or nickel dichloride was added into the reaction system as catalysts. The double bond was not affected by Ce^{3+} but was reduced when there was Ni^{2+} in the system (Scheme

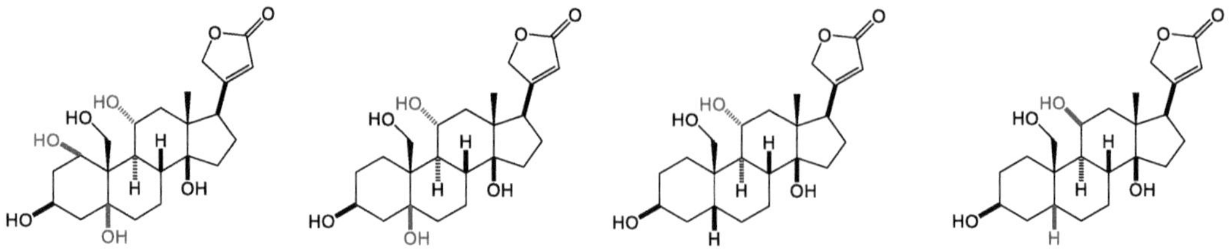

Ouabagenin (**139**) Sarmentologenin (**140**) 19-Hydroxy-sarmentogenin (**141**) 5-*epi*-Panogenin (**142**)

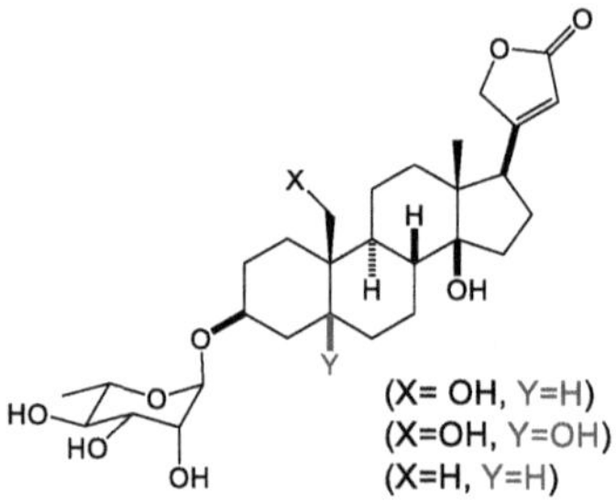

(X= OH, Y=H) Cannogenol-3-O-α-*L*-rhamnoside (**143**)
(X=OH, Y=OH) Strophanthidol-3-O-α-*L*-rhamnoside (**144**)
(X=H, Y=H) Digitoxigenin-3-O-α-*L*-rhamnoside (**145**)

FIGURE 4.6 Cardiotonic steroids [51].

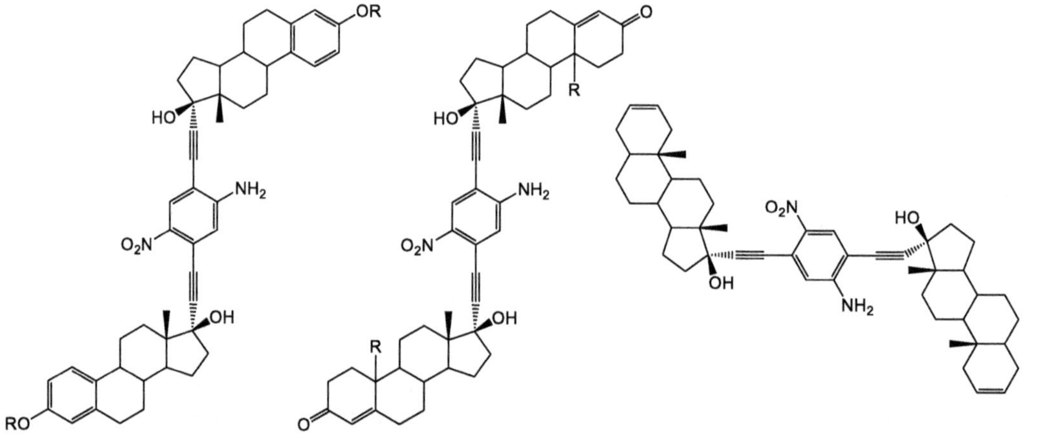

146 R = H (92 %) from Ethynylestradiol
147 R = CH_3 (93 %) from Mestranol
148 R = $^{19}CH_3$ (88 %) from Ethisterone
149 R = H (90 %) from Norethisterone
150 (89 %) from 17α-ethynyl-5α-androst-2-en-17β-ol

FIGURE 4.7 Synthesis of steroidal molecular compasses (146–150) [52].

4.25). The authors also measured their inhibitory activities against AKR1B10 and AKR1B1, and the most active compound, (152a), showed an IC_{50} of 0.50 μM for AKR1B10, and the most AKR1B10-selective compound's (151a) IC_{50} was 0.81 μM with AKR1B1/AKR1B10 selectivity of 195. In addition, the binding modes of (151–152a) in the active site of human AKR1B10 were identified by docking [53].

Researchers also reported on the synthesis of highly functionalized core skeletons of estrogenic and cardiotonic (159,161) steroids. The key steps for this synthesis were Mizoroki-Heck and intramolecular Diels–Alder (IMDA) reactions. The diastereoselectivity of the IMDA reaction used to access the cardiotonic steroidal skeleton was found to be significantly enhanced by performing the

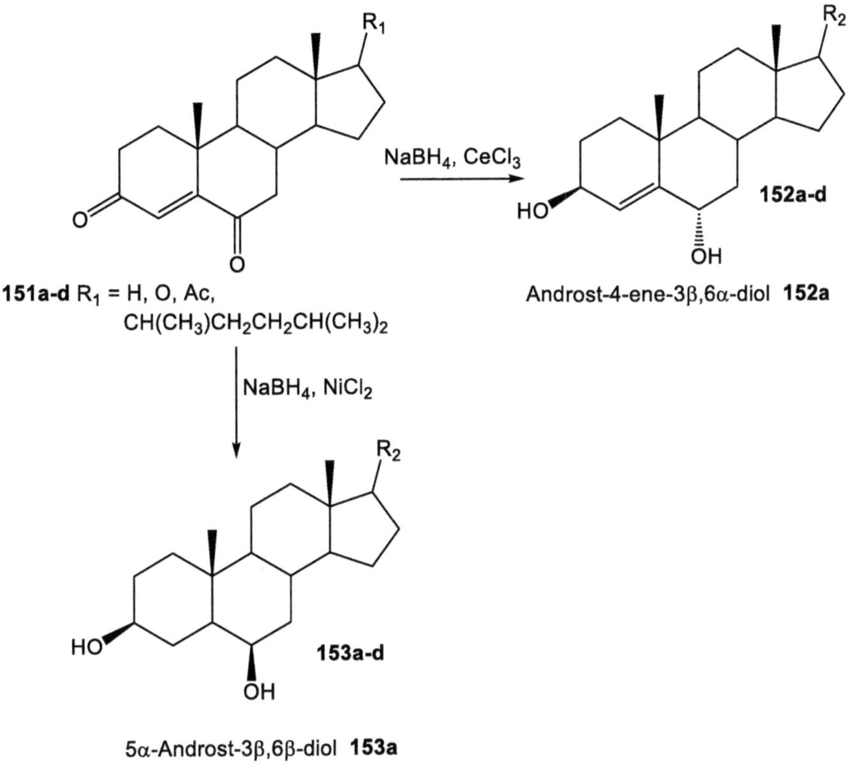

NaBH$_4$, CeCl$_3$

151a-d R$_1$ = H, O, Ac, CH(CH$_3$)CH$_2$CH$_2$CH(CH$_3$)$_2$

152a-d

Androst-4-ene-3β,6α-diol **152a**

NaBH$_4$, NiCl$_2$

153a-d

5α-Androst-3β,6β-diol **153a**

152a-d and **153a-d:** R$_2$ = H, OH, CH(OH)CH$_3$, CH(CH$_3$)CH$_2$CH$_2$CH(CH$_3$)$_2$

SCHEME 4.25 Reduction of keto-steroids (151a-d) with NaBH$_4$ and different catalysts [53].

reaction in water. The Mizoroki–Heck reaction was used in the first step of the synthesis and the IMDA in the final step (Scheme 4.26) [54].

In 2015, Baran and group improved the reaction conditions initially reported by Schönecker and demonstrated that this method could be applied to the synthesis of a variety of polyhydroxylated derivatives of natural steroids. They studied the C-H oxidation protocol mechanistically and applied the method to a range of additional substrates. For example, dehydro-*epi*-androsterone and 3-methylestrone were oxidized to their corresponding C12-hydroxylated analogues 164 and 165 in good yield (Scheme 4.27) [55].

Beletskaya and group developed a highly efficient procedure for the synthesis of azolyl-substituted steroids by utilizing a catalyst system comprising CuI and dipivaloylmethane as a ligand. A number of nitrogen heterocycles were used in the vinylation process, affording the corresponding coupling products in good to excellent yields. Their protocol was also applicable for a limited number of secondary amides (Scheme 4.28) [56].

Ollevier presented a review of uses of bismuth as a catalyst highlighting several organic reactions in the presence of this metal [11]. Specifically in oxidations, Salvador and Silvestre demonstrated the use of bismuth(III) salts as an efficient catalyst for selective allylic oxidation using *tert*-butylhydroperoxide. BiCl$_3$ was especially effective, and it could be recovered and reused as BiOCl. Using BiCl$_3$/K-10 as catalyst increased the reaction rate (Scheme 4.29) [11,57].

In 2017, Baran *et al.* achieved a decarboxylative cross-coupling method that afforded terminal and substituted alkynes from various carboxylicacids. They used both nickel- and iron-based

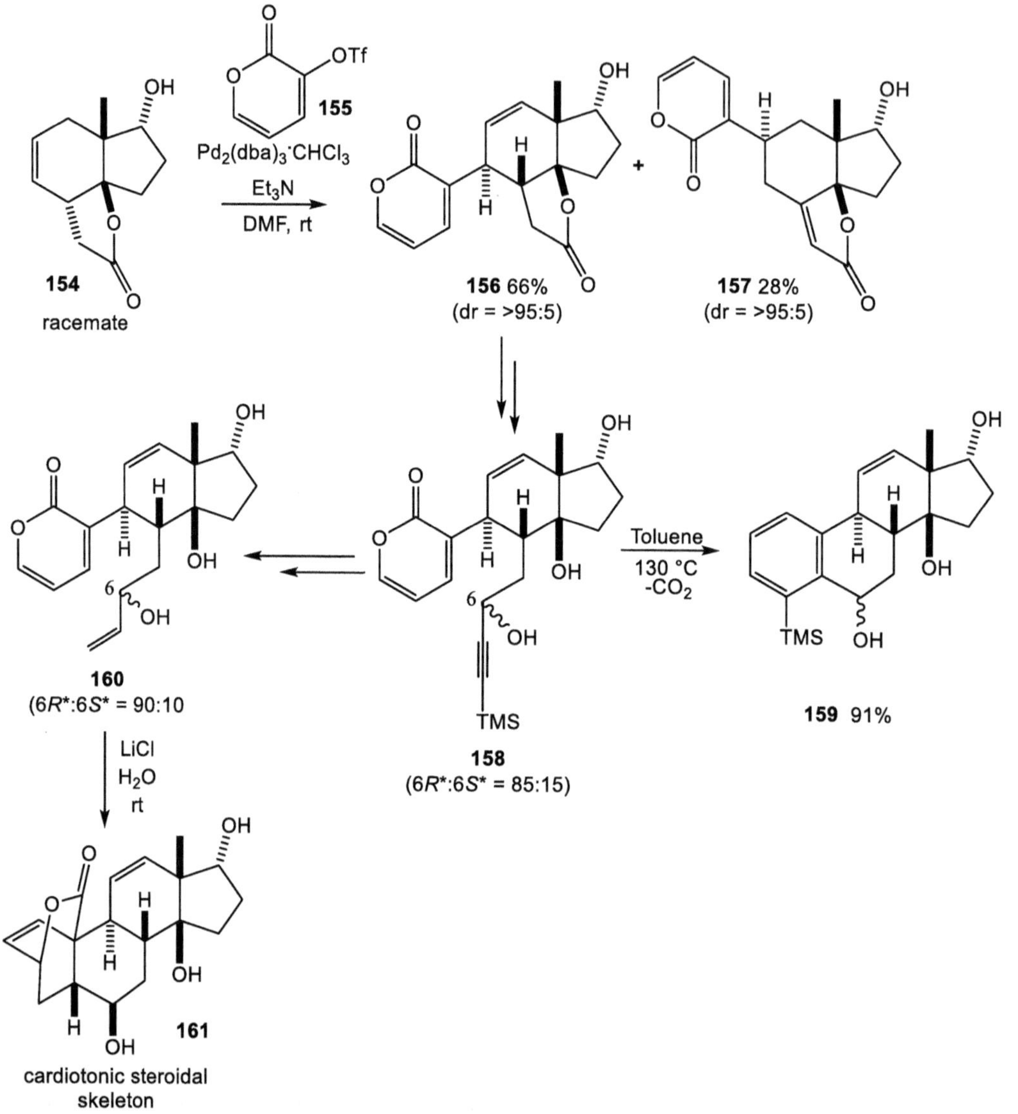

154 racemate
155
$Pd_2(dba)_3 \cdot CHCl_3$, Et_3N, DMF, rt
156 66% (dr = >95:5)
157 28% (dr = >95:5)
158 (6R*:6S* = 85:15)
Toluene, 130 °C, $-CO_2$
159 91%
160 (6R*:6S* = 90:10)
LiCl, H_2O, rt
161 cardiotonic steroidal skeleton

SCHEME 4.26 Synthesis of precursors 158 and 160 for the IMDA reaction [54].

catalysts, and in this strategy, *N*-hydroxytetrachlorophthalimide (TCNHPI) esters was crucial to the success of the transformation; the reaction was amenable to *in situ* carboxylic acid activation (Scheme 4.30) [58].

Heretsch *et al.* gave another example of using copper in steroid chemistry; they prepared B-nor-steroids using a Cu-mediated one-pot oxidation/benzilic acid rearrangement (Scheme 4.31) [59].

Jones and Stewart [47] reported on the total synthesis of cannogenol-3-*O*-*α*-*L*-rhamnoside (143) [60a] *via* key enantioselective CuII-catalyzed Michael-intramolecular double aldol cyclization, as well as the total synthesis of *ent*-pregnanolone (182) [60b] by key tandem copper-catalyzed conjugate addition-oxygenation reaction as an example of constructing a steroid ring system in their publication "A tribute to Sir Derek Barton". Others reported on the total synthesis of limonin (183) [60c] using $Mn(OAc)_3$ and $(Ph_3P)_3RuCl_2$ as metallic catalysts (Figure 4.8).

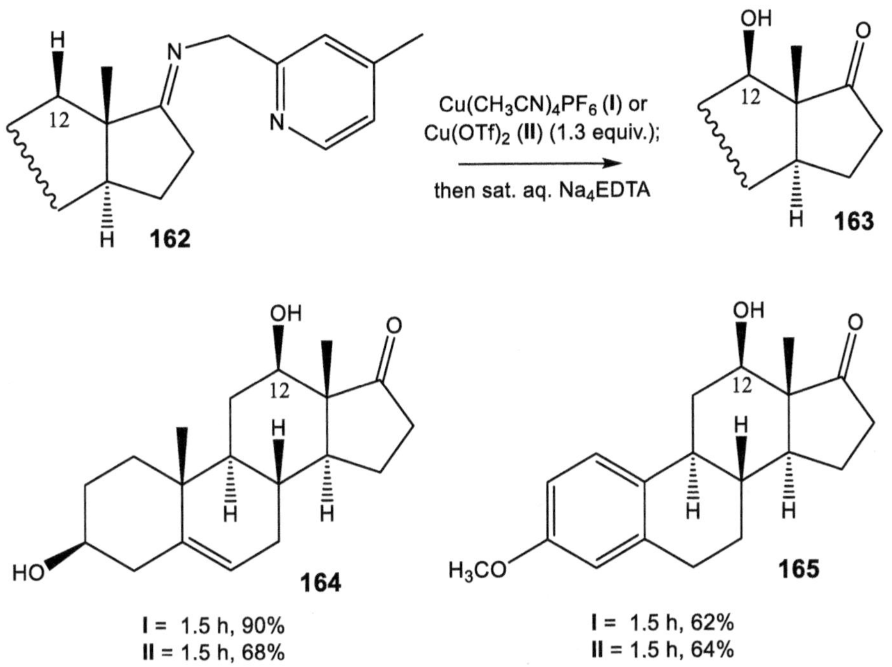

Cu(CH$_3$CN)$_4$PF$_6$ (I) or
Cu(OTf)$_2$ (II) (1.3 equiv.);
then sat. aq. Na$_4$EDTA

162 163 164 165

I = 1.5 h, 90%
II = 1.5 h, 68%

I = 1.5 h, 62%
II = 1.5 h, 64%

SCHEME 4.27 CH-12 hydroxylation of steroids with Cu(CH$_3$CN)$_4$PF$_6$ (I) or Cu(OTf)$_2$ (II) [55].

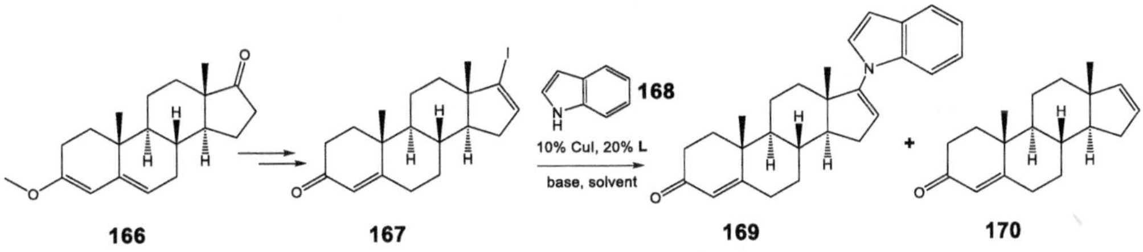

168
10% CuI, 20% L
base, solvent

166 167 169 170

SCHEME 4.28 Amination of 17-iodosteroid (167) with indole (168) [56].

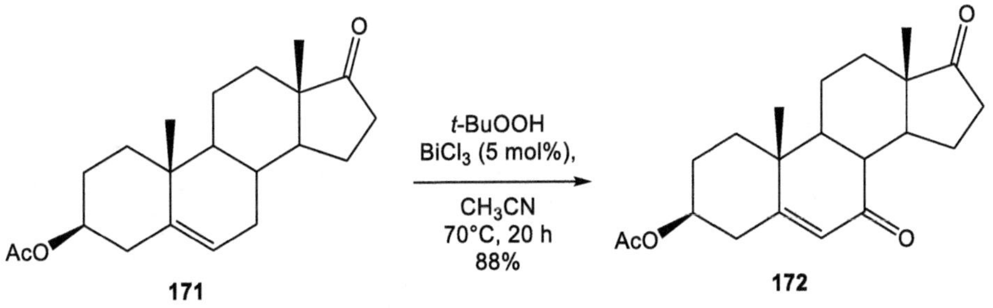

t-BuOOH
BiCl$_3$ (5 mol%),
CH$_3$CN
70°C, 20 h
88%

171 172

SCHEME 4.29 Allylic oxidation of unsaturated steroids.

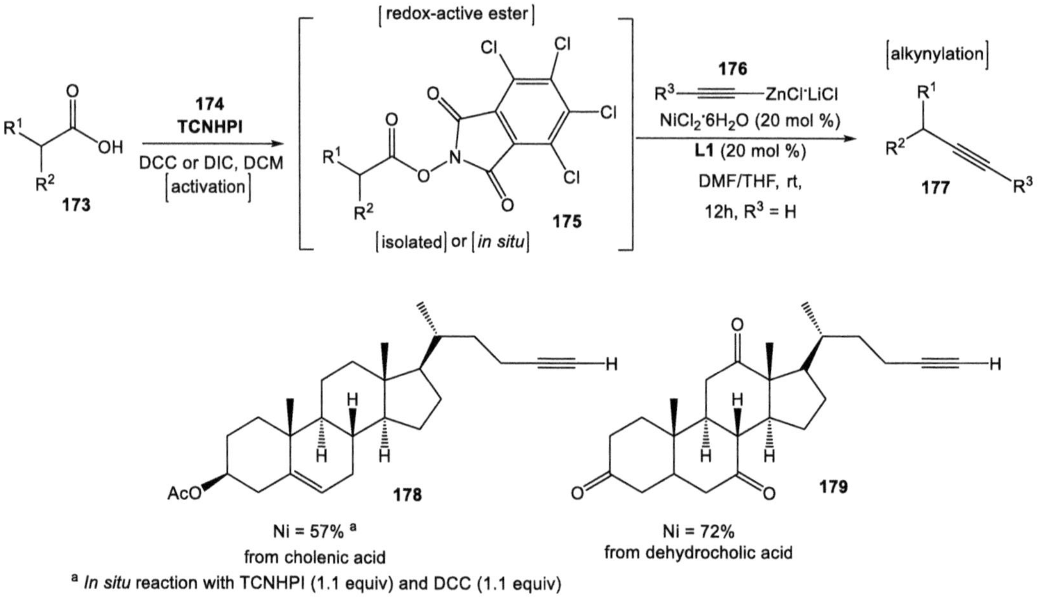

[redox-active ester]

174
TCNHPI
DCC or DIC, DCM
[activation]

173 175 [isolated] or [in situ]

176
R^3—≡—ZnCl·LiCl
$NiCl_2 \cdot 6H_2O$ (20 mol %)
L1 (20 mol %)
DMF/THF, rt,
12h, R^3 = H

[alkynylation]
177

178
Ni = 57% [a]
from cholenic acid

179
Ni = 72%
from dehydrocholic acid

[a] *In situ* reaction with TCNHPI (1.1 equiv) and DCC (1.1 equiv)

SCHEME 4.30 Nickel-catalyzed decarboxylative alkynylation with TCNHPI (174) redox-active esters [58].

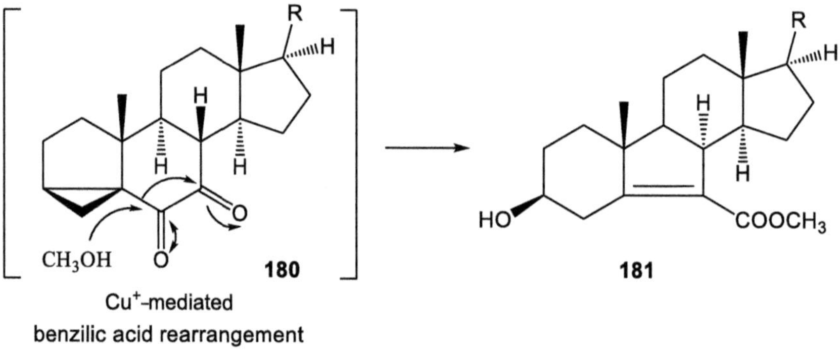

CH_3OH **180**
Cu^+-mediated
benzilic acid rearrangement

HO **181** $COOCH_3$

SCHEME 4.31 Benzilic acid rearrangement of *i*-steroid ketones and their subsequent opening giving access to 5(6→7) *abeo*-steroids [59].

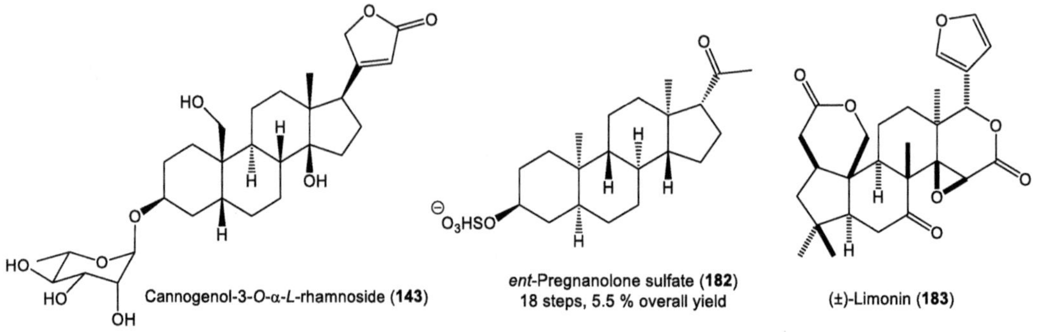

Cannogenol-3-*O*-α-*L*-rhamnoside (**143**)

ent-Pregnanolone sulfate (**182**)
18 steps, 5.5 % overall yield

(±)-Limonin (**183**)

FIGURE 4.8 Steroid synthesis *via* different catalytic methods.

Wicha *et al.* described other advances in the field of cross-coupling reactions to synthesize cardenolide and bufadienolide aglycones [61], and others described using various metal catalysts in Diels–Alder reactions with application to steroid synthesis [62]. In this context, Mahajan and Gupta provide diverse examples of heterogeneous catalysis used in organic steroidal reactions using microwave synthesis and microbial transformations [63]. More recently, Xun and Xu summarized the contributions of diverse research groups in the synthesis of aplykurodinone-1 in either racemic or enantiomeric form [64]. Although research into the construction of the steroids core has been ongoing for over 70 years, there is still much interest in expanding this methodology owing to the underlying biology of this system. As this highlight shows, steroids with all their complexity still test the creativity of the synthetic chemist even after all these years [47].

4.4.2 Acid Catalysis

In 2016, Fernández-Herrera *et al.* described the synthesis of dihydropyran (DHP) derivatives from diosgenin, hecogenin, and sarsasapogenin by regioselective opening of the ring E of spiroketal sapogenins in excellent yields and one step; they used Lewis acid (boron trifluoride etherate)-mediated acetolysis followed by a basic workup. Additionally, they analyzed the reaction mechanism *via* density functional theory computations and found that the formation of DHP derivatives proceeds through an intermediary oxacarbenium ion (Scheme 4.32) [65].

One research group synthesized and screened a series of novel D-ring-substituted oxazoline and oxazoline derivatives of dehydroepiandrosterone and pregnenolone, respectively, for anticancer activity against a panel of human prostate cancer cell lines and found that all the compounds showed promising anticancer activity especially against LNCaP and DU-145 cell lines. Compound (188) (Ar = $C_6H_4OCH_3$) was found to be the most active in this study. For the synthesis of oxazolines, in one of the last steps, they used $BF_3.OEt_2$ as Lewis acid catalyst, which activated α,β-azidoalcohol (187) with the appropriately substituted aromatic aldehydes to give the corresponding acetylated product. Upon deacetylation in the presence of methanolic $NaOCH_3$, the acetylated product yielded deacetylated D-ring steroidal oxazolines (188) containing aryl groups with different substituents (Scheme 4.33) [66].

Scientists described preparing novel steroidal heterocycles containing 4,6-diaryl substituted pyridine moiety fused to the 2,3- and 16,17-positions of the steroid nucleus. The Michael reaction of steroidal ketones (189) with *in situ* generated chalcones yielded intermediates of 3,5-diaryl-1,5-dicarbonyl steroidal derivatives (190). Subsequently, the intermediates (190) were converted to pyridine derivatives (192) by solid phase reaction with urea in the presence of $BF_3.OEt_2$ as the catalyst under microwave irradiation (Scheme 4.34) [67].

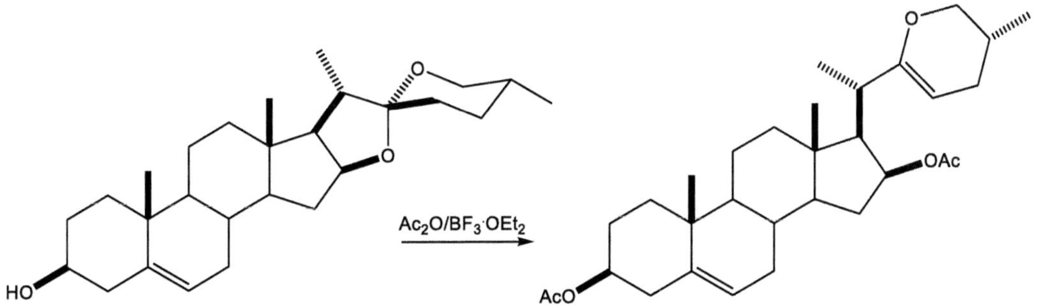

HO 184 $Ac_2O/BF_3 \cdot OEt_2$ AcO OAc 185

SCHEME 4.32 Regioselective opening of ring E of spiroketal sapogenins [65].

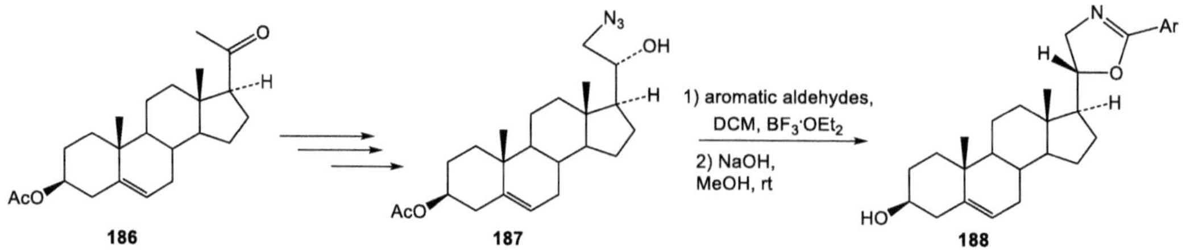

SCHEME 4.33 Synthesis of D-ring-substituted pregnenolone oxazolines (188) [66].

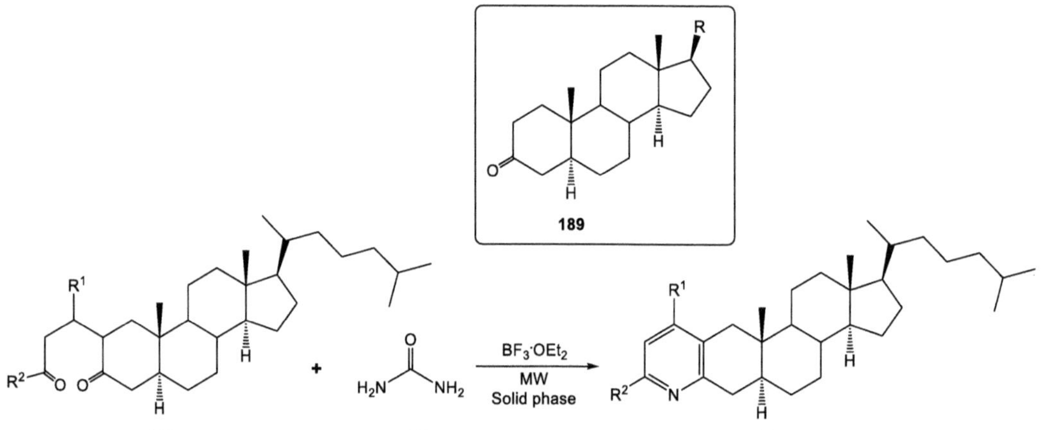

SCHEME 4.34 Synthesis of 4',6'-diaryl-chloest[3,2-b]pyridine derivatives (192) [67].

Others used acid catalysis to synthesize steroidal lactams (193) and (194) (Figure 4.9) [68]. Zavarzin and group reported that the reaction of thiohydrazides of oxamic acids with 16-hydroxymethylidene derivatives of androstane and estrone (195, 197) involving the hydroxymethylidine group led to thiohydrazones (200). These underwent heterocyclization in acidic medium (TsOH or acetic acid) to form 16-(1,3,4-thiadiazol-2-yl)-substituted steroids (201) [69]. When acetic acid was used, 3-OH acetylation of the steroid part took place along with the cyclization (Scheme 4.35).

The reactivity of spirostanic steroids 22-oxo-23-spiroketals (207a-e) against several acidic conditions was studied. Different rearrangements in the presence of HCl(g), $BF_3.OEt_2$ and $TiCl_4$ were reported [70]. The Lewis acid $BF_3.OEt_2$-catalyzed cleavage of 22-oxo-23-spiroketals with Ac_2O in DCM provided novel cholestanic frame works with pyranone E ring on the sidechain (208a-d) in higher yields than in the presence of protic acid HCl(g) [70b–c]. These mild reaction conditions promoted the regio- and stereoselective opening of the tetrahydrofuran F ring in the 23-spiroketals (207a-d). The X-ray analysis of (208c) confirmed the *Z* configuration of the double bond C23–C24. Moreover, when 22-oxo-23-spiroketals were treated with $TiCl_4/Ac_2O$ and DCM, the cleavage of the E/F rings selectively yielded furostanols (211a-d) containing a carbonyl group at C-23 and cholestanic derivatives with pyranone E ring (208a-d) [70a]. The α-orientation of the 22-hydroxyl group in compound (211a) was proved by X-ray diffraction analysis and by comparison of the ^{13}C chemical shifts of the hemiketal carbon atom within the family of compounds (Scheme 4.36).

More recently, researchers reported that the acetolysis of smilagenin (212) (25*R*) using boron trifluoride diethyl etherate in acetic anhydride yielded 20-α-methylepoxycholestene (213) as the major product, two furostene derivatives (214, 215), and the β-oriented derivative (216) (20*R*)

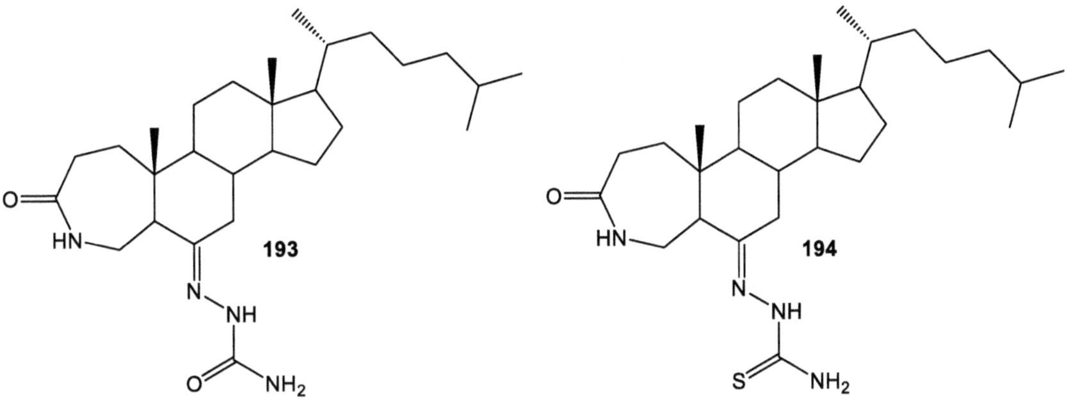

FIGURE 4.9 Steroid derivatives (193, 194) obtained *via* their lactam precursor with thiosemicarbazide or semicarbazide adding a few drops of glacial acetic acid as a catalyst [68].

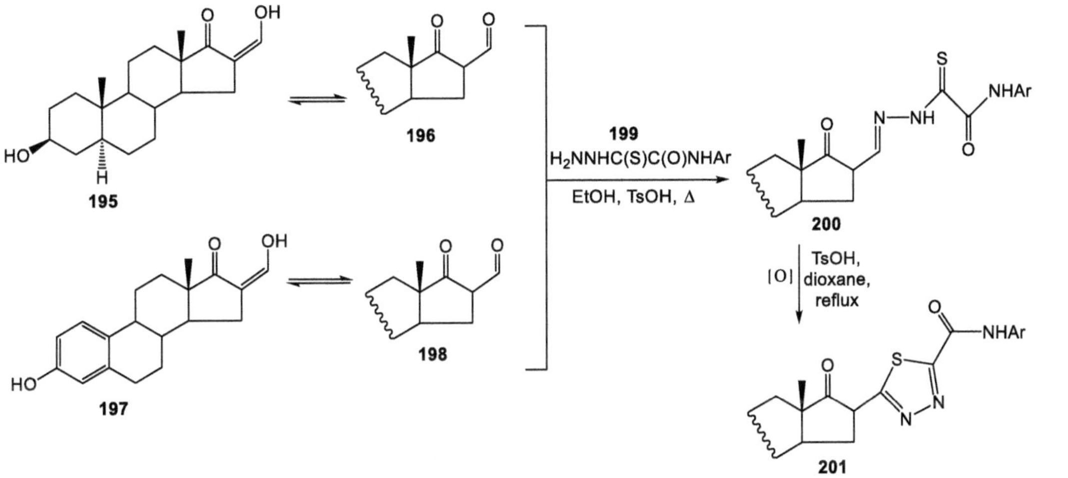

SCHEME 4.35 Cyclization of the thiohydrazone fragment in dioxane under reflux in the presence of TsOH. The cyclization is oxidative and may be air promoted [69].

(Scheme 4.37). They observed higher regioselectivity of the acetolysis of (212) using $ZnCl_2$ in acetic anhydride and obtained 20-α-methylepoxycholestene (213) in 90% yield. Additionally, the treatment of each epimeric epoxycholestene (213) and (216) under basic hydrolysis yielded two new diastereomeric 23-acetyl-spirostanols (217) and (218) stereospecifically (Scheme 4.38). The structure and stereochemistry for the iso-type 23-acetyl-spirostanol (218) was confirmed by X-ray analysis and by comparison with the crystallographic structures of the spirostanic sapogenins of the 25*S* series and smilagenin (212) [71].

4.4.3 Organocatalysis

Organocatalysis has become a major category of catalysis, and it has been successfully employed in the synthesis of steroids. The fundamental studies that revolutionized the synthesis of steroids and terpenoids helping to reveal the power of organocatalysis for the synthesis of complex natural products are well documented [72], but the development of new organic catalysts and organocatalyzed

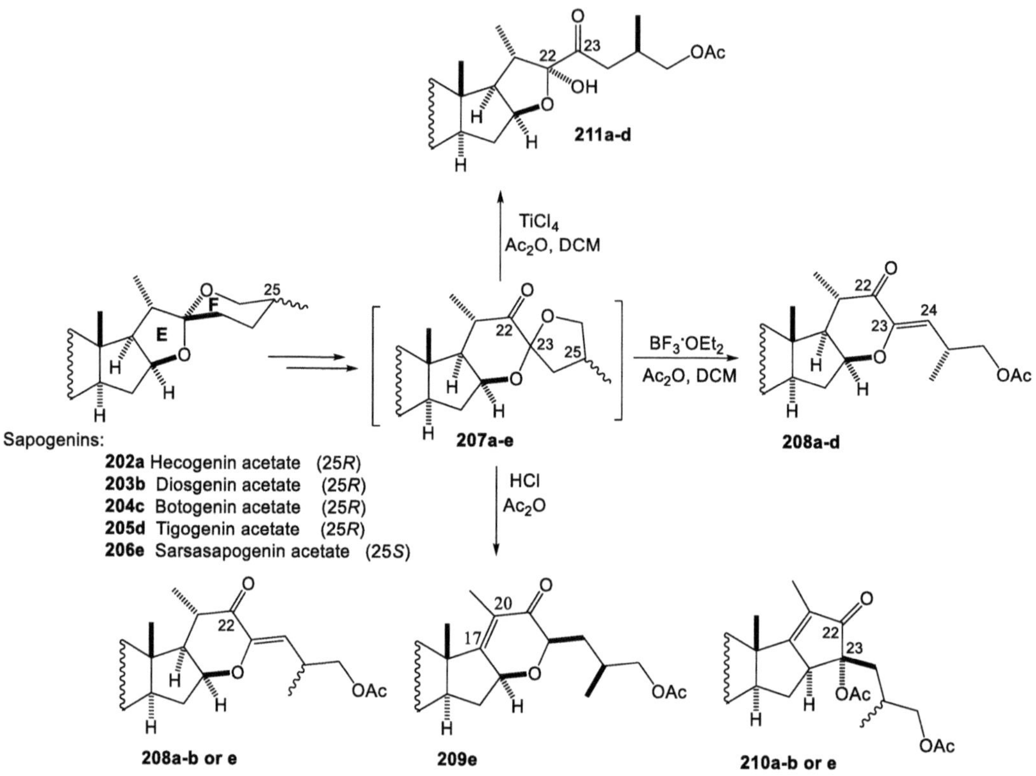

211a-d
TiCl4
Ac2O, DCM
207a-e
BF3·OEt2
Ac2O, DCM
208a-d
Sapogenins:
202a Hecogenin acetate (25R)
203b Diosgenin acetate (25R)
204c Botogenin acetate (25R)
205d Tigogenin acetate (25R)
206e Sarsasapogenin acetate (25S)
HCl
Ac2O
208a-b or e
209e
210a-b or e

SCHEME 4.36 Steroidal frameworks derived from the opening of E/F rings [70].

Smilagenin 212
BF3·OEt2
Ac2O
213
214
215
216

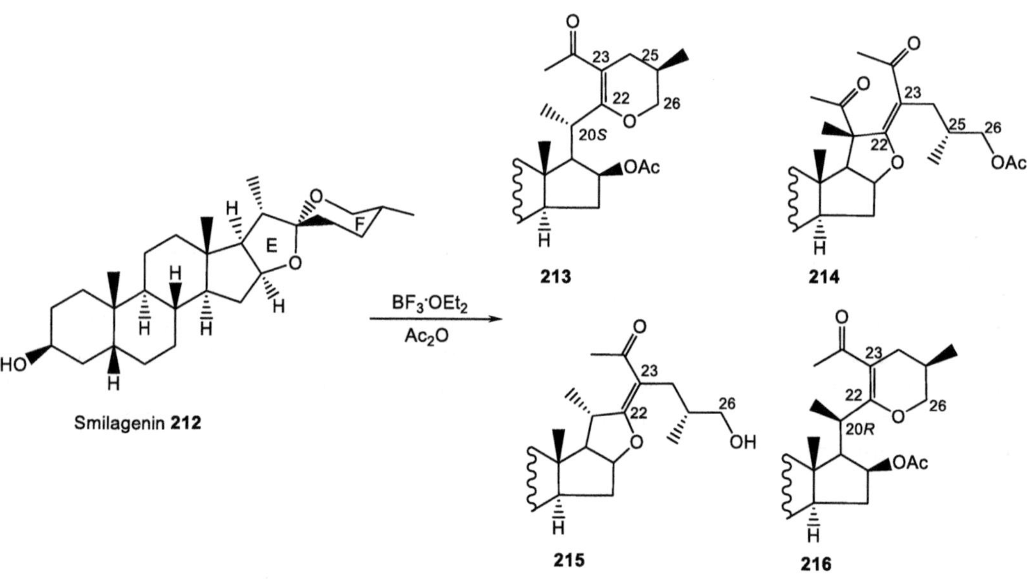

SCHEME 4.37 Steroidal derivatives from cleavage of rings E/F of smilagenin (212) using $BF_3.OEt_2$ [71].

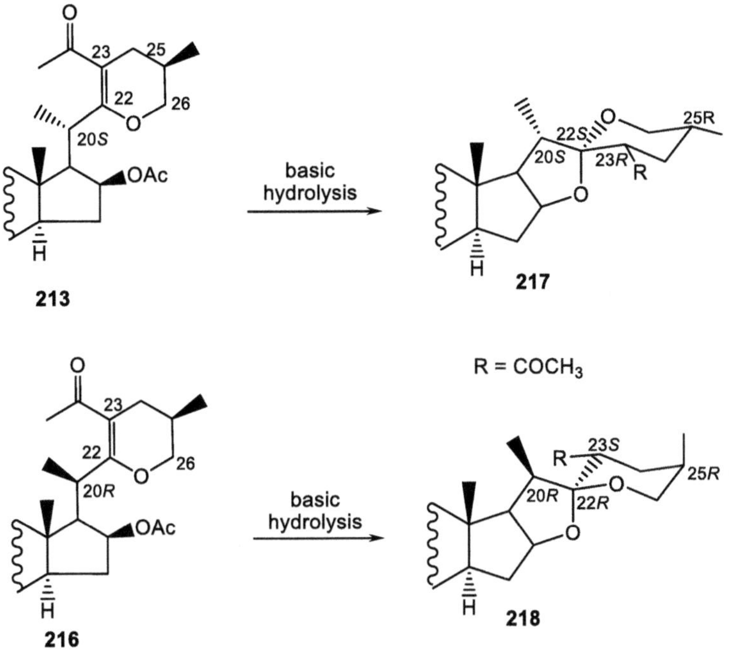

SCHEME 4.38 23-Acetyl-spirostanols (217) and (218) obtained from compounds (213) and (216), respectively [71].

tandem reactions is still of great importance to the synthesis of steroids, including major contributions by Hong [73], Hayashi [74], Jørgensen [75], and List [76] inspired by the application of organocatalysis to form biologically relevant molecules and in the total synthesis of complex natural products.

In 2014, Jørgensen reported a simple and highly stereoselective organocatalytic approach to 14*β*-steroids. The methodology showed a broad generality that generated a wide range of variations in the formed products. These included several modifications on rings A, B and C, as well as the introduction of various substituents in the angular position at C-13 (Scheme 4.39) [75].

In 2018, Pinho e Melo and group developed a pyrrolidine (225)-catalyzed diastereoselective reaction of a steroidal *N*-sulfonyl-1-azadiene (224) with carbonyl compounds (226) that led to novel chiral steroids (227a-c). They found that the use of cyclic ketones (226), allowed the synthesis of hexacyclic steroids, whereas acyclic ketones produced pentacyclic steroids (Scheme 4.40). The annulation reactions they described represented a simple, versatile, and highly selective approach to structurally diverse steroids with potential applications in medicinal chemistry. They prepared the *N*-sulfonyl-1-azadienesteroidal (224) from 16-dehydropregnenolone acetate 16-DPA (223) [77].

Hayashi *et al.* achieved the enantioselective total synthesis of estradiol methyl ether (231) in five reaction vessels with four purification steps (Scheme 4.41). The starting point was the A-ring system in the form of *p*-methoxycinnamaldehyde (229), and when this was combined with the organocatalyst diphenylprolinol silyl ether (221), the resulting minimum activation allowed an *in situ* reaction with nitroalkanediketones (228) (D-ring) leading to the bicyclo[4.3.0]nonane derivatives (230) with A, C, and D rings of the steroids as a single isomer with excellent enantioselectivity. Significantly, this domino Michael–aldol reaction sequence, where five contiguous stereogenic centers were created, provides a single isomer (230) with the required configuration in almost optically pure form (>99% enantiomeric excess) [74].

Several publications related to the application of different catalytic methods in steroid synthesis have been described in the last ten years [78], most related to homogeneous, heterogeneous catalysis, organocatalysis, and enzymatic catalysis [79]. Indeed, although research into the construction

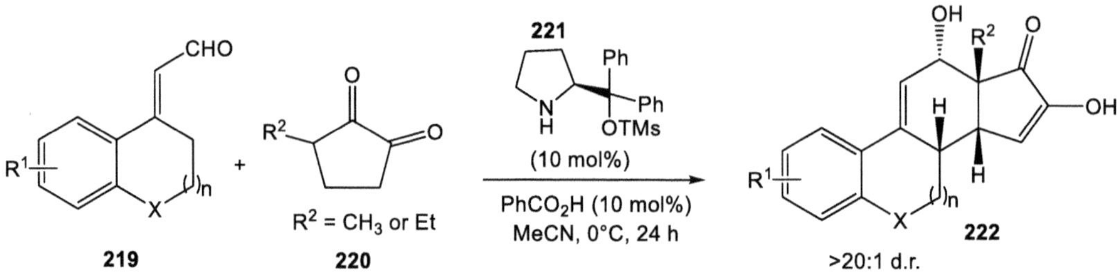

SCHEME 4.39 Synthesis of 14*β*-steroids *via* the organocatalytic reaction of enals (219) with diketones (220) [75].

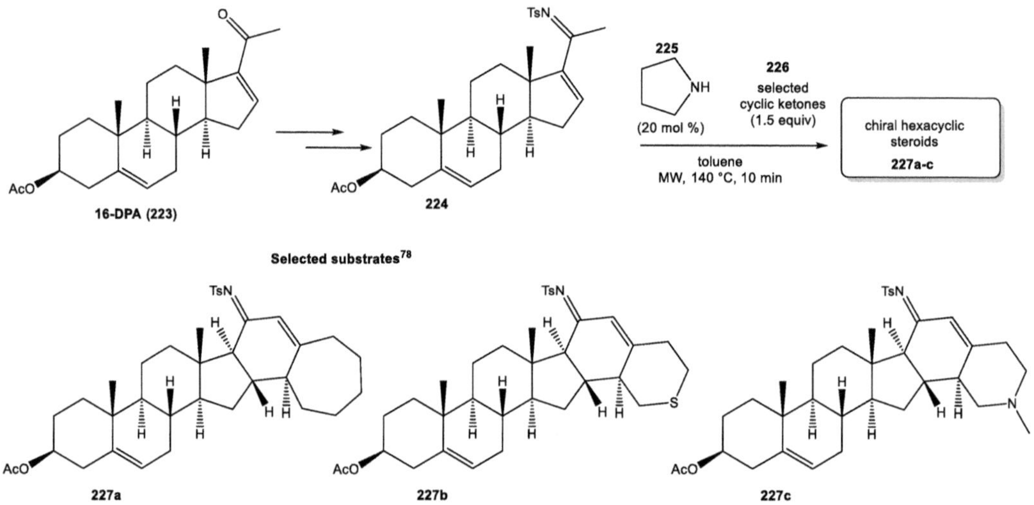

SCHEME 4.40 Synthesis of chiral hexacyclic steroids [77].

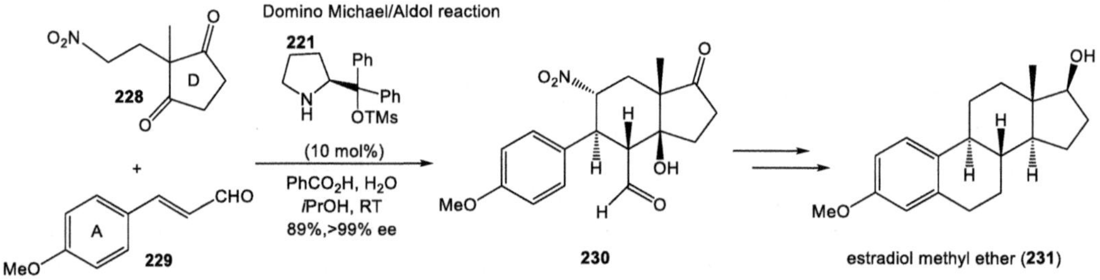

SCHEME 4.41 Enantioselective total synthesis of estradiol methyl ether (221) [74].

of the steroids core has been carried out for over 70 years, there is still much interest in expanding further methodologies, as a consequence of the importance of these compounds in medicinal chemistry and also inspired by its complex steroidal structures.

To summarize, in this chapter, we have highlighted some recent progress in the synthesis of carbohydrates, lipids, and steroids using different catalytic methods. These approaches provide a new variety of organic molecules for subsequent explorations in areas such as, medicinal chemistry, molecular pharmacology, materials science, polymer chemistry, theoretical and computational chemistry.

REFERENCES

[1] (a) S.S. Wang, G.Y. Yang. Recent advances in polyoxometalate-catalyzed reactions. Chem. Rev. 115(11) (2015) 4893–4962. https://doi.org/10.1021/cr500390v; (b) H. Zheng, Z. Sun, X. Yi, S. Wang, J. Li, X. Wang, Z. Jiang. A water-tolerant C16H3PW11CrO39 catalyst for the efficient conversion of monosaccharides into 5-hydroxymethylfurfural in a micellar system. RSC. Adv. 45(3) (2013) 23051–23056. https://doi.org/10.1039/C3RA43408G

[2] Q. Zhao, L. Wang, S. Zhao, X. Wang, S. Wang. High selective production of 5-hydroymethylfurfural from fructose by a solid heteropolyacid catalyst. Fuel 90(6) (2011) 2289–2293. https://doi.org/10.1016/j.fuel.2011.02.022

[3] C. Fan, H. Wang, H. Zhang, J. Wang, S. Wang, X. Wang. Conversion of fructose and glucose into 5-hydroxymethylfurfural catalyzed by a solid heteropolyacid salt. Biomass Bioenergy. 35(7) (2011) 2659–2665. https://doi.org/10.1016/j.biombioe.2011.03.004

[4] Y. Qu, C. Huang, J. Zhang, B. Chen. Efficient dehydration of fructose to 5-hydroxymethylfurfural catalyzed by a recyclable sulfonated organic heteropolyacid salt. Bioresour. Technol. 106 (2012) 170–172. https://doi.org/10.1016/j.biortech.2011.11.069

[5] M.J. Climent, A. Corma, S. Iborra, M.J. Sabater. Heterogeneous catalysis for tandem reactions. ACS Catal. 4(3) (2014) 870–891. https://doi.org/10.1021/cs401052k

[6] (a) N. Villandier, A. Corma. One pot catalytic conversion of cellulose into biodegradable surfactants. Chem. Comm. 46(24) (2010) 4408–4410. https://doi.org/10.1039/C0CC00031K; (b) K. Tomishige, Y. Nakagawa, M. Tamura. Selective hydrogenolysis of C-O bonds using the interaction of the catalyst surface and OH Groups. Top Curr. Chem. 353 (2014) 127–162. https://doi.org/10.1007/128_2014_538

[7] K. Chen, M. Tamura, Z. Yuan, Y. Nakagawa, K. Tomishige. One-pot conversión of sugar and sugar polyols to n-Alkanes without C-C dissociation over the Ir-ReOx/SiO2 catalyst combined with H-ZSM-5. Chem. Sus. Chem. 6(4) (2013) 613–621. https://doi.org/10.1002/cssc.201200940

[8] S. Matthies, D.T. McQuade, P.H. Seeberger. Homogeneous gold-catalyzed glycosylations in continuous flow. Org. Lett. 17(15) (2015) 3670–3673. https://doi.org/10.1021/acs.orglett.5b01584

[9] S.A. Shahzad, M.A. Sajid, Z.A. Khan, D.C. Gonzalez. Gold catalysis in organic transformations: A review. Synth. Commun. 47(8) (2017) 735–755. https://doi.org/10.1080/00397911.2017.1280508

[10] B.N. Harris, P.P. Patel, C.P. Gobble, M.J. Stark, C. De Meo. C-5 modified *S*-benzoxazolyl sialyl donors: Towards more efficient selective sialylations. Eur. J. Org. Chem. 2011(20–21) (2011) 4023–4027. https://doi.org/10.1002/ejoc.201100539

[11] T. Ollevier. New trends in bismuth-catalyzed synthetic transformations. Org. Biomol. Chem. 11(17) (2013) 2740–2755. https://doi:10.1039/C3OB26537D

[12] H.Y. Choi, J.H. Bae, Y. Hasegawa, S. An, I.S. Kim, H. Lee, M. Kim. Thiol-functionalized cellulose nanofiber membranes for the effective adsorption of heavy metal ions in water. Carbohyd. Polym. 234 (2020) 115881. https://doi.org/10.1016/j.carbpol.2020.115881

[13] F. Rol, M.N. Belgacem, A. Gandini, J. Bras. Recent advances in surface-modified cellulose nanofibrils. Prog. Polym. Sci. 88 (2019) 241–264. https://doi.org/10.1016/j.progpolymsci.2018.09.002

[14] Q. Wang, Q. Yao, J. Liu, J. Sun, Q. Zhu, H. Chen. Processing nanocellulose to bulk materials: A review. Cellulose 26 (2019) 7585–7617. https://doi.org/10.1007/s10570-019-02642-3

[15] (a) H. Xu, T. Bronner, M. Yamamoto, H. Yamane. Regeneration of cellulose dissolved in ionic liquid using laser-heated melt-electrospinning. Carbohyd. Polym. 201 (2018) 182–188. https://doi.org/10.1016/j.carbpol.2018.08.062; (b) H. Lee, M. Nishino, D. Sohn, J.S. Lee, I.S. Kim. Control of the morphology of cellulose acetate nanofibers via electrospinning. Cellulose 25 (2018) 2829–2837. https://doi.org/10.1007/s10570-018-1744-0

[16] S. Kang, J. Ye, Y. Zhang, J. Chang. Preparation of biomass hydrochar derived sulfonated catalysts and their catalytic effects for 5-hydroxymethylfurfural production. RSC Adv. 3(20) (2013) 7360–7366. https://doi.org/10.1039/C3RA23314F

[17] Y. Lin, J. Yu, X. Zhang, J. Fang, G.P. Lu, H. Huang. Carbohydrate-derived porous carbon materials: An ideal platform for green organic synthesis. Chin. Chem. Lett. 33(1) (2022) 186–196. https://doi.org/10.1016/j.cclet.2021.06.045

[18] X. Xie, J. Shi, Y. Pu, Z. Wang, L.L. Zhang, J.X. Wang, D. Wang. Cellulose derived nitrogen and phosphorus co-doped carbon-based catalysts for catalytic reduction of *p*-nitrophenol. J. Colloid Interface Sci. 571(1) (2020) 100–108. https://doi.org/10.1016/j.jcis.2020.03.035

[19] M.S. Mirhosseyni, F. Nemati, A. Elhampour. Pyrolysis of functional cellulose by ionothermal method to synthesis of mesoporous triazine carbon for supporting of Pd and its application. Carbohyd. Polym. 217 (2019) 199–206. https://doi.org/10.1016/j.carbpol.2019.04.005

[20] B. Sahoo, A.E. Surkus, M.M. Pohl, J. Radnik, M. Schneider, S. Bachmann, M. Scalone, K. Junge, M. Beller. A biomass-derived non-noble cobalt catalyst for selective hydrodehalogenation of alkyl and (hetero) aryl halides. Angew. Chem. Int. Ed. 56(37) (2017) 11242–11247. https://doi.org/10.1002/anie.201702478

[21] R. Zhong, Y. Liao, L. Peng, R.I. Iacobescu, Y. Pontikes, R. Shu, L. Ma, B.F. Sels. Silica-carbon nanocomposite acid catalyst with large mesopore interconnectivity by vapor-phase assisted hydrothermal treatment. ACS Sustain. Chem. Eng. 6(6) (2018) 7859–7870. https://doi.org/10.1021/acssuschemeng.8b01003

[22] N.P. Bizier, S.R. Atkins, L.C. Helland, S.F. Colvin, J.R. Twitchell, M.J. Cloninger. Indium triflate catalyzed peracetylation of carbohydrates. Carbohydr. Res. 343(10–11) (2008) 1814–1818. https://doi.org/10.1016/j.carres.2008.04.009

[23] E.A. Mensah, F.R. Reyes, E.S. Standiford. Highly efficient cationic palladium catalyzed acetylation of alcohols and carbohydrate-derived polyols. Catalysts 6(2) (2016) 27. https://doi.org/10.3390/catal6020027

[24] T.Y. Chiu, W. Chin, J.R. Guo, C.F. Liang, P.H. Lin. A dinuclear dysprosium complex as an air-stable and recyclable catalyst: Applications in the deacetylation of carbohydrate, aliphatic, and aromatic molecules. Chem. Asian J. 14(5) (2019) 627–633. https://doi.org/10.1002/asia.201801652

[25] (a) Á. Magyar, K. Juhász, Z. Hell. The application of 4Å molecular sieves in organic chemical syntheses: An overview. Synthesis 53 (2) (2021) 279–295. https://doi.org/10.1055/s-0040-1706535; (b) Y. H. Liu, L. H. Cao. An expedient one-step synthesis of polysubstituted guanidinoglucosides using HgO-4 Å molecular sieves as catalyst. Carbohydr. Res. 343(14) (2008) 2376–2383. https://doi.org/10.1016/j.carres.2008.07.016

[26] (a) M. Heuckendorff, J. Bendix, C. M. Pedersen, M. Bols. *β*-Selective mannosylation with a 4,6-silylene-tethered thiomannosyl donor. Org. Lett. 16(4) (2014) 1116–1119. https://doi.org/10.1021/ol403722f (b) B. Dhakal, L. Bohé, D. Crich. Trifluoromethanesulfonate anion as nucleophile in organic chemistry. J. Org. Chem. 82(18) (2017) 9263–9269. https://doi.org/10.1021/acs.joc.7b01850; (c) M. Huang, P. Retailleau, L. Bohé, D. Crich. Cation clock permits distinction between the mechanisms of *α*- and *β-O*- and *β-C*-glycosylation in the mannopyranose series: Evidence for the existence of a mannopyranosyl oxocarbenium ion. J. Am. Chem. Soc. 134(36) (2012) 14746–14749. https://doi.org/10.1021/ja307266n; (d) E. Wojaczyńska, F. Steppeler, D. Iwan, M.C. Scherrmann, A. Marra. Synthesis and applications of carbohydrate-based organocatalysts. Molecules 26(23) (2021) 7291. https://doi.org/10.3390/molecules26237291

[27] H.B. Jadhav, U. Annapure. Designer lipids-synthesis and application: A review. Tr. Food Sci. Technol. 116 (2021) 884–902. https://doi.org/10.1016/j.tifs.2021.08.020

[28] C.C. Akoh. Food lipids chemistry, nutrition and biotechnology. Taylor and Francis (2017).

[29] (a) P.K. Mukherjee, V.L. Marcheselli, C.N. Serhan, N.G. Bazan. Neuroprotectin D1: A docosahexaenoic acid-derived docosatriene protects human retinal pigment epithelial cells from oxidative stress. Proc. Natl. Acad. Sci. USA. 101(22) (2004) 8491–8496. https://doi.org/10.1073/pnas.0402531101; (b) A. Ariel, P.L. Li, W. Wang, W.X. Tang, G. Fredman, S. Hong, K.H. Gotlinger, C.N. Serhan. The docosatriene protectin D1 is produced by TH2 skewing and promotes human T cell apoptosis via lipid raft clustering. J. Biol. Chem. 280(52) (2005) 43079–43086. https://doi.org/10.1074/jbc.M509796200; (c) K. Hamidzadeh, J. Westcott, N. Wourms, A.E. Shay, A. Panigrahy, M.J. Martin, R. Nshimiyimana, C.N. Serhan. A newly synthesized 17-epi-neuroprotectin D1/17-epi-protectin D1: Authentication and functional regulation of inflammation-resolution. Biochem. Pharmacol. 203 (2022) 115181. https://doi.org/10.1016/j.bcp.2022.115181

[30] M. Aursnes, J.E. Tungen, A. Vik, J. Dallib, T.V. Hansen. Stereoselective synthesis of protectin D1: A potent anti-inflammatory and proresolving lipid mediator. Org. Biomol. Chem. 12(3) (2014) 432–437. https://doi.org/10.1039/C3OB41902A

[31] A. de La Torre, Y.Y. Lee, C. Oger, P.T. Sangild, T. Durand, J.C.Y. Lee, J.M. Galano. Synthesis, discovery, and quantitation of dihomo-isofurans: Biomarkers for in vivo adrenic acid peroxidation. Angew. Chem. 126 (2014) 6363–6366. https://doi.org/10.1002/anie.201402440

[32] C.D. Hunter, T. Guo, G. Daskhan, M.R. Richards, C.W. Cairo. Synthetic strategies for modified glycosphingolipids and their design as probes. Chem. Rev. 118(17) (2018) 8188–8241. https://doi.org/10.1021/acs.chemrev.8b00070

[33] G. van Meer, D.R. Voelker, G.W. Feigenson. Membrane lipids: Where they are and how they behave. Nat. Rev. Mol. Cell Biol. 9(2008) 112–124. https://doi.org/10.1038/nrm2330

[34] R.L. Schnaar, R. Sandhoff, M. Tiemeyer, T. Kinoshita. Glycosphingolipids. In Essentials of glycobiology, 3rd ed. A. Varki, R.D. Cummings, J.D. Esko, P. Stanley, G.W. Hart, M. Aebi, A.G. Darvill, T. Kinoshita, N.H. Packer, J.H. Prestegard, et al., Eds., Cold Spring Harbor Laboratory Press: Cold Spring Harbor, NY (2017).

[35] Y.A. Hannun, L.M. Obeid. Sphingolipids and their metabolism in physiology and disease. Nat. Rev. Mol. Cell Biol. 19(2018) 175–191. https://doi.org/10.1038/nrm.2017.107
[36] L. Opálka, A. Kováčik, M. Sochorová, J. Roh, J. Kuneš, J. Lenčo, K. Vávrová. Scalable synthesis of human ultralong chain ceramides. Org. Lett. 17(21) (2015) 5456–5459. https://doi.org/10.1021/acs.orglett.5b02816
[37] (a) P. Wisse, M.A. de Geus, G. Cross, A.M. van den Nieuwendijk, E.J. van Rooden, R.J. van den Berg, J.M. Aerts, G.A. van der Marel, J.D. Codée, H.S. Overkleeft. Synthesis of 6-hydroxysphingosine and α-hydroxy ceramide using a cross-metathesis strategy. J. Org. Chem. 80(14) (2015) 7258–7265. https://doi.org/10.1021/acs.joc.5b00823; (b) T. Yamamoto, H. Hasegawa, T. Hakogi, S. Katsumura. Versatile synthetic method for sphingolipids and functionalized sphingosine derivatives via olefin cross metathesis. Org. Lett. 8(24) (2006) 5569–5572. https://doi.org/10.1021/ol062258l
[38] N.R. Yepuri, S.A. Holt, G. Moraes, P.J. Holden, K.R. Hossain, S.M. Valenzuela, M. James, T.A. Darwish. Stereoselective synthesis of perdeuterated phytanic acid, its phospholipid derivatives and their formation into lipid model membranes for neutron reflectivity studies. Chem. Phys. Lipids 183 (2014) 22–33. https://doi.org/10.1016/j.chemphyslip.2014.04.004
[39] X.G. Yin, J. Lu, J. Wang, R.Y. Zhang, X.F. Wang, C.M. Liao, X.P. Liu, Z. Liu, J. Guo. Synthesis and evaluation of liposomal anti-GM3 cancer vaccine candidates covalently and noncovalently adjuvanted by αGalCer. J. Med. Chem. 64 (2021) 1951–1965. https://doi.org/10.1021/acs.jmedchem.0c01186
[40] J. Sibold, S. Ahadi, D.B. Werz, C. Steinem. Chemically synthesized Gb 3 glycosphingolipids: Tools to access their function in lipid membranes. Eur. Biophys. J. 50 (2021)109–126. https://doi.org/10.1007/s00249-020-01461-w
[41] Y.H. Tsai, S. Götze, I. Vilotijevic, M. Grube, D. Varon Silva, P.H. Seeberger. A general and convergent synthesis of diverse glycosylphosphatidylinositol glycolipids. Chem. Sci. 4(1) (2013) 468–481. https://doi.org/10.1039/C2SC21515B
[42] J.W. Winkler, J. Uddin, C.N. Serhan, N.A. Petasis. Stereo controlled total synthesis of the potent anti-inflammatory and pro-resolving lipid mediator resolvin D3 and its aspirin-triggered 17*R*-epimer. Org. Lett. 15(7) (2013) 1424–1427. https://doi.org/10.1021/ol400484u
[43] M. Rosell, M. Villa, T. Durand, J.M. Galano, J. Vercauteren, and C. Crauste. Total syntheses of two bis-allylic-deuterated DHA analogues. Asian J. Org. Chem. 6(3) (2017) 322–334. https://doi.org/10.1002/ajoc.201600565
[44] D. Listunov, V. Maraval, R. Chauvin, Y. Genisson. Chiral alkynyl carbinols from marine sponges: Asymmetric synthesis and biological relevance. Nat. Prod. Rep. 32(1) (2015) 49–75. https://doi.org/10.1039/C4NP00043A
[45] W. Wei, Y. Feng, X. Zhang, X. Cao, F. Feng. Synthesis of structured lipid 1,3-dioleoyl-2-palmitoylglycerol in both solvent and solvent-free system. LWT Food Sci. Technol. 60(2) (2015) 1187–1194. https://doi.org/10.1016/j.lwt.2014.09.013
[46] C.O. Pérez-Gómez, J. Vazquez-Chavez, R. Yepéz, J.P. García-Merinos, M.I. Ramírez-Díaz, R.E. del Río, R. Santillan, Y. López. Synthesis and structural characterization of an oxaziridine derived from 6-azadiosgenin. J. Mol. Struct. 1255 (2022) 132386. https://doi.org/10.1016/j.molstruc.2022.132386
[47] K.D. Jones, S.G. Stewart. Recent advances in steroid synthesis: A tribute to Sir Derek Barton. Aust. J. Chem. 71 (2018) 627–633. https://doi.org/10.1071/CH18256
[48] E.S. Pintér, Z. Csók, Z. Berente, L. Kollár, R. Skoda-Földes. Synthesis of novel 13*a*-18-nor-16-carboxamido steroids via a palladium-catalyzed aminocarbonylation reaction. Steroids 78 (2013) 1177–1182. https://doi.org/10.1016/j.steroids.2013.08.011
[49] K.P. Labrada, C. Morera, I. Brouard, R. Llerena, D.G. Rivera. Synthesis and conformational study of triazole-linked bis-spirostanic conjugates. Tetrahedron Lett. 54 (2013) 1602–1606. https://doi.org/10.1016/j.tetlet.2013.01.058
[50] N.V. Sokolova, V.G. Nenajdenko. Azidoisocyanides, new bifunctional reagents for multicomponent reactions and biomolecule modifications. Chem Nat. Compd. 50(2) (2014) 197–213. https://doi.org/10.1007/s10600-014-0914-z
[51] H.R. Khatri, B. Bhattarai, W. Kaplan, Z. Li, M.J.C. Long, Y. Aye, P. Nagorny. Modular total synthesis and cell-based anticancer activity evaluation of ouabagenin and other cardiotonic steroids with varying degrees of oxygenation. J. Am. Chem. Soc. 141(12) (2019) 4849–4860. https://doi.org/10.1021/jacs.8b12870
[52] N. Aguilar-Valdez, M. Maldonado-Domínguez, R. Arcos-Ramos, M. Romero-Ávila, R. Santillan, N. Farfán. Synthesis of steroidal molecular compasses: Exploration on the controlled assembly of solid organic materials. Cryst. Eng. Comm. 19(13) (2017) 1771–1777. https://doi.org/10.1039/C7CE00157F

[53] W. Zhang, L. Wang, L. Zhang, W. Chen, X. Chen, M. Xie, G. Yan, X. Hu, J. Xu, J. Zhang. Synthesis and biological evaluation of steroidal derivatives as selective inhibitors of AKR1B10. Steroids 86 (2014) 39–44. https://doi.org/10.1016/j.steroids.2014.04.010

[54] S. Watanabe, T. Nishikawa, A. Nakazaki. Synthesis of oxy-functionalized steroidal skeletons via Mizoroki-Heck and intramolecular Diels-Alder reactions. Org. Lett. 21(18) (2019) 7410–7414. https://doi.org/10.1021/acs.orglett.9b02716

[55] (a) R.R. Karimov, J.F. Hartwig. Transition-metal-catalyzed selective functionalization of C(sp3)-H bonds in natural products. Angew. Chem. Int. Ed. 57(16) (2018) 4234–4241. https://doi.org/10.1002/anie.201710330; (b) Y.Y. See, A.T. Herrmann, Y. Aihara, P.S. Baran. Scalable C-H oxidation with copper: Synthesis of polyoxypregnanes. J. Am. Chem. Soc. 137(43) (2015) 13776–13779. https://doi.org/10.1021/jacs.5b09463

[56] Y.N. Kotovshchikov, G.V. Latyshev, N.V. Lukashev, I.P. Beletskaya. An efficient approach to azolyl-substituted steroids through copper-catalyzed Ullmann C-N coupling. Eur. J. Org. Chem. 2013(34) (2013) 7823–7832. https://doi.org/10.1002/ejoc.201300719

[57] J.A.R. Salvador, S.M. Silvestre. Bismuth-catalyzed allylic oxidation using *t*-butyl hydroperoxide. Tetrahedron Lett. 46(15) (2005) 2581–2584. https://doi.org/10.1016/j.tetlet.2005.02.080

[58] J.M. Smith, T. Qin, R.R. Merchant, J.T. Edwards, L.R. Malins, Z. Liu, G. Che, Z. Shen, S.A. Shaw, M.D. Eastgate, P.S. Baran. Decarboxylative alkynylation. Angew. Chem. Int. Ed. 56(39) (2017) 11906–11910. https://doi.org/10.1002/anie.201705107

[59] F. Noack, B. Hartmayer, P. Heretsch. Access to 5 (6→7) *abeo*-Steroids through benzylic acid rearrangement of *i*-steroids. Synthesis 50(4) (2018) 809–820. https://doi.org/10.1055/s-0036-1591883

[60] (a) B. Bhattarai, P. Nagorny. Enantioselective total synthesis of cannogenol-3-*O*-α-*L*-rhamnoside via sequential Cu(II)-catalyzed Michael addition/intramolecular aldol cyclization reactions. Org. Lett. 20(1) (2018) 154–157. https://doi.org/10.1021/acs.orglett.7b03513; (b) V. Kapras, V. Vyklicky, M. Budesinsky, I. Cisarova, L. Vyklicky, H. Chodounska, U. Jahn. Total synthesis of *ent*-pregnanolone sulfate and its biological investigation at the NMDA Receptor. Org. Lett. 20(4) (2018) 946–949. https://doi.org/10.1021/acs.orglett.7b03838; (c) S. Yamashita, A. Naruko, Y. Nakazawa, L. Zhao, Y. Hayashi, M. Hirama. Total synthesis of Limonin. Angew. Chem. Int. Ed. 54(29) (2015) 8538–8541. https://doi.org/10.1002/anie.201503794

[61] M. Michalak, K. Michalak, J. Wicha. The synthesis of cardenolide and bufadienolide aglycones, and related steroids bearing a heterocyclic subunit. Nat. Prod. Rep. 34(4) (2017) 361–410. https://doi.org/10.1039/C6NP00107F

[62] E.G. Mackay, M.S. Sherburn. The Diels–Alder reaction in steroid synthesis. Synthesis 47(1) (2015) 1–21. https://doi.org/10.1055/s-0034-1378676

[63] P. Gupta, A. Mahajan. Sustainable approaches for steroid synthesis. Environ. Chem. Lett. 17(2) (2019) 879–895. https://doi.org/10.1007/s10311-018-00845-x

[64] W. Xun, B. Xu. Synthetic approaches of aplykurodinone-1: A minireview. Asian J. Org. Chem. 11(2) (2022) e202100695. https://doi.org/10.1002/ajoc.202100695

[65] J.C. Hilario-Martínez, R. Zeferino-Díaz, M.A. Muñoz-Hernañdez, M.G. Hernández-Linares, J.L. Cabellos, G. Merino, J. Sandoval-Ramírez, Z. Jin, M.A. Fernández-Herrera. Regioselective spirostan E-ring opening for the synthesis of dihydropyran steroidal frameworks. Org. Lett. 18(8) (2016) 1772–1775. https://doi.org/10.1021/acs.orglett.6b00492

[66] A.H. Banday, S.M.M. Akram, R. Parveen, N. Bashir. Design and synthesis of D-ring steroidal isoxazolines and oxazolines as potential antiproliferative agents against LNCaP, PC-3 and DU-145 cells. Steroids 87 (2014) 93–98. https://doi.org/10.1016/j.steroids.2014.05.009

[67] M. Dutta, P. Saikia, S. Gogoi, R.C. Boruah. Microwave-promoted and Lewis acid catalysed synthesis of steroidal A- and D-ring fused 4,6-diarylpyridines. Steroids 78 (2013) 387–395. https://doi.org/10.1016/j.steroids.2013.01.006

[68] M. Ibrahim-Ouali, F. Dumur. Steroidal lactams: A review. ARKIVOC—Online J. Org. Chem. 2022(1) (2022) 262–284. https://doi.org/10.24820/ark.5550190.p011.773

[69] I.V. Zavarzin, Y.S. Antonov, E.I. Chernoburova, M.A. Shchetinina, N.G. Kolotyrkina, A.S. Shashkov. Interaction of 16-hydroxymethylidene derivatives of androstane and estrone with thiohydrazides of oxamic acids. Russ. Chem. Bull. Int. Ed. 62(12) (2013) 2603–2608. https://doi.org/10.1007/s11172-013-0379-4

[70] (a) A. Corona-Díaz, J.P. García-Merinos, M.E. Ochoa, R.E. del Río, R. Santillan, S. Rojas-Lima, J.W. Morzycki, Y. López. $TiCl_4$ catalyzed cleavage of (25*R*)-22-oxo-23-spiroketals: Synthesis of sapogenins with furostanol and pyranone E rings on the side chain. Steroids 152 (2019) 108488. https://doi.org/10.1016/j.steroids.2019.108488; (b) Y. López, L. Rodríguez, R.E. del Río, N. Farfán, J.W. Morzycki, R.

Santillan. Regioselective cleavage of 22-oxo-23-spiroketals. Novel cholestanic frameworks with pyranone and cyclopentenone E rings on the side chain. Steroids 77 (2012) 534–541. https://doi.org/10.1016/j.steroids.2012.01.018; (c) A. Corona-Díaz, J.P. García-Merinos, Y. López, J.B. González-Campos, R.E del Río, R. Santillan, N. Farfán, J.W. Morzycki. Regio- and stereoselective cleavage of steroidal 22-oxo-23-spiroketals catalyzed by BF3·Et2O. Steroids 100 (2015) 36–43. https://doi.org/10.1016/j.steroids.2015.04.004

[71] J.P. García-Merinos, R. Yépez, C.M. Ramírez-Lozano, S. Rincón, M.E. Ochoa, Y. López, N. Farfán, R. Santillan. Smilagenin transformation products under Lewis acid catalysis in acetic anhydride and synthesis of 23-acetyl-spirostanols. Nat. Prod. Commun. 18(11) (2023) 1–14. https://doi.org/10.1177/1934578X231212341

[72] H.R. Khatri, N. Carney, R. Rutkoski, B. Bhattarai, P. Nagorny. Recent progress in steroid synthesis triggered by the emergence of new catalytic methods. Eur. J. Org. Chem. 2020(7) (2020) 755–776. https://doi.org/10.1002/ejoc.201901466

[73] (a) D.H. Jhuo, B.C. Hong, C.W. Chang, G.H. Lee. One-pot organocatalytic enantioselective Michael-Michael-Aldol-Henry reaction cascade: A facile entry to the steroid system with six contiguous stereogenic centers. Org. Lett. 16(10) (2014) 2724–2727. https://doi.org/10.1021/ol501011t; (b) C.H. Peng, B.C. Hong, A. Raja, C.W. Chang, G.H. Lee. Constructing densely functionalized Hajos-Parrish-type ketones with six contiguous stereogenic centers and two quaternary carbons in a formal [2+2+2] cycloaddition cascade. RSC Adv. 6(97) (2016) 95314–95319. https://doi.org/10.1039/C6RA22430J

[74] (a) Y. Hayashi, S. Koshino, K. Ojima, E. Kwon. Pot economy in the total synthesis of Estradiol methyl ether by using an organocatalyst. Angew. Chem. Int. Ed. 56(39) (2017) 11812–11815. https://doi.org/10.1002/anie.201706046; (b) S. Koshino, E. Kwon, Y. Hayashi. Total synthesis of Estradiol Methyl Ether and its five-pot synthesis with an organo catalyst. Eur. J. Org. Chem. 2018(41) (2018) 5629–5638. https://doi.org/10.1002/ejoc.201800910

[75] K.S. Halskov, B.S. Donslund, S. Barfüsser, K.A. Jørgensen. Organocatalytic asymmetric formation of steroids. Angew. Chem. Int. Ed. 53(16) (2014) 4137–4141. https://doi.org/10.1002/anie.201400203

[76] (a) S. Prévost, N. Dupré, M. Leutzsch, Q. Wang, V. Wakchaure, B. List. Catalytic asymmetric torgov cyclization: A concise total synthesis of (+)-Estrone. Angew. Chem. Int. Ed. 53(33) (2014) 8770–8773. https://doi.org/10.1002/anie.201404909; (b) S. Prévost, N. Dupré, M. Leutzsch, Q. Wang, V. Wakchaure, B. List. Die katalytische asymmetrische Torgov-cyclisierung: einekurze total synthese von (+)-Estron. Angew. Chem. 126(33) (2014) 8915–8918. https://doi.org/10.1002/ange.201404909

[77] S.M.M. Lopes, C.S.B. Gomes, T.M.V.D. Pinho e Melo. Reactivity of steroidal 1-azadienes toward carbonyl compounds under enamine catalysis: Chiral penta- and hexacyclic steroids. Org. Lett. 20(14) (2018) 4332–4336. https://doi.org/10.1021/acs.orglett.8b01783

[78] (a) N. Arichi, K. Hata, Y. Takemoto, K.I. Yamada, Y. Yamaoka, K. Takasu. Synthesis of steroidal derivatives bearing a small ring using a catalytic [2+2] cycloaddition and a ring-contraction rearrangement. Tetrahedron 71 (2015) 233–242. https://doi.org/10.1016/j.tet.2014.11.065; (b) D. Czajkowska-Szczykowska, J.W. Morzycki, A. Wojtkielewicz. Pd-catalyzed steroid reactions. Steroids 97 (2015) 13–44. https://doi.org/10.1016/j.steroids.2014.07.018; (c) M. Ibrahim-Ouali, F. Dumur. Recent syntheses of steroid derivatives using the CuAAC "click" reaction. ARKIVOC—Online J. Org. Chem. 2021(9) (2021) 130–149. http://dx.doi.org/10.24820/ark.5550190.p011.543; (d) P. Borah, V.D. Shivling, B.K. Banik, B.M. Sahoo. An overview on steroids and microwave energy in multi-component reactions to wards the synthesis of novel hybrid molecules. Curr. Org. Synth. 17(8) (2020) 594–609. https://doi.org/10.2174/1570179417666200503050106; (e) N. Liaba, J. Lee, C.H. Oh. Access toward steroid skeleton and its derivatives via copper-catalyzed intramolecular cyclization: A short synthesis of estradiol and estrone. Asian J. Org. Chem. 12(5) (2023) 93–100. https://doi.org/10.1002/ajoc.202300062; (f) D.B. Ramachary, R. Sakthidevia, P.S. Reddya. Direct organocatalytic stereoselective transfer hydrogenation of conjugated olefins of steroids. RSC Adv. 3(32) (2013) 13497–13506. https://doi.org/10.1039/C3RA41519H

[79] (a) E. Kozłowska, N. Hoc, J. Sycz, M. Urbaniak, M. Dymarska, J. Grzeszczuk, E. Kostrzewa-Susłow, Ł. Stępień, E. Pląskowska, T. Janeczko. Biotransformation of steroids by entomopathogenic strains of *Isaria farinosa*. Microb. Cell Fact. 17(71) (2018) 1–11. https://doi.org/10.1186/s12934-018-0920-0; (b) F. Zappaterra, S. Costa, D. Summa, V. Bertolasi, B. Semeraro, P. Pedrini, R. Buzzi, S. Vertuani. Biotransformation of cortisone with *Rhodococcusrhodnii*: Synthesis of new steroids. Molecules. 26(5) (2021)1352. https://doi.org/10.3390/molecules26051352; (c) Z. Liu, R. Zhang, W. Zhang, Y. Xu. Structure-based rational design of hydroxysteroid dehydrogenases for improving and diversifying steroid synthesis. Crit. Rev. Biotechnol. 43(5) (2023) 770–786. https://doi.org/10.1080/07388551.2022.2054770

5 The Chemical Synthesis of Oligosaccharides in the Enterobacteriaceae Family

Mana Mohan Mukherjee, Nabamita Basu, Susmita Poddar, and Rina Ghosh

5.1 INTRODUCTION

Throughout human history, we have been engaged in a constant struggle for survival against the most treacherous of enemies: microorganisms, in particular the mighty bacteria. Bacteria are ancient single-celled organisms that are widely distributed in the air, soil, and water. They can form symbiotic relationships with other organisms, and their infections are evidence of long-lasting conflict between human culture and microorganisms ever since there was written record [1].

These pathogenic bacteria (bacteria that can cause human diseases are called pathogenic bacteria) show toxigenesis by producing toxins, generally categorized into two classes: exotoxins and endotoxins [2]. Exotoxin-secreting bacteria (Gram-positive or Gram-negative) release polypeptides or proteins into the surrounding medium ultimately poisoning the host cell [3]. Bacterial endotoxins are lipopolysaccharides that constitutes a major component of outer membrane of Gram-negative bacteria and release toxin when they are lysed or phagocytized by host cells and are then transported through blood and lymph; this can cause meningococcemia or sepsis [4].

Gram-negative bacteria are coated with impervious and fortified outer membrane (OM) that allows them to resist stress like antibiotics, other bacteria, and environmental factors. Lipopolysaccharide (LPS) is the major component for the outer leaflet of the Enterobacteriaceae OM. LPS can be subdivided into three regions: i) lipid A, the hydrophobic membrane anchor; ii) an oligosaccharide core (core OS); and iii) a repetitive polymer of glycosyl units known as O polysaccharide (O-PS) (Figure 5.1) [5].

Diverse environmental conditions on Earth (e.g., heat, pH, salinity, pressure, and osmotic activity) immensely affect the function of the cell, necessitating adaptation through structural modification. Beyond its role as protective shield against chemical and environmental stresses, the OM plays an acute role in controlling the admission of molecules such as nutrients, water, ions, and small hydrophilic antibiotics into the cell. The Gram-negative OM is coated with highly variable polymer molecules that can cause immune activation, known as antigens. Bacteria are divided into serotypes based on different antigen combinations. The three major types of antigens present on the cell surface are O (somatic), K (capsular), and H (flagellar) [6]. These antigens play roles in motility (H-antigen), interaction with the environment (K-antigens and O-antigen), protection from an antagonistic environment (K-antigens and O-antigen) and increasing the ability of the OM to provide structural support to the cell (O-antigen) [7].

K-antigens are the capsule coated on the surface of bacteria outside the cell envelope and the H-antigen is a protein antigen based on the flagellar structure [8]. On the other hand, the O-antigen is serotype specific and consists of a highly variable chain of high-molecular-weight polysaccharides with some exceptions (e.g., K-88 and K-99 of E. coli, which are protein antigens) [9]. Enterobacterales is a bacterial order that is defined in part by the presence of a carbohydrate antigen

DOI: 10.1201/9781003437413-5

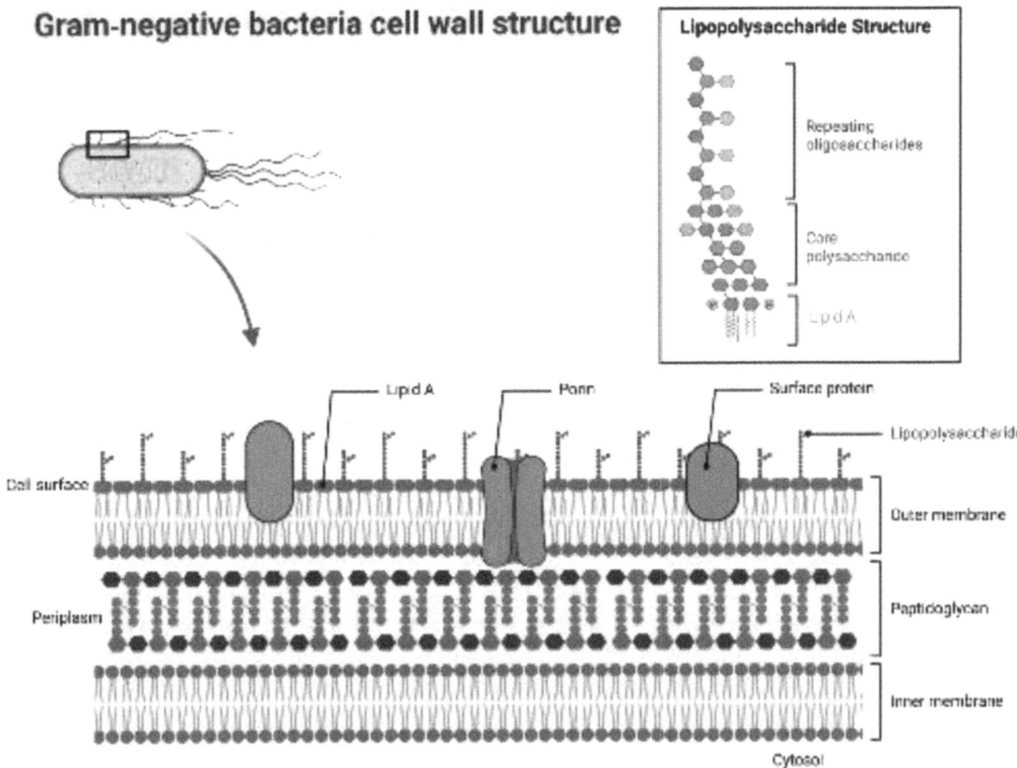

FIGURE 5.1 Cartoon picture of Gram-negative bacteria (created in Biorender.com).

in the outer leaflet of the OM and in the periplasm known as enterobacterial common antigen (ECA) [10]. Although Enterobacterales express various antigens (e.g., K, O, and H), ECA is unique in that it is restricted to one order and it is invariant, allowing cross-reactivity among the members of Enterobacterales [11].

It has been a dogma that mutants that are not capable to synthesize a minimal core structure are not viable, thus, the oligosaccharide core and lipid A represent the common structural unit occurring in all LPS and important for viability and membrane function of Gram-negative bacteria. All enterobacteria possess LPS in the outer membranes of their cell walls. Apart from representing components highly important for the bacteria (e.g., protective and permeability barrier functions, interacting with the environment), LPS are potent virulence factors in pathogenic strains. Thus, the structures and functions of many enterobacterial lipopolysaccharides have been intensively investigated. Thus, the solution to this worldwide health problem has had little progress due to the lack of efficient diagnostics and increasing antibiotic resistance of certain bacteria [12].

Significant studies have been achieved over the past centuries leading to emerging field of modern vaccinology and most importantly the discovery of the 'magic bullets' named antibiotics. Despite these advances, bacterial infections remain a leading cause of death and sickness, particularly in the developing world. The main reason behind this is that the bacteria is conceivably becoming antimicrobial resistant (AMR) and multidrug resistant. These AMR bacteria can be gained through processes like i) restricting the entry of the antimicrobial agent by changing the entryway, ii) forming an antibiotic pump at the bacterial cell wall to remove the antibiotic drug, iii) secreting enzymes that can destroy the antibiotic drug, iv) changing the antibiotic target part so that the drug can no longer fit into the site, or v) developing new cell processes that avoid using the antibiotic's target.

The introduction of penicillin in the 1940s allowed for the treatment of often fatal infections such as sepsis, cellulitis, pneumonia, and endocarditis caused by organisms such as streptococcal species or *S. aureus*. *S. pneumoniae* was first noted to develop resistance to penicillin, by producing β-lactamases that render penicillin inactive, in South Africa in 1977. Among the WHO list of priority pathogens [13], defined as the antibiotic-resistant pathogens posing the greatest threat to human health and for which the development of new antibiotics is urgently needed, are the Gram-negative Enterobacteriaceae, organisms such as *Escherichia coli* (EC), *Klebsiella pneumoniae* (KP), *Salmonella, Shigella, Enterobacter spp., Providencia spp.* And other species. Enterobacteriaceae cause a wide range of infections, including pneumonia, urinary tract infection (UTI), bacteremia, and surgical site infection. Even the clinically ubiquitous Gram-negative EC, an organism associated with routine outpatient UTI and bacteremia, has developed resistance to multiple classes of antibiotics, including most oral agents used for common UTI treatment. It has been estimated that if the AMR trend continues, the cumulative loss to world economies might be as high as $100 trillion by 2050, and 10 million lives a year may be lost to these AMR bacteria, exceeding the 8.2 million lives a year currently lost to cancer. To put this number in perspective, currently, at least 700,000 people die of resistant infections every year globally, more than the combined number of deaths caused by tetanus, cholera, and measles [14]. Perhaps more perplexing is that the most vulnerable portion of our population to the high bacterial morbidity and mortality is children aged less than five years. Researchers estimated that nearly 4.95 million (3.62–6.57) deaths occurring in 2019 worldwide can be connected with infections by AMR bacterial strains. This suggests prioritizing preventive measures toward this end [15]. It is also worthy of mentioning here that unlike drug resistance, vaccine resistance doesn't progress [16]. Development of antibacterial vaccines is thus considered one of the important measures that could prevent or even eradicate the corresponding bacterial diseases [17].

Surface-located bacterial carbohydrates are promising candidates to be explored as vaccines to prevent bacterial infections [18]. Carbohydrate antigens were recognized early in the twentieth century as potential vaccine determinants when capsular polysaccharide isolated from *Streptococcus pneumoniae* were shown to be immunogenic in both rodents and humans [18c] Although these capsular polysaccharides (CPS) can be harvested by isolation from bacteria, it is nearly impossible to achieve the required homogeneity for licensed vaccine [19]. Additional difficulty arises when other polysaccharides from the cell surface mix with the desired CPS in the isolation process. The role of polysaccharide impurities in the immune system is still unclear, but they lead to decreased immunogenicity. Multiple conjugating sites present in the isolated polysaccharides lead to a variety of undesired conjugations [20].

With recent determinations of structural motifs of bacterial polysaccharides, chemically synthesized polysaccharides provide an alternative way to develop structurally well-defined polysaccharide-based vaccines [21]. Moreover, pure chemical entities can enable the investigation of structure-activity relationships between a polysaccharide's structural motif and antibody immune responses [22]. The immunogenicity of bacterial polysaccharides is well studied, and their conjugation to carrier proteins has been widely used in industry to produce potent polysaccharide-based vaccines [23].

Keeping all these in mind, synthetic carbohydrate chemists are relentlessly trying to synthesize polysaccharides related to the LPS of the enterotoxigenic bacteria. But the chemical synthesis of this complex oligosaccharide core has long been a challenge for carbohydrate chemists because of its prolong reaction sequences and difficulty. In addition to that, successful oligosaccharide synthesis requires significant knowledge on protecting group manipulation [24], stereo controlled glycosylation reaction [25], and late-stage modifications [26] where there is every possibility of failure even at the very final stage of synthesis.

Despite all the bitterness, there has been a significant development in the total synthesis of oligosaccharides/polysaccharides (OSs/PSs) [27], and there is at least hope for the ultimate development of a 'sweet vaccine' based on carbohydrates. In this chapter, we review the chemical synthesis of the OSs/PSs, present in enterotoxigenic bacteria along with those of AMR strains of the Enterobacteriaceae family.

5.1.1 Enterobacteriaceae

The family Enterobacteriaceae belongs to the *Gammaproteobacteria* (order: *Enterobacteriales*) and comprises a quite high number of species that are Gram-negative and facultative aerobe and oxidase negative. Several species include serovars or strains that are pathogenic for humans (responsible for 1.5 million death per year), animals, or plants, like *Salmonella enterica*, EC, *Shigella dysenteriae* (generally pathogenic in humans), *Yersinia pestis* (pathogenic in animal and human), or *Erwinia carotovora* (in plant). Other species may be opportunistic/nosocomial human pathogens like KP and *Serratia marcescens* [28].

A polysaccharide composed of repeating unit (RU) α-D-Fucp4NAc-(1→4)-β-D-ManpNAcA-(1→4)-α-D-GlcpNAc [29], known as enterobacterial common antigen (ECA), is present in all members of Enterobacteriaceae. Boons' research group synthesized a tri- and a hexa-saccharide related to ECA toward uncovering the corresponding immunodominant moiety. They synthesized the target compounds (Scheme 5.1) by overcoming several hurdles and then carefully planning suitable orthogonal glycan donors and acceptors, along with introducing carboxylic acid in the appropriate step and choosing suitable differential protecting groups at different stages and also a suitable

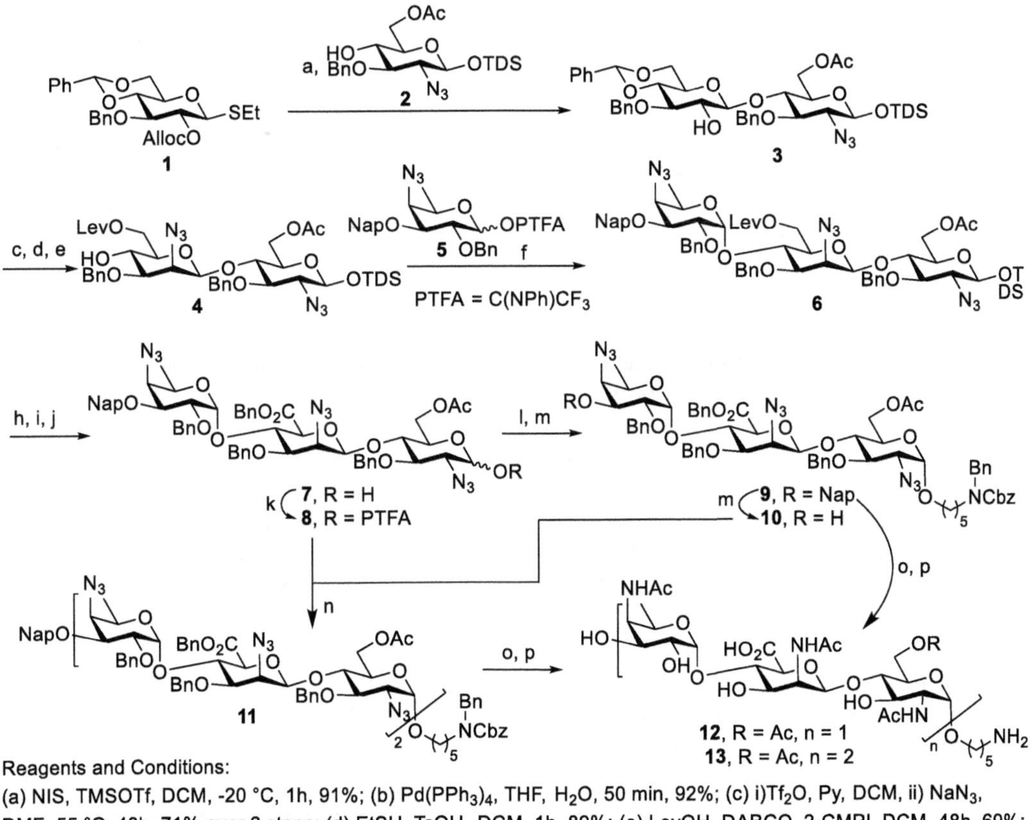

Reagents and Conditions:
(a) NIS, TMSOTf, DCM, -20 °C, 1h, 91%; (b) $Pd(PPh_3)_4$, THF, H_2O, 50 min, 92%; (c) i)Tf_2O, Py, DCM, ii) NaN_3, DMF, 55 °C, 48h, 71% over 2 steps; (d) EtSH, TsOH, DCM, 1h, 89%; (e) LevOH, DABCO, 2-CMPI, DCM, 48h, 69%; (f) TfOH, Et_2O, -20 °C, 30 min, α:β = 9:1, 82%; (g) $N_2H_4.H_2O$, Py, AcOH, 2h, (h) TEMPO, BAIB, DCM, H_2O, 18 h; (i) BnBr, Cs_2CO_3, DMF, 3 h; (j) HF/Py, THF, 48 h, 88% over 4 steps; (k) DBU, $CF_3C(NPh)Cl$, DCM, 1 h, 90%; (l) $HO(CH_2)_5NBnCbZ$, TfOH, dioxane, Toluene, rt, 45 min, α:β = 3:1, 62%; (m) DDQ, DCM, 100 mM pH 7.4 PBS buffer, rt, 4 h, 77%; (n) TfOH, Et_2O, -20 °C to 0 °C, 43%; (o) $SnCl_2$, PhSH, Et_3N, MeCN, THF, Ac_2O, MeOH, 71%; (p) H_2, $Pd(OH)_2$, ^tBuOH, H_2O, AcOH, 85% for **12**, 71% for **13**.

SCHEME 5.1 Synthesis of trisaccharide **12** and hexasaccharide **13** related to ECA.

reagent for crucial reduction of azido moiety. The glycosylation reaction of glycosyl acceptor **2** with thioglycoside donor **1** using NIS-TMSOTf activator followed by sequential i) deprotection of allyloxycarbonyl (Alloc) in excellent yield using $Pd(PPh_3)_4$ in aqueous THF, ii) OH triflylation and S_N2 inversion at C-2' using NaN_3 (combined yield 71%), iii) EtSH-TfOH-mediated 4,6-O-benzylidene group deprotection (89% yield), and then iv) regioselective levulinoyl (Lev) ester protection (69% yield) furnished the 4′-hydroxy group free disaccharide acceptor **4**. This on further glycosylation with rare-sugar *N*-phenyltrifluoromethylacetimidate (PTFA) donor **5** using TfOH activator in Et_2O solvent generated trisaccharide **6** in 82% yield and high anomeric selectivity (α/β = 9:1). Lev ester group deprotection with hydrazine hydrate followed by TEMPO-BAIB mediated oxidation of the resulting primary OH group and subsequent carboxylic acid-benzylation utilizing BnBr, Cs_2CO_3 and then anomeric TDS-deprotection with HF/pyridine afforded trisaccharide **7** at 88% yield over the last four steps. The glycosyl hemiacetal **7** was then converted to its corresponding PTFA donor **8** at 90% yield with DBU and *N*-phenyltrifluoroacetimidyl chloride. A TfOH-promoted glycosylation reaction of donor **8** with 5-*N*-benzyl-*N*′-carboxybenzylpentanol in a mixture of toluene and dioxane gave the corresponding spacer linked trisaccharide **9** at 62% yield (α/β = 3:1), and this on DDQ-mediated naphthylmethyl (Nap) group deprotection in alkaline buffer afforded the trisaccharide acceptor **10** at 77% yield. After several trials, the final glycosylation of acceptor **10** with donor **8** furnished the corresponding hexasaccharide derivative **11** at 43% yield. Careful reduction of azide to its corresponding amine and its simultaneous conversion to NHAc using $SnCl_2$/PhSH/Et_3N/Ac_2O/MeOH followed by global deprotection by $H_2/Pd(OH)_2$ of **11** and **9** generated the corresponding target trisaccharide **12** and hexasaccharide **13** as their 5-aminopentyl glycoside at 85% and 71% yields, respectively. Both of these compounds were finally converted to their corresponding BSA glycoconjugates and also deacetylated BSA glycoconjugates. Immunochemical experiments were then carried out to unveil the immunodominant part of ECA. Moreover, a developed monoclonal antibody that bound to ECA also recognized a wide number of strains belonging to the Enterobacteriaceae family [30].

5.1.2 *Klebsiella pneumonia*

KP is a Gram-negative encapsulated bacterium causing infections like pneumonia, liver abscess, UTI, and bacteremia, particularly in newborn and immune-compromised patients [31]. Due to the emergence of strains with hypervirulence and multidrug resistance, treatment of pneumonia is becoming highly challenging [32], and incidents of outbreaks with mortality and morbidity are increasing [33]. Both CPS and the O-antigenic part of the cell wall LPS are known as virulence factors [34].

Researchers recently gave a comprehensive report on the diversity and distribution of O- and K-antigens with a collection of blood from invasive KP-infected patients from 13 different countries. They established that among 645 KP isolates, there were four major O-antigens accounting for 90.1% of strains; among 519 strains, 62.1% were multidrug resistant, and there were 49 varieties of K-antigens, with the most common type being K2. The limited number of O-antigen types points logically toward a preferred consideration of O-antigens for KP vaccine development [35]. Several immune-prophylactic and immune-therapeutic approaches have been exploited in vaccine development against KP [36] including broad-spectrum multivalent glycoconjugate-vaccine candidates via the conjugation of the most common O-antigens (O1, O2, O3, and O5) of KP at the reducing end, forming 2',5'-anhydromannose aldehyde, to the lysine residues of purified proteins of *Pseudomonas aeruginosa* flagellins (FlaA and FlaB) [37], but a valid licensed vaccine for KP is yet to be developed and marketed.

Type O1 is the most predominant serotype of KP [38]. LPS of KP type O1 (KP-O1) contains two O-antigens, inner Galactan I [→3)-β-Galf-(1→3)-α-Galp-(1→]$_n$ linked to core lipid A and outer Galactan II [→3)-β-Galp-(1→3)-α-Galp-(1→]$_n$ [39] Zhu and Yang investigated and published the synthesis of a tetra- and a hexasaccharide corresponding to the O-antigenic Galactan I of the cell-wall polysaccharide of KP of serotype O1 (KP-O1), bearing RU [→3)-β-Galf-(1→3)-α-Galp-(1→] [40].

Antigenic components of KP of types O1 and O2 correspond to Galactan I-III, which are the most prevalent types of carbapenem-resistant strains and are responsible for 50%–68% of all KP infections. Galactan I is located in the inner part of the O-chains. Nifantiev et al. synthesized Galactan I and II-related oligosaccharides using a novel technique involving pyranoside into furanoside rearrangement [41]. They applied their novel process to synthesize a disaccharide [41a] and oligosaccharides (Scheme 5.2) of Galactan I of KP [42a]; the oligosaccharides corresponding to the Galactan II of KP were synthesized as 3-aminopropyl glycosides.

Researchers adopted [2+2+2] iterative convergent glycosylation to synthesize the oligosaccharides corresponding to the core Galactan I [42a]. They obtained β-allyl galactofuranoside-based acceptor **15** from galactopyranose derivative **14** in three steps using a pyranose-furanose rearrangement in the first step [41b]. The researchers strategically applied temporary *p*-methoxybenzyl (PMB) protection to the C-3-O position of the galactopyranosyl PTFA donor **16**. This strategic protection facilitated the glycosylation of **15**, followed by consecutive steps to produce the next disaccharide donor **18**, featuring a 3′-OPMB as a temporary protection. Glycosylation with **18** of acceptor 3-trifluoroacetamidopropanol followed by de-*p*-methoxybenzylation of **19a** yielded the next disaccharide acceptor **19b**. Their process took three successive steps: i) glycosylation with common donor **18**, ii) de-*p*-methoxybenzylation, and iii) similar glycosylation that ultimately furnished hexasaccharide derivative **21** in addition to a intermediate tetrasaccharide derivative **20** (Scheme 5.2).

The targeted oligosaccharides corresponding to Galactan II was also achieved based on [2+2+2] convergent glycosylation. For these, the starting disaccharide donor **25** was planned and synthesized with a 65% yield based on an orthogonal technique in ether, with the aim of controlling [1,2]-*cis* anomeric selectivity. This process involved the corresponding monomer armed-donor **23** having non-participating protection on C-2-O as well as disarmed-donor **24**, and both **23** and **24** had PMB temporary protection on C-3. Donor **25** was converted to acceptor **26b** in two steps, and then its glycosylation with donor **25** generated tetrasaccharide derivative **27**. Removal of PMB followed by iterative glycosylation with disaccharide trichloroacetimidate donor **29** using NIS-TMSOTf as activator yielded the desired hexasaccharide derivative **30** at 85%. In the last two glycosylation steps, the β-anomeric selectivity was based on NGP [43] (Scheme 5.2).

Global deprotection of glycan derivatives (**19a**, **20**, and **21** related to Galactan I and **26a**, **27**, and **30** related to Galactan II) finally furnished the respective oligosaccharides **22a-22c** and **31a-31c** related to Galactan I and Galactan II of KP. The research group then converted the Galactan I- and Galactan II-related spacer-armed free oligosaccharides to the corresponding biotinylated derivatives to select antibodies for sera of anti-KP O1. These experiments indicated their potential applicability (the smaller ones for detection kit or the larger ones for development of vaccine against KP type O1 infections) [44].

A common tetrasaccharide related to the O-antigen of KP-O3, EC-O9 and *Hafnia alvei* PCM 1223 was synthesized by Ghosh *et al.* utilizing reactivity based [1+1+2] one-pot glycosylation techniques (Scheme 5.3) [45]. Disaccharide building block **36** was obtained by *N*-(*p*-methylphenylthio)-ε-caprolactam (NMPTC)-TMSOTf-mediated glycosylation of D-Man-based acceptor **33** with the thiomannoside donor **32** followed by Zemplén deacetylation. Then, iterative, NMPTC-TMSOTfp-mediated, three-component, one-pot glycosylation utilizing monosaccharides **32** and **35** and disaccharide **36** generated the desired tetrasaccharide derivative **37** at 54% yield. On global deprotection, this yielded the KP-O3 tetrasaccharide **38** as methyl glycoside (Scheme 5.3).

Bacterial isolate with sequence type 258 expressing carbapenemase enzyme makes this strain carbapenem resistant and responsible for the spread of this β-lactam-resistant strain globally [46]. Seeberger et al. synthesized the branched repeating hexasaccharide unit as its 5-aminopentyl glycoside (**52**) related to carbapenem-resistant K-antigen of type 258 KP following a [3+3]-convergent glycosylation strategy (Scheme 5.4) [47]. *De novo* synthesis from **39** and **40** of the differentially protected D-GalpA-based thioglycoside building block **41** followed by its orthogonal glycosylation with α-L-Rha-based TCA-donor **42** furnished disaccharide derivative **43**, which was gradually converted to trisaccharide glycosyl donor **47** via the generation of chain-extended trisaccharide thioglycoside **46**.

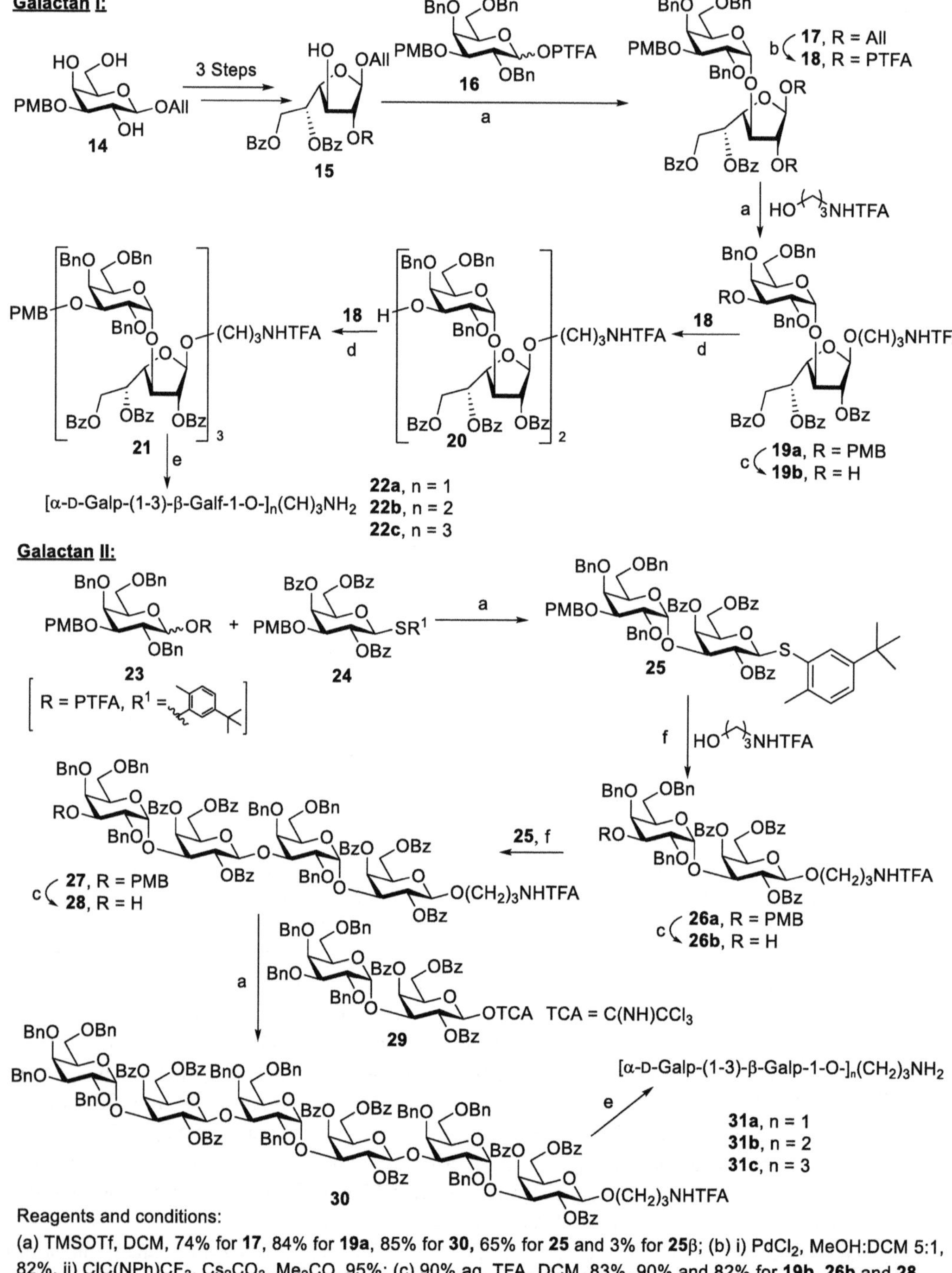

Reagents and conditions:

(a) TMSOTf, DCM, 74% for **17**, 84% for **19a**, 85% for **30**, 65% for **25** and 3% for **25β**; (b) i) $PdCl_2$, MeOH:DCM 5:1, 82%, ii) $ClC(NPh)CF_3$, Cs_2CO_3, Me_2CO, 95%; (c) 90% aq. TFA, DCM, 83%, 90% and 82% for **19b**, **26b** and **28**, respectively; (d) TMSOTf, DCM, -50 °C to -15 °C, 76% for **20** and 81% for **21**; (e) i) H_2, $Pd(OH)_2$-C (20%), EtOAc:MeOH 1:1, ii) NaOMe, MeOH, H_2O, 71% for **Galactan I** and 70% for **Galactan II**; (f) NIS, TfOH, Et_2O 85% for **26a** and **90%** for **27**.

SCHEME 5.2 Convergent synthesis of oligosaccharides related to Galactan I and Galactan II of KP.

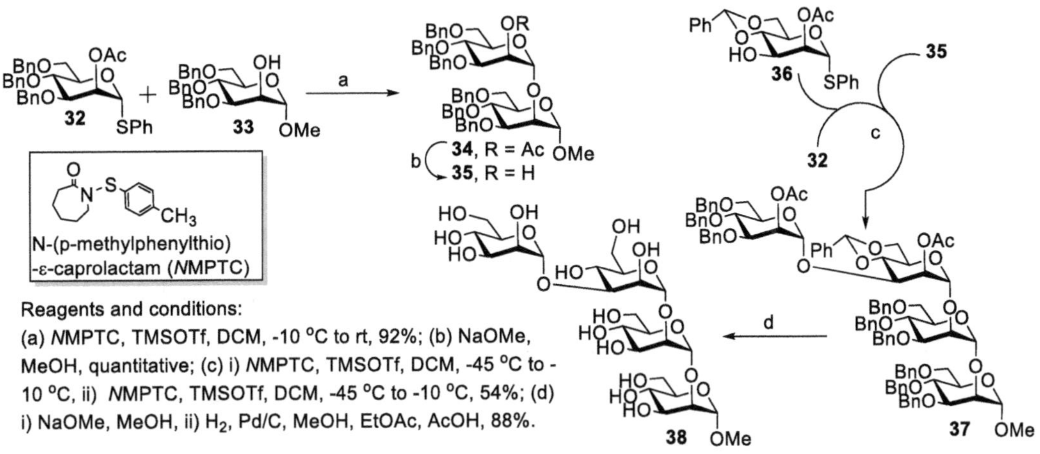

N-(p-methylphenylthio)-ε-caprolactam (*N*MPTC)

Reagents and conditions:
(a) *N*MPTC, TMSOTf, DCM, -10 °C to rt, 92%; (b) NaOMe, MeOH, quantitative; (c) i) *N*MPTC, TMSOTf, DCM, -45 °C to -10 °C, ii) *N*MPTC, TMSOTf, DCM, -45 °C to -10 °C, 54%; (d) i) NaOMe, MeOH, ii) H_2, Pd/C, MeOH, EtOAc, AcOH, 88%.

SCHEME 5.3 Synthesis of tetrasaccharide **38** related to LPS of EC-O9, KP-O3, and *Hafnia alvei* PCM 1223.

On the other hand, the other trisaccharide-derived glycosyl acceptor **50** was obtained employing two successive sequential glycosylation-deprotection steps starting with monomer donor **42**. Union of **50** and **47** in the presence of TBSOTf activator gave the α-anomer of hexasaccharide derivative **51** at high yield. Finally, global deprotection steps afforded the desired hexasaccharide **52** (Scheme 5.4). Immunization based on semisynthetic **52**-CRM197 glycoconjugate elicited cross-reactive antibodies with high titer value against CPS of a carbapenem-resistant, highly contagious KP strain in mice and rabbits; moreover, the antibodies supported phagocytosis. All these indicated the semi-synthetic glycoconjugate to be a potential vaccine candidate.

Among a number of K-antigens, only a few are responsible for pneumonia [48]. Some researchers choose K-antigens, particularly those of drug-resistant ones for development of vaccine candidates. Mukhopadhyay's research group synthesized a hexasaccharide RU of the K-antigen corresponding to carbapenem-resistant strains KP 2796 and 3264 [49]. A sequential stepwise stitching of monomeric units **53**, **54**, and **61** that formed the higher glycosides **55**, **57**, **59**, and **62** and the subsequent removal of chloroacetic functionalities resulted in glycoside acceptors **56**, **58**, **60**, and **63**, respectively (required for the next-step glycosylation reactions). Final coupling of thiorhamnoside donor **64** and pentasaccharide acceptor **63** yielded hexasaccharide **65** at 84%. Then, PMB deprotection and the regioselective oxidation of primary hydroxy group followed by global deprotection furnished the targeted hexasaccharide **66** at 64% yield (Scheme 5.5).

Very recently, researchers activated thioglycoside donors with NIS-$HClO_4$-SiO_2 toward the synthesis of disaccharide donor **78** and tetrasaccharide acceptor **75** using orthogonal glycosylation and then [4+2] convergent glycosylation; then they introduced the carboxylic acid group at the late stage of the total synthesis. Shit et al. achieved the synthesis of the target hexasaccharide **80** related to *Klebsiella* serotype K-34 [50]. The glycosylation of L-rhamno-1-thiopyranoside donor **67** with *O*-PMP-glucopyranoside acceptor **68** using the aforesaid promoter system yielded 78% of corresponding disaccharide **69**. Selective removal of O-acetyl protection quantitatively produced the corresponding disaccharide acceptor **70**, which was then coupled with 3-OPMB protected L-rhamno-1-thiopyranoside donor **71** to yield corresponding trisaccharide **72** at 73%. Functional group alteration of this trisaccharide **72** generated 3″-OH group free trisaccharide acceptor **73**, which was then glycosylated with the first L-rhamno-1-thiopyranoside donor **67** to yield tetrasaccharide **74** at 74%. Separately, glycosylation of L-rhamnopyranosyl TCA donor **76** with D-galactopyrano thioglycoside acceptor **77** yielded corresponding disaccharide **78** at 71%. Coupling of disaccharide donor **78** with tetrasaccharide acceptor **75** in DCM:Et_2O (3:1) solvent generated hexasaccharide **79** at 65% yield.

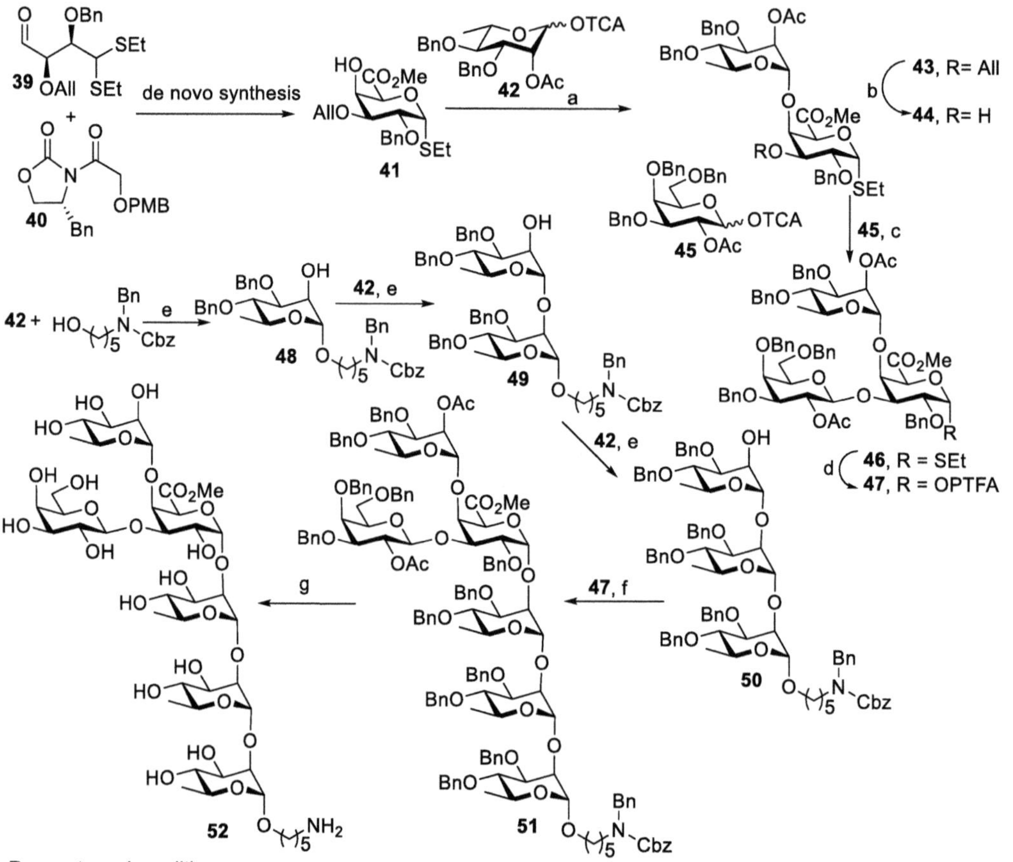

Reagents and conditions:
(a) TBSOTf, DCM, 0 °C, 98%; (b) i) Ir[COD(PCH_3Ph_2)]PF_6, H_2, THF, rt, ii) pTsOH, MeOH, 68%; (c) TBSOTf, DCM, -30 °C, 82%; (d) (i) TCCA, Me_2CO-H_2O, 92% ii) PhN=C(Cl)CF_3, K_2CO_3, rt, 94%; (e) i) TBSOTf, DCM, 0 °C, ii) NaOMe, DCM, MeOH, rt, 53% over 2 steps for **48**, 86% for **49** and 87% for **50**; (f) TBSOTf, Tol:Et_2O 10:1, -70 °C to rt [α 81%, β 17%]; (g) i) H_2, 10% Pd/C, MeOH, H_2O-AcOH, rt, 77%, ii) LiOH, 30% H_2O_2, MeOH, H_2O, 0 °C to rt, 99%.

SCHEME 5.4 Synthesis of acidic hexasaccharide **52** related to carbapenem-resistant KP antigen.

Final deprotection and oxidation of primary hydroxy group produced the desired hexasaccharide acid **80** in 43% overall yield (Scheme 5.6).

The structure of the RU of the CPS isolated from primary pyrogenic liver abscess of patients caused by KP infection, determined by Yang et al. is [→3)-β-D-Glc-(1→4)-[2,3-(*S*)-pyruvate]-β-D-GlcA-(1→4)-α-L-Fuc-(1→] having either C-2 or C-3 O of L-Fuc unit acetylated and an unusual 2,3-*O*-pyruvylation on D-GlcA [51]. It was also reported that the CPS could induce macrophage activation through Toll-like receptor 4. With a view to check if the repeating trisaccharide is having immunomodulatory properties, later authors synthesized the same as an attempt to isolate the trisaccharide by partial hydrolysis, causing loss of acetyl and pyruvyl protections [52].

Apart from CPS and LPS, extracellular biofilms secreted by KP toward their defense from external agents can also cause pathogenicity of the particular strain [53]. *Klebsiella* strain Ts113 isolated from a UTI produces biofilms in cellulose medium, and the structure of the biofilm-PS is similar to that of *Klebsiella* K24 [54]. Recently, Misra's research group synthesized the pentasaccharide RU

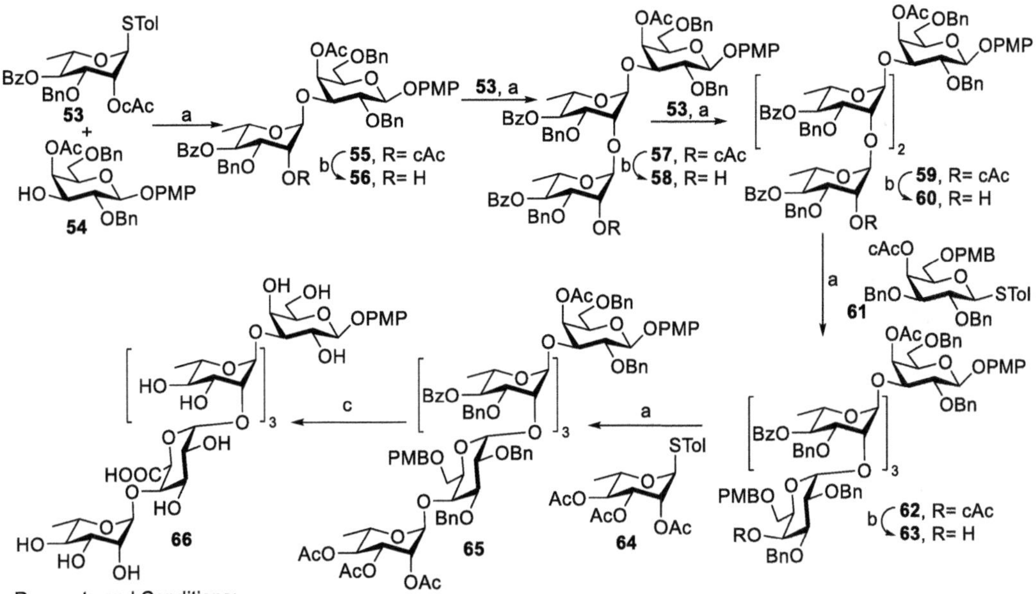

Reagents and Conditions:
(a) NIS, HSO_4-SiO_2, 0 °C, 91% for **55**, 90% for **57**, 89% for **59** and 84% for **65**, respectively, -50 °C, 82% for **62**; (b) Thiourea, Colidine, DCM, MeOH, 87% for **56**, 88% for **58**, 90% for **60** and 85% for **63**, respectively; (c) i) DDQ, DCM, H_2O, 78%, ii) TEMPO, BAIB, iii) H_2, Pd-C, iv) NaOMe, MeOH, 64%.

SCHEME 5.5 Synthesis of the hexasaccharide related to the repeating unit of the capsular polysaccharide from carbapenem-resistant KP-2796 and -3264.

of the acidic polysaccharide of the mentioned biofilm utilizing convergent synthesis and orthogonal glycosylation with excellent stereochemical outcome from the glycosylation steps [55].

Stereoselective glycosylation reaction of D-mannopyrano thioglycoside donor **81** with D-glucopyranoside acceptor **82** using NIS-TMSOTf promoter produced the corresponding 1→3 linked disaccharide **83** at 80% yield, which on treatment with sodium methoxide in methanol produced the disaccharide acceptor **84** in 98% yield. BSP, TTBP-mediated preactivation-based glycosylation reaction of 4,6-O-benzylidene-derived thiomannopyranoside donor **85** with 1-*O*-allyl glycopyranoside acceptor **86** yielded β-linked manno (1→3) glucopyrano disaccharide **87** at 67%. *O*-Allyl disaccharide **87** was then converted to its corresponding TCA donor **88**, which was then coupled with 3-hydroxy free thiomannopyranoside acceptor **89** to produce trisaccharide **90** at satisfactory yield (45%). Further glycosylation of trisaccharide donor **90** with disaccharide acceptor **84** generated pentasaccharide **91** at moderate yield (41%); then global deprotection and oxidation of primary hydroxy group produced the fully deprotected acidic pentasaccharide **92** in 61% overall yield (Scheme 5.7).

5.1.3 *Salmonella*

Salmonella serovars are a group of Gram-negative Enterobacteriacea that include *Salmonella* Typhi (ST) and *Salmonella* Paratyphi (SP-A) as more common; these cause enteric fever, typhoid, and paratyphoid in humans. Vaccines for ST have been in use for decades including the formulated purified capsular polysaccharide Vi-antigen; recently, a conjugate of Vi and CRM197 was established as prominently immunogenic in an animal model without the use of adjuvant [56], but there are no licensed vaccines for SP-A or B; some SP-A vaccines are in development and under phase trials. A compiled update was published on the status of the development on SP vaccine [57]. The capsular

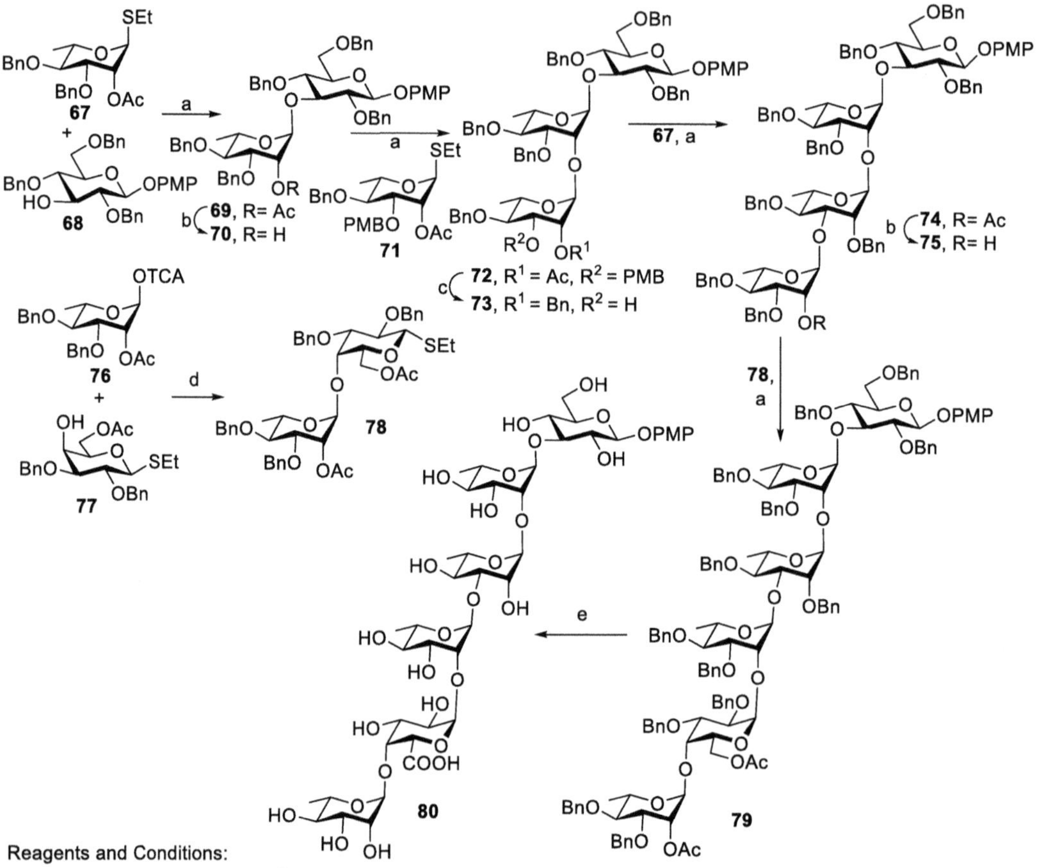

Reagents and Conditions:
(a) NIS, $HClO_4$-SiO_2, DCM, -20 °C, 1 h, 78% for **69**, 73% for **72** and 74% for **74**, respectively and DCM:Et_2O 3:1, 65% for **79**; (b) NaOMe, MeOH, 3 h, 96% for **70** and 93% for **75**; (c) i) BnBr, NaOH, TBAB, DMF, rt, 2 h, ii) DDQ, DCM:H_2O 5:1, rt, 2 h, 72%; (d) $HClO_4$-SiO_2, DCM, 0 °C, 1 h, 71%; (e) i) NaOMe, MeOH, rt, 3 h, ii) BAIB, TEMPO, DCM:H_2O 2:1, rt, 2 h, iii) H_2, $Pd(OH)_2$/C, MeOH, rt, 24 h, 43%.

SCHEME 5.6 Synthesis of hexasaccharide repeating-unit **80** of the capsular polysaccharide of *Klebsiella* serotype K-34.

polysaccharide of ST, a Vi-antigen, has an RU consisting of α-(1→4)-linked D-GalpNAc, primarily acetylated at C3-O, and is a major virulent factor of ST [58].

Using *N*-acetyl-2,3-oxazolidinone protected building blocks and a preactivation protocol, Ye et al. achieved a short synthesis of oligosaccharide fragments related to ST Vi having α-(1→4)-linked *N*-acetyl-D-galacturonic acid of different chain lengths (I–V, inset, Scheme 5.8) [59]. ELISA assays using di-, tri-, and tetrasaccharides established the latter two as having higher antigenic activities than the native polysaccharides; moreover, the presence of O-acetyl group on the oligosaccharides increased the antigenic activities. The authors further synthesized the oligosaccharide mimics tetra-, hexa-, and octa-pseudosaccharides corresponding to Vi-PS of ST following preactivation. Scheme 5.8 displays the synthesis of tetrasaccharide derivative **99** through the successive use of 2,3-*N*,*O*-oxazolidinone-protected thiogalactopyranoside donor **93**; pTolSCl/AgOTf as thiophilic activator; and allyl glycoside acceptor building blocks **93, 96,** and **98**. Ruthenium-based olefin self-cross-metathesis of the allyl glycosides generated the corresponding carbon-carbon coupled glycosides **100, 101,** and **102** at excellent yields under properly optimized condition. The glycomimics were

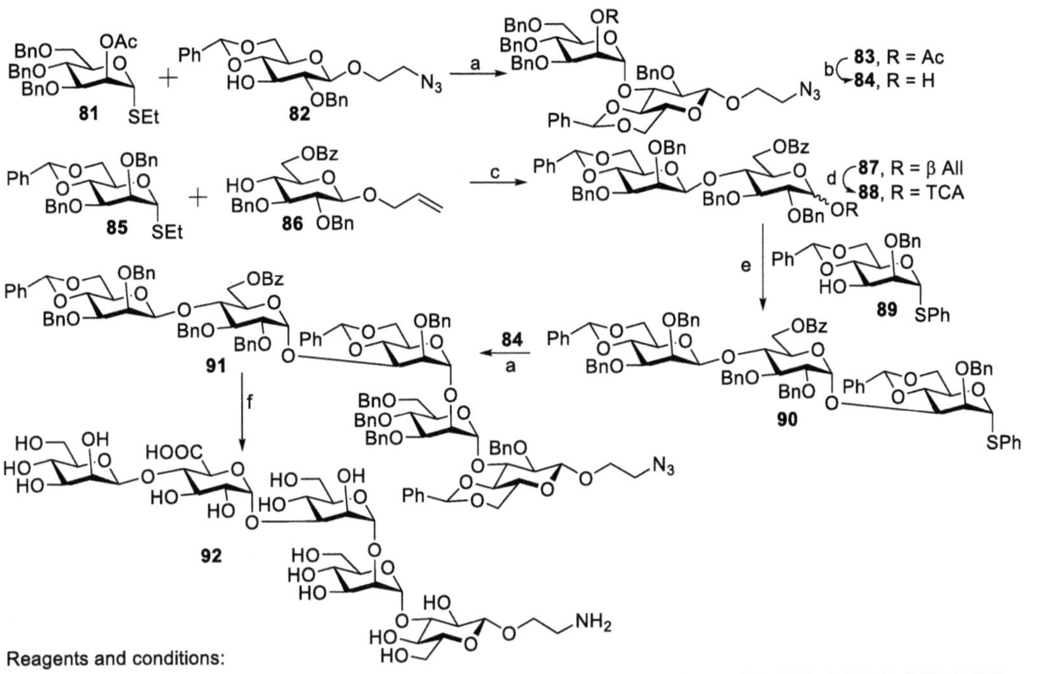

Reagents and conditions:
(a) NIS, TMSOTf, -10 °C, 30 min, 80% for **83** and 40% for **91**; (b) NaOMe, MeOH, rt, 2 h, 98%; (c) Tf_2O, BSP, TTBP, DCM, -60 °C, 1 h then compound **86**, -78 °C, 2h, 65%; (d) i) $PdCl_2$, MeOH, 0 °C to rt, 3 h, 75%, ii) CCl_3CN, DBU, DCM, -10 °C, 90%; (e) TMSOTf, DCM, -10 °C 30 min, 45%; (f) i) MeOH, NaOMe, rt, 4 h, ii) TEMPO, NaOCl, NaBr, TBAB, $NaHCO_3$, 5 °C, 3 h then $NaClO_2$, NaH_2PO_4, ^tBuOH, 2-methyl but-2-ene, rt, 3 h, iii) $Pd(OH)_2/C$, H_2, MeOH, rt, 24 h, 61%.

SCHEME 5.7 Synthesis of pentasaccharide repeating unit **92** of the biofilms produced by KP.

prepared using carbon chain (glycosides **103, 104,** and **105**) or triazole (the latter based on CuAAC click chemistry) as the linker between glycoside **V**. They also investigated the binding affinities of all these to anti-Vi-antibodies [60].

Ye's research group further employed their *N*-acetyl-2,3-*N,O*-oxazolidin-based preactivation technique for obtaining higher homologous oligosaccharides like penta-, hepta- and octa-saccharide fragments. Their comparative ELISA experiments toward vaccine development established that the hexasaccharide is the minimum epitope for Vi-antigen [61].

Lay et al. also synthesized α-selective neutral di- and trisaccharide and zwitterionic analogues to Ye et al.'s structure followed by evaluating their antigenic properties [62] in competitive ELISA experiments; all these showed specificity to anti-Vi polyclonal antibodies, and the binding was not affected by zwitterionic synthetic fragment.

SP-A is mainly responsible for paratyphoid in humans, infecting approximately six million people annually worldwide [63]. Although the disease is treated with antibiotics, but there is still no licensed para typhi vaccine available. The Strategic Advisory Group of Experts on Immunization felt and expressed the need for combined typhoid-paratyphoid vaccine [64] Due to the emergence of drug-resistant paratyphoid, although one conjugate vaccine is under phase I and phase II trials [65], development of a suitable paratyphoid or typhoid–paratyphoid vaccine is still warranted.

With this end in view, Misra, Tennet, and Huang very recently stereoselectively prepared tetra- and pentasaccharides corresponding to the O-antigen of SP-A, utilizing convergent [2+2] (Scheme 5.9) and [2+3] glycosylation reactions, respectively [66]. The disaccharide building donor block **109** was prepared by armed-disarmed technique starting with NIS-TfOH-mediated glycosylation of D-Man acceptor **107** with armed D-Glc donor **106** followed by the conversion of the resulting

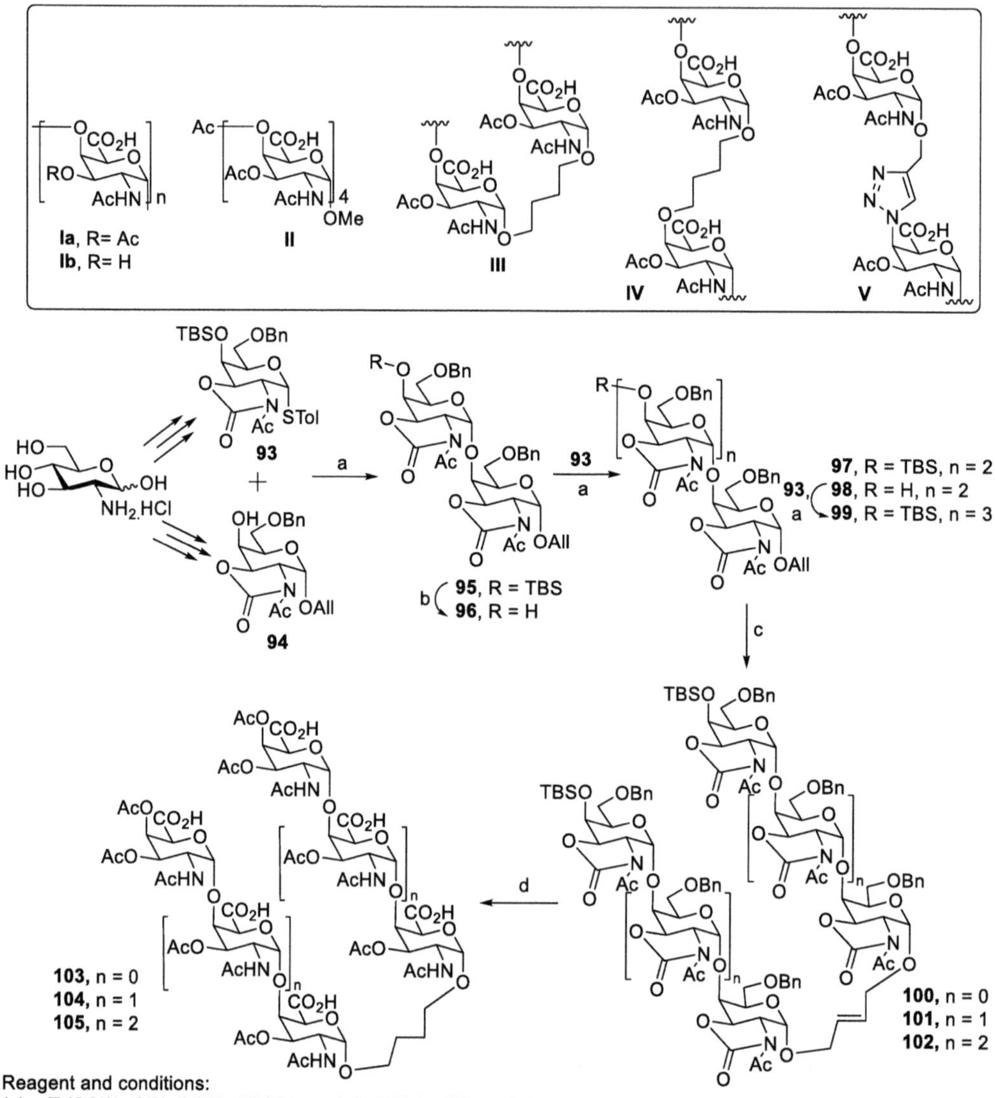

Reagent and conditions:
(a) *p*-TolSCl/AgOTf, DCM, -72 °C to rt, 3 h, 52% for **95**, 51% for **97**, 64% for **99**; (b) TBAF, THF, rt, 10 min, 90% for **96**, 77% for **98**; (c) Ru-based catasyst; (d) i) aq NaOH, 1,4-dioxane 1:1, 40 °C, 24 h, ii) Ac_2O, DMAP, Py, 0 °C to rt, 1 h, iii) H_2, Pd/C, THF:AcOH:H_2O 4:2:1, 12 h, iv) $NaIO_4$, $RuCl_3.xH_2O$, CCl_4:CH_3CN:H_2O 2:2:3, 24 h.

SCHEME 5.8 Synthesis of oligosaccharides and their carbon chain mimics related to Vi Antigen from ST.

disaccharide **108** to the corresponding paratose bearing disaccharide donor **109** through a xanthate deoxygenation pathway.

On the other hand, the disaccharide acceptor building block **113** was obtained by first coupling L-Rha donor **110** with D-Gal acceptor **111** followed by Zemplén deacetylation of disaccharide **112**. Then, [2+2] convergent glycosylation of **109** with **113** afforded tetrasaccharide derivative **114** that after global deprotection steps generated target tetrasaccharide **115**. For the synthesis of the pentasaccharide, tailor-made trisaccharide **119** was synthesized by coupling donor **116** with acceptor **117** forming initially disaccharide acceptor **118** followed by second glycosylation with L-rhamnopyranoside donor **110**. Reaction with NaOMe resulted in trisaccharide acceptor **120**, which was then further glycosylated with paratose-bearing disaccharide donor **109** to yield

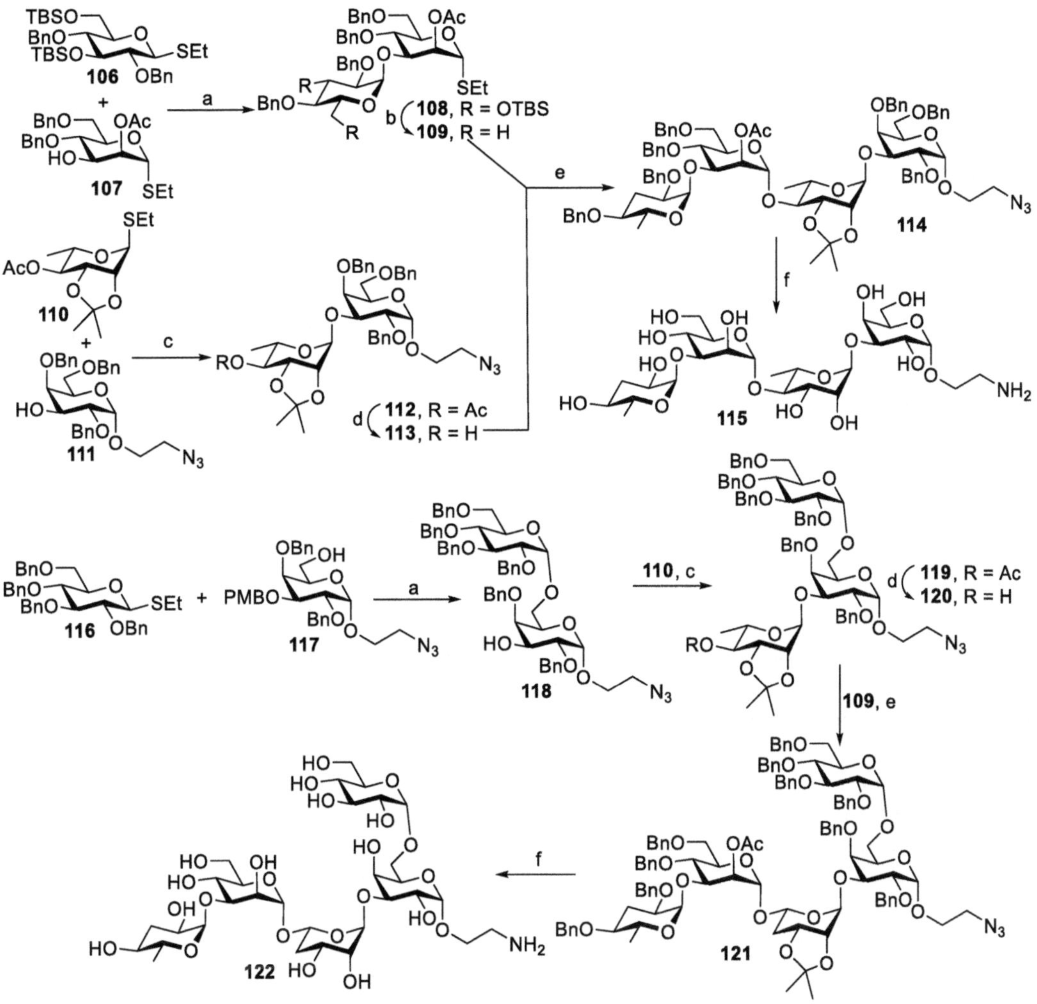

Reagents and conditions:
(a) NIS, TfOH, DCM, -40 °C, 15 min, 72% for **108** and DCM:Et_2O 1:4, -20 °C, 30 min then 0 °C, 1h, 72% for **118**; (b) i) p-TSA, MeOH, rt, 2 h, 79%, ii) PhOCSCl, Py, rt, 4 h, 91%, iii) TBTH, AIBN, Toluene, reflux, 4 h, 55%; (c) NIS, TMSOTf, DCM, -20 °C, 30 min, 76% for **112** and 40 min, 76% for **119**; (d) NaOMe, MeOH, rt, 2 h, 92% for both **113** and **120**; (e) NIS, AgOTf, DCM, -20 °C, 45 min, 71% for **114** and 74% for **121**; (f) i) NaOMe, MeOH, rt, 2 h, ii) NaH, DMF, BnBr, rt, 2 h, iii) 80% aq AcOH, 80 °C, 2 h, iv) Ac_2O, Py, rt, 3 h, v) 20% $Pd(OH)_2$-C, H_2, MeOH, rt, 24 h, overall 47% for **115** and 53% for **122**.

SCHEME 5.9 Synthesis of tetrasaccharide **115** and pentasaccharide **122** related to SP.

pentasaccharide **121** at 74%; global deprotection of **121** finally generated pentasaccharide **122** at 53% overall yield.

Of the tailor-made oligosaccharides, to develop the corresponding potential vaccine, researchers conjugated tetrasaccharide **115** via a bifunctional linker with bacteriophage Qβ, a powerful carrier for bioconjugate-based vaccine [63a, 67]. The resulting glycoconjugate showed higher antibody response than observed earlier based on keyhole limpet hemocyanin carrier [68].

Salmonella enterica (SE) is known to cause foodborne gastrointestinal infection in humans and animals [69]. The repeating tetrasaccharide [→4)-α-L-Rhap-(1→3)-β-D-GlcpNAc-(1 →2)-β-D-Galp-(1 →3)-α-D-GlcpNAc-(1→] of SE type 59 O-antigen contains one α-L-Rhap, two D-GlcNAc (one α and another β-linked) and one β-D-Galp units. Misra et al. synthesized tetrasaccharide derivative **130** of the RU by a [2+2] block glycosylation reaction of **126** and **129** (Scheme 5.10) [70]. One of the disaccharide-derived building blocks **129** was prepared from reactivity-based glycosylation reaction of the thioglycoside monomeric building units **127** and **128**. Interestingly, during the synthesis of the other non-reducing end disaccharide building block **125** unusual α-selectivity was obtained in spite of having a participating group (OAc) on C-2 of the donor ethyl 2-*O*-acetyl-3-*O*-benzyl-4,6-*O*-benzylidene-1-thio-β-D-galactopyranoside **124** in coupling with acceptor **123**.

Misra's group further investigated the convergent synthesis of the pentasaccharide RU corresponding to SE-O44 using NIS-TfOH as the activator system (Scheme 5.11) [71]. Glycosylation of thioglucopyranoside donor **132** with acceptor **133** and subsequent deprotection of PMB group resulted in disaccharide acceptor **134** at 74% yield. Reactivity-based coupling between thiogalactoside donor **135** and glucosamine derived acceptor **128** generated disaccharide thioglycoside **136** at 72% yield. Coupling disaccharides **134** and **136** produced tetrasaccharide acceptor **137** at 69% yield, which was further glycosylated with glucosamine-derived donor **138** yielded pentasaccharide derivative **139** at 67%. Global deprotection produced the final pentasaccharide **140** as its 2-amino ethyl glycoside at 57% overall yield.

In different experimental work, Das and Mukhopadhyay reported on the straightforward linear synthesis of the repeating pentasaccharide (in the form of its PMP-glycoside **152**) related to the O-antigen of SE type 44 by activating thioglycoside donor using NIS-H_2SO_4-SiO_2 and that of TCA-donor with H_2SO_4-SiO_2 (Scheme 5.12) [72]. Coupling thiogalactoside donor **142** with acceptor **141** using the before mentioned thiophilic activator system yielded disaccharide **143** at 91%. Removal of acetate protection under base mediation generated the corresponding 3′,4′-diol acceptor **144**, which was then regioselectively glycosylated at its 3′-position with thioglycoside donor **145**, yielding trisaccharide **146** at 68%. Glycosylation at 4′-position of this trisaccharide acceptor **146** with TCA donor **147** produced tetrasaccharide **148** at 72% yield. Removal of chloroacetyl functionality generated tetrasaccharide acceptor **149**, which was further glycosylated with thioglucoside donor **150**, yielding pentasaccharide **151** at 89%; conversion of NPhth to NHAc followed by global deprotection finally afforded target pentasaccharide **152** (Scheme 5.12).

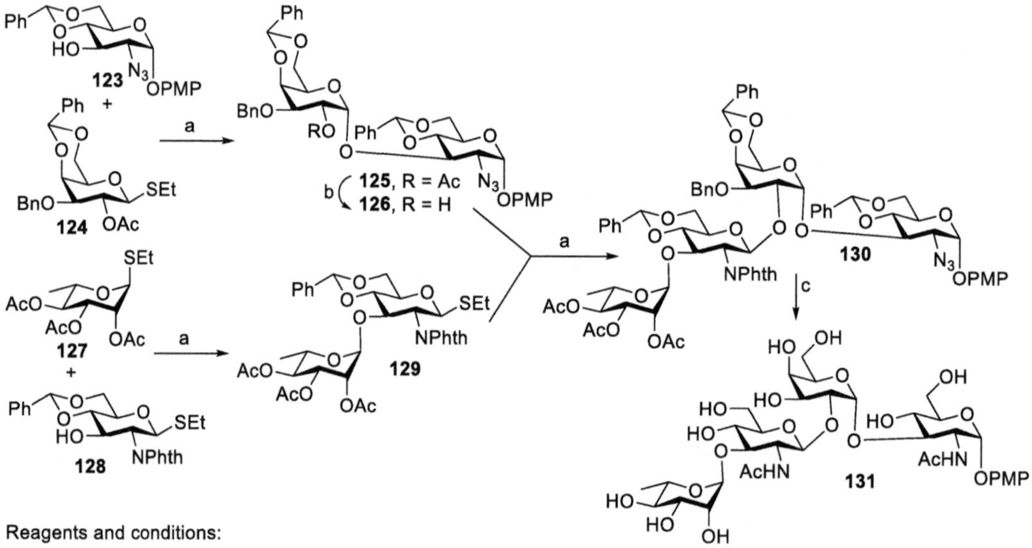

Reagents and conditions:
(a) NIS, TfOH, DCM, -30 °C, 45 min, 77% for **125**, 81% for **129** and 74% for **130**, respectively; (b) NaOMe, MeOH, rt, 2 h, quantitative; (c) i) $N_2H_4.H_2O$, EtOH, 80 °C, 8 h, ii) Ac_2O, Py, rt, 1 h, iii) H_2, 20% $Pd(OH)_2$-C, MeOH, rt, 24 h, iv) Ac_2O, Py, rt, 1 h, v) NaOMe, MeOH, rt, 2 h, 56% overall.

SCHEME 5.10 Synthesis of tetrasaccharide **131** of the O-polysaccharide of SE-O59.

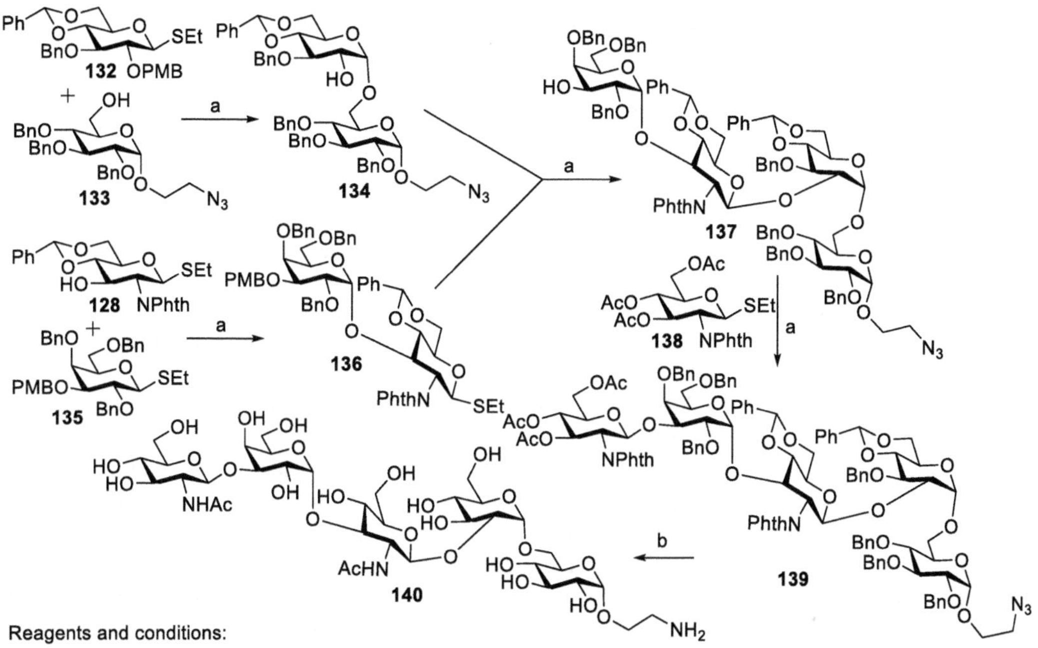

Reagents and conditions:
(a) NIS, TfOH, DCM:Et_2O 1:1, -45 °C, 45 min then 0 °C, 1 h, 74% for **134**, -35 °C, 30 min, 72% for **136**, -30 °C, 30 min then 0 °C, 1 h, 69% for **137** and -25 °C, 20 min, 67% for **139**, respectively; (b) i) $N_2H_4.H_2O$, EtOH, 80 °C, 8 h, ii) Ac_2O, Py, rt, 1 h, iii) NaOMe, MeOH, rt, 2 h, iv) Et_3SiH, 20% $Pd(OH)_2$-C, MeOH:EtOAc 3:1, rt, 10 h, 57%.

SCHEME 5.11 Synthesis of pentasaccharide **140** corresponding to the cell wall O-antigen of SE-O44.

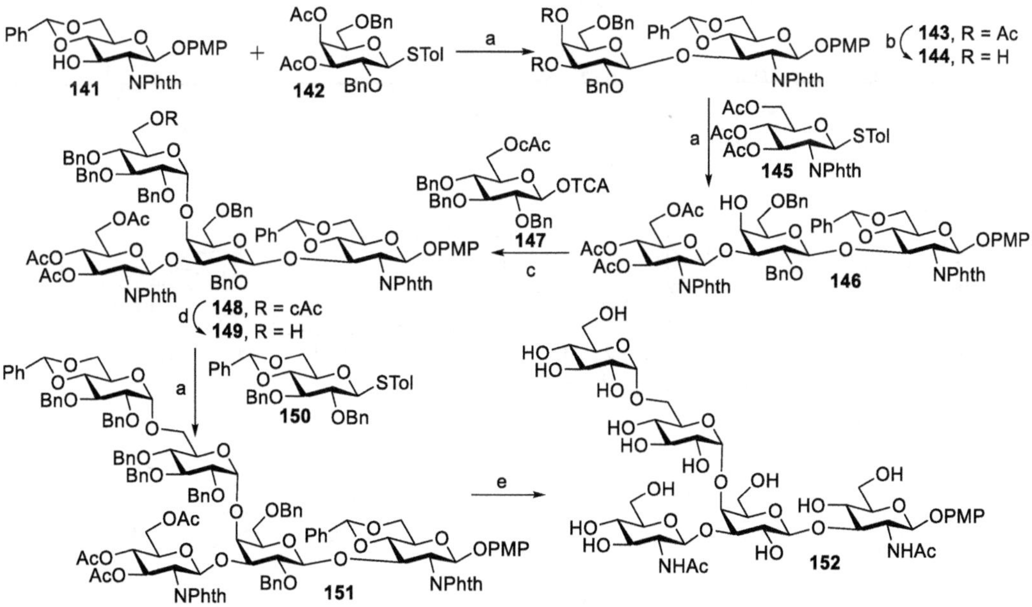

Reagents and conditions:
(a) NIS, H_2SO_4-SiO_2, -40 °C, 91% for **143**, -20 °C, 68% for **146** and 89% for **151**, respectively; (b) NaOMe, MeOH, 82%; (c) H_2SO_4-SiO_2, -55 °C, 72%; (d) Thiourea, 2,6-colidine, 87%; (e) i) 80% aq. AcOH, 80 °C, ii) Ethylenediamine, n-butanol, reflux, iii) Ac_2O, Py, iv) H_2, Pd-C, MeOH, v) NaOMe, MeOH, rt, 58% overall yield.

SCHEME 5.12 Synthesis of pentasaccharide **152** related to the repeating unit of the O-antigen from SE-O4.

Using thioglycosides as the glycosyl donors and NIS-$HClO_4$-SiO_2 as the corresponding activator system, Mishra's research group used linear stereoselective glycosylation along with PMB-deprotection in each glycosylation reaction pot to concisely synthesize the pentasaccharide corresponding to SE-O51 antigen [73].

Misra's research group also undertook the synthesis of the tetrasaccharide RU related to SE-O53 using convergent [2+2] block glycosylation and also following one-pot [1+1+2] glycosylation (Scheme 5.13) [74]. After synthesizing the required monomeric building blocks **153**, **154**, **158**, **161**, and **162** from the corresponding native monosaccharides, they prepared 2-aminoehanol glycoside disaccharide **155** by first coupling (71% yield) glycosyl acceptor **153** with thiorhamnoside donor **154** building blocks in the presence of NIS and $HClO_4$-SiO_2 followed by $PdCl_2$-mediated deallylation (67%). Compound **156** after deprotection of acetyl as well as benzylidene and NPhth-to-NHAc conversion generated disaccharide **157**. Acceptor **156** on glycosylation with thioglycoside donor **158** yielded trisaccharide derivative **159** at 70%. This on functional group deprotection and transformation provided the trisaccharide **160**.

Separately, disaccharide derivative **163**, comprising galactofuranoside moiety, was prepared at 65% yield by $NOBF_4$-mediated coupling of TCA-furanoside donor **162** with thioglycoside acceptor **161**. Disaccharide-thioglycoside donor **163** after activation with NIS and $HClO_4$-SiO_2 and reaction with disaccharide acceptor **156** afforded tetrasaccharide **164** at 66% yield. Required deprotection and conversion of NPhth to NHAc of **164** finally gave target tetrasaccharide **165**. Other researchers used a one-pot approach to build the same tetrasaccharide **164** by initial coupling **166** with **162**

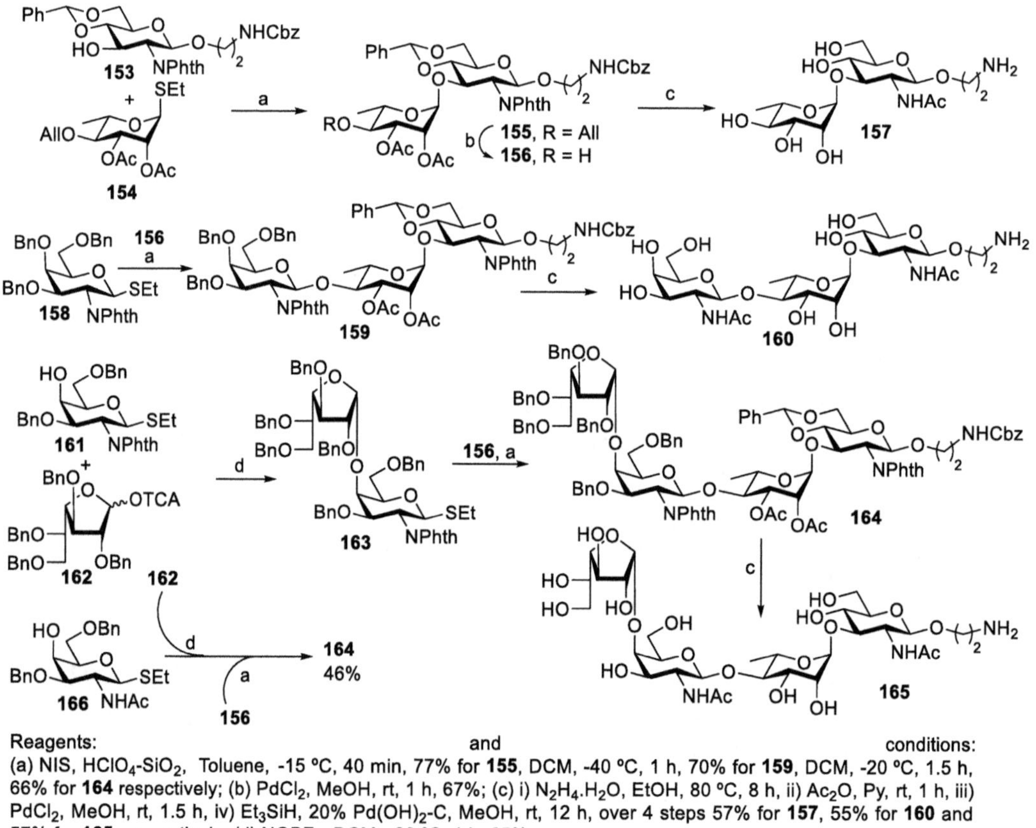

Reagents: and conditions:
(a) NIS, $HClO_4$-SiO_2, Toluene, -15 °C, 40 min, 77% for **155**, DCM, -40 °C, 1 h, 70% for **159**, DCM, -20 °C, 1.5 h, 66% for **164** respectively; (b) $PdCl_2$, MeOH, rt, 1 h, 67%; (c) i) $N_2H_4.H_2O$, EtOH, 80 °C, 8 h, ii) Ac_2O, Py, rt, 1 h, iii) $PdCl_2$, MeOH, rt, 1.5 h, iv) Et_3SiH, 20% $Pd(OH)_2$-C, MeOH, rt, 12 h, over 4 steps 57% for **157**, 55% for **160** and 57% for **165**, respectively; (d) $NOBF_4$, DCM, -20 °C, 1 h, 65%.

SCHEME 5.13 Synthesis of SE-O53 antigen-related tetrasaccharide **166**.

using $NOBF_4$. This was followed by NIS and $HClO_4$-SiO_2-mediated in situ glycosylation of that thioglycoside with disaccharide acceptor **156** to generate tetrasaccharide **164** at 46% yield.

Thoughtfully planning and preparing the rare sugar-based donor **171** from D-Fuc and the other required monomeric building blocks **168, 169, 170** and **143** from the corresponding native monosaccharides, Misra's research group synthesized pentasaccharide **167** as a PMP glycoside corresponding to the O-antigen of SE-O55 (Scheme 5.14) following linear progressive stereoselective glycosylation. For this, they used thioglycosides as donors and NIS and $HClO_4$-SiO_2 as the corresponding activator, benzyl as the permanent protecting group, and acetyl as the temporary protecting group as well as 1,2-*trans* glycosylation modulating group [75].

A sequential stepwise glycosylation strategy was utilized by Misra's research group [76] to prepare the tetrasaccharide RU corresponding to the cell wall PS of SE-O60 employing thioglycoside donors (one of which was a tailor-made rare sugar-based thioglycoside), NIS-H_2SO_4-SiO_2, or NIS-TMSOTf as the corresponding activators. To generate the crucial β-mannoside linkage and, in situ, the next acceptor, they judiciously used a differentially protected thiomannoside holding an in situ removable temporary PMB on C3-O to direct H-bond-mediated β-mannosylation [77].

The first potential synthetic glycoconjugate vaccine corresponding to non-typhoidal SE sero group D was recently developed by the collaborative effort of Simon, Misra, and Huang [78]. They synthesized tetrasaccharide glycoside capped with aminopropyl chain related to this sero group and then conjugated it with carrier bacteriophage Qβ to develop a non-paratyphoid *Salmonella* vaccine; approximately 300 copies of the tetrasaccharide were immobilized per Qβ particle. The resulting glycoconjugate elicited IgG antibody responses in both rabbit and mice. The antisera from rabbit raised against the said glycoconjugate protected mice challenged using lethal dose of SE.

Scheme 5.15 schematically depicts the synthesis of required tetrasaccharide **179**. Disaccharide donor building block **174** was obtained by selective armed–disarmed coupling of D-Tyv-based armed thioglycoside donor **172** with D-Man-derived disarmed thioglycoside **173** in the presence of NIS-TMSOTf. Separately, disaccharide acceptor **177** was obtained efficiently by NIS-TMSOTf-promoted glycosylation of D-Gal acceptor **175** with L-Rha-based thioglycoside donor **110** followed by a Zemplén deacetylation of the generated disaccharide **176**. Tetrasaccharide derivative **178** was then synthesized at 78% yield by NIS-TMSOTf-mediated [2+2] convergent glycosylation of disaccharide building blocks. Global deprotection of **178** in three steps ultimately gave target tetrasaccharide **179** at 57% combined yield.

5.1.4 *Shigella*

Shigella are a group of Gram-negative entero-invasive bacteria invading the epithelium of the ileum, colon, and rectum, causing Shigellosis in humans with dysentery and strong diarrhea that is sometimes bloody. *Shigella* are also responsible for morbidity and mortality in children of age group 1 to 4, making it a public health challenge, especially in low- and middle-income communities. *S. dysenteriae* (SD), *S. boydii* (SB), *S. flexneri* (SF), and *S. sonnei* (SS) can bring about Shigellosis

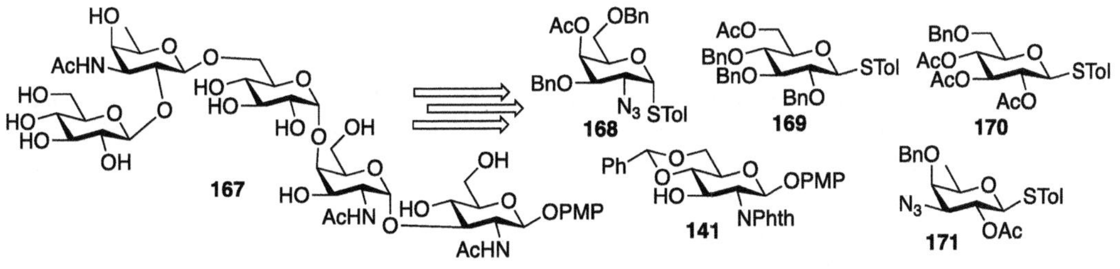

SCHEME 5.14 Retrosynthetic plan towards synthesis of SE-O55 related pentasaccharide **167**.

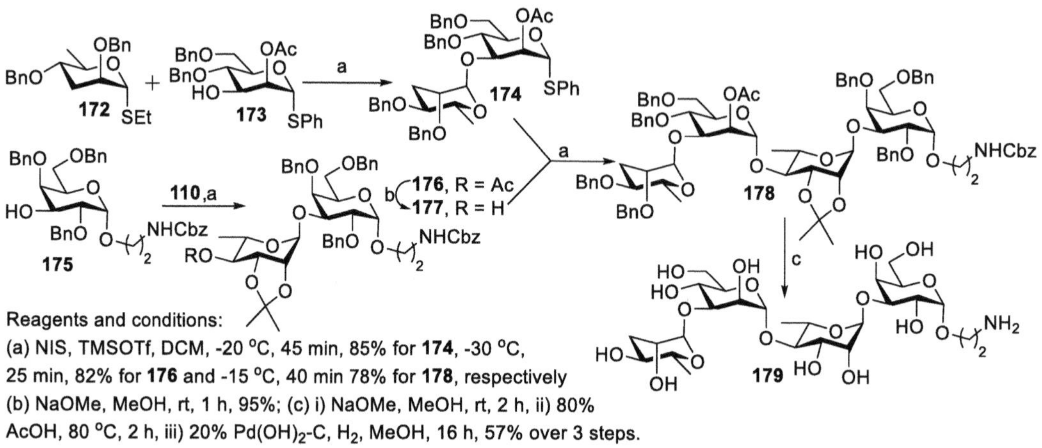

Reagents and conditions:
(a) NIS, TMSOTf, DCM, -20 °C, 45 min, 85% for **174**, -30 °C, 25 min, 82% for **176** and -15 °C, 40 min 78% for **178**, respectively
(b) NaOMe, MeOH, rt, 1 h, 95%; (c) i) NaOMe, MeOH, rt, 2 h, ii) 80% AcOH, 80 °C, 2 h, iii) 20% $Pd(OH)_2$-C, H_2, MeOH, 16 h, 57% over 3 steps.

SCHEME 5.15 Synthesis of target tetrasaccharide **179** related to SE-O-antigen.

and are the four most common sources [79]. Although these are curable by antibiotic treatment, but increasing instances of drug-resistant *Shigella* strains make the curative treatment difficult [79b,c]. Although there have been several efforts to develop *Shigella*-related vaccines, there is to date no licensed effective marketed *Shigella* vaccine; however, some glycoconjugate-based vaccine candidates are under either phase or preclinical trials [80]. In their recent review, Mulard et al. discussed current developments of *Shigella*-related glycoconjugate vaccines [80a], as have other researchers [81]. Thus, our discussion will only include research carried out during the past ten years and focusing on chemical syntheses of oligosaccharides as potential *Shigella* vaccine candidates.

Among the *Shigella* species, SD1 and SF2a are the most-researched strains for glycoconjugate vaccine development [80a] and are under preclinical or phase trials. Of these, SD type 1 is of more concern because of its ability to express Shiga toxin [82]. We will not here discuss Pozsgay et al.'s remarkable achievement in semi-synthetic glyco-HSA conjugates with varying OS chain length related to SD1 toward vaccine development before 2010; some of these glycoconjugates are under phase I clinical trial [83].

Researchers have investigated SDs other than SD1 for related OS syntheses. Misra et al. reported the synthesis of an acidic pentasaccharide corresponding to the O-antigen of SD4. They first prepared the corresponding neutral pentasaccharide by [3+2] convergent glycosylation and introduced the acid function at the later stage of the total synthesis by oxidation of a primary OH in buffer medium; they then utilized thioglycosides as glycosyl donors with their activation using NIS-TfOTf activator in all glycosylation steps [84]. They also synthesized a di- and a trisaccharide related to the O-polysaccharide of SD8 [85].

Species/sub-group SF accounts for 60% of Shigellosis, most cases of which most are suffered by children in developing countries. Except serotype 6, all known serotypes of SF O-antigens share the following RU: [→2)α-L-Rhap-(1→2)-α-L-Rhap-(1→3)-α-L-Rhap-(1→3)-β-D-GlcpNAc(1→] and about 95% of serotypes 1a, 1b, and 2a contain 3-/4-O-acetylation on the non-reducing end Rha [86].

Mulard's research group extensively studied SF type 2a (SF2a). Following on their earlier research on development of glycoconjugate vaccines for this strain [87], his group next examined the effect of site-selective non-stoichiometric acetylation on the antigenicity [88]. They also utilized a partly enzymatic and mostly chemical approach to synthesize several oligosaccharides including a decasaccharide **190** and several diversely site-selective internal O-acetylated decasaccharides corresponding to serotype 2a based on disaccharides as transglucosylase acceptors in one of the steps [89]. The differentially protected decasaccharide derivative **189** was prepared by a combination of linear and [5+5] convergent glycosylation. Immunological binding studies of the tailor-made decasaccharides to five murine protective antibodies mIgGs by ELISA-inhibition against SF2a-LPS resulted in one typical mIgG having a unique binding pattern. Their work also indicated how such modifications including acetylation help towards a better understanding of vaccine development programs. The synthesis of representative decasaccharide **190** as 2-aminoethyl glycoside is depicted in Scheme 5.16.

To achieve the desired target decasaccharide, they judiciously chose the protecting groups in the glycosyl donors or acceptors and corresponding suitable chemoselective deprotection methods and used TCA- or PTFA-based glycosyl donors along with their TMSOTf-mediated activation in the high-yielding glycosylation steps. They enzymatically converted related disaccharide **180** to corresponding trisaccharide **181**. Compound **181** was then transformed to partially protected trisaccharide **182** by a series of multistep protection–deprotection profiling of the hydroxy groups. This on glycosylation with monomer donor **183** yielded tetrasaccharide **184** at 81% [89].

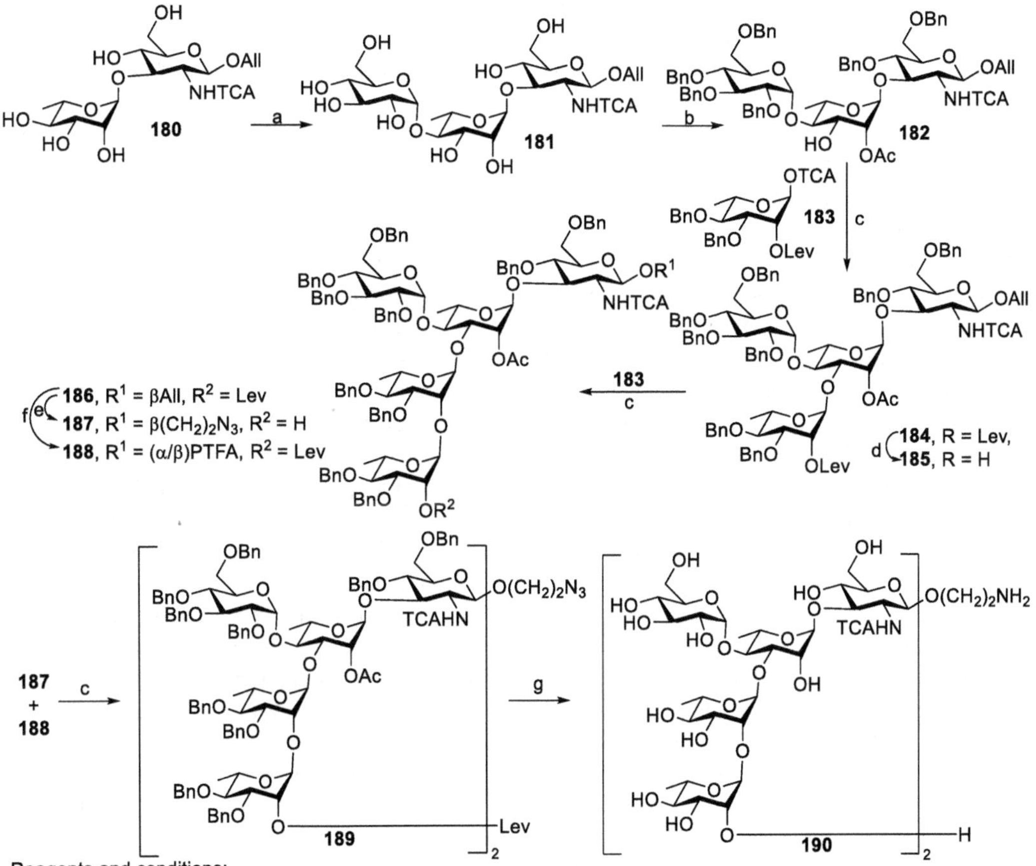

Reagents and conditions:

(a) Sucrose, GBD-CD , NaOAc Buffer, 30 °C, 24 h, 87%; (b) i) BzCl, Sym-Collidine, MeCN:Me_2CO 1:1, -40 °C, 46 h, 65%, ii) DMP, CSA, Me_2CO, 2 h, 92%, iii) NaOMe, MeOH, 3.3 h, 94%, iv) NaH, BnBr, DMF, -10 °C, 4 h, 78%, v) 50% aq. TFA, DCM, 75 min, 94%, vi) MeC(OMe)3, p-TSA, MeCN, 35 min, then 80% aq AcOH, 0 °C, 25 min; (c) TMSOTf, Et_2O, -15 °C, 40 min, 81% for **184**, 86% for **186**, toluene, -10 °C, 2 h, 82% for **189**; (d) N_2H_4, Py:AcOH 3:2, 0 °C,70 min, 92%; (e) i) [Ir], H_2, THF, 2.5 h, then I_2, THF:H_2O 3:1, 4.5 h, 93%, ii) CCl_3CN, DBU, DCE, -10 °C, 40 min, 97%, iii) Bromoethanol, TMSOTf, DCE, 0 °C, 25 min, then NaI, NaN_3, DMF, 80 °C, 2 h, 78%, iv) N_2H_4, AcOH/Py 2:3, 0 °C, 90%; (f) i) [Ir], H_2, THF, 2.5 h, then I_2, THF:H_2O 3:1, 4.5 h, 93%, ii) $CF_3C(NPh)Cl$, Cs_2CO_3, Me_2CO, rt, 2 h, 83%; (g) i) NaOMe, MeOH, 21 h, 71%, ii) H_2, $Pd(OH)_2/C$, ^tBuOH, DCM, H_2O, 3 d, 58%.

SCHEME 5.16 Synthesis of representative decasaccharide **190** related to SF-2a.

Removal of the Lev group generated tetrasaccharide acceptor **185**, which was further glycosylated with the same monomeric donor **183**, producing 86% of the desired pentasaccharide **186.** Transformation of the anomeric *O*-allyl to the corresponding PTFA glycoside through the specific removal of allyl group followed by reaction with $CF_3C(NPh)Cl$ under basic condition formed the pentasaccharide glycosyl donor **188**. Corresponding pentasaccharide acceptor **187** was obtained by glycosylation with pentasaccharide TCA donor with bromoethanol acceptor followed by in situ azidation and then hydrazine-mediated deprotection of the Lev group. Final [5+5] convergent glycosylation of acceptor **187** with donor **188** afforded the desired decasaccharide derivative **189** at high yield (82%). Finally, two-step global deprotection ultimately furnished the target decasaccharide **190** at 58% overall yield [89].

Mulard et al. also showed that a synthesized glycoconjugate equipped with a thiol-maleimide-anchored pentadecasaccharide (of chain extended **190**) conjugated with TT as a carrier protein was a promising vaccine candidate against SF-2a. The hapten loading was critical: The best immunogenicity was obtained based on a hapten at a carrier ratio of (17± 5); moreover, the authors demonstrated for the first time a strong anti-SF2a antibody response with an increased titer sustained for more than one year using alum as adjuvant [90].

In their earlier immunochemical studies [87, 90] with the partially acetylated and non-acetylated variants of SF 2a-related oligosaccharides, Mulard et al. demonstrated that a nonacetylated pentadecasaccharide was the best hapten for glycoconjugate vaccine development in the process of very recently synthesizing a glycoconjugate vaccine candidate, SF2a-TT15 (stability, 72 months). They standardized an improved method of bioconjugation fulfilling Good Manufacturing Practice criteria for obtaining high yields of glycoconjugate with controllable and consistent glycan-TT ratio. They obtained the planned conjugate vaccine candidate by scaling up their earlier reported syntheses [91] after a 40-step synthesis of pentadecasaccharide [nonacetylated, as $-(CH_2)_2NHC(O)CH_2SAc$ glycoside] followed by its reaction with tetanus toxoid-maleimide. Using alum as an adjuvant, the tested SF2a-TT15 vaccine candidate fulfilled the toxicity criteria, showed immunogenicity to rabbits, and produced bactericidal antibodies in mice and thus was a probable candidate for phase I trial [92].

Apart from sero type 2a, researchers have synthesized oligosaccharides of the O-antigen of other serotypes, including branched pentasaccharide RU of serotypes 1a, 1b (Mulard et al. [93]), hexasaccharide RU corresponding to serotype 1d (Mukhopadhyay et al. [94]), and type 1d (Misra et al.95). Mulard's research group also worked on *S. flexneri* type 3a. Utilizing iterative glycosylation, they effectively synthesized several oligosaccharides of different chain lengths (three to fifteen) related to SF type 3a O-antigen from the corresponding building blocks [96].

One of the main objectives of any semi-synthetic vaccine development program is efficient preparation of the hapten at multigram level followed by its conversion to an effective glycoconjugate. With this end in view, Mulard's research group developed and reported on the effective synthesis of an orthogonally protected pentasaccharide **201** at multigram scale by [3+2] glycosylation and also its conversion to a decasaccharide **205**; the synthesis is depicted in Scheme 5.17 [97].

To synthesize trisaccharide acceptor **196**, researchers initially combined TCA-rhamnose donor **183** bearing a temporary protecting Lev group on C2-O to promote 1,2-*trans* anomeric selectivity and rhamnose-acceptor **191** with C2-OAc as a permanent protection in the presence of TMSOTf; this yielded disaccharide **192** at 93%. Its $PdCl_2$-iodine mediated anomeric deallylation followed by usual trichloroacetylation yielded the next disaccharide donor **193** in 89%. Glycosylation of glycosamine-derived acceptor **194** with **193** in the presence of TMSOTf at 0 °C generated the corresponding trisaccharide **195** at 86% yield (45 g scale); its delevoloylation using hydrazine hydrate yielded trisaccharide acceptor **196** at 92%.

In other work, glycosylation of rhamnose acceptor **191** with glucoside TCA donor **197**, modulated by DMF and activated by TfOH, followed by DMAP-promoted debenzoylation generated the corresponding disaccharide **198** at 90% yield. Iridium–iodine-mediated anomeric deallylation (85%, 29 g scale) and then anomeric trichloroacetimidate formation generated the required disaccharide donors **199** (93%, 23 g scale) and **200** (90%, 34 g scale). Its coupling with earlier-made trisaccharide acceptor **196** mediated by TMSOTf yielded pentasaccharide derivative **201** (94% from **200**, 12 g scale; 94% from **199**, 27 g scale). This on iridium-mediated anomeric deallylation followed by C1-O-*N*-phenyltrifluoroacetimidylation (93% yield) gave pentasaccharide donor **202**. Reaction of **202** with 2-azidoethanol followed by debenzoylation with ethylene diammine-pyridine-acetic acid afforded 2-azido ethyl glycoside acceptor **203**. Then, TMSOTf-promoted [5+5]-glycosylation of acceptor **203** and donor **202** finally provided the desired decasaccharide derivative **204** at 91% yield (1.6 g scale). Compound **204** on functional group deprotections and transformation generated the target decasaccharide **205** (Scheme 5.17).

Shigella sonnei causes diarrheal diseases in industrial belts and in developing countries. A zwitterionic (1→2)-*trans*-linked disaccharide [→4)-α-L-Alt*p*NAcA-(1→3)-β-D-Fuc*p*-NAc4N-(1→] composed of two rare sugars (L-altruonate and L-FucNAc) is the RU of the cell wall polysaccharide of this strain [98].

Mulard et al. first reported [99] the synthesis of this rare sugar-based zwitterionic disaccharide **210**, and its two frame-shifted trisaccharides, **217** and **218** (Scheme 5.18), by chain elongation of **209** at either end. TMSOTf-mediated glycosylation of glycosyl acceptor **207** with donor **206** formed the corresponding disaccharide derivative **208** (78%), which was then converted via sequential debenzylidenation, selective oxidation, and then benzyl ester formation to

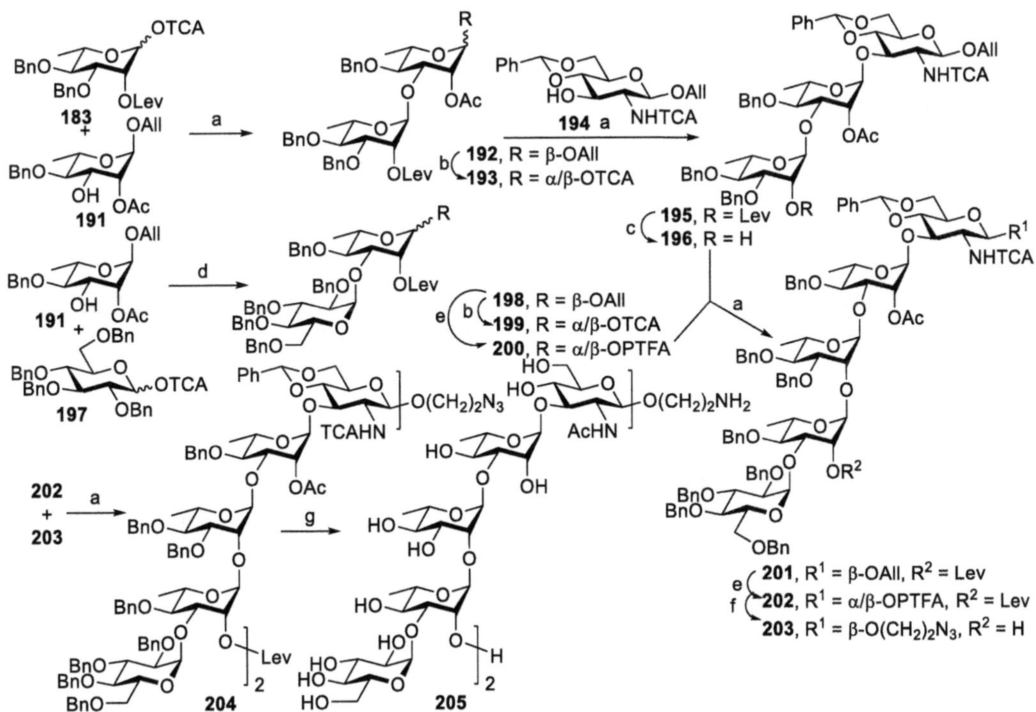

Reagents and conditions:

(a) TMSOTf, Toluene, -78 °C to rt, 93% for **192**, DCM, 0 °C to rt, 86% for **195** (45 g), toluene 94% for **201** and toluene, -78 °C 88% for **204**, respectively; (b) i) $PdCl_2$, DCM, H_2O, 50 °C then I_2, THF, H_2O, ii) CCl_3CN, DBU, 1,2-DCE, 89% for **193**, 93% for **199**, respectively; (c) $N_2H_4.H_2O$, Py/AcOH 3:2, 92%; (d) i) TfOH, DMF, DCM, -78 °C to rt ii) LevOH, EDC, DMAP, DCM, 90% (23 g); (e) i) $[Ir(COD)\{PCH_3Ph_2\}_2]^+PF_6^-$, THF then I_2, THF, H_2O, ii) PTFACl, K_2CO_3, Me_2CO, 90% for **200** (34 g) and 93% for **202**, respectively; (f) i) 2-azidoethanol, TMSOTf, toluene, -40 °C, 78%, ii) $N_2H_4.H_2O$, Py/AcOH (3:2), 91%; (g) i) $N_2H_4.H_2O$, Py/AcOH 3:2, 92%, ii) $Pd(OH)_2$, H_2, ^tBuOH, DCM, H_2O, 52%.

SCHEME 5.17 Synthesis of SF-3a related decasaccharide derivative **205**.

intermediate disaccharide ester **209**. Hydrogenation with Pearlman catalyst produced target disaccharide **210**, which in separate work was converted to the next glycosyl donor **212** following a three-step functional group transformation: i) chloroacetylation of 3′-free hydroxy group forming **211**, ii) iridium-mediated deprotection of anomeric O-allyl, and then iii) formation of anomeric *N*-phenyltrifluoromethyltrilylamide. TMSOTf-mediated glycosylation of monomer-acceptor **213** with donor **212** afforded at 69% yield the corresponding desired trisaccharide derivative **214** after taking proper care during quenching with base to avoid C-5′ epimerization. Trisaccharide derivative **216** was obtained similarly after TMSOTf-based glycosylation of glycosyl disaccharide acceptor **209** with monosaccharide donor **215**. Target trisaccharides **217** and **218** were obtained, respectively, from their protected precursors **214** and **216** after proper hydrolysis of benzyl ester followed by H_2, $Pd(OH)_2$/C-based reduction or deprotection of residual protecting groups (Scheme 5.18).

Mulard further explored a different route for synthesizing orthogonally protected repeating disaccharide unit **210**, which is ready to be elongated at either ends toward oligosaccharides, starting from L-Glc and D-GlcNAc acetate. Here, the most challenging was a late-stage selective installment of azido group on C-4 at the reducing end residue masking its amino moiety, through O-triflation followed by displacement of the resulting triflate by azide nucleophile [100].

SB is one of the causes of diarrheal infection and bacillary dysentery in developing countries. Misra et al. synthesized tri- and pentasaccharide related to the O-antigen of SB6 following a linear iterative glycosylation approach [101]. Misra's group further developed a convergent stereoselective synthesis of the tetrasaccharide RU as its PMP glycoside sodium salt, corresponding to SB-O9 [102] This group also reported the block synthesis [103] of a common tetrasaccharide shared by SB-O8 and EC-O143.

Sequential and iterative one-pot synthesis of the pentasaccharide RU corresponding to the cell wall O-antigen SB type 18 was reported by Shit et al. [104] They used the acid lability of

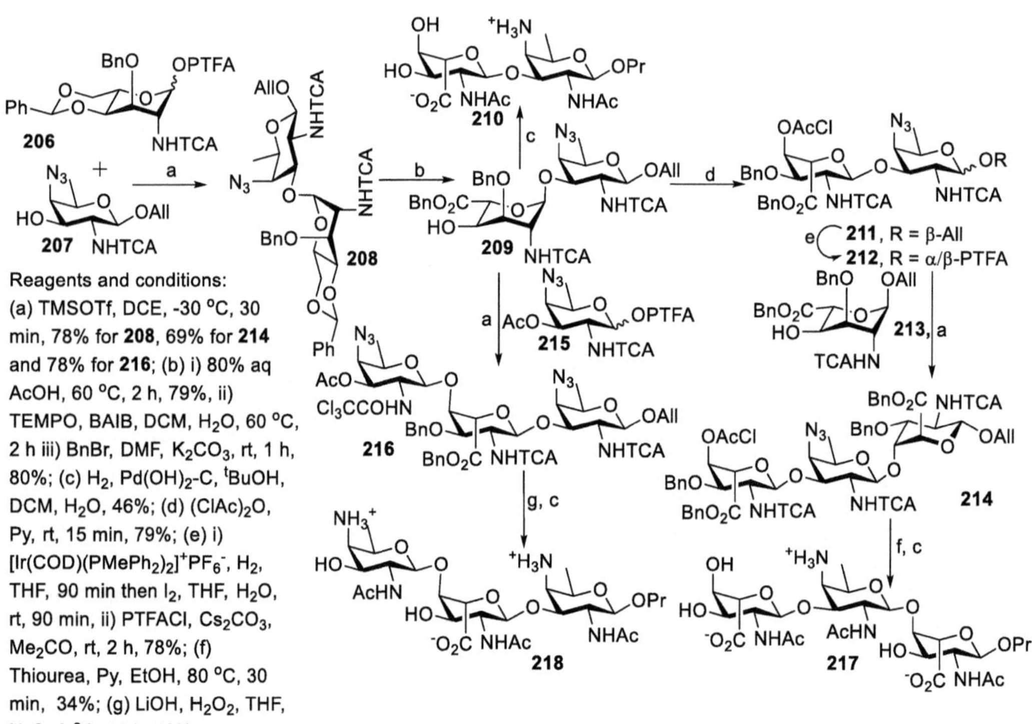

SCHEME 5.18 Synthesis of the zwitterionic repeating unit of the O-antigen from *Shigella sonnei*.

PMP group as a key feature for this synthesis, where this functionality was removed soon after the glycosylation step to generate the corresponding glycoside acceptors in situ. Another notable feature of this synthesis was the construction of β-L-rhamnosidic linkage from the corresponding L-rhamnopyranosyl thioglycoside donor using remote picoloyl protection at C-3 influencing the glycosylation reaction toward desired stereoselectivity. D-Galactosyl thioglycoside donor **220** was coupled with *p*-methoxyphenyl 4,6-di-*O*-acetyl-2-azido-2-deoxy-α-D-galactopyranoside **219** in the presence of NIS/$HClO_4$-SiO_2 at −30 °C for 45 minutes and then at 0 °C for 30 minutes resulting in the stereoselective α-disaccharide in O-PMB deprotected derivative **221** in 75% yield (α:β = 19:1).

Similar coupling of this disaccharide acceptor **221** with L-rhamnopyranosyl thioglycoside donor **222** under similar glycosylation reaction condition yielded trisaccharide acceptor **223** at 71%. Stereoselective 1,2-*trans* glycosylation of L-rhanmopyranosyl thioglycoside donor **224** and trisaccharide acceptor **223** and removal of the O-PMB protection resulted in 74% of tetrasaccharide acceptor **225**. Final β-linked glycosylation of the L-rhamnopyranosyl thioglycoside donor **226** was achieved with the help of remote picoloyl protection at C-3, producing the final pentasaccharide derivative **227** at 72% yield with the formation of a minor amount (5%) of the undesired isomer. Similar reaction sequences were performed without isolation of the intermediate glycoside to achieve the one-pot iterative synthesis of the pentasaccharide to result in the final compound **227** at 48% overall yield [104].

Here it is worth mentioning that despite the formation of an anomeric mixture (α:β = 19:1) of the very first disaccharide **221** in the reaction sequence, which was later glycosylated, the final β-linked L-rhamnosylation resulted in a 19:1 mixture of glycosides along with an additional side reaction with deprotected PMB-alcohol, and the one-pot total synthesis was ultimately quite interesting. Removal of the benzylidene and isopropylidene acetals with acetic acid followed by the selective TEMPO/BAIB-mediated oxidation of primary hydroxy group to carboxylic acid, then hydrogenolysis with

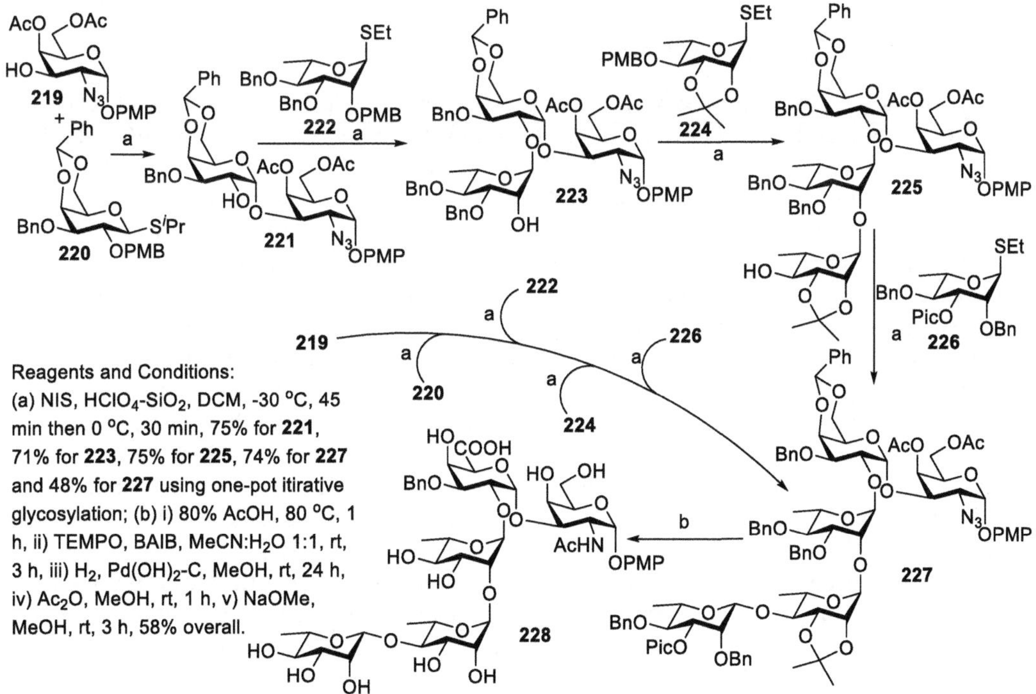

SCHEME 5.19 Synthesis of SB-18-related pentasaccharide **228** by both stepwise glycosylations and one-pot iterative glycosylation reactions.

$Pd(OH)_2$-C, acetylation using acetic anhydride in methanol, and final deacetylation with NaOMe/MeOH resulted in deprotected pentasaccharide **228** in 58% yield after five steps [104].

Plesiomonas shigelloides, a Gram-negative anaerobic bacterium, causes diarrhea via consumption of raw seafood and high mortality among children and immune-compromised patients. The O-antigen is composed of a branched tetrasaccharide RU consisting of D-GlcNAc, D-GalNAc, and two L-Rha [105]. Das and Mukhopadhyay developed a synthesis protocol for this branched tetrasaccharide corresponding to *P. shigelloides* strain AM36565 [106].

5.1.5 *Escherichia coli*

Although Gram-negative anaerobic bacteria EC present in the intestine of humans and animals are usually symbiotic to the host, there are several modified clones of EC attributing to some virulence factors which cause a broad spectrum of disease like enteric disease, UTI, and even sepsis and meningitis [107]. There are six well defined categories of intestinal pathogens of EC, like enteropathogenic, enterohaemorrhagic, enterotoxigenic, enteroaggregative, entero-invasive, and diffusively adherent EC, and both commensal and pathogenic EC strains are identified by their capsular (K), somatic (O), or flagellar (H) antigens [108]· Extra-intestinal pathogenic EC is the primary cause of both UTI and blood-stream infections [109], often causing infections among health care personnel [110]. About 180 serotypes of EC have been identified so far [111]; this diversification is due to the genetic variation between *galF* and *gnd* genes on the EC chromosome that code for O-antigen [112]. Increased incidences of hospitalization, treatment failure, and mortality have been witnessed due to surges in multidrug-resistant extra-intestinal pathogenic EC. This along with inadequate antibiotic pipelines [113] warrant the promotion of a suitable vaccine against EC. Some efforts have been already made toward developing glycoconjugate vaccines against different serotypes of EC [114] including a tetravalent vaccine candidate (ExPEC4V) containing four O-antigens of EC causing UTI, developed based on a bioconjugate approach [115].

5.1.6 EHEC, STEC, and VTEC Strains

EHEC strains are also known as vero toxin or Shiga toxin producing EC (VTEC or STEC) as they produce Shiga-like toxin during early stage of infection to host. EC O4, O5, O16, O26, O41, O46, O48, O55, O59, O91, O98, O111ab, O113, O117, O118, O119, O120, O125, O126, O128, O145, O157, O171, EC O181, etc. belong to the STEC family [116]. The seriousness of EHEC and STEC infections is reflected in incidents of outbreak [117]. The most significant consequences of STEC infections are complications of the central nervous system and hemolytic uremic syndrome, which gets even complicated by use of antibiotics resulting in ultimate renal failure. Several non-glycoconjugate vaccines have also been developed through preclinical or clinical trials [118].

Other researchers synthesized a common pentasaccharide related to EC-O4:K3, O4:K6 and O4:K12 involving [3+2] convergent glycosylation in his laboratory [119]. A nonreducing end-sialic acid holding pentasaccharide RU of STEC strain EC-O171 antigen was synthesized by Mandal and Misra involving [2+2] block glycosylation followed by conversion of the resulting tetrasaccharide to its corresponding glycosyl acceptor and then glycosylation using sialic acid-derived donor and finally global deprotection [120]. Misra's group also synthesized tetrasaccharide RU of verotoxin producing EC-O176 antigen following a [2+2] convergent pathway [121].

Si and Misra accomplished the synthesis of the pentasaccharide RU of EC-O166:H15; the strain belongs to EHEC and caused diarrheal outbreak in Japan in 1996 [122]. Later, Misra et al. also synthesized an acidic pentasaccharide corresponding to EC-O113-PS [123] and an acidic hexasaccharide related to EC-O41-PS [124] employing [3+2] and [3+3] block glycosylation, respectively, for these two total syntheses. The $-CO_2H$ function in each case was installed by TEMPO-mediated oxidation of a CH_2OH group at a late stage of the total syntheses.

Misra also accomplished the synthesis of a pentasaccharide related to the PS of another STEC strain, EC-O117:K98:H4. After linear synthesis of a trisaccharide building block **235** by two glycosylation reactions based on TCA donors (Gal-TCA donor **230** and acceptor **229** for the first glycosylation resulting in disaccharide **231** and Glc-TCA donor **233** and acceptor **232** for the second one resulting in trisaccharide **234**). Then they prepared the repeating pentasaccharide derivative **239** of EC-O117:K98:H4-PS utilizing [3+1+1] one-pot glycosylation reactions at overall of 65% one-pot yield via in situ generation of an intermediate glycosyl acceptor by removal of a temporary protecting PMB group at an elevated temperature after the first glycosylation with thio-L-rhamnoside donor **236** followed by addition of the second donor **237**. The synthesis is of pentasaccharide derivative **238** along with its conversion in five steps to target pentasachharide **239** is shown in Scheme 5.20 [125].

They further adopted linear glycosylation and used thioglycosides as glycosyl donors for the synthesis of EC-O166-related pentasaccharide [126]. A pentasaccharide corresponding to the PS of ETEC strain EC-O11 was synthesized by Si and Misra following [3+2] block glycosylation from the corresponding di- and tri-saccharide building blocks; they prepared these through glycosylation from the corresponding thioglycoside building blocks using NIS, $HClO_4$-SiO_2 as the activator system in all glycosylation steps [127]. EC-O139 is also an ETEC strain. Misra's group achieved a linear synthesis of the repeating hexasaccharide of EC-O139 as its PMP glycoside. For the glycosylation, they depended on thioglycoside donors and NIS and $HClO_4$-SiO_2 as activator. The acid group was generated at a late stage of the total synthesis, after synthesis of a neutral heptasaccharide derivative [128].

EC-O59 is another EHEC strain that was the cause of several diarrheal outbreaks in European countries [129]. Si and Misra accomplished the synthesis of EC-O59-related pentasaccharide following [1+2+2] one-pot glycosylation reactions using orthogonal glycosylation based on one mono- **246** and two disaccharide building blocks **242** and **245**. The disaccharide building block **242** was prepared by β-mannosylation at 55% yield on acceptor **241** through activation of the 4,6-O-benzylidenated thiomannoside donor **240** using BSP-Tf_2O followed by several functional group transformations with anomerization and introduction of azido function at C2 and then $PdCl_2$-mediated deallylation.

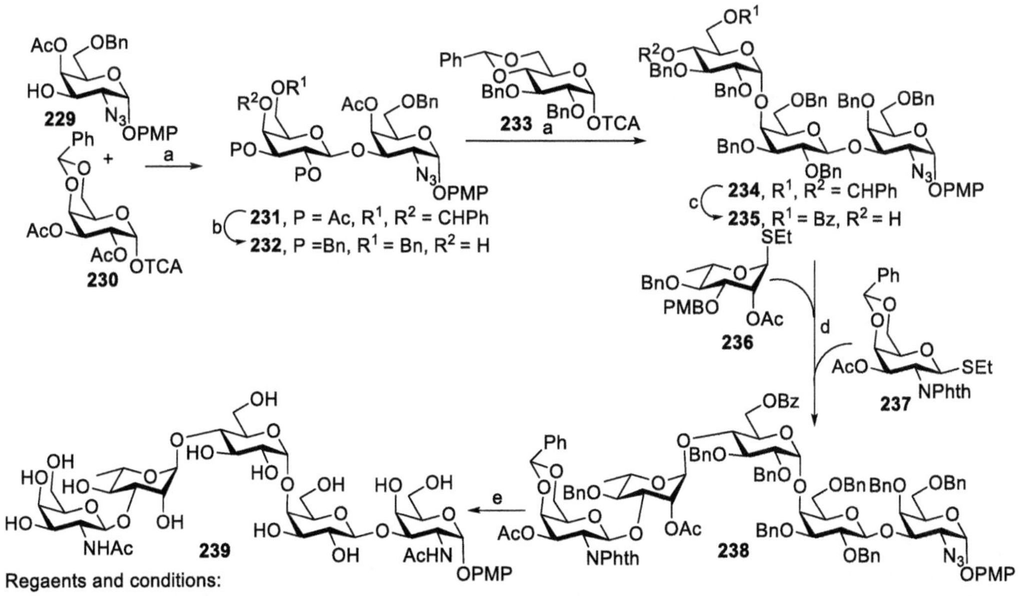

Regaents and conditions:
(a) $NOBF_4$, DCM, -20 °C, 45 min, 79% for **231**, DCM-Et_2O 1:4, -10 °C, 30 min, 75% for **234**, (b) i) BnBr, NaOH, TBAB, THF, rt, 5 h, 90%, ii) $NaBH_3CN$, HCl-Et_2O, THF. 5 °C, 1 h, 75%; (c) i) $HClO_4$-SiO_2, MeCN, rt, 25 min, ii) BzCN, Py, DCM, 0 °C, 4 h, 76%; (d) **236**, NIS, $HClO_4$-SiO_2, DCM, -45 °C, 30 min then 10 °C, 30 min then **237**, NIS, -30 °C, 1 h, 65%; (e) i) $N_2H_4.H_2O$, EtOH, 80 °C, 8 h, ii) Ac_2O, Py, rt, 1 h, iii) CH_3COSH, Py, rt, 18 h, iv) Et_3SiH, 20% $Pd(OH)_2$-C, MeOH, rt, 24 h, v) NaOMe, MeOH, rt, 4 h, overall 52%.

SCHEME 5.20 Synthesis of repeating pentasaccharide **239** corresponding to EC-O117:K98:H4 antigen.

The other disaccharide building block **245** was obtained by NIS and $HClO_4$-SiO_2-mediated glycosylation of **244** with thiogalactoside donor **243** followed by conversion of the anomeric PMP to TCA. Pentasaccharide derivative **247** was obtained at 64% yield following a [1+2+2] one-pot glycosylation reaction using orthogonal glycosylation based on activators C2$HClO_4$-SiO_2 for first activation of TCA-disaccharide donor **245;** this they reacted with thiomannoside acceptor **246**, and then in the same pot, they added $HClO_4$-SiO_2 for an NIS-mediated second activation of an in situ formed intermediate trisaccharide thioglycoside to react with the second disaccharide building block **242**. Finally, **247** was converted to the target pentasaccharide as aminoethyl glycoside **248** through a series of functional group transformations including TEMPO-BAIB mediated selective primary hydroxy oxidation to carboxylic acid of the Gal-moiety in one of these transformation steps and final global deprotection (Scheme 5.21) [130].

In 2017, following a convergent glycosylation approach, Si and Misra performed the synthesis of the pentasaccharide RU as its 2-aminoethyl glycoside containing nonreducing 4,6-O-pyruvic acid acetalated D-Gal, corresponding to EC-O156 antigen [131]. In 2020, they used linear synthetic to prepare the repeating pentasaccharide unit of EC-O43 antigen as their attempt for an initially planned convergent [3+2] glycosylation step failed to generate the corresponding pentasaccharide [132].

Other researchers synthesized oligosaccharides of several other enterotoxicogenic EC strains. Several OSs (tetra- to dodeca-saccharides) and their BSA and CRM197 glycoconjugates corresponding to O-antigens of EC-O148 having similarity in PS structural pattern with *S. dysentriae* type 1 (SD-O1) were synthesized. Immunological evaluation of these glycoconjugates demonstrated that these elicited anti-EC-O-PS specific antibodies. Crossreactivity of EC-O148 and SD-O1 PS with the antisera and inhibition of binding between antisera and O-antigens of both the bacterial strains were also witnessed [133].

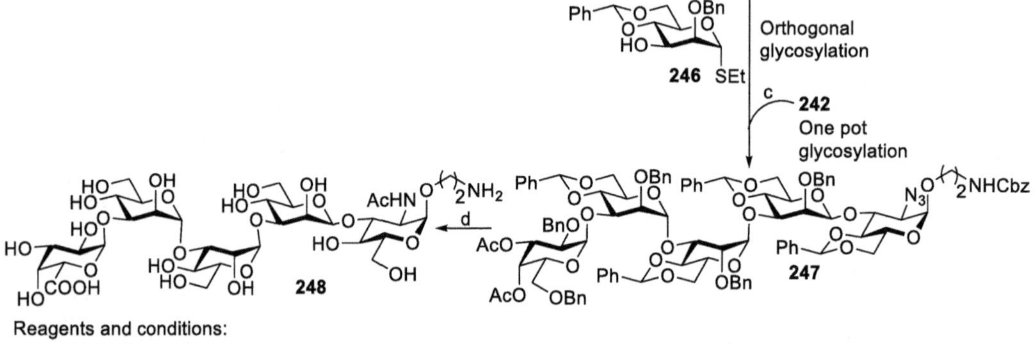

Reagents and conditions:

(a) i) BSP, TTBP, Tf2O, -60 °C, then **241**, -78 °C, 2 h, 55%, β:α = 4:1, ii) NaOMe, MeOH, 1h, 92%, iii) Tf2O, Py, DCM, 1 h, -10 °C, iv) NaN_3, DMF, TBAB, 70 °C, 37% in 2 steps, v) $PdCl_2$, MeOH, 0 °C - rt 3 h, 76%; (b) i) NIS, $HClO_4$-SiO_2, Et_2O-DCM 3:1, -10 °C, 1 h, 68%, ii) CAN, MeCN-H_2O 4:1, 0 °C, 2 h, iii) CCl_3CN, DBU, DCM, -10 °C, 1 h, 58%; (c) $HClO_4$-SiO_2, DCM, -10 °C, 20 min then **242** followed by NIS, -10 °C, 30 min; (d) i) CH_3COSH, Py, rt, 24 h, ii) 20% $Pd(OH)_2$-C, H_2, MeOH, rt, 6 h, iii) NaBr, TBAB, TEMPO, $NaHCO_3$, NaOCl, DCM, H_2O, 0 - 5 °C, 3 h, ^{t}BuOH, 2-methylbut-2-ene, $NaClO_2$, NaH_2PO_4, rt, 3 h, iv) H_2, $Pd(OH)_2$-C, MeOH, rt, 24 h, overall 50% in 4 steps.

SCHEME 5.21 Synthesis of repeating acidic pentasaccharide **248** related to avian EC-O59.

EC-O111 shows a different virulence mechanism and can be categorized as an EAEC/EPEC/STEC strain [134]. O-111 contains a rare sugar, L-colitose. Seeberger et al. achieved the de novo synthesis of L-colitose-based glycosyl donor by starting with commercially available *S*-ethyl lactate. They employed sequential linear glycosylation to synthesize a trisaccharide diol acceptor and double glycosylation of the latter with colitose-based donor for the first synthesis of the pentasaccharide as its 5-aminopentyl glycoside related to EC-O111 antigen [135].

O-antigenic PS of EC-O120, an STEC strain, has a repeating hexasaccharide unit [α-L-Rhap-(1→4)-β-D-GlcAp-(1→2)-α-L-Rhap-(1→2)-α-L-Rhap-(1→2)-α-D-Galp-(1→3)-β-D-GalNAcp] [136]. Budhadev and Mukhopadhyay achieved the synthesis of the repeating hexasaccharide as its PMP glycoside related to EC-O120 by introducing the carboxylic acid function at a late stage of the total synthesis by primary hydroxy oxidation on a precursor neutral hexasaccharide intermediate. For the glycosylation steps, they relied on thioglycoside donors and one trichloroacetimidate donor, activating the latter with $HClO_4$-SiO_2 and the former with NIS, $HClO_4$-SiO_2 [137].

Later, Mukherjee and Ghosh developed efficient one-pot synthetic routes for the synthesis of EC-O120 PS-related acidic pentasaccharide **257** separately utilizing sequential three-component (3C) [1+ 2+2] one-pot and four-component (4C) [1+2+1+1] one-pot glycosylation with late-stage development of the acid function (Scheme 5.22) [138]. Thus, they synthesized the thiodisaccharide acceptor **251** by initial orthogonal α-selective TMSOTf-mediated glycosylation of TCA donor **249** and thiorhamnoside acceptor **250** under optimized condition followed by selective O-deacetylation. In other work, the other disaccharide acceptor **254** was obtained by initial NIS-$FeCl_3$-mediated efficient α-glycosylation of acceptor **252** with thioglycoside donor **253** followed by Zemplén deacetylation. They effectively synthesized neutral pentasaccharide derivative **256** at 78% and 72% yields following [3+1+1] 3C one-pot (using $FeCl_3$ and NIS as activators) glycosylation based on building blocks **255**, **251**, and **254** and 4C [1+2+1+1] one-pot glycosylation (using sequentially semiorthogonal activation of TCA donor using $FeCl_3$; preactivation of the in situ formed trisaccharide thioglycoside with Ph_2SO, TTBP, Tf_2O and final activation of the in situ formed thioglycoside tetrasaccharide using $FeCl_3$-NIS activator systems) based on building blocks **255**, **251**, **250**, and **252** under controlled conditions as depicted in the one-pot schemes. Debenzylation of **256** followed by selective primary OH oxidation, benzyl ester formation, and global debenzylation finally afforded the target acidic pentasaccharide **257** (Scheme 5.22).

EC-O132 is an STEC strain. Mukhopadhyay et al. synthesized EC-O132 PS-related pentasaccharide based on [3+2] convergent glycosylation, thioglycosides as glycosyl donors and NIS-TMSOTf as the activator group [139].

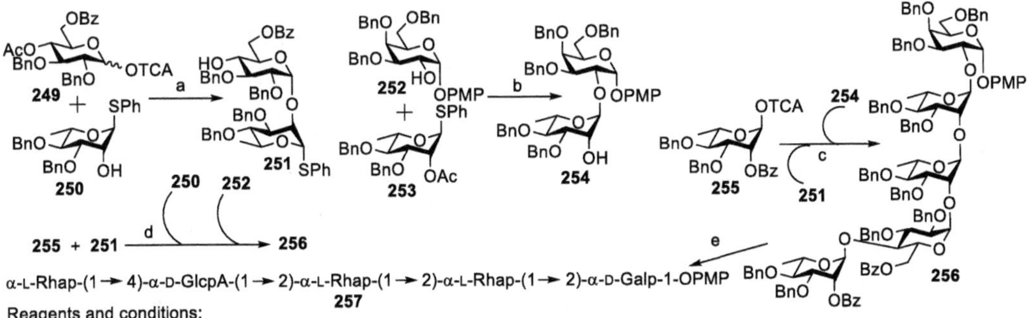

Reagents and conditions:
(a) (i) TMSOTf, 0 °C, DCM/Et_2O 3:2, 90%, α:β 9:1, ii) CH_3COCl, MeOH, 0 °C, 2 h, 88%; (b) i) NIS, $FeCl_3$, DCM, 0 °C, 15 min; 92%, ii) NaOMe, MeOH, DCM, quantitative; (c) $FeCl_3$, -60 °C - rt, 45 min then **254**, NIS, $FeCl_3$, 0 °C to rt, 10 min, 78%; (d) $FeCl_3$, -60 °C - rt, 45 min then Ph_2SO, Tf_2O, TTBP, -60 °C, 10 min then -40 °C and add **250** raise temerature to rt through 2 h, then **252**, NIS, $FeCl_3$, 0 °C to rt, 10 min, 72% for [1+2+1+1] four component one-pot reaction; (e) i) NaOMe, MeOH, DCM 1:1, 3 h, ii) BAIB, TEMPO, DCM, H_2O 2:1, 8 h, iii) K_2CO_3, BnBr, DMF, 79% over 3 steps, iv) Pd-C, H_2, 3 h, 92%.

SCHEME 5.22 Synthesis of EC-O120-related pentasaccharide derivative **256** utilizing one-pot [1+2+2] 3C and one-pot [1+2+1+1] 4C glycosylation and synthesis of pentasaccharide **257**.

Mandal's group also pursued STEC strains EC-O163 and EC-O181; the former strain causes renal failure in children and renal/nonrenal consequences in other survivors. Mandal et al. carried out synthesis of the EC-O163 RU tetrasaccharide and its analogues as their PMP glycosides (**265** and **266**) utilizing one-pot glycosylation reactions of two mono and one disaccharide building blocks (Scheme 5.23). They initially prepared a disaccharide derivation (**260**) at 87% yield by NIS, H_2SO_4–silica-mediated initial glycosylation reaction based on two glucose units (**258** and **259**), and then three-step C2 epimerization of the reducing end D-glucose generated the disaccharide building block **261**. One-pot glycosylations of this with other two monosaccharide building blocks (**262** and **263**) afforded the desired tetrasaccharide derivative **264**. Following a series of functional group transformations, **264** yielded the tetrasaccharide **265** related to EC-O163 and its analogue **266** [140].

The structure of EC-O181 antigen is as follows: [→4)-{α-L-QuipNAc-(1→3)}-α-D-GalpNAc-(1→6)-α-D-Glcp-(1→P-4)-α-L-QuipNAc-(1→3)-β-D-GlcpNAc-(1→]n [141]. Mandal's group recently reported the synthesis of the EC-O181-related pentasaccharide as its PMP glycoside via synthesis of a trisaccharide block using one-pot glycosylation reactions followed by [3+2] block glycosylation and utilizing NIS-H_2SO_4-silica as the activator in all glycosylation steps based on thioglycoside donors (Scheme 5.24) [142].

After preparing the QuiNAc-related rare sugar building block **268** starting with L-Rha through rhamnal triacetate **267** and the disaccharide building block **269** by armed–disarmed glycosylation reaction of **268** and **128**, they set forward for 3C one-pot glycosylation using building blocks **268**, **270** and **271** to generate high yields of trisaccharide derivative **272**, which upon selective debenzylidenation afforded the trisaccharide acceptor block **273**. Next, [3+2] block glycosylation of the disaccharide donor **269** and trisaccharide acceptor **273** yielded the desired pentasaccharide derivative **274** at 77% yield; several functional group transformation reactions finally gave the target pentasaccharide **275** (Scheme 5.24) [142].

The structure of the PS of the LPS of EC of type 74 belonging to STEC group, reported by Wildmaim et al. [143], shows the presence of rare sugar FucNAc. It has structural similarity with that of EC-O2, and the two strains exhibit crossreactivity. Recently, Bera and Mukhopadhyay synthesized the tetrasaccharide RU of EC-O74 antigenic PS [144]. The challenging donor FucNAc-based

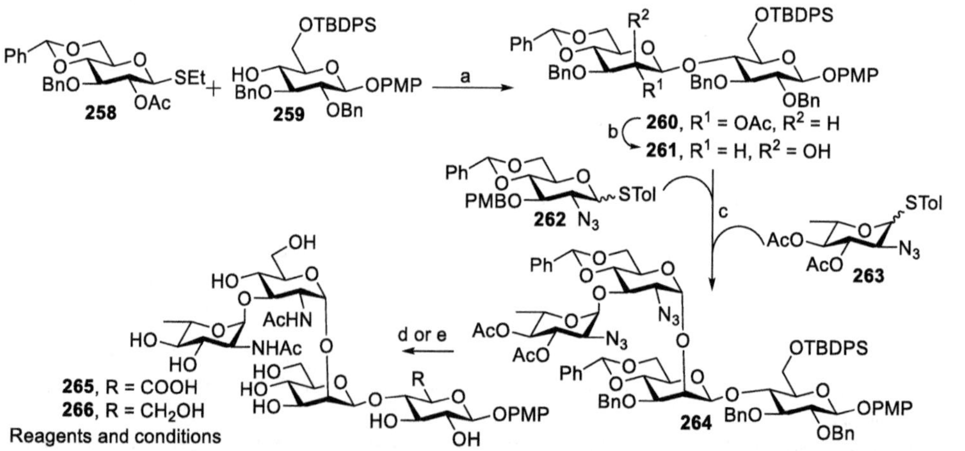

Reagents and conditions

(a) NIS, H_2SO_4-SiO_2, DCM, -20 °C, 30 min, 87%; (b) i) NaOMe, MeOH, rt. 1 h, 93%, ii) Dess-Martin periodinane, DCM, rt, 3 h, iii) NaBH4, MeOH, rt, 10 h, 74%; (c) NIS, H_2SO_4-SiO_2, DCM-Et_2O 1:2, -20 °C, 30 min then 10 °C, 30 min then **263**, NIS, -20 °C, 30 min, 61%; (d) i) CH_3COSH, Py, rt, 10 h, ii) Bu_4NF-THF, AcOH, 0 °C to rt, 6 h, iii) TEMPO, NaBr, $NaOCl_2$, 2-methylbut-2-ene, NaH_2PO_4, $NaHCO_3$, DCM, 0 °C to rt, 5 h, iv) H_2, 10% $Pd(OH)_2$-C MeOH, rt, 24 h, v) NaOMe, MeOH, rt, 1 h, 54%; (e) steps for (d) except oxidation step iii), 62%.

SCHEME 5.23 Synthesis of tetrasaccharides **265** and **266** related to the O-antigen of EC-O163.

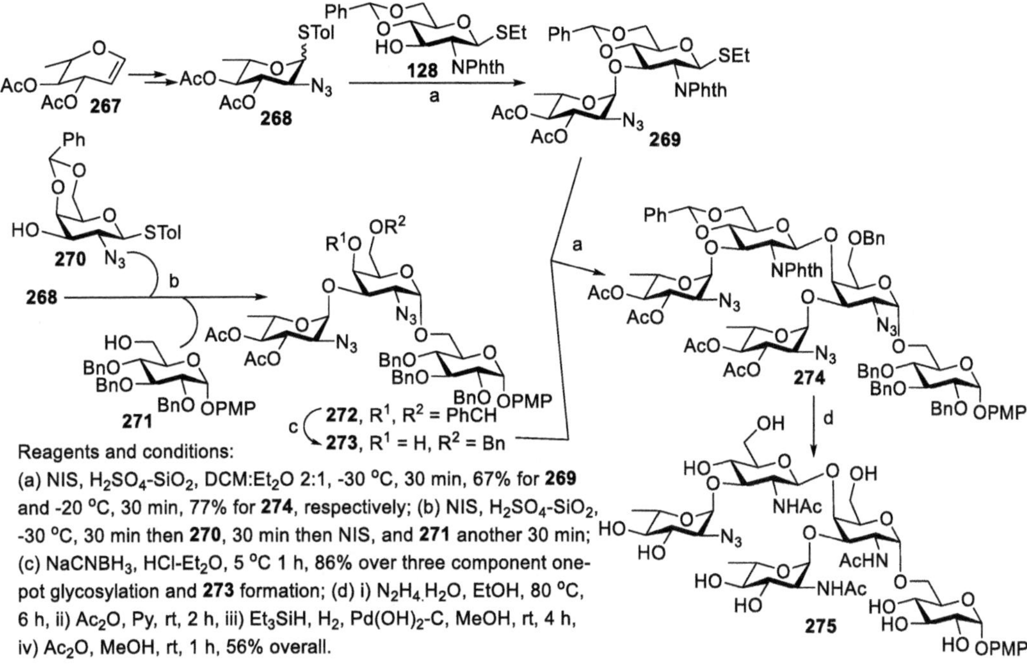

SCHEME 5.24 Synthesis of pentasaccharide repeating unit **275** related to the O-antigen of EC-O181.

building block **277** was prepared from **276** in seven steps using DTBP-TIPST for 6-deoxygenation in one of the steps. To achieve anomeric selectivity, 2-OBz was used in all glycosyl donors for NGP and NIS-TfOTf as the thiophilic activator. Thus, they initially coupled donor **277** with acceptor **278** to yield the disaccharide derivative **279** (70%); on selective 4,6-*O*-debenzylidenation using cyanuric chloride -$NaBH_4$ in MeCN, **279** gave the next acceptor **280** at 78%. Separately they obtained disaccharide-derived donor building block **283** through glycosylation of monosaccharide acceptor **282** with galactose TCA donor **281** through activation with TMSOTf. With this donor **283**, glycosylation of the previously prepared disaccharide acceptor **280** afforded the desired tetrasaccharide derivative **284** at 82% yield. This on acid promoted debenzylidenation followed by deprotection of NPhth group using ethylene diamine and acetylation, then azido conversion to NHAc and finally global deprotection ultimately provided the target tetrasaccharide **285** of EC O74 (Scheme 5.25).

The core structure of LPS in EC has limited distinction with only five different cores: R1, R2, R3, R4, and R12 [145]. Of these, R3 is particularly important as it is found in most of the VTEC isolates such as EC-O157:H7. Nearly 210 million diarrheal infections with 380,000 deaths are observed each year from this strain [146]. Following linear glycosylation, Li et al. synthesized the outer core pentasaccharide as 5-aminopentyl glycoside related to EC-R3 [147]. After preparing all the required monosaccharide building blocks for the target synthesis, they glycosylated thioglucoside acceptor **286** with orthogonal TCA-donor **197** using TMSOTf as a mediator in 50:50 Et_2O-DCM solvent to modulate the α-anomeric selectivity; this was followed by debenzylidenation, and then 4,6-di-O-benzylation to give Glc-disaccharide as thioglycoside **287** at 53% yield over these three steps.

Anomeric SPh deprotection with subsequent conversion to next TCA donor **288** was achieved with NBS-aqueous acetone and then DBU and Cl_3CCN at 68% yield in two steps. Glycosylation catalyzed by TMSOTf between donor **288** and Gal acceptor **289** yielded 55% of trisaccharide **290**, which after Zemplén debenzoylation gave trisaccharide acceptor **291**. Next, 2-azidoglucosylation (76% yield) using donor **292** and acceptor **291** afforded tetrasaccharide **293** after conversion in two steps (reduction using $NaBH_4$-$NiCl_2$ and then N-acetylation with Ac_2O) of the azido to acetamido (94% yield). CAN-mediated anomeric PMP deprotection followed by conversion of the resulting free anomeric OH to TCA using DBU, Cl_3CCN (53% in two steps) resulted in tetrasaccharide donor **294**. TMSOTf catalyzed the final glycosylation of acceptor **295** with tetrasaccharide TCA donor **294** generated pentasaccharide derivative **296** at 42% yield; on global deprotection in two steps, this ultimately accomplished the synthesis of target pentasaccharide **297** (Scheme 5.26) [147].

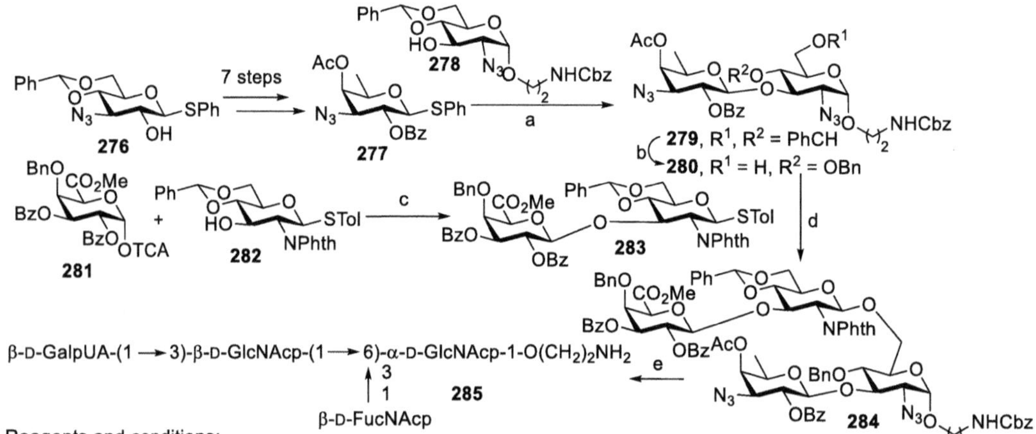

Reagents and conditions:
(a) NIS, THSOTf, DCM, -4 °C, 10 min, 70%; (b) TCT, $NaBH_4$, MeCN, rt, 8 h, 78%; (c) TMSOTf, DCM, 12 h, 75%; (d) NIS, TMSOTf, DCM, -4 °C, 10 min, 82%; (e) i) 80% Acetic acid, 80 °C, 78%, ii) Ethylene diamine, Ac_2O, Py, iii) Thioacetic acid, iv) NaOMe, MeOH then H_2O, v) H_2, Pd-C, 63%.

SCHEME 5.25 Synthesis of repeating tetrasaccharide **285** of EC-O74 antigen.

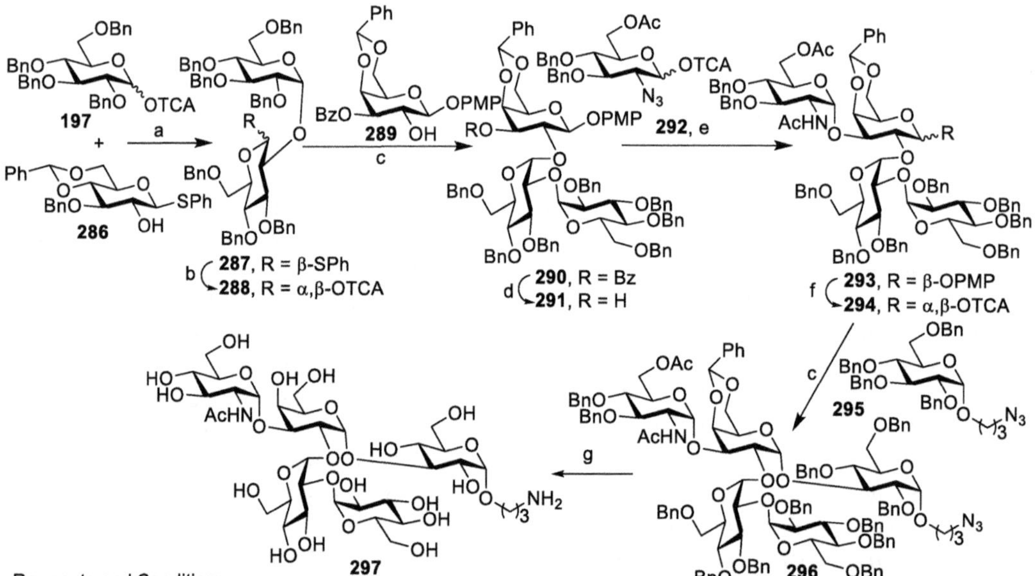

Reagents and Conditions:
(a) i) TMSOTf, Et_2O:DCM 1:1, -40 °C to rt, ii) DCM:TFA:H_2O 10:1:0.1, ii) NaH, BnBr, DMF, 53% (3 steps); (b) i) NBS, Me_2CO:H_2O 9:1, ii) CCl_3CN, DBU, DCM, 68% (2 steps); (c) TMSOTf, Et_2O:DCM 1:1, -20 °C to rt, 55% for **290**, and -25 °C to rt, 42% for **296**, respectively; (d) NaOMe, MeOH, 45 °C, 96%; (e) i) TMSOTf, Et_2O:DCM 1:1, -20°C to rt, 76%, ii) $NaBH_4$, $NiCl_2.H_2O$, DCM:MeOH 1:1.5, iii) Ac_2O, 94% over 2 steps; (f) i) CAN, Toluene:MeCN:H_2O 1:1.5:1, ii) CCl_3CN, DBU, DCM, 53% (2 steps); (g) i) NaOMe, MeOH, ii) H_2, $Pd(OH)_2/C$, 72% (2 steps).

SCHEME 5.26 Synthesis of outer core oligosaccharide **297** of EC-R3.

They also prepared the corresponding CRM197-glycoconjugate. Immunological evaluation of these revealed that the semisynthetic glycoconjugate could induce specific antibodies with significant in vitro opsonophagocytic activity against EC-O157:H7.

5.1.7 Enteropathogenic EC (EPEC) and Enteroaggressive EC (EAEC)

Epidemiological studies indicated that atypical EPEC is prevalent in both advanced and developing countries and is one of the causes of endemic diarrhea, particularly in infants and related outbreaks [148].

As mentioned earlier, Ghosh et al. synthesized a common tetrasaccharide as methyl glycoside related to LPS of enteropathogenic EC-O9, KP-O3 and *H. alvei* PCM 1223 utilizing 3C one-pot glycosylation employing NMPTC/TMSOTf as promoting agents for reactivity-based thioglycoside donors relying on successive one-pot glycosylations (Scheme 5.3) [45]. Her research group next used inexpensive and readily available nontoxic trichloroisocyanuric acid as an activator in combination with TMSOTf for thioglycoside activation during sequential one-pot glycosylation reactions using orthogonal glycosylation to synthesize a tri- and a tetra- mannoside; these motifs were observed in outer PS of HIV-1, *M. tuberculosis*, and EC-O9 [149]. Following glycosylation comprising iterative as well as convergent [3+2] block synthesis, Mandal carried out the synthesis of the pentasaccharide RU of enteropathogenic EC-O117:K98:H4 using TCA as well as thioglycoside donors [150].

EC-O158 is an EPEC strain. Perepelov et al. established the structure of EC-O158 PS with a pentasaccharide RU to contain a doubly branched β-ManNAc sugar [151]. Mitra and Mukhopadhyay reported in 2016 the synthesis of the repeating pentasaccharide as its PMP-derived triethylene glycol-based glycoside. They employed a [3+2] convergent approach and thioglycosides as the glycosyl donor and NIS with H_2SO_4-SiO_2 as the activator system in all glycosylation steps. β-D-ManNAc unit was achieved from a β-D-Glc-derived unit by inversion with N-function at C2; then, they obtained the α-D-GalpA unit from the corresponding α-D-Gal unit by TEMPO-mediated oxidation at a late stage of the total synthesis [152].

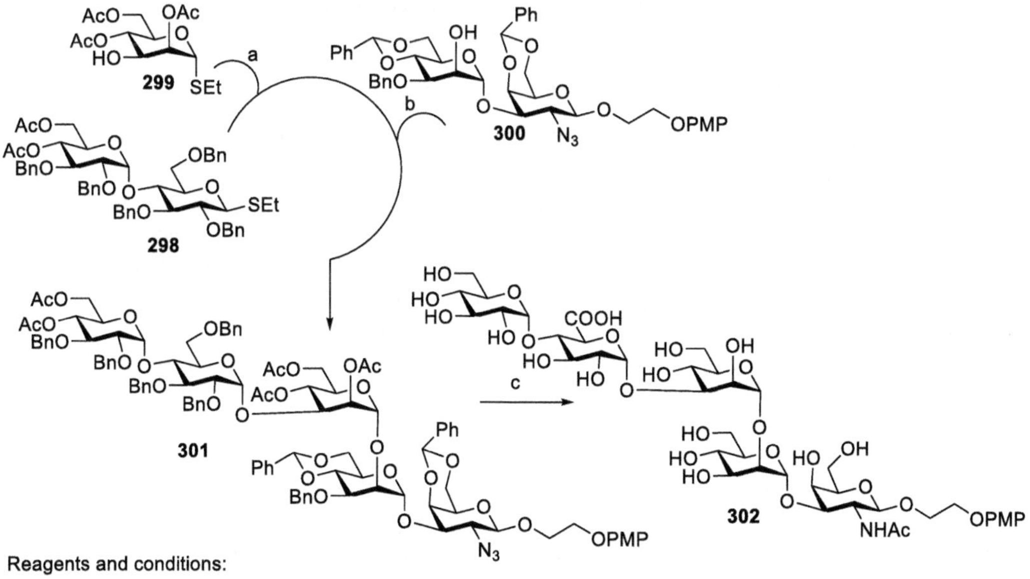

Reagents and conditions:
(a) NIS, TfOH, Et_2O, 10 °C, 1 h; (b) NIS, TfOH, Et_2O:DCM 1:1, -10 °C, 1 h, 54% over two steps; (c) i) H_2, Pd-C, MeOH:DCM 2:1, rt, 4 h, ii) Ac_2O, MeOH, rt, 2 h, iii) NaBr, TBAB,TEMPO, NaOCl, DCM, H_2O, 0 - 5 °C, 4 h, iv) ^{t}-BuOH, 2-methyl-but-2-ene, $NaClO_2$, NaH_2PO_4, rt, 4 h, v) H_2, Pd-C, MeOH, rt, 12 h, vi) NaOMe, MeOH, rt, 5 h, 49% overall.

SCHEME 5.27 Synthesis of pentasaccharide **302** corresponding to the EC-O175 antigen.

Apart from synthesizing oligosaccharides related to ETEC strains, Misra's group also carried out syntheses of oligosaccharides corresponding to EAEC and EPEC strains. His group reported on the synthesis of the pentasaccharide RU of EAEC strain EC-O175 and carried out conformational analysis of the same by ROSEY-NMR (nuclear magnetic resonance) and simulation. They synthesized the pentasaccharide derivative as 2-(*p*-methoxyphenoxy)ethyl glycoside **301** at 54% yield from the corresponding building blocks **298**, **299**, and **300** following armed–disarmed glycosylation based on thioglycosides in one-pot as shown in Scheme 5.27. A series of multistep functional group transformations including selective removal of O-benzyl group in presence of 4,6-O-benzylidine protection followed by acetylation of ammine group in presence of free hydroxy group and then oxidation of the primary hydroxy in one of the late steps finally yielded the target pentasaccharide **302** [153].

To synthesize a pentasaccharide related to EPEC strain EC-O115, researchers employed [3+2] convergent glycosylation using thioglycosides as the common glycosyl donor except in one step and NIS-$HClO_4$-SiO_2 as the corresponding activator [154]. To generate disaccharide **311**, β-rhamnosylation was carried out at 70% yield using L-Rha-TCA donor **310** bearing a remote picoloyl (Pico) group on C3-O for H-bond-assisted aglycon delivery and a thioglycoside acceptor **309**. $Cu(OAc)_2$-based deprotection Pico-group of **311** followed by acetylation using Ac_2O gave disaccharide thiglycoside building block **312**.

Separately, to prepare trisaccharide acceptor **308**, the first glycosylation was performed based on acceptor **303** and Man-thioglycoside donor **304** with PMB-protected C3-O as temporary protection for getting directly disaccharide acceptor **305** in one-pot (71% yield). Next, glycosylation in 73% on disaccharide acceptor **305** using Gal-thioglycoside **306** bearing a temporary levoloyl group on C4-O and temporary benzoyl group on C6-O generated trisaccharide **307**, which after delevoloylation using hydrazine hydrate resulted in trisaccharide acceptor **308**. Although [3+2] convergent glycosylation utilizing **308** and 3′-O-Pico donor **311** failed, but successful [3+2] glycosylation using 3′-OAc donor **312** and acceptor **308** afforded the desired neutral pentasaccharide derivative **313** at 71% yield. Zemplén debenzylation of **313** followed by TEMPO-DAIB-mediated oxidation of the exposed primary OH group and then after four subsequent deprotection steps: i) phthalimido deprotection by $N_2H_4.H_2O$, ii) N-acetylation using Ac_2O, iii) Zemplén deacetylation, and iv) hydrogenative deprotection of benzyl and benzylidene groups using $H_2/Pd(OH)_2/C$; this finally provided the target pentasaccharide as its aminoethyl glycoside **314** at 43% over the last six steps (Scheme 5.28).

5.1.8 Other Diarrhea-Causing *Escherichia* Strains

Sau and Misra accomplished the synthesis of the tetrasaccharide RU of Shiga toxin producing EPEC-O40-PS following [2+2] glycosylation using thioglycoside donors and NIS-$HClO_4$-SiO_2 as the activator system [155]. Misra et al. achieved the linear iterative synthesis of a pentasaccharide related to EC-O13-PS and performed conformational analysis by ROSEY-NMR and molecular dynamics [156]. Mandal used a [2+3] convergent approach to synthesize a pentasaccharide of EC-O36-PS [157].

Mukhopadhyay's research group accomplished the synthesis of an acidic tetrasaccharide corresponding to the O-antigen of mastitis-causing EC-O174:H28 involving linear iterative glycosylation based on thioglycosides as glycosyl donors and NIS–H_2SO_4-SiO_2 as the activator; the acid function was introduced at a late stage of the total synthesis [158]. Later they used glycosylation to synthesize two hexasaccharides of EC-TD2158 O-antigen [159].

Very recently, Mukhopadhyay's group and Misra's group separately synthesized the repeating tetrasaccharide as its aminopropyl glycoside related to EC-O131 antigen [160]. The RU of EC-O131 antigen is a tetrasaccharide composed of (2→6)-linked-α-D-NeupA along with two D-Gal and one-D-GalNAc monomer units. Mukhopadhyay's group used differentially protected acceptors, thioglycosides as glycosyl donors, and NIS-TMSOTf or NIS-TfOH as the thioglycoside activators were employed in the high-yielding stereoselective linear sequential glycosylation steps using 4,6-O-benzylidene and chloroacetyl as temporary protecting groups at suitable positions of the building blocks [160a].

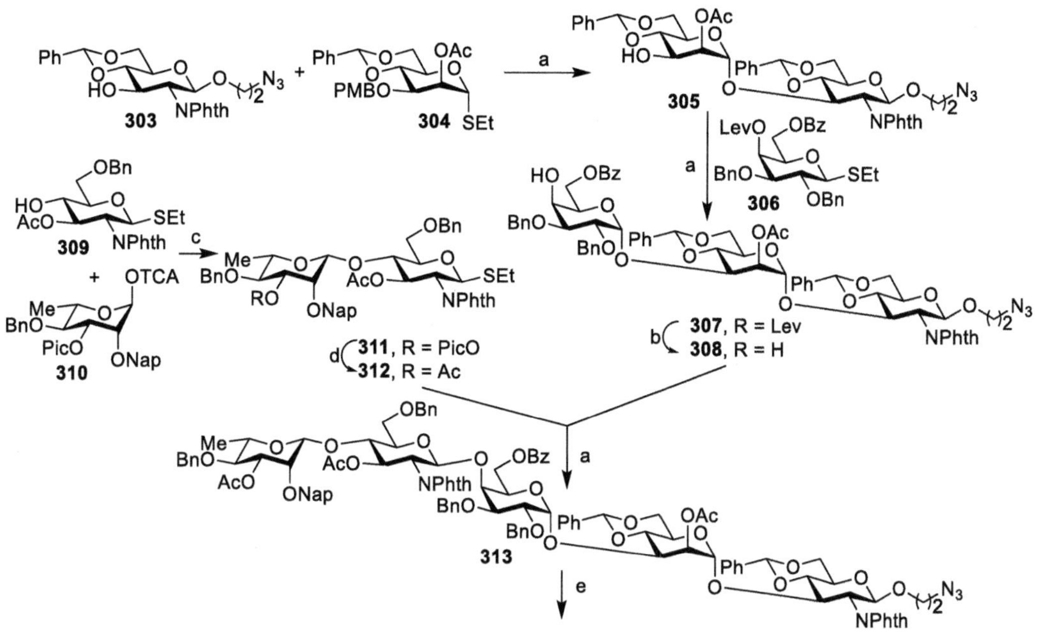

β-L-Rhap-(1→4)-β-D-GlcNAcp-(1→4)-α-D-GalpA-(1→3)-α-D-Manp-(1→3)-β-D-GlcNAcp-1-O-$(CH_2)_3NH_2$
314

Reagents and conditions:
(a) NIS, $HClO_4$-SiO_2, DCM, -30 °C, 45 min then 15 °C, 30 min, 71% for **305**, -15 °C, 45 min, 73% for **307** and -40°C, 1 h, 71% for **313**, respectively; (b) $N_2H_4.H_2O$, AcOH, DCM-MeOH 1:1, rt, 5 h, 76%; (c) $HClO_4$-SiO_2, DCM, -45 °C, 70%; (d) i) $Cu(OAc)_2$, DCM:MeOH 4:1, rt, 30 min, ii) Ac_2O, Py, rt, 1 h, 75%, overall 53%; (e) i) NaOMe, MeOH, rt, 30 min, ii) TEMPO, BAIB, MeOH:H_2O 1:1, rt, 3h, iii) $N_2H_4.H_2O$, EtOH, 80 °C, 15 h, iv) Ac_2O, Py, rt, 1 h, v) NaOMe, MeOH, rt, 3 h, vi) H_2, $Pd(OH)_2$-C, MeOH, rt, 24 h, 43% over 6 steps.

SCHEME 5.28 Convergent synthesis of pentasaccharide **314** corresponding to the EC-O115 antigen.

Disaccharide acceptor **317** was obtained at 76% yield in two steps by glycosylation of acceptor **315** with donor **316** using NIS-TMSOTf activator system followed by debenzylidenation with 80% aqueous acetic acid. Acceptor **317** was further glycosylated using glycosyl donor **318** under the earlier-mentioned reaction condition to generate trisaccharide **319** in high yield (85%), which after thiourea-mediated chloroacetyl deprotection gave trisaccharide acceptor **320** at 86% yield. On glycosylation with neuraminic acid-based thioglycoside donor **321** through activation with NIS-TfOH, this furnished the desired acidic tetrasaccharide derivative **322** at 78% yield with high α selectivity. Finally, required protection and deprotections were made in five sequential steps: i) Zemplén deacylation, ii) phthlimido deprotection using ethylene diamine, iii) usual N-acetylation, iv) deacetylation, and finally v) hydrogenative deprotection, giving an overall 49% yield of target 3-aminopropyl tetrasaccharide **323** (A in Scheme 5.29).

Misra's group adopted a similar approach with a different protection group profile on monosaccharide units and glycosidic reagent to synthesize the target tetrasaccharide **131** as PMP glycoside [160b]. Disaccharide diol acceptor **326** was achieved by glycosylation of thioglycoside donor **325** with acceptor **324** using NIS–H_2SO_4-SiO_2 followed by removal of the 4,6-O-benzylidene protection using H_2SO_4-SiO_2. Coupling of this diol acceptor with the same donor **325** using the same activator system produced corresponding 1→6 trisaccharide **327** at 70% yield. Acetylation of free 4′-OH group followed by 4,6-O-benzylidene deprotection using H_2SO_4-SiO_2 resulted in trisaccharide diol acceptor **228** at 76% yield; this was further glycosylated with neuraminic acid-based thioglycoside donor **329** producing tetrasaccharide derivative **330** at 71% yield. Global deprotection of this tetrasaccharide furnished 46% of the desired deprotected tetrasaccharide **331** as its PMP glycoside (B in Scheme 5.29).

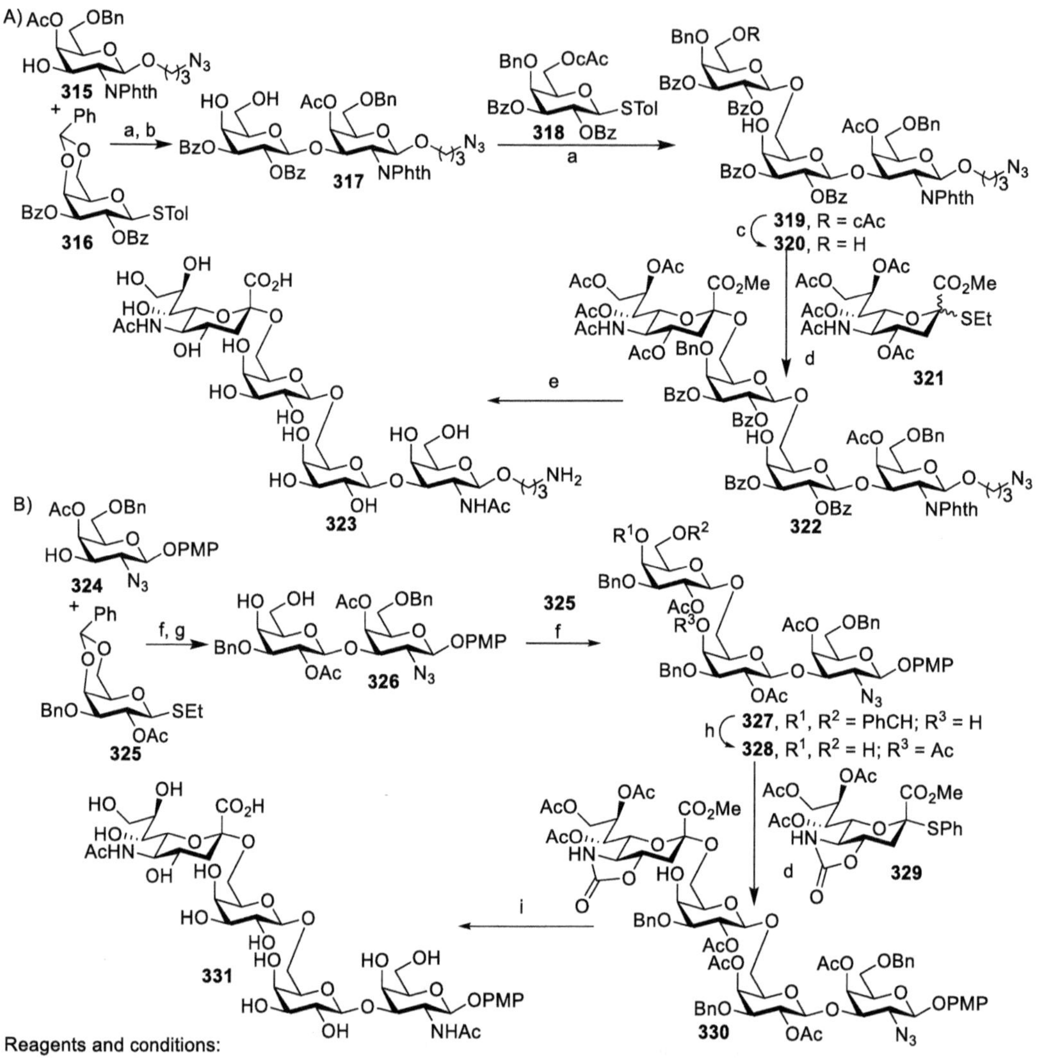

Reagents and conditions:
(a) NIS, TMSOTf, DCM, 0 °C; (b) 80% aq AcOH, 80 °C, 76% over 2 steps; (c) thiourea, 2,4,6-Collidine, rt, 86%; (d) NIS, TfOH, DCM:MeCN 3:2, -60 °C, 78% for **322** and DCM, -35 °C, 2 h, 71% for **330**; (e) i) MeONa, MeOH, then H_2O, 65 °C, ii) Ethylenediammine, n-BuOH, reflux, iii) Ac_2O, Py, rt, iv) MeONa, MeOH, rt, v) H_2, Pd/C, MeOH, rt, 49% over 5 steps; (f) NIS, $HClO_4$-SiO_2, DCM, -25 °C, 1 h, 78% for **326** precursor and -40 °C, 1 h, 70% for **327**; (g) $HClO_4$-SiO_2, MeCN, rt, 30 min, 85%; (h) i) Ac_2O, Py, rt, 2 h, ii) $HClO_4$-SiO_2, MeCN, rt, 30 min,76% in two steps; (i) i) MeCOSH, Py, rt, 24 h, ii) $LiOH.H_2O$, $EtOH:H_2O$ 3:1, 80 °C, 16 h, (ii) $NaHCO_3$, Ac_2O, H_2O, rt, 5 h, iv) NaOMe, MeOH, rt, 5 h, v) H_2, $Pd(OH)_2$-C, MeOH, rt, 24 h, 46% over 4 steps.

SCHEME 5.29 Synthesis of tetrasaccharide related to EC-O131 containing *N*-acetyl neuraminic acid.

EC-O44:H18 is an EAEC strain that caused several diarrhetic outbreaks in Europe [161]. Recently, Misra et al. used convergent glycosylation to synthesize the pentasaccharide RU as its PMP glycoside corresponding to the cell wall PS of EC-O44:H18 [162]. A linear sequential glycosylation pathway proved to be much more effective than [3+2] convergent synthesis of a branched pentasaccharide related to this strain. To generate the challenging β-mannosidic linkage, they selected a 2-OPMB and 4,6-O-benzylidene protected mannosyl thioglycoside donor **333**. In all glycosylation steps, they used thioglycoside donors and for their activation NIS–H_2SO_4-SiO_2. They were fortunate enough to get the desired β-linked disaccharide **334** at 74% yield without generating any detectable α-mannoside-linked disaccharide through glycosylation of 2-azido-2-deoxy -D-Glc acceptor **332** with thiomannoside donor.

NMR characterization of **334** by 1H- and 13C-NMR, δ values for H-1 and C1 of the mannoside (4.54, bs, and 102.1 ppm) including $J_{C1\text{-}H1}$ coupling at a constant value of 157.5 Hz for the Man unit confirmed the β-mannoside anomeric stereochemistry of compound **334**. An attempt for β-mannosylation under similar condition based on the corresponding 4,6-di-O-benzyl derivative analogous to the acceptor **332** failed. Conversion of **334** to the next glycosyl acceptor took place in four steps: i) excellent direct benzylation of 4,6-di-OAc, ii) DDQ-mediated deprotection of PMB, iii) usual acetylation by acetic anhydride-pyridine followed by iv) selective debenzylidenation-6′-O-benzylation afforded acceptor **335** in 68% yield over the last three steps. On next glycosylation with thioglucoside donor **336**, this generated trisaccharide **337**, which on deacetylation gave trisaccharide acceptor **338** at overall high yield. After this, their attempt to generate the pentasaccharide through a [3+2]-convergent glycosylation using a planned disaccharide donor failed completely and prompted them to instead use linear sequential glycosylation [162].

The next glycosylation was based on acceptor **338** and thiomannoside donor **339**, which afforded the corresponding tetrasaccharide-based acceptor **340** after the deacetylation step in high α-anomeric selectivity. The last α-mannosylation reaction was then carried out on **340** with common donor **339** to afford pentasaccharide derivative **341** in 78% yield. This after thioacetic acid mediated direct conversion of azido to acetamido, then deacylation followed by hydrogenative deprotection of benzyl and benzylidene groups finally furnished the target pentasaccharide **342** as its PMP glycoside (Scheme 5.30) [162].

EC O20:K83:H26 belongs to EPEC and has caused multiple epidemics [163]. Misra's group recently reported the synthesis of the repeating pentasaccharide of this strain [164]. They used thioglycoside as the glycosyl donor unit and NIS-$HClO_4$-SiO_2 as the thiophilic promoter in all

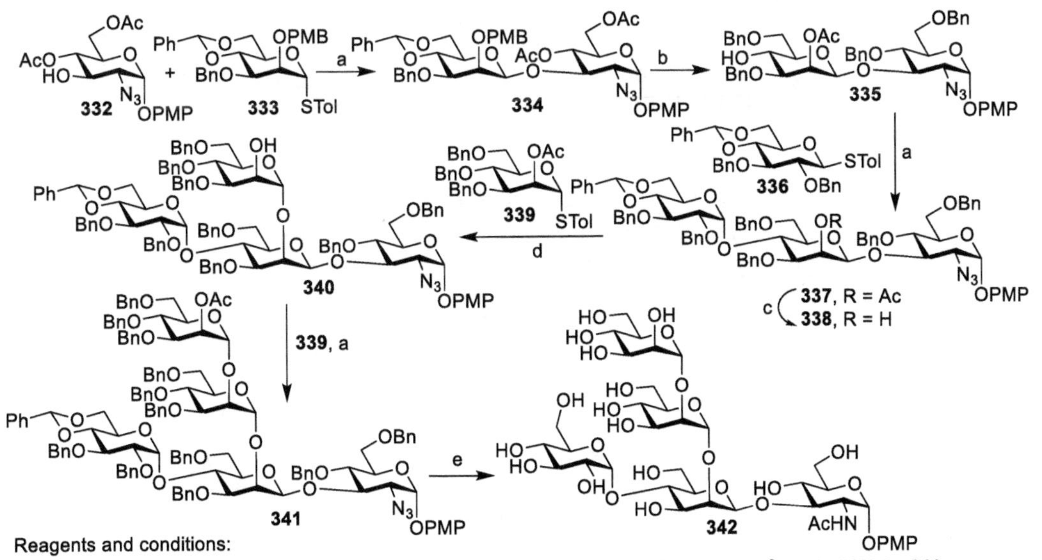

Reagents and conditions:
(a) NIS, $HClO_4$-SiO_2, DCM, -45 °C, 1 h, 74% for **334**, -15 °C, 1 h, 72% for **337**, and -15 °C, 1 h, 78% for **341** respectively; (b) i) BnBr, NaOH, TBAB, THF, rt, 3 h, 90%, ii) DDQ, DCM:H_2O 9:1, rt, 2 h, iii) Ac_2O, Py, rt, 1 h, iv) Et_3SiH, BF_3.Et_2O, DCM, 0 °C, to rt, 2 h, 68% in three steps; (c) NaOMe, MeOH, rt, 2 h, 92%; (d) i) NIS, $HClO_4$-SiO_2, DCM, -20 °C, 1 h, ii) NaOMe, MeOH, rt, 1 h, 70% in two steps; (e) i) CH_3COSH, Py, rt, 24 h, ii) NaOMe, MeOH, rt, 2 h, iii) H_2, $Pd(OH)_2$-C, MeOH, rt, 24 h, 51%.

SCHEME 5.30 Synthesis of the pentasaccharide **342** related to EC-O44:H18 antigen.

glycosylation steps (Scheme 5.31). They employed Lev as the temporary protecting group in the Manp-based donor unit. After synthesis of a neutral pentasaccharide by a [3+2] convergent strategy, they generated the galactopyranosyl uronic acid unit to synthesize the pentasaccharide RU of the O-antigen of this strain as its PMP glycoside. Coupling **343** with **344** in a dichloromethane–ether mixture using NIS-$HClO_4$-SiO_2 as the activator furnished disaccharide derivative **345**; after subsequent de-O-benzylidenation followed by regioselective 6-O-acetylation, **345** generated disaccharide acceptor **346**. Glycosyl acceptor **346** on further glycosylation with donor **347** utilizing the same activator and then delevoloylation by hydrazine hydrate afforded the required trisaccharide acceptor building block **348** at good yield [164].

In separate work, they obtained disaccharide donor **350** by reaction of partners **344** and **349**, again activating the donor with NIS-$HClO_4$-SiO_2. Coupling trisaccharide acceptor **348** and disaccharide donor **350** based on NIS-$HClO_4$-SiO_2 generated the desired neutral pentasaccharide **351** at 65% yield. On conversion of azido to acetamido by thioacetic acid–pyridine followed by sequential Zemplén deacetylation, BAIB-TEMPO-mediated selective oxidation of primary OH, and finally debenzylidenation–debenzylation by transfer hydrogenation, this ultimately afforded the target pentasaccharide **352** as its PMP glycoside (Scheme 5.31).

E. albertii (EA) is a new *Escherichia* strain that causes diarrhea. Mukhopadhyay et al. recently synthesized the pentasaccharide repeated unit of EA-O2 utilizing a [3+2] convergent pathway [165]. Misra et al. synthesized a pentasaccharide related to another strain, EA-O4 PS, in 2020 based on linear glycosylation [166]. They used [3+2] convergent armed–disarmed synthesis using thioglycosides as the anomeric leaving group for total synthesis as 2-aminoethyl glycoside of the pentasaccharide RU of EA-O2 antigen. They achieved 1,2-*trans* anomeric selectivity through NGP by the 2-phthlimido

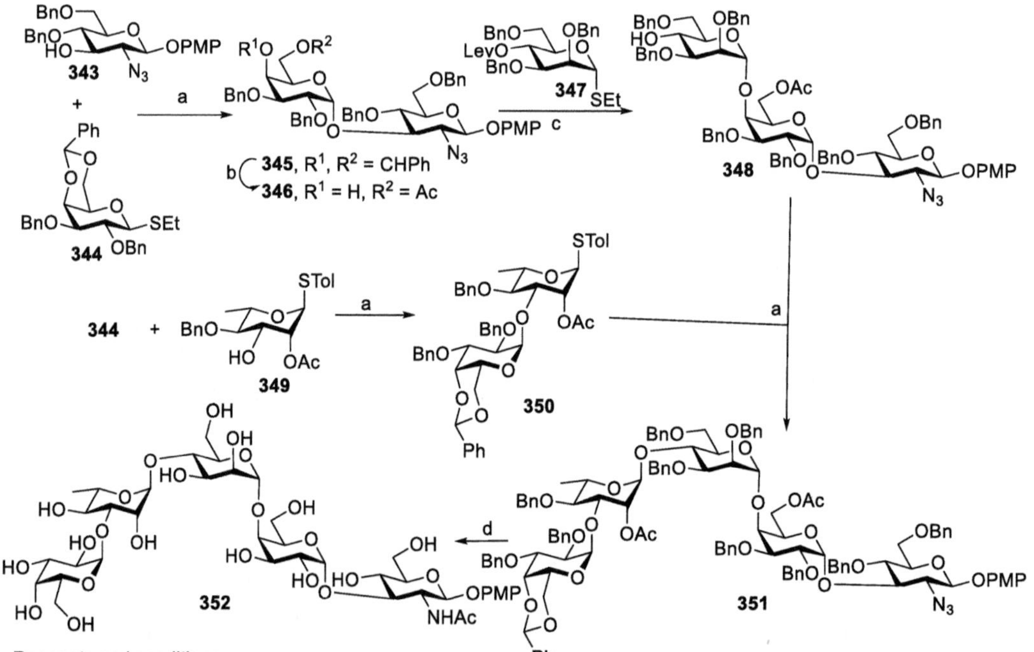

Reagents and conditions:
(a) NIS, $HClO_4$-SiO_2, DCM:Et_2O 3:1, -15 °C, 1 h, 67% for **345**, 68% for **350** and DCM, -15 °C, 2 h, 65% for **351**, respectively; (b) i) $HClO_4$-SiO_2, MeCN, rt, 30 min, ii) Ac_2O, Py, DCM, 0 °C, 2 h, 74% in 2 steps; (c) i) NIS, $HClO_4$-SiO_2, DCM, -25 °C, 1h, ii) AcOH, N_2H_4, H_2O, MeOH:DCM 1:1, 0 °C, 8 h, 72%; (d) i) CH_3COSH, Py, rt, 12 h, ii) NaOMe, MeOH, rt, 6 h, iii) BAIB, TEMPO, DCM;H_2O 2:1, rt, 15 h, iv) Et_3SiH, 10% Pd/C, MeOH, rt, 24 h, 52% in 4 steps.

SCHEME 5.31 Synthesis of repeating pentasaccharide **352** corresponding to EC-O20:K83:H26 antigen.

group during glycosylation to incorporate the GlcNAc moieties in it [167]. However, Manna et al. had to follow a linear pathway to obtain a tailor-made petasaccharide related to EA-O4 as their attempt to generate a tetrasaccharide-intermediate by a convergent [2+2] pathway didn't work well [168].

5.1.9 EC Strains Causing UTI

Only a limited number of EC strains—EC-O4, EC-O6, EC-O14, EC-O22, EC-O69, EC-O75, and EC-O83—are responsible for UTI in humans [169]. Misra et al. reported the synthesis of the tetrasaccharide RU as its 2-aminoethyl glycoside of O-antigen of UTI strain EC-O69 utilizing one-pot [1+1+2] iterative glycosylation. They also performed conformational analysis based on NOE and ROESY-NMR and performed molecular dynamics simulation studies [170].

The structure of the RU of UTI-causing EC-O75 was elucidated by Erbing et al. [171] Sau and Misra first synthesized the tetrasaccharide and its one-anomeric analogue as their PMP glycosides related to EC-O75 employing [2+2] block glycosylation and generating β-mannosyl moiety in the tetrasaccharide derivative from its corresponding β-epimer precursor D-glycosyl moiety by C2-epimerization [172]. Later, Toshima et al. developed an effective stereo- and regioselective β-mannosylation reaction based on protected 1,2-anhydromannose as the glycosyl donor and boronic acid catalyst. Glycosyl diol acceptor-derived boronate ester, generated in situ from catalytic 4-nitrophenylboronic acid and a 4,6-diol acceptor, facilitated regioselective glycosylation based on 1,2-anhydro mannosyl donor. They applied this methodology for the concise synthesis of a tetrasaccharide related to EC-O75 antigen as shown in Scheme 5.32 [173].

The reducing end galactosyl acceptor **354** was selectively glycosylated at C3-O with disaccharide donor **356** at 96% yield through an in situ temporary blocking of **354** with PMP boronic acid (**353a**) as 4,6-boronate ester **355**. The resulting trisaccharide derivative **357** was again regioselectively glycosylated with 1,2-anhydromannosyl donor **358** and catalytic *p*-nitrophenylboronic acid (**353b**) to give the tetrasaccharide derivative **359** at excellent yield and with exclusive β-anomeric selectivity. Compound **359** after a series of protection-deprotection finally afforded the target tetrasaccharide **360** as its octyl glycoside (Scheme 5.32).

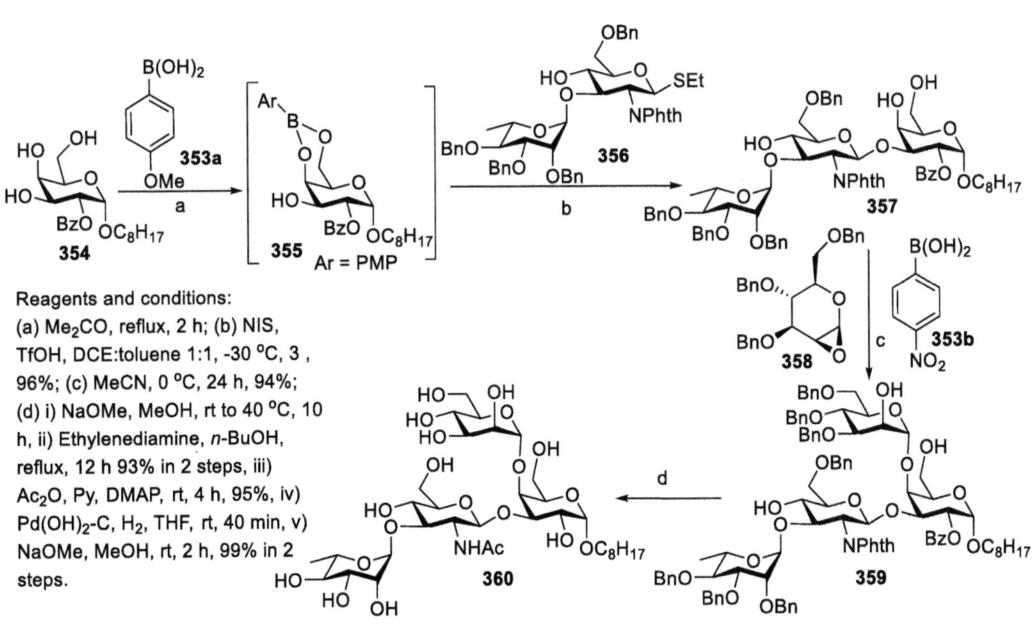

Reagents and conditions:
(a) Me_2CO, reflux, 2 h; (b) NIS, TfOH, DCE:toluene 1:1, -30 °C, 3 , 96%; (c) MeCN, 0 °C, 24 h, 94%; (d) i) NaOMe, MeOH, rt to 40 °C, 10 h, ii) Ethylenediamine, *n*-BuOH, reflux, 12 h 93% in 2 steps, iii) Ac_2O, Py, DMAP, rt, 4 h, 95%, iv) $Pd(OH)_2$-C, H_2, THF, rt, 40 min, v) NaOMe, MeOH, rt, 2 h, 99% in 2 steps.

SCHEME 5.32 Synthesis of tetrasaccharide **360** related to EC-O75 antigen.

5.1.10 EC Strains with Human Blood Group Antigens

EC-O86, EC-O90, EC-O127, and EC-O128 have human blood group antigens, and EC-O86 is an EPEC strain. In 2014, Mong's research group published the synthesis of EC-O86-related pentasaccharide RU utilizing *N*-morpholine-modulated [1+1+2] one-pot glycosylation to generate a tetrasaccharide-derived intermediate ***368*** at 56% yield from corresponding thiotolyl building blocks ***361*** and ***362*** and a disaccharide acceptor building block ***367*** (Scheme 5.33) [174]. Disaccharide acceptor ***367*** was prepared by coupling donor ***364*** with acceptor ***365*** followed by DDQ-mediated selective removal of the naphthylmethyl protection from disaccharide ***366***. Compound ***368*** on selective debenzoylation followed by NIS-TMSOTf-mediated glycosylation of the resulting tetrasaccharide acceptor ***369*** with thiofucoside donor ***370*** generated pentasaccharide derivative ***371*** at 87% yield; after several functional group transformations, this finally afforded the target pentasaccharide as methyl glycoside ***372*** (Scheme 5.33) [175].

Later, Misra's research group accomplished the synthesis of the same pentasaccharide as its conjugation-ready 2-aminoethyl glycoside following a linear glycosylation strategy; they also studied its conformational behavior with 2D-NMR and molecular dynamics simulation [176]. Misra et al. also synthesized the repeating pentasaccharide as its PMP glycoside corresponding to EPEC strain EC-O127 following linear glycosylation mediated by NIS–$HClO_4$-SiO_2 using thioglycoside

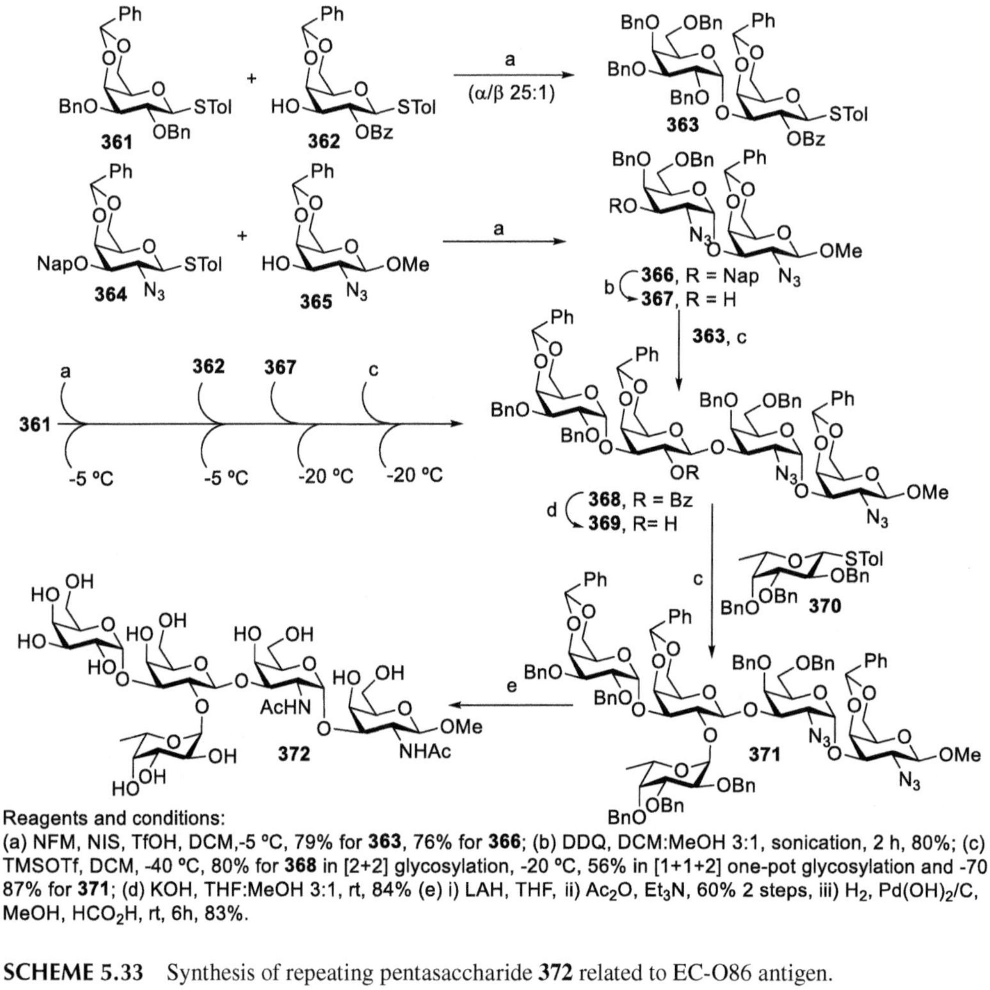

Reagents and conditions:
(a) NFM, NIS, TfOH, DCM,-5 °C, 79% for **363**, 76% for **366**; (b) DDQ, DCM:MeOH 3:1, sonication, 2 h, 80%; (c) NIS, TMSOTf, DCM, -40 °C, 80% for **368** in [2+2] glycosylation, -20 °C, 56% in [1+1+2] one-pot glycosylation and -70 °C, 87% for **371**; (d) KOH, THF:MeOH 3:1, rt, 84% (e) i) LAH, THF, ii) Ac_2O, Et_3N, 60% 2 steps, iii) H_2, $Pd(OH)_2/C$, MeOH, HCO_2H, rt, 6h, 83%.

SCHEME 5.33 Synthesis of repeating pentasaccharide **372** related to EC-O86 antigen.

donors [177]. To synthesize the pentasaccharide RU (as its PMP glycoside) of O-antigen of EC-O142, Misra's research group recently adopted a linear pathway for regioselective, 1,2-*cis* stereoselective, and two 1,2-*trans*selective glycosylation based on thioglycoside donors and NIS-$HClO_4$-SiO_2 as the thiophilic activator [178].

5.2 MISCELLANEOUS

Avian pathogenic EC strains cause colibacillosis in poultry farms, affecting poultry-related economy. Synthesis of a hexasaccharide corresponding to avian (duck) pathogenic EC-O133-PS was reported from Mukhopadhyay's laboratory [179]. The research group employed [2+2+1+1] convergent synthesis of a related suitable neutral hexasaccharide derivative. They achieved late-stage installation of the carboxylic acid function by the TEMPO-mediated oxidation of a primary OH group of the mentioned hexasaxccharide (Scheme 5.34). In the process, they obtained disaccharide-derived acceptor **376** and donor **380** in good yields, first mediated by H_2SO_4-SiO_2 and then by NIS–H_2SO_4-SiO_2 -mediated high-yielding glycosylation using TCA monomeric donor **373** and thiorhamnoside acceptor **374**. For the second method, they used thioglycoside donor **378** and rhamnopyranoside acceptor **377** followed by the required functional group transformations of corresponding disaccharide derivatives **375** and **379**. NIS, H_2SO_4-SiO_2 mediated [2+2] convergent glycosylation of disaccharide-based building blocks **376** and **380** furnished the tetrasaccharide derivative **381** at 83% yield. After conversion to corresponding tetrasaccharide-derived acceptor **382** and glycosylation with donor **383**, followed by DDQ-mediated deprotection, the process resulted in tetrasaccharide acceptor **385**. Subsequent glycosylation with thioglucoside donor **386** afforded desired hexasaccharide derivative **387**. After a series of functional group transformations including TEMPO-mediated oxidation of CH_2OH, this finally yielded the target hexasaccharide as conjugation-ready aminoethyl glycoside **388** (Scheme 5.34).

To advance a semisynthetic vaccine against avian EC-virulent O1 strain, Toshima's group recently synthesized a pentasaccharide related to this strain [180]. To prepare trisaccharide building block **393**, the authors prepared corresponding disaccharide **391**, first by the regio- and stereoselective rhamosylation of glycosyl acceptor **390** with 1,2-epoxy rhamno-donor **389** mediated by *p*-nitrophenylboronic acid, which furnished exclusively the β-(1-4)-linked product **391** at 92% yield, and then by di-O-benzylation followed by deprotection of PMB group by HCl-triethylsilane in DCM-hexafluoroisopropanol. The next step was TfOH-mediated glycosylation of the resulting disaccharide acceptor with the monomer TCA donor **392** bearing 3-OPMB, and then PMB deprotection furnished the trimeric acceptor building block **393**.

In separate work, in synthesizing disaccharide **395**, Toshima et al. faced the primary challenge of generating a β-mannosaminidic linkage of the pentasaccharide that had been installed by first forming a β-glucosidic linkage followed by C2-inversion-amination. Initially, the researchers synthesized disaccharide donor **396** starting with orthogonal β-selective glucosylation with donor **394** of monomeric building block **395**; this was followed by 2-O-debenzoylation under Zemplén condition and then 2′-O-triflylation for the next S_N2 reaction using tetrabutylammonium azide as the azide-nucleophile and then subsequent hydrolysis of the resulting disaccharide based thioglycoside and its anomeric trifluoroacetimidate protection [180].

Then, in [3+2] convergent glycosylation with disaccharide donor **396** and trisaccharide acceptor **393** afforded pentasaccharide derivative **397** at 80% yield. Compound **397** ultimately generated the target pentasaccharide **398** after the following high-yielding process: i) azido reduction under Staudinger condition; ii) *N*-acetylation; iii) Zn-Cu-Ac_2O-AcOH-mediated transformation of NTroc to NHAc; iv) Zemplén diacylation; and v) hydrogenative deprotection of benzyl, benzylidene, and Cbz with $Pd(OH)_2/C$-H_2]. The group then prepared the corresponding glycoconjugates of different chain lengths. Immunochemical studies with these glycoconjugates established pentasaccharide **398** as the required glycotope for future synthetic/semisynthetic vaccine development (Scheme 5.35).

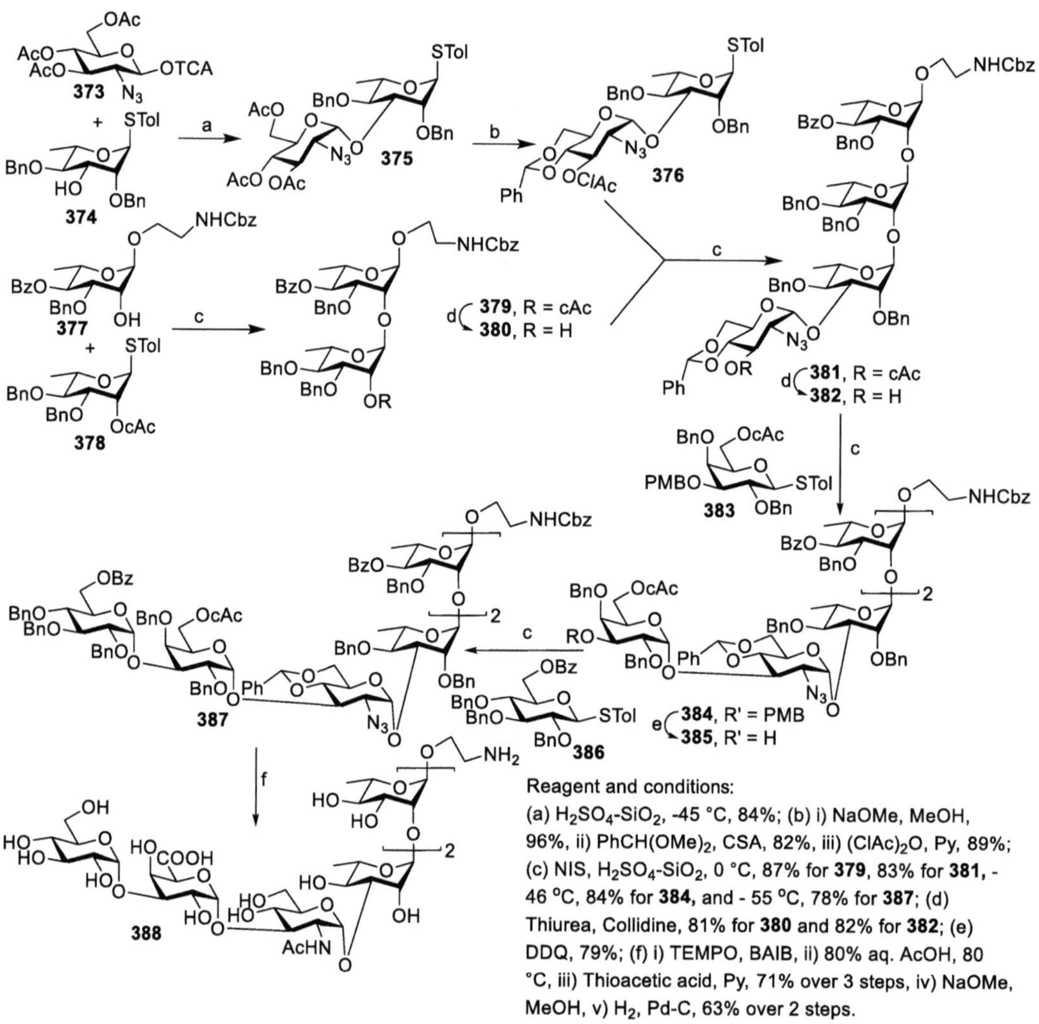
373
374
375
376
377
378
379, R = cAc
380, R = H
381, R = cAc
382, R = H
383
384, R' = PMB
385, R' = H
386
387
388
Reagent and conditions:
(a) H2SO4-SiO2, -45 °C, 84%; (b) i) NaOMe, MeOH, 96%, ii) PhCH(OMe)2, CSA, 82%, iii) (ClAc)2O, Py, 89%; (c) NIS, H2SO4-SiO2, 0 °C, 87% for 379, 83% for 381, -46 °C, 84% for 384, and - 55 °C, 78% for 387; (d) Thiurea, Collidine, 81% for 380 and 82% for 382; (e) DDQ, 79%; (f) i) TEMPO, BAIB, ii) 80% aq. AcOH, 80 °C, iii) Thioacetic acid, Py, 71% over 3 steps, iv) NaOMe, MeOH, v) H2, Pd-C, 63% over 2 steps.

SCHEME 5.34 Synthesis of hexasaccharide **388** related to EC-O133 antigen.

To synthesize glycoconjugate toward the development of extraintestinal pathogenic EC-strain (ExPEC), an EC-O25B-based vaccine candidate, Naini et al. planned to synthesize three glycans of different chain lengths comprising one to three RU of EC-O25B natural antigen: antigen 1, 1RU; antigen 2, 2RU; antigen 3, 3RU (Schemes 5.36a and 5.36b) and two other frame shifted RUs (not shown) [181]. The main hurdles of their syntheses were the presence of a highly branched (1,3,4,6-branching with other glycose units) core D-Glc, glycosylation on the less nucleophilic C4-O of the core Glc unit, and α-glucosylation using an orthogonally protected core D-Glcp unit of the synthetic target molecules.

The 1,2-*trans* linkages were developed through NGP by a participating group on C2-O of the glycosyl donor building blocks via use of a temporary protecting 2-azidomethylbenzoyl (AZMB) group and acetyl as a permanent protecting group, except for one α-rhamnopyranosylation that used a Bn on C2-O of rhamnosyl donor. They chose AZMB protection as a participating and 1,2-*trans* glycosylation-modulating group because of its selective deprotection under mild conditions at the late stage of the total target synthesis in the presence of permanent protecting OAc group. Based on carefully manipulated building units, the authors achieved improved synthetic pathways of the

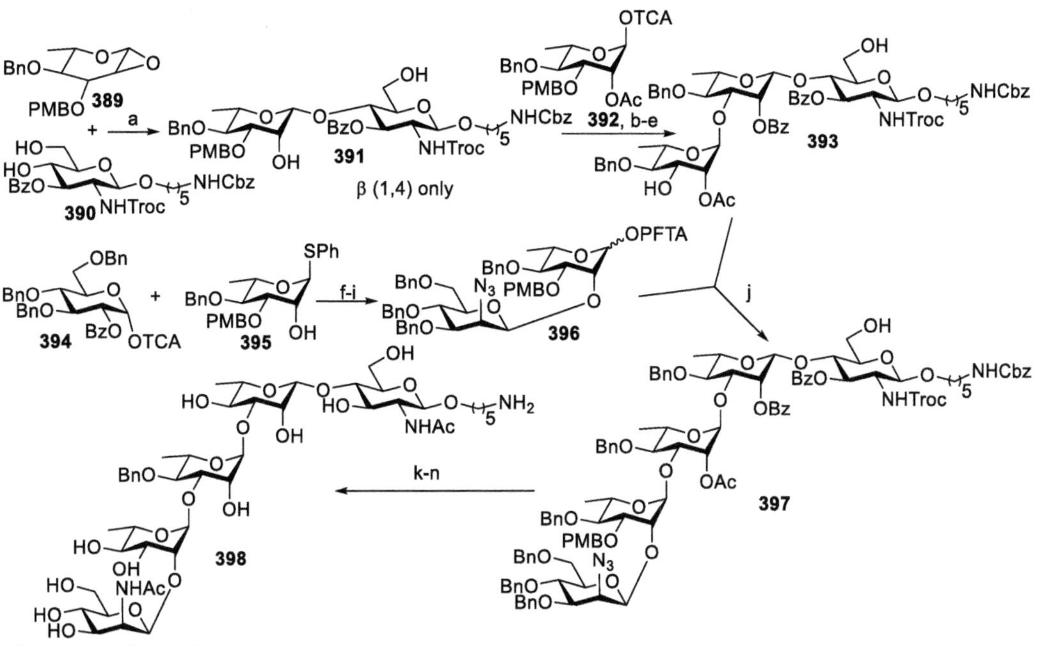

Reagents and conditions:
(a) *p*-Nitrophenyl boronic acid, THF, 0 °C, 1 h, 92%; (b) BzCl, Py, rt, 10 h, 97%; (c) HCl, TESH, DCM, HFIP, 0 °C, 3 h, 98%; (d) TfOH, DCM, -20 °C, 2h, 77%; (e) HCl, TESH, DCM HFIP, 0 °C, 3 h, 96%; (f) i) TMSOTf, DCM, -80 °C, 1.5 h, ii) NaOMe, THF, MeOH, rt to 50 °C, 24 h, 86% over two steps; (g) i) Tf_2O, Py, DCM, -20 °C to 0 °C, 4 h, ii) $TBAN_3$, Toluene, 60 °C, 24 h, 87%, over two steps; (h) NBS, H_2O, MeCN, THF, -40 °C to -30 °C, 4 h, quantitative; (i) N-phenyltrifluoroacetimidoyl chloride, Cs_2CO_3, DCM, rt, 17h, 97% (α:β = 23:77); (j) TfOH, Toluene, -40 °C, 2 h, 80%; (k) i) PPh_3, THF, rt to 40 °C, 8 h then H_2O, reflux, 9h, ii) Ac_2O, DCM, AcOH, rt, 2 h, 87% over two steps; (l) Zn-Cu, Ac_2O, AcOH, 1,4-dioxane, rt, 2 h, 80%; (m) NaOMe, THF, MeOH, rt, 24h, 84%; (n) $Pd(OH)_2/C$, H_2, MeOH, rt, 3 h, 86%.

SCHEME 5.35 Synthesis of pentasaccharide **398** related to virulent avian EC-O1.

required penta- and hexa-saccharide building blocks (**410** and **413**, respectively); Scheme 5.36a shows part of the process. Scheme 5.36b shows the syntheses of the required tetrasaccharide building block **407** and of target 1RU using stepwise sequential glycosylation: 2U following [5+1+4] glycosylation and 3RU by [6+5+4] building. They used either *N*-phenyltrifluoroacetimidates (NPTFA) or thioglycosides as glycosyl donors; for activation of the first type, TMSOTf was used and except in one case, all thioglycosides were activated using NIS-TMSOTf [181].

Initial glycosylation of rhamnopyranoside acceptor **400** (bearing a temporary protecting PMP on C1-O) using core thioglucoside donor **399** containing 3-O-Nap and 4,6-O-benzylidene as temporary protecting groups generated the corresponding inseparable disaccharide mixture **401** at 71% yield but with poor anomeric selectivity. Then PTSA-EtSH-mediated cleavage of benzylidene group generated **402** followed by separation of anomers; then, 6-O selective β-glucosylation of resulting **402** with donor **403** directly gave trisaccharide acceptor **404** at 70% yield. This on further β-glycosylation yielded **406** as the only specific β-anomer at 68% based on 2-NTroc-protected thioglucoside donor **405**; on Nap deprotection with DDQ, this yielded tetrasaccharide acceptor **407** at 86%. Its glycosylation with Rha-donor **408** yielded pentasaccharide **409** (68%); successive anomeric PMP deprotection of **409** and then anomeric OH protection resulted in corresponding pentasaccharide-NPTFA-donor **410** at 84% yield over the last two steps. This was used for glycosylation of next acceptor **411** to generate 84% of **412**; after TBS deprotection using HF, pyridine afforded hexasaccharide acceptor **413** [181].

For the synthesis of the tetrasaccharide donor building block **419**, Naini et al. first prepared disaccharide **415** as an inseparable mixture at 53% yield and 6:1 ratio in favor of the α-anomer by glycosylation of rhamnopyranoside acceptor **400** with differentially protected core thioglucoside donor **414**, activated by NIS-TBSOTf. Regioselective reductive deprotection of the Nap group by BH_3.THF-TMSOTf gave disaccharide acceptor **416** (60% after separation). This on combination with glycosyl donor **403** in the presence of TMSOTf generated trisaccharide **417** (in 75%), which after de-chloroacetylation utilizing DABCO gave acceptor **418** (80%). Further glycosylation with armed Rha-donor **408** then gave tetrasachharide **419** at 77% yield. This was then transformed to corresponding donor **420** (52% over two steps) by anomeric deprotection followed by protection of C1-O as NPTFM. Chain extension of **420** by nonreducing end acceptor **411**, sequential selective removal under mild condition of AZMB group, functional group manipulation of NHTroc to NHAc, and global hydrogenative deprotection furnished target pentasaccharide **421** (1RU, antigen 1) as its aminoethyl glycoside [181].

They then prepared other glycans. To achieve the target decasaccharide (2RU), Naini et al. reacted tetrasaccharide donor **420** with hexasaccharide acceptor **413** in the presence of TMSOTf to give corresponding decasaccharide derivative **422** at 69% yield. On functional group manipulation and global deprotection, this afforded decasaccharide **423** (2RU, antigen 2) as aminoethyl glycoside. They separately generated pentadecasaccharide **426** as its aminoethyl glycoside by the following sequence of reactions: i) chain extension of pentasaccharide donor **410** by hexasaccharide acceptor

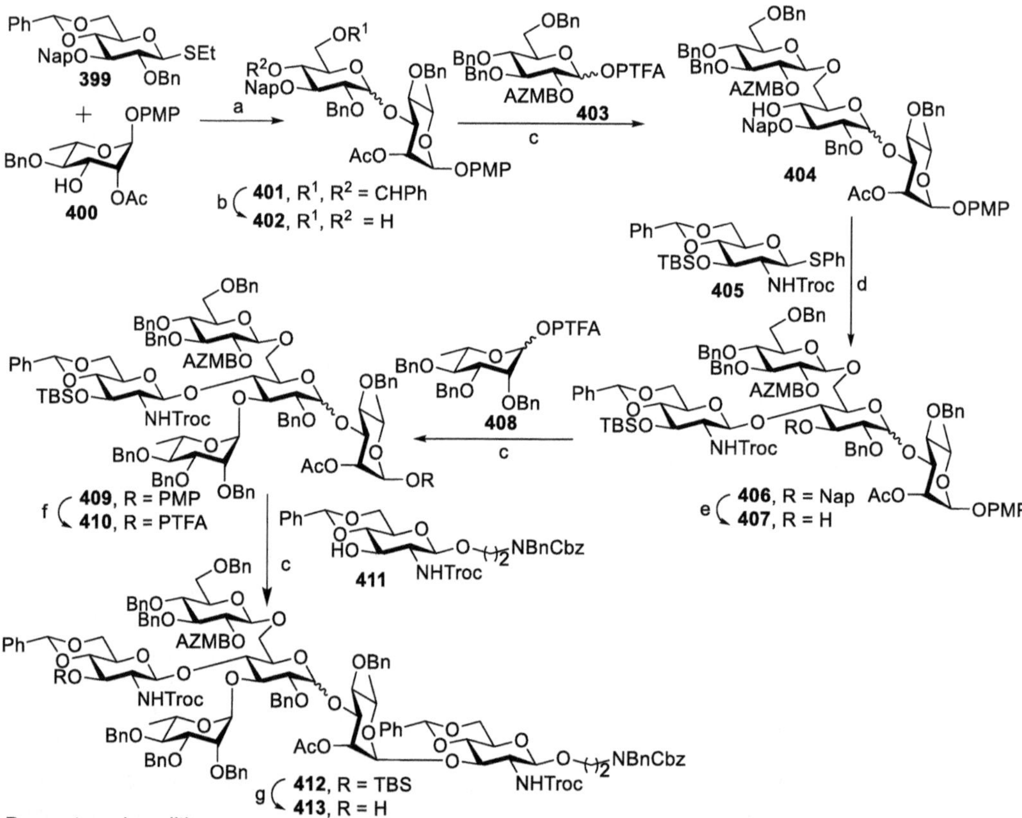

SCHEME 5.36a Syntheses of pentasaccharide donor **410** and hexasaccharide acceptor **413** toward the synthesis of 1RU, 2RU, and 3RU of EC-O25B antigen.

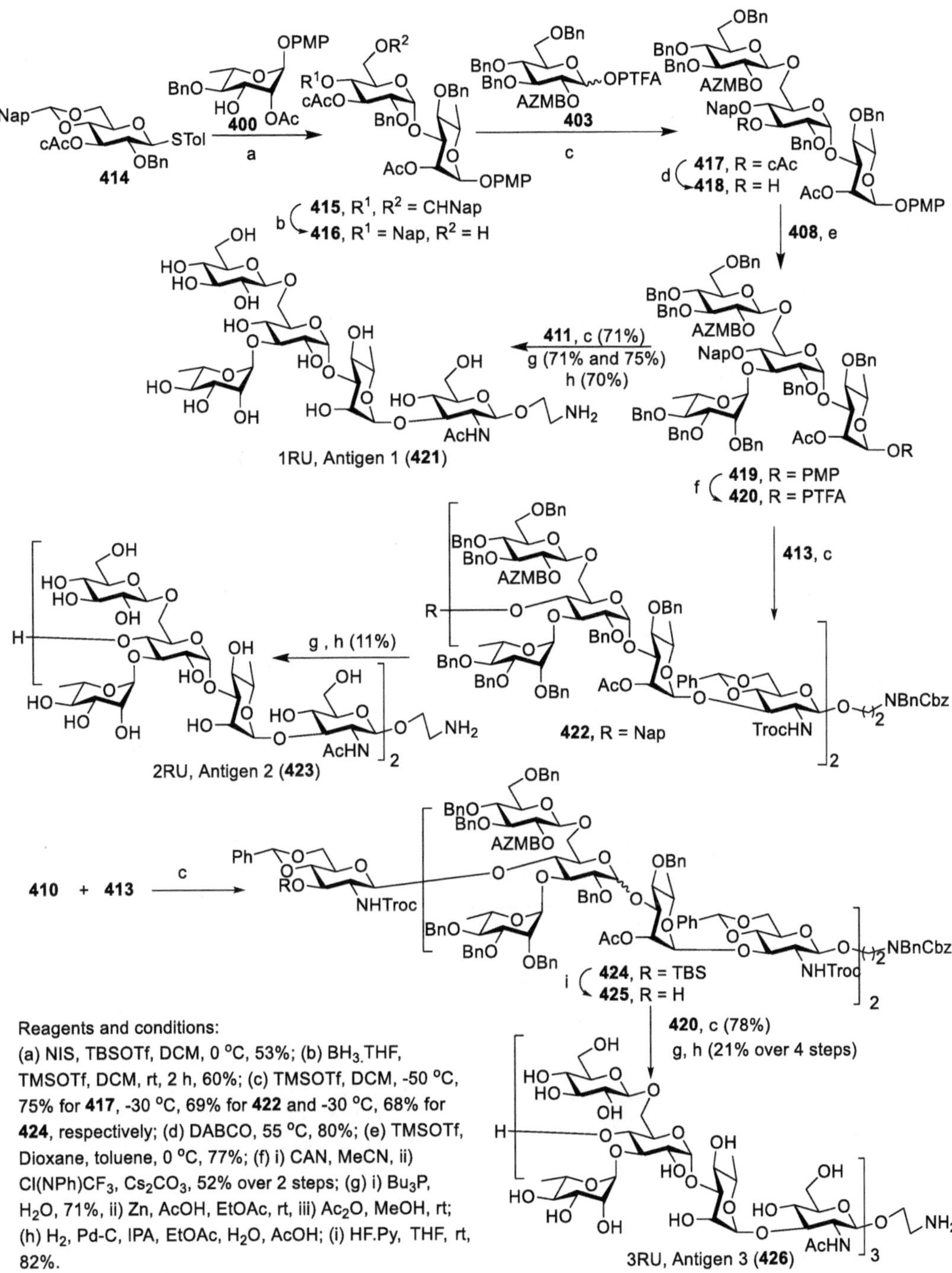

SCHEMES 5.36b Syntheses of **421** (1RU, antigen 1), **423** (2RU, antigen 2), and **426** (3RU, antigen 3) of EC-O25B antigen.

413 (yield of this step 68%), ii) HF-mediated cleavage of TBS group from **424** to give acceptor **425** (82%), iii) glycosylation with tetrasaccharide donor **420** (78%), and iv) functional group deprotection and manipulation in four steps (overall 21% of **426**, 3RU, antigen 3) (Scheme 5.36b). They then prepared the CRM_{197} glycoconjugates of 1RU, 2RU, and 3RU along with two other frame-shifted pentasaccharides (syntheses not shown) and performed immunochemical studies of all three. These experiments indicated 1RU, 2RU, and 3RU to be promising vaccine candidates, and longer glycan chains improved the immune response [181].

5.2.1 *Providencia*

Providencia, a rod-shaped bacterium, is a Gram-negative member of the *Proteeae* tribe of Enterobacteriaceae and is responsible for causing several types of enteric infections including wound and urinary tract infections. *Providencia* is subdivided into eight species: *P. alcalifaciens, P. stuartii, P. rettgeri, P. rustigianii, P. heimbachae, P. vermicola, P. sneebia*, and *P. burhodogranariea*. These species have been identified and isolated from urine, stool, sputum, perineum, axilla, blood, and wound specimens of infected patients as well as from polluted soil and wastewater. Of these species, clinically important ones are *P. alcalifaciens* (PA), *P. stuartii* (PS), and *P. rettgeri* (PRe), which are particularly responsible for antibiotic-resistant infections in hospitalized patients with long-term urinary catheters, particularly immuno-compromised patients. PA mainly causes

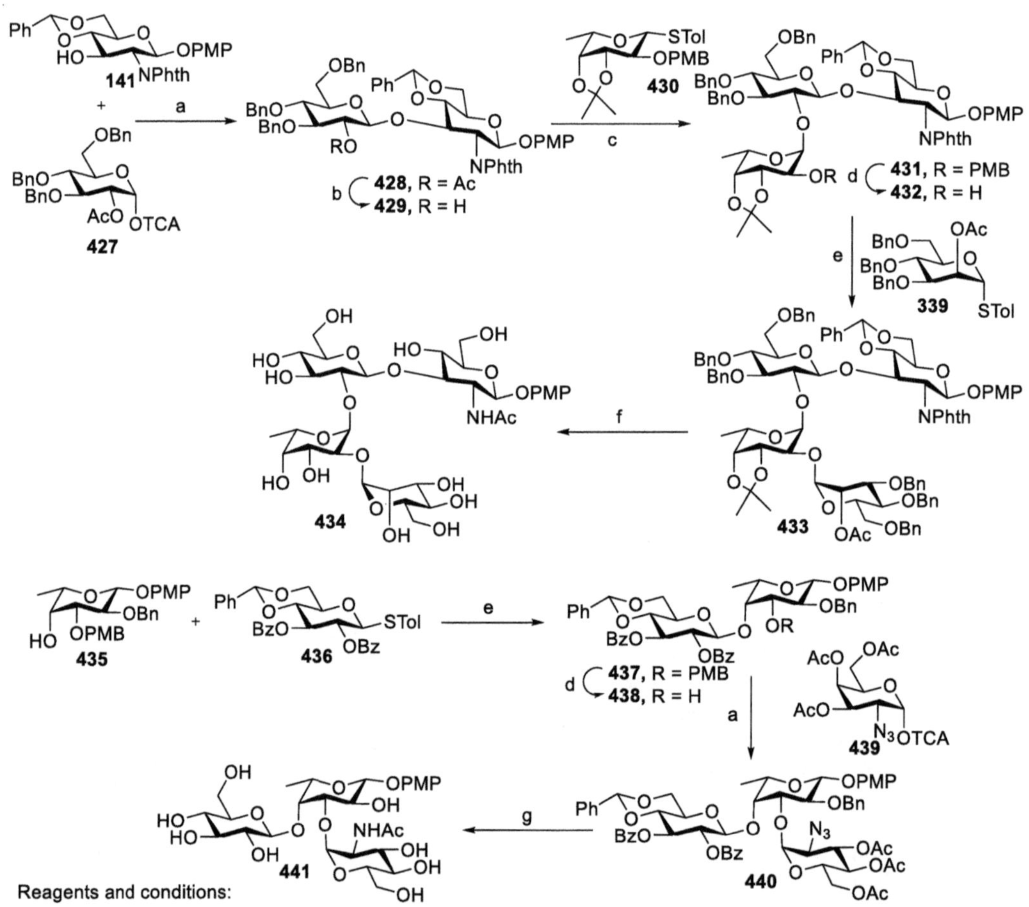

Reagents and conditions:

(a) H_2SO_4-SiO_2, DCM, 5 -10 °C; 91% for **428** and 80% for **440** respectively; (b) NaOMe, MeOH, 96%; (c) NIS, H_2SO_4-SiO_2, DCM, -40 °C, 83%; (d) DDQ, DCM:H_2O 4:1, 81% for both **432** and **438**; (e) NIS, H_2SO_4-SiO_2, DCM, 5 -10 °C; 84% **433** and 85% for **437** respectively; (f) i)Ethylene diamine, n-BuOH then Ac_2O, Py, ii) 80% AcOH, 80 °C, H_2, Pd-C, MeOH, iii) NaOMe, MeOH, 63%; (g) i)Thioacetic acid, ii) 80% AcOH, 80 °C, H_2, Pd-C, MeOH, iii) NaOMe, MeOH, 67%.

SCHEME 5.37 Syntheses of trisaccharide **441** and tetrasaccharide **434** related to PR-O34 antigen.

diarrhea in children and in developing countries [182]. In this section, we discuss the recent chemical syntheses of oligosaccharides related to *P. rustigianni* O34 (PR-O34), *P.* stuartii O49 (PS-O49), and *P. alcalifaciens* O28 (PA-O28) antigens.

Verma et al. reported on concise syntheses of tetra- and trisaccharides related to the RU of the O-antigen from PR-O34 as PMP glycosides **434** and **441**, respectively [183]. The glycosylation reactions were achieved by either H_2SO_4-silica for TCA donors or NIS/H_2SO_4-silica for thioglycoside donors at different temperatures. To synthesize tetrasaccharide **434**, Verma et al. first coupled known glucosamine-derived acceptor **141** with glucopyranosyl TCA donor **427** at 5 °C –10 °C; this gave 91% of corresponding disaccharide **428**. Removal of acetate protection from this disaccharide produced disaccharide acceptor **429**, which was further coupled with suitably protected L-fucosyl thioglycoside donor **430** at –40 °C to produce 83% of trisaccharide **431**.

Selective removal of PMB with DDQ gave acceptor **432**, which on final glycosylation reaction with known Man-derived thioglycoside donor **339** at 5 °C –10 °C afforded the fully protected tetrasaccharide **433** at 80% yield. Reaction of PMP β-L-fucopyranoside accepter **435** with Glcp donor **436** at 5 °C–10 °C produced disaccharide **437** at 85% yield. Selective removal of PMB protection from **437** resulted in disaccharide acceptor **438**, which was further glycosylated with galactosamine-derived TCA donor **439** at 5 °C–10 °C; this generated 80% of the trisaccharide **440**. A final series of deprotection reactions on both tetra- and trisaccharide derivatives (**433** and **440**, respectively) afforded their corresponding fully deprotected glycosides **434** and **441** at respective yields of 63% and 67% (Scheme 5.38) [183].

Werz and coworkers employed [4+2+1] glycosylation for the synthesis of the seven different monosaccharide building blocks containing the heptasaccharide substructure of the lipopolysaccharide of PR-O34 as pentenyl glycoside [184]. Their convergent strategies focused on [4+3], [3+4] and [4+2+1] coupling reactions, among which the first two strategies failed under a plethora of different glycosylation conditions, but [4+2+1] was successful. Glycosylation of pentenyl fucoside acceptor **452** and D-galactosamine-derived TCA donor **453** using TMSOTf at –30°C produced corresponding α-disaccharide **454** at 87% yield; this was then converted to corresponding TCA donor **455** at 90% yield.

In a different test, Werz et al. coupled Galp acceptor **442** with Glcp phosphate donor **443** using TMSOTf promoter and produced corresponding 1→3 lined disaccharide **444** at 80% yield. Deprotection of the acid labile Fmoc protection produced disaccharide acceptor **445**, which was then coupled with fucosyl phosphate donor **446** to yield 90% of trisaccharide **447** with complete α selectivity; to some extent, however, acid-labile PMB protection was removed under glycosylation. Complete deprotection of the PMB protection with CAN produced 76% of trisaccharide acceptor **448**, which was then glycosylated with Manp phosphate donor **449**, producing tetrasaccharide **450** at 80% yield. Subsequent saponification removed the acetate protection resulting in tetrasaccharide acceptor **451** [184].

Glycosylation of disaccharide TCA donor **455** and tetrasachharide acceptor **451** using TMSOTf as promoter produced 67% of hexasaccharide **456**. Subsequent transesterification under basic condition removed the acetate protection, generating hexasaccharide acceptor **457**, which was subsequently glycosylated with glucuronic acid ester TCA donor **458** using TESOTf (slow addition and dilution was the key for efficient reaction) as promoter at 0 °C for 24 hours; this procedure afforded final heptasaccharide **459** at 54% yield. Sequential deprotection of heptasaccharide derivative **459** afforded the target acidic heptasaccharide **460** at 18% yield (Scheme 5.39) [184].

To synthesize a trisaccharide RU of the O-antigen polysaccharide of PS-O49, Halder et al. utilized both stepwise and one-pot approaches [185], coupling D-galactosamine donor **461** with D-Galp acceptor **462** using NIS/TMSOTf to afford disaccharide **463** at 85% yield. Selective removal of chloroacetyl protecting group using thiourea produced disaccharide acceptor **464**, which they then coupled with Gal donor **465** producing the trisaccharide **466** in 89% yield. In an alternative [1+1+1] one-pot synthesis, they coupled Galp TCA donor **468** with galactosamine thioglycoside acceptor **469** using TMSOTf alone followed by activating the in situ-generated thioglycoside with NIS/TMSOTf

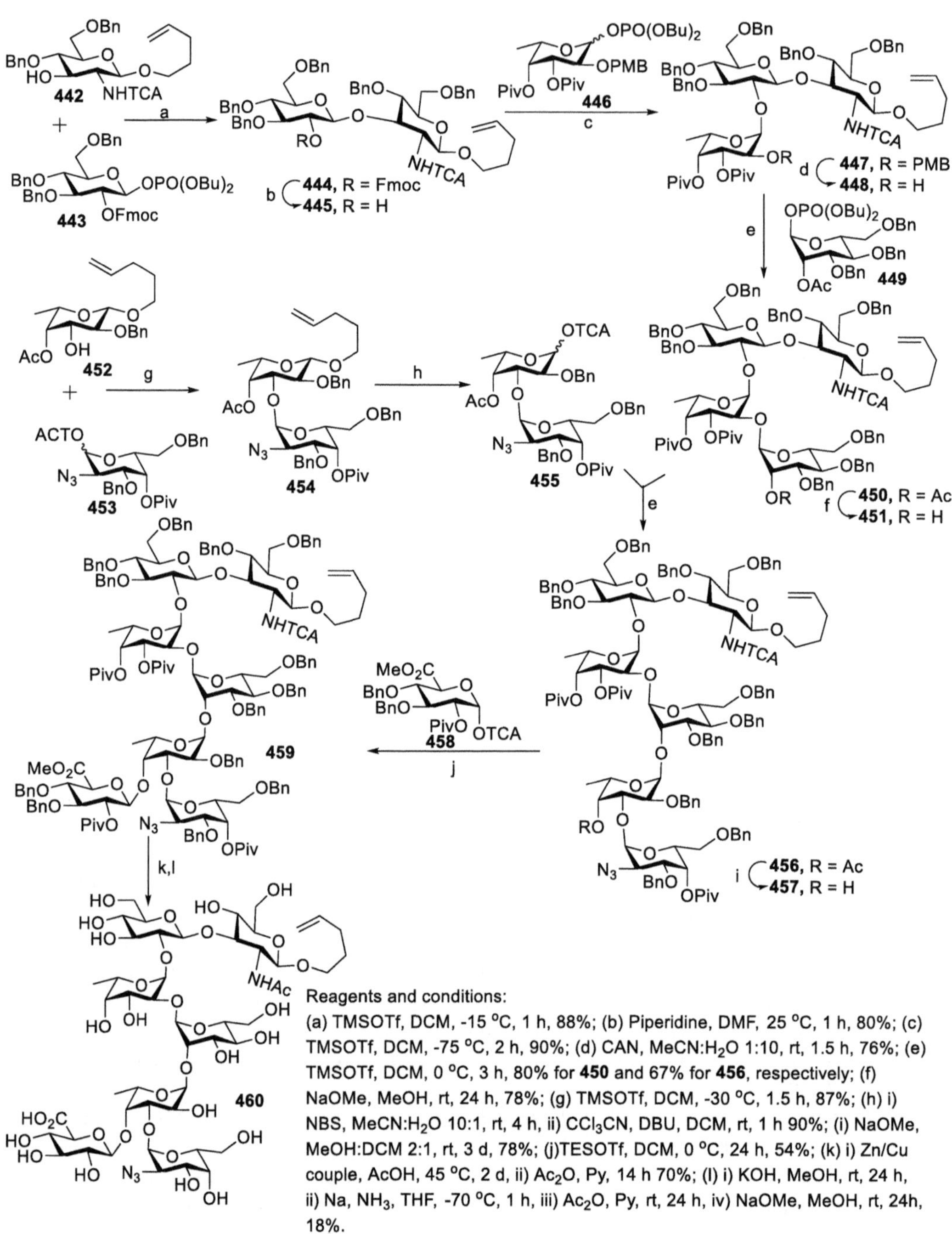

SCHEME 5.38 Synthesis of heptasaccharide **460** related to PR-O34 antigen.

to couple with acceptor **462** to produce trisaccharide **466** at 73% overall yield. Deprotection of the trisaccharide derivative **466** resulted in the final trisaccharide **467** as its PMP glycoside at 68% overall yield (Scheme 5.40).

Mandal and Chheda reported a straightforward convergent [3+2] block synthesis of the pentasaccharide RU of the O-specific lipopolysaccharide of PA-O28 [186]. The known 6-*O*-acetyl-2,3,4-tri-*O*-benzyl-α-D-glucopyranosyl trichloroacetimidate donor **470** was allowed to stereoselectively

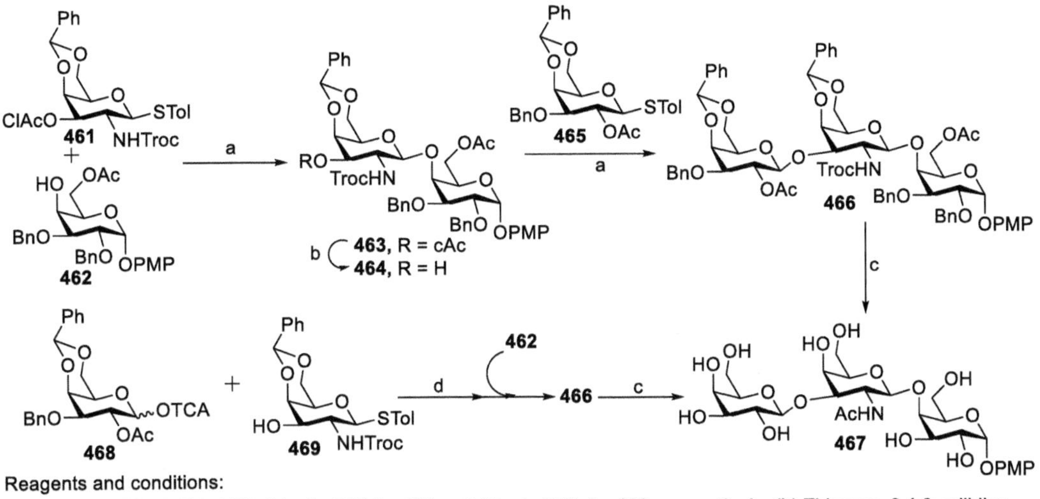

Reagents and conditions:
(a) NIS, TMSOTf, DCM, 0 °C, 20 min 85% for **463** and 15 min 89% for **466**, respectively; (b) Thiourea, 2,4,6-collidine, DCM:MeOH 2:1, 12 h, reflux, 87%; (c) i) Zn, THF:AcOH:Ac_2O 3:2:1, 3 h, 0 °C to rt, ii) NaOMe, MeOH, iii) H_2, $Pd(OH)_2$-C, MeOH, 24 h, rt, 68%; (d) TMSOTf, DCM, -20 °C, 2 h, rt, 30 min then acceptor **462** and again rt, 30 min, 0 °C, 10 min, NIS, TMSOTf, DCM, 0 °C, 15 min 73% overall yield.

SCHEME 5.39 Synthesis of the O-antigen repeating unit as aminoethyl glycoside **467** of PS-O49.

couple with the known L-fucopyranoside accepter **471** using H_2SO_4-silica activator to produce the disaccharide thioglycoside **472** at 79% yield. Selective activation of disaccharide thiothycoside **473** over thioglycoside motif **474** using armed–disarmed Fraser-Reid furnished the trisaccharide **475** at 72% yield with complete desired stereoselectivity. Thiofucoside donor **476** was glycosylated with known glucosamine derived acceptor **303** using a mixed solvent DCM: Et_2O 1:4 in the presence of H_2SO_4-silica activator to yield 83% of disaccharide **477** (with formation of 5% of the minor isomer). Functional group manipulations of this disaccharide **476** generated disaccharide acceptor **477** at 90% yield. Its consecutive coupling with trisaccharide donor **475** in the presence of NIS/H_2SO_4-silica activator produced desired pentasaccharide **478** at 83% yield. Sequential late-stage functional group modifications resulted in target pentasaccharide **479** as the sodium salt of the 2-aminoethyl glycoside at 57% yield (Scheme 5.41).

Kulkarni and Podilapu reported on the total synthesis of the phosphorylated trisaccharide RU of PA-O22 containing rare sugar unit AAT using one-pot glycosylation and late-stage phosphorylation [187]. The key rare sugar building block **481** was synthesized from known D-rhamnosyl 2,4-diol **480** through several functional group modifications including S_N2 displacement of sulfonate. Similar S_N2 displacement of sulfonate was also utilized to prepare D-galactosamine donor **483** starting from an inexpensive glucosamine derivative. Glycosylation of 4-OH thiophenyl galactoside donor **482** and 3-OH AAT acceptor **481** using NIS/TMSOTf cleanly afforded the 4′-OH disaccharide in one hour. Further addition of glycosyl donor **483** in the same reaction vessel using the same activator combination produced 3″-OH free trisaccharide **484** at 72% yield.

One interesting outcome of this synthesis approach was that the reaction sequence that produced TfOH as a byproduct of the glycosylation removed the required Fmoc protection to afford 3″-OH trisaccharide **484**. Mannitol derivative **485** was used to prepare 2-OH D-glyceramide derivative **486** in five steps, which was then phosphorylated using imidazole, PCl_3, and NEt_3 to produce *H*-phosphonate **487** at 89% yield. Phosphonate **487** was coupled with trisaccharide **484** in the presence of pivaloyl chloride and pyridine followed by oxidation with iodine to furnish the phosphorylated trisaccharide derivative **488** ats 64% yield. Global deprotection of **488** using Zn, AcOH, and

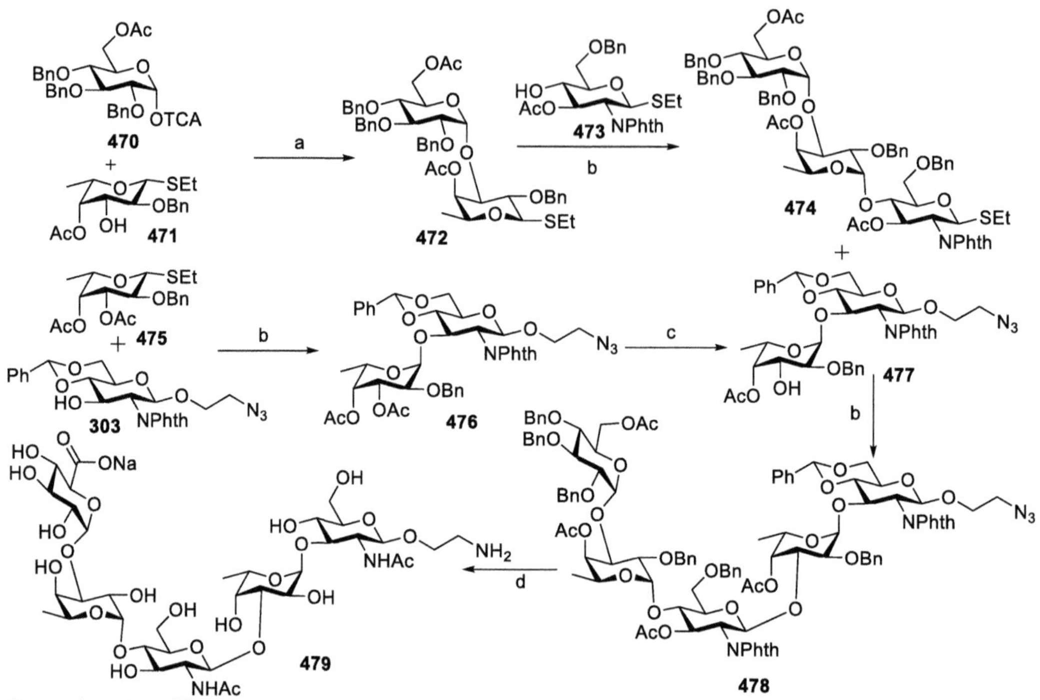

Reagents and conditions:

(a) H_2SO_4-SiO_2, DCM, -40 °C, 1 h, 79%; (b) NIS, H_2SO_4-SiO_2, DCM, -40 °C, 30 min, 72% for **474**, -25 °C, 1 h, 83% for **478** and Et_2O:DCM 4:1, -30 °C, 30 min, 83% for **476**; (c) i) NaOMe, MeOH, rt, 1 h, ii) triethylorthoacetate, pTSA, DMF, 2 h then 80% AcOH, rt, 1 h, 90%; (d) i) Ethylene diamine, EtOH, 80 °C, 6 h then Ac_2O, Py, rt, 3h, ii) NaOMe, MeOH, rt, 1 h, iii) TEMPO, NaBr, NaOCl, TBAB, $NaClO_2$, 2-methyl-but-2-ene, NaH_2PO_4, $NaHCO_3$, DCM, 0 °C to rt, 5 h, iv) H_2, 20% $Pd(OH)_2$-C, MeOH, rt, 24 h, 57%.

SCHEME 5.40 Synthesis of repeating pentasaccharide **479** of PA-O28 antigen.

Ac_2O first, followed by ester hydrolysis using Et_3N, MeOH, and water at 60 °C and final hydrogenolysis with H_2 and $Pd(OH)_2$/C with a drop of AcOH in MeOH produced target zwitterionic trisaccharide **489** at 64% yield in three steps (Scheme 5.41) [187].

5.2.2 *Enterobacter*

E. cloacae complex is a group of Gram-negative anaerobic bacteria of the family Enterobacteriaceae, of which *E. cloacae* and *E. hormaechei* are the most common nosocomial pathogens infecting immunocompromised patients. These are also reported to be antibiotic resistant [188], and special attention is needed for developing new effective antibiotics and vaccines. Chaudhury and Mukhopadhyay gave a detailed description of the synthesis of a repeating pentasaccharide bearing a rare sugar corresponding to *E. cloacae* C4115 [189].

After they synthesized all the monosaccharide building blocks, Chaudhury and Mukhopadhyay fully synthesized target pentasaccharide **498** following linear glycosylation (Scheme 5.42). L-Rha-TCA donor **490** activated with TMSOTf was reacted with thiorhamnoside acceptor **491** furnishing 86% disaccharide **492**; after Zemplén deacetylation, this generated disaccharide acceptor **493** (95%). Its glycosylation with donor **490** generated trisaccharide thioglycoside **494**, which was then transformed to corresponding TCA donor **495**. After TMSOTf-activation and coupling with thioglucoside acceptor **496**, this gave tetrasaccharide derivative **497** at 75% yield. In a separate process, a

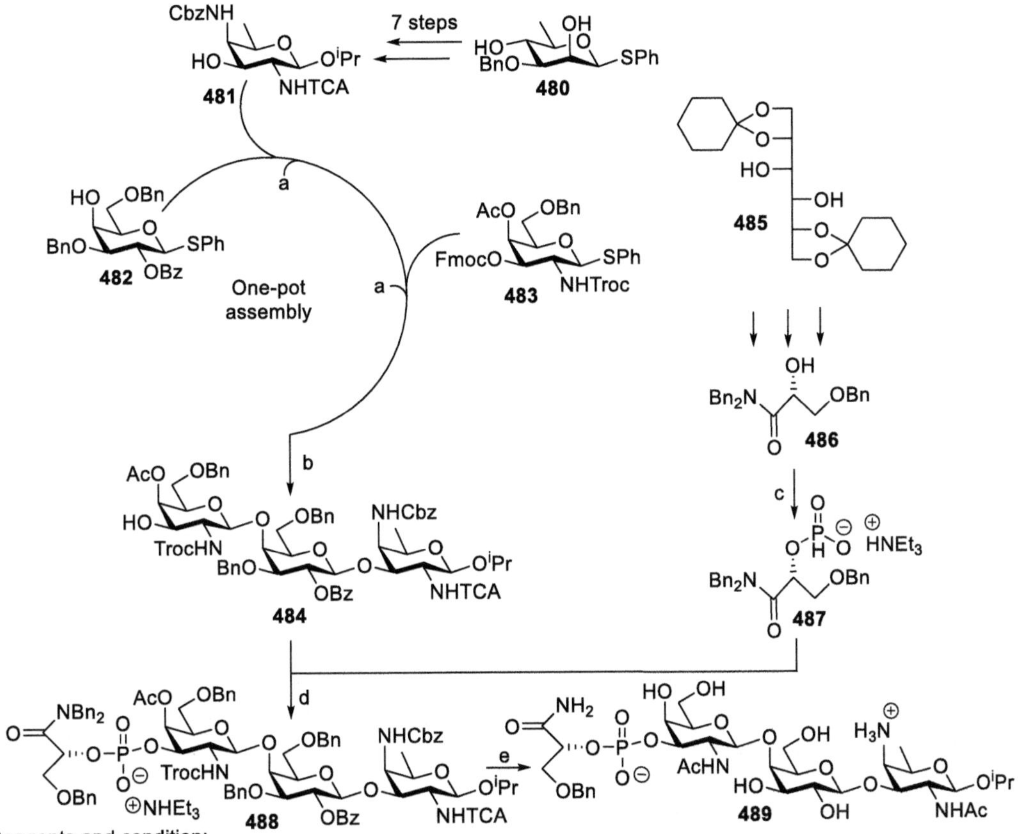

Reagents and condition:
(a) NIS, TMSOTf, DCM, 1 h; (b) Et_3N, 10 min, 72% over 3 steps; (c) Im, PCl_3, Et_3N, toluene, DCM, 3 h, 89%; (d) i) PivCl, Py, 12 h, ii) I_2, Py, H_2O, 2 h, 64% over 2 steps; (e) i) Zn, AcOH, Ac_2O, 24 h, ii) Et_3N, MeOH, H_2O, 60 °C, 6 h, iii) $Pd(OH)_2/C$, H_2, MeOH, AcOH, 36 h, 64% over 3 steps.

SCHEME 5.41 Synthesis of a phosphorylated trisaccharide related to PA-O22 utilizing one-pot glycosylation.

known 2-azido-2,6-dideoxyglucose derivative **498** was converted in five steps to rare sugar acceptor **499**. On glycosylation with tetrasaccharide donor **497** in the presence of NIS, TMSOTf generated desired pentasaccharide derivative **500** at 75% yield. Compound **500**, after deisopropylidenation followed by conversion of azido to acetamido and final global deprotection, this furnished target pentasaccharide **501** [189].

5.3 CONCLUSION

The COVID-19 pandemic triggered expedited research on antibacterial vaccines. Although AMR bacterial strains were known even in the last millennium, increasing numbers of AMR strains in different bacteria families, along with the corresponding considerable motility and morbidity particularly in low- and middle-income, underdeveloped and developing countries, the WHO, apart from other preventive measures, prioritized developing corresponding anti-AMR-drugs or vaccines against the Enterobacteriaceae.

Carbohydrates in the cell wall or capsule play a crucial role in bacterial infections and the corresponding immune responses in the host. Keeping in mind of the limitations of earlier generation carbohydrate-vaccines (attenuated bacteria-based, isolated bacterial PS/LPS-based, and glycoconjugate-based using isolated-bacterial PS), synthesizing antimicrobial glycoconjugate vaccines via

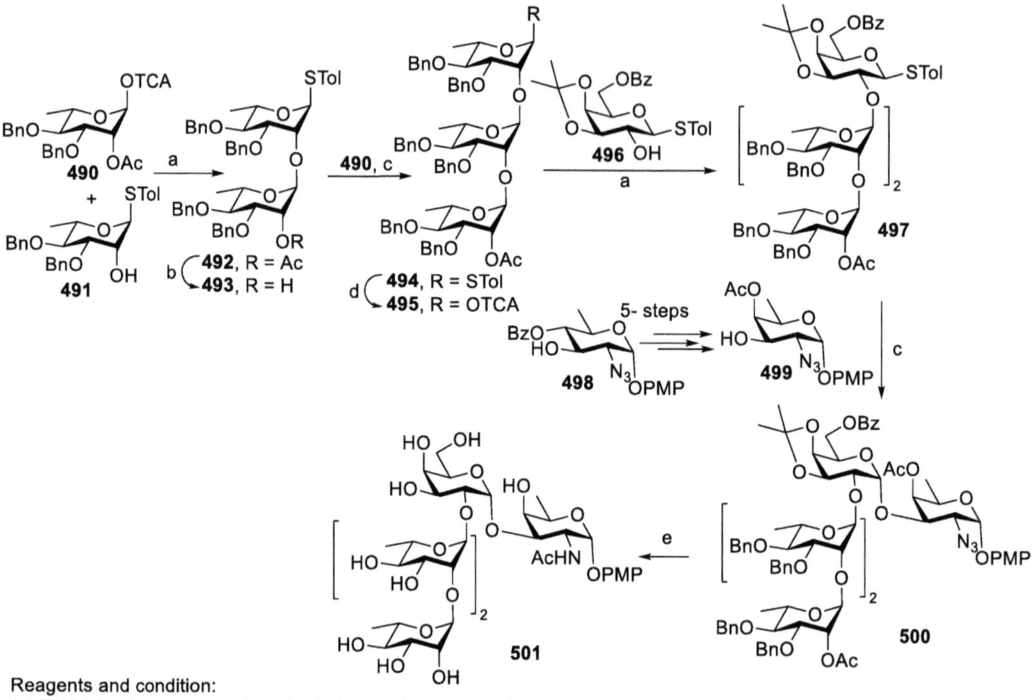

Reagents and condition:
(a) TMSOTf, DCM, -15 °C, 86% for **492**, 75% for **497**; (b) NaOMe, MeOH, 95%; (c) NIS, TMSOTf, CH_2Cl_2, -15°C, 90% for **498** and -30°C, 75% for **500**; (d) i) TCCA, Me_2CO, H_2O, 85% ii) CCl_3CN, DBU, DCM; (e) i) 80% aq. AcOH, 80°C, ii) AcSH, 85% over 2 steps, iii) H_2, Pd-C, MeOH, iv) MeONa, MeOH, 90% over 2 steps.

SCHEME 5.42 The linear synthesis of repeating pentasaccharide **501** of *Enterobacter cloacae* antigen.

synthesis of homogeneous OS/PS has gained much attention in the scientific community working with carbohydrates around the world. In this chapter, we have discussed syntheses of OSs/PSs corresponding to some bacteria including AMR Enterobacteriaceae bacteria.

Although during total syntheses of OS/PS haptens, failure is possible even at the very final stage; however, researchers have reported on chemoselective glycosylation including one-pot iterations and removing some of the bottlenecks for machine-driven automated syntheses of OS/PS [190]. It is our hope that these techniques along with improved methods of glycoconjugation with protein, larger-scale preparation of haptens and vaccines could be possible on glycan synthesizer, paving the way toward the ultimate development of FDA-approved sweet-vaccines.

REFERENCES

[1] Gyles, C. L.; Prescott, J. F.; Boerlin, P. Themes in bacterial pathogenic mechanisms. *Pathogenesis of Bacterial Infections in Animals*, **2004**, 3–12. https://doi.org/10.1002/9780470344903.ch1.

[2] Erkmen, O.; Bozoglu, T. F., Eds. Bacterial pathogenicity and microbial toxins. In *Food microbiology: principles into practice*, 1st Edition, Wiley, **2016**, 126–137. https://doi.org/10.1002/9781119237860.ch7.

[3] Boquet, P.; Ricci, V. Bacterial exotoxins. In *Reference module in biomedical science*, Elsevier, **2014**. https://doi.org/10.1016/B978-0-12-801238-3.00134-3.

[4] (a) Sampath, V. Bacterial endotoxin-lipopolysaccharide; structure, function, and its role in immunity in vertebrates and invertebrates. *Agric. Nat. Resour.* **2018**, *52*, 115–120. https://doi.org/10.1016/j.anres.2018.08.002; (b) Fransen, F.; Heckenberg, G. B.; Hamstra, H. J.; Feller, M.; Boog, C. J.; van Putten, J. P. M.; van de Beek, D.; van der Ende, A.; van der Ley, P. Naturally occurring lipid A mutants in *Neisseria meningitidis* from patients with invasive meningococcal disease are associated with reduced coagulopathy. *PLOS Pathog.* **2009**, *5*, e1000396. https://doi.org/10.1371/journal.ppat.1000396.

[5] Raetz, C. R. H.; Whitfield, C. Lipopolysaccharide endotoxins. *Annu. Rev. Biochem.* **2002**, *71*, 635–700. https://doi.org/10.1146/annurev.biochem.71.110601.135414.
[6] (a) Whitfield, C.; Roberts, I. S. Structure, assembly, and regulation of expression of capsules in *Escherichia coli. Mol. Microbiol.* **1999**, *31*, 1307–1319. https://doi.org/10.1046/j.1365-2958.1999.01276.x; (b) Browne, R. M.; Hartland, E. L. *Escherichia coli* as a cause of diarrhea. *J. Gastroenterol. Hepatol.* **2002**, *17*, 467–475. https://doi.org/10.1046/j.1440-1746.2002.02769.x.
[7] (a) Berry, J.; Rajaure, M.; Pang, T.; Young, R. The spanin complex is essential for lambda lysis. *J. Bacteriol.* **2012**, *194*, 5667–5674. https://doi.org/10.1128/JB.01245-12; (b) Lerouge, I.; Vanderleyden, J. O-antigen structural variation: mechanisms and possible roles in animal/plant-microbe interactions. *FEMS Microbiol. Rev.* **2002**, *26*, 17–47. https://doi.org/10.1111/j.1574-6976.2002.tb00597.x; (c) Wang, L.; Rothemund, D.; Curd, H.; Reeves, P. R. Species-wide variation in the Escherichia coli flagellin (H-antigen) gene. *J. Bacteriol.* **2003**, *185*, 2936–2943. https://doi.org/10.1128/JB.185.9.2936-2943.2003; (d) Williams, P.; Lambert, P. A.; Brown, M. R.; Jones, R. J. The role of the O and K antigens in determining the resistance of *Klebsiella* aerogenes to serum killing and phagocytosis. *Microbiol.* **1983**, *129*, 2181–2191. https://doi.org/10.1099/00221287-129-7-2181.
[8] Ørskov, F.; Ørskov, I. Serotyping of Enterobacteriaceae, with special emphasis on K antigen determination. In *Methods in microbiology*, Vol. 11, Elsevier, **1978**, 1–77, Chapter I. https://doi.org/10.1016/S0580-9517(08)70486-0.
[9] (a) Ørskov, I.; Ørskov, F.; Sojka, W.; Leach, J. Simultaneous occurrence of *E. coli* B and L antigens in strains from diseased swine: influence of cultivation temperature two new E. coli K antigens: K 87 and K 88. *Acta Pathol. Microbiol. Scand.* **2009**, *53*, 404–422. https://doi.org/10.1111/j.1699-0463.1961.tb00424.x-24; (b) Ørskov, I.; Ørskov, F.; Smith, H. W.; Sojka, W. The establishment of K99, a thermolabile, transmissible *Escherichia coli* K antigen, previously called "Kco", possessed by calf and lamb enteropathogenic strains. *Acta Pathol. Microbiol. Scand. B Microbiol.* **2009**, *83B*, 31–36. https://doi.org/10.1111/j.1699-0463.1975.tb00066.x.
[10] Rai, A. K.; Mitchell, A. M. Enterobacterial common antigen: synthesis and function of an enigmatic molecule. *mBio*.**2020**, *11*, e01914–e01920. https://doi.org/10.1128/mBio.01914-20.
[11] Kunin, C. M.; Beard, M. V.; Halmagyi, N. E. Evidence for a common hapten associated with endotoxin fractions of E. coli and other Enterobacteriaceae. *Proc. Soc. Exp. Biol. Med.* **1962**, *111*, 160–166. https://doi.org/10.3181/00379727-111-27734.
[12] (a) Ali, Y. M.; Hayat, A.; Saeed, B. M.; Haleem, K. S.; Alshamrani, S.; Kenawy, H. I.; Ferreira, V. P.; Saggu, G.; Buchberger, A.; Lachmann, P. J.; Sim, R. B.; Goundis, D.; Andrew, P. W.; Lynch, N. J.; Schwaeble, W. Low-dose recombinant properdin provides substantial protection against *Streptococcus pneumoniae* and *Neisseria meningitidis* infection. *J. Proc. Natl. Acad. Sci. USA.* **2014**, *111*, 5301–5306. https://doi.org/10.1073/pnas.1401011111; (b) Amani, J.; Mirhosseini, S. A.; Fooladi, A. A. I. A review approaches to identify enteric bacterial pathogens. *Jundishapur J. Microbiol.* **2015**, *8*, e17473. https://doi.org/10.5812/jjm.17473.
[13] http://www.who.int/mediacentre/news/releases/2017/bacteria-antibiotics-needed/en/.
[14] (a) O'Neill, J. Tackling drug-resistant infections globally: final report and recommendations. *The Review on Antimicrobial Resistance*, **2015**. https://amr-review.org/sites/default/files/160518_Final%20paper_with%20cover.pdf; (b) Jansen, K.; Knirsch, C.; Anderson, A. The role of vaccines in preventing bacterial antimicrobial resistance. *Nat. Med.* **2018**, *24*, 10–19. https://doi.org/10.1038/nm.4465.
[15] Laxminarayan, R. The overlooked pandemic of antimicrobial resistance. *Lancet.* **2022**, *399*, 606–607. https://doi.org/10.1016/S0140-6736(22)00087-3.
[16] Kennedy, D. A.; Read, A. F. Why does drug resistance readily evolve but vaccine resistance does not? *Proc. Biol. Sci.* **2017**, *284*, 2016–2562. https://doi.org/10.1098/rspb.2016.2562.
[17] (a) Rosini, R.; Nicchi, S.; Pizza, M.; Rappuoli, R. Vaccines against antimicrobial resistance. *Front. Immunol.* **2020**, 11, 1048. https://doi.org/10.3389/fimmu.2020.01048; (b) Seeberger, P. H. Discovery of semi- and fully-synthetic carbohydrate vaccines against bacterial infections using a medicinal chemistry approach. *Chem. Rev.* **2021**, 121(7), 3598–3626. https://doi.org/10.1021/acs.chemrev.0c01210.
[18] (a) Slovin, S. F.; Ragupathi, G.; Musselli, C.; Olkiewicz, K.; Verbel, D.; Kuduk, S. D.; Schwarz, J. B.; Sames, D.; Danishefsky, S.; Livingston, P. O. Fully synthetic carbohydrate-based vaccines in biochemically relapsed prostate cancer: clinical trial results with α-*N*-acetylgalactosamine-*O*-serine/threonine conjugate vaccine. *J. Clin. Oncol.* **2003**, *21*, 4292–4298. https://doi.org/10.1200/JCO.2003.04.112; (b) Cavallari, M.; Stallforth, P.; Kalinichenko, A.; Rathwell, D. C.; Gronewold, T. M.; Adibekian, A.; Mori, L.; Landmann, R.; Seeberger, P. H.; De Libero, G. A semisynthetic carbohydrate-lipid vaccine that protects against *S. pneumoniae* in mice. *Nat. Chem. Biol.* **2014**, *10*, 950–956. https://doi.org/10.1038/nchembio.1650; (c) Tillett, W. S.; Francis Jr., T. Cutaneous reactions to the polysaccharides and proteins of pneumococcus in lobar pneumonia. *J. Ex. Med.* **1929**, *50*, 687–701. https://doi.org/10.1084/jem.50.5.687.

[19] Nativi, C.; Renaudet, O. Recent progress in antitumoral synthetic vaccines. *ACS Med. Chem. Lett.* **2014**, *5*, 1176–1178. https://doi.org/10.1021/ml5003794.

[20] (a) Talaga, P.; Bellamy, L.; Moreau, M. Quantitative determination of C-polysaccharide in *Streptococcus pneumoniae* capsular polysaccharides by use of high-performance anion-exchange chromatography with pulsed amperometric detection. *Vaccine* **2001**, *19*, 2987–2994. https://doi.org/10.1016/s0264-410x(00)00535-1.

[21] Richichi, B.; Thomas, B.; Fiore, M.; Bosco, R.; Qureshi, H.; Nativi, C.; Renaudet, O.; BenMohamed, L. A cancer therapeutic vaccine based on clustered Tn-antigen mimetics induces strong antibody-mediated protective immunity. *Angew. Chem. Int. Ed.* **2014**, *53*, 11917–11920. https://doi.org/10.1002/anie.201406897.

[22] (a) Fernandez-Tejada, A.; Chea, E. K.; George, C.; Pillarsetty, N.; Gardner, J. R.; Livingston, P. O.; Ragupathi, G.; Lewis, J. S.; Tan, D. S.; Gin, D. Y. Development of a minimal saponin vaccine adjuvant based on QS-21. *Nat. Chem.* **2014**, *6*, 635–643. https://doi.org/10.1038/nchem.1963; (b) Zhou, Z.; Mondal, M.; Liao, G.; Guo, Z. Synthesis and evaluation of monophosphoryl lipid A derivatives as fully synthetic self-adjuvanting glycoconjugate cancer vaccine carriers. *Org. Biomol. Chem.* **2014**, *12*, 3238–3245. https://doi.org/10.1039/C4OB00390J; (c) Adamo, R.; Hu, Q.-Y.; Torosantucci, A.; Crotti, S.; Brogioni, G.; Allan, C.; Chiani, M.; Bromuro, C.; Quinn, D.; Tontini, M.; Berti, F. Deciphering the structure-immunogenicity relationship of anti-*Candida* glycoconjugate vaccines. *Chem. Sci.* **2014**, *5*, 4302–4311. https://doi.org/10.1039/C4SC01361A; (d) Crotti, S.; Zhai, H.; Zhou, J.; Allan, M.; Proietti, D.; Pansegrau, W.; Hu, Q. Y.; Berti, F.; Adamo, R. Defined conjugation of glycans to the lysines of CRM197 guided by their reactivity mapping. *ChemBioChem.* **2014**, *15*, 836–843. https://doi.org/10.1002/cbic.201300785; (e) McLellan, J. S.; Chen, M.; Joyce, M. G.; Sastry, M.; Stewart-Jones, G. B.; Yang, Y.; Zhang, B.; Chen, L.; Srivatsan, S.; Zheng, A.; Zhou, T.; Graepel, K. W.; Kumar, A.; Moin, S.; Boyington, J. C.; Chuang, G. Y.; Soto, C.; Baxa, U.; Bakker, A. Q.; Spits, H.; Beaumont, T.; Zheng, Z.; Xia, N.; Ko, S. Y.; Todd, J. P.; Rao, S.; Graham, B. S.; Kwong, P. D. Structure-based design of a fusion glycoprotein vaccine for respiratory syncytial virus. *Science* **2013**, *342*, 592–598. https://doi.org/10.1126/science.1243283.

[23] (a) Zhu, J.; Warren, J. D.; Danishefsky, S. J. Synthetic carbohydrate-based anticancer vaccines: the memorial Sloan-Kettering experience. *Expert Rev. Vaccines.* **2009**, *8*, 1399–1413. https://doi.org/10.1586/erv.09.95; (b) Adamo, R. Advancing homogeneous antimicrobial glycoconjugate vaccines. *Acc. Chem. Res.* **2017**, *50*, 1270–1279. https://doi.10.1021/acs.accounts.7b00106.

[24] (a) Zhang, Z.; Ollmann, I. R.; Ye, X.-S.; Wischnat, R.; Baasov, T.; Wong, C.-H. Programmable one-pot oligosaccharide synthesis. *J. Am. Chem. Soc.* **1999**, *121*, 734–753. https://doi.org/10.1021/ja982232s; (b) Kulkarni, S. S.; Wang, C.-C.; Sabbavarapu, N. M.; Podilapu, A. R.; Liao, P.-H.; Hung, S.-C. "One-pot" protection, glycosylation, and protection-glycosylation strategies of carbohydrates. *Chem. Rev.* **2018**, *118*, 8025–8104. https://doi.10.1021/acs.chemrev.8b00036; (c) Cheng, C.-W.; Wu, C.-Y.; Hsu, W. L.; Wong, C.-H. Programmable one-pot synthesis of oligosaccharides. *Biochem.* **2020**, *59*, 3078–3088. https://doi.org/10.1021/acs.biochem.9b00613.

[25] (a) Nigudkar, S. S.; Demchenko, A. V. Stereo controlled 1, 2-*cis* glycosylation as the driving force of progress in synthetic carbohydrate chemistry. *Chem. Sci.* **2015**, *6*, 2687–2704. https://doi.org/10.1039/C5SC00280J; (b) Mensink, R. A.; Boltje, T. J. Advances in stereoselective 1,2-*cis* glycosylation using C-2 auxiliaries. *Chem. Eur. J.* **2017,** *23*, 17637–17653. https://doi.org/10.1002/chem.201700908; (c) Bennett, C. S.; Galan, M. C. Methods for 2-deoxyglycoside synthesis. *Chem. Rev.* **2018**, *118*, 7931–7985. https://doi.org/10.1021/acs.chemrev.7b00731; (d) Ling, J.; Bennett, C. S. Recent developments in stereoselective chemical glycosylation. *Asian J. Org. Chem.* **2019**, *8*, 802–813. https://doi.org/10.1002/ajoc.201900102; (e) Ishiwata, A.; Tanaka, K.; Ao, J.; Ding, F.; Ito, Y. Recent advances in stereoselective 1,2-cis-O-glycosylations. *Front. Chem.* **2022**, *10*, 972429. https://doi.org/10.3389/fchem.2022.972429; (f) Mukherjee, M. M.; Ghosh, R.; Hanover, J. A. Recent advances in stereoselective chemical O-glycosylation reactions. *Front. Mol. Biosci.* **2022**, *9*, 896187. https://doi.org/10.3389/fmolb.2022.896187.

[26] (a) Mukherjee, M. M.; Xu, P.; Stevens, E. D.; Kováč, P. Towards the complete synthetic *O*-antigen of vibrio cholerae O1, serotype Inaba: improved synthesis of the conjugation-ready upstream terminal hexasaccharide determinant. *RSC Adv.* **2019**, *9*, 36440–36454. https://doi.10.1039/c9ra08232h; (b) Dhara, D.; Dhara, A.; Murphy, P. V.; Mulard, L. A. Protecting group principles suited to late-stage functionalization and global deprotection in oligosaccharide synthesis. *Carbohydr. Res.* **2022**, *521*, 108644–108662. https://doi.org/10.1016/j.carres.2022.108644.

[27] (a) Yu, B.; Yang, Z.; Cao, H. One-pot glycosylation (OPG) for the chemical synthesis of oligosaccharides. *Curr. Org. Chem.* **2005**, *9*, 179–194. https://doi.org/10.1016/j.anres.2018.08.002; (b) Das, R.;

Mukhopadhyay, B. Chemical O-glycosylations: an overview. *ChemistryOpen* **2016**, *5*, 401–433. https://doi.org/10.1002/open.201600043; (c) Yang, W.; Yang, B.; Ramadan, S.; Huang, X. Preactivation-based chemoselective glycosylations: a powerful strategy for oligosaccharide assembly. *Beilstein J. Org. Chem.* **2017**, *13*, 2094–2114. https://doi.org/10.3762/bjoc.13.207; (d) Cheng, C.-W.; Zhou, Y.; Pan, W.-H.; Dey, S.; Wu, C.-Y.; Hsu, W.-L.; Wong, C.-H. Hierarchical and programmable one-pot synthesis of oligosaccharides. *Nat. Commun.* **2018**, *9*, Article 5202. https://doi.org/10.1038/s41467-018-07618-8; (e) Pardo-Vargas, A.; Delbianco, M.; Seeberger, P. H. Automated glycan assembly as an enabling technology. *Curr. Opin. Chem. Biol.* **2018**, *46*, 48–55. https://doi.org/10.1016/j.cbpa.2018.04.007; (f) Panza, M.; Pistorio, S. G.; Stine, K. J.; Demchenko, A. V. Automated chemical oligosaccharide synthesis: novel approach to traditional challenges. *Chem. Rev.* **2018**, *118*, 8105–8150. https://doi.org/10.1021/acs.chemrev.8b00051; (g) Nielsen, M. M.; Pedersen, C. M. Catalytic glycosylations in oligosaccharide synthesis. *Chem. Rev.* **2018**, *118*, 8285–8358. https://doi.org/10.1021/acs.chemrev.8b00144; (h) Bauer, B. Transition metal catalyzed glycosylation reactions: an overview. *Org. Biomol. Chem.* **2020**, *18*, 9160–9180. https://doi.org/10.1039/D0OB01782E.

[28] Qadri, F.; Svennerholm, A-M.; Faruque, A. S. G.; Sack, R. B. Enterotoxigenic *Escherichia coli* in developing countries: epidemiology, microbiology, clinical features, treatment, and prevention. *Clin. Microbiol. Rev.* **2005**, *18*, 465–483. https://doi.org/10.1128/CMR.18.3.465-483.2005.

[29] (a) Dell, A.; Oates, J.; Lugowski, C.; Romanowska, E.; Kenne, L.; Lindberg, B. The enterobacterial common-antigen, a cyclic polysaccharide. *Carbohydr. Res.* **1984**, *133*, 95–104. https://doi.org/10.1016/0008-6215(84)85186-1; (b) Lugowski, C.; Romanowska, E.; Kenne, L.; Lindberg, B. Identification of a trisaccharide repeating-unit in the enterobacterial common-antigen. *Carbohydr. Res.* **1983**, *118*, 173–181. https://doi.org/10.1016/0008-6215(83)88045-8.

[30] Liu, L.; Zha, J.; DiGiandomenico, A.; McAllister, D.; Kendall Stover, C.; Wang, Q.; Boons, G. J. Synthetic enterobacterial common antigen (ECA) for the development of a universal immunotherapy for drug-resistant Enterobacteriaceae. *Angew. Chem. Int. Ed.* **2015**, *54*, 10953–10957. https://doi.org/10.1002/anie.201505420.

[31] (a) Podschun, R.; Ullmann, U. *Klebsiella* spp. as nosocomial pathogens: epidemiology, taxonomy, typing methods, and pathogenicity factors. *Clin. Microbiol. Rev.* **1998**, *11*, 589–603. https://doi.org/10.1128/CMR.11.4.589; (b) Fang, C. T.; Chuang, Y. P.; Shun, C. T.; Chang, S. C.; Wang, J. T. A novel virulence gene in *Klebsiella pneumoniae* strains causing primary liver abscess and septic metastatic complications. *J. Exp. Med.* **2004**, *199*, 697–705. https://doi.org/10.1084/jem.20030857.

[32] Paczosa, M. K.; Mecsas, J. *Klebsiella pneumoniae*: going on the offense with a strong defense. *Microbiol. Molecular Biol. Rev.* **2016**, *80*, 629–661. https://doi.org/10.1128/MMBR.00078-15.

[33] (a) Kaur, J.; Sheemar, S.; Chand, K.; Chopra, S.; Mahajan, G. Outbreak of carbapenemase producing *Klebsiella pneumoniae* BSI in NICU unit. *Int. J. Curr. Microbiol. Appl. Sci.* **2016**, *5*, 727–733. http://dx.doi.org/10.20546/ijcmas.2016.501.073; (b) Yu, J.; Tan, K.; Rong, Z.; Wang, Y.; Chen, Z.; Zhu, X.; Wu, L.; Tan, L.; Xiong, W.; Sun, Z.; Chen, L. Nosocomial outbreak of KPC-2- and NDM-1-producing *Klebsiella pneumoniae* in a neonatal ward: a retrospective study. *BMC Infect. Dis.* **2016**, *16*, 563. https://doi.org/10.1186/s12879-016-1870-y; (c) Kola, A.; Piening, B.; Pape, U. F.; Schlieker, W. V.; Kaase, M.; Geffers, C.; Wiedenmann, B.; Gastmeier, P. An outbreak of carbapenem-resistant OXA-48—producing *Klebsiella pneumonia* associated to duodenoscopy. *Antimicrob. Resist. Infect. Control.* **2015**, *4*, 8. https://doi.org/10.1186/s13756-015-0049-4; (d) Stillwell, T.; Green, M.; Barbadora, K.; Ferrelli, J. G.; Roberts, T. L.; Weissman, S. J.; Nowalk, A. Outbreak of KPC-3 producing carbapenem-resistant *Klebsiella pneumoniae* in a US pediatric hospital. *J. Pediatr. Infect. Dis.* **2014**, *4*, 330–338. https://doi.org/10.1093/jpids/piu080; (e) Ben-David, D.; Kordevani, R.; Keller, N.; Tal, I.; Marzel, A.; Gal-Mor, O.; Maor, Y.; Rahav, G. Outcome of carbapenem resistant *Klebsiella pneumoniae* bloodstream infections. *Clin. Microbiol. Infect.* **2012**, *18*, 54–60. https://doi.org/10.1111/j.1469-0691.2011.03478.x.

[34] (a) Zamze, S.; Martinez-Pomares, L.; Jones, H.; Taylor, P. R.; Stillion, R. J.; Gordon, S.; Wong, S. Y. Recognition of bacterial capsular polysaccharides and lipopolysaccharides by the macrophage mannose receptor. *J. Biol. Chem.* **2002**, *277*, 41613–41623. https://doi.10.1074/jbc.M207057200; (b) Cortes, G.; Borrell, N.; de Astorza, B.; Gomez, C.; Sauleda, J.; Alberti, S. Molecular chain to the virulence of *Klebsiella pneumoniae* in a murine model of pneumonia. *Infect. Immun.* **2002**, *70*, 2583–2590. https://doi.org/10.1128/IAI.70.5.2583-2590.2002; (c) Shankar-Sinha, S.; Valencia, G. A.; Janes, B. K.; Rosenberg, J. K.; Whitfield, C.; Bender, R. A.; Standiford, T. J.; Younger, J. G. The *Klebsiella pneumoniae* O antigen contributes to bacteremia and lethality during murine pneumonia. *Infect. Immun.* **2004**, *72*, 1423–1430. https://doi.org/10.1128/IAI.72.3.1423-1430.2004.

[35] Choi, M.; Hegerle, N.; Nkeze, J.; Sen, S.; Jamindar, S.; Nasrin, S.; Sen, S.; Permala-Booth, J.; Sinclair, J.; Tapia, M. D.; Johnson, J. K.; Mamadou, S.; Thaden, J. T.; Fowler Jr, V. G.; Aguilar, A.; Terán, E.;

Decre, D.; Morel, F.; Krogfelt, K. A.; Brauner, A.; Protonotariou, E.; Christaki, E.; Shindo, Y.; Lin, Y. -T.; Kwa, A. L.; Shakoor, S.; Singh-Moodley, A.; Perovic, O.; Jacobs, J.; Lunguya, O.; Simon, R.; Cross, A. S.; Tennant, S. M. The diversity of lipopolysaccharide (O) and capsular polysaccharide (K) antigens of invasive *Klebsiella pneumoniae* in a multi-country collection. *Front. Microbiol.* **2020**, *11*, Article 1249. https://doi.org/10.3389/fmicb.2020.01249.

[36] (a) Ahmad, T. A.; Haroun, M.; Hussein, A. A.; Ashry, E. S. H. E.; El-Sayed, L. H. Development of a new trend conjugate vaccine for the prevention of *Klebsiella pneumonia. Infect. Dis. Rep.* **2012**, *4*, e33, 128–133. https://doi.org/10.4081/idr.2012.e33; (b) Malachowa, N.; Kobayashi, S. D.; Porter, A. R.; Freedman, B.; Hanley, P. W.; Lovaglio, J.; Saturday, G. A.; Gardner, D. J.; Scott, D. P.; Griffin, A.; Cordova, K.; Long, D.; Rosenke, R.; Sturdevant, D. E.; Bruno, D.; Martens, C.; Kreiswirth, B. N.; DeLeo, F. R. Vaccine protection against multidrug-resistant *Klebsiella pneumoniae* in a nonhuman primate model of severe lower respiratory tract infection. *mBio* **2019**, *10*, e02994.19. https://doi.org/10.1128/mBio.02994-19; (c) Feldman, M. F.; Bridwell, A. E. M.; Scott, N. E.; Vinogradov, E.; McKee, S. R.; Chavez, S. M.; Twentyman, J.; Stallings, C. L.; Rosen, D. A.; Harding, C. M. A promising bioconjugate vaccine against hypervirulent *Klebsiella pneumonia. PNAS.* **2019**, *116*, 18655–18663. https://doi.org/10.1073/pnas.1907833116; (d) Patro, L. P. P.; Rathinavelan, T. Targeting the sugary armor of *Klebsiella* species. *Front. Cell. Infect. Microbiol.* **2019**, *9*, Article 367. https://doi.org/10.3389/fcimb.2019.00367; (e) Opoku-Temeng, C.; Kobayashi, S. D.; DeLeo, F. R. *Klebsiella pneumoniae* capsule polysaccharide as a target for therapeutics and vaccines. *Comput. Struc. Biotechnol. J.* **2019**, *17*, 1360–1366. https://doi.org/10.1016/j.csbj.2019.09.011; (f) Ahmad, T. A.; El-Sayed, L. H.; Haroun, M.; Hussein, A. A.; El Ashry, E. S. H. Development of immunization trials against *Klebsiella pneumonia. Vaccine* **2012**, *30*, 2411–2420. https://doi.org/10.1016/j.vaccine.2011.11.027; (g) Choi, M.; Tennant, S. M.; Simon, R.; Cross, A. S. Progress towards the development of *Klebsiella* vaccines. *Exp. Rev. Vaccines* **2019**, 681–691. https://doi.org/10.1080/14760584.2019.1635460.

[37] Hegerle, N.; Choi, M.; Sinclair, J.; Amin, M. N.; Ollivault-Shiflett, M.; Curtis, B.; Laufer, R. S.; Shridhar, S.; Brammer, J.; Toapanta, F. R.; Holder, I. A.; Pasetti, M. F.; Lees, A.; Tennant, S. M.; Cross, A. S.; Simon, R. Development of a broad spectrum glycoconjugate vaccine to prevent wound and disseminated infections with *Klebsiella pneumonia* and *Pseudomonas aeruginosa. PLOS One* **2018**, *13*, e0203143. https://doi.org/10.1371/journal.pone.020.

[38] Alberti, S.; Hernandez-Alles, S.; Gil, J.; Reina, J.; Martinez-Beltran, J.; Camprubi, S.; Tomas, J. M.; Benedi, V. J. Development of an enzyme-linked immunosorbent assay method for typing and quantitation of *Klebsiella pneumoniae* lipopolysaccharide: application to serotype 01. *J. Clin. Microbiol.* **1993**, *31*, 1379–1381. https://doi.org/10.1128/jcm.31.5.1379-1381.1993.

[39] Vinogradov, E.; Frirdich, E.; MacLean, L. L.; Perry, M. B.; Petersen, B. O.; Duus, J. O.; Whitfield, C. Structures of lipopolysaccharides from *Klebsiella pneumonia*, eluicidation of the structure of the linkage region between core and polysaccharide O chain and identification of the residues at the non-reducing termini of the O chains. *J. Biol. Chem.* **2002**, *277*, 25070–25081. https://doi.org/10.1074/jbc.M202683200.

[40] Zhu, S.-Y.; Yang, J.-S. Synthesis of tetra- and hexasaccharide fragments corresponding to the O-antigenic polysaccharide of *Klebsiella pneumonia. Tetrahedron* **2012**, *68*, 3795–3802. https://doi.org/10.1016/j.tet.2012.03.074.

[41] (a) Krylov, V. B.; Argunov, D. A.; Vinnitskiy, D. Z.; Verkhnyatskaya, S. A.; Gerbst, A. G.; Ustyuzhanina, N. E.; Dmitrenok, A. S.; Huebner, J.; Holst, O.; Siebert, H.-C.; Nifantiev, N. E. Pyranoside-into-furanoside rearrangement: new reaction in carbohydrate chemistry and its application in oligosaccharide synthesis. *Chem. Eur. J.* **2014**, *20*, 16516–16522. https://doi.org/10.1002/chem.201405083; (b) Krylov, V. B.; Argunov, D. A.; Vinnitskiy, D. Z.; Gerbst, G.; Ustyuzhanina, N. E.; Dmitrenok, A. S.; Nifantiev, N. E. The Pyranoside-into-furanoside rearrangement of alkyl glycosides: scope and limitations. *Synlett.* **2016**, *27*, 1659–1664. https://doi.org/10.1055/s-0035-1561595.

[42] (a) Stella, A.; Verkhnyatskaya, S. A.; Krylov, V. B.; Nifantiev, N. E. Pyranoside-into-furanoside rearrangement of 4-pentenyl glycosides in the synthesis of a tetrasaccharide-related to Galactan I of *Klebsiella pneumonia. Eur. J. Org. Chem.* **2017**, 710–718. https://doi.org/10.1002/ejoc.201601413; (b) Argunov, D. A.; Trostianetskaia, A. S.; Krylov, V. B.; Kurbatova, E. A.; Nifantiev, N. E. Convergent synthesis of oligosaccharides structurally related to Galactan I and Galactan II of *Klebsiella pneumoniae* and their use in screening of antibody specificity. *Eur. J. Org. Chem.* **2019**, 4226–4232. https://doi.org/10.1002/ejoc.201900389.

[43] (a) Argunov, D. A.; Krylov, V. B.; Nifantiev, N. E. The use of pyranoside-into-furanoside rearrangement and controlled O (5)→ O (6) benzoyl migration as the basis of a synthetic strategy to assemble (1→ 5)-and (1→ 6)-linked galactofuranosyl chains. *Org. Lett.* **2016**, *18*, 5504–5507. https://doi.org/10.1021/acs.orglett.6b02735; (b) Gerbst, A. G.; Krylo, V. B.; Argunov, D. A.; Dmitrenok, A. S.; Nifantiev, N. E.

Driving force of the pyranoside-into-furanoside rearrangement. *ACS Omega* **2019**, *4*, 1139–1143. https://doi.org/10.1021/acsomega.8b03274.

[44] Argunov, D. A.; Trostianetskaia, A. S.; Krylov, V. B.; Kurbatova, E. A.; Nifantiev, N. E. Convergent synthesis of oligosaccharides structurally related to Galactan I and Galactan II of *Klebsiella Pneumoniae* and their use in screening of antibody specificity. *Eur. J. Org. Chem.* **2019**, 4226–4232. https://doi.org/10.1002/ejoc.201900389.

[45] Maity, S. K.; Ghosh, R. A three-component, one-pot, sequential synthesis of a common tetrasaccharide block related to the lipopolysaccharide of the *Escherichia coli* O9, *Klebsiella pneumonia* O3, and *Hafnia alvei* PCM 1223. *Synlett.* **2012**, *23*, 1919–1922. https://doi.org/10.1055/s-0032-1316685.

[46] Kitchel, B.; Rasheed, J. K.; Patel, J. B.; Srinivasan, A.; Venezia, S. N.; Carmeli, Y.; Brolund, A.; Giske, C. G. Molecular epidemiology of KPC-producing *Klebsiella pneumoniae* isolates in the United States: clonal expansion of multi locus sequence type 258. *Antimicrob. Agents Chemother.* **2009**, *53*, 3365–3370. https://doi.org/10.1128/AAC.00126-09.

[47] Seeberger, P. H.; Pereira, C. L.; Khan, N.; Xiao, G.; Diago-Navarro, E.; Reppe, K.; Opitz, B.; Fries, B. C.; Witzenrath, M. A semi-synthetic glycoconjugate vaccine candidate for carbapenem-resistant *Klebsiella pneumonia*. *Angew. Chem. Int. Ed.* **2017**, *56*, 13973–13978. https://doi.org/10.1002/anie.201700964.

[48] Cryz Jr, S. J.; Mortimer, P.; Cross, A. S.; Fürer, E.; Germanier, R. Safety and immunogenicity of a polyvalent *Klebsiella* capsular polysaccharide vaccine in humans. *Vaccine* **1986**, *4*, 15–20. https://doi.org/10.1016/0264-410x(86)90092-7.

[49] Sarkar, V.; Mukhopadhyay, B. Chemical synthesis of the hexasaccharide related to the repeating unit of the capsular polysaccharide from carbapenem resistant *Klebsiella pneumonia* 2796 and 3264. *RSC Adv.* **2016**, *6*, 40147–40154. https://doi.org/10.1039/C6RA07351D.

[50] Shit, P.; Sahaji, S.; Misra, A. K. Convergent synthesis of the hexasaccharide repeating unit of the capsular polysaccharide of *Klebsiella* serotype K-34. *Tetrahedron* **2022**, *129*, 133159. https://doi.org/10.1016/j.tet.2022.133159.

[51] Yang, F. L.; Yang, Y. L.; Liao, P. C.; Chou, J. C.; Tsai, K. C.; Yang, A. S.; Sheu, F.; Lin, T. L.; Hsieh, P. F.; Wang, J. T.; J. T.; Hua, J. T.; Wu, S. H. Structure and immunological characterization of the capsular polysaccharide of a pyrogenic liver abscess caused by *Klebsiella pneumoniae*: activation of macrophages through Toll-like receptor 4. *J. Biol. Chem.***2011**, *24*, 21041–21051. https://doi.org/10.1074/jbc.M111.222091.

[52] Susanto, W.; Kong, K. -H.; Hua, K. -F.; Wu, S. -H.; Lam, Y. Synthesis of the trisaccharide repeating unit of capsular polysaccharide from *Klebsiellapneumonia*. *Tetrahedron Lett.* **2019**, *60*, 288–291. https://doi.org/10.1016/j.tetlet.2018.12.031.

[53] Flemming, H. -C.; Wingender, J.; Szewzyk, U.; Steinberg, P.; Rice, S. A.; Kjelleberg, S. Biofilms: an emergent form of bacterial life. *Nat. Rev. Microbiol.* **2016**, *14*, 563–575. https://doi.org/10.1038/nrmicro.2016.94.

[54] Cescutti, P.; De Benedetto, G.; Rizzo, R. Structural determination of the polysaccharide isolated from biofilms produced by a clinical strain of *Klebsiella pneumoniae*. *Carbohydr. Res.* **2016**, *430*, 29–35. https://doi.org/10.1016/j.carres.2016.05.001.

[55] Gucchait, A.; Ghosh, A.; Misra, A. K. Convergent synthesis of the pentasaccharide repeating unit of the biofilms produced by *Klebsiella pneumoniae*. *Beilstein J. Org. Chem.* **2019**, *15*, 431–436. https://doi.org/10.3762/bjoc.15.37.

[56] Micoli, F.; Rondini, S.; Pisoni, I.; Proietti, D.; Berti, F.; Costantino, P.; Rappuoli, R.; Szu, S.; Saul, A.; Martin, L. B. Vi-CRM197 as a new conjugate vaccine against *Salmonella Typhi*. *Vaccine* **2011**, *29*, 712–720. https://doi.org/10.1016/j.vaccine.2010.11.022.

[57] Martin, L. B.; Simon, R.; MacLennan, C. A.; Tennant, S. M.; Sahastrabuddhe, S.; Khan, M. I. Status of paratyphoid fever vaccine research and development. *Vaccine* **2016**, *34*, 2900–2902. https://doi.org/10.1016/j.vaccine.2016.03.106.

[58] Santhanam, S. K.; Dutta, D.; Parween, F.; Qadri, A. The virulence polysaccharide Vi released by *Salmonella Typhi* targets membrane prohibitin to inhibit T-cell activation. *J. Infect. Dis.* **2014**, *210*, 79–88. https://doi.org/10.1093/infdis/jiu064.

[59] Yang, L.; Zhu, J.; Zheng, X.-J.; Tai, G.; Ye, X.-S. A highly α-stereoselective synthesis of oligosaccharide fragments of the Vi antigen from *Salmonella typhi* and their antigenic activities. *Chem. Eur. J.* **2011**, *17*, 14518–14526. https://doi.org/10.1002/chem.201102615.

[60] Zhang, G.-L.; Yang, L.; Zhu, J.; Wei, M.; Yan, W.; Xiong, D.-C.; Ye, X.-S. Synthesis and antigenic evaluation of oligosaccharide mimics of Vi antigen from *Salmonella typhi*. *Chem. Eur. J.* **2017**, *23*, 10670–10677. https://doi.org/10.1002/chem.201702114.

[61] Zhang, G.-L.; Wei, M.-M.; Song, C.; Ma, Y.-F.; Zheng, X.-J.; Xionga, D.-C.; Ye, X.-S. Chemical synthesis and biological evaluation of penta- to octa- saccharide fragments of Vi polysaccharide from Salmonella typhi. *Org. Chem. Front.* **2018**, *5*, 2179–2188. https://doi.org/10.1039/C8QO00471D.

[62] Fusari, M.; Fallarini, S.; Lombardi, G.; Lay, L. Synthesis of di- and tri-saccharide fragments of *Salmonellatyphi* Vi capsular polysaccharide and their zwitterionic analogues. *Bioorg. Med. Chem.* **2015**, *23*, 7439–7447. https://doi.org/10.1016/j.bmc.2015.10.043.

[63] (a) Yin, Z.; Comellas-Aragones, M.; Chowdhury, S.; Bentley, P.; Kaczanowska, K.; BenMohamed, L.; Gildersleeve, J. C.; Finn, M. G.; Huang, X. Boosting immunity to small tumor-associated carbohydrates with bacteriophage Qβ capsids. *ACS Chem. Biol.* **2013**, 1253–1262. https://doi.org/10.1021/cb400060x; (b) Whitaker, J. A.; Franco-Paredes, C.; del Rio, C.; Edupuganti, S. Rethinking typhoid fever vaccines: implications for travelers and people living in highly endemic areas. *J. Travel Med.* **2009**, *16*, 46–52. https://doi.org/10.1111/j.1708-8305.2008.00273.x.

[64] Duclos, P.; Okwo-Bele, J.; Salisbury, D. Establishing global policy recommendations: the role of the strategic advisory group of experts on immunization. *Expert Rev. Vaccines* **2011**, *10*, 163–73. https://doi.org/10.1586/erv.10.171.

[65] Martina, L. B.; Simon, R.; MacLennan, C. A.; Tennant, S. M.; Sahastrabuddhe, S.; Khan, M. I. Status of paratyphoid fever vaccine research and development. *Vaccine* **2016**, *34*, 2900–2902. https://doi.org/10.1016/j.vaccine.2016.03.106.

[66] Dhara, D.; Baliban, S. M.; Huo, C.-X.; Rashidijahanabad, Z.; Sears, K. T.; Nick, S. T.; Misra, A. K.; Tennant, S. M.; Huang, X. Syntheses of *Salmonella paratyphi* A associated oligosaccharide antigens and development towards anti-paratyphoid fever vaccines. *Chem. Eur. J.* **2020**, *26*, 15953–15968. https://doi.org/10.1002/chem.202002401.

[67] Bachmann, M. F.; Jennings, G. T. Vaccine delivery: a matter of size, geometry, kinetics and molecular patterns. *Nat. Rev. Immunol.* 2010, *10*, 787–796. https://doi.org/10.1038/nri2868.

[68] Dhara, D.; Baliban, S. M.; Huo, C.-X.; Rashidijahanabad, Z.; Sears, K. T.; Nick, S. T.; Misra, A. K.; Tennant, S. M.; Huang, X. Syntheses of *Salmonella Paratyphi* A associated oligosaccharide antigens and development towards anti-paratyphoid fever vaccines. *Chem. Eur. J.* **2020**, *26*, 15953–15968. https://doi.org/10.1002/chem.202002401.

[69] (a) Grassel, G. A.; Finlay, B. B. Pathogenesis of enteric Salmonella infections. *Curr. Opin. Gastroenterol.* **2008**, *24*, 22–26. https://doi.org/10.1097/MOG.0013e3282f21388; (b) Cobern, B.; Grassel, G. A.; Finlay, B. B. *Salmonella*, the host and disease: a brief review. *Immuno*: a brief review. *Immunol. Cell. Biol.* **2007**, *85*, 112–118. https://doi.org/10.1038/sj.icb.7100007.

[70] Sau, A.; Panchadhayee, R.; Ghosh, D.; Misra, A. K. Synthesis of a tetrasaccharide analog corresponding to the repeating unit of the O-polysaccharide of *Salmonella enterica* O59: unexpected stereo outcome in glycosylation. *Carbohydr. Res.* **2012**, *352*, 18–22. https://doi.org/10.1016/j.carres.2012.01.026.

[71] Dhara, D.; Mandal, P. K.; Misra, A. K. Convergent synthesis of a pentasaccharide repeating unit corresponding to the cell wall O-antigen of *Salmonella enterica* O44. *Tetrahedron: Asym.* **2014**, *25*, 263–267. https://doi.org/10.1016/j.tetasy.2013.12.005.

[72] Das, R.; Mukhopadhyay, B. Chemical synthesis of the pentasaccharide related to the repeating unit of the O-antigen from *Salmonella enterica* O4. *J. Carbohydr. Chem.* **2015**, *34*, 247–262. https://doi.org/10.1080/07328303.2015.1047503.

[73] Bhaumik, I.; Misra, A. K. Concise synthesis of a pentasaccharide repeating unit corresponding to the O-antigen of *Salmonella enterica* O51. *ChemistrySelect* **2017**, *2*, 937–939. https://doi.org/10.1002/slct.201601918.

[74] Dhara, D.; Misra, A. K. Convergent synthesis of oligosaccharide fragments corresponding to the cell wall O-polysaccharide of *Salmonella enterica* O53. *ChemistryOpen* **2015**, *4*, 768–773. https://doi.org/10.1002/open.201500102.

[75] Gucchait, A.; Kundu, M.; Misra, A. K. Synthesis of pentasaccharide repeating unit corresponding to the cell wall O-polysaccharide of *Salmonella enterica* O55 strain containing a rare sugar 3-acetamido-3-deoxy-D-fucose. *Synthesis* **2021**, *53*, 3613–3620. https://doi.org/10.1055/s-0037-1610777.

[76] Gucchait, A.; Shit, P.; Misra, A. K. Concise synthesis of a tetrasaccharide related to the repeating unit of the cell wall O-antigen of *Salmonella enterica* O60. *Tetrahedron* **2020**, *76*, 131412. https://doi.org/10.1016/j.tet. 2020.131412.

[77] Gucchait, A.; Misra, A. K. Influence of remote functional groups towards the formation of 1,2-*cis* glycosides: special emphasis on β-mannosylation. *Org. Biomol. Chem.* **2019**, *17*, 4615–4610. https://doi.org/10.1039/C9OB00670B.

[78] Huo, C.-X.; Dhara, D.; Baliban, S. M.; Nick, S. T.; Tan, Z.; Simon, R.; Misra, A. K.; Huang, X. Synthetic and immunological studies of *Salmonella Enteritidis* O-antigen tetrasaccharides as potential anti-*Salmonella* vaccines. *Chem. Commun.* **2019**, *55*, 4519–4522. https://doi.org/10.1039/C8CC08622B.

[79] (a) Kotloff, K. L.; Riddle, M. S.; Platts-Mills, J. A.; Pavlinac, P.; Zaidi, A. K. M. Shigellosis. *Lancet.* **2018**, *391*, 801–812. https://doi.org/10.1016/S0140-6736(17)33296-8. (b) Camacho, A. I.; Irache, J. M.; Gamazo, C. Recent progress towards development of a *Shigella* vaccine. *Expert Rev. Vaccines* **2013**, *12*, 43–55. https://doi.org/10.1586/erv.12.135; (c) Kosek, M.; Yori, P. P.; Olortegui, M. P. Shigellosis update: advancing antibiotic resistance, investment empowered vaccine development and green bananas. *Curr. Opin. Infect. Dis.* **2010**, *23*, 475–480. https://doi.org/10.1097/QCO.0b013e32833da204.

[80] (a) Barel, L.-A.; Mulard, L. A. Classical and novel strategies to develop a *Shigella* glycoconjugate vaccine: from concept to efficacy in human. *Hum. Vac. Immunotherapeutics* **2019**, *15*, 1338–1356. https://doi.org/10.1080/21645515.2019.1606972; (b) Launay, O.; Lewis, D. J. M.; Anemona, A.; Loulergue, P.; Leahy, J.; Sciré, A. S.; Maugard, A.; Marchetti, E.; Zancan, S.; Huo, Z.; Rondini, S.; Marhaba, R.; Finco, O.; Martin, L. B.; Auerbach, J.; Cohenf, D.; Saul, A.; Gerke, C.; Podda, A. Safety profile, immunologic responses of a novel vaccine against *Shigella sonnei* administered intramuscularly, intradermally and intranasally: results from two parallel randomized phase 1 clinical studies in healthy adult volunteers in Europe. *EBioMed.* **2017**, *22*, 164–17. https://doi.org/10.1016/j.ebiom.2017.07.013; (c) Obiero, C. W.; Ndiaye, A. G. W.; Sciré, A. S.; Kaunyangi, B. M.; Marchetti, E.; Gone, A. M.; Schütte, L. D.; Riccucci, D.; Auerbach, J.; Saul, A.; Martin, L. B.; Bejon, P.; Njuguna, P.; Podda, A. A phase 2a randomized study to evaluate the safety and immunogenicity of the 1790GAHB generalized modules for membrane antigen vaccine against *Shigella sonnei* administered intramuscularly to adults from a shigellosis-endemic country. *Front. Immunol.* **2017**, *8*, 1884. https://doi.org/10.3389/fimmu.2017.01884.

[81] (a) Mani, S.; Wierzba, T.; Walker, R. I. Status of vaccine research and development for *Shigella. Vaccine* **2016**, *34*, 2887–2894. https://doi.org/10.1016/j.vaccine.2016.02.075; (b) Ashkenazi, S.; Cohen, D. An update on vaccines against *Shigella. Ther. Adv. Vaccines* **2013**, *1*, 113–123. https://doi.org/10.1177/2051013613500428; (c) Kosek, M.; Yori, P. P.; Olortegui, M. P. Shigellosis update: advancing antibiotic resistance, investment empowered vaccine development and green bananas. *Curr. Opin. Infect. Dis.* **2010**, *23*, 475–480. https://doi.org/10.1097/QCO.0b013e32833da204.

[82] Lindberg, A. A.; Browns, J. E.; Strombergli, N.; Westling-Ryd, M.; Schultz, J. E.; Karlsson, K.-A. Identification of the carbohydrate receptor for Shiga toxin produced by *Shigella dysenteriae* type 1. *J. Biol. Chem.* **1987**, *262*, 1779–1785. https://doi.org/10.1016/S0021-9258(19)75706-8.

[83] (a) Pozsgay, V. Synthesis of glycoconjugate vaccines against *Shigella dysenteriae* type1. *J. Org. Chem.* **1998**, *63*, 5983–5999. https://doi.org/10.1021/jo980660a; (b) Pozsgay, V.; Chu, C.; Pannell, L.; Wolfe, J.; Robbins, J. B.; Schneerson, R. Protein conjugates of synthetic saccharides elicit higher levels of serum IgG lipopolysaccharide antibodies in mice than do those of the O-specific polysaccharide from *Shigella dysenteriae* type 1. *Proc. Natl. Acad. Sci. USA.* **1999**, *96*, 5194–5197. https://doi.org/10.1073/pnas.96.9.5194; (c) Pozsgay, V.; Ekborg, G.; Sampathkumar, S. G. Synthesis of hexa- to tridecasaccharides related to *Shigella dysenteriae* type 1 for incorporation in experimental vaccines. *Carbohydr. Res.* **2006**, *341*, 1408–1427. https://doi.org/10.1016/j.carres.2006.04.006; (d) Pozsgay, V. Towards a synthetic carbohydrate vaccine against *Shigella dysenteriae* type 1. In Nifantiev, N. E., Ed. *Progress in the synthesis of complex carbohydrate chains of plant and microbial polysaccharides*, Transworld Research Network, **2009**, 327–344.

[84] Guchhait, G.; Misra, A. K. Convergent synthesis of a common pentasaccharide corresponding to the O-antigen of *Escherichia coli* O168 and *Shigella dysenteriae* type 4. *Glycoconj. J.* **2011**, *28*, 11–19. https://doi.org/10.1007/s10719-010-9317-y.

[85] Santra, A.; Ghosh, T.; Misra, A. K. Synthesis of di- and trisaccharides related to the *O*-polysaccharide of *Shigella dysenteriae* type 8. *J. Carbohydr. Chem.* **2015**, *34*, 501–513. https://doi.org/10.1080/07328303.2015.1108424.

[86] (a) Knirel, Y. A.; Sun, Q.; Senchenkova, S. N.; Perepelov, A. V.; Shashkov, A. S.; Xu, J. O-Antigen modifications providing antigenic diversity of *Shigella flexneri* and underlying genetic mechanisms. *Biochem. Moscow.* **2015**, *80*, 901–914. https://doi.org/10.1134/S0006297915070093; (b) Perepelov, A. V.; Shekht, M. E.; Liu, B.; Shevelev, S. D.; Ledov, V. A.; Senchenkova, S. N.; L'vov, V. L.; Shashkov, A. S.; Feng, L.; Aparin, P. G.; Wang, L.; Knirel, Y. A. Shigella flexneri O-antigens revisited: final elucidation of the O-acetylation profiles and a survey of the O-antigen structure diversity. *FEMS Immunol. Med. Microbiol.* **2012**, *66*, 201–210. https://doi.org/10.1111/j.1574-695X.2012.01000.x.

[87] (a) Wright, K.; Guerreiro, C.; Laurent, I.; Baleux, F.; Mulard, L. A. Preparation of synthetic glycoconjugates as potential vaccines against *Shigella flexneri* serotype 2a disease. *Org. Biomol. Chem.* **2004**, *2*, 1518–1527. https://doi.org/10.1039/B400986J; (b) Bélot, F.; Guerreiro, C.; Baleux, F.; Mulard, L. A. Synthesis of two linear PADRE conjugates bearing a deca- or pentadecasaccharide B epitope as potential synthetic vaccines against *Shigella flexneri* serotype 2a infection. *Chem. Eur. J.* **2005**, *11*, 1625–1635. https://doi.org/10.1002/chem.200400903; (c) Vulliez-Le Normand, B.; Saul, F. A.; Phalipon, A.; Be'lot,

F.; Guerreiro, C.; Mulard, L. A.; Bentley, G. A. Structures of synthetic O-antigen fragments from serotype 2a *Shigella flexneri* in complex with a protective monoclonal antibody. *PNAS.* **2008**, *105*, 9976–9981. https://doi.org/10.1073/pnas.0801711105; (d) Phalipon, A.; Tanguy, M.; Grandjean, C.; Guerreiro, C.; Belot, F.; Cohen, D.; Sansonetti, P. J.; Mulard, L. A. A synthetic carbohydrate-protein conjugate vaccine candidate against Shigella flexneri 2a infection. *J. Immunol.* **2009**, *182*, 2241–2247. https://doi.org/10.4049/jimmunol.0803141; (e) Said Hassane, F.; Phalipon, A.; Tanguy, M.; Guerreiro, C.; Belot, F.; Frisch, B.; Mulard, L. A.; Schuber, F. Rational design and immunogenicity of liposome-based diepitope constructs: application to synthetic oligosaccharides mimicking the *Shigella flexneri* 2a O-antigen. *Vaccine* **2009**, *27*, 5419–5426. https://doi.org/10.1016/j.vaccine.2009.06.031; (f) Vulliez-Le Normand, B.; Saul, F. A.; Phalipon, A.; Belot, F.; Guerreiro, C.; Mulard, L. A.; Bentley, G. A. Structures of synthetic O-antigen fragments from serotype 2a *Shigella flexneri* in complex with a protective monoclonal antibody. *Proc. Natl. Acad. Sci. U. S. A.* **2008**, *105*, 9976–9981. https://doi.org/10.1073/pnas.0801711105; (g) Phalipon, A.; Tanguy, M.; Grandjean, C.; Guerreiro, C.; Belot, F.; Cohen, D.; Sansonetti, P. J.; Mulard, L. A. A synthetic carbohydrate-protein conjugate vaccine candidate against *Shigella flexneri* 2a infection. *J. Immunol.* **2009**, *182*, 2241–2247. https://doi.org/10.4049/jimmunol.0803141.

[88] Gauthier, C.; Chassagne, P.; Theillet, F.-X.; Guerreiro, C.; Thouron, F.; Nato, F.; Delepierre, M.; Sansonetti, P. J.; Phalipon, A.; Mulard, L. A. Non-stoichiometric O-acetylation of *Shigella flexneri* 2a O-specific polysaccharide: synthesis and antigenicity. *Org. Biomol. Chem.* **2014**, *12*, 4218–4232. https://doi.org/10.1039/c3ob42586j.

[89] Salamone, S.; Guerreiro, C.; Cambon, E.; Hargreaves, J. M.; Tarrat, N.; Remaud-Siméon, M.; André, I.; Mulard, L. A. Investigation on the synthesis of *Shigella flexneri* specific oligosaccharides using disaccharides as potential transglucosylase acceptor substrates. *J. Org. Chem.* **2015**, *80*, 11237–11257. https://doi.org/10.1021/acs.joc.5b01407.

[90] van der Put, R. M. F.; Kim, T. H.; Guerreiro, C.; Thouron, F.; Hoogerhout, P.; Sansonetti, P. J.; Westdijk, J.; Stork, M.; Phalipon, A.; Mulard, L. A. A synthetic carbohydrate conjugate vaccine candidate against Shigellosis: improved bioconjugation and impact of alum on immunogenicity. *Bioconjugate Chem.* **2016**, *27*, 883–892. https://doi.org/10.1021/acs.bioconjchem.5b00617.

[91] Bélot, F.; Guerreiro, C.; Baleux, F.; Mulard, L. A. Synthesis of two linear PADRE conjugates bearing a deca- or pentadecasaccharide B epitope as potential synthetic vaccines against *Shigella flexneri* serotype 2a infection. *Chem. Eur. J.* **2005**, *11*, 1625–1635; *Bioconjugate Chem.* **2016**, *27*, 883–892. https://doi.org/10.1021/acs.bioconjchem.5b00617.

[92] (a) van der Put, R. M. F.; Smitsman, C.; de Haan, A.; Hamzink, M.; Timmermans, H.; Uittenbogaard, J.; Westdijk, J.; Stork, M.; Ophorst, O.; Thouron, F.; Guerreiro, C.; Sansonetti, P. J.; Phalipon, A.; Mulard, L. A. The first-in-human synthetic glycan-based conjugate vaccine candidate against *Shigella*. *ACS Cent. Sci.* **2022**, *8*, 449–460. https://doi.org/10.1021/acscentsci.1c01479; (b) Cohen, D.; Atsmon, J.; Artaud, C.; Meron-Sudai, S.; Gougeon, M.-L.; Bialik, A.; Goren, S.; Asato, V.; Ariel Cohen, O.; Reizis, A.; Dorman, A.; Hoitink, C. W. G.; Westdijk, J.; Ashkenazi, S.; Sansonetti, P. J.; Mulard, L. A.; Phalipon, A. Safety and immunogenicity of a synthetic carbohydrate conjugate vaccine against *Shigella flexneri* 2a in healthy adult volunteers: a phase 1, dose escalating, single-blind, randomised, placebo-controlled study. *Lancet Infect. Dis.* **2020**, *21*, 546–558. https://doi.org/10.1016/S1473-3099(20)30488-6.

[93] Hargreaves, J. M.; Le Guen, Y.; Guerreiro, C.; Descroix, K.; Mulard, L. A. Linear synthesis of the branched pentasaccharide repeats of O-antigens from *Shigella flexneri* 1a and 1b demonstrating the major steric hindrance associated with type-specific glucosylation. *Org. Biomol. Chem.* **2014**, *12*, 7728–7749. https://doi.org/10.1039/C4OB01200C.

[94] Mitra, L. A.; Mukhopadhyay, B. Linear synthesis of the hexasaccharide related to the repeating unit of the O-antigen from *Shigella flexneri* serotype 1d. *Carbohydr. Res.* **2016**, *426*, 1–8. https://doi.org/10.1016/j.carres.2016.03.012.

[95] Dhara, D.; Kar, R. K.; Bhunia, A.; Misra, A. K. Convergent synthesis and conformational analysis of the hexasaccharide repeating unit of the *O*-antigen of *Shigella flexneri* serotype 1d. *Eur. J. Org. Chem.* **2014**, 4577–4584. https://doi.org/10.1002/EJOC.201402392.

[96] Hu, Z.; White, A. F. B.; Mulard, L. A. Efficient Iterative Synthesis of O-acetylated tri- to pentadecasaccharides related to the lipopolysaccharide of *Shigella flexneri* type 3 a through di- and trisaccharide glycosyl donors. *Chem. Asian J.* **2017**, *12*, 419–439. https://doi.org/10.1002/asia.201600819.

[97] Cornil, J.; Hu, Z.; Bouchet, M.; Mulard, L. A. Multigram synthesis of an orthogonally-protected pentasaccharide for use as a glycan precursor in a *Shigella flexneri* 3a conjugate vaccine: application to a ready-for-conjugation decasaccharide. *Org. Chem. Front.* **2021**, *8*, 6279–6299. https://doi.org/10.1039/D1QO00761K.

[98] Kenne, L.; Lindberg, B.; Petersson, K.; Katzenellenbogen, E.; Romanowska, E. Structural studies of the O-specific side-chains of the *Shigella Sonnei* phase I lipopolysaccharide. *Carbohydr. Res.* **1980**, *78*, 119–126. https://doi.org/10.1016/S0008-6215(00)83665-4.

[99] Pfister, H. B.; Mulard, L. A. Synthesis of the zwitterionic repeating unit of the O-antigen from *Shigella Sonnei* and chain elongation at both ends. *Org. Lett.* **2014**, *16*, 4892–4895. https://doi.org/10.1021/ol502395k.

[100] Pfister, H. B.; Paoletti, J.; Poveda, A.; Jimenez-Barbero, J.; Mulard, L. A. Zwitterionic polysaccharides of *Shigella sonnei*: synthetic study toward a ready-for-oligomerization building block made of two rare amino sugars. *Synthesis* **2018**, *50*, 4270–4282. https://doi.org/10.1055/s-0037-1609719.

[101] Santra, A.; Misra, A. K. Synthesis of tri- and pentasaccharide fragments corresponding to the O-antigen of *Shigella boydii* type 6. *Tetrahedron Asym.* **2010**, *21*, 2612–2618. https://doi.org/10.1016/j.tetasy.2010.10.032.

[102] Santra, A.; Misra, A. K. Convergent synthesis of the tetrasaccharide repeating unit of the *O*-antigen of *Shigella boydii* type 9. *Beilstein J. Org. Chem.* **2011**, *7*, 1182–1188. https://doi.org/10.3762/bjoc.7.137.

[103] Santra, A.; Misra, A. K. Synthesis of a common tetrasaccharide repeating unit of the *O*-antigen of enteroinvasive *Escherichia coli* O143 and *Shigella boydii* type 81. *Curr. Org. Syn.* **2013**, *10*, 178–182. https://doi.org/10.2174/157017913804810997.

[104] Shit, P.; Gucchait, A.; Misra, A. K. Straightforward sequential and one-pot synthesis of a pentasaccharide repeating unit corresponding to the cell wall O-antigen of *Shigella boydii* type 18. *Tetrahedron* **2019**, *75*, 130697–130703. https://doi.org/10.1016/j.tet.2019.130697.

[105] Säwén, E.; Östervall, J.; Landersjö, C.; Edblad, M.; Weintraub, A.; Ansaruzzaman, M.; Widmalm, G. Structural studies of the O-antigenic polysaccharide from *Plesiomonas shigelloides* strain AM36565. *Carbohydr. Res.* **2012**, *348*, 99–103. https://doi.org/10.1016/j.carres. 2011.10.021.

[106] Das, R.; Mukhopadhyay, B. Chemical synthesis of the tetrasaccharide repeating unit of the O-antigenic polysaccharide from *Plesiomonas shigelloides* strain AM36565. *Carbohydr. Res.* **2013**, *376*, 1–6. https://doi.org/10.1016/j.carres.2013.04.033.

[107] Kaper, J. B.; Nataro, J. P.; Mobley, H. L. T. Pathogenic *Escherichia coli*. *Nat. Rev. (Microbiol.)* **2004**, *2*, 123–140. https://doi.org/10.1038/nrmicro818.

[108] Nataro, J. P.; Kaper, J. B. Diarrheagenic *Escherichia coli*. *Clin. Microbiol. Rev.* **1998**, *11*, 142–201. https://doi.org/10.1128/CMR.11.1.142.

[109] (a) Laupland, K. B.; Church, D. L. Population-based epidemiology and microbiology of community-onset bloodstream infections. *Clin. Microbiol. Rev.* **2014**, *27*, 647–664. https://doi.org/10.1128/CMR.00002-14; (b) Flores-Mireles, A. L.; Walker, J. N.; Caparon, M.; Hultgren, S. J. Urinary tract infections: epidemiology, mechanisms of infection and treatment options. *Nat. Rev. Microbiol.* **2015**, *13*, 269–284. https://doi.org/10.1038/nrmicro3432.

[110] Weiner, L. M.; Webb, A. K.; Limbago, B.; Dudeck, M. A.; Patel, J.; Kallen, A. J.; Edwards, J. R.; Sievert, D. M. Antimicrobial-resistant pathogens associated with healthcare-associated infections: summary of data reported to the national healthcare safety network at the Centers for disease control and prevention, 2011–2014. *Infect. Control Hosp. Epidemiol.* **2016**, *37*, 1288–1301. https://doi.org/10.1017/ice.2016.174.

[111] Huttner, A.; Gambillara, V. The development and early clinical testing of the ExPEC4V conjugate vaccine against uropathogenic *Escherichiacoli*. *Clin. Microbiol. Infect.* **2018**, *24*, 1046e1050. https://doi.org/10.1016/j.cmi.2018.05.009.

[112] (a) Wang, L.; Wang, Q.; Reeves, P. R. The variation of O antigens in gram-negative bacteria. *Subcell. Biochem.* **2010**, *53*, 123–152. https://doi.org/10.1007/978-90-481-9078-2_6; (b) Scotland, S. M.; Rowe, B.; Smith, H. R.; Willshaw, G. A.; Gross, R. J. Vero cytotoxin-producing strains of *Escherichia coli* from children with haemolytic uraemic syndrome and their detection by specific DNA probes. *J. Med. Microbiol.* **1988**, *25*, 237–243. https://doi.org/10.1099/00222615-25-4-237.

[113] (a) Boucher, H. W.; Talbot, G. H.; Benjamin, Jr., D. K.; Bradley, J.; Guidos, R. J.; Jones, R. N.; Murray, B. E.; Bonomo, R. A.; Gilbert, D. 10 x '20 progress-development of new drugs active against gram-negative bacilli: an update from the infectious diseases society of America. *Clin. Infect. Dis.* **2013**, *56*, 1685–1694. https://doi.org/10.1093/cid/cit152; (b) Lee, J. H. Antimicrobial resistance of *Escherichia coli* O26 and O111 isolates from cattle and their characteristics. *Vet. Microbiol.* **2009**, *135*, 401–405. https://doi.org/10.1016/j.vetmic. 2008.09.076.

[114] (a) Huttner, A.; Gambillara, V. The development and early clinical testing of the ExPEC4V conjugate vaccine against uropathogenic *Escherichiacoli*. *Clin. Microbiol. Infec.* **2018**, *24*, 1046–1050. https://doi.org/10.1016/j.cmi.2018.05.009. (b) Ihssen, J.; Kowarik, M.; Dilettoso, S.; Tanner, C.; Wacker, M.;

Thöny-Meyer, L. Production of glycoprotein vaccines in *Escherichia coli*. *Microb. Cell Fact.* **2010**, *9*, 61. https://doi.org/10.1186/1475-2859-9-61; (c) Ma, Z.; Z.; Zhang, Z.; Shang, W.; Zhu, F.; Han, W.; Zhao, X.; Han, D.; Wang, P. G.; Chen, M. Glycoconjugate vaccine containing *Escherichia coli* O157:H7 O-antigen linked with maltose-binding protein elicits humoral and cellular responses. *PLOS One* **2014**, *9*, e105215. https://doi.org/10.1371/journal.pone.0105215; (d) Andrade, G. R.; New, R. R. C.; Sant'Anna, O. A.; Williams, N. A.; Alves, R. C. B.; Pimenta, D. C.; Vigerelli, H.; Melo, B. S.; Rocha, L. B.; Piazza, R. M. F.; Mendonca-Previato, L. M.; Domingos, M. O. A universal polysaccharide conjugated vaccine against O111 *E. coli*. *Human Vac. Immunother.* **2014**, *10*, 2864–2874. http://dx.doi.org/10.4161/21645515.2014.972145.

[115] Huttner, A.; Hatz, C.; van den Dobbelsteen, G.; Abbanat, D.; Hornacek, A.; Frölich, R.; Dreyer, A. M.; Martin, P.; Davies, T.; Fae, K.; van den Nieuwenhof, I.; Thoelen, S.; de Vallière, S.; Kuhn, A.; Bernasconi, E.; Viereck, V.; Kavvadias, T.; Kling, K.; Ryu, G.; Hülder, T.; Gröger, S.; Scheiner, D.; Alaimo, C.; Harbarth, S.; Poolman, J.; Fonck, V. G. Safety, immunogenicity, and preliminary clinical efficacy of a vaccine against extraintestinal pathogenic *Escherichia coli* in women with a history of recurrent urinary tract infection: a randomised, single-blind, placebo-controlled phase 1b trial. *Lancet Infect. Dis.* **2017**, *17*, 528–537. https://doi.org/10.1016/S1473-3099(17)30108-1.

[116] Stenutz, R.; Weintraub, A.; Widmalm, G. The structures of *Escherichia coli* O-polysaccharide antigens. *FEMS Microbiol. Rev.* **2006**, *30*, 382–403. https://doi.10.1111/j.1574-6976.2006.00016.x.

[117] (a) Riley, L. W.; Remis, R. S.; Helgerson, S. D.; McGee, H. B.; Wells, G. J.; Herbert, R. J.; Olcott, E. S.; Johnson, L. M.; Hargrett, N. T.; Blake, P. A.; Cohen, M. L. Hemorrhagic colitis associated with a rare *Escherichia coli* serotype. *New Engl. J. Med.* **1983**, *308*, 681–685. https://doi.org/10.1056/NEJM198303243081203; (b) Werber, D.; Krause, G.; Frank, C.; Fruth, A.; Flieger, A.; Mielke, M.; Schaade, L.; Stark, K. Outbreaks of virulent diarrhoeagenic *Escherichia coli*—are we in control? *BMC Med.* **2012**, *10*, 11. https://doi.org/10.1186/1741-7015-10-11.

[118] Goldwater, P. N.; Bettelheim, K. A. Treatment of enterohemorrhagic *Escherichia coli* (EHEC) infection and hemolytic uremic syndrome (HUS). *BMC Med.* **2012**, *10*, 12. https://doi.org/10.1186/1741-7015-10-12.

[119] Panchadhayee, R.; Misra, A. K. Convergent synthesis of a common pentasaccharide-repeating unit corresponding to the O-specific polysaccharide of *Escherichia coli* O4:K3, O4:K6, and O4:K12. *Tetrahedron Asym.* **2010**, *21*, 859–863. https://doi.org/10.1016/j.tetasy.2010.04.056.

[120] Mandal, P. K.; Misra, A. K. Concise synthesis of the pentasaccharide O-antigen corresponding to the Shiga toxin producing *Escherichia coli* O171. *Bioorg. Chem.* **2010**, *38*, 56–61. https://doi.org/10.1016/j.bioorg.2010.01.001.

[121] Guchhait, G.; Misra, A. K. Convergent synthesis of the tetrasaccharide repeating unit corresponding to the O-antigen of the verotoxin-producing *Escherichia coli* O176. *Glycoconj. J.* **2011**, *28*, 519–524. https://doi.org/10.1007/s10719-011-9351-4.

[122] Zhou, Z.; Ogasawara, J.; Nishikawa, Y.; Seto, Y.; Helander, A.; Hase, A.; Iritani, N.; Nakamura, H.; Arikawa, K.; Kai, A.; Y. Kamata, Hoshi, H.; Haruki, K. An outbreak of gastroenteritis in Osaka, Japan due to *Escherichia coli* serogroup O166:H15 that had a coding gene for enteroaggregative *E. coli* heat-stable enterotoxin 1 (EAST1). *Epidemiol. Infect.* **2002**, *128*, 363–371. https://doi.org/10.1017/s0950268802006994.

[123] Santra, A.; Misra, A. K. Convergent synthesis of the pentasaccharide repeating unit of the O-antigenic polysaccharide of enterohaemorrhagic *Escherichia coli* O113. *Glycoconj. J.* **2012**, *29*, 181–188. https://doi.org/10.1007/s10719-012-9383-4.

[124] Ghosh, T.; Santra, A.; Misra, A. K. Convergent synthesis of a hexasaccharide corresponding to the cell wall O-antigen of *Escherichia coli* O41. *RSC Adv.* **2014**, *4*, 54–60. https://doi.org/10.1039/C3RA45493B.

[125] Bhaumik, I.; Misra, A. K. Expedient synthesis of a pentasaccharide related to the O-specific polysaccharide of *Escherichia coli* O117:K98:H4 strain. *RSC Adv.* **2014**, *4*, 61589–61595. https://doi.org/10.1039/C4RA10538A.

[126] Si, A.; Misra, A. K. Efficient synthesis of the pentasaccharide repeating unit of the O-antigenic polysaccharide of *Escherichia coli* O166 strain. *Synthesis* **2015**, *47*, 83–88. https://doi.org/10.1002/open.201500129.

[127] Si, A.; Misra, A. K. Expedient synthesis of the pentasaccharide repeating unit of the polysaccharide O-antigen of *Escherichiacoli* O11. *ChemistryOpen* **2016**, *5*, 47–50. https://doi.org/10.1002/open.201500129.

[128] Ghosh, T.; Misra, A. K. Synthesis of the heptasaccharide repeating unit of the cell wall O-polysaccharide of enterotoxigenic *Escherichiacoli* O139. *ChemistryOpen* **2016**, *5*, 43–46. https://doi.org/10.1002/open.201500164.

[129] Guo, H.; Kong, Q.; Cheng, J.; Wang, L.; Feng, L. Characterization of the *Escherichia coli* O59 and O155 O-antigen gene clusters: the atypical wzx genes are evolutionary related. *FEMS Microbiol. Lett.* **2005**, *248*, 153–161. https://doi.10.1016/j.femsle.2005.05.036.
[130] Si, A.; Misra, A. K. Synthesis of a pentasaccharide repeating unit corresponding to the cell wall O-antigen of *Escherichia coli* O59 using iterative glycosylations in one pot. *Tetrahedron* **2016**, *72*, 4435–4441. https://doi.org/10.1016/j.tet.2016.06.025.
[131] Si, A.; Misra, A. K. Concise synthesis of the pyruvic acid acetal containing pentasaccharide repeating unit of the cell wall O-antigen of *Escherichia coli* O156. *RSC Adv.* **2017**, *7*, 49903–49909. https://doi.org/10.1039/C7RA07567G.
[132] Shit, P.; Misra, A. K. Straightforward synthesis of the pentasaccharide repeating unit of the cell wall O-antigen of *Escherichia coli* O43 strain. *Glycoconjugate J.* **2020**, *37*, 647–656. https://doi.org/10.1007/s10719-020-09933-z.
[133] Pozsgay, V.; Kubler-Kielb, J.; Coxon, B.; Santacroce, P.; Robbins, J. B.; Schneerson, R. Synthetic oligosaccharides as tools to demonstrate cross-reactivity between polysaccharide antigens. *J. Org. Chem.* **2012**, *77*, 5922–5941. https://doi.org/10.1021/jo300299p.
[134] Campos, L. C.; Whittam, T. S.; Gomes, T. A.; Andrade, J. R.; Trabulsi, L. R. *Escherichia coli* serogroup O111 includes several clones of diarrheagenic strains with different virulence properties. *Infect. Immun.* **1994**, *62*, 3282–3288. https://doi.org/10.1128/iai.62.8.3282-3288.1994.
[135] Calin, O.; Eller, S.; Hahm, H. S.; Seeberger, P. H. Total synthesis of the *Escherichia coli* O111 O-specific polysaccharide repeating unit. *Chem. Eur. J.* **2013**, *19*, 3995–4002. https://doi.org/10.1002/chem.201204394.
[136] Perepelov, A. V.; Wang, Q.; Senchenkova, S. N.; Gong, Y.; Shaskov, A. S.; Wang, L.; Knirel, Y. A. Structure and gene cluster of the O-antigen of *Escherichia coli* O120. *Carbohydr. Res.* **2012**, *353*, 106–110. https://doi.org/10.1016/j.carres.2012.03.022.
[137] Budhadev, D.; Mukhopadhyay, B. Convergent synthesis of the hexasaccharide related to the repeating unit of the O-antigen from *E. coli* O120. *RSC Adv.* **2015**, *5*, 98033–98040. https://doi.org/10.1039/C5RA20363E.
[138] Mukherjee, M. M.; Ghosh, R. Synthetic routes toward acidic pentasaccharide related to the O-antigen of *E. coli* 120 using one-pot sequential glycosylation reactions. *J. Org. Chem.* **2017**, *82*, 5751–5760. https://doi.org/10.1021/acs.joc.7b00561.
[139] Pal, D.; Mukhopadhyay, B. Chemical synthesis of the pentasaccharide repeating unit of the *O*-specific polysaccharide from *Escherichia coli* O132 in the form of its 2-aminoethyl glycoside. *Beilstein J. Org. Chem.* **2019**, *15*, 2563–2568. https://doi.org/10.3762/bjoc.15.249.
[140] Karki, G.; Kumar, H.; Rajan, R.; Mandal, P. K. Expeditious synthesis of a tetrasaccharide repeating unit of the O-antigen of *Escherichia coli* O163. *Synlett.* **2016**, *27*, 2581–2586. https://doi.org/10.1055/s-0035-1562606.
[141] Fontana, C.; Weintraub, A.; Widmalm, G. Structural Elucidation of the O-Antigen Polysaccharide from *Escherichia coli* O181. *ChemistryOpen* **2015**, *4*, 47–55. https://doi.org/10.1002/open.201402068.
[142] Kumar, H.; Mandal, P. K. Synthetic routes toward pentasaccharide repeating unit corresponding to the O-antigen of *Escherichia coli* O181. *Tetrahedron Lett.* **2019**, *60*, 860–863. https://doi.org/10.1016/j.tetlet.2019.02.027.
[143] Hanna, K.; Jonsson, M.; Weintraub, A.; Widmalm, G. Structural determination of the O-antigenic polysaccharide from *Escherichia coli* O74. *Carbohydr. Res.* **2009**, *344*, 1592–1595. https://doi.org/10.1016/j.carres.2009.03.023.
[144] Bera, A.; Mukhopadhyay, B. Chemical synthesis of the rare D-Fuc3NAc containing tetrasaccharide repeating unit of the O-antigenic polysaccharide from *E. coli* O74. *Carbohydr. Res.* **2021**, *506*, 108366. https://doi.org/10.1016/j.carres.2021.108366.
[145] Heinrichs, D. E.; Yethon, J. A.; Whitfield, C. Molecular basis for structural diversity in the core regions of the lipopolysaccharides of *Escherichia coli* and *Salmonella enteric*. *Mol. Microbiol.* **1998**, *30*, 221–232. https://doi.org/10.1046/j.1365-2958.1998.01063.x.
[146] Gupta, S. K.; Keck, J.; Ram, P. K.; Crump, J. A.; Miller, M. A.; Mintz, E. D. Part III. Analysis of data gaps pertaining to enterotoxigenic *Escherichia coli* infections in low and medium human development index countries, 1984–2005. *Epidemiol. Infect.* **2008**, *136*, 721–738. https://doi.org/10.1017/S095026880700934X.
[147] Shang, W.; Xiao, Z.; Yu, Z.; Wei, N.; Zhao, G.; Zhang, Q.; Wei, M.; Wang, X.; Wang, P. G.; Li, T. Chemical synthesis of the outer core oligosaccharide of *Escherichia coli* R3 and immunological evaluation. *Org. Biomol. Chem.* **2015**, *13*, 4321–4330. https://doi.org/10.1039/c5ob00177c.

[148] (a) Ochoa, T. J.; Barletta, F.; Contreras, C.; Mercado, E. New insights into the epidemiology of enteropathogenic *Escherichia coli* infection. *Trans. R Soc. Trop. Med. Hyg.* **2008**, *102*, 852–856. https://doi.org/10.1016/j.trstmh.2008.03.017; (b) Yatsuyanagi, J.; Saito, S.; Sato, H.; Miyajima, Y.; Amano, K.-I.; Enomoto, K. Characterization of enteropathogenic and enteroaggregative *Escherichia coli* isolated from diarrheal outbreaks. *J. Clin. Microbiol.* **2002**, *40*, 294–297. https://doi.org/10.1128/JCM.40.1.294-297.2002; (c) Dulguer, M. V.; Fabbricotti, S. H.; Bando, S. Y.; Moreira-Filho, C. A.; Fagundes-Neto, U.; Scaletsky, I. C. A. Atypical enteropathogenic *Escherichia coli* strains: phenotypic and genetic profiling reveals a strong association between enteroaggregative *E. coli* heat-stable enterotoxin and diarrhea, *J. Infect. Dis.* **2003**, *188*, 1685–1694. https://doi.org/10.1086/379666.

[149] Basu, N.; Mukherjee, M. M.; Ghosh, R. Synthesis of tri- and tetramannosides based on sequential one-pot three component glycosylation reactions. *J. Indian Chem. Soc.* **2013**, *90*, 1815–1820. https://doi.org/10.5281/zenodo.5792055.

[150] Mandal, P. K. Synthesis of the pentasaccharide repeating unit of the *O*-antigen of *E. coli* O117:K98:H4. *Beilstein J. Org. Chem.* **2014**, *10*, 2724–2728. https://doi.org/10.3762/bjoc.10.287.

[151] Perepelov, A. V.; Wang, Q.; Shashkov, A. S.; Wen, L.; Knirel, Y. A. Structure and genetics of the O-antigen of *Escherichia coli* O158. *Carbohydr. Res.* **2011**, *346*, 2274–2277. https://doi.10.1016/j.carres.2011.07.021.

[152] Mitra, A.; Mukhopadhyay, B. Convergent chemical synthesis of the pentasaccharide repeating unit of the O-antigen from *E. coli* O158. *RSC Adv.* **2016**, *6*, 85135–85141. https://doi.org/10.1039/C6RA13909D.

[153] Ghosh, T.; Kar, R. K.; Bhunia, A.; Misra, A. K. Synthesis of the pentasaccharide repeating unit of the O-antigen of *Escherichia coli* O175 using one-pot glycosylations and its conformational analysis. *Tetrahedron* **2014**, *70*, 9262–9267. https://doi.org/10.1016/j.tet.2014.09.080.

[154] Kundu, M.; Gucchait, A.; Misra, A. K. Convergent synthesis of a pentasaccharide corresponding to the cell wall O-polysaccharide of enteropathogenic *Escherichia coli* O115. *Tetrahedron* **2020**, *76*, 130952. https://doi.org/10.1016/j.tet.2020.130952.

[155] Sau, A.; Misra, A. K. Convergent synthesis of the tetrasaccharide repeating unit of the cell wall lipopolysaccharide of *Escherichia coli* O40. *Beilstein J. Org. Chem.* **2012**, *8*, 2053–2059. https://doi.org/10.3762/bjoc.8.230.

[156] Santra, A.; Si, A.; Kar, R. K.; Bhunia, A.; Misra, A. K. Linear synthesis and conformational analysis of the pentasaccharide repeating unit of the cell wall O-antigen of *Escherichia coli* O13. *Carbohydr. Res.* **2014**, *391*, 9–15. https://doi.org/10.1016/j.carres.2014.03.012.

[157] Mandal, P. K. Convergent synthesis of the pentasaccharide repeating unit of the *O*-antigen of *Escherichia coli* O36. *Synthesis* **2015**, *47*, 836–844. https://doi.org/10.1055/s-0034-1379952.

[158] Mitra, A.; Mukhopadhyay, B. Concise chemical synthesis of the tetrasaccharide repeating unit of the O-antigen from bivine mastitis isolate *E. coli* serotype O174:H28. *Synthesis* **2015**, *47*, 3061–3066. https://doi.org/10.1055/s-0034-1380922.

[159] Pal, K. B.; Mukhopadhyay, B. Synthesis of two hexasaccharides related to the repeating unit of the O-antigen from *Escherichia coli* TD2158. *ChemistrySelect* **2017**, *2*, 7378–7381. https://doi.org/10.1002/slct.201701082.

[160] (a) Adak, A.; Balasaria, S.; Mukhopadhyay, B. Concise synthesis of the tetrasaccharide repeating unit of the O-antigen from *Escherichia coli* O131 containing N-acetyl neuraminic acid. *Carbohydr. Res.* **2022**, *521*, 108654. https://doi.org/10.1016/j.carres.2022.108654; (b) Manna, T.; Misra, A. K.; Synthesis of the sialic acid-containing tetrasaccharide repeating unit corresponding to the cell wall *O*-antigen of *Escherichia coli* O131 strain. *Carbohydr. Res.* **2022**, *521*, 108668–108674. https://doi.org/10.1016/j.carres.2022.108668.

[161] (a) Smith, H. R.; Scotland, S.M.; Willshaw, G.A.; Rowe, B.; Craviato, A.; Eslava, Isolates of *Escherichia coli* O44:H18 of diverse origin are enteroaggregative. *J Infect Dis.* **1994**, *170,* 1610–1613. https://doi.org/10.1093/infdis/170.6.1610; (b) Wallace-Gadsden, F.; Johnson, J. R.; Wain, J.; Okeke, I. N. Enteroaggregative *Escherichia coli* related to uropathogenic clonal group A. *Emerg. Infect. Dis.***2007**, *13*, 757–760. https://doi.org/10.3201/eid1305.061057.

[162] Kundu, M.; Gucchait, A.; Misra, A. K. Synthesis of the pentasaccharide repeating unit of the O-antigenic polysaccharide of enteroaggregative *Escherichia coli* O44:H18 strain. *Tetrahedron* **2021**, *91*, 132245. https://doi.org/10.1016/j.tet.2021.132245.

[163] Garabal, J. I.; Gonzalez, E. A.; Vazquez, F.; Blanco, J.; Blanco, M.; Blanco, J. E. Serogroups of *Escherichia coli* isolated from piglets in Spain. *Vet. Microbiol.* **1996**, *48*, 113–123. https://doi.org/10.1016/0378-1135(95)00150-6.

[164] Shit, P.; Misra, A. K. Synthesis of the pentasaccharide repeating unit corresponding to the capsular polysaccharide of *Escherichia coli* O20:K83:H26. *Tetrahedron* **2022**, *122*, 132948. https://doi.org/10.1016/j.tet.2022.132948.

[165] Bera, M.; Adak, A.; Mukhopadhyay, B. Concise chemical synthesis of the pentasaccharide repeating unit of the O-antigen from *Escherichia albertii* O2. *Carbohydr. Res.* **2019**, *485*, 107817. https://doi.org/10.1016/j.carres.2019.107817.
[166] Manna, T.; Gucchait, A.; Misra, A. K. Convenient synthesis of the pentasaccharide repeating unit corresponding to the cell wall O-antigen of *Escherichia albertii* O4. *Beilstein J. Org. Chem.* **2020**, *16*, 106–110. https://doi.org/10.3762/bjoc.16.12.
[167] Bera, M.; Adak, A.; Mukhopadhyay, B. Concise chemical synthesis of the pentasaccharide repeating unit of the O-antigen from *Escherichia albertii* O2. *Carbohydr. Res.* **2019**, *485*, 107817. https://doi.org/10.1016/j.carres.2019.107817.
[168] Manna, T.; Gucchait, A.; Misra, A. K. Convenient synthesis of the pentasaccharide repeating unit corresponding to the cell wall O-antigen of *Escherichia albertii* O4. *Beilstein J. Org. Chem.* **2020**, *16*, 106–110. https://doi.org/10.3762/bjoc.16.12.
[169] Stenutz, R.; Weintraub, A.; Widmalm, G. The structures of *Escherichia coli* O-polysaccharide antigens. *FEMS Microbiol. Rev.* **2006**, *30*, 382–403. https://doi.org/10.1111/j.1574-6976.2006.00016.x.
[170] Jana, M.; Kar, R. K.; Bhunia, A.; Misra, A. K. Synthesis of the tetrasaccharide repeating unit of the O-antigen of the *Escherichia coli* O69 strain and its conformational analysis. *RSC Adv.* **2014**, *4*, 37079–37084. https://doi.org/10.1039/C4RA06989G.
[171] Erbing, C.; Kenne, L.; Lindberg, B.; Hammarstrom, S. Structure of the O specific side chains of the Escherichia coli O75 lipopolysaccharide: a revision. *Carbohydr. Res.* **1978**, *60*, 400–403. https://doi.org/10.1016/S0008-6215(78)80049-4.
[172] Sau, A.; Misra, A. K. Synthesis of the tetrasaccharide motif and its structural analog corresponding to the lipopolysaccharide of *Escherichia coli* O75. *PLOS One* **2012**, *7*, e37291, 1–8. https://doi.org/10.1371/journal.pone.0037291.
[173] Nishi, N.; Nashida, J.; Kaji, E.; Takahashi, D.; Toshima, K. Regio- and stereoselective β-mannosylation using a boronic acid catalyst and its application in the synthesis of a tetrasaccharide repeating unit of lipopolysaccharide derived from E. coli O75. *Chem. Commun.* **2017**, *53*, 3018–3021. https://doi.org/10.1039/c7cc00269f.
[174] Springer, G. F.; Williamson, P.; Brandes, W. C. Blood group activity of gram-negative bacteria. *J. Exp. Med.* **1961**, *113*, 1077–1093. https://doi.org/10.1084/jem.113.6.1077.
[175] Ingle, A. B.; Chao, C.-S.; Hung, W.-C.; Mong, K.-K. T. Chemical synthesis of the O-antigen repeating unit of *Escherichiacoli* O86 by an N-formylmorpholine-modulated one-pot glycosylation strategy. *Asian J. Org. Chem.* **2014**, *3*, 870–876. https://doi.org/10.1002/ajoc.201402057.
[176] Bhaumik, I.; Kar, R. K.; Bhunia, A.; Misra, A. K. Expedient synthesis of the pentasaccharide repeating unit of the O-antigen of *Escherichia coli* O86 and its conformational analysis. *Glycoconj. J.* **2016**, *33*, 887–896. https://doi.org/10.1007/s10719-016-9687-x.
[177] Santra, A.; Ghosh, T.; Misra, A. K. Expedient synthesis of two structurally close tetrasaccharides corresponding to the O-antigens of *Escherichia coli* O127 and *Salmonella enterica* O13. *Tetrahedron Asym.* **2012**, *23*, 1385–1392. https://doi.org/10.1016/j.tetasy.2012.08.006.
[178] Jana, S. K.; Shit, P.; Misra, A. K. Straightforward synthesis of the pentasaccharide repeating unit of the O-antigenic polysaccharide from the enteropathogenic *Escherichia coli* O142. *Synthesis* **2023**, *54*, 773–778. https://doi.org/10.1055/s-0041-1738428.
[179] Mitra, A.; Mukhopadhyay, B. Convergent synthesis of the hexasaccharide repeating unit of the *O*-antigenic OPS of *Escherichia coli* O133. *Eur. J. Org. Chem.* **2019**, 4869–4878. https://doi.org/10.1002/ejoc.201900815.
[180] Nishi, N.; Seki, K.; Takahashi, D.; Toshima, K. Synthesis of a pentasaccharide repeating unit of lipopolysaccharide derived from virulent E. coli O1 and identification of a glycotope candidate of avian pathogenic *E. coli* O1. *Angew. Chem. Int. Ed.* **2021**, *60*, 1789–1796. https://doi.org/10.1002/anie.202013729.
[181] Naini, A.; Bartetzko, M. P.; Rao Sanapala, S.; Broecker, F.; Wirtz, V.; Lisboa, M. P.; Parameswarappa, S. G.; Knopp, D.; Przygodda, J.; Hakelberg, M.; Pan, R.; Patel, A.; Chorro, L.; Illenberger, A.; Ponce, C.; Kodali, S.; Lypowy, J.; Anderson, A. S.; Donald, R. G. K.; von Bonin, A.; Pereira, C. L.; Semisynthetic glycoconjugate vaccine candidates against *Escherichia coli* O25B induce functional IgG antibodies in mice. *JACS Au.* **2022**, *2*, 2135–2151. https://doi.org/10.1021/jacsau.2c00401.
[182] (a) Wie, S.-H. Clinical significance of *Providencia* bacteremia or bacteriuria. *Korean J. Intern. Med.* **2015**, *30*, 167–169. https://doi.org/10.3904/kjim.2015.30.2.167; (b) Shin, S.; Jeong, S. H.; Lee, H.; Hong, Park, J. S.; Song, W. Emergence of multidrug-resistant *Providencia rettgeri* isolates co-producing NDM-1 carbapenemase and PER-1 extended-spectrum β-lactamase causing a first outbreak in Korea. *Annals Clin. Microbiol. Antimicrobials* **2018**, *17*, Article 20. https://doi.org/10.1186/s12941-018-0272-y; (c) Liu, J.; Wang, R.; Fang, M. Clinical and drug resistance characteristics of *Providencia stuartii* infections in 76

patients. *J. Int. Med. Res.* **2020**, *48*, 0300060520962296. https://doi.org/10.1177/0300060520962296; (d) Carlson, C. D; Senti, E.; Oostema, J. *Providencia alcalifaciens* is a highly antimicrobial resistant bacteria found in a suburban creek. *FASEB J. Biochem. Mol. Biol.* **2022**, *36*. https://doi.org/10.1096/fasebj2022.36.S1.0R432.

[183] Verma, P. R.; Mukhopadhyay, B. Concise synthesis of a tetra- and a trisaccharide related to the repeating unit of the O-antigen from *Providencia rustigianni* O34 in the form of their p-methoxyphenyl glycosides. *RSC Adv.* **2013**, *2*, 201–207. https://doi.org/10.1039/c2ra22407k.

[184] Ahadi, S.; Awan, S. I.; Werz, D. B. Total synthesis of tri-, hexa- and heptasaccharidic substructures of the O-polysaccharide of *Providencia rustigianii* O34. *Chem. Eur. J.* **2020**, *26*, 6264–6270. https://doi.org/10.1002/chem.202000496.

[185] Halder, T.; Yadav, S. Total synthesis of the O-antigen repeating unit of *Providencia stuartii* O49 serotype through linear and one-pot assemblies. *Beilstein J. Org. Chem.* **2021**, *17*, 2915–2921. https://doi.org/10.3762/bjoc.17.199.

[186] Mandal, P. K.; Chheda, P. R. Synthesis of a pentasaccharide repeating unit of the O-specific polysaccharide from the lipopolysaccharide of *Providencia alcalifaciens* O28. *Tetrahedron Lett.* **2015**, *56*, 900–902. https://doi.org/10.1016/j.tetlet.2014.12.143.

[187] Podilapu, A. R.; Kulkarni, S. S. Total synthesis of repeating unit of O-polysaccharide of *Providencia alcalifaciens* O22 via one-pot glycosylation. *Org. Lett.* **2017**, *19*, 5466–5469. https://doi.org/10.1021/acs.orglett.7b02791.

[188] (a) Davin-Regli, A.; Pages, J. M. *Enterobacter aerogenes* and *Enterobacter cloacae*; versatile bacterial pathogens confronting antibiotic treatment. *Front. Microbiol.* **2015**, *6*, Article 392. https://doi.org/10.3389/fmicb.2015.00392; (b) Mezzatesta, M. L.; Gona, F.; Stefani, S. Enterobacter cloacae complex: clinical impact and emerging antibiotic resistance. *Future Microbiol.* **2012**, *7*, 887–902. https://doi.org/10.2217/fmb.12.61.

[189] Chaudhury, A.; Mukhopadhyay, B. Synthesis of the pentasaccharide repeating unit of the O-antigen from *Enterobacter cloacae* C4115 containing the rare α-D-FucNAc. *RSC Adv.* **2020**, *10*, 4942–4948. https://doi.org/10.1039/c9ra09807k.

[190] (a) Lin, M.-H.; Wolf, J. B.; Sletten, E. T.; Cambié, D.; Danglad-Flores, J.; Seeberger, P. H. Enabling technologies in carbohydrate chemistry: automated glycan assembly, flow chemistry and data science. *ChemBioChem* **2023**, *24*, e202200607. https://doi.org/10.1002/cbic.202200607; (b) Yao, W.; Xiong, D.-C.; Yan, Y.; Geng, C.; Cong, Z.; Li, F.; Li, B.-H.; Qin, X.; Wang, L.-N.; Xue, W.-Y.; Yu, N.; Zhang, H.; Wu, X.; Liu, M.; Ye, X.-S. Automated solution-phase multiplicative synthesis of complex glycans up to a 1,080-mer. *Nat. Syn.* **2022**, *1*, 854–863. https://doi.org/10.1038/s44160-022-00171-9; (c) Joseph, A.; Pardo-Vargas, A.; Seeberger, P. H. Total synthesis of polysaccharides by automated glycan assembly A. *J. Am. Chem. Soc.* **2020**, *142*, 8561–8564. https://dx.doi.org/10.1021/jacs.0c00751.

6 Synthesis of Selected Nucleoside Analogues against RNA Viruses

Evan Saillard, Mathias Brouillard, Vincent Roy, and Luigi A. Agrofoglio

6.1 INTRODUCTION

The continuous growth of the human population as well as the human interaction with wild environments have favored several emerging and re-emerging RNA viruses that are responsible for highly fatal viral diseases, which threaten the public health with deadly outbreaks of global concern [1–4]. They are the most common class of pathogens (two or thow novel viruses discovered each year) and they represent a challenge for global disease control [5] because of their fast evolutionary rates and increased mutability compared with other DNA viruses [6].

Hemorrhagic fever virus (HFV) [7], which includes Dengue (DENV), Ebola (EBV), Lassa fever (LASV), Yellow fever (YFV), Crimean-Congo (CCHFV), and Hantavirus (HANTV) fevers, are highly infectious viruses belonging to different families and orders that cause life-threatening diseases. They represent real concerns due to poor prognosis and lack of specific vaccines or drugs for effective treatment. Owing to multiple deadly epidemics, the WHO classifies them as "high-impact, high-threat diseases" with resurgent epidemic potential. Beside vaccine developments (SARS-CoV-2, YFV, EBV, DENV), the biological diversity and rapid adaptive rates of RNA viruses make drug development necessary. In particular, there is currently no approved commercial antiviral drug against HFV. All these viruses are enveloped negative strand RNA (except *Flaviviridae* which is positive stranded), undergo cytoplasmic replication and code for a relatively conserved *RNA-dependent RNA polymerase (RdRp)* (L-protein for (-) RNA viruses and NS5 protein for (+) RNA viruses) [8–10].

RdRp is essential for the replication and the expression of the viral genome, and it is a target for novel therapeutics, especially for nucleoside analogues [11, 12]. In all RNA viral classes, RdRps share multiples structures. The core structure of RdRp has a right-hand shape complete with palm, thumb, and finger domains. Five of the seven classical RdRp catalytic motifs (A–E) are located within the most conserved palm domain, while the other two (F and G) are found in the finger domains [13, 14]. The structurally conserved RdRp core and the related motifs are essential for viral RdRp catalytic function and there by represent key therapeutic targets [15–17].

However, the majority of RNA viral replicases lack proofreading activity (i.e. an exonuclease) which leads to high error rates (2.64 x 10^5 mutations per site per generation for DENV) [18] and low replicative fidelity [19–20]. RNA viruses can thus become resistant but also escape vaccine-induced immunity [21]. Flaviviruses possess a viral methyltransferase [22], involved in viral mRNA capping and favoring the evasion of the host innate system, which may be also targeted. Nucleoside analogues, with more than 40 derivatives marketed to treat viral DNA and RNA infections [23], represent some of the most promising and leading drugs [24] with Sofosbuvir (HCV) or Remdesivir (EBV) being recent examples. These analogues are the most promising broad-spectrum antivirals against emerging RNA viruses.

However, the development of ribonucleoside analogues capable of inhibiting HFV viral RdRp is still in its infancy. The first (and only) reported effective antiviral drug against LASV is the nucleoside analogue ribavirin (RBV, Virazole), although favipiravir (a nucleobase converted intracellularly

DOI: 10.1201/9781003437413-6

into its active nucleotide) is in clinical trial against LASV (Figure 6.1). Other viral RdRp nucleosides include *N*-nucleosides such as ribavirin and 4'-azidocytidine and C-nucleosides such as galidesivir and remdesivir.

6.2 MAIN MECHANISM OF ACTION OF NUCLEOSIDE INHIBITORS OF VIRAL POLYMERASES

Nucleoside analogues (and acyclic nucleoside phosphonates) generally act in their 5'-triphosphorylated form and target viral DNA/RNA polymerase and HIV reverse transcriptase, thereby preventing the formation of viral nucleic acid. As the nucleoside triphosphate is the active form, it cannot cross the cell membrane because it is negatively charged; one thus uses parent nucleosides that after penetration in the cell are then transformed into analogues of the nucleoside triphosphate by three successive phosphorylations (Nu -> NuMP -> NuDP -> NuTP) catalyzed by various nucleoside and nucleotide kinases present in the host cell or coming from some viruses [25].

The phosphorylation of a nucleoside into a nucleoide triphosphate follows this order: step 1, the deoxynucleoside kinases, e.g. deoxyguanosine kinase, deoxycytidine kinase, and thymidine kinases 1 and 2; step 2, the nucleoside monophosphate kinases adenylate kinases 1 to 6, guanylate kinase, thymidylate kinase, and uridylate-cytidylate kinase; and step 3, nucleoside diphosphate kinase, phosphoglycerate kinase, and creatine kinase. However, both the first phosphorylation step and the penetration of Nu or NuMP into the cell remain limiting steps: some nucleosides appear inactive while their triphosphates inhibit the viral polymerase. To address these limitations, kinase bypass strategies have been developed that involve the direct delivery of phosphorylated nucleosides into cells [26–29]. To mask the negative charges of the phosphate moiety and increase cellular penetration while maintaining a good solubility in physiological fluids, researchers have developed biolabile phosphate protecting groups. The nucleoside monophosphate is then released by enzymatic or intracellular chemical degradation of the biolabile groups by various enzymes such as reductases, carboxylesterases, cytochrome P450, allowing for targeting specific organs. Many biolabile groups have been developed to date, such as cycloSal, Hept-direct, nitrofuranylmethyl amidate, *bis*dithioethanole, *bis*(*S*-acyloylthioethyl) (Figure 6.2). They have their own characteristics (stability, mechanism of release, polarity, solubility, etc.) that guide their use. Although some of these nucleoside prodrugs went to clinical trials, none has been approved to date.

Other prodrugs such as *bis*(POM), *bis*(POC), which were applied to ANP, led to marketed antiviral nucleosides (Figure 6.4). Furthermore, ProTide technology, invented by C. Mc Guigan in 1990 based on triester aryloxy phosphoramidate prodrugs, has been successfully applied to various nucleoside analogues including the marketed antiviral drugs sofosbuvir, tenofovir alafenamide, and remdesivir, with increased antiviral activity compared with the parent nucleoside. This is a very versatile method since variations can be made at the ester (R), aminoacid (R') and aryl moieties. In addition, the chirality at the phosphorus (*R*p or *S*p) is also of great importance for the antiviral activity [30–33].

Once incorporated into the growing viral nucleic acid, if a nucleoside lacks the 3'-OH group (such as AZT, D4T, 3TC, PMEA, PMPA), chain elongation is limited, and the nucleoside acts as a

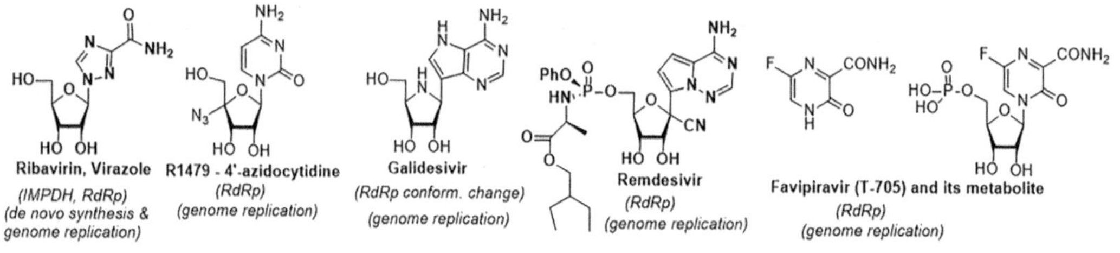

FIGURE 6.1 Chemical structures of some nucleosides and pronucleotides that target RdRp of selected viruses

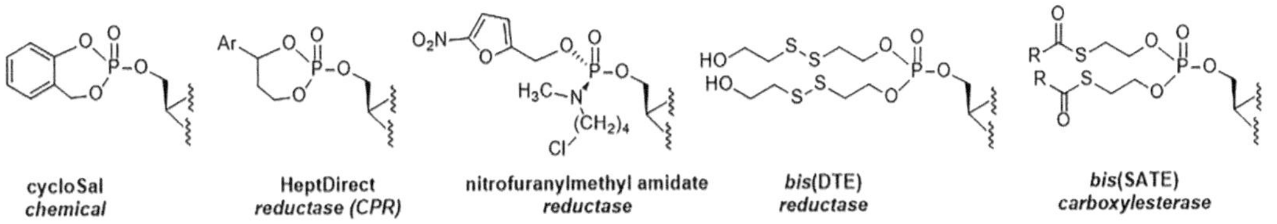

FIGURE 6.2 Chemical structure of some biolabile groups found in pronucleotides

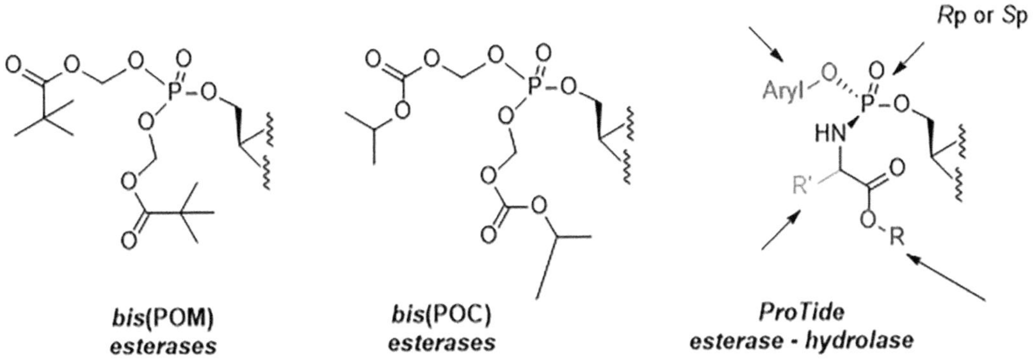

FIGURE 6.3 Chemical structure of biolabile groups found in FDA approved pronucleotides

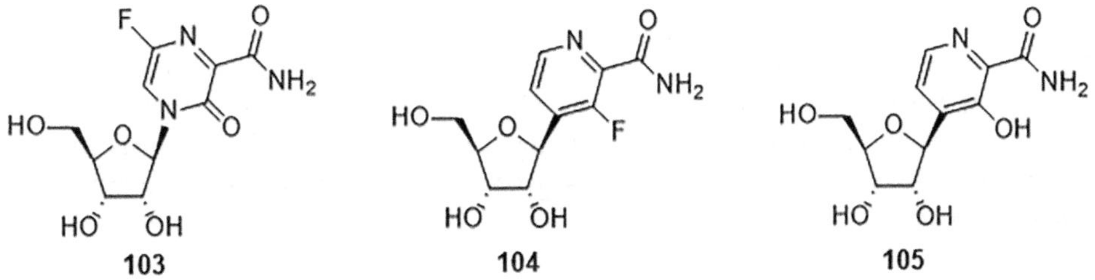

FIGURE 6.4 C-Nucleosides as favipiravir (T-705) analogues

(obligate) chain terminator. However, some nucleosides that still have a 3'-OH group (such as entecavir, L-dT, IDU, HPMPC) can also inhibit the chain elongation; after several incorporations that change the structures of viral nucleic acids and halt synthesis, these compounds are referred to as delayed chain terminators.

6.3 RIBAVIRIN AND ITS ANALOGS

Ribavirin **4** (Virazole, RBV) discovered more than 40 years ago, has a broad-spectrum activity profile on several DNA and RNA viruses [34]. It is a unique ribonucleoside that was synthesized in 1970 at ICN Pharmaceuticals, and it bears a *1H*-1,2,4-triazole-3-carboxamide moiety as nucleobase. RBV, 1-β-D-ribofuranosyl-1*H*-1,2,4-triazol-3-carboxamide, can be easily synthesized from peracylated β-D-ribofuranose **2** in presence of the 1,2,4-triazol-3-carboxylic acid methyl ester **1** to intermediate **3**, which, when treated with MeOH/NH_3 gives ribavirin **4** (Scheme 6.1) [34–36].

An RBV prodrug bearing various vinyl esters was developed and showed improved bioavailability and reduced side effects [37]. The synthesis starts by the regioselective acylation of unprotected

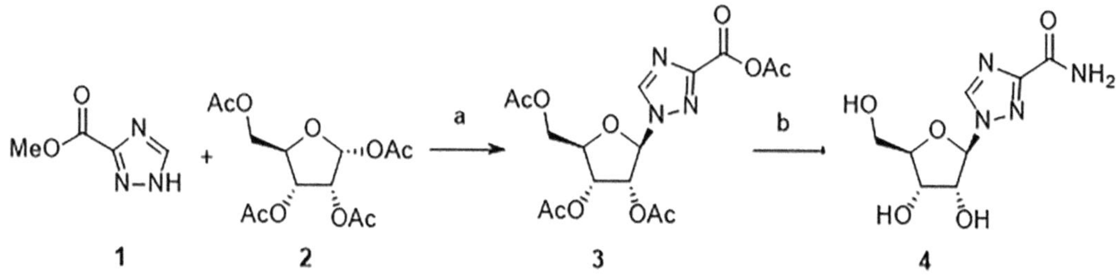

SCHEME 6.1 *Reagents and conditions:* (a) acid-catalyzed fusion; (b) $MeOH/NH_3$

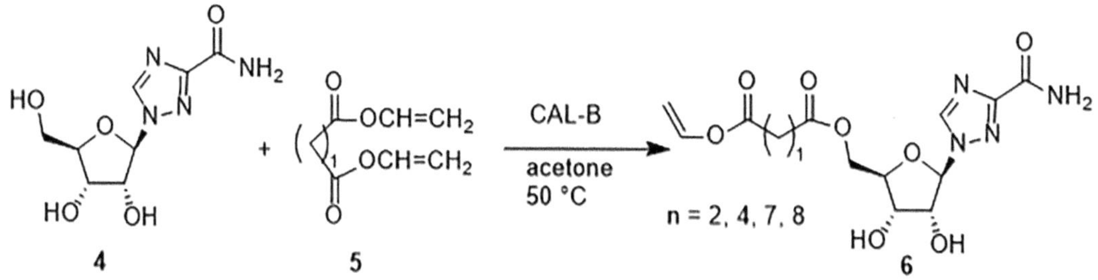

SCHEME 6.2 Synthesis of ribavirin prodrug bearing vinyl esters.

RBV with different divinyl dicarboxylates as acyl donors by lipase immobilized on acrylic resin from *Candida antarctica* (CAL-B) in acetone (Scheme 6.2).

This "old" antiviral compound is a broad-spectrum agent, active against various DNA and RNA viruses [38–41]. It has been clinically approved together with IFN-α [42] (but with side effects) in HCV treatment but also for treating infections by respiratory syncytial virus, adenovirus, hantavirus, LASV, and CCHFV. RBV has several mechanisms of action; RBV monophosphate first inhibits *inosine monophosphate dehydrogenase*, which is involved in the *de novo* synthesis of purine nucleotides (IMP, GTP); this depletes intracellular GTP, a direct impact on both cell and viral replication. RBV also shows immunomodulatory activity, specifically increased Th lymphocytes activity. The 5'-triphosphate form of RBV directly inhibits the RdRp of RNA viruses. RBV also interferes with the formation of the 5' cap structure of viral mRNA, probably by inhibiting guanyl and methyl transferase. Finally, RBV enhances viral mutagenesis by substitution of RTP for GTP, as almost RdRp lack proofreading, which is potentially why RBV does not see more widespread clinical use [42].

6.4 SELECTED *N*-NUCLEOSIDES

6.4.1 Sofosbuvir

Sofosbuvir was approved by FDA in 2013 for the treatment of chronic HCV infection in 2007 by M. Sofia *et al.* [43] through in-depth investigation of the impact of structural modifications at the C2' position of ribofuranose [e.g., by 2'-methyl and 2'-fluoro modifications (direct impact on the 3'-endo conformation)] on antiviral activity. The synthesis occurred from the commercially available *iso*propylidene D-glyceraldehyde **7** which, by treatment with (carbethoxyethylidene)-triphenylmethylphosphorane afforded the chiral ester **8** in 79% yield a mixture of *E/Z* 97:3 (Scheme 6.3). Various selective oxidation of the double bond gave the diol **9** with the desired stereochemistry. The cyclic sulfate **10** was obtained from oxidation with catalytic TEMPO and sodium hypochlorite of the cyclic sulfite made with thionyl chloride. The tetraethylammonium fluoride hydrate in acetonitrile or dioxane was used as fluorinated agent. Of interest, no C-3 regioisomer could be

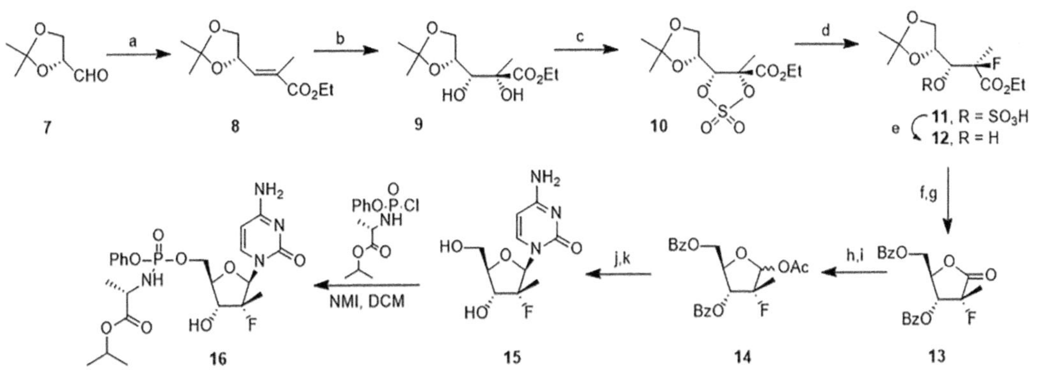

SCHEME 6.3 *Reagents and conditions*: (a) $Ph_3PC(Me)CO_2Et$,DCM; (b) $KMnO_4$, acetone; (c) i) $SOCl_2$/TEA/DCM; ii) aq. NaOCl/TEMPO/$NaHCO_3$/MeCN; (d) TEAF, dioxane; (e) $(MeO)_2C(Me)_2$ conc. HCl; (f) EtOH, con. HCl; (g) BzCl, pyr.; (h) $Li(OtBu)_3AlH$, THF; (i) Ac_2O/DMAP; (j) silylated N^4Bz-cytosine, $SnCl_4$, PhCl; (k) $MeOH/NH_3$.

observed. Selective hydrolysis of the sulfate ester **11** was done by treating the crude with concentrated hydrochloric acid in 2,2-dimethoxypropane. Treatment of the resulting acyclic **12** in EtOH with concentrated HCl, afforded the lactone which was benzoylated to **13**. After reduction, the ribonolactol obtained was acylated to **14** and condensed with silylated N^4-benzoylcytosine using $SnCl_4$. Subsequent deprotection in methanolic ammonia gave the desired nucleoside **15,** which was then converted to its phosphoramidate derivative **16**.

Sofosbuvir was approved by FDA in 2013 for the treatment of chronic HCV infection [44–46]. Sofia determined that the *Sp* isomer (EC_{90} = 0.42 μM) was tenfold more active than the *Rp* isomer (EC_{90} = 7.5 μM) with no cytotoxicity (up to 100 μM). After cell penetration, the prodrug is cleaved by host enzymes and chemical hydrolysis, releasing sofosbuvir-5'-monophosphate, which is then converted by various host kinases to its active metabolite, resulting in high levels of the triphosphate analogue in the liver. Sofosbuvir acts as a chain terminator.

6.4.2 '-C-and 3'-C-Alkylated Nucleosides

In 2009, Hong *et al.* [47] reported the first synthesis of (±)-1'(α), 2'(β)-C-dimethyl carbocyclic adenosine analogue **23** starting from aldehyde **18** that can be synthesized from 1,4-dihydroxy-2-butene **17** [48], (Scheme 6.4). Grignard addition of *iso*propenyl magnesium bromide on **18** followed by its further oxidation and a second Grignard addition furnished advance intermediate **19.** Diene **19** was then subjected to a RCM to yield a 1:1 mixture of cyclopentenols **20a** and **20b**. The 6-chloropurine was coupled on **20a** under Mitsubonu condition to **21.** Dihydroxylation of the alkene with catalytic amounts of OsO_4 gave **22** and subsequent transformations gave the (±)-1,2-dimethyl carboxylic adenine analogue **23**.

Audran *et al.* attempted to synthesize methyl-carbanuclesides, such as the chemoenzymatic synthesis of **28a** and **28b**, (Scheme 6.5) [49]. In this approach, enantiopure building block **24** was obtained using an enzymatic kinetic resolution of its racemate [50]. After inversion of the alcohol at the C1' position was achieved using a Mitsunobu reaction and further reduction of the ester moiety and global deprotection, diol **25** was obtained. The biocatalytic acetylation of **25** using *porcine pancreas lipase* furnished desired primary acetate **26** as a single product. After stereoselective epoxidation on **26**, the resulting **27** was coupled to purines to give carbanucleosides **28a** and **28b**.

Cho *et al.* [51] reported the synthesis of 2'-C-Me branched C-nucleosides, starting from the *O*-benzyl protected 2'-C-Me ribonolactone **29** [52] and a fully elaborated bromo heterocycle [53] **30** (Scheme 6.6). Addition of around 3 equivalents of *n*-BuLi to a cold suspension of **30** in THF in the presence of the wanted base resulted in both protection of the free amino group and exchange

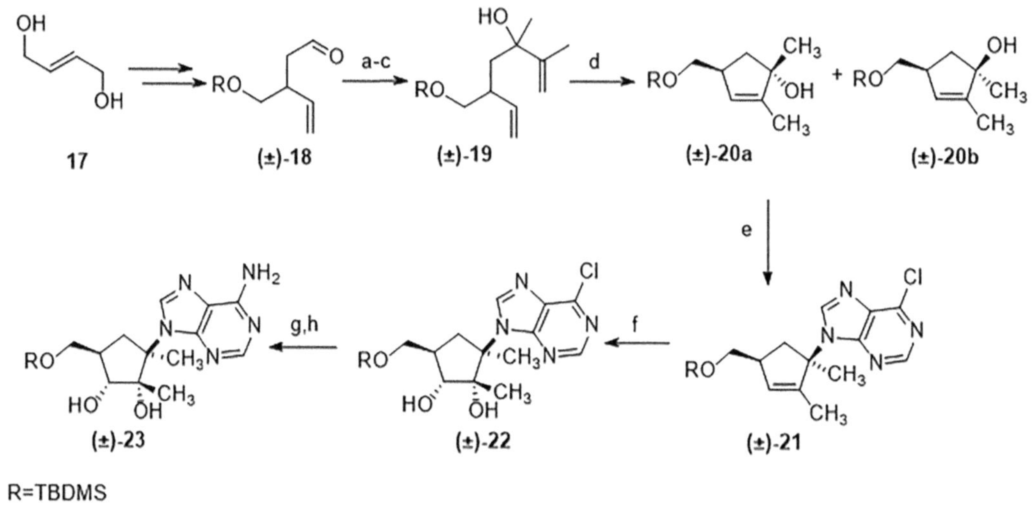

R=TBDMS

SCHEME 6.4 *Reagents and conditions:* (a) *iso*propenyl magnesium bromide, THF; (b) MnO_2, CCl_4, 60 °C; (c) methyl magnesium bromide, THF; (d) Grubbs catalyst, benzene, reflux; (e) 6-chloropurine, DIAD, PhP_3, Dioxane, DMF; (f) OsO_4, NMO; (g) NH_3, methanol, 90 °C; (h) TBAF, THF, CH_3CN.

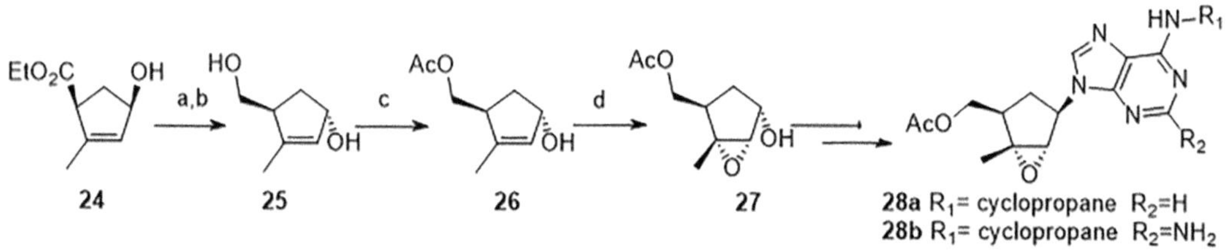

28a R_1= cyclopropane R_2=H
28b R_1= cyclopropane R_2=NH_2

SCHEME 6.5 *Reagents and conditions*: (a) PPh_3I, DIAD, AcOH, THF, 0 °C; (b) $LiAlH_4$, ether, –20 °C; (c) vinyl acetate, *porcine pancreas lipase*; (d) m-CPBA, CH_2Cl_2, 0 °C

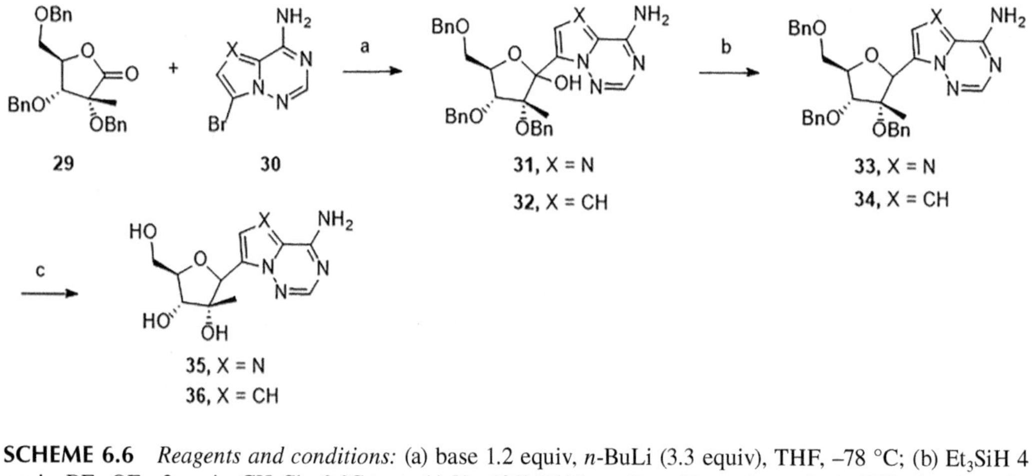

SCHEME 6.6 *Reagents and conditions:* (a) base 1.2 equiv, *n*-BuLi (3.3 equiv), THF, –78 °C; (b) Et_3SiH 4 equiv, BF_3-OEt_2 2 equiv, CH_2Cl_2, 0 °C to rt; (c) H_2, 10% Pd/C, acetic acid, rt, alternatively BCl_3 4 to 8 equiv, CH_2Cl_2, –78 °C.

of lithium-bromine, which was then reacted with the lactone **29** to make a 1,2-addition, affording adducts **31** or **32**, respectively. These compounds appear to exist in equilibrium with the two anomers in solutions. Remarkably, subsequent reduction of **31** and **32** with triethylsilane and boron trifluoride etherate was highly stereoselective, leading to the formation of **33** and **34**, respectively, as an exclusive product for both of them. Final deprotection of hydroxyl group, *O*-debenzylation via either hydrogenolysis or BCl_3 provided the desired nucleosides **35** and **36**, respectively.

Synthesized C-nucleosides **35** and **36** were evaluated together with the corresponding known *N*-nucleosides for their anti-HCV activity (EC_{50}) using a subgenomic GT1b replicon and cytotoxicity (CC_{50}) from Huh-7 cells. These compounds inhibited HCV; for **35**, the activity is GT1b EC_{50} = 1.28 μM, Huh-7 CC_{50} = >89 μM, GT1b NS5B TP IC_{50} = 2.10 μM and for the molecule **36**, the activity of it is GT1b EC_{50} = 1.98 μM, Huh-7 CC_{50} = 85 μM, GT1b NS5B TP IC_{50} = 0.31 μM, but they showed poorer EC_{50} values than their analogue corresponding N-nucleosides. Analogue **36** was observed to be stable against ADA degradation, while **35** was not; **36** had a favorable pharmacokinetic profile and potent anti-HCV activity in chimpanzees. Therefore, it was decided to characterize it for its potential as an orally administered therapeutic agent for chronic HCV infections. To conclude, C-nucleosides (**35** and **36**) show selective replicon activity and potent inhibition of NS5B as their triphosphates [54].

Furthermore, **36** was found to have a favorable pharmacokinetic profile and *in vivo* potential for enhanced potency over the corresponding *N*-nucleoside. In search of inhibitors of HCV NS5B polymerase, Wang et al. synthesized another 2'-C-methyl nucleosides, AL-611 [55]. The amino group of the nucleoside **37** was monomethoxytritylated to protect it, and the obtained product was treated with NaOEt in ethanol to remove both the benzoyl protecting group and make a SNAr with the chorine group on the base to afford **38**, (Scheme 6.7). Then an iodination was performed to obtain the iodine group instead of the primary alcohol to obtain **39**, which was reduced with DBU to obtain the corresponding alkene **40**. Iodofluorination of **40** with triethylamine trihydrofluoride and NIS provided nucleoside **41** as a single isomer in a good yield. Nucleophilic substitution of the iodo moiety with OBz, and the secondary alcohol was monomethoxytritylated to afford the tri-protected compound **42.** After that, they performed a selective deprotection to give debenzoylated **43,** which was converted to 5′-phosphoroamidate prodrugs with **44** to yield the AL-611 **45**.

Novel compound AL-611 was tested against the HCV Replicon and showed good inhibition. With the phosphoramidate prodrug, it showed a very good EC_{50} (= 0.005 μM) and an acceptable CC_{50} (>100 μM). The resulting SI is >20,000. Furthermore, the Sp-isomer 9 was 10 times more

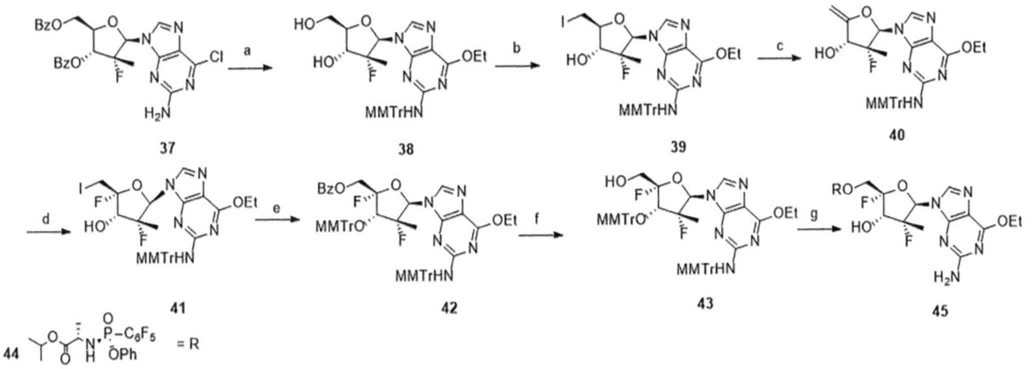

SCHEME 6.7 *Reagents and conditions:* (a) (i) MMTrCl, $AgNO_3$, collidine, DCM, rt; (ii) NaOEt, EtOH, rt; (b) I_2, PPh_3, imidazole, THF, rt; (c) DBU, THF, 75 °C; (d) NEt_3.3HF, NIS, DCM, 0 °C; (e) (i) MMTrCl, $AgNO_3$, collidine, DCM, rt; (ii) NaOBz, DMF, 15-crown-5, 95 °C; (f) *n*-butylamine, rt; (g) (i) 44, tert-BuMgCl, THF, 5 °C; (ii) HCl/CH_3CN

active than corresponding *Rp*-isomer of **45** (0.005 and 0.057 μM, respectively) with SI > 20,000. This Sp-isomer **45** (AL-611) was selected for further evaluation. Its analogue guanine triphosphate showed IC_{50} = 0.16 μM for 6-TP and appreciable inhibition of human DNA pol-α with 79% and 73% inhibition, respectively, at 100 μM. Overall, the triphosphate analogue of AL-611 tested did not show any alarming inhibition of the human mitochondrial polymerases and showed low inhibition of DNA pol-α. NTP levels in the target tissue are important for *in vivo* efficacy of an antiviral nucleoside polymerase inhibitor. Compound **45** was tested *in vitro* with hepatocytes with a concentration of 321 pmol/million cells at 24 h, after *in vivo* testing with dog liver NTP formation, and had a concentration of 22.3 μM post dose at 4 h. Because of these good results, the Sp-diastereomer prodrug **45** (AL-611) was selected for preclinical toxicology studies [55].

6.4.3 4′-Substituted Nucleosides

The 4′-azidocytidine **50** (R1479), developed by Roche [56], is an inhibitor of the HCV-RdRp that is also active against DENV, henipaviruses, and RSV. Roche synthetized this molecule by starting with commercial adenosine **46**, which was iodinated with iodine and triphenylphosphine, (Scheme 6.8). After that, this iodine group was eliminated with sodium methanolate (NaOMe) to give the corresponding alkene compound **47**. For the next step, the author performed simultaneous iodination and azidation with $[Bn(Et)_3]N_3{:}I_2$ in NMO to afford intermediate **48**. This intermediate was protected with a solution of benzoyl chloride; then, the iodine group changed into the primary alcohol group, and the benzoyl protected group was removed with methanolic ammonia to yield 4′-azidoadenine **49**. Finally, the base was chlorinated with $POCl_3$, and an amine was substituted by SnAr to give desired product **50**.

The antiviral activity of 4'-azidocytidine **50** (R1479) was tested against henipaviruses Nipah virus and Hendra, highly pathogenic zoonotic paramyxoviruses [57] as well as against a potent and selective inhibitor of HCV replication targeting the RNA-dependent RNA polymerase of hepatitis C virus, NS5B (IC_{50} = 1.28 μM and CC_{50} > 2000 μM in the HCV replicon system) [58]. For NIV, EC_{50} = 1.12 μM for NCI-H358 and 13.55 μM for HeLa depending on the assay; CC_{50} > 100 μM. For HeV, EC_{50} = 1.75 μM for NCI-H358 and 9.85 μM for HeLa depending on the assay, and CC_{50} > 100 μM.R1479 was also tested against MV (rMV^{EZ}GFP(3)), hPIV3 (hPIV3-GFP), MuV (rMuV-EGFP), and Pneumovirus RSV (rgRSV224) for its inhibition of infection-induced GFP expression, with EC_{50} < 5 μM for each these tests. All results showed that R1479 is efficient against henipaviruses with low micromolar EC_{50} values and has minimal cellular cytotoxicity (CC_{50} > 100 mM), against paramyxoviruses but also for flaviviruses and a pneumovirus [59].

Other 4'-substituted ribonucleoside cytidine were described [60]. Selective protection of the 4'-α-hydroxymethyl substituent of compound **51** was achieved by treatment with dimethoxytrityl chloride in pyridine followed by protection of the 4'-β-hydroxymethyl group with TBDMS chloride, (Scheme 6.9). Removal of the dimethoxytrityl group in acidic condition with acetic acid, followed by Swern oxidation of the free primary alcohol, gave the aldehyde derivative **52**. Then, compound **52**

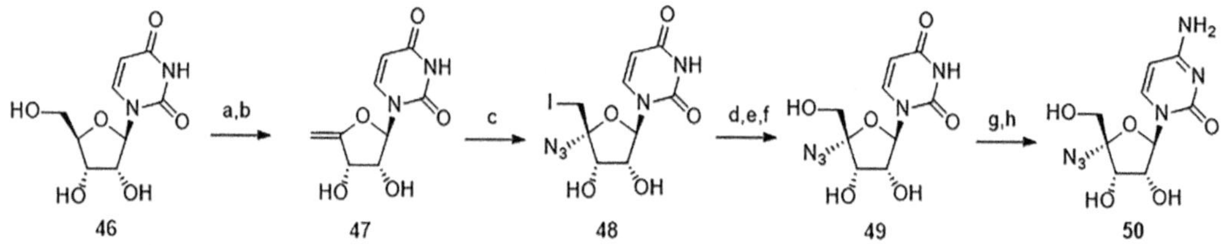

46 —a,b→ 47 —c→ 48 —d,e,f→ 49 —g,h→ 50

SCHEME 6.8 *Reagents and conditions:* (a) I_2, PPh_3, imidazole, THF; (b) NaOMe, MeOH; (c) $[Bn(Et)_3]N_3{:}I_2$, NMO; (d) BzCl, DMAP, NMP, THF; (e) m-CPBA, m-CBA, $(NH_4)HSO_4$, CH_2Cl_2; (f) NH_3, MeOH; (g) triazole, $POCl_3$, Et_3N, MeCN; (h) NH_3, MeOH; (i) BzCl, pyridine

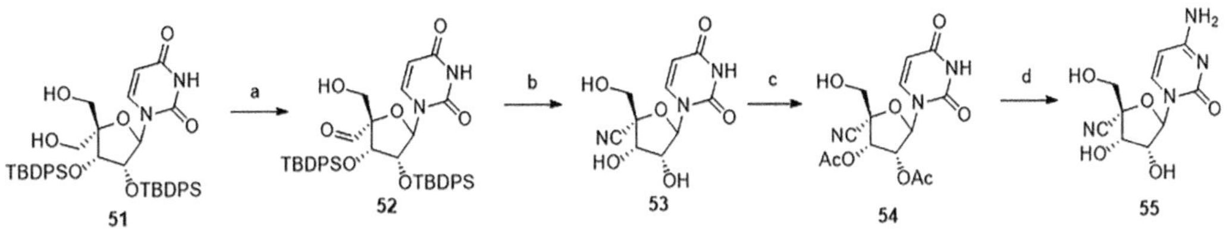

SCHEME 6.9 *Reagents and reaction conditions:* (a) (i) DMTCl, pyridine, (ii) TBDMSCl, pyridine, (iii) AcOH, (iv) $(COCl)_2$, DMSO, Et_3N, DCM; (b) (i) $NH_2OH.HCl$/pyr, Ac_2O, NaOAc, n-$BuNF_4$; (c) Ac_2O, pyridine; (d) (i) 1,2,4-triazole, Et_3N, $POCl_3$, (ii) NH_4OH, (iii) NH_3/MeOH

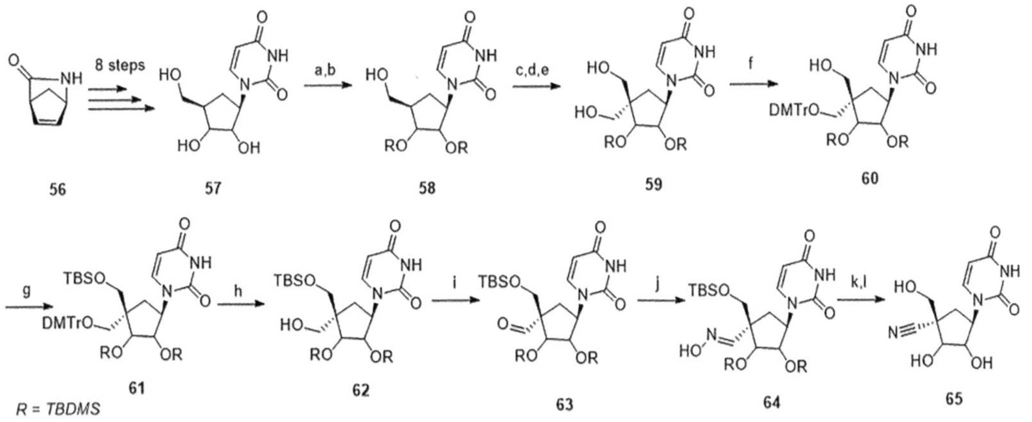

SCHEME 6.10 *Reagents and reaction conditions:* (a) TBDMSCl, imidazole, DMF, 50 °C; (b) TFA/H_2O/THF (1:1:2), 0 °C; (c) DMP, pyridine, DCM, rt; (d) 37% CHO aq. 2 M NaOH aq. 1,4-dioxane, rt; (e) $NaBH_4$, 0 °C to rt; (f) DMTrCl, DCM/pyridine (3:1), rt; (g) TBDMSCl, imidazole, DMF, 50 °C; (h) AcOH 80% aq. THF, rt; (i) DMP, pyridine, DCM, rt; (j) $NH_2OH.HCl$, pyridine, rt; (k) Burgess reagent, toluene, 100 °C; (l) TBAF 1M, THF, rt.

was used to condense the aldehyde function with hydroxylamine, giving the corresponding oxime, which was dehydrated with acetic anhydride in the presence of sodium acetate to give the nitrile group. Global deprotection with TBAF gave 4'-α-cyanouridine **53**. This cyanouridine was protected peracetylated with acetic anhydride in pyridine to give the corresponding protected derivative **54**. Activation of the 4-keto group using phosphorus oxychloride and triazole gave the chloride group, followed by treatment with aqueous ammonia to afford the corresponding triacetylated cytidine derivative, which after treatment with methanolic ammonia gave the desired unprotected cyanocytidine **55** by removing the acetyl group in good yield.

Amblard *et al.* [61] also reported on 4'-substituted carbocyclic uracil derivatives (Scheme 6.10). The carbocyclic uridine precursor was synthetized from the (-)-Vince lactam **56** in eight steps following reported procedures [62]. Compound **57** was first per-silylated with *tert*-butyldimethylsilyl chloride (TBDMSCl) in presence of imidazole before selective 5'-deprotection in presence of trifluoroacetic acid (TFA) and water to afford **58**. Then, they performed Dess Martin oxidation **58** using Dess–Martin periodinane (DMP) to form the corresponding aldehyde intermediate followed by further reaction with 37% in water paraformaldehyde (CHO) in presence of sodium hydroxide (NaOH) and final reduction with $NaBH_4$ to afford intermediate 4'-diol **59**. Compound **62** was synthesized from **59** through a three-step process by selective protection of the 4'-α-hydroxymethyl group with 4,4'-dimethoxytrityl chloride (DMTrCl) to give **60**, followed by protection of the 4'-β-hydroxymethyl group with TBDMSCl to **61** and then selective 5'-DMTr deprotection under acidic conditions.

Compound **62** was first oxidized in Dess-Martin conditions with DMP to form corresponding 5′-aldehyde intermediate **63**, which was reacted with hydroxylamine hydrochloride in pyridine to form the oxime, followed by dehydration of intermediate **64** in the presence of Burgess reagent; the final deprotection with TBAF afforded **65**.

These compounds and their monophosphate analogues were evaluated for antiviral activity against SARS-CoV-2, influenza A/B, norovirus and also evaluated for cytotoxicity. The cytotoxicity was determined *in vitro* using the CellTiter 96 nonradioactive cell proliferation colorimetric assay in primary human peripheral blood mononuclear (PBM), human T lymphoblast (CEM), human hepatocellular carcinoma (Huh7) and kidney epithelial (Vero) cell lines. None displayed toxicities up to 100 μM in primary human PBM cells, CEM cells, Vero cells, and Huh7. Furthermore, none inhibited SARS-CoV-2, influenza A/B, or norovirus at concentration up to 10 μM; none of these compounds or their corresponding monophosphate prodrugs displayed significant antiviral activity. However, they can be tried against other RNA viruses.

6.5 GALIDESIVIR

Galidesivir ***70***, a broad-spectrum RNA antiviral, is in advanced development for the treatment of viral infections, including SARS-CoV-2, EBV, Marburg, YFV and Zika [63–65]. After triphosphorylation, it binds to the active site of the viral RdRp, leading to conformational modifications and premature chain termination. Galidesivir (BCX4430), developed by BioCryst Pharmaceuticals, was first reported by Evans *et al.* in 2001 [66].

This antiviral agent is an adenosine analogue in which the oxygen in the ribose ring has been replaced by nitrogen and the nitrogen at position 9 has been replaced by carbon. Galidesivir has broad-spectrum antiviral properties and can be synthesized from forodesine (BCX1177), also known as Immucillin H, a selective inhibitor of human T-lymphocytes. Galidesivir's synthetic route involves seven steps to reach the desired compound with a 25% overall yield, (Scheme 6.11). To substitute the ketone with an amine, the 4'aza and the 2',3',5'-hydroxyl groups must be protected (from **66** to **67** then **70**). However, this approach requires a series of tedious protection and deprotection steps. An alternative method involves three steps beginning with the acetylation of BCX1177 **66** using acetic anhydride to **68**, followed by chlorination to **69** and finally ammonolysis to BCX4430 **70** [67].

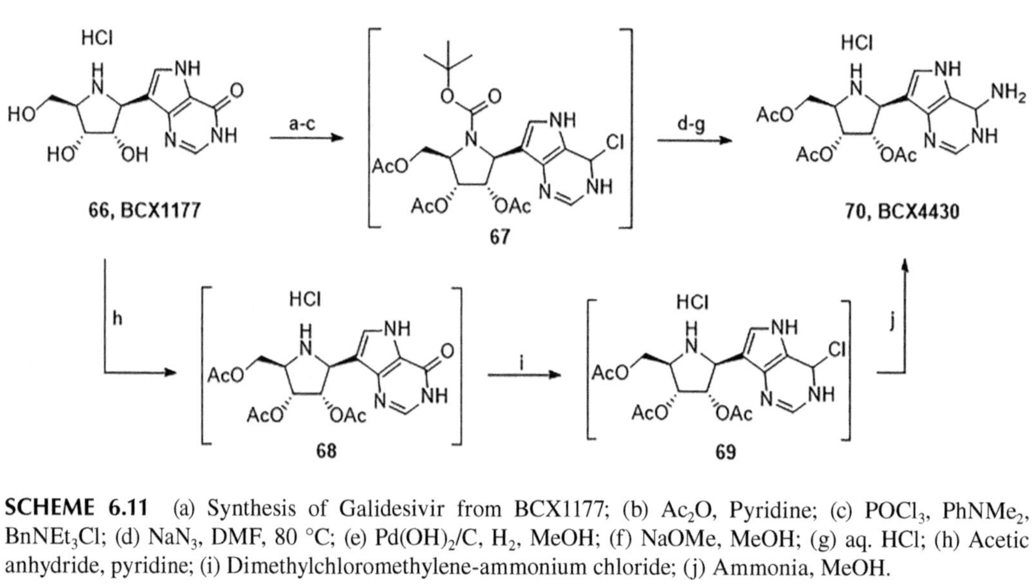

SCHEME 6.11 (a) Synthesis of Galidesivir from BCX1177; (b) Ac_2O, Pyridine; (c) $POCl_3$, $PhNMe_2$, $BnNEt_3Cl$; (d) NaN_3, DMF, 80 °C; (e) $Pd(OH)_2/C$, H_2, MeOH; (f) NaOMe, MeOH; (g) aq. HCl; (h) Acetic anhydride, pyridine; (i) Dimethylchloromethylene-ammonium chloride; (j) Ammonia, MeOH.

For BCX4430 to become active, it needs to be phosphorylated by cellular kinases until it reaches its triphosphate form. Once it becomes active, it functions as a nucleotide analogue of adenosine triphosphate and is integrated into the viral RNA. This integration leads to premature chain termination, which is highly efficient due to the preference of viral RNA polymerase over host polymerase. Galidesivir has been tested *in vitro* against various viruses and found to have limited to weak activity against Togaviridae and Arenaviridae. However, it has shown stronger activity against Filoviridae, Flaviviridae, Phenuiviridae, Paramyxoviridae, Orthomyxoviridae, Pneumoviridae, and Picornaviridae families [68]. Furthermore, it demonstrated effectiveness in animal models of viral diseases such as EBV, Marburg, YFV, Zika, and Rift Valley fever virus. Interestingly, galidesivir was more effective in animal models than what cell culture activity would suggest, indicating that the *in vivo* process is more efficient in animal tissues than cell culture, for instance by reducing lung infection 24 h after galidesivir treatment [69]. Originally used in Japan to treat T-cell lymphoma, forodesine was repurposed due to the pandemic. Computational molecular modeling revealed its potency against the SARS-CoV-2 omicron variant, and further biological assays confirmed its potential as a lead compound in the fight against COVID-19 [70]. In fact, forodesine exhibited an EC_{50} of 0.70 μM, which is more efficient than remdesivir and molnupiravir, two antiviral nucleosides.

The synthesis of forodesine **77** starts with the addition of lithiated acetonitrile to imine **72** to form cyanomethyl C-glycoside derivative **73**, (Scheme 6.12) [71]. The nitrogen atom was protected as the Boc carbamate. Bredereck's reagent was used to yield the enamine, which upon mild acid hydrolysis gave **74**. Compound **74** then reacted with ethyl glycinate to form enamine **75**. To facilitate cyclization to the pyrrole, the nitrogen atom of the enamine was protected by treatment with excess benzyl chloroformate and DBU. Pyrrole **76** was readily obtained by hydrogenolysis. To generate the target nucleoside **77**, formamidine acetate treatment was carried out followed by full deprotection.

Riboprine is a galidesivir analogue that occurs naturally and has recently drawn attention due to its potential in fighting the global SARS-CoV-2 pandemic. It is present in yeast and mammalian t-RNA, and has been found to inhibit cell multiplication in different types of human epithelial cancers [72,73]. Riboprine differs from BCX1177 and BCX4430 in its N6 substitution, where instead of a ketone or an amine, it presents an isopentenyl group. This group is of particular interest due to its strong *in vitro* activity against the SARS-CoV-2 omicron variant, with an EC_{50} of 0.40 μM.

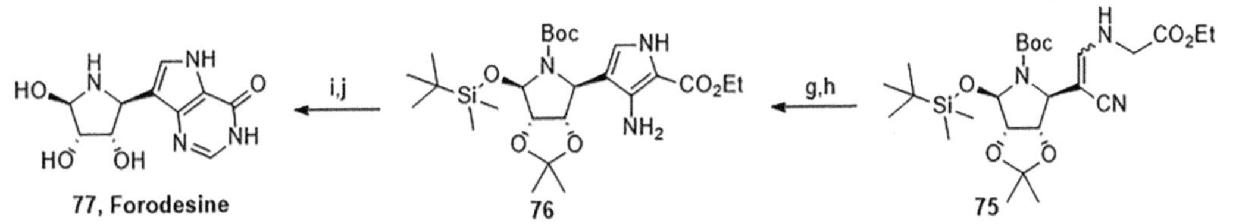

SCHEME 6.12 *Reagents and conditions:* (a) i. NCS, pentane; ii. LiTMP, −78 °C; (b) CH_2CNMgX or CH_2CNLi; (c) $(Boc)_2O$, CH_2Cl_2; (d) $tBuOCH(NMe)_2$, DMF, 70 °C; (e) THF, Acetic acid, H_2O; (f) $H_2NCH_2CO_2Et.HCl$, NaOAc, MeOH; (g) $ClCO_2Bn$, DBU, CH_2Cl_2, reflux; H_2, Pd/C, EtOH; (h) $H_2NCH{=}NH.HOAc$, EtOH, reflux; (i) TFA.

6.6 REMDESIVIR

Remdesivir **85** (GS-5734) is a prodrug of an adenosine analogue that has broad-spectrum antiviral activity against (+)RNA and (-)RNA viruses.74 Originally developed for the treatment of EBV infection, it is also active against RSV, Junin, LASV, Nipah, Hendra, and coronaviruses [75]. The presence of a cyano group in the C1'-position is critical for its activity and selectivity toward viral RNA polymerases [76]. This adenosine analogue can exist in *Sp* or *Rp* conformations, but chiral HPLC can isolate the more selective *Sp* isomer, which has the best therapeutic window. Siegel *et al.* synthesized GS-5734 from the glycosylation of 2,3,5-tri-*O*-benzyl-D-ribono-1,4-lactone ***78*** with bromo heterocycle **79** to nucleoside **80** (Scheme 6.13) [77]. The hydroxyl nucleoside of **80** was then cyanated before the deprotection of the hydroxyl groups to **81**. Next, coupling with phosphoramidyl chloride **83** (as a diastereomeric mixture) achieved ~1:1 diastereomeric mixture **84**. Chiral HPLC afforded remdesivir **85**.

In vitro remdesivir exhibited anti-Ebola activity on HeLa and TERT-immortalized human foreskin microvascular endothelial cell, with an EC_{50} of 0.10 μM and 0.053 μM, respectively. The molecule also showed activity against HCV from the Flaviviridae and RSV from *Pneumoviridae* (EC_{50} of 0.057 μM and 0.015 μM, respectively). GS-5734 was tested *in vivo* using a rhesus monkey model that had contracted the fatal Ebola virus disease [78]. The best results showed that the medication had antiviral effects, reduced the infection signs, and provided a significant survival benefit, even when treatment was initiated on the third day after infection when the systemic viral RNA was detectable. These findings suggest that GS-5734 could be used as a viable postexposure protection against Ebola virus disease. During the COVID-19 pandemic, remdesivir was approved as a therapeutic alternative, but clinical trials do not agree on the possible benefits of its use [79, 80]. Ongoing clinical trials are still in progress. Second-generation synthesis was developed to avoid chiral chromatography, operating a selective crystallization of *Sp* phosphorus isomer partner before its

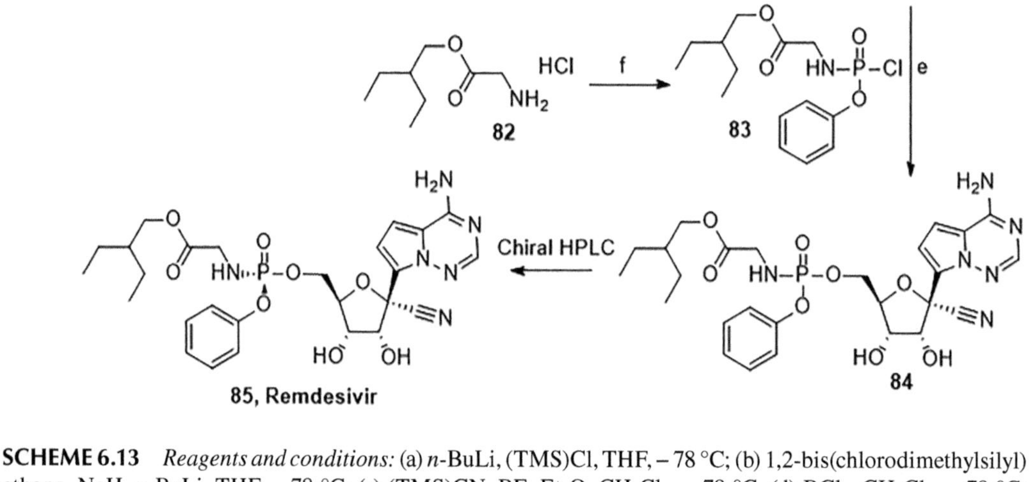

SCHEME 6.13 *Reagents and conditions:* (a) *n*-BuLi, (TMS)Cl, THF, – 78 °C; (b) 1,2-bis(chlorodimethylsilyl) ethane, NaH, *n*-BuLi, THF, – 78 °C; (c) (TMS)CN, $BF_3{\cdot}Et_2O$, CH_2Cl_2, – 78 °C; (d) BCl_3, CH_2Cl_2, – 78 °C; (e) **83**, NMI, $OP(OMe)_3$; (f) $OP(OPh)Cl_2$, Et_3N, CH_2Cl_2, 0 °C.

combination with an *iso*propylidene derivative nucleoside. A third route has been described from **86** to achieve the desired isomer through a catalytic asymmetric synthesis of *P* stereogenic phosphoric acid derivatives (Scheme 6.14) [81].

Remdesivir is a versatile drug with a wide range of activities. When it enters the cell, it is converted into an alanine metabolite that is then transformed into a triphosphate nucleoside analogue through three successive phosphorylations. This analogue is highly effective in inhibiting the RdRp enzyme of viruses and acts as a delayed chain terminator. The cyano group in the C1' position is essential for its selectivity toward viral RNA polymerases and activity [82]. Specifically, this group initiates steric clash that blocks the virus transcription process, making remdesivir an effective treatment for various viral infections. Analogues were synthesized to enhance the antiviral efficiency by emphasizing the crucial role of C1'-substitution.

Recently, Cardoza *et al.* incorporated triazoles at C1' to optimize this pivotal sterically conflicted area [83]. To obtain the most effective nucleosides against SARS-CoV-2, they substituted the C1' position with a propargyl. Then, a click reaction was performed on **88** before the deprotection of the three hydroxyls to give **89**. After persilylation, selective 5'-hydroxy deprotection was carried out to obtain **90**. Finally, a tough phosphoramidation was conducted on the protected final compound (Scheme 6.15). Due to the impossible silyl deprotection, they compared **91** with remdesivir **85**. Both compounds exhibited the same SARS-CoV-2 infection inhibition at micromolar levels.

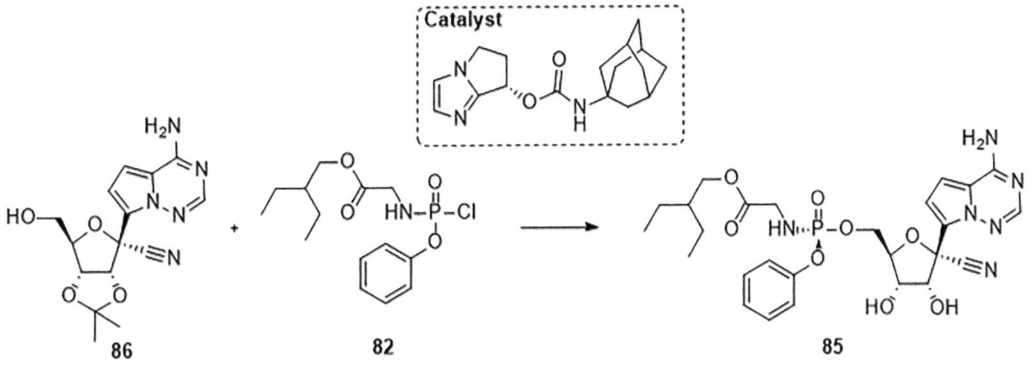

SCHEME 6.14 *Reagents and conditions:* 2,6-lutidine, Cat., DCM 4Å MS, −40 °C, 48 h.

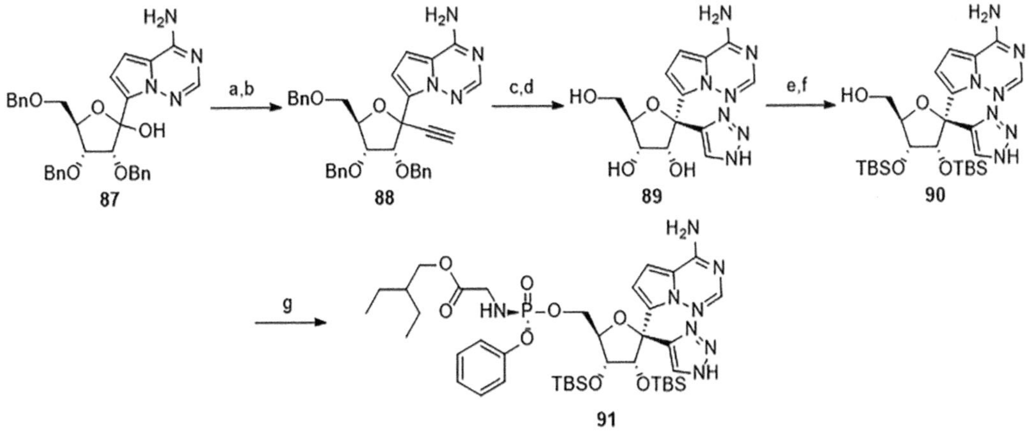

SCHEME 6.15 *Reagents and conditions*: (a) Ethynylmagnesium chloride, THF, 0 °C to rt; (b) CH_3SO_3H, DCM; (c) NaN_3, $CuSO_4.5H_2O$, Sodium ascorbate, *t*-BuOH:H_2O, 80 °C; (d) BCl_3, CH_2Cl_2, −40 °C; (e) TBSOTf, Lutidine, DMF, 55 °C; (f) TFA:Water, THF, −7 °C; (g) $MgCl_2$, (*i*-Pr)$_2$Net, CH_3CN, rt to 50 °C.

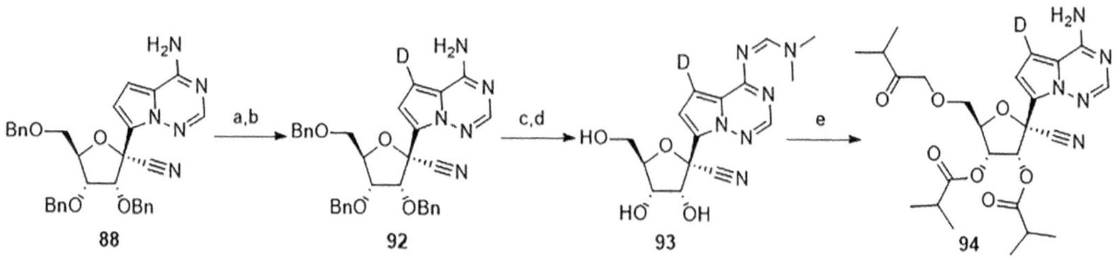

SCHEME 6.16 *Reagents and conditions:* (a) *N*-Iodosuccinimide, DMF, 50 °C; (b) $PdCl_2(dppf)CH_2Cl_2$, Tetramethylethyldiamine, $NaBD_4$, THF/D_2O; (c) BCl_3, CH_2Cl_2; (d) DMF-DMA, MeOH, DMF, 60 °C; (e) Isobutyryl chloride, Et_3N, DMAP, CH_2Cl_2, then AcOH, MeOH, 50 °C.

In a recent report, Xie *et al.* developed a tri-*iso*butyrate ester prodrug of remdesivir **94** called VV116 that showed effectiveness against SARS-CoV-2 [84]. The team inserted a deuterium at a particular position to prevent ring enzymatic inhibition. They also performed a tri-*iso*butyrate ester functionalization to improve its *in vivo* pharmacokinetics. Deuteration was carried out from the commercially available compound **88** through an iodination step to produce compound **92**. Then, the hydroxyls were deprotected, and the amine was converted into its protected form, **93**, allowing the *O*-acylation. Finally, the amine was deprotected to obtain the desired compound **94** (Scheme 6.16).

Remdesivir, VV116 is administered as a ProTide to prevent it from being metabolized by the liver. Once absorbed, it is converted to **94a** before reaching its active triphosphate form, **94b** (Scheme 6.17). Recent research showed that VV116 was effective against the SARS-CoV-2 delta and omicron BA.1 variants in Vero E6 cells, with EC_{50} values of 0.45 μM and 0.17 μM, respectively. VV116 was more effective than remdesivir (EC_{50} = 2.2 μM and 0.54 μM, respectively) [85]. Against omicron BA.5, VV116 and remdesivir displayed similar activities: EC_{50} = 0.15 μM and 0.13 μM, respectively. Moreover, when combined with mirmatrelvir, an antiviral drug developed by Pfizer, VV116 **94** exhibited a synergistic effect on coronaviruses (HCoV-OC43 and SARS-CoV-2), as confirmed *in vivo* with enhanced antiviral efficiency in mice.

6.7 FAVIPIRAVIR

Favipiravir ***98*** (T-705, Avigan®), developed by Toyama Chemical Industry, targets a broad spectrum of viral RdRps from filo-, arena-, noro-, bunya-, toga-, hanta- and flavi-viruses [86]. It is currently in clinical phase II against LASV. Once converted to favipiravir ribofuranosyl 5'-triphosphate, it is incorporated selectively into the growing RNA by viral RdRp and acts as a lethal antiviral mutagen [87]. Favipiravir was obtained through a seven-step process [88]. First, the researchers methylated 3-aminopyrazine-2-carboxylic acid ***95***, then brominated the pyrazine moiety to ***96***. Next, the amino group was substituted with a methoxy group, followed by amination under Pd-catalyzed conditions to ***97***. The amino group was then converted into a fluoro group, and the hydroxy deprotection delivered favipiravir ***98***. In 2014, Shi *et al.* reported a highly improved protocol [89] that involved the amidation of 3-hydroxypyrazine-2-carboxylic acid ***99*** to ***100*** followed by nitration to ***101*** and reduction to ***102***. Fluorination was then used to isolate T-705 (Scheme 6.18).

Avigan® was first developed as an anti-influenza drug, but it was also effective against EBV, LASV, norovirus, and rabies virus [90]. Once converted by hypoxanthine-guanine phosphoribosyl transferase and subsequently phosphorylated by cell enzymes, it acts as a chain terminator. Its integration into the viral RNA allows it to bind to RdRp domains and inhibit virus replication. Favipiravir was evaluated against SARS-CoV-2 in response to the COVID-19 pandemic and improved clinical recovery in various clinical trials [91]. For instance, Zhao *et al.* investigated favipiravir's interaction with RNA viral synthesis. This purine nucleotide analogue mimics nucleotide incorporation, but T-705 did not significantly inhibit RNA production. This suggests that its

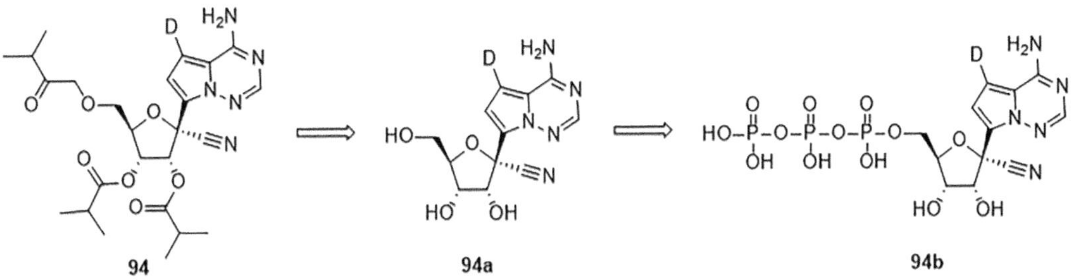

SCHEME 6.17 VV116 and its metabolized forms.

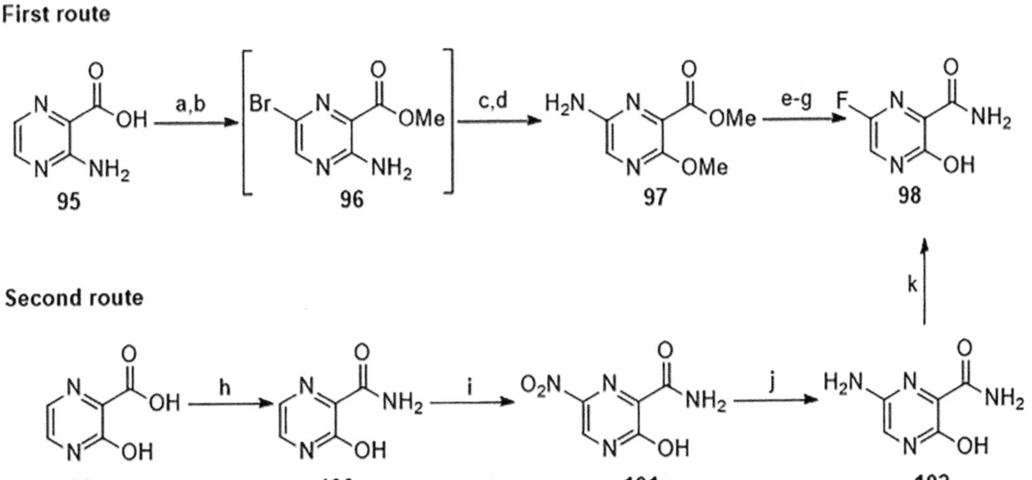

SCHEME 6.18 *Reagents and conditions:* (a) MeOH, H_2SO_4; (b) NBS, CH_3CN; (c) H_2SO_4, $NaNO_2$, MeOH; (d) Benzophenone imine, (S)-BINAP, $Pd_2(dba)_3$; (e) $NH_3.H_2O$; (f) $NaNO_2$, HF.Py; (g) NaI, TMSCl; (h) 1. $SOCl_2$, MeOH, 40 °C, 2. $NH_3.H_2O$; (i) H_2SO_4, KNO_2, 40 °C; (j) Raney Ni, $N_2H_4.H_2O$, MeOH; (k) $NaNO_2$, HF-Py, –20 °C then rt.

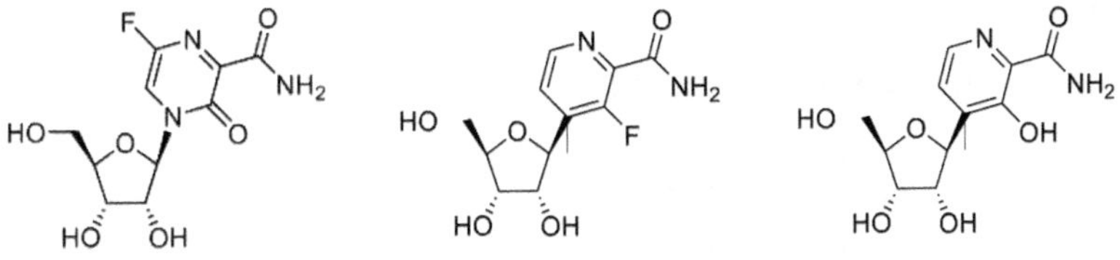

FIGURE 6.5 C Nucleosides as favipiravir (T-705)

integration is likely to induce mutations. To study their activities and to optimize their antiviral efficiency, favipiravir analogues have been reported and evaluated (Figure 6.5) [92].

Several analogues have been synthesized and assessed, but T-205 and Avigan® were selected because of their superior inhibitory activity against influenza A polymerase *in vitro*. Both these analogues are derived from **38** (Scheme 6.19). To obtain analogue **104**, compound **106** was coupled with the pyridine derivative to **107** before the 1'-hydroxyl elimination and removal of *iso*propylidene **108**. Oxidative hydrolysis isolated thc amide β-isomer, and final compound **104** was obtained

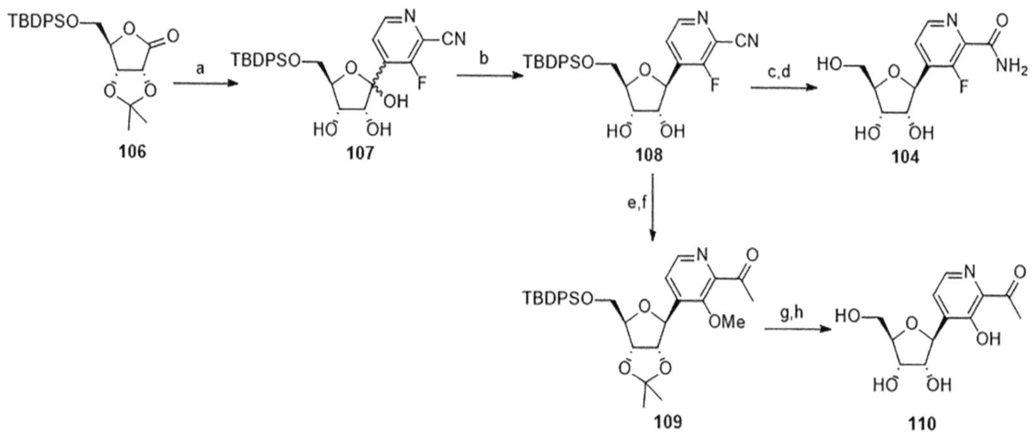

SCHEME 6.19 *Reagents and conditions:* (a) 3-Fluoro-2-pyridinecarbonitrile, lithium di*iso*propylamide, THF, –78 °C; (b) BF_3-Et_2O, Et_3SiH, 0 °C to rt; (c) NH_4OH, H_2O_2, MeOH; (d) NH_4F, MeOH; (e) i. 2,2-dimethoxypropane, TsOH, acetone; ii. NaOMe, dioxane; (f) H_2O_2,NH_3, MeOH/H_2O; (g) EtSNa, DMF, 40 °C; (h) HCl, MeOH.

through desilylations. Compound **110** was obtained from intermediate **108**. After a tough substitution of the 3-fluorine with methoxy, oxidative hydrolysis followed. Then 2'-3'-deprotection followed the demethylation and the 5'-desilylation to isolate the desired final molecule.

Compound **103** showed an EC_{50} of 2.2 μM, in the same range of favipiravir (EC_{50} = 2.7 μM). Compounds **104** and **105** showed lower EC_{50} values of 1.9 μM and 1.3 μM, respectively. However, this last compound appeared to be cytotoxic, with a CC_{50} of 2 μM. In 2022, *in vivo* evaluations in macaques demonstrated for the first-time antiviral activity of favipiravir on Zika virus viremia [93]. Favipiravir significantly reduced Zika viral load in plasma, in contrast with the results in untreated animals. These results indicate the molecule's potential for use in combination with other antiviral drugs such as galidesivir, which has also shown effectiveness in non-human primate models. However, it is worth noting that using favipiravir against SARS-CoV-2 infection did not yield any positive results and may have even worsened the disease.

6.8 MISCELLANEOUS

6.8.1 3-Deazaneplanocin A Analogues

The precursor of this synthesis [94] comes from a nine-step synthesis from D-ribose, this one modified from a previous method, (Scheme 6.20) [95]. First, D-ribose was treated with a catalytic amount of H_2SO_4 in acetone to protect it by forming an *iso*propylidene at the 3' and 4' positions. Then the primary alcohol was protected with TBDPSCl in DCM to give reactional intermediate in very good yield. Next, a Grignard reaction using excess vinyl-magnesiumbromide at –78 °C successfully gave selectively β-allylic hydroxyl group **111**. Next, selective protection of **111** with TBDMSCl primarily yielded **112** due to an allylic position with less steric hindrance than the secondary; then, **112** was retaken in DCM, and Swern oxidation of the secondary hydroxyl group was carried out to subsequently be able to carry out a Wittig reaction using triphenilphosphine methyl bromine to obtain **114**. The two-silyl groups on dienes **114** were removed with ammonium fluoride (NH_4F) in methanol to obtain diene/diol **115**. The RCM reaction of **115** with the second Grubbs catalyst successfully provided a cyclopentenyl derivative **116**. Finally, the reaction of dihydroxyl cyclopentene ***116*** with TBDPSCl gave chemoselective precursor **117** with 5-OTBDPS. For the glycosylation, a Mitsunobu reaction was carried out in the presence of triphenylphosphine (Ph_3P) and di*iso*propyl

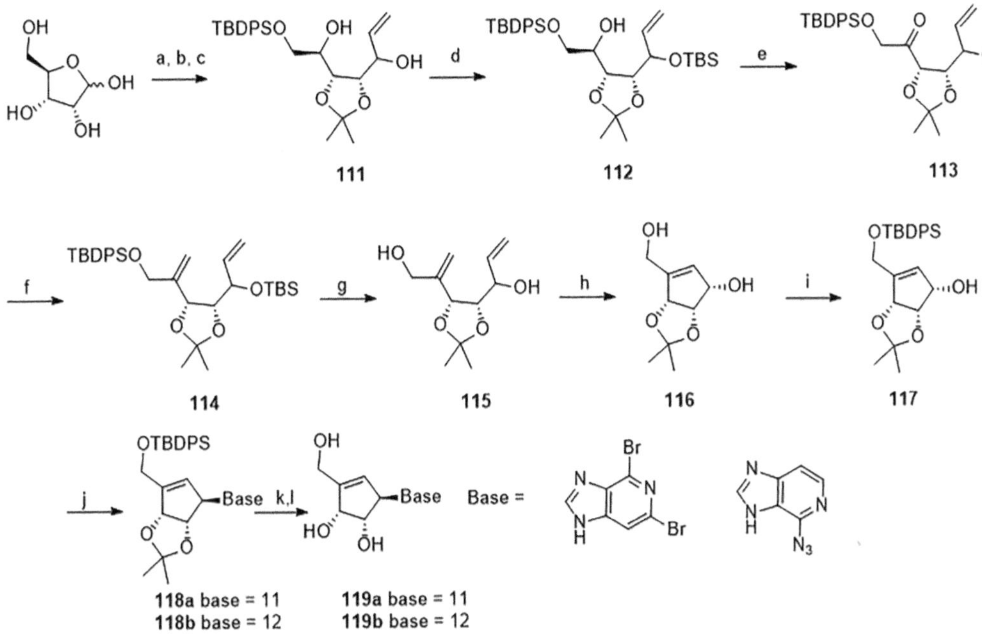

SCHEME 6.20 *Reagents and reaction conditions:* (a) acetone, H_2SO_4, 0 °C then rt; (b) TBDPSCl, imidazole, CH_2Cl_2, 0 °C then rt; (c) vinyl-magnesiumbromide, THF, –78 °C then rt; (d) TBDMSCl, imidazole, CH_2Cl_2, DMF, 0 °C then rt; (e) $(COCl)_2$, DMSO, Et_3N, CH_2Cl_2, –78 °C then rt; (f) Ph_3PCH_3Br, *n*-BuLi, THF, –20 °C then rt; (g) NH_4F, MeOH, 0 °C then 40 °C; (h) H-G 2nd cat, CH_2Cl_2, rt; (i) TBDPSCl, imidazole, CH_2Cl_2, 0 °C then rt; (j) Ph_3P, DIAD, THF, 0 °C then rt; (k) TBAF, THF, rt, 3 h for **10a** and **10b**; (l) HCl/MeOH, 0 °C then rt.

azodicarboxylate (DIAD) in THF provided compound **118a** and **118b** which were deprotected with TBAF and HCl to afford compounds **119a** and **119b**.

Deazanucleoside such as 3-deazaneplanocin A analogues showed a broad spectrum of biological activity against certain DNA/RNA viruses as a promising inhibitor of *S*-adenosyl-L-homocysteine hydrolase (SAH) and enhancer of zest homolog 2 (EZH2). They were investigated *in vitro* for antiviral activity against influenza viruses H1N1 and H3N2 and dengue virus-2 replicon (DENV-2). The novel *N9*-2,6-dibromo-3-deazaneplanocin (**119a**) exposed the most potent activity (EC_{50} = 7.0 μM) against H1N1, CC_{50} >100 *in vitro,* which was comparable with that of ribavirin (EC_{50} = 11.1 μM). Unexpectedly, *N7*-6-azido-3-deazaneplanocin (**119b**) displayed EC_{50} values of 22.5 μM and 25.3 μM against H1N1 and H3N2, respectively, with CC_{50} > 100 for both. Other 3-DZNep analogues showed no significant antiviral activity against influenza A and B in vitro or against DENV-2. *N9*-2,6-Dibromo 3-deazaneplanocin (**119a**) has been discovered as a novel hit compound against H1N1, H3N2, H5N1, and Flu B in MDCK cells [96]. In addition, 3-fluorodeazaneplanocin A showed potent activity against H5N1 and Flu B as well as dengue virus in Vero cells [97]. Compound **119a** was developed as a new promising anti-H1N1 agent with expected SAH hydrolase or a polymerase inhibition. Both compounds have no activity against DENV-2.

6.8.2 6-Methyl-7-Deaza Analogue of Adenosine

Glycosylation was employed to synthesize the target product **127** (Scheme 6.21) [98]. The nucleobase was synthetized in five steps from 2-amino-4-picoline **120**. The first step was bromination at C-5 position to afford **121**. Then, nitration with nitric acid in sulfuric acid gave nitro group **122**, which was reduced with dichlorotin in acidic condition and yield the diamine pyridine

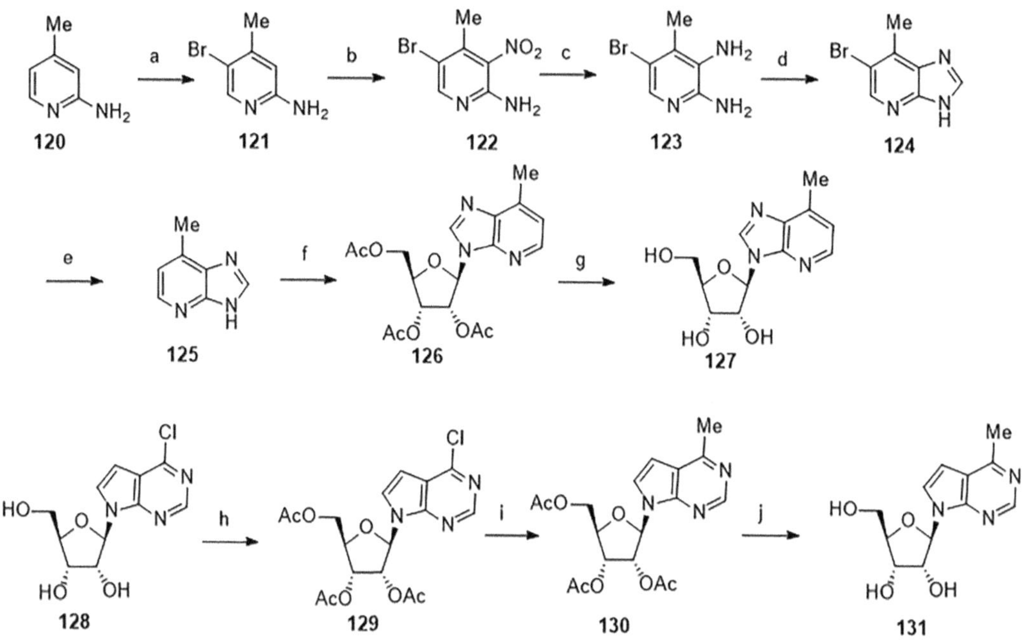

SCHEME 6.21 *Reagents and conditions:* (a) 1,3-Dibromo-5,5-dimethylhydantoin, CH_2Cl_2; (b) (i) HNO_3, H_2SO_4, (ii) H_2SO_4; (c) $SnCl_2$, HCl; (d) HCOOH; (e) PtO_2, Pd/C, NaOH; (f) p-TsOH, β-D-ribofuranose 1,2,3,5-tetraacetate; (g) NH_3, MeOH; (h) Ac_2O, pyridine; (i) $Pd(PPh_3)_4$, $AlMe_3$, THF; (j) NH_3, MeOH

derivate **123**. After that, formic acid was used to give **124**, and the bromine group was reduced by catalytic oxide platinum and palladium on charcoal in basic condition to afford base **125**. Glycosylation was carried out with the *tetra*-protected ribofuranose to synthetize protected compound **126**, which was deprotected with ammonia in methanol to **127**. The second target compound, 7-deaza-6-methyl-9-β-D-ribofuranosylpurine **131**, employed the commercially available nucleoside **128**, which was triacetyled for protecting group with Ac_2O in pyridine to give compound **129**. Then, the base was methylated to **130**, and the resulting product was deprotected with methanolic ammonia to afford the desired product **131**.

Compound **131** exhibits highly potent activity against PV and DENV-2 in cell culture assays and possesses a better cytotoxicity profile than those of 6-methyl-9-β-D-ribofuranosylpurine and tubercidin. Compounds **127** and **131** were evaluated against infectious PV in a whole-cell assay. HeLa S3 cells were pretreated with **127** and **131** for 1 h before being exposed to a high multiplicity of infection with PV, and there efficacies were compared with tubercidin and 6-methyl-9-β-D-ribofuranosylpurine, both of which are very effective against these cells: IC_{50} (6-methyl-9-β-D-ribofuranosylpurine) = 1.01 μM; IC_{50} (tubercidin) = 0.030 μM). Novel nucleoside **131** was tested as the most potent inhibitor of PV replication in the series of its analogues (IC_{50} = 0.011 μM). For the PV assays, **127** and **131** were treated for 7, 24, and 48 h. Molecule **127** showed no cytotoxic effects to HeLa cells after treatment for 7 h; it was also tested with live dengue virus DENV-2 (IC_{50} = 0.877 μM (replicon), 0.062 μM (plaque), and 0.039 μM (PCR) and demonstrated highly potent inhibition activity.

6.8.3 6′,6′-Difluoroaristeromycins

In 2019, Jeong *et al.* [99] reported an ambitious synthesis of 6′-fluorinated aristeromycins, which show broad-spectrum antiviral properties and is analogue of AT-9010. Synthesis of the precursor ***138*** began with the 1,4-conjugate addition of ***132*** with Gilman's reagent, which gave derivative

133 (Scheme 6.22). Treatment of ***133*** with lithium hexamethyldisilazide (LiHMDS) followed by trapping with triethylsilyl chloride (TESCl) gave silylenol ether ***134***. This ether was treated with selectfluor in dimethylformamide (DMF) at 0 °C to give ***135*** in fairly good yield. These two steps were repeated to give difluoro analogue ***136*** in equilibrium with ***137***. Reduction of ***136*** with lithium borohydride ($LiBH_4$) in MeOH produced 1-hydroxyl derivate ***138***. After the synthesis of the precursor, ***138*** was treated with triflic anhydride (Tf_2O), followed by treatment with NaN_3 to yield azido derivate **139**. The catalytic hydrogenation of **139** gave amino derivative **140**. This compound was treated with 5-amino-4,6 dichloropyrimidine in the presence of *N,N*-di*iso*propylethylamine (DIPEA) under microwave radiation, afforded **141** with a good yield. This was then cyclized with diethoxymethyl acetate under microwave radiation to produce 6-chloropurine derivative **142**. Treatment of **142** with *t*-butanolic ammonia followed by the removal of the protecting groups under acidic conditions yielded 6′,6′-difluoroaristeromycin **143**.

The 6′-fluorinated aristeromycins were designed as dual-targeted antiviral compounds aimed at inhibiting both the viral RdRp and the SAH in the host cell [100]. The introduction of a fluorine group at the 6′ position improved SAH hydrolase inhibition and activity against RNA viruses. Viral load reduction assays were performed with compound **13** by infecting cells with CHIKV, ZIKV, SARS-CoV, and MERS-CoV, followed by treatment with different concentrations of **143**; **13**

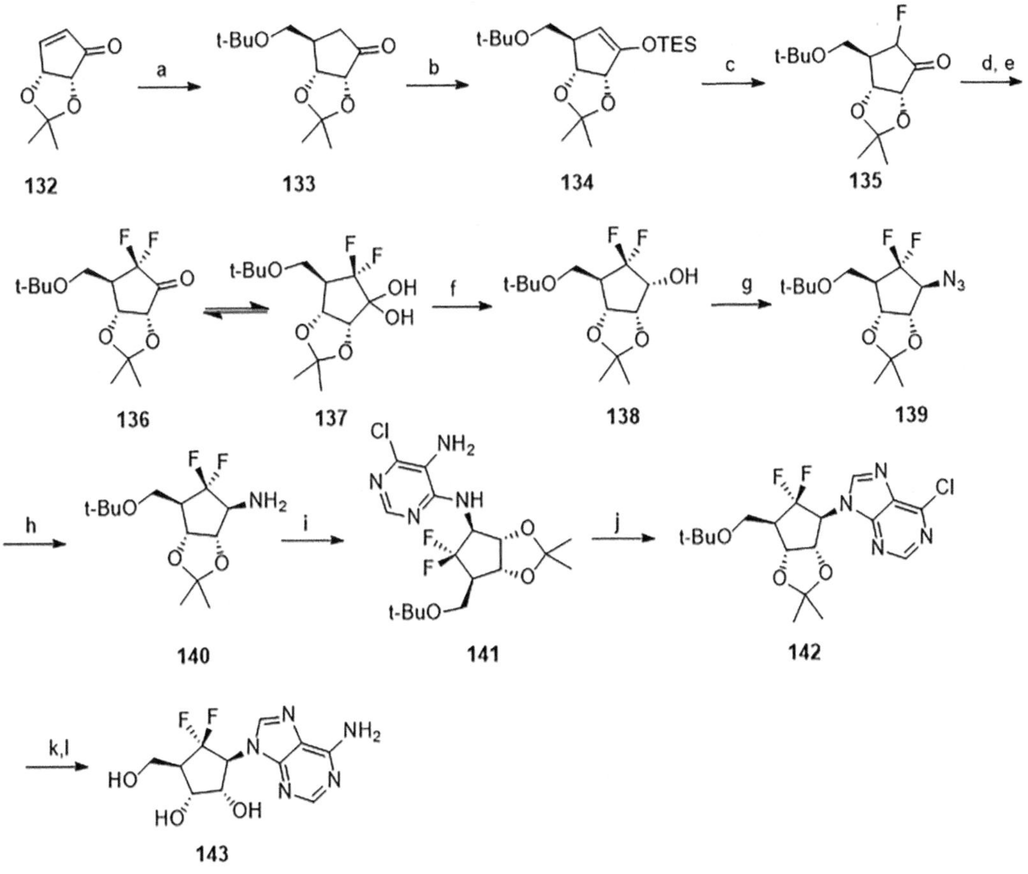

SCHEME 6.22 *Reagents and conditions:* (a) $LiCu(CH_2Ot\text{-}Bu)_2$; (b) TESCl, LiHMDS, THF, –78 °C; (c) Selectfluor, DMF, 0 °C; (d) TESCl, LiHMDS, THF, –78 °C; (e) Selectfluor, DMF, 0 °C; (f) $LiBH_4$, MeOH, 0 °C; (g) (i) Tf_2O, pyridine, 0 °C; (ii) NaN_3, DMF, 60–100 °C; (h) Pd/C, H_2, MeOH, rt; (i) 5-amino-4,6-dichloropyrimidine, DIPEA, *n*-BuOH, 170–200 °C, MW; (j) $CH_3C(O)OCH(OEt)_2$, 140 °C, MW; (k) NH_3/*t*-BuOH, 120 °C; (l) 67 % aq TFA, 50 °C

showed interesting properties against SARS-CoV (EC_{50} = 0.5 μM), MERS-CoV (EC_{50} = 0.2 μM), ZIKV (EC_{50} = 0.26 μM), and CHIKV (EC_{50} = 0.13 μM), and adding a phophoramidate increased the effectiveness of this molecule on the targets by a factor of 10 against SARS-CoV as a spectrum coronavirus inhibitor.

6.8.4 5′-Nor-3-Deaza-1′,6′-Isoneplanocin

The synthesis of 5'-nor-3-deaza-1',6'-*iso*neplanocin began with the iodination of protected (−)-cyclopentenone ***132*** to ***144*** with diiodine in tetrachloride methane, (Scheme 6.23) [101]. Luche reduction of ***144*** to allylic alcohol ***145*** was carried out with $NaBH_4$ and $CeCl_3.7H_2O$ to avoid the alkene reduction. The mixture was subjected to Ullmann conditions with 3-deazaadenine and gave a low yield of ***146*** occurred. Acidic deprotection with HCl in methanol gave *(4'R)-**147***. Obtaining epimer ***152*** began with ***145***, which upon acid catalyzed *iso*propylidene rearrangement was expected to provide ***148***; this followed by a Mitsunobu C-4' inversion to benzoylated ***150***. Basic removal of the benzoate of ***150*** to obtain ***151*** and subsequent acidic condition deprotection yielded *(4'S)-**152***.

Initially, 5′-norcarbocyclic nucleoside analogues were created as inhibitors of S-adenosylhomocysteine hydrolase (*SAHase*), a cellular enzyme involved in many biological methylations [102]. Recently it was reported that 5′-norcarbocyclic nucleoside analogues could serve as HIV non-nucleoside reverse transcriptase inhibitors [103]. Compounds **147** and **152** showed potential activity and non-cytotoxic antiviral profiles, which is why they tested in the antiviral assays [104]. Compound **152** displayed potent activity (EC_{50} = 1.12 μM) against the polyomavirus JC virus; **147** had much lower activity (EC_{50} = 59.14 μM). Both epimers showed no cytotoxicity (CC_{50} >

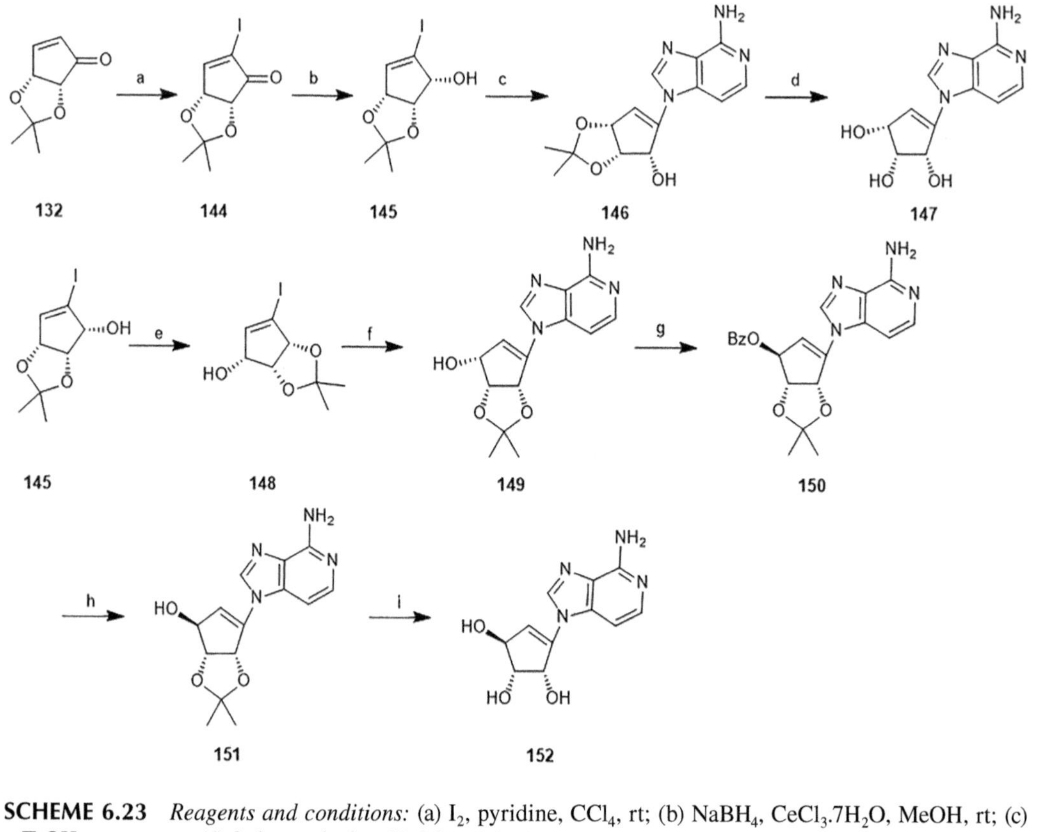

SCHEME 6.23 *Reagents and conditions:* (a) I_2, pyridine, CCl_4, rt; (b) $NaBH_4$, $CeCl_3.7H_2O$, MeOH, rt; (c) p-TsOH, acetone, rt; (d) 3-deazaadenine, K_2CO_3, dipivaloylmethane, CuI, 120 °C; (e) 2 M HCl/MeOH, rt; (f) HCl, MeOH, rt; (g) Ph_3P, DIAD, benzoic acid, THF, rt; (h) LiOH, THF-H_2O (1:1), rt; (i) HCl, MeOH, rt.

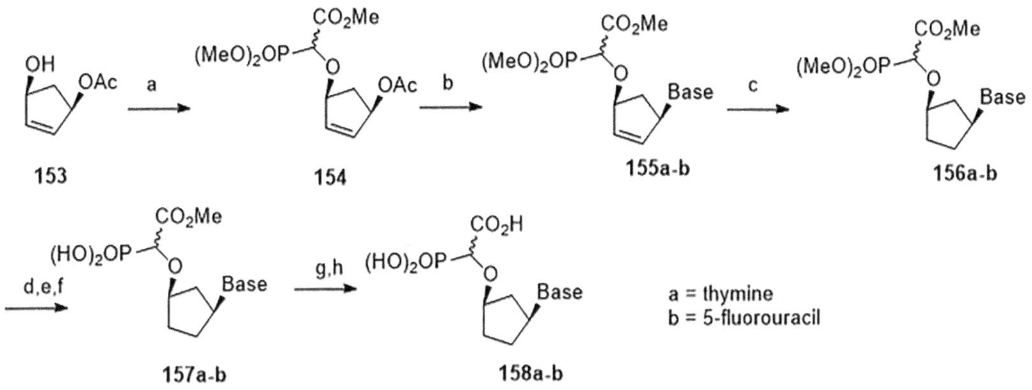

SCHEME 6.24 *Reagents and conditions:* (a) $(MeO)_2OPC(N_2)CO_2Me$, $Rh_2(OAc)_4$ or $Cu(OTf)_2$, C_6H_6, 80 °C; (b) base, $Pd(PPh_3)_4$, Na_2CO_3, aq MeCN, 55 °C; (c) H_2, Pd/C, MeOH; (d) TMSBr, MeCN microwave 50 °C; (e) H_2O; (f) concentration in vacuo; (g) aq NaOH, 50 °C; (h) charcoal chromatography

150 μM). These molecules were also tried and showed no activity against human cytomegalovirus, adenovirus, vaccinia virus, Epstein–Barr virus and human norovirus, that is, no cytotocicity.

6.8.5 α-Carboxy Nucleoside Phosphonate [105]

The key O-H insertion reaction was carried out using the racemic acetoxy alcohol **153** and trimethyl phosphonodiazoacetate in the presence of rhodium(II) acetate or copper(II) triflate to afford desired product **154** (Scheme 6.24). Rhodium-catalyzed O–H insertion provides a mild and neutral way of attaching the phosphonate group to suitably protected nucleosides. The introduction of the nucleobases onto allylic acetate **154** was next undertaken by a Tsuji–Trost-type palladium(0)-catalyzed allylic substitution to afford compound **155a-b**. These compounds were used to make a hydrogenation reaction over palladium on carbon catalyst to afford the saturated derivatives **156a-b**. The derivatives **156a-b** were deprotected using the trimethyl derivatives were treated with excess TMSBr, followed by addition of water and then saponification treatment with aqueous NaOH at 50 °C, giving the fully deprotected unsaturated thymine and 5-fluorouracil derivatives ***158a-b***, after using charcoal chromatography purification.

McClure *et al.* evaluated novel compounds **158a** and **158b** for their inhibitory activity against a broad range of DNA and RNA viruses, herpes simplex virus type 1 (HSV-1), herpes simplex virus type 2 (HSV-2), vaccinia virus, respiratory syncytial virus (RSV), vesicular stomatitis virus (VSV), coxsackie virus B4, parainfluenza virus 3, influenza virus A, influenza virus B, reovirus-1, Sindbis virus, Reovirus-1, Punta Toro virus, HIV-1, and HIV-2. Both **158a** and **158b** showed some inhibitory activity against HIV-1 reverse transcriptase, with IC_{50} = 0.41 μM for **158a** and IC_{50} = 3.8 μM for **158b** in [^{3}H]dTTP/poly rA-dT and IC_{50} = 293 μM for **158a** in [^{3}H]dATP/poly rU-dA. Compound **158a** possessed greater anti-HIV RT activity than all of the tested NRTIs (IC_{50} = 0.316–10 μM) [106].

6.8.6 5′-Norcarbanucleosides

Hong's group reported a series of racemic 5'-norcarbanucleosides [107–109] including a phosphonate analogue of 2'-modified 5'-norcarbocyclic adenine [110]. But-3-en-1-ol ***159*** was transformed into three steps to (±)-**160** (Scheme 6.25). After selective deprotection of PMB, the resulting hydroxyl group was oxidized to aldehyde (±)-**161**. Adding vinyl-magnesiumbromide and subsequent RCM afforded the isomers (±)-**262a** and (±)-**262b**. Cyclopentenol (±)-**262a** was subjected to Mitsunobu reaction in the presence of *N6-bis*-Boc-adenine to give (±)-**163**, which was then converted to its phosphonate analogue (±)-**164**.

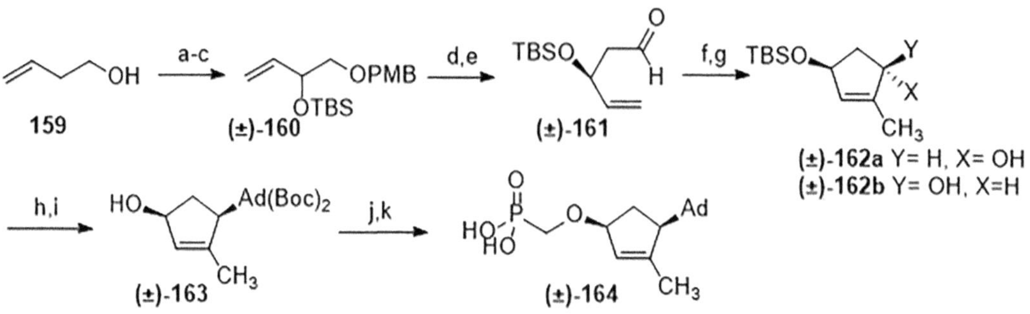

SCHEME 6.25 *Reagents and conditions*: (a) PMBCl, NaH, DMF, 0 °C; (b) (i) O_3, DMS, –78 °C; (ii) vinyl-MgBr, THF, –78 °C; (c) TBDMSCl, imidazole, CH_2Cl_2, 0 °C; (d) DDQ, CH_2Cl_2/H_2O, rt; (e) PCC, 4MS, CH_2Cl_2; (f) *iso*propenyl MgBr, THF, –78 °C; (g) Grubbs catalyst, benzene, 60 °C, reflux; (h) PPh_3, DIAD, *N6-bis*-Boc-adenine, THF, –20 °C; (i) TBAF, THF, rt; (j) di*iso*propyl bromomethyl phosphonate, LiO-tBu, LiI, DMF, 60 °C; (k) TMSBr, CH_3CN.

Schneller's group reported the synthesis of various 5'-nor-carbanucleosides modified at either 4'- or 1'- and also at the base moiety [111, 112]. Silylation and deacylation of chiral acetate **163**, (Scheme 6.26), afforded **164**, which by oxidation gave the cyclopentenone derivative **165**. Stereoselective addition of methyllithium on **165** afforded **166**, the key intermediate for obtaining both series 1'-methyl and 4'-methyl-5'-norcarbanucleosides. Under Mitsunobu reaction in the presence of chloropurine, **166** gave coupling product **167**. Deprotection of silyl group, followed by the inversion of the C4' center through a second Mitsunobu reaction provided **168**. Stereoselective dihydroxylation (*trans* to the heterocycle base) followed by an ammonolysis gave the desired 5'-*nor*carbanucleoside **169**.

6.8.7 5′-Norphosphonomethoxy Carbocyclic Nucleosides

Recently, Maguire *et al.* [113] reported the synthesis of racemic guanine α-carboxy nucleoside phosphonate (±)-***174*** (Scheme 6.27). The nucleobase construction of (±)-***171*** was performed from *cis*-3-aminocyclopentanol (±)-***170*** in presence of 2-amino-4,6-dichloro-5-formamidopyrimidine and DIPEA under microwave activation. Silylation of the alcohol followed by Boc protection of amine and subsequent desilylation with TBAF gave alcohol (±)-***172***, which in presence of rhodium acetate and trimethylphosphodiazoacetate gave a mixture of diastereomers at α-position of phosphate (±)-***173.*** Removal of protecting groups afforded the desired (±)-***174***.

6.8.8 Carbocyclic 5′-nor "Reverse" Fleximers

Construction of targets ***179a–c*** started with commercial diacetylated carbopentene, (Scheme 6.28) [114]. To synthetize the target compounds, enzymatic resolution of the *meso*-diacetate ***175*** with *Pseudomonas cepacia* lipase was used to give desired enantiomer ***176.*** This enantiomer was the intermediate and was needed for Tsuji-Trost coupling to the pyrimidine ring system. With ***176***, the *N*3-benzoyl protected pyrimidine base was added using a palladium catalyzed Tsuji-Trost coupling and gave the second protected intermediate ***177***. Intermediate ***177*** was then deprotected in mild methanolic ammonia to yield compound ***178***, which was then subjected to Stille coupling with the organotin reagent and either $PdCl_2(PPh_3)_2$ or $Pd(PPh_3)_4$ to yield compounds ***179a–c***.

The final compounds were evaluated for their potential inhibitory activity against different abroad antiviral aspect. They were evaluated for their potential inhibitory activity against different types of viruses including HSV-1, HSV-2, HSV TK, vaccina virus, vesicular stomatitis virus, respiratory syncytial virus, Sindbis virus, Punta Toro virus, Reovirus-1, Coxsackie virus B4, parainfluenza-3 virus, influenza virus types A and B, HIV-1, and HIV-2. Compounds **179a–c** showed no activity at subtoxic concentrations, but **179a** showed inhibitory activity against HCMV (AD-169 & Davis) and

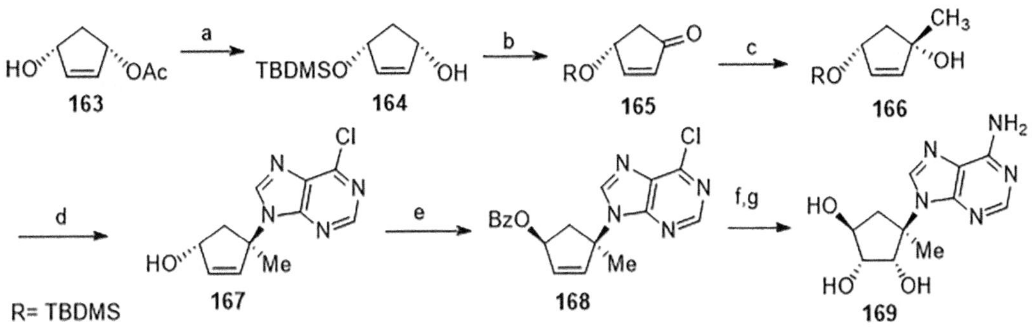

SCHEME 6.26 *Reagents and conditions:* (a) (i)TBDMSCl, imidazole (ii) K_2CO_3 in MeOH; (b) pyridinium chloroformate (PCC), CH_2Cl_2 (c) MeLi, Et_2O, −40 °C (d) (i) Ph_3P/DIAD, chloropurine,THF; (ii) K-10 clay, EtOAc/MeOH/H_2O (1:6:2); (e) $PhCO_2H$, DIAD, PPh_3; (f) OsO_4, NMO; (g) NH_3 in MeOH.

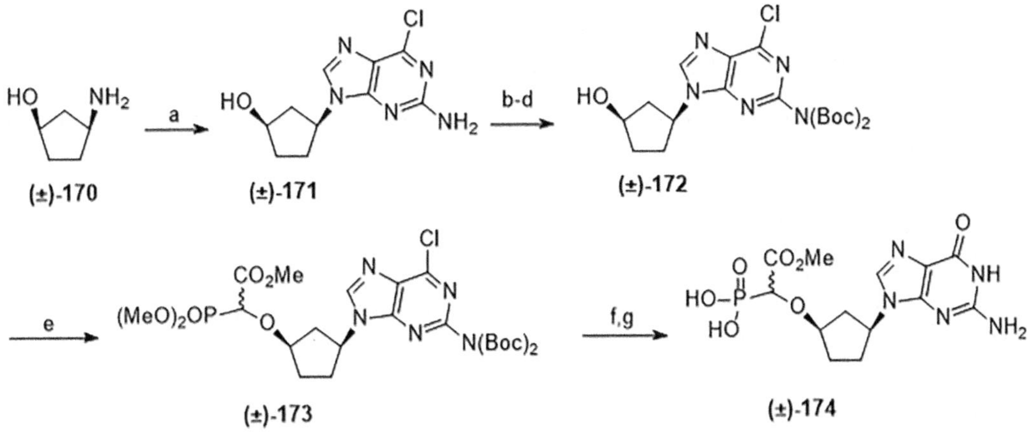

SCHEME 6.27 *Reagents and conditions*: (a) 2-amino-4,6-dichloro-5-formamidopyrimidine, DIPEA, MW, EtOH/H_2O; (b) TBDMSCl, imidazole; (c) Boc_2O, DMAP; (d) TBAF, THF; (e) $(MeO)_2OPC(N_2)CO_2Me$, $Rh_2(OAc)_4$, C_6H_6; (f) TMSBr, CH_3CN, MW; (g) 1M NaOH.

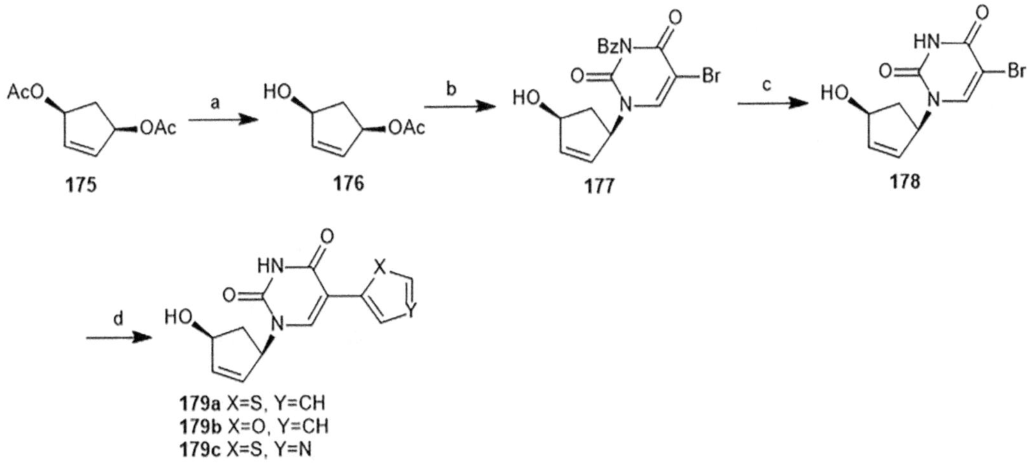

SCHEME 6.28 *Reagents and conditions:* (a) *Pseudomonas cepacia* lipase, potassium phosphate buffer pH 7.4, acetone, 1 N NaOH; (b) NaH, DMF, $Pd_2(dba)_3$, DPPP, N^3-benzoylated-5-bromouracil, 55 °C; (c) methanolic ammonia, rt; (d) for **5a**, tributylstannyl thiophene, dioxane, $(PPh_3)_2PdCl_2$; for **5b**, tributylstannyl furan, dioxane, $(PPh_3)_2PdCl_2$; for **5c**, tributylstannyl thiazole, THF, $Pd(PPh_3)_4$.

VZV (OKA & 07/1): EC_{50} = 1.8–2.0 μM and 1.6–1.7 μM respectively. However, **179a** also showed global cytotoxicity in the lower micromolar concentration of 4 μM against two different cell lines (HEL, MDCK) and 20 μM against HeLa and Vero. Their effects were compared with RBV.

6.9 CONCLUDING REMARKS

More than 70 nucleoside and their analogues are licensed as direct-acting antiviral drugs mainly against DNA viruses. Recently, RNA viruses, which have caused various pandemics, are the targets of new antivirals, which prompted the search of more complex analogues with efficient synthetic routes. Lessons learned from DNA viruses as well as recent structural findings regarding RdRp in RNA viruses could help design new broad-spectrum nucleosides analogues through practical guidelines and facilitate their clinical development. However, small modifications of the nucleoside scaffold (sugar, nucleobase, or prodrug) can impact their biological activities (gain and loss) in terms of biological and pharmacokinetics parameters, cellular uptake, and so on. Altogether, new nucleoside drugs will be discovered in next future.

6.10 DECLARATION OF COMPETING INTEREST

The authors declare that they have no known competing financial interests or personal relationships that could have appeared to influence the work reported in this paper.

ACKNOWLEDGMENTS

L.A.A. is grateful to the participants of GAVO consortium for fruitful discussions on emerging viruses as well as the members of FEDER FérI consortium for sharing their points of view concerning 'one-health' and zoonotic viruses.

REFERENCES

[1] Roychoudhury S, Das A, Sengupta P, Dutta S, Roychoudhury S, Choudhury AP, *et al.* Viral pandemics of the last four decades: Pathophysiology, health impacts and perspectives. Int J Environ Res Public Health **2020**; 17: 9411.

[2] WHO. Report of the WHO/FAO/OIE joint consultation on emerging zoonotic diseases (World Health Organization, **2004**).

[3] Nichol ST, Arikawa J, Kawaoka Y. Emerging viral diseases. Proc Natl Acad Sci USA **2000**; 97: 12411–12412.

[4] Choi YK. Emerging and re-emerging fatal viral diseases. Exp Mol Med **2021**; 53: 711–712.

[5] Carrasco-Hernandez R, Jacome R, Lopez Vidal Y, Ponce de Leon S. Are RNA viruses candidate agents for the next global pandemic? A Review. ILAR J **2017**; 58: 343–358.

[6] Duffy S, Schackelton L, Holmes EC. Rates of evolutionary change in viruses: Patterns and determinants. Nat Rev Genet **2008**; 9: 267–276.

[7] Mariappan V, Pratheesh P, Shanmugan L, Rao SR, Pillai AB. Viral hemorrhagic fever: Molecular pathogenesis and current trends of disease management-an update. Curr Res Virol Sci **2021**; 2: 100009.

[8] Jacome R, Becerra A, de Leon SP, Lazcano A. Structural analysis of monomeric RNA-dependent polymerases: Evolutionary and therapeutic implications. PLoS One **2015**; 10: e0139001.

[9] Ferrero D, Ferrer-Orta C, Verdaguer N. Viral RNA-Dependent RNA polymerases: A structural overview. Subcell Biochem **2018**; 88: 39–71.

[10] Te Velthuis AJ. Common and unique features of viral RNA-dependent polymerases. Cell Mol Life Sci **2014**; 71: 4403–4420.

[11] Debing Y, Neyts J, Delang L. The future of antivirals: Broad-spectrum inhibitors. Curr Opin Infect Dis **2015**; 28: 596–602.

[12] Meganck RM, Baric RS. Developing therapeutic approaches for twenty-first-century emerging infectious viral diseases. Nat Med **2021**; 27: 401–410.

[13] Gorbalenya AE, Pringle FM, Zeddam JL, Luke BT, Cameron CE, Kalmakoff J, *et al.* The palm subdomain-based active site is internally permuted in viral RNA-dependent RNA polymerases of an ancient lineage. J Mol Biol **2002**; 324: 47–62.

[14] Venkataraman S, Prasad BVLS, Selvarajan R. RNA Dependent RNA Polymerases: Insights from structure, function and evolution. Viruses **2018**; 10: 76.
[15] Pathania S, Rawal RK, Singh PK. RdRp (RNA-dependent RNA polymerase): A key target providing anti-virals for the management of various viral diseases. J Mol Struct **2022**; 1250: 131756.
[16] Picarazzi F, Vicenti I, Saladini F, Zazzi M, Mori M. Targeting the RdRp of emerging RNA viruses: The structure-based drug design challenge. Molecules **2020**; 25: 5695.
[17] Huchting J. Targeting viral genome synthesis as broad-spectrum approach against RNA virus infections. Antivir Chem Chemother **2020**; 28: 1–27.
[18] Afreen N, Naqvi IH, Broor S, Ahmed A, Kazim SN, Dohare R, *et al.* Evolutionary analysis of Dengue serotype 2 viruses using phylogenetic and Bayesian methods from New Delhi, India. PLoS Negl Trop Dis. **2016**: 10: e0004511.
[19] Lauring AS, Andino R. Quasispecies theory and the behavior of RNA viruses. PLoS Pathog **2010**; 6: e1001005.
[20] Novella IS, Presloid JB, Taylor RT. RNA replication errors and the evolution of virus pathogenicity and virulence. Curr Opin Virol **2014**; 9: 143–147.
[21] Duffy S. Why are RNA virus mutation rates so damn high? PLoS Biol **2018**; 16: e3000003.
[22] Ramdhan P, Li C. Targeting viral methyltransferases: An approach to antiviral treatment for ssRNA viruses. Viruses **2022**; 14: 379.
[23] Slusarczyk M, Serpi M, Pertusati F. Phosphoramidates and phosphonamidates (ProTides) with antiviral activity. Antivir Chem Chemother **2018**; 26: 1–31.
[24] Roy V, Agrofoglio LA. Nucleosides and emerging viruses: A new story. Drug Discov Today **2022**; 27: 1945–1953.
[25] Deville-Bonne D, El Amri C, Meyer P, Chen Y, Agrofoglio LA, Janin Y. Human and viral nucleoside/nucleotide kinases involved in antiviral drug activation: Structural and catalytic properties. Antivir Res **2010** 86: 101–120.
[26] De Clercq E, Field HJ. Antiviral prodrugs – The development of successful prodrug strategies for antiviral chemotherapy. Br J Pharmacol **2006**; 147: 1–11.
[27] Gosselin G, Girardet JL, Perigaud C, Benzaria S, Lefebvre I, Schlienger N, *et al.*, New insights regarding the potential of the pronucleotide approach in antiviral chemotherapy. Acta Biochim Pol **1996**; 43: 196–208.
[28] Ray S, Hostetler KY. Application of kinase bypass strategies to nucleoside antivirals. Antivir Res **2011**; 92: 277–291.
[29] Pradere U, Garnier-Amblard EC, Coats SJ, Amblard F, Schinazi RF. Synthesis of nucleoside phosphate and phosphonate prodrugs. Chem Rev **2017**; 114: 9154–9218.
[30] McGuigan C, Bellevergue P, Sheeka H, Mahmood N, Hay AJ. Certain phosphoramidate derivatives of dideoxy uridine (ddU) are active against HIV and successfully by-pass thymidine kinase. FEBS Lett **1994**; 351: 11–14.
[31] Mehellou Y, Balzarini J, McGuigan C. Aryloxy phosphoramidate triesters: A technology for delivering monophosphorylated nucleosides and sugars into cells. ChemMedChem **2009**; 4: 1779–1791.
[32] Serpi M, Pertusati F. An overview of ProTide technology and its implications to drug discovery. Expert Opin Drug Discov **2021**; 16: 1149–1161.
[33] Mehellou Y, Rattan HS, Balzarini J. The ProTide prodrug technology: From the concept to the clinic. J Med Chem **2018**; 61: 2211–2226.
[34] Witkowski JT, Robins RK, Sidwell RW, Simon LN. Design, synthesis, and broad spectrum antiviral activity of 1-β-D-ribofuranosyl-1,2,4-triazole-3-carboxamide and related nucleosides. J Med Chem **1972**; 15: 1150–1154.
[35] Huffman JH, Sidwell RW, Khare GP, Witkowski JT, Allen LB, Robins RK. In vitro effect of 1-beta-D-ribofuranosyl-1,2,4-triazole-3-carboxamide (virazole, ICN 1229) on deoxyribonucleic acid and ribonucleic acid viruses. Antimicrob Agents Chemother. **1973**; 3: 235–241.
[36] Ito Y, Nii Y, Kobayashi S, Ohno M. Regioselective synthesis of virazole using benzyl cyanoformate as a synthon. Tetrahedron Lett **1979**; 20: 2521.
[37] Liu BK, Wang N, Wu Q, Xie CY, Lin XF. Regioselective enzymatic acylation of ribavirin to give potential multifunctional derivatives. Biotechnol Lett **2005**; 27: 717–720.
[38] Sidwell RW, Huffman JH, Khare GP, Allen LB, Witkowski JT, Robins RK. Broad-spectrum antiviral activity of virazole: 1-β-D-ribofuranosyl-1,2,4-triazole-3-carboxamide. Science **1972**; 177: 705–706.
[39] Graci JD, Cameron CE. Mechanisms of action of ribavirin against distinct viruses. Rev Med Virol **2006**; 16: 37–48.
[40] Ramirez-Olivencia G, Estebanez M, Membrillo FJ, Ybarra MC. Use of ribavirin in viruses other than hepatitis C. A review of the evidence. Enferm Infect Microbiol Clin **2019**; 37: 602–608.
[41] Liatsos GD. Controversies' clarification regarding ribavirin efficacy in measles and coronaviruses: Comprehensive therapeutic approach strictly tailored to COVID-19 disease stages. World J Clin Cases **2021**; 9: 5235–5278.

[42] Loustaud-Ratti V, Carrier P, Rousseau A, Maynard M, Babany G, Alain S, *et al.* Pharmacological exposure to ribavirin: A key player in the complex network of factors implicated in virological response and anaemia in hepatitis C treatment. Dig Liver Dis **2011**; 43: 850–855.
[43] Sofia MJ, Bao D, Chang W, Du J, Nagarathnam D, Rachakonda S, *et al.* Discovery of a β-D-2'-deoxy-2'-α-fluoro-2'-β-C-methyluridine nucleotide prodrug (PSI-7977) for the treatment of hepatitis C virus. J Med Chem. **2010**; 53: 7202–7218.
[44] Sofia MJ, Furman PA. HCV: The journey from discovery to a cure. In: Sofia, M. (eds), Topics in Medicinal Chemistry, **2018**, vol 32. Springer, Cham.
[45] Murakami E, Tolstykh T, Bao H, Niu C, Steuer HM, Bao D, Chang W, Espiritu C, Bansal S, Lam AM, Otto MJ, Sofia MJ, Furman PA. Mechanism of activation of PSI-7851 and its diastereoisomer PSI-7977. J Biol Chem. **2010**; 285: 34337–34347.
[46] Lam AM, Murakami E, Espiritu C, Steuer HM, Niu C, Keilman M, *et al.* PSI-7851, a pronucleotide of beta-D-2'-deoxy-2'-fluoro-2'-C-methyluridine monophosphate, is a potent and pan-genotype inhibitor of hepatitis C virus replication. Antimicrob Agents Chemother **2010**; 54: 3187–3196.
[47] Li H, Lee W, Yoo JC, Hong JH. Selective synthesis of 1'(α), 2'(β)-C-dimethyl carbocyclic adenosine analogue as potential anti-HCV agent. Bull Korean Chem Soc **2009**; 30: 2039–2042.
[48] Hong JH. Synthesis of novel 2'-methyl carbovir analogues as potent antiviral agents. Arch. Pharmacal Res **2007**; 30: 131–137.
[49] Aubin Y, Audran G, Monti H, De Clercq E. Chemoenzymatic synthesis and antiviral evaluation of conformationally constrained and 3'-methyl-branched carbanucleosides using both enantiomers of the same building block. Bioorg Med Chem **2008**; 16: 347–381.
[50] Audran G, Acherar S, Monti H. Enantioselective synthesis of 3-methylcarbapentofuranose derivatives, based on a chemoenzymatic procedure. Eur J Org Chem **2003**; 92–98.
[51] Cho A, Zhang L, Xu J, Babusis D, Butler T, *et al.* Synthesis and characterization of 2'-C-Me branched C-nucleosides as HCV polymerase inhibitors. Bioorg Med Chem Lett **2012**; 22: 4127–4132.
[52] Butler T, Cho A, Kim CU, Saunders OL, Zhang L. 1'-substituted carba-nucleoside analogs for antiviral treatment. Patent WO 2009132135, **2009.**
[53] Lindell SD, Maechling S, Sabina RL. Synthesis and biochemical testing of 3-(carboxyphenylethyl) imidazo[2,1-f][1,2,4]triazines as inhibitors of AMP deaminase. ACS Med Chem Lett **2010**; 1: 286.
[54] Carroll SS, Ludmerer S, Handt L, Koeplinger K, Zhnag NR, Graham D, *et al.* Robust antiviral efficacy upon administration of a nucleoside analog to hepatitis C virus-infected chimpanzees. Antimicrob Agents Chemother **2009**; 53: 926.
[55] Wang G, Dyatkina N, Prhavc M, Williams C, Serebryany V, Hu Y, Huang Y, *et al.* Synthesis and anti-HCV activity of sugar-modified guanosine analogues: Discovery of AL-611 as an HCV NS5B polymerase inhibitor for the treatment of chronic Hepatitis C. J Med Chem **2020**; 63: 10380–10395.
[56] Smith DB, Kalayanov G, Sund C, Winqvist A, Pinho P, Maltseva T, Morisson V, Leveque V, *et al.* The design, synthesis, and antiviral activity of 4'-azidocytidine analogues against hepatitis C virus replication: The discovery of 4'-azidoarabinocytidine. J Med Chem **2009**; 52: 219–223.
[57] Hotard AL, He B, Nichol ST, Spiropoulou CF, Lo MK. 4'-Azidocytidine (R1479) inhibits henipaviruses and other paramyxoviruses with high potency. Antiviral Res **2017**; 144: 147–152.
[58] Klumpp K, Leveque V, Le Pogam S, Ma H, Jiang WR, Kang H, Granycome C, Singer M, Laxton C., *et al.* The novel nucleoside analog R1479 (4'-azidocytidine) is a potent inhibitor of NS5B-dependent RNA synthesis and hepatitis C virus replication in cell culture. J Biol Chem **2006**; 281: 3793–3799.
[59] Wang G, Deval J, Hong J, Dyatkina N, Prhavc M, Taylor J, *et al.* Discovery of 4'-chloromethyl-2'-deoxy-3',5'-di-O-isobutyryl-2'-fluorocytidine (ALS-8176), a first-in-class RSV polymerase inhibitor for treatment of human respiratory syncytial virus infection. J Med Chem **2015**; 58: 1862–1878.
[60] Smith DB, Martin JA, Klumpp Kl, Baker SJ, Blomgren PA, Devos R, Granycome C, Hang J, Hobbs CJ, *et al.* Design, synthesis, and antiviral properties of 4'-substituted ribonucleosides as inhibitors of hepatitis C virus replication: The discovery of R1479. Bioorg Med Chem Lett **2007**; 17: 2570–2576.
[61] Biteau NG, Amichai SA, Azadi N, De R, Downs-Bowen J, Lecher JC, *et al.* Synthesis of 4'-substituted carbocyclic uracil derivatives and their monophosphate prodrugs as potential antiviral agents. Viruses, **2023**; 15: 544.
[62] Liu J, Du J, Wang P, Nagarathnam D, Espiritu CL, Bao H, *et al.* A 2'-deoxy-2'-fluoro-2'-C-methyl uridine cyclopentyl carbocyclic analog and its phosphoramidate prodrug as inhibitors of HCV NS5B polymerase. Nucleosides Nucleotides Nucleic Acids **2012**; 31: 277–285.
[63] Julander JG, Demarest JF, Taylor R, Gowen B, Walling DM, Mathis A, *et al.* An update on the progress of galidesivir (BCX4430), a broad-spectrum antiviral. Antiviral Res **2021**; 195: 105180.
[64] Warren TK, Wells J, Panchal RG, Stuthman KS, Garza NL, Van Tongeren SA, *et al.* Protection against filovirus diseases by a novel broad-spectrum nucleoside analogue BCX4430. Nature **2014**; 508: 402–405.

[65] Julander JG, Bantia S, Taubenheim BR, Minning DM, Kotian P, Morrey JD, *et al.* BCX4430, a novel nucleoside analog, effectively treats yellow fever in a Hamster model. Antimicrob Agents Chemother **2014**; 58: 6607–6614.
[66] Evans GB, Furneaux RH, Hutchison TL, Kezar HS, Morris PE, Schramm VL, Tyler PC. Addition of lithiated 9-deazapurine derivatives to a carbohydrate cyclic imine: Convergent synthesis of the aza-C-nucleoside immucillins. J Org Chem **2001**; 66: 5723–5730.
[67] Montgomery JA, Morris PE. Deazaguanine analog, preparation thereof and use thereof. US6458799B1, **2002**
[68] Julander JG, Furuta Y, Shafer K, Sidwell RW. Activity of T-1106 in a hamster model of yellow Fever virus infection. Antimicrob Agents Chemother. **2007**; 51: 1962–1966.
[69] Taylor R, Bowen R, Demarest JF, DeSpirito M, Hartwig A, Bielefeldt-Ohmann H, *et al.* Activity of galidesivir in a hamster model of SARS-CoV-2. Viruses **2021**; 14: 8.
[70] Abdalla M, Rabie AM. Dual computational and biological assessment of some promising nucleoside analogs against the COVID-19-Omicron variant. Comput Biol Chem **2023;** 104: 107768.
[71] Evans GB, Furneaux RH, Gainsford GJ, Schramm VL, Tyler PC. Synthesis of transition state analogue inhibitors for purine nucleoside phosphorylase and N-riboside hydrolases. Tetrahedron **2000**; 56: 3053–3062.
[72] Spinola M, Colombo F, Falvella FS, Dragani TA. N6-isopentenyladenosine: A potential therapeutic agent for a variety of epithelial cancers. Int J Cancer **2007**; 120: 2744–2748.
[73] Rabie AM, Abdalla M. Forodesine and riboprine exhibit strong anti-SARS-CoV-2 repurposing potential: *In silico* and *in vitro* studies. ACS Bio Med Chem Au **2022**; 2: 565–585.
[74] Cho A, Saunders OL, Butler T, Zhang L, Xu J, Vela JE, *et al.* Synthesis and antiviral activity of a series of 1'-substituted 4-aza-7,9-dideazaadenosine C-nucleosides. Bioorg Med Chem Lett **2012**; 22: 2705–2707.
[75] Taha HR, Keewan N, Slati F, Al-Sawalha NA. Remdesivir: A closer look at its effect in COVID-19 pandemic. Pharmacology. **2021**; 106: 462–468.
[76] Gordon CJ, Lee HW, Tchesnokov EP, Perry JK, Feng JY, Bilello JP, *et al.* Efficient incorporation and template-dependent polymerase inhibition are major determinants for the broad-spectrum antiviral activity of remdesivir. J Biol Chem **2022**; 298: 101529.
[77] Siegel D, Hui HC, Doerffler E, Clarke MO, Chun K, Zhang L, Neville S, *et al.* Discovery and synthesis of a phosphoramidate prodrug of a pyrrolo[2,1-f][triazin-4-amino] adenine C-nucleoside (GS-5734) for the Treatment of Ebola and Emerging Viruses. J Med Chem **2017**; 60: 1648–1661.
[78] Warren TK, Jordan R, Lo MK, Ray AS, Mackman RL, Soloveva V, *et al.* Therapeutic efficacy of the small molecule GS-5734 against Ebola virus in rhesus monkeys. Nature. **2016**; 531: 381–385.
[79] Wang Y, Zhang D. *et al.* Remdesivir in adults with severe COVID-19: A randomised, double-blind, placebo-controlled, multicentre trial. Lancet **2020**; 395: 1569–1578.
[80] Rubin D, Chan-Tack K, Farley J, Sherwat A. FDA approval of remdesivir – a step in the right direction. N. Engl. J. Med. **2020**; 383: 2598–2600.
[81] Wang M, Zhang L, Huo X, Zhang Z, Yuan Q, Li P, Chen J, Zou Y, *et al.* Catalytic asymmetric synthesis of the anti-COVID-19 drug remdesivir. Angew Chem Int Ed Engl **2020**; 59: 20814–20819.
[82] Gordon CJ, Lee HW, Tchesnokov EP, Perry JK, Feng JY, Bilello JP *et al.* Efficient incorporation and template-dependent polymerase inhibition are major determinants for the broad-spectrum antiviral activity of remdesivir. J Biol Chem **2022**; 298: 101529.
[83] Cardoza S, Shrivash MK, Riva L, Chatterjee AK, Mandal A, Tandon V. Multistep synthesis of analogues of remdesivir: Incorporating heterocycles at the C-1' position. J Org Chem **2023**, 88, 9105.
[84] Xie Y, Yin W, Zhang Y, Shang W, Wang Z, Luan X, *et al.* Design and development of an oral remdesivir derivative VV116 against SARS-CoV-2. Cell Res. **2021**; 31: 1212–1214.
[85] Zhang Y, Sun Y, Xie Y, Shang W, Wang Z, Jiang H, *et al.* A viral RNA-dependent RNA polymerase inhibitor VV116 broadly inhibits human coronaviruses and has synergistic potency with 3CLpro inhibitor nirmatrelvir. Sig Transduct Target Ther **2023**; 8: 360.
[86] Furuta Y, Komeno T, Nakamura T. Favipiravir (T-705), a broad spectrum inhibitor of viral RNA polymerase. Proc Jpn Acad Ser B Phys Biol Sci **2017**; 93: 449–463.
[87] Baranovich T, Wong SS, Armstrong J, Marjuki H, Webby RJ, Webster RG, Govorkova EA. T-705 (favipiravir) induces lethal mutagenesis in influenza A H1N1 viruses in vitro. J Virol **2013**; 87: 3741–3751.
[88] Furuta Y, Takahashi K. Nitrogenous heterocyclic carboxamide derivatives or salts thereof and antiviral agents containing both. WO 00/10569, **2001**.
[89] Shi F, Li Z, Kong L, Xie Y, Zhang T, Xu W. Synthesis and crystal structure of 6-fluoro-3-hydroxypyrazine-2-carboxamide. Drug Discov Ther **2014**; 3: 117–120.
[90] Shirakia K, Daikoku T. Favipiravir, an anti-influenza drug against life-threatening RNA virus infections. Pharmacol. Ther. **2020**; 209: 107512.

[91] Zhao L, Zhong W. Mechanism of action of favipiravir against SARS-CoV-2: Mutagenesis or chain termination? Innovation (Camb). **2021**; 2: 100165.

[92] Wang G, Wan J, Hu Y, Wu X, Prhavc M, Dyatkina N, Rajwanshi VK, *et al.* Synthesis and anti-influenza activity of pyridine, pyridazine, and pyrimidine C-nucleosides as Favipiravir (T-705) analogues. J Med Chem **2016**; 59: 4611–4624.

[93] Marlin R, Desjardins D, Contreras V, *et al.* Antiviral efficacy of favipiravir against Zika and SARS-CoV-2 viruses in non-human primates. Nat Commun **2022**; 13: 5108.

[94] Song GY, Paul V, Choo H, Morrey J, Sidwell RW, Schinazi RF, Chu CK. Enantiomeric synthesis of D- and L-cyclopentenyl nucleosides and their antiviral activity against HIV and West Nile virus. J Med Chem. **2001**; 44: 3985–3993.

[95] Cho JH, Bernard DL, Sidwell RW, Kern ER, Chu CK. Synthesis of cyclopentenyl carbocyclic nucleosides as potential antiviral agents against orthopoxviruses and SARS. J Med Chem **2006**; 49: 1140–1148.

[96] Liu C, Chen Q, Schneller SW. 3-Bromo-3-deazaneplanocin and 3-bromo-3-deazaaristeromycin: Synthesis and antiviral activity. Bioorg Med Chem Lett **2012**; 22: 5182–5184.

[97] Liu C, Chen Q, Gorden JD, Schneller SW. 2- and 3-Fluoro-3-deazaneplanocins, 2-fluoro-3-deazaaristeromycins, and 3-methyl-3-deazaneplanocin: Synthesis and antiviral properties. Bioorg Med Chem **2015**; 23: 5496–5501.

[98] Wu R, Smidansky ED, Oh HS, Takhampunya R, Padmanabhan R, Cameron CE, Peterson BR. Synthesis of a 6-methyl-7-deaza analogue of adenosine that potently inhibits replication of polio and dengue viruses. J Med Chem **2010**; 53: 7958–7966.

[99] Yoon JS, Kim G, Jarhad DB, Kim HR, Shin YS, Qu S, *et al.* Design, Synthesis, and anti-RNA virus activity of 6'-fluorinated-aristeromycin analogues. J Med Chem. **2019**; 62: 6346–6362.

[100] Turner MA, Yang X, Yin D, Kuczera K, Borchardt RT, Howell PL. Structure and function of S-adenosylhomocysteine hydrolase. Cell Biochem Biophys **2000**; 33: 101–125.

[101] Qi C, Schneller SW, Liu C, Jones KL, Singer T. 5'-Nor-3-Deaza-1',6'-Isoneplanocin, the Synthesis and Antiviral Study. Molecules **2020**; 25: 3865.

[102] Matyugina ES, Khandazhinskaya AL. 5′-Norcarbocyclic nucleoside analogs. Russ Chem Bull **2014**; 63: 1069–1080.

[103] Matyugina ES, Valuev-Elliston VT, Babkov DA, Novikov MS, Ivanov AV, Kochetkov SN, Balzarini J, Seley-Radtke KL, Khandazhinskaya AL. 5'-Nor carbocyclic nucleosides: Unusual nonnucleoside inhibitors of HIV-1 reverse transcriptase. Med Chem Comm **2013**; 4: 741–748.

[104] Liu C, Coleman R, Archer A, Hussein I, Bowlin TL, Chen Q, Schneller SW. Enantiomeric 4'-Truncated 3-deaza-1',6'-isoneplanocins: Synthesis and antiviral properties including Ebola. Bioorg Med Chem Lett **2019**; 29: 2480–2482.

[105] Keane SJ, Ford A, Mullins ND, Maguire NM, Legigan T, Balzarini J, Maguire AR. Design and synthesis of α-carboxy nucleoside phosphonate analogues and evaluation as HIV-1 reverse transcriptase-targeting agents. J Org Chem **2015**; 80: 5, 2479–2493.

[106] Rosenblum LL, Patton G, Grigg AR, Frater AJ, Cain D, Erlwein O, Hill CL, Clarke JR, McClure MO. Differential susceptibility of retroviruses to nucleoside analogues. Antiviral Chem Chemother **2001**; 12: 91–97.

[107] Shen GH, Hong JH. Recent advances in the synthesis of cyclic 5'-nornucleoside phosphonate analogues. Carbohydr Res. **2018**; 463: 47–106.

[108] Liu LJ, Kim E, Hong JH. Synthesis and some properties of 4'-phenyl-5'-norcarbocyclic adenosine phosphonic acid analogues. Bull. Korean Chem Soc **2011**; 32: 1662–1668.

[109] Kim S, Kim E, Yoo JC, Hong JH. Efficient synthesis of novel 4'-trifluoromethyl-5'-norcarbocyclic purine phosphonic acid analogs by using the Ruppert-Prakash reaction. Bull Korean Chem Soc **2014**; 35: 2743–2748.

[110] Li H, Kim W, Hong JH. Antiviral activity enhancement through the SATE prodrug of a 2'-modified 5'-norcarbocyclic adenine analogue. Bull Korean Chem Soc **2010**; 31: 2180–2184.

[111] Roy A, Schneller SW. 4'- and 1'-methyl-substituted 5'-norcarbanucleosides. J Org Chem **2003**; 68: 9269–9273.

[112] Yang M, Serbessa T, Schneller SW. The 3-deaza and 7-deaza derivatives of 5'-amino-5'-deoxy-5'-noraristeromycin. Collect Czech Chem Commun **2006**; 71: 1122–1129.

[113] Maguire NM, Ford A, Balzarini J, Maguire AR. Synthesis of guanine α-carboxy nucleoside phosphonate (G-α-CNP), a direct inhibitor of multiple viral DNA polymerases. J Org Chem **2018**; 83: 10510–10517.

[114] Zimmermann SC, Sadler JM, Andrei G, Snoeck R, Balzarinib J, Seley-Radtke KL. Carbocyclic 5'-nor "reverse" fleximers. Design, synthesis, and preliminary biological activity. Med Chem Commun **2011**; 2: 650–654.

7 Synthesizing Nanocatalysts with Carbohydrates

Farahnaz K. Behbahani

7.1 INTRODUCTION

Researchers began investigating nanocatalysis using nanoparticles (NPs) as homogeneous or heterogeneous catalysts within the last two decades [1–5]. The large surface-to-volume ratio of NPs compared with bulk materials makes them attractive candidates for use as catalysts. An important challenge has been the preparation and characterization of NPs from noble metals, which constitutes an important branch of heterogeneous catalysis in the chemical industry. Their very high superficial area make them excellent catalysts, reducing the amount of catalyst per gram of product and making the process more sustainable. These NPs are synthesized by chemical methods in good amounts with controlling their sizes and shapes. However, in most cases, synthesis uses hazardous conditions, toxic solvents, and high amounts of energy (200 °C), which reduce possible industrial applications. Most methods are still in the development stage, and problems are often experienced with stability of the prepared NPs, such as controlling crystal growth and particle aggregation [6–8]. Therefore, designing new synthesis approaches that are easy, rapid, and sustainable is critical.

In recent years, magnetic nanoparticles (MNPs), have proven to be valuable assets in organic chemistry such as for the synthesis of N-substituted pyrroles [9]. benzo [c] acridine-8(9 H)-ones and 2-amino-4H-chromenes [10], 8-aryl-7H-acenaphtho [1, 2-d]imidazole's [11], 2-substituted benzimidazoles [12], 2-arylbenzimidazoles [13], 2,6-diamino-4-arylpyridine–3,5-dicarbonitriles [14], and 2-amino-4H-chromene [15].

Carbohydrates, also known as saccharides, are one of the most abundant biomolecules. They are poly-hydroxylated aldehydes or ketones and can exist as monosaccharides, disaccharides, oligosaccharides, and polysaccharides. The mono- and disaccharides, which are smaller carbohydrates, are generally referred to as sugars. Carbohydrates are synthesized by plants through photosynthesis (Scheme 7.1); in the presence of light, CO_2 reacts with water to give oxygen and carbohydrates, with the general formula $C_m(H_2O)n$ [16].

Carbohydrates contain a large variety of monosaccharide units with backbones of varying length. The monosaccharides have carbonyl groups, either as aldehydes or ketones, as well as primary and secondary alcohols. The sequence of stereogenic centers in the monosaccharide influences how it occupies a 3-dimensional space. Each monosaccharide has its own enantiomer with respect to the most distant stereogenic center from the carbonyl group. The prefixes D and L are assigned to distinguish between the two enantiomers (Scheme 7.2). In the D form, the hydroxyl group (the higher-priority group) on the last chiral carbon of the open chain Fischer projection is on the right side, whereas in the L form, the hydroxyl group is on the left [17].

The open chain form of monosaccharides can undergo intramolecular reaction; a nucleophilic attack of one of the hydroxyl groups at the intrachain aldehyde or ketone results in the formation of a cyclic hemiacetal or hemiketal, respectively. The most commonly formed ring structures are furanoses (five-membered rings) and pyranoses (six-membered rings), as shown in Scheme 7.3. The newly formed stereocenter is called the anomeric center. The two possible stereo chemistries formed at the anomeric center are called anomers, which are diastereoisomers of each other. The configuration at the anomeric carbon is termed alpha- (α-) or beta- (β-) with respect to the stereocenter that

DOI: 10.1201/9781003437413-7

$$CO_2 + H_2O \xrightarrow{\text{Light}} Cm(H_2O)n + O_2$$

SCHEME 7.1 Photosynthesis.

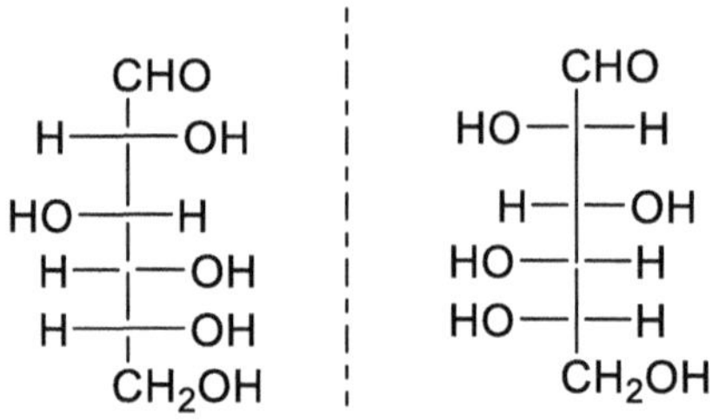

SCHEME 7.2 Fisher projections of D-glucose (left) and L-glucose (right).

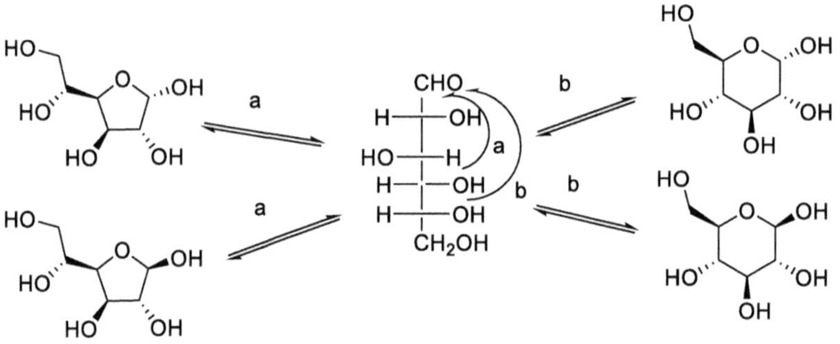

SCHEME 7.3 Formation of furanose and pyranose from D-glucose open chain form.

decides the absolute configuration (D or L form). As shown in Scheme 7.3, cyclic sugars can be represented in different ways such as Mills, Fischer, Haworth, and chair representations. In a Fischer projection, if the hydroxyl group at the anomeric center is on the same side as the hydroxyl group of the configurational (D- or L-) carbon, it is the α-anomer of the sugar. If it is in the opposite side, then it is the β-anomer. In general, the functionalization of the anomeric hydroxyl group blocks ring opening and closing mechanisms, locking the ring in its cyclic conformation [17].

In other works, researchers have synthesized a variety of nanocatalysts in the presence of carbohydrates, and I focus on such applications to synthesize nanocatalysts in this chapter. The requirement of green and sustainable materials to prepare heterogeneous catalysts has intensified for practical reasons in recent decades. Carbohydrates are possibly the most plentiful and renewable organic materials in nature, with inimitable physiochemical properties, plausible low-cost and large-scale production, and sustainability features, all of which could be exploited in the generation of nanostructured heterogeneous catalysts based on the capacity of biomolecules to form these nanoparticles, control particle size and structural form, and avoid aggregation [18–28]. Researchers used small molecules such as monosaccharides (glucose or galactose), amino acids, and short peptides as reducing agents for in situ creation of these metallic NPs [18–22].

In one study, zinc oxide (ZnO) NPs were synthesized using maltose, dextrose, and fructose extract for the first time by precipitation; the different carbohydrate sources were effective reducing and stabilizing agents. The biologically synthesized ZnONPs were characterized using scanning electron microscope (SEM), UV–visible spectrophotometer (UV), Fourier transform infrared (FTIR) spectra, X-ray diffraction (XRD), zeta potential, and particle size analysis. The UV results indicated the formation of ZnONPs around 200 nm–400 nm. FTIR studies showed the presence of various functional groups responsible for the synthesis of ZnONPs. Morphological analysis using SEM revealed spherical and cube-shaped ZnONPs, and XRD analysis revealed the high crystalline nature of the synthesized nanomaterials; zeta potential showed the high stability of the synthesized nanomaterials. To determine the antimicrobial activity, the synthesized ZnONPs were treated against Gram-positive and Gram-negative *Staphylococcus epidermidis* and *Enterobacter aerogenes,* and the results showed high activity of ZnONPs synthesized from fructose. The study highlighted the high antimicrobial activity of biologically synthesized ZnONPs [29].

Ecofriendly biosynthesis of silver NPs (AgNPs) [30] using plant extracts is also an exciting advancement in bio-nanotechnology and was successfully attempted in nearly 41 plant species; however, an established model plant system for systematically unraveling the biochemical components required for AgNP production is lacking. Researchers used *Arabidopsis thalianaas*, the model plant for in vitro AgNP biosynthesis. Employing biochemistry and spectroscopy to analyze selected mutants and overexpressor *Arabidopsis* plants involved in pleotropic functions and sugar homeostasis showed that carbohydrates, polyphenolics, and glycoproteins were essential for AgNP synthesis. Molecular genetics data enforced sugar-conjugated proteins as essential for AgNP, in contrast with protein alone. Additionally, researchers performed a comparative analysis of AgNP synthesis using the aqueous extracts of some of the plant species found in a brackish water ecosystem (*Gracilaria*, *Potamogeton*, Enteromorpha, and *Scenedesmus*). Plant extract of *Potamogeton* showed the highest NP production, comparable with that of *Arabidopsis*; producing AgNPs in Arabidops opens the possibility of using molecular genetics to understand biochemical pathways and components in detail.

7.2 SYNTHESIS OF METAL NPs WITH GLUCOSIDESIS

Researchers explored the green synthesis of metal NPs; using glycosides to synthesize metal NPs avoids the need for toxic materials [31]. Specifically, they used glucose and different glycopyranosides to successfully synthesize gold NPs (AuNPs) and found very interesting results using eight different glucose derivatives. The scientists developed an easy, room temperature method of synthesizing the AuNPs using auric acid, sodium hydroxide, and different glycosides as reducing agents (Scheme 7.4).

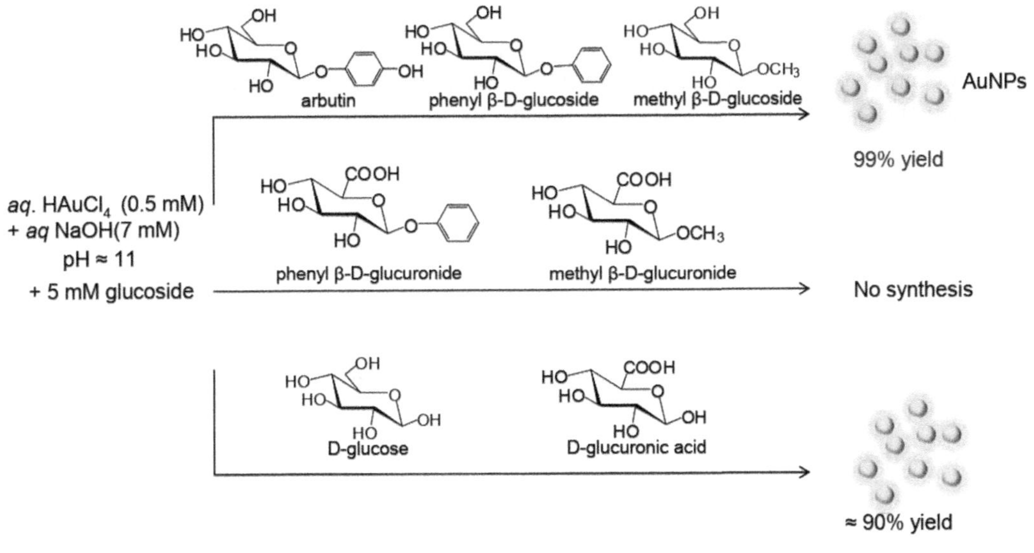

SCHEME 7.4 Synthesis of AuNPs induced by different glucose derivatives.

The researchers used eight sugar-containing reductants for comparison, oxidizing the C-6 positions of the glycosides to carboxylic acid during the reduction of auric acid in forming the AuNPs and substituting sugars at the anomeric carbon position (Scheme 7.4). With glucose and glucuronic-acid (COOH in C-6), NPs formed due to the oxidation of an aldehyde generated via nanomerization; this explains why the synthesis in the presence of the glucuronic acid derivatives substituted at the anomeric position did not work (Scheme 7.5). Furthermore, significant differences in the final yield in AuNPs, especially in NP form and size, were observed on high-resolution transmission electron microscopy (HRTEM; 8 nm to 27 nm) and depended on the sugar derivative. The use of 1-phenyl or 1-methyl-β-glucopyranoside gave the highest synthetic yield (>99%) of homogeneous monodispersed round AuNPs (13.15 nm and 10.95 nm, respectively), whereas with glucose or glucuronic acid, the synthesized AuNPs showed multiple forms (16 nm and 8.8 nm). In the presence of arbutin, the NPs were amorphous (27.4 nm) (Figure 7.1) [31].

The authors of this work showed how AuNP size and form can be controlled using different sugars as additives. This aspect is of special interest for example in the case of ultrasmall NPs, where their colloidal stability can be affected by these small size differences, underscoring the importance of particle uniformity in nanomedicine. These strategies can also be extended with other metals and also using other sugar derivatives with different degrees of hydrophobicity by substitution in the anomeric and other positions [31]. They synthesized AgNPs using aqueous extract of *Clerodendron serratum* leaves, which had high polyphenol glycoside and quercetin 3-**O**-β-D glucoside content. This glucose derivative reduced the Ag ions of silver nitrate in 30 min at room temperature, forming spherical AgNPs with the size range of 5 nm–30 nm (Figure 7.2A). In addition, the researchers successfully used glycosides to prepare mesoporous silica-coated AgNPs (Ag@MSN) in two steps. Glucose was used as reducing agent for Ag ions, and arginine was used for the formation of silica. The nanostructure presented single AgNPs as cores of diameter ca. 30 nm surrounding a silica shell (thickness 30 nm–40 nm) with a total average particle size of around 110 nm (Figure 7.2B).

Another interesting example was the in situ creation of glycosylated functionalized AuNPs [31]. In this case, the glycoside acted as a reducing agent but also was a specific moiety for particle functionalization. A glycopolymer (cellulose) was activated in the anomeric position by thiosemicarbazide, producing glucoside thiosemicarbazone (Figure 7.3A). This activated sugar combined with aqueous N-methylmorpholine N-oxide (NMMO)-a permitted the solubilization of the water-insoluble cellulose and a dilute aqueous HAuCl solution, finally producing glyco-AuNPs. TEM analysis confirmed the formation of AuNP aggregates around 10 nm–20 nm in diameter (Figure 7.3B). The same protocol was used to successfully synthesize glyco-AgNPs around 5 nm–30 nm in diameter. The combination of NMMO-mediated GNP synthesis and immobilization of sugar reducing ends to an Au matrix allowed the design of a diverse array of carbohydrate-GNP conjugates by tailoring the functional sugars cellulose, chitin, chitosan, maltose, and lactose.

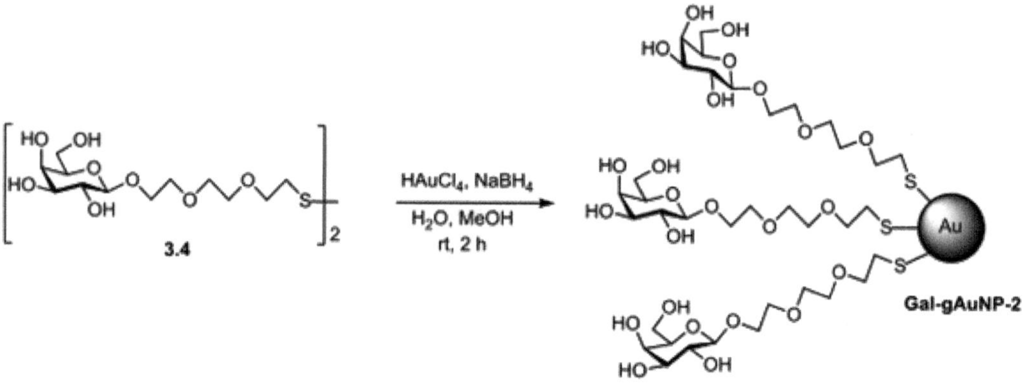

3.4

$HAuCl_4$, $NaBH_4$

H_2O, MeOH

rt, 2 h

Gal-gAuNP-2

SCHEME 7.5 Synthesis of Gal.gAuNP-2 following a modified version of BSM.

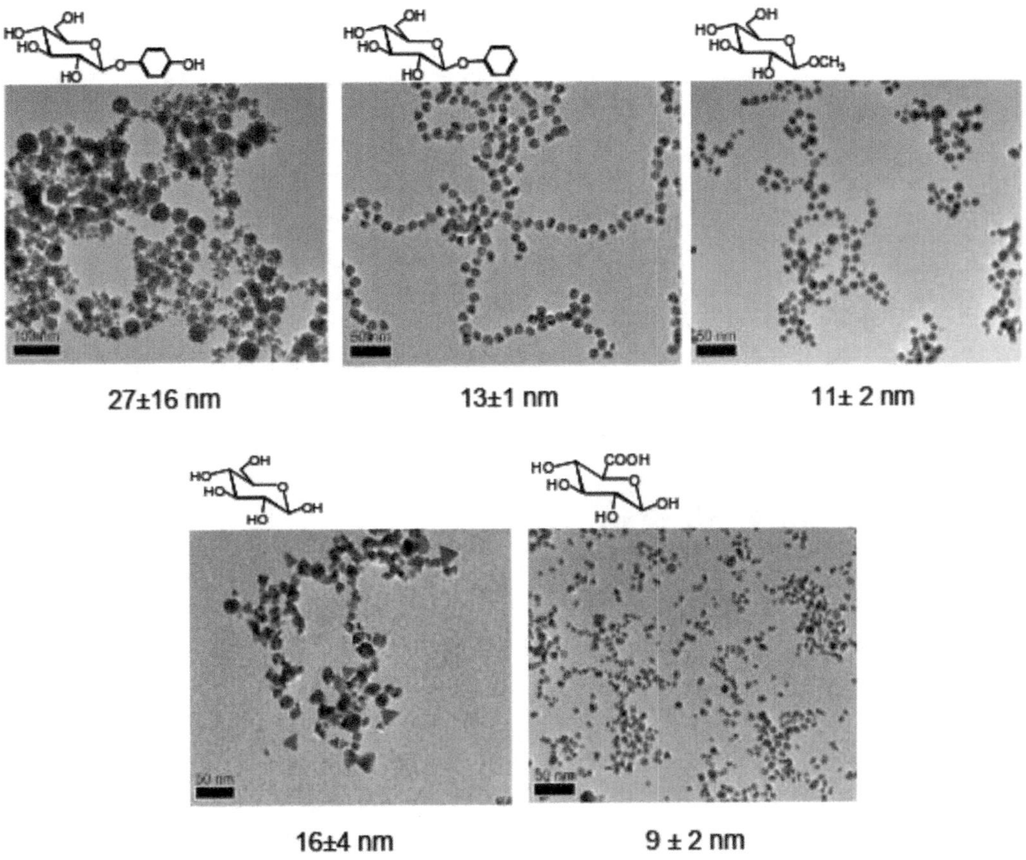

27±16 nm 13±1 nm 11± 2 nm

16±4 nm 9 ± 2 nm

FIGURE 7.1 Characterization of glucoside-induced AuNPs by HRTEM.

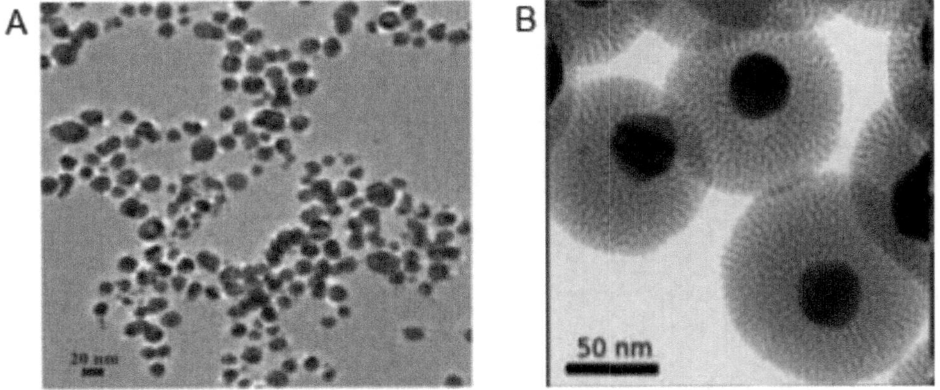

FIGURE 7.2 TEM images of (A) AgNPs and (B) Ag@MSN coated on a silicon substrate.

In another study, glyco-AuNPs [32] combined in a single entity the peculiar properties of AuNPs with the biological activity of carbohydrates. The result was an exciting nanosystem that mimicked the natural multivalent presentation of saccharide moieties and exploited the peculiar optical properties of the metallic core. In short, advances in glyco-AuNP applications in different biological fields are highlighting the key parameters that inspired the glycol-NP design (Figure 7.4). Another

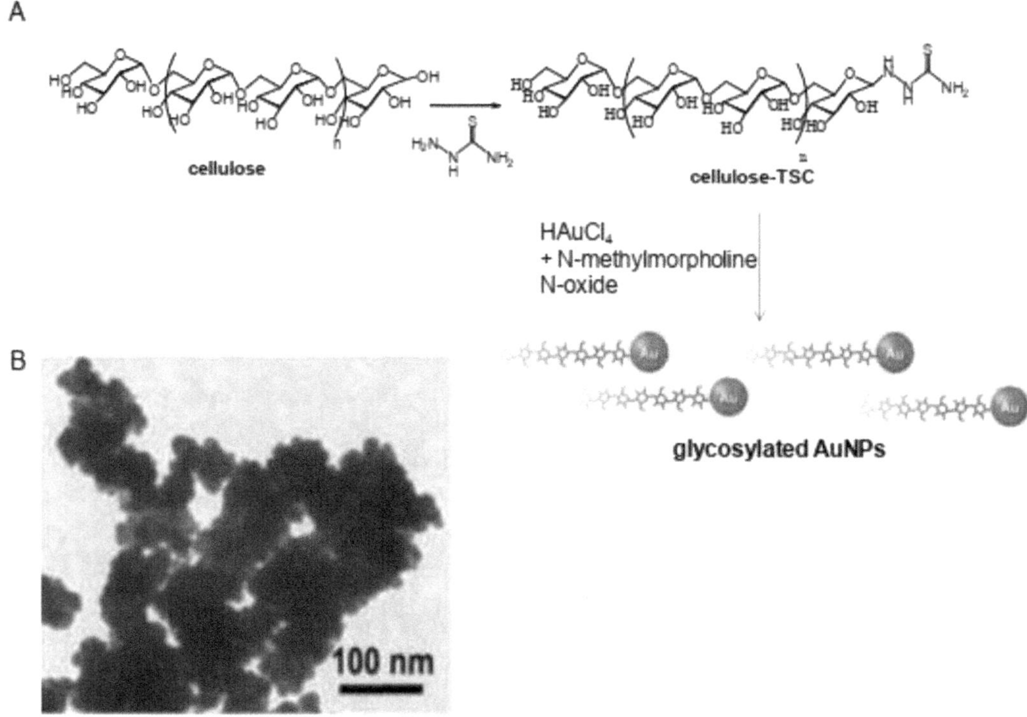

FIGURE 7.3 Synthesis of glycosylated AuNPs: (A) synthesis scheme; (B) TEM image of glyco-AgNPs.

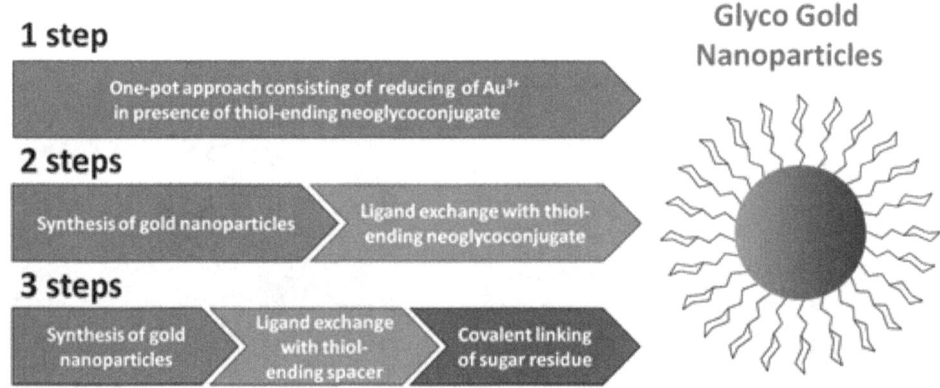

FIGURE 7.4 Schematic overview of synthesizing glyco-AuNPs as nanosensors.

study focused on the synthesis of AuNPs coated with carbohydrates. The authors first introduce the main protocols for producing glycol-AuNPs and then the main parameters that should be taken into account for designing the optimal nanosystems. Then, three sections summarize the achievements and ongoing research published from 2013 to 2016 with a special focus on glyco-AuNPs applications (in biosensing, as drug carriers or contrast agents, and as generally effective immunology tools) (Figure 7.5).

Compostella et al. [32]. used plasmon resonance spectroscopy and the naked eye to study the interactions of galactoside-AuNPs with peanut agglutinin (Figure 7.6); the glycosyl residue was not coated on the gold surface by a thiol moiety but exploited strain-promoted azide–alkyne

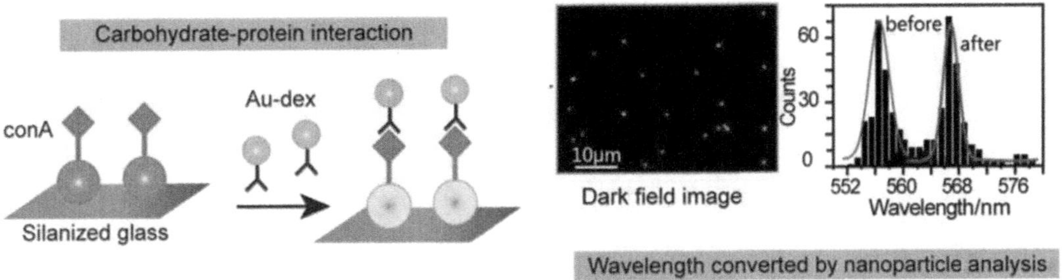

FIGURE 7.5 Protein–carbohydrate interactions between ConA-functionalized AuNPs (60 nm) and dextran-coated AuNPs (20 nm) analyzed by conventional dark field microscopy.

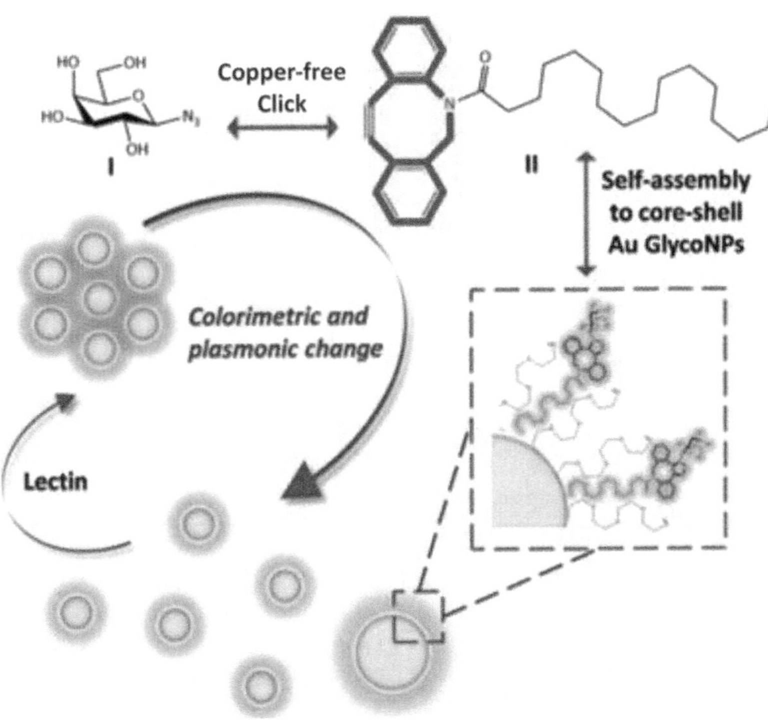

FIGURE 7.6 Copper-free cycloaddition of azido galactoside on cyclooctyne and schematic depiction of Au surface coating. When galactoside-coated AuNPs interact with lectin, the NPs aggregate, inducing a plasmonic band shift. Reprinted with permission from Compostella and coworkers [32].

cycloaddition to covalently link an azido galactoside on a lipid cyclooctyne. In this manner, the amphiphilic glycolipid could be embedded on PEGylated AuNPs, and the sugar moiety displayed on the outer shell of the nanosystem was available for interaction with lectin.

In 2016, Compostella and coworkers [32] combined the multiple display of the carbohydrate moiety on AuNPs with the photoaffinity labeling to concentrate and purify proteins by covalent crosslinking (Figure 7.7).

A very recent and interesting application showed for the first time the immobilization of AuNPs on the end face of an optical fiber by means of thiochemistry. A second layer of glycosyl residues (maltose or lactose) containing thioctic acid was covalently linked on supported AuNPs, affording

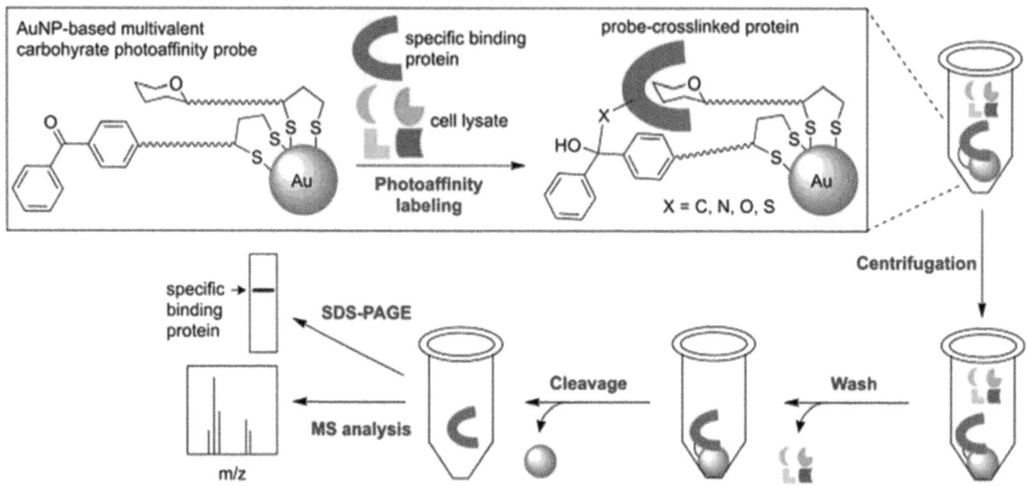

FIGURE 7.7 Photoaffinity labeling based on glyco-AuNPs is able to recognize and isolate only the desired protein. After washing and cleavage, the protein can be fully analyzed. Reprinted from Compostella and coworkers [32].

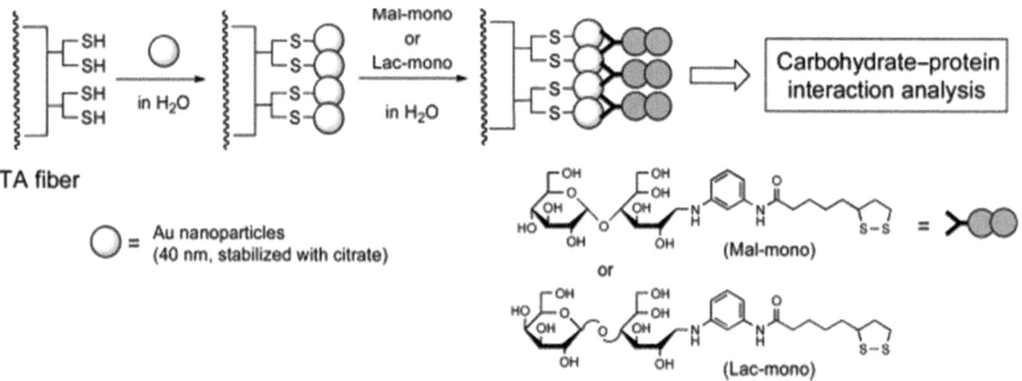

FIGURE 7.8 Fiber-type sugar chips. AuNPs were first immobilized on an optical fiber functionalized with α-lipoic acid; successively, glycosil-thioctic residues were grafted on AuNPs, forming the sugar chips.

fiber-type "sugar chips" (Figure 7.8) that allowed for analyzing and quantifying the carbohydrate–protein interactions in very small volumes with results confirmed by conventional localized surface plasmon resonance biosensors.

Glyco-AuNPs as drug carriers and contrast agents. Anticancer therapy based on platinum compounds such as cisplatin, carboplatin, and oxaliplatin induces the regression of cancer cell proliferation. Unfortunately, these compounds are poorly tolerated; therefore, new metal-based anticancer drugs need to be developed. Recently, researchers reported on the interesting synthesis of AuNPs coated with glycopolymers and functionalized with gold(I) triphenylphosphine (Figure 7.9).

Glyco-AuNPs are typically synthesized by attaching carbohydrates to the NP surfaces via thiol-terminated linkers. They were used to detect concanavalin A, *Ricinus communis* agglutinin 120, cholera toxin, and viral hemagglutinin [33]. Researchers are actively seeking facile methods of synthesizing AuNPs and glyco-AuNPs and demonstrations of their effective use as colorimetric sensors for toxic proteins. This group used the Turkish reaction to synthesize AuNPs of various shapes

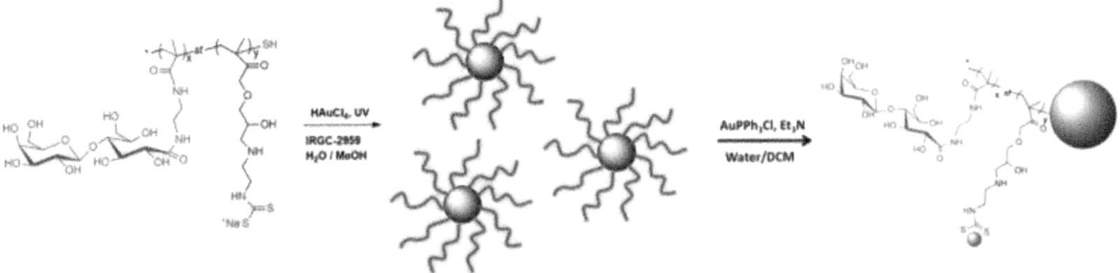

FIGURE 7.9 Glyco-AuNPs displaying both sugar residues and anticancer Au(I)PPh_3. Reprinted with permission from Compostella and coworkers [32].

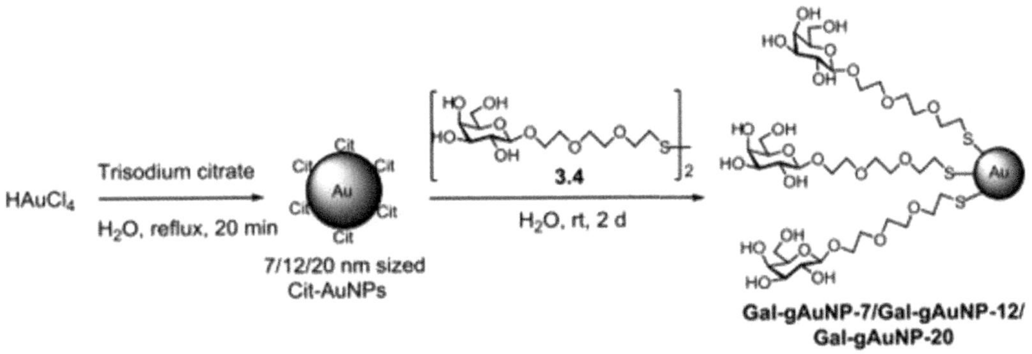

SCHEME 7.6 Synthesis of Gal.gAuNP-7/Gal-gAuNP-12/Gal-gAuNP-20 by ligand exchange.

and sizes, and for glyco-AuNPs, they used Brust-Schiffrin method, ligand exchange, and copper (I)-catalyzed azide alkyne Huisgen cycloaddition. The authors presented galactose-capped AuNPs and full-length sialic acid-terminated complex bi-antennary N-glycan-capped AuNPs as colorimetric sensors for detecting lectin heat-labile enterotoxin and influenza viral particles (Schemes 7.5 and 7.6).

Increasing use of AgNPs) in various products is resulting in a greater likelihood of human exposure to these materials. Nevertheless, little is still known about the influence of carbohydrates on the toxicity and cellular uptake of nanoparticles. AgNPs functionalized with three different monosaccharides and ethylene glycol were synthesized and characterized. Oxidative stress and toxicity were evaluated by protein carbonylation and MTT assay, respectively. Cellular uptake was evaluated by confocal microscopy and inductively coupled plasma-mass spectrometry. AgNPs coated with galactose and mannose were considerably less toxic to neuronal-like cells and hepatocytes than particles functionalized by glucose, ethylene glycol, or citrate; additionally, toxicity correlated to oxidative stress but not to cellular uptake. Carbohydrate coating on AgNPs modulated both oxidative stress and cellular uptake, but the first had the main impact on toxicity. These findings provide new perspectives on modulating the bioactivity of AgNPs using carbohydrates (Figure 7.10) [34].

Researchers [35] outlined the organic transformations catalyzed by diverse carbohydrate-based nanostructured catalysts in green, environmentally friendly processes. They highlighted selected examples for a variety of organic reactions exploiting the proposed catalysts' reactivity and reusability, presenting the intrinsic natures of the applied carbohydrate supports as well as the advantages and challenges of the introduced catalysts (Scheme 7.7).

Starch NPs have been demonstrated as a highly efficient, ecofriendly, and heterogenous organocatalyst for the synthesis of diheteroaryl thioethers via one-pot reaction of methylcarbonyls, thiourea, and iodine in DMSO [36]. This method offered significant advantages such as available

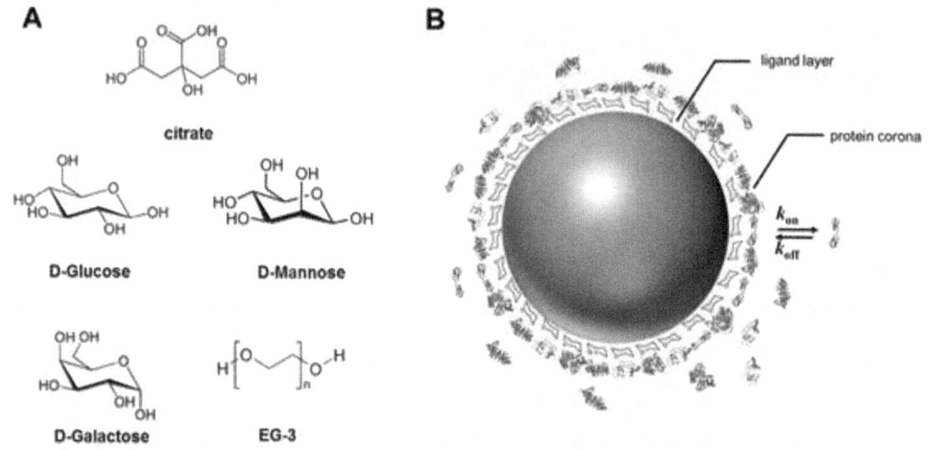

FIGURE 7.10 AgNPs functionalized with three different monosaccharides and ethylene glycol were synthesized and characterized.

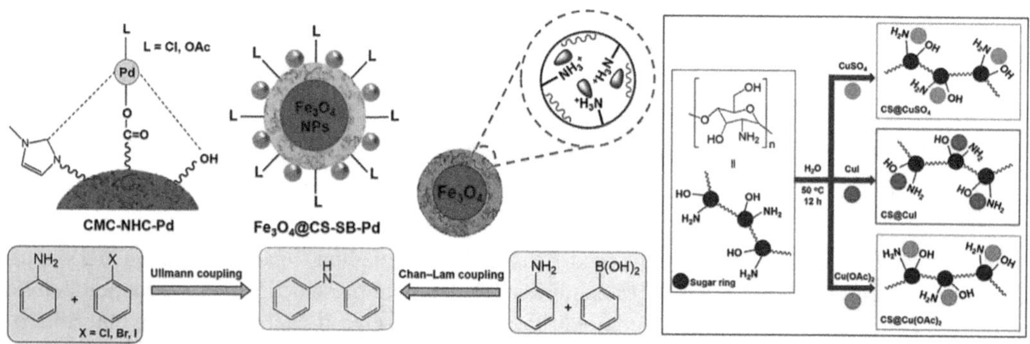

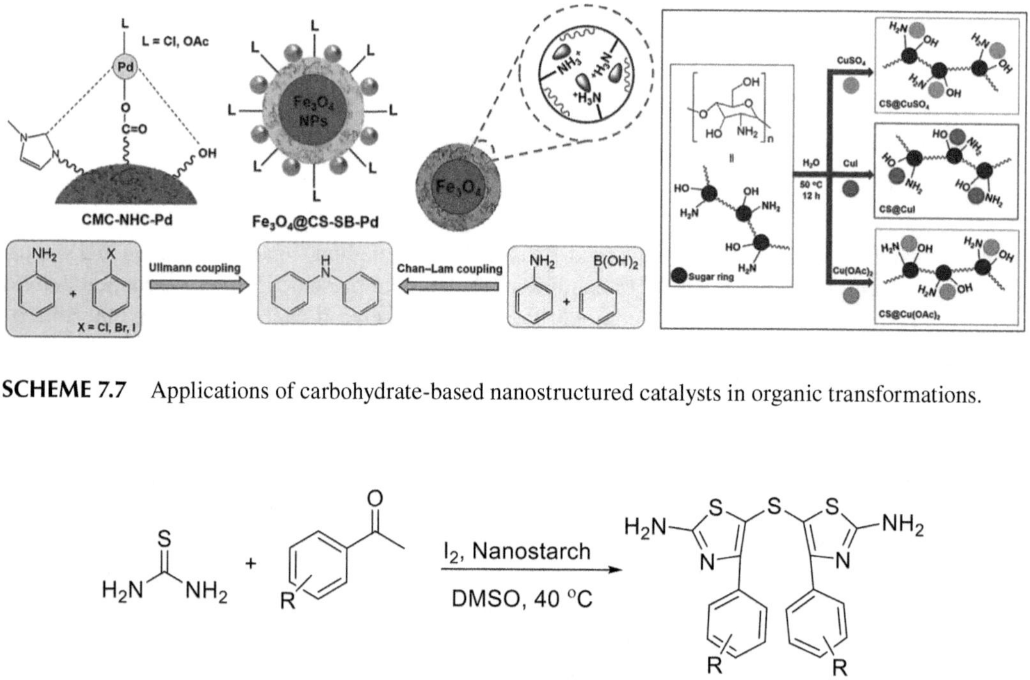

SCHEME 7.7 Applications of carbohydrate-based nanostructured catalysts in organic transformations.

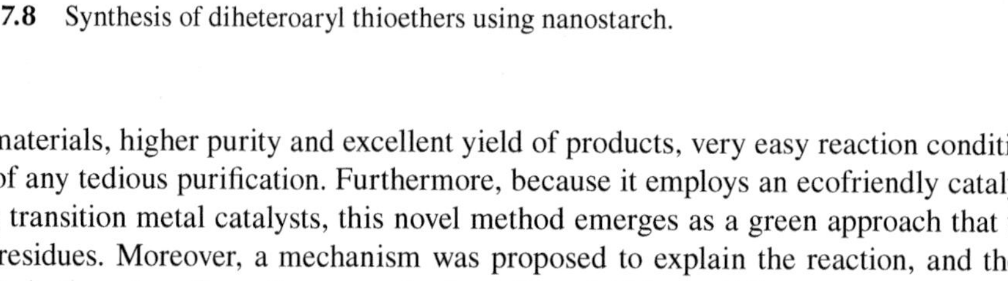

SCHEME 7.8 Synthesis of diheteroaryl thioethers using nanostarch.

starting materials, higher purity and excellent yield of products, very easy reaction conditions, and absence of any tedious purification. Furthermore, because it employs an ecofriendly catalyst without using transition metal catalysts, this novel method emerges as a green approach that uses less harmful residues. Moreover, a mechanism was proposed to explain the reaction, and the role of starch NPs in these transformations was also investigated (Scheme 7.8).

7.3 CONCLUSION

Nanoparticles have drawn significant interest in recent years due to their unique optical, electronic, and molecular recognition properties. Nanoparticles are three-dimensional multivalent systems based on carbohydrate decorated NPs; glyco-NPs are typically synthesized by attaching carbohydrates to the surface of NPs via terminated linkers. In this chapter, I highlighted the synthesis of NPs of various shapes and sizes using saccharides, which I hope will be of benefit for all readers of this book.

REFERENCES

[1] Dong, X.Y., Gao, Z.W., Yang, K.F., Zhang, W.Q. and Xu, L.W., 2015. Nanosilver as a new generation of silver catalysts in organic transformations for efficient synthesis of fine chemicals. *Catalysis Science & Technology*, *5*(5), pp.2554–2574.

[2] Filice, M. and Palomo, J.M., 2014. Cascade reactions catalyzed by bionanostructures. *ACS Catalysis*, *4*(5), pp.1588–1598.

[3] Serna, P. and Corma, A., 2015. Transforming nano metal nonselective particulates into chemoselective catalysts for hydrogenation of substituted nitrobenzenes. *ACS Catalysis*, *5*(12), pp.7114–7121.

[4] Mishra, K., Basavegowda, N. and Lee, Y.R., 2015. Biosynthesis of Fe, Pd, and Fe-Pd bimetallic nanoparticles and their application as recyclable catalysts for [3+ 2] cycloaddition reaction: a comparative approach. *Catalysis Science & Technology*, *5*(5), pp.2612–2621.

[5] Aditya, T., Pal, A. and Pal, T., 2015. Nitroarene reduction: a trusted model reaction to test nanoparticle catalysts. *Chemical Communications*, *51*(46), pp.9410–9431.

[6] MubarakAli, D., Gopinath, V., Rameshbabu, N. and Thajuddin, N., 2012. Synthesis and characterization of CdS nanoparticles using C-phycoerythrin from the marine cyanobacteria. *Materials Letters*, *74*, pp.8–11.

[7] Saldan, I., Semenyuk, Y., Marchuk, I. and Reshetnyak, O., 2015. Chemical synthesis and application of palladium nanoparticles. *Journal of Materials Science*, *50*, pp.2337–2354.

[8] Gutiérrez, L., Costo, R., Grüttner, C., Westphal, F., Gehrke, N., Heinke, D., Fornara, A., Pankhurst, Q.A., Johansson, C., Veintemillas-Verdaguer, S. and Morales, M.D.P., 2015. Synthesis methods to prepare single-and multi-core iron oxide nanoparticles for biomedical applications. *Dalton Transactions*, *44*(7), pp.2943–2952.

[9] Shokri, F. and Behbahani, F.K., 2022. Synthesis of Fe3O4@ L-proline@ SO3H as a novel and reusable acidic magnetic nanocatalyst and its application for the synthesis of N-substituted pyrroles at room temperature under ultrasonic irradiation and without solvent. *Inorganic and Nano-Metal Chemistry*, *52*(8), pp.1143–1152.

[10] Karimirad, F. and Behbahani, F.K., 2020. γ-Fe2O3@ Si-(CH2) 3@ mel@(CH2) 4SO3H as a magnetically bifunctional and retrievable nanocatalyst for green synthesis of benzo [c] acridine-8 (9 H)-ones and 2-amino-4 H-chromenes. *Inorganic and Nano-Metal Chemistry*, *51*(5), pp.656–666.

[11] Hasanzadeh, F. and Behbahani, F.K., 2020. Synthesis of 8-Aryl-7 H-acenaphtho [1, 2-d] imidazoles using Fe 3 O 4 NPs@ GO@ C 4 H 8 SO 3 H as a green and recyclable magnetic nanocatalyst. *Russian Journal of Organic Chemistry*, *56*, pp.1070–1076.

[12] Behbahani, F.K., Rezaee, E. and Fakhroueian, Z., 2014. Synthesis of 2-substituted benzimidazoles using 25% Co/Ce-ZrO 2 as a heterogeneous and nanocatalyst. *Catalysis Letters*, *144*, pp.2184–2190.

[13] Behbahani, F.K., Ziaei, P., Fakhroueian, Z. and Doragi, N., 2011. An efficient synthesis of 2-arylbenzimidazoles from o-phenylenediamines and arylaldehydes catalyzed by Fe/CeO 2-ZrO 2 nano fine particles. *MonatsheftefürChemie-Chemical Monthly*, *142*, pp.901–906.

[14] Sadri, Z., Behbahani, F.K. and Keshmirizadeh, E., 2022. One-pot multicomponent synthesis of 2, 6-diamino-4-arylpyridine-3, 5-dicarbonitriles using prepared nanomagnetic Fe3O4@ SiO2@(CH2) 3NHCO-adenine sulfonic acid. *Inorganic and Nano-Metal Chemistry*, pp.1–10.

[15] Ferdousian, R., Behbahani, F.K. and Mohtat, B., 2022. Synthesis and characterization of Fe3O4@ Sal@ Cu as a novel, efficient and heterogeneous catalyst and its application in the synthesis of 2-amino-4H-chromenes. *Molecular Diversity*, *26*(6), pp.3295–3307.

[16] Pigman, W. and Horton, D., 1972. *The carbohydrates: chemistry and biochemistry*. Academic Press: San Diego.

[17] Gold, V., Loening, K., McNaught, A. and Shemi, P., 1997. *IUPAC Compendium of Chemical Terminology*. Blackwell Science: Oxford.

[18] Shervani, Z. and Yamamoto, Y., 2011. Carbohydrate-directed synthesis of silver and gold nanoparticles: effect of the structure of carbohydrates and reducing agents on the size and morphology of the composites. *Carbohydrate Research*, *346*(5), pp.651–658.
[19] Yokota, S., Kitaoka, T., Opietnik, M., Rosenau, T. and Wariishi, H., 2008. Synthesis of gold nanoparticles for in situ conjugation with structural carbohydrates. *Angewandte Chemie International Edition*, *47*(51), pp.9866–9869.
[20] Engelbrekt, C., Sørensen, K.H., Zhang, J., Welinder, A.C., Jensen, P.S. and Ulstrup, J., 2009. Green synthesis of gold nanoparticles with starch-glucose and application in bioelectrochemistry. *Journal of Materials Chemistry*, *19*(42), pp.7839–7847.
[21] Care, A., Bergquist, P.L. and Sunna, A., 2015. Solid-binding peptides: smart tools for nanobiotechnology. *Trends in Biotechnology*, *33*(5), pp.259–268.
[22] Tan, Y.N., Lee, J.Y. and Wang, D.I., 2010. Uncovering the design rules for peptide synthesis of metal nanoparticles. *Journal of the American Chemical Society*, *132*(16), pp.5677–5686.
[23] Filice, M., Marciello, M., del Puerto Morales, M. and Palomo, J.M., 2013. Synthesis of heterogeneous enzyme-metal nanoparticle biohybrids in aqueous media and their applications in C-C bond formation and tandem catalysis. *Chemical Communications*, *49*(61), pp.6876–6878.
[24] Mittal, A.K., Chisti, Y. and Banerjee, U.C., 2013. Synthesis of metallic nanoparticles using plant extracts. *Biotechnology Advances*, *31*(2), pp.346–356.
[25] Chinnadayyala, S.R., Santhosh, M., Singh, N.K. and Goswami, P., 2015. Alcohol oxidase protein mediated in-situ synthesized and stabilized gold nanoparticles for developing amperometric alcohol biosensor. *Biosensors and Bioelectronics*, *69*, pp.155–161.
[26] Hulkoti, N.I. and Taranath, T.C., 2014. Biosynthesis of nanoparticles using microbes—a review. *Colloids and Surfaces B: Biointerfaces*, *121*, pp.474–483.
[27] Mashwani, Z.U.R., Khan, T., Khan, M.A. and Nadhman, A., 2015. Synthesis in plants and plant extracts of silver nanoparticles with potent antimicrobial properties: current status and future prospects. *Applied Microbiology and Biotechnology*, *99*, pp.9923–9934.
[28] Sousa, A.A., Hassan, S.A., Knittel, L.L., Balbo, A., Aronova, M.A., Brown, P.H., Schuck, P. and Leapman, R.D., 2016. Biointeractions of ultrasmall glutathione-coated gold nanoparticles: effect of small size variations. *Nanoscale*, *8*(12), pp.6577–6588.
[29] Naveenkumar, S., Chandramohan, S. and Muthukumaran, A., 2021. A novel synthesis of zinc oxide nanoparticles using various carbohydrate sources and its antimicrobial effects. *Materials Today: Proceedings*, *36*, pp.520–525.
[30] Kumar, A., Kumar, A.A., Nayak, A.P., Mishra, P., Panigrahy, M., Sahoo, P.K. and Panigrahi, K.C., 2019. Carbohydrates and polyphenolics of extracts from genetically altered plant acts as catalysts for in vitro synthesis of silver nanoparticle. *Journal of Biosciences*, *44*, pp.1–10.
[31] Palomo, J.M. and Filice, M., 2016. Biosynthesis of metal nanoparticles: novel efficient heterogeneous nanocatalysts. *Nanomaterials*, *6*(5), p.84.
[32] Compostella, F., Pitirollo, O., Silvestri, A. and Polito, L., 2017. Glyco-gold nanoparticles: synthesis and applications. *Beilstein Journal of Organic Chemistry*, *13*(1), pp.1008–1021.
[33] Poonthiyil, V., 2016. *Synthesis and application of gold and glycogold nanoparticles*, Thesis, University of Canterbury Christchurch, New Zealand.
[34] Kennedy, D.C., Orts-Gil, G. and Lai, C.H., 2014. LM ller, A. Haase, A. Luch and PH Seeberger. *Journal of Nanobiotechnology*, *12*(59), pp.1–8.
[35] Khalilzadeh, M.A., Kim, S.Y., Jang, H.W., Luque, R., Varma, R.S., Venditti, R.A. and Shokouhimehr, M., 2022. Carbohydrate-based nanostructured catalysts: applications in organic transformations. *Materials Today Chemistry*, *24*, p.100869.
[36] Monjezi, H.R., Zarnegar, Z. and Safari, J., 2019. Starch nanoparticles as a bio-nanocatalyst in synthesis of diheteroarylthioethers. *Journal of Saudi Chemical Society*, 23(7), pp.973–979.

8 The Photochemistry of Carbohydrates and Steroids

Angelo Albini

8.1 INTRODUCTION

Sugars, due to their polymers, constitute by far the most abundant class of products in living matter. Because of their low molecular weight compounds and the large number of asymmetric centers present in a 1:1 ratio with the total number of carbon atoms, they represent an unparalleled source of chemical messengers among water-soluble compounds. Their importance cannot be underestimated, as evidenced by the prominent role they have always played in chemistry.

Figure 8.1 shows the number of publications over the years, highlighting that although it started somewhat slowly, interest in this discipline has steadily grown to reach about one percent of the entire photochemical literature. There are clearly few patents filed in relation to the total number of publications; the ratio of 0.1 to 0.2 does not increase and is much lower than the proportions of other areas of interest in the chemical industry; companies in the sector are not yet ready to invest in photochemistry despite the increasing number of photochemical processes that operate effectively in industry. A number of chemistry practitioners still feel that photochemistry is a province of physical chemistry, useful for acquiring academic renown but unsuitable for a utilitarian and industrial approach. This state of affairs should change thanks to the changing sensitivity (and legislation) that will impose more punctual attention on the environment, with the adoption of innovative chemical methods capable of operating without aggressive reagents or the need for high temperatures and pressures.

Ciamician opposed nonrenewable energy derived from fossil fuels, a nonrenewable energy source, and proclaimed the much greater amount of energy provided by the sun, which would have been usable as soon as scientists discovered how to harness it; as Ciamician reminded, society—unlike nature—is in a hurry [1]. This brief writing aims to illustrate some elements of the photochemistry of steroids and carbohydrates, two very different fields. The former, when sufficient spectroscopic and chromatographic support was obtained for the attribution of structure to photoproducts, reached its peak around 1990. However, the latter has also been thoroughly studied, and new aspects of photochemistry have been revealed.

8.2 THE PHOTOCHEMISTRY OF CARBOHYDRATES

It is appropriate to present this topic by referring to the glycosylation reaction because glycosides are low-molecular-weight substances in which sugars are present and, as already mentioned, they form the most numerous class of chemical messengers present in the cell. They also exert a wide range of vital actions, as is presumed by drugs that attempt to imitate their action when natural mechanisms malfunction.

As often happens when venturing into the roots of contemporary science, photochemistry is the right place to find its beginning. Here, glycosidation under solar irradiation was first reported by Ciamician and Ravenna in 1911 [2]. These scientists discovered that when a xenobiotic substance is injected into a plant and then exposed to sunlight, the corresponding glycosides are formed, and when a preformed glycoside is injected, part of the aglycone is released. This discovery could have sparked rapid development in the field if experimentation on these specific compounds had not been

DOI: 10.1201/9781003437413-8

extremely difficult because carbohydrates are soluble in water and colorless and absorb only in the UV-C region. This is not a limitation today, but the instrumentation and knowledge of the time did not yet allow definitive results to be obtained [3–5].

In modern times, the problem instead is determining whether the mechanism is primary or secondary, as the use of high-energy photons often forms photoproducts of species that still have high energy, even if they are the lowest-energy configuration of the considered structure (typically species with an odd number of electrons, such as radicals or radical ions) and capable of overlaying their selectivity on that of the starting molecule. In summary, it seems reasonable to assume that the primary step with glycosides involves homolytic fragmentation, but this leads to species that are still highly reactive and, as mentioned, can further react to give rise to secondary processes that can obscure the selectivity in the first step [3–5].

This fact, combined with the notorious difficulty of separating pure compounds from mixtures of (photo)products and their tendency to "caramelize," explains why the results presented in the early publications were often contradictory (the exception for the structure of sugars is naturally represented by the work of Emil Fischer, whose nomination for the Nobel Prize nomination was indeed supported successfully by Ciamician in 1902; in general, however, solving the problem of the photochemistry of such particular molecules was impossible with the means of the time). In such circumstances, the only progress can come from experiments resolved over time, in which primary photoproducts are identified and calculated. Fortunately, substantial progress has been made in both cases, and there is now much better knowledge of sugars. In the case of glucose, accessible chemical pathways include hydrogen detachment processes for each of the five O-H groups and a process of C-O bond cleavage and ring opening. Both types of reactions are guided by repulsive $n\sigma^*$ states easily accessed from the Franck–Condon region and lead to barrierless conical intersections. The calculation of vertical singlet excitation energies predicts absorption from 6.0 eV up to a densely populated zone of weakly absorbing states (lifetime 200 fs) [6, 7].

All modern experiments on this topic are carried out using germicidal lamps (discharge lamps at low mercury pressure that emit at 254 nm and 184 nm; in the latter case, given that oxygen also absorbs, the contribution is taken into consideration only for experiments performed in vacuum), with laser or LED, or for photosensitization with benzene or acetone [8].

After optimization, the conical intersection for O-H bond cleavage was localized around 5.0 eV, and that for C-O bond cleavage was around 4.0 eV. Radicals can be further mediated, for example by substitutive reduction elimination after electron transfer from a species such as hydroxide anion,

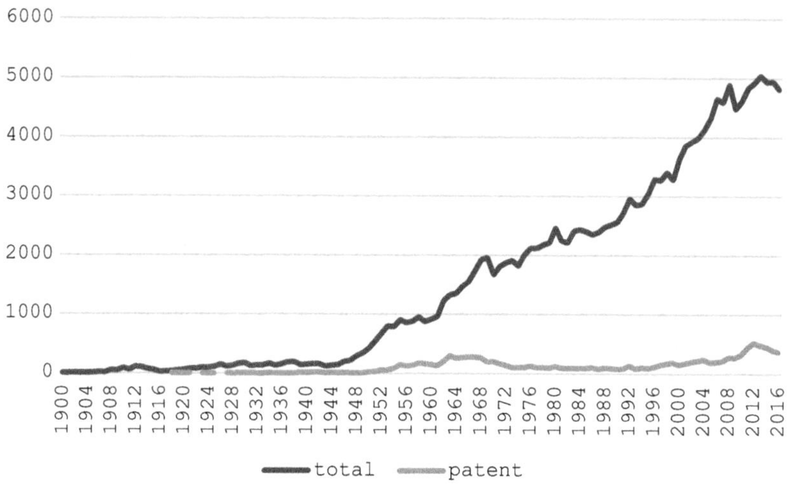

FIGURE 8.1 Carbohydrate photochemistry proportional to patents deposited, in red, compared with the total number of documents in the same time interval.

dimethylformamide, or hexamethylphosphoramide [9] or by substituent change, as in the case of the preparation of α-bromo [10, 11] and α-fluoro sugars [12, 13], which are interesting synthetic intermediates. Finally, it should be mentioned that polymeric carbohydrates are by far the most abundant class among natural compounds, and their role in photochemistry is quite complex and depends on structural elements different from phenols or other aromatic fragments [14]. From a technological point of view, at least the photochemically generated arrays [15–17] should be remembered, still exploiting the multiplicity of asymmetric centers for various industrial applications. [18, 19] and, given the relationship between glucose and diabetes, the use of noninvasive systems including some photochemicals for medical tests [20]. See Schemes 8.1 to 8.3).

Other photolabile molecules are nucleic acids, where intramolecular hydrogen atom exchange from the sugar portion to the present (and absorbing) heterocycle usually occurs. A key point for glycosidation is that unlike other natural polymers, each step adds an asymmetric center with a covalent bond (αβ). Glycosidation is therefore inherently more difficult for stereochemical control (Scheme 8.4).

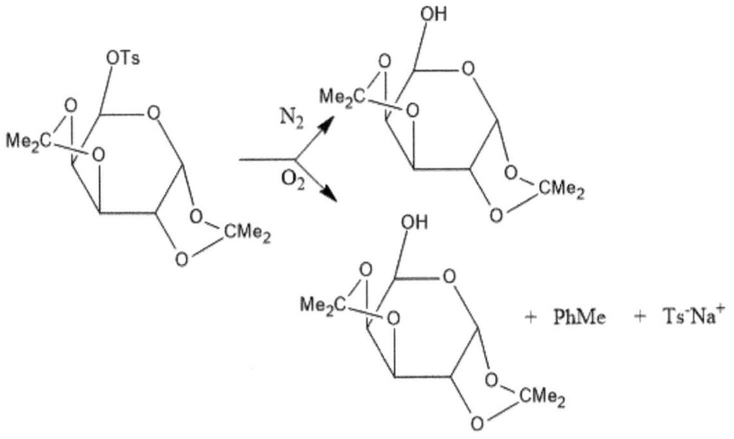

SCHEME 8.1 Substituent modification under reducing conditions.

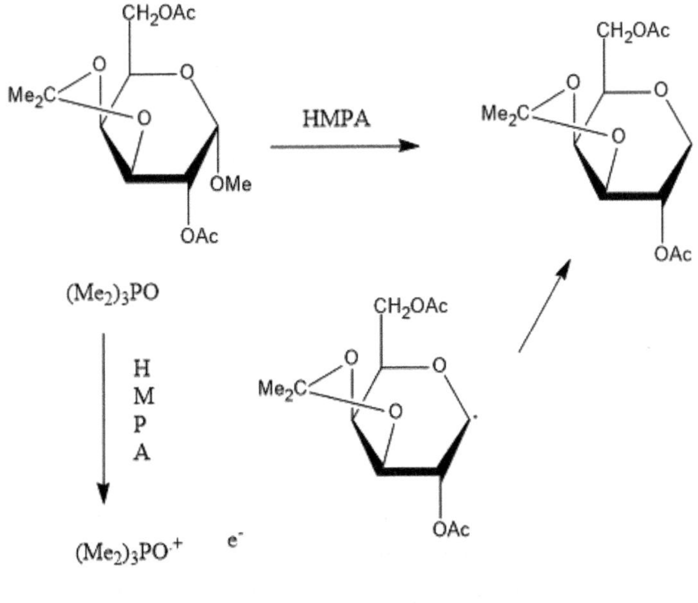

SCHEME 8.2 Substituent elimination under reducing conditions.

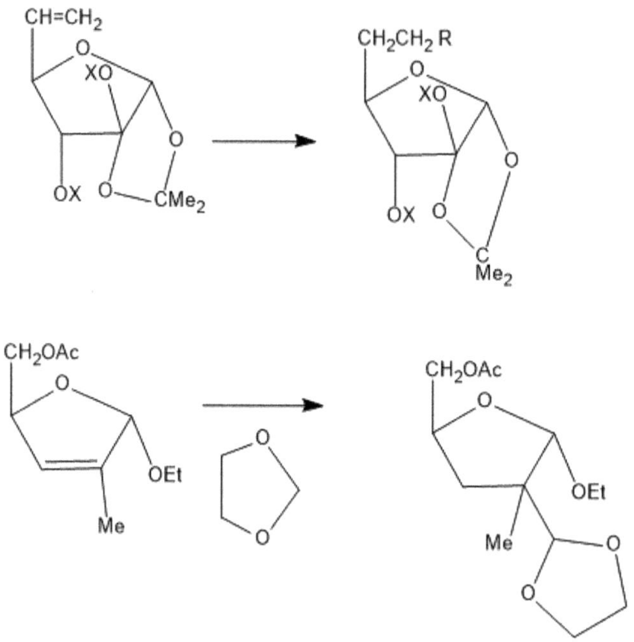

SCHEME 8.3 Other trapping pathways via radical with formation of carbon-carbon bonds.

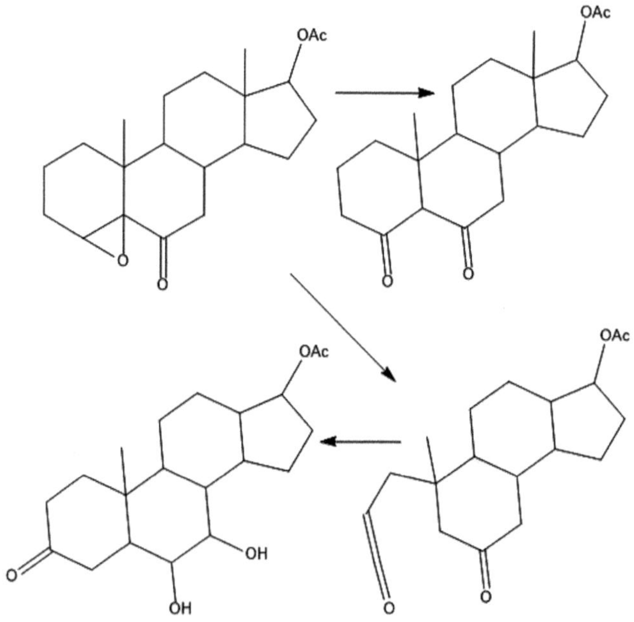

SCHEME 8.4 Exemplative photoreaction of steroidal epoxyketones.

An indication of potential value of this further aspect is the cross-pinacolization reactions involving sugars reported in a few cases. As we progress in the study of photochemical applications, it should be noted that excited states can be generated not only via direct light absorption but also through photosensitization, as when the reacting molecule differs from the one that absorbs light. The concept is very similar to that of catalysis, and indeed the terms photosensitization and photocatalysis are interchangeable in the literature. There are two ways of photosensitization: energy transfer, which results in nonspectroscopic excited states called triplets, and electron transfer for generating radical ions. The latter is very important for carbohydrates since covalent molecules only access redox processes under drastic conditions and absorb only in the UV. Notably, however, this double limitation becomes a doubly favored process in the excited state because the thermodynamic disadvantage is compensated by the excitation energy (see the Weller equation). In fact, metallic sodium is needed to reduce naphthalene, but triethylamine is enough to reduce its excited state. Speaking of carbohydrates, it should never be forgotten that these compounds in solution exist in equilibrium between the semiacetal cyclic form and the open chain form, in which the carbonyl function is present [21].

8.3 THE PHOTOCHEMISTRY OF STEROIDS

Steroids are secondary metabolites of the isoprenoid class, based on the cyclopentanoperhydrophenanthrene skeleton. As I described, these compounds are similar to carbohydrates due to the negligible absorption of the aliphatic skeleton, although the lack of thermal analogy to compare with makes it difficult to assign structure to photoproducts. Furthermore, all steroids of any relevance have (at least) two chromophores that, together with their mutual interaction, determine their photochemistry. Consider a simple model that uses a pair of test tubes that each contain a compound that carries a chromophore; study how the two chromophores react differently when both are present compared with when they are irradiated individually, and relate the results to how they are held in mutual position by the 77 aliphatic skeleton, in this case symbolized by a test tube holder (Scheme 8.5).

The data in the literature actually support this classification. In siloxy steroids, the siloxy group acts as an antenna, but only for photophysical processes such as absorption and energy transfer, rather than as chemical reactions. The photoreactivity is carried out by the ketonic functions present in the molecules, which reach both singlet and triplet states through through-bond interactions [22]. In contrast, in some systems, two independent groups behave differently: One group leads to a Norrish type II product via UV-A and UV-C excitation on ring D, while the other leads to rearrangement to a 'lumi'ketone in ring A (Schemes 8.6 and 8.7) [23].

Rearranged compounds are of great interest because they lead to profound modifications of molecular structures in a single step and under mild conditions that are often of industrial interest because they are high yielding. Other widely studied steroid derivatives include α,β-conjugated ketones with three-membered rings, such as epoxides or cyclopropanes (Scheme 8.8) [24–26].

Another example of an industrially important reaction is the steroid class of nitrite homolysis [27]. These compounds also weakly absorb in the visible spectrum and undergo homolytic fragmentation of the nitrite function, releasing NO and selectively forming a methyl radical from the closest methyl at position 20 (Barton reaction). This then formed a nitroso derivative that isomerized to the oxime. Taken together, the sequence corresponds to the transfer of chemical activation from one position to another with chemo- and regioselective reaction. Although the mechanism has been contested, it remains an example of creative science [27].

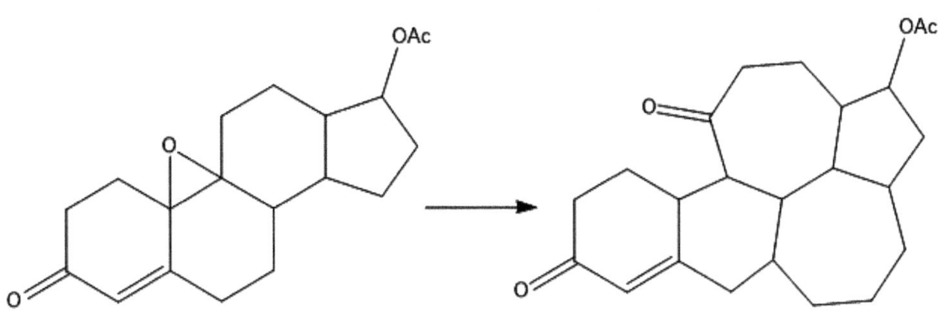

OAc O O → OAc O O

SCHEME 8.5 Photochemical reaction of a steroidal epoxyketone.

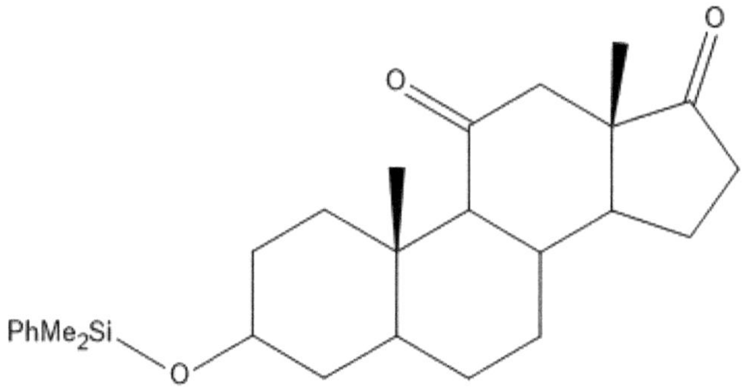

SCHEME 8.6 In a group of steroids, absorption (or photosensitization) occurs on the siloxy group, but through-bond interaction brings energy to the ketonic functions (antenna effect).

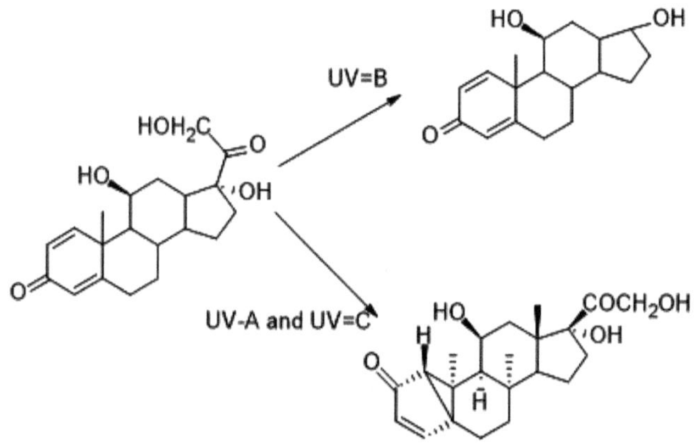

SCHEME 8.7 UV-A and UV-C excitation produces a Norrish type II reaction, while UV-A and UV-B excitation leads to the luminal rearrangement.

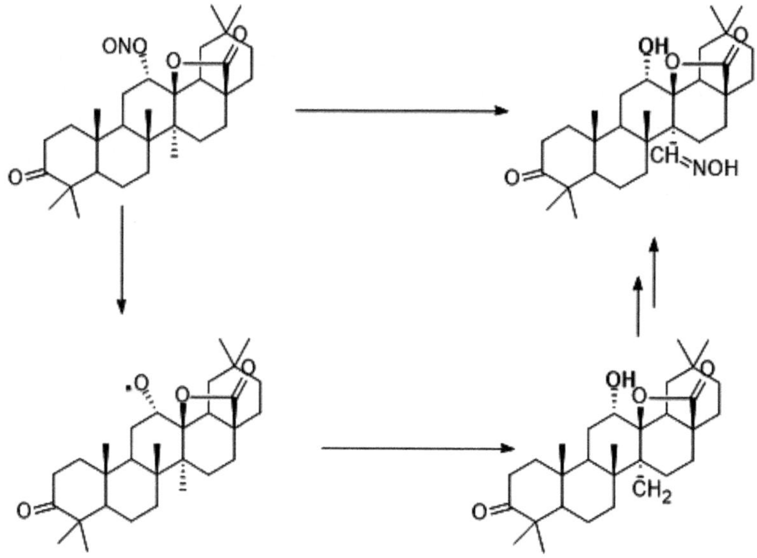

SCHEME 8.8 Photochemistry of some steroid cyclopropane ketones.

8.4 CONCLUSIONS AND OUTLOOK

Broadly speaking, we can say that photochemistry has gone through three important eras. The first was the pioneering era, in which this discipline, created thanks to the intuition of a few individuals, was developed only in a small number of laboratories worldwide. Those people, endowed with excellent foresight, knew how to foresee a large booty of discoveries that if valued could change the future of human society. In particular, they glimpsed the possibility of finding alternative sources of energy before the accumulation of fossil fuels was exhausted. They realized that the only viable path was that traced by nature in ancient times: harnessing solar energy.

Then as now, fossil fuels provide much more energy than that obtainable from wind, tides, and waterfalls, and it was not possible to assess the actual contributions of nuclear energy; on the contrary, the sun made available a very powerful energy source, so much so that the sunlight hitting an area the size of Lazio, Italy, would have been sufficient to power the whole planet in 1912. Moreover, the use of light as an activator offered the possibility of obtaining in the laboratory chemical substances identical to natural ones but prepared in less polluting conditions. At the time, however, without the support of sufficient experiments for their verification and discussion, these intuitions remained interesting speculations. Things only changed after 1950 with the availability of modern analytical techniques [28–31].

The second era, called by some the renaissance of organic photochemistry, was a time of great fervor. Thanks to the development of suitable analytical methods, it was enough to take a sample from the reagent bottle and by irradiating it be certain that something new and unexpected would come out, in contrast with nonphotochemical reactions. Finally, we have the third era, in which this discipline has been widely developed, and thanks to the consequent availability of many examples, it has been possible to propose general mechanisms comparable with those of any other field of chemistry and even surpassing them; now, reactions are explained in terms of electronic structure, and what interpretation can be more appropriate than that referring to an electronic isomer of the ground states? [28–31].

Comparing the electronic configurations of excited states enriches the description of both. Thanks to the general schemes in the literature, the choice of compounds to irradiate is no longer left to chance, and the wealth of known reactions allows for setting up any project. Two research themes that have recently been fully developed are particularly related to the topics considered in this chapter: photosensitization for electron transfer [28–31] and flow photochemistry (Scheme 8.9) [32].

The first theme fits well with aliphatic compounds because the double limitation due to the intrinsic low propensity for redox passages and the high excitation energy becomes a double facilitation (Eox, Ered, Eex are all large in the Weller equation) for the excited state, revealing chemistry different from that of the ground or excited states. Many systems suitable for playing the role of single electron transfer photosensitizer, often referred to as photocatalysts, have been described including organic molecules or their salts, metal ions, or, in heterogeneous version, semiconductor powders.

In conclusion, photochemistry is at a turning point. The current situation shows the application of concepts from molecular photochemistry and the virtual disappearance of the dichotomy that

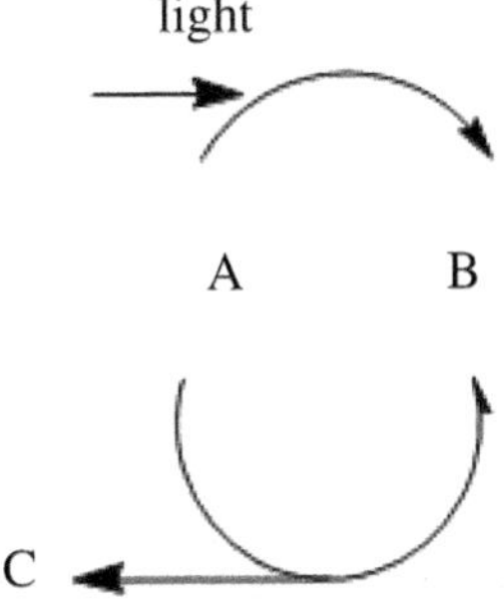

SCHEME 8.9 Single electron transfer sensitization reaction.

FIGURE 8.2 Giacomo Ciamician on the roof of the Chemistry Department in Bologna observing his flasks exposed to solar light.

once narrowed the rich, pervasive discipline that to a handful of large-scale reactions (halogenation, sulfonation, chlorosulfonation, oxime rearrangement). Additionally, there's significant development in synthetic methods in flow. If photochemists can direct their discipline towards fine chemistry and take advantage of the green chemistry train, which requires a reassessment of reaction conditions, then a general photochemical reactor can be developed in collaboration with chemists and engineers, avoiding the pitfall of exclusively engineering competence in large-scale reactions. For reactions in test tubes and for those on a small scale, the peculiarities of each method can be exploited and, en passant, contribute to environmental disaster control. An indication of the fruitfulness of this approach is evident in the development of flow reactions and modular reactors reported recently.

Reflecting on the data, which show few publications on aliphatic compounds and even fewer patents in the field, we can arrive at two opposing conclusions. In the first, more pessimistic scenario, the particular difficulties in managing this discipline combined with the fact that most systems of interest have already been thoroughly investigated predict that the space for future developments is small if not nonexistent; the second scenario, on the contrary, outlines a future in which a radical change of perspective frees the space occupied by old chemistry and gives the green light to innovative projects of great interest. Ingenuity will be needed, but here photochemists have a bright light house to follow: Giacomo Ciamician (Figure 8.2).

REFERENCES

[1] Ciamician, G, 1912, The Photochemistry of the Future. *Science*, 36, 385–394

[2] Ciamician, G, Ravenna, C, 1911, Genesis of alkaloids in plants. *C Reali Acad Lincei*, 20, 614

[3] Brinkley, RW, 1981, Photochemical reactions of carbohydrates. *Adv Carbohydr Chem Biochem*, 38, 102–113

[4] Bundle, DR, Descotes, G, Gigg, J, Gigg, R, Klaffke, W, Meyer, B, Descotes, G, 1990, Synthesis of Oligosaccharides Related to Bacterial *O*-Antigens. *Top Curr Chem*, 154, 285–332

[5] Brinkley, ER, Brinkley, RW, eds, 1999, *Carbohydrates Photochemistry*, American Chemical Society, Washington, DC

[6] Nag, L, Lukas, A, Vos, MH, 2019, Short-lived radical intermediates in the photochemistry of glucose oxidase. *ChemPhysChem*, 20, 1793–1798

[7] Tuma, D, Sobolewski, AL, Domcke, W, 2014, Electronically excited states and photochemical reaction mechanisms of b-glucose. *Phys Chem Chem Phys*, 16, 38–47

[8] Schönberg, A, 1968, *Preparative Organic Photochemistry, Preparative Organic Photochemistry*, Springer, Berlin

[9] Elliott, Q, Dos Passos Gomes, G, Evoniuk, CJ, Alabugin, IV, 2020, Testing the limits of radical-anionic CH-amination: a 10-million-fold decrease in basicity opens a new path to hydroxyisoindolines via a mixed C–N/C–O-forming cascade. *Chem Sci*, 11, 6539–6555

[10] Yang, Y, Pignatello, JJ, 2017, Participation of the Halogens in Photochemical Reactions in Natural and Treated Waters. *Molecules*, 13(22), 1684

[11] Ortner, K, Buchberger, W, Himmelsbach, M, 2009, Capillary electrokinetic chromatography of insulin and related synthetic analogues. *J Chromatogr A*, 1216(14), 2953–2957

[12] Schmid, T, Baumann, B, Himmelsbach, M, 2016, Analysis of saccharides in beverages by HPLC with direct UV detection. *Anal Bioanal Chem*, 408, 1871–1878

[13] Hinge, C, Henriksen, O, Lindberg, UL, Hasselbalch, S, Højgaard, L, Law, I, Andersen, F, Ladefoged, C, 2022, A zero-dose synthetic baseline for the personalized analysis of [18F]FDG-PET: Application in Alzheimer's disease. *Front Neurosci*, 16, 1053783

[14] Zhang, J, Luo, XZ, Wu, X, Gao, CF, Wang, PY, Chai, JZ, Liu, M, Ye, W, Xi, DC, 2023, Photosensitizer-free visible-light-promoted glycosylation enabled by 2-glycosyloxy tropone donors. *Nature Commun*, 14, 8025

[15] Salvadori, G, Macaluso V, Pellicci G, Cupellini L, Granucci G, Mennucci B, 2022, Protein control of photochemistry and transient intermediates in phytochromes. *Nat Commun*, 13(1), 6838

[16] Eisch JJ, Munson PR, Gitua JN, 2004, Observed Conversion of Methanol into Ethylene Glycol as Possible Prototype for Sugar Alcohol Formation. *Orig Life Evol Biosph*, 34, 441–454

[17] Kommedal, EG, Angeltveit, CF, Klau, LJ, Ayuso-Fernández, I, Arstad, B, Antonsen, SG, Stenstrøm, Y, Ekeberg, D, Gírio, F, Carvalheiro, F, Horn, SJ, 2023, Visible light-exposed lignin facilitates cellulose solubilization by lytic polysaccharide monooxygenases. *Nat Commun*, 14, 1063

[18] Goddard, ED, Turro, NJ, Kuo, PL, 1985, Fluorescence probes for critical micelle concentration determination. *Langmuir*, 1, 352–355

[19] Maric, T, Mikhaylov, G, Khodakivskyi, P, Bazhin, A, Sinisi, R, Bonhoure, N, Yevtodiyenko, A, Jones, A, Muhunthan, V, Abdelhady, G, Shackelford, D, 2019, Bioluminescent-based imaging and quantification of glucose uptake in vivo. *Nat Methods*, 16, 526–532

[20] Mara, MW, Tatum, AD, March, AM, Doumy, G, Moore, EG, Raymond, KN, 2019, Energy Transfer from Antenna Ligand to Europium(III) Followed Using Ultrafast Optical and X-ray Spectroscopy. *J Am Chem Soc*, 141(28), 11071

[21] Ricci, A, Fasani, E, Mella, M, Albini, A, 2003, General Patterns in the Photochemistry of Pregna-1,4-dien-3,20-diones. *J Organic Chem*, 68, 4361–4366

[22] Waters, JA, Kondo, Y, Witkop, B, 1972, Photochemistry of steroids. *J Pharm Sci*, 61, 321–334

[23] Lehmann, C, Schaffner, K, Jeger, O, 1962, PhotochemischeReaktionen. 13. Mitteilung. ZurphotochemischenUmlagerung von 3-Oxo-4,5-oxido-Steroiden in 10(5→4)-abeo-Steroide. *Helv Chim Acta*, 45, 1031

[24] Wekrli, H, Lehmann, C, Lizuka, T, Schaffner, K, 1967, PhotochemischeReaktionen 42. Mitteilung [1]. Photoisomerisierung von α, β-Epoxyketonen II Der sterischeVerlauf der Umlagerung von 3-Oxo-4, 5-oxido-Steroiden. *Helm Chim Acta*, 2403

[25] Wang, Z, 2010, *Comprehensive Organic Name Reactions and Reagents*, Wiley, Berlin

[26] Grossi, L, 2005, The Barton Reaction: Can a More Tenable Pathway Be Hypothesized for the Formation of Nitrosoalkyl Derivatives? *Chemistry Eur J*, 18, 5419–5425

[27] Sugimoto, A, Sumino, Y, Takagi, M, Fukuyama, T, Ryu, I, 2006, The Barton reaction using a microreactor and black light. Continuous-flow synthesis of a key steroid intermediate for an endothelin receptor antagonist. *Tetrahedron Lett*, 47, 6197–6200

[28] Escudero, D, 2016, Revising Intramolecular Photoinduced Electron Transfer (PET) from First-Principles. *Acc Chem React*, 49, 1812–1824

[29] Melchiorrer, P, 2022, Introduction: Photochemical Catalytic Processes. *Chem Rev*, 122, 3183

[30] Bauer, A, Westkämper, F, Grimme, S, Bach, T, 1995, Carbon-carbon bond cleavage reactions of 1, 2-diamines initiated by photoinduced electron transfe. *Res Chem Intermed*, 21, 587–611

[31] Protti, S, Fagnoni, M, Albini, A, Zhang, W, Cue, BW, Eds, 2018, in *Green Techniques for Organic Synthesis and Medicinal Chemistry*, Spinger, Berlin

[32] Rehm, T, 2020, Flow Photochemistry as a Tool in Organic Synthesis. *Chemistry Eur J*, 16952–16974

9 Carbohydrates as Biosurfactants

Ahindra Nag and Dipak Patil

9.1 INTRODUCTION OF SURFACTANTS

The term surfactant is a shortening of surface active agent. Surfactants are organic compounds that contain both hydrophobic and hydrophilic groups; therefore, a surfactant molecule contains both water-soluble and water-insoluble components. Surfactant molecules migrate to the water surface, where the insoluble hydrophobic group may extend out of the bulk water phase and mix with oil phase while the water-soluble head group remains in the water phase. This alignment and aggregation of surfactant molecules at the surface alters the surface properties of water at the water/oil interface. Ultimately, surfactants are compounds that lower the surface tension of a liquid, allowing easier spreading, and the interfacial tension between two liquids or between a liquid and a solid.

Surfactants play an important role as cleaning, wetting, dispersing, lubricant, and emulsifying agents in many practical applications. They are useful in the production and processing of foods, agrochemicals, pharmaceuticals, personal care and laundry products, petroleum, fuel additives, and photographic films. In addition, surfactants can be found throughout a broad spectrum of biological systems and medical applications, in soil remediation, and in other environmental, health, and safety applications.

9.2 CLASSIFICATION OF SURFACTANTS

Commercially, surfactants are classified according to their use. However, this is not that useful because many surfactants have several uses. The most accepted classification of surfactants is based on their dissociation in water.

9.2.1 Anionic Surfactants

Anionic surfactants dissociate in water as amphiphilic anions and cations, which are generally alkaline metals (Na^+, K^+) or a quaternary ammonium. These include alkylbenzene sulphonates (detergents), (fatty acid) soaps, lauryl sulfate (foaming agent), and di-alkyl sulfosuccinate (wetting agent).

9.2.2 Cationic Surfactants

Cationic surfactants dissociate in water as amphiphilic cations and anions, generally of the halogen type. A very large proportion of this class corresponds to nitrogen compounds such as fatty amine salts and quaternary ammoniums, with one or several long alkyl chains that often come from natural fatty acids. These surfactants are in general more expensive than anionic surfactants because of the high-pressure hydrogenation reaction to be carried out during their synthesis.

9.2.3 Nonionic Surfactants

As the name itself states, these surfactants cannot be ionized in aqueous solution because their hydrophilic group is a nondissociable type, such as alcohol, phenol, ether, ester, or amide. Large

DOI: 10.1201/9781003437413-9

proportions of nonionic surfactants are made hydrophilic by the presence of a polyethylene glycol chain, obtained by the polycondensation of ethylene oxide, and are called polyethoxylated nonionics. In the past decade, glucoside (sugar-based) head groups have been introduced in the market because of their low toxicity. The lipophilic group is often of the alkyl group from fatty acids of natural origin. These surfactants are biodegradable and hence called biosurfactants.

9.2.4 Biosurfactants

Fatty acid esters of polyol are non-ionic surfactants classically known as biosurfactants. Combination of polar and non-polar group in a single molecule turns it to an amphiphilic or surface active compound. These are generally prepared by enzyme catalyst and are biodegradable. These are widely used in foodstuff, cosmetics, and medicine preparations. When polar part is ascorbic acid the resulting fatty acid ester may act as surfactant cum antioxidant.

Biosurfactants have many advantages over their chemically synthesized counterparts, for instance, their biodegradability; biological surfactants are easily degraded by microorganism [1]. Biosurfactants are also less toxic than chemical surfactants and show higher EC_{50} (the effective concentration to decrease 50% of the test population) values than synthetic dispersants [2]. In terms of availability, biosurfactants can be produced from very cheap, very widely available raw materials, hydrocarbons, carbohydrates, and lipids separately or in combinations [3]. Physically, many biosurfactants are not affected by environmental [3] factors such as temperature, pH, and ionic strength. Lichenysin produced by *Bacillus licheniformis* strain was not affected by temperatures of up to 50 °C, pH range of 4–9.0, and NaCl concentration of 50g/l and Ca concentration of 25g/l. Mulligan stated that a good surfactant can lower the surface tension [4] of water from 75 to 35 mN/m and the interfacial tension between water/hexadecane from 40 to 1 mN/M [5]. Surfactin possesses the ability to reduce the surface tension of water to 25m N/M and the interfacial tension of water/hexadecane to <1 mN/M. Surfactants that show high biocompatibility and digestibility are effectively applied in the cosmetics and pharmaceutical industries and as functional food additives [6].

9.3 ENZYMATIC ESTERIFICATION

Enzymatic esterification of fatty acids with carbohydrates such as sucrose, glucose, fructose, and sorbitol has been reported using different lipases [7]. Ter Haar et al. [8] studied the lipase-catalyzed synthesis of fructose oligoglycoside lauryl esters. Cao et al. [9] reported on the lipase-catalyzed, solid-phase synthesis of sugar ester using immobilized enzyme on various carriers. Many reports are available on parameters of enzymatic sugar ester synthesis such as carbon-chain length of the fatty acid [10]. Scheckermann et al. [11] found that an excess of fatty acid gave better esterification of fructose using lipase catalyst in 2-methyl-2-butanol. Sabeder et al. [12] studied the influence of different reaction parameters on the enzymatic esterification of fructose in organic solvents.

Fatty acid esters of polyols have important applications due to their surface active properties resulting from the combination of hydrophilic and hydrophobic properties in a single molecule. Incorporation of the polar moiety to fatty acid converts it to an amphiphilic or surface-active compound. These biosurfactants are widely used as in the food, detergent, pharmaceutical, and cosmetics industries [13]. Ascorbyl ester can be used as a biosurfactant as well as a fat-soluble antioxidant for the prevention of fat oxidation in foods.

Enzymatic synthesis in organic media provides higher selectivity and purer products due to enzyme specificity [14]. The catalysis is conducted under mild temperature and pH, conditions that have fewer side reactions than the chemical process [15]. Several chemical methodologies applied to making sugar esters showed low selectivity and consequently gave rise to different compounds [9, 16].

Lipases (triacylglycerol hydrolases, EC 3.1.1.3) are characterized by their ability to catalyze hydrolysis, esterification, and transesterification. Researchers have recently presented a number of reports on the lipase-catalyzed synthesis of monosaccharide fatty acid esters in organic solvents.

Different methods for removing the water produced during esterification were suggested such as molecular sieves [17, 18]; azeotropic distillation [19]; and pervaporation, vapor permeation, or membrane separation [20]. There are reports on the synthesis of sugar ester by direct esterification in the presence of a cosolvent [21, 22], and this reaction produces water, which promotes reverse hydrolysis. In addition, when a cosolvent is used, a complementary step is required to remove it.

9.4 ENZYMATIC ACYLATION OF FRUCTOSE

A reaction mixture that consisted of 2 mmol of fructose, 10 mmol of acyl donor, and 300 mg of lipase in 80 mL of 2-methyl-2-butanol was prepared in a 250 mL three-necked, round-bottom flask with continuous argon flow over a magnetic stirrer (100 rpm) kept in temperature-controlled oil bath at 65 °C for 12 h (Scheme 9.1-a). Sugar ester was confirmed by TLC, performed on silica gel plates $60F_{254}$ (Merck) using a developing solvent of chloroform:methanol:acetic acid:water (80:15:8:2) or ethyl acetate:hexane (1:1). The plates were sprayed with 50% sulfuric acid and heated at 110 °C for 5 min, and the presence of sugar ester was confirmed [24, 25].

The concentrated reaction mass was pre-adsorbed on silica gel, and the product fructose oleate was separated by silica gel column chromatography. Unreacted oleic acid or OAME was removed by eluting 0%–5% ethyl acetate in hexane and with gradually increasing the solvent polarity (20%–25%) fructose oleate was obtained. The yield was calculated on the basis of fructose (Figure 9.1). The product was confirmed by ^{1}H NMR (nuclear magnetic resonance) spectroscopy (400 MHz,

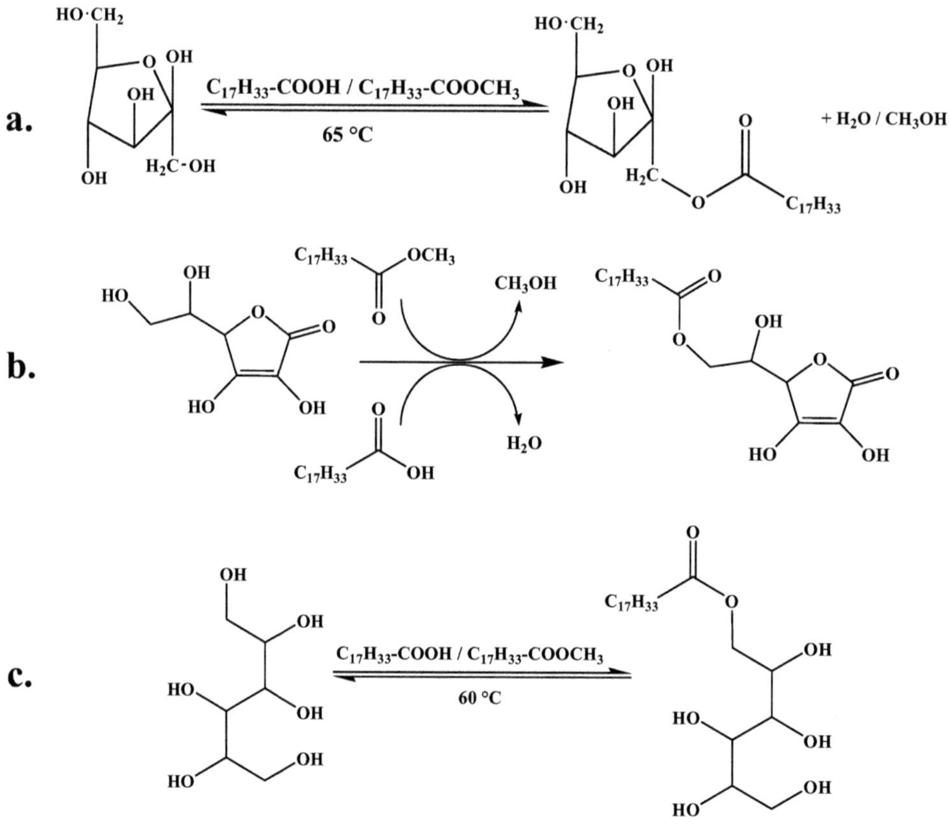

FIGURE 9.1 Lipase catalyzed acylation of d-fructose (a), ascorbic acid (b) and sorbitol (c) with oleic acid/OAME.

CD_3OD): 5.34 (m,2H), 4.28 (td,1H), 4.12 (m,3H), 3.99 (m,2H), 3.85 (ddd,1H), 2.34 (t,2H), 2.02 (m,4H), 1.61 (t,2H), 1.30 (b,20H), and 0.90 (t,3H). Previous studies on enzymatic reaction of oleic acid with fructose using *Candida Antarctica* showed the formation of two products: monoacetylated esters of C-1 of fructofuranose and fructopyranose. The protons of sugar moiety were observed as multiplates at 4.28(1H), 4.12 (3H), 3.99 (2H), and 3.85 (1H). These multiples were overlapped while equating the protons of both β-1-fructopyranose-oleate and 1-fructofuranose-oleate esters.

9.5 ENZYMATIC ACYLATION OF ASCORBIC ACID

Reaction was carried out with 2 mmol of ascorbic acid, 8 mmol of acyl donor, 200 mg of lipase, and 12 mL 2-methyl-2-butanol (Scheme 9.1b) in 20 mL screw-capped glass vials. Molecular sieves (250 mg) were added to control the water generated in the reaction (esterification). The reaction was carried out at 55 °C in an incubator shaker for 24 h and monitored by silica gel TLC plates ($60F_{254}$) using ethyl acetate/hexane (90%) as developing solvent; the products were purified using flash chromatography [28, 29]. Each column was eluted by medium pressure gas flow, which gave rapid separation and minimized the time of contact with silica gel. The columns were prepared by making a 5% slurry, and products were isolated by eluting 40%–50% of eluting solvent. Yield of isolated esters was calculated on the basis of ascorbic acid. Purified ascorbyl oleate was confirmed by ^1H NMR spectroscopy (400 MHz, $CDCl_3$): 5.35(m,2H), 4.78(s,1H), 4.39(s,1H), 4.24(d,2H), 2.36(t,2H), 2.05(m,4H), 1.62(b,2H), 1.27(b,20H), and 0.8(t,3H).

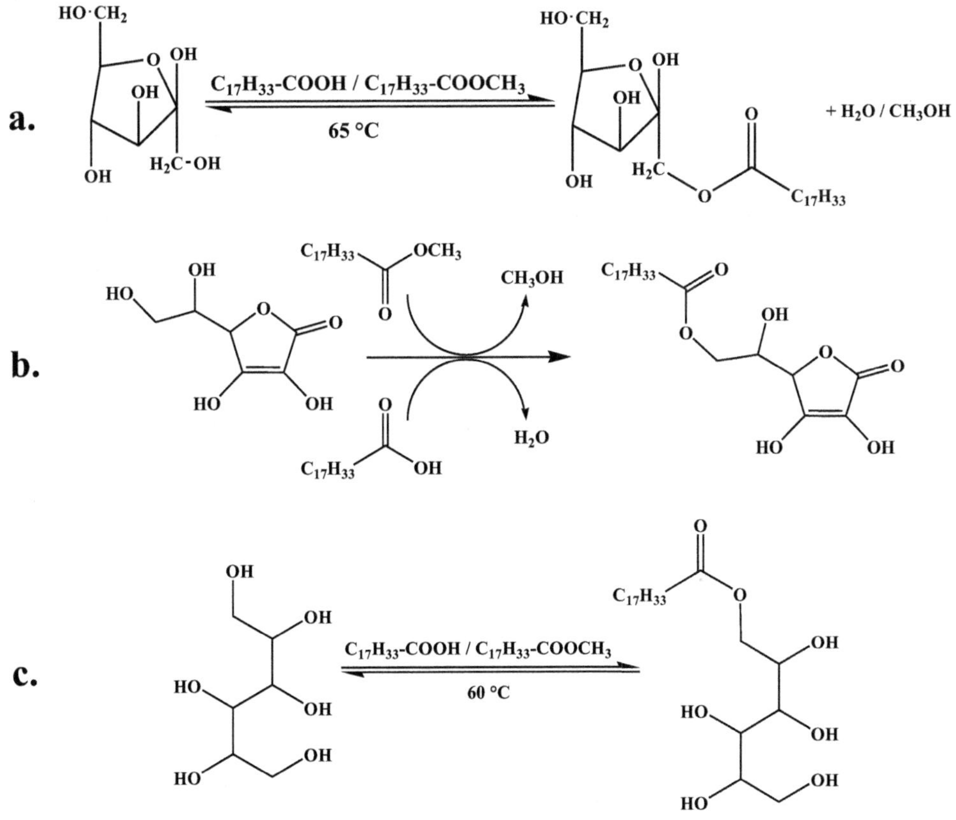

SCHEME 9.1 Lipase-catalyzed acylation of ascorbic acid (a), d-fructose (b), and sorbitol (c) with oleic acid/OAME.

9.6 ENZYMATIC ACYLATION OF SORBITOL

In a 20 mL screw-caped glass vial, 2 mmol of sorbitol was reacted with 6 mmol of acyl donor in the presence of 200 mg of lipase and molecular sieves (250 mg) in 12 mL 2-methyl-2-butanol (Scheme 9.1c). The reaction was carried out at 60 °C in an incubator shaker for 12 h and monitored by silica gel TLC plates ($60F_{254}$) using ethyl acetate as the developing solvent. The product was purified by washing the crude product with hexane and then centrifugation (4000 rpm) to remove excess of oleic acid, similarly dissolving the product in diethyl ether. Yield of isolated esters was calculated on the basis of sorbitol and was confirmed by NMR spectroscopy (^1H 400 MHz, CD_3OD): 5.33(m, 2H), 4.34(d, 0.5H), 4.20(dd, 0.5H), 4.14(m, 1H), 3.90(m, 2H), 3.77(m, 1H), 3.67(dd,1H), 3.61(d, 1H), 2.36(q, 2H), 2.04(m, 4H), 1.62(b, 2H), 1.30(b, 20H), and 0.89(t, 3H).

9.7 MEASUREMENT OF ANTIOXIDANT PROPERTIES

The stable DPPH radical was used to determine radical-scavenging activity by spectrophotometric measurement [27]. After scanning the wavelength between 200 nm and 800 nm in UV-vis double beam spectrophotometer 6.59, the maximum absorbance of DPPH (100 μM) in methanol was found at 515 nm.

Different concentrations of each test sample (30 μL) were added to 3 mL of DPPH solution. These samples were kept in dark for 30 min. at room temperature, and then the decrease in absorption was measured at 515 nm. Blank sample (A_0) was prepared by adding the same amount of methanol as in the absence of any sample. All measurements were made in triplicate. Radical scavenging was calculated using the following equation:

$$\%Inhibition = \frac{A_0 - A_C}{A_0} \times 100$$

where $A0$ = absorbance of blank sample and AC = absorbance at concentration C [18].

9.8 RESULTS AND DISCUSSION

The weight of the dried seed kernel powder of Bahera fruit was 54 g. The oil was extracted from the powdered meal with hexane six times, using a Soxhlet apparatus. The oil was recovered from hexane by distillation (23.4 g, 3.9% w/w of Bahera fruits). The yield of methylated oil was 62% (2.48 g) based on the starting weight of Bahera oil. The fatty acid profile of chicken and Bahera oil is shown in Table 9.1. Chicken viscera and Bahera oil contain both saturated and unsaturated fatty acid (FA) and fatty acid methyl esters (FAME). These were separated by low-temperature crystallization using acetone treatment (12%, w/v) followed by vacuum distillation (yield: FA, 34% and FAME, 45%). Separated oleic acid and OAME were confirmed by 1H NMR spectroscopy [27].

TABLE 9.1
Percentage Composition of FFA in Chicken Viscera and Bahera Oil

Fatty Acids	Chicken Viscera	Bahera Oil
Oleic (18:1)	38.4	50.2
Linoleic (18.2)	17.8	10.8
Palmitic (16.0)	16.5	18.2
Eicosenoic (22:1)	6.4	-
Stearic (18:0)	5.0	8.2
Others	15.9	12.6

9.9 FRUCTOSE OLEATE

Acylation took place using oleic acid and OAME by esterification and transesterification. During transesterification, methanol was generated as a byproduct; due to its high polarity, the enzyme activity was affected by removing the hydrated layer of enzyme [26]. Elimination of methanol significantly shifted the equilibrium toward the ester synthesis. The reaction was carried out at the boiling temperature of methanol (65 °C) under continuous argon flow, which removed the methanol generated in the reaction mixture.

The product fructose oleate synthesized from oleic acid and OAME was confirmed by ^{1}H NMR spectroscopy. In acylation using oleic acid and methyl oleate, two spots of fructose ester were found on TLC. Previous studies [27, 28] on the enzymatic acylation of fructose with stearic acid using *Mucor miehei* and oleic acid using *C. antarctica* showed the formation of two monoacylated esters at C-1: fructofuranose and fructpyranose. These multiplets were overlapped while equating the protons of both β-1-fructopyranose-oleate and 1-fructofuranose-oleate esters. The reaction equilibrium shifted in favor of ester formation with more transesterification than esterification because methanol elimination was favored by the temperature of reaction mixture.

9.10 SORBITOL OLEATE

The inhibition study of corbyl esters for same concentrations had shown slightly higher inhibition than Vitamin C as a control showed slightly higher inhibition than vitamin C as a control (Figure 9.2), 26.36% inhibition for ascorbyl oleate in contrast with 21.69% for vitamin C after 30 min, when 30 μL (0.1 mmol) was added in 3 mL of 100 μM DPPH solution. Ascorbyl oleate showed the better antioxidant activity.

9.10.1 Antioxidant Activity

Esterification of sorbitol showed trace of diester on TLC with agreement with the previously studied monoesterification of sorbitol [27]. Monoester was separated from the unreacted starting material by chemical treatment and was confirmed by NMR spectroscopy.

9.11 ASCORBYL OLEATE

Earlier researchers reported that silica gel chromatography is unsuitable for purifying ascorbyl oleate because considerable losses of product occur; this could be due to oxidation and discoloration of product [30].

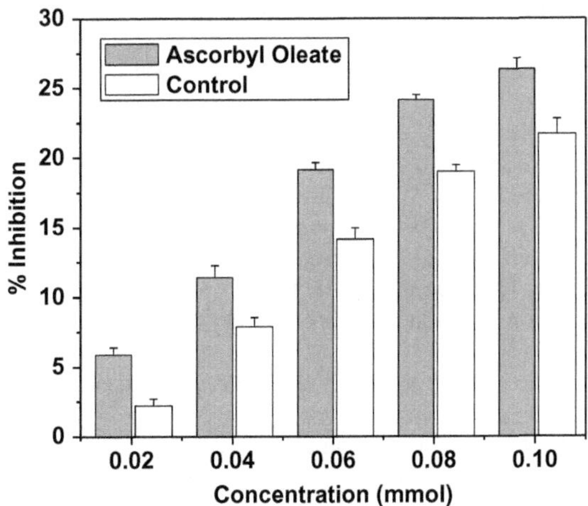

FIGURE 9.2 Scavenging effect of ascorbyl oleate in methanol.

TABLE 9.2
Yield of Isolated Esters

Ester	Esterification	Transesterification
Fructose oleate	16	21
Ascorbyl oleate	10	12
Sorbitol oleate	19	27

In the present work, ascorbyl oleate was successfully purified using flash chromatography. The column was eluted with the help of medium pressure gas flow, which gave rapid separation and minimized the time of contact with silica gel. Purified ascorbyl oleate was confirmed by NMR spectroscopy. On comparison, synthesized esters gave better yield through transesterification than through esterification [29] (Table 9.2).

Oxygen radical absorbance capacity has become the current industry standard for assessing antioxidant strength of foods, juices, and food additives [30, 31]. During food processing and storage, oxidation deteriorates food quality; degradation affects mostly lipids, carbohydrates, and proteins. Usually, synthetic antioxidants [32] such as butylhydroxyanisole or butylhydroxytoluene are used to delay this degradation; these antioxidants are volatile and easily decomposed at high temperatures and can produce toxic components [32]. Vitamin C [33], a well-known natural antioxidant, cannot be used directly in food due to its highly hydrophilic nature [32]. Therefore, ascorbyl fatty acid esters are very interesting because they can be synthesized for use as surfactant cum antioxidant. Lipase plays an important role in catalyzing fatty acid esterification and transesterification to produce biosurfactants or emulsifier/antioxidant in food products.

REFERENCES

[1] Mohan, P. K., Nakhla, G. & Yanful, E. K. Biokinetics of biodegradation of surfactants under aerobic, anoxic and anaerobic conditions. *Water Research* 40, 533–540 (2006).
[2] Desai, J. D. & Banat, I. M. Microbial production of surfactants and their commercial potential. *Microbiology and Molecular Biology Reviews* 61, 47–64 (1997).
[3] Kosaric, N. Biosurfactants and their applications for soil bioremediation. *Food Technology and Biotechnology* 39, 295–304 (2001)
[4] Md, F. Biosurfactant: production and application. *Journal of Petroleum & Environmental Biotechnology* 3(4), 124 (2012).
[5] Mulligan, C. N. Environmental applications for biosurfactants. *Environmental Pollution* 133, 183–198 (2005).
[6] Md, F. Biosurfactant: production and application. *Journal of Petroleum & Environmental Biotechnology* 3(4), 124 (2012)
[7] Seino, H., Uchibori, T., Nishitani, T. & Inamasu, S. Enzymatic synthesis of carbohydrate esters of fatty acid (I) esterification of sucrose, glucose, fructose and sorbitol. *Journal of the American Oil Chemists' Society* 61, 1761–1765, doi:10.1007/bf02582144 (1984).
[8] ter Haar, R. et al. Molecular sieves provoke multiple substitutions in the enzymatic synthesis of fructose oligosaccharide-lauryl esters. *Journal of Molecular Catalysis B: Enzymatic* 62, 183–189 (2010).
[9] Cao, L., Bornscheuer, U. T. & Schmid, R. D. Lipase-catalyzed solid-phase synthesis of sugar esters. Influence of immobilization on productivity and stability of the enzyme. *Journal of Molecular Catalysis B: Enzymatic* 6, 279–285 (1999).
[10] Khaled, N., Montet, D., Farines, M., Pina, M. & Graille, J. Synthesis of sugar monoesters by biocatalysis . *Oléagineux* 47, 181–189 (1992).
[11] Scheckermann, C., Schlotterbeck, A., Schmidt, M., Wray, V. & Lang, S. Enzymatic monoacylation of fructose by two procedures. *Enzyme and Microbial Technology* 17, 157–162 (1995).

[12] Sabeder, S., Habulin, M. & Knez, Z. Lipase-catalyzed synthesis of fatty acid fructose esters. *Journal of Food Engineering* 77, 880–886 (2006).
[13] Fiechter, A. Biosurfactants: Moving towards industrial application. *Trends in Biotechnology* 10, 208–217 (1992).
[14] Klibanov, A. M. Asymmetric transformations catalyzed by enzymes in organic-solvents. *Accounts of Chemical Research* 23, 114–120 (1990).
[15] Tarahomjoo, S. & Alemzadeh, I. Surfactant production by an enzymatic method. *Enzyme and Microbial Technology* 33, 33–37, doi:10.1016/S0141-0229(03)00085-1 (2003).
[16] Plat, T. & Linhardt, R. Syntheses and applications of sucrose-based esters. *Journal of Surfactants and Detergents* 4, 415–421, doi:10.1007/s11743-001-0196-y (2001).
[17] Chamouleau, F., Coulon, D., Girardin, M. & Ghoul, M. Influence of water activity and water content on sugar esters lipase-catalyzed synthesis in organic media. *Journal of Molecular Catalysis B: Enzymatic* 11, 949–954 (2001).
[18] Adamczak, M., Bornscheuer, U. T. & Bednarski, W. Synthesis of ascorbyl oleate by immobilized Candida antarctica lipases. *Process Biochemistry* 40, 3177–3180 (2005).
[19] Yan, Y. et al. Production of sugar fatty acid esters by enzymatic esterification in a stirred-tank membrane reactor: Optimization of parameters by response surface methodology. *Journal of the American Oil Chemists' Society* 78, 147–153, doi:10.1007/s11746-001-0235-x (2001).
[20] Sakaki, K. et al. Enzymatic synthesis of sugar esters in organic solvent coupled with pervaporation. *Desalination* 193, 260–266 (2006).
[21] Janssen, A. E. M., Klabbers, C., Franssen, M. C. R. & van't Riet, K. Enzymatic synthesis of carbohydrate esters in 2-pyrrolidone. *Enzyme and Microbial Technology* 13, 565–572 (1991).
[22] Oguntimein, G. B., Erdmann, H. & Schmid, R. D. Lipase catalysed synthesis of sugar ester in organic solvents. *Biotechnology Letters* 15, 175–180, doi:10.1007/bf00133019 (1993).
[23] Cao, L. Q., Fischer, A., Bornscheuer, U. T. & Schmid, R. D. Lipase-catalyzed solid phase synthesis of sugar fatty acid esters . *Biocatalysis and Biotransformation* 14, 269–283 (1997).
[24] Schlotterbeck, A., Lang, S., Wray, V. & Wagner, F. Lipase-catalyzed monoacylation of fructose. *Biotechnology Letters* 15, 61–64, doi:10.1007/bf00131554 (1993).
[25] Coulon, D., Girardin, M., Rovel, B. & Ghoul, M. Comparison of direct esterification and transesterification of fructose by Candida Antartica lipase. *Biotechnology Letters* 17, 183–186, doi:10.1007/bf00127985 (1995).
[26] Dubreucq, E., Ducret, A. & Lortie, R. Optimization of lipase-catalyzed sorbitol monoester synthesis in organic medium. *Journal of Surfactants and Detergents* 3, 327–333, doi:10.1007/s11743-000-0136-x (2000).
[27] Brandwilliams, W., Cuvelier, M. E. & Berset, C. Use of a free-radical method to evaluate antioxidant activity. *Food Science and Technology-Lebensm* 28, 25–30 (1995).
[28] Gorman, L. A. & Dordick, J. S. Organic solvents strip water off enzymes. *Biotechnology and Bioengineering* 39, 392–397, doi:10.1002/bit.260390405 (1992).
[29] Patil, D., De Leonardis, A. & Nag, A. Synthesis of biosurfactants from natural resources. *Journal of Food Biochemistry* 35(3), 747–758 (2011)
[30] Viklund, F., Alander, J. & Hult, K. Antioxidative properties and enzymatic synthesis of ascorbyl FA esters . *Journal of the American Oil Chemists' Society* 80, 795–799, doi:10.1007/s11746-003-0774-1 (2003).
[31] Cao, G., Alessio, H. M. & Cutler, R. G. Oxygen-radical absorbance capacity assay for antioxidants. *Free Radical Biology and Medicine* 14, 303–311 (1993).
[32] Halliwell, B. *Advances in Pharmacology*, Volume 38 (ed Sies Helmut), 3–20, Academic Press (1996).
[33] Yan, Y., Bornscheuer, U. T. & Schmid, R. D. Lipase-catalyzed synthesis of vitamin C fatty acid esters. *Biotechnology Letters* 21, 1051–1054 (1999), doi:10.1023/a:1005620125533.

10 Chemical Tools for the Synthesis and Application of Artificial Lipids

Federica A. Souto-Trinei, Brenda Portasany, and Roberto J. Brea

10.1 INTRODUCTION

Lipids, which are extensively distributed among fauna and flora, constitute a heterogeneous family of biomolecules that possess the shared attribute of having little or no solubility in water but considerable solubility in nonpolar substances [1]. They are essential for maintaining cell structure and function [2]. and play important roles in many biological processes, including membrane formation, cell signaling, and energy storage [3, 4]. Moreover, lipids are related to numerous diseases (Figure 10.1) [5]. For instance, they can be excessively stored, causing cellular and tissue damage that results in metabolic and neurodegenerative disorders such as Gaucher and Niemann–Pick diseases [6, 7].

Although the importance of lipids is well-acknowledged, deciphering their biosynthesis and precise roles has proven elusive. Research on them is challenging and limited in scope, not only because of their structural diversity but also because of the difficulty in reconstructing their natural synthetic pathways [2]. To better understand the structure and function of lipids, there is enormous interest in developing simple methodologies to generate artificial lipids in a controlled manner.

Noncanonical lipids are becoming more popular as research tools to overcome limitations and gain a better understanding of the production, trafficking, and functions of lipids [2]. In particular, artificial phospholipids capable of self-assembling into functional synthetic cells serve as models for explaining the origin of life, fabricating efficient bioreactors, and building smart drug delivery vectors [8, 9]. While most recent research has focused on developing in situ artificial membrane-forming phospholipids [10], it is worth mentioning that such synthesis is also important for the production of lipid species with specific properties, such as pharmacologic, fluorescence-labelled lipid derivatives and reversible lipids [11, 12]. Furthermore, artificial lipids produced in situ help explain important biological processes such as cell growth and division [13]. This type of synthesis leads to a biomimetic approach that is simpler and more practical than conventional synthesis, thus facilitating the study of the fundamental biology of lipids. Moreover, cellulosic synthesis of artificial lipids may provide insight into how natural lipids influence fundamental cellular properties such as membrane curvature, signaling, and communication. Additionally, current chemoselective methodologies enable the precise construction of synthetic lipids, thus answering fundamental questions about life functions and the biological effects of lipids.

10.2 CLASSIFICATION OF LIPIDS

Lipids are usually classified as simple or complex according to their intricacy [14]. Acylglycerols and waxes fall under the category of simple lipids, while complex lipids include phospholipids, glycolipids, and lipoproteins. Interestingly, compounds such as sterols, vitamins, and terpenes are soluble in organic solvents and are regularly associated with lipids.

DOI: 10.1201/9781003437413-10

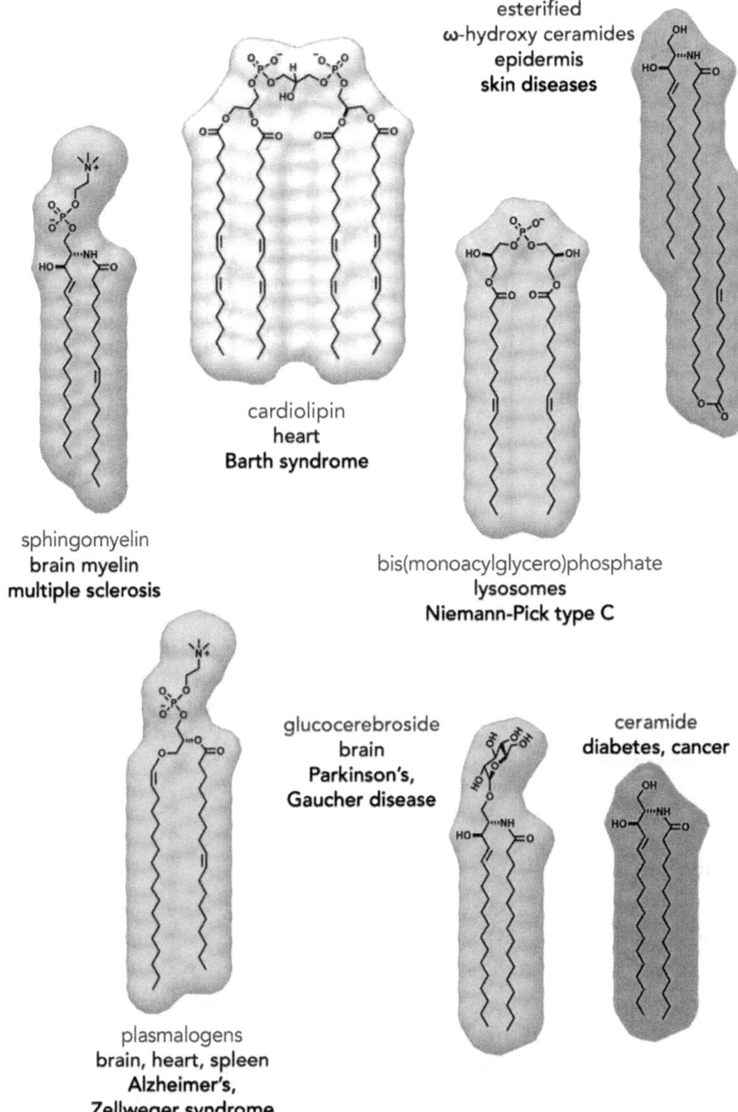

FIGURE 10.1 Representation of relevant lipid species, indicating where they display their main functions (in black) and their associated diseases (in red).

10.2.1 Simple Lipids

10.2.1.1 Acylglycerols

The majority of the body's fatty acids are combined with various alcohols, with a preference for glycerol, which produces acylglycerols (also called acylglycerides) (Figure 10.2) [14, 15]. The glycerol molecule is covalently bonded to one or two fatty acid chains via ester linkages, affording monoacylglycerol or diacylglycerol species, respectively, that are generally very hydrophobic [15].

10.2.1.2 Waxes

Essentially, waxes consist of a long-chain fatty acid linked to a long-chain alcohol through an ester bond (Figure 10.3) [14]. Therefore, they are usually defined as esters derived from monohydric

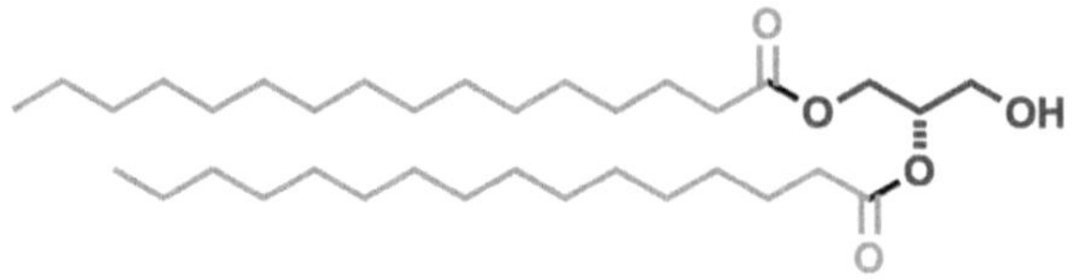

FIGURE 10.2 Chemical structure of dipalmitoylglycerol. This acylglycerol consists of two palmitic acid chains (in grey) covalently bonded to a glycerol molecule (in blue) through ester linkages.

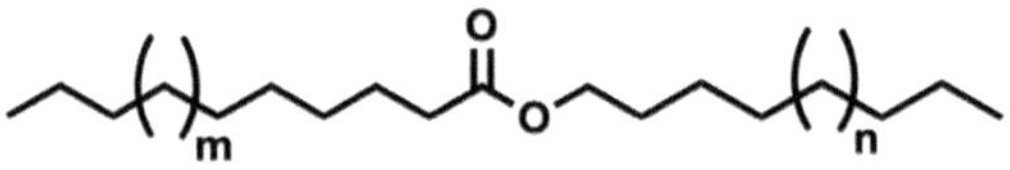

FIGURE 10.3 Schematic representation of waxes.

alcohols with long chain structures and more fatty acids. For instance, one of the most essential constituents found in beeswax is the ester formed from a 30-carbon alcohol and palmitic acid.

Waxes exhibit solidity at ambient temperature and are completely insoluble in aqueous solutions. These properties make them key biomolecules due their structural and energy-storage applicability. Typically, waxes contribute to skin lubrication and provide water-resistant shielding to hair and feathers; bees employ them in the construction of honeycombs. In plants, waxes create a protective coating on leaves and fruits. Plankton organisms possess a substantial amount of waxes, and cold-region marine animals consume and accumulate waxes that serve as energy reserves [14].

10.2.2 Complex Lipids

10.2.2.1 Phospholipids

Phospholipids are esters characterized by the presence of an alcohol (usually glycerol), phosphoric acid, and fatty acids (Figure 10.4) [14]. Certain tissues exhibit a high abundance of phospholipids that afford them specific chemical and physical properties as well as well-defined functions.

Phospholipids can be further categorized into glycerophospholipids (Figure 10.5), wherein the alcohol component is glycerol, and sphingophospholipids, wherein the alcohol component is sphingosine [14]. Glycerophospholipids, the most prevalent phospholipids, are primarily located in cellular membranes, although they are also found in minimal amounts within adipose tissues. Furthermore, they serve as reservoirs for bioactive agents, contribute to cellular communication networks, and function as anchoring mechanisms for proteins residing in cellular membranes.

The most abundant sphingophospholipid is sphingomyelin (Figure 10.6), which consists of an alcohol known as sphingosine, fatty acids, phosphoric acid, and choline [14]. The fatty acids are connected to sphingosine amine through an amide linkage, forming a fundamental structure called ceramide. Sphingomyelin plays a crucial role as a constituent of cell membranes, specifically in myelin sheaths in nervous tissue. Like glycerophosphates, sphingomyelin possesses a polar head and two nonpolar tails.

10.2.2.2 Glycolipids

Glycolipids consist of one or more monosaccharide components connected to a hydrophobic fragment through a glycosidic bond (Figure 10.7) [16, 17]. Various classes of glycolipids exist, each one characterized by different molecular structures in their backbone, including acylglycerols, sphingoids, ceramides, and sterols [18]. It is important to note that these classes are not uniform and exhibit distinct variations in their composition, such as the number and type of carbohydrate moieties or fatty acid residues. Due to their amphiphilic nature, these lipids display both biological and technological functionality.

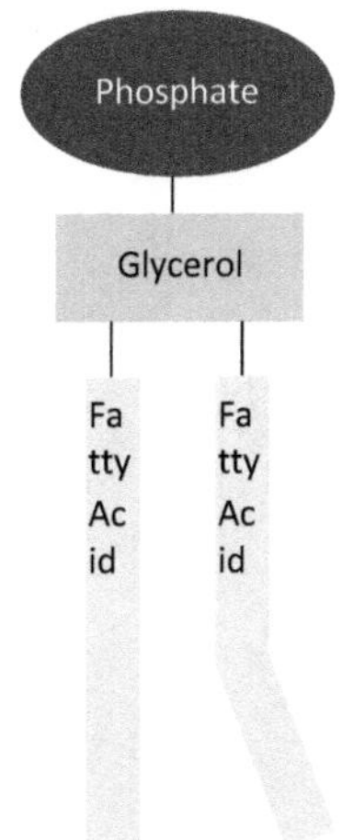

FIGURE 10.4 Schematic representation of a phospholipid.

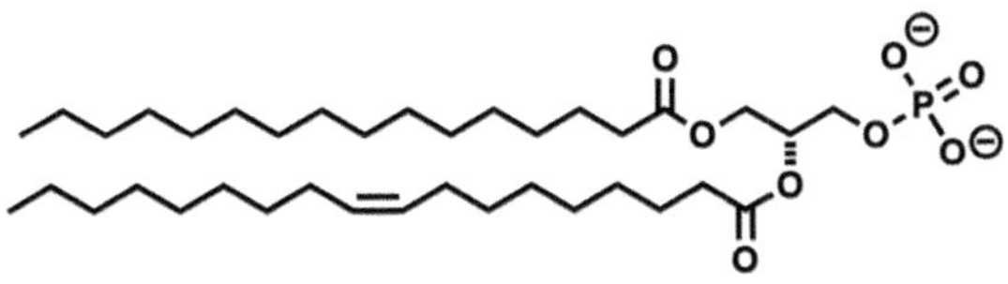

FIGURE 10.5 Chemical structure of a glycerophospholipid.

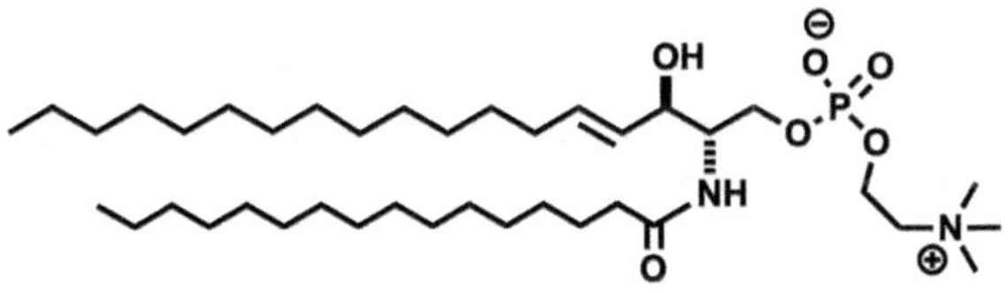

FIGURE 10.6 Chemical structure of a sphingomyelin.

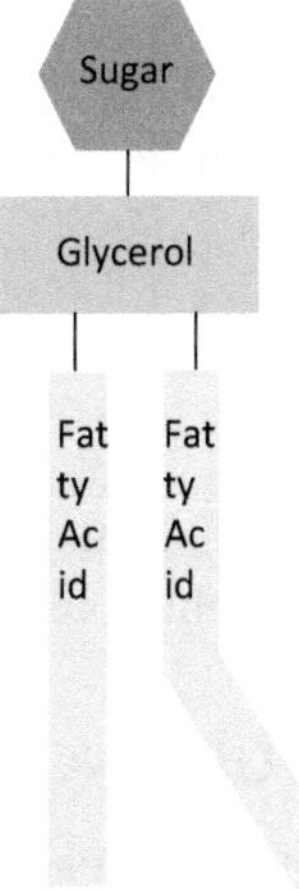

FIGURE 10.7 Schematic representation of a glycolipid.

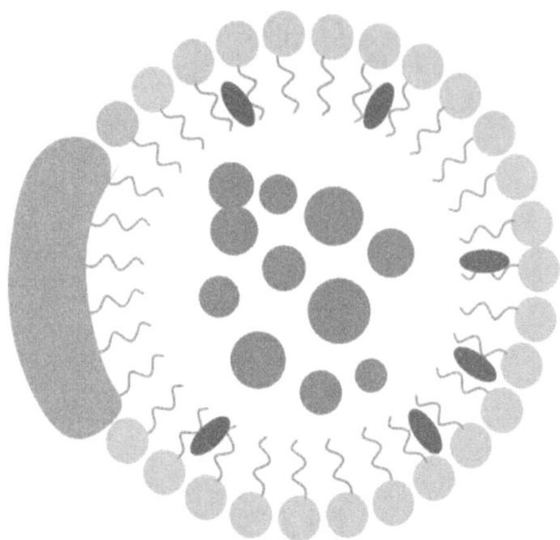

FIGURE 10.8 General structure of a lipoprotein.

10.2.2.3 Lipoproteins

Lipoproteins consist of spherical macromolecular complexes that encompass hydrophobic lipids, specifically triglycerides and cholesterol esters (Figure 10.8) [19]. These complexes are enveloped by a hydrophilic layer composed of phospholipids, free cholesterol, and apoproteins, which are specialized proteins exhibiting amphipathic characteristics [20].

10.3 IMPORTANCE OF ARTIFICIAL LIPIDS

Artificial lipids can be intentionally engineered with distinct chemical compositions and properties to satisfy a wide range of application requirements [11, 12]. Scientists can successfully modify the lipid's constitution to attain desired attributes such as stability, drug-carrying ability, and release kinetics [12]. In particular, synthetic lipids have a crucial role in the delivery of medication systems, specifically liposomes and lipid nanoparticles [12]. By customizing the lipid composition, researchers are capable of regulating drug encapsulation, release rates, and target specificity. This customization facilitates the creation of more efficacious and precisely targeted methods for drug delivery, reducing undesirable effects and improving therapeutic outcomes.

Noncanonical lipids are also integral constituents in the formulation of liposomal vaccines, gene therapies, and nucleic acid delivery systems [12]. These lipids perform the crucial functions of safeguarding and conveying delicate biomolecules such as mRNA and siRNA to their designated cellular destinations. Thus, they are necessary in advancing state-of-the-art biotechnological initiatives. Interestingly, biomimetic lipids can be used as key building blocks for the precise construction of cell analogues in the laboratory [10, 12, 13]. These artificial assemblies can give us a deeper understanding of relevant biological processes and shed light on how life emerged on Earth.

In brief, synthetic lipids provide considerable adaptability and flexibility that render them essential in diverse domains encompassing pharmaceuticals and materials science. Their capacity to imitate and regulate biological mechanisms and their contribution to the development of sophisticated drug transportation systems make them invaluable tools for scientific and technological progress.

10.4 SYNTHESIS OF MEMBRANE-FORMING ARTIFICIAL LIPIDS

There has been recent great interest in mimicking natural lipid synthesis pathways using simple precursors in aqueous media [10, 13]. In particular, several research groups have recently described chemoselective reactions for the de novo formation of synthetic membranes [21–23]. These robust

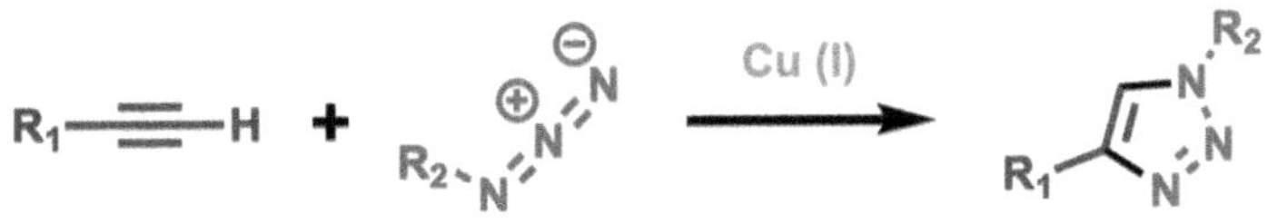

FIGURE 10.9 Schematic representation of a CuAAC reaction.

methods have offered a powerful toolbox to prepare artificial lipids chemically. In this section, we highlight relevant strategies that have been recently employed for the in situ generation of noncanonical lipids capable of driving the self-assembly of biomimetic membranes.

10.4.1 Copper-Catalyzed Click Chemistry

Copper-catalyzedazide-alkyne cycloaddition (CuAAC) is a highly suitable reaction for obtaining small molecules in a biorthogonal and regioselective way [24]. CuAAC is based on the cycloaddition between a terminal alkyne and an aliphatic azide in the presence of copper, which affords a 1,4-disubstituted-1,2,3-triazole (Figure 10.9); this cycloaddition was classified as a click chemistry reaction [25] that was highly efficient and stereospecific and that generates stable products under mild conditions without forming byproducts. Moreover, it is suitable in both organic and aqueous solvents, being compatible with a broad range of functional groups.

Taking this into account, Budin et al. employed a CuAAC reaction for the straightforward preparation of noncanonical phospholipids at physiological conditions (Figure 10.10) [26], thus exploring the use of bioorthogonal reactions for mimicking natural lipid generation. In their seminal study, an alkyne-modified lysophospholipid and an oleyl azide were coupled using a copper(I) catalyst and producing a biomimetic triazole-containing phospholipid (Figure 10.10) [26]. Interestingly, it was demonstrated that phospholipids obtained using this method share physical properties as natural ones, as the triazole does not interfere with the membrane-forming capabilities.

More recently, a self-reproducing system that can propel the repetitive synthesis and expansion of artificial phospholipid membranes was built using CuAAC [27]. Triazole phospholipids are continually formed by the regeneration of membrane-bound autocatalysts, thus simulating the spontaneous creation of membranes.

Hardy et al. further demonstrated that the CuAAC reaction could be light-promoted by ruthenium complexes under aqueous conditions [28]. Specifically, the ruthenium–copper electron transport chain drives the generation of membrane-forming artificial phospholipids. Taking this into consideration, Konetski et al. corroborated that the reaction could be spatiotemporally controlled, enabling the photoinduced construction of synthetic membranes [29]. Using a similar approach, Enomoto et al. were also able to spatiotemporally control phospholipid synthesis by generating copper(I) from a photosensitizer dyad, promoting CuAAC and subsequently driving spontaneous phospholipid membrane generation [30].

10.4.2 Native Chemical Ligation

Native chemical ligation (NCL) is one of the most powerful techniques for the synthesis and derivatization of polypeptides, small proteins, and nucleic acids [31, 32]. Brea et al. used this nonenzymatic and chemoselective approach to couple long-chain acylthioesters to cysteine-modified lysophospholipids, thus producing membrane-forming artificial phospholipids (Figure 10.11) [21]. Further studies demonstrated the powerful use of the NCL approach for the functional reconstitution of integral membrane proteins into synthetic membranes [33, 34]. Interestingly, Brea et al. also described a reversible NCL reaction that enabled the in situ remodeling of synthetic lipids, thus modulating changes in vesicle spatial organization, composition, and morphology [35].

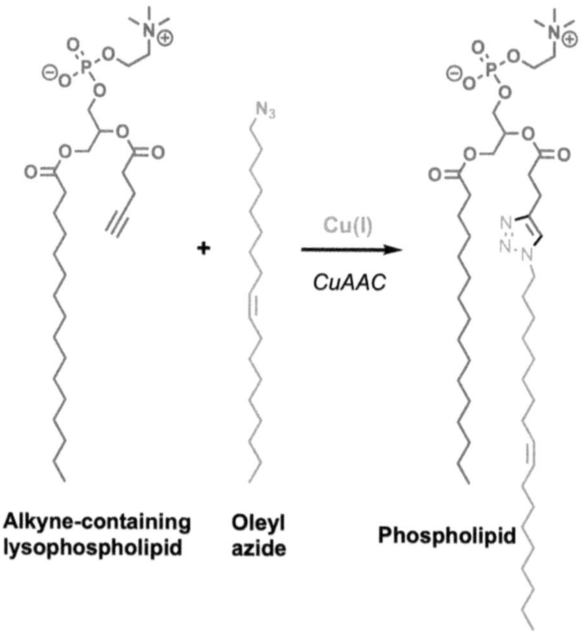

Cu(I)

CuAAC

Alkyne-containing lysophospholipid

Oleyl azide

Phospholipid

FIGURE 10.10 Synthesis of biomimetic phospholipids driven by a copper(I) CuAAC reaction.

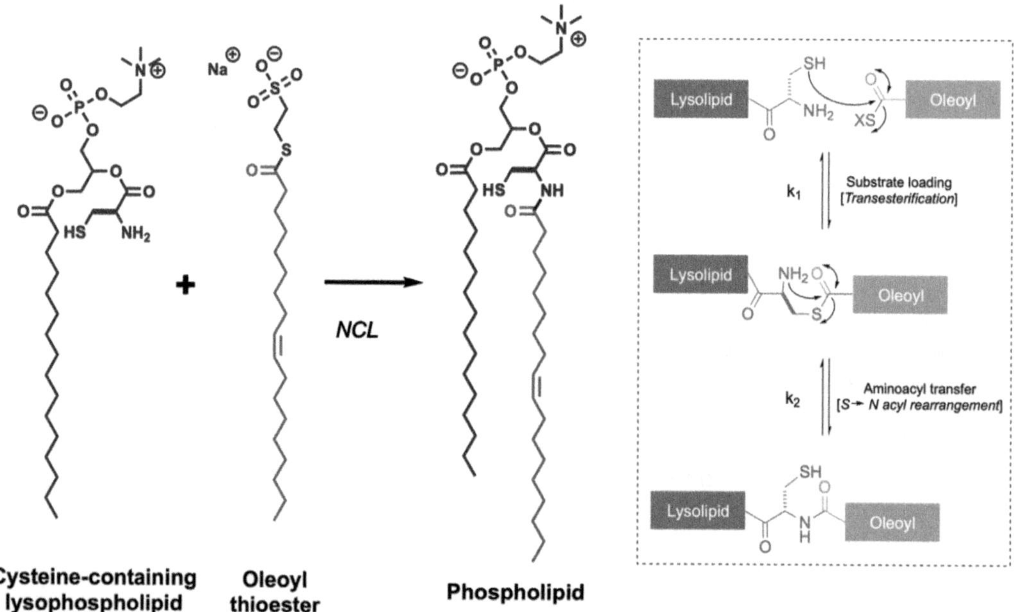

FIGURE 10.11 Synthesis of noncanonical phospholipid by NCL. The mechanism of the reaction is highlighted.

10.4.3 Direct Aminolysis

Direct aminolysis creates amide bonds by utilizing the natural nucleophilicity of *N*-terminal amines through reactions with *N*-acyl donors, most frequently C-terminal thioesters [36, 37]. Although these ligations are significantly slower than NCL [38], their reaction rates can be increased by the

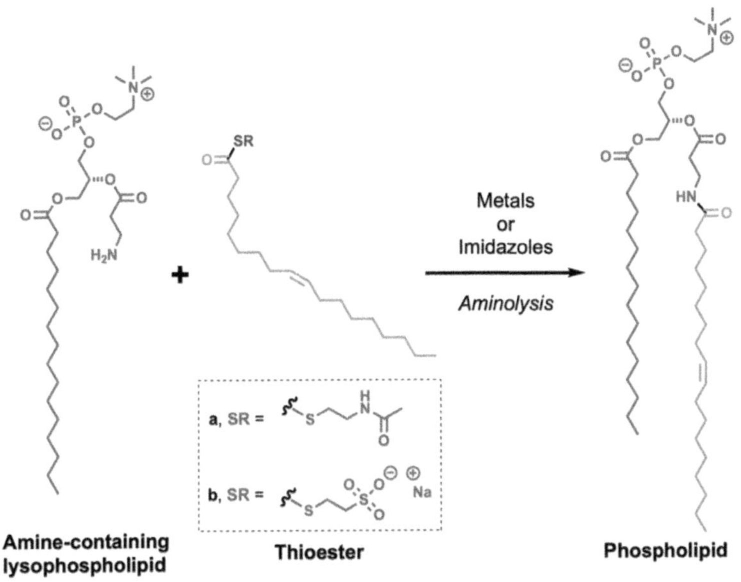

FIGURE 10.12 In situ formation of phospholipids by metal- or imidazole-promoted aminolysis ligations between an amine-functionalized lysophospholipid and oleoyl thioesters.

addition of catalysts such as metal ions [39] or imidazole [40]. Given the significance of this cysteine-free ligation strategy, novel approaches have recently emerged to explore its use for generating biomimetic phospholipids.

For instance, Souto-Trinei et al. described a methodology that employs metal and imidazole to promote in situ generation of membrane-forming artificial phospholipids via direct aminolysis between amine-modified lysophospholipids and fatty acyl thioesters (Figure 10.12) [41]. In brief, aminolysis has been a powerful nonenzymatic methodology for forming amidophospholipid membrane from simple water-soluble precursors. Moreover, their studies also showed the suitability of the approach for driving the formation of native phospholipids via direct esterification.

10.4.4 KAHA and KAT Ligation

Devaraj et al. recently described novel bioconjugation strategies to reproduce lipid synthesis by using KAHA and KAT ligation (Figure 10.13) [42], and Bode et al. developed amide bond-forming methodologies [43, 44]. These chemoselective ligations were based on the reaction between *O*-substituted hydroxylamines (HAs) and α-ketoacids (KAs) [43] or potassium acyltrifluoroborates (KATs) [44]. Applying both methods to the ligation of lipid fragments allowed Chen et al. [42] to rapidly synthesize noncanonical phospholipids in aqueous solution at physiological pH. These enzyme-free methodologies generate membranes from biocompatible water-soluble precursors, affording cell-like compartments. Interestingly, they also showed that KAT ligations facilitated the in situ formation of phospholipids in living cells.

10.4.5 Traceless Ceramide Ligation

Rudd et al. recently developed traceless ceramide ligation (TCL) for synthesizing natural ceramides in live cells [45]. This nonenzymatic methodology uses fatty acid salicylaldehyde esters and sphingosine as simple precursors, leading to in situ formation of ceramides (Figure 10.14). Ceramides are a relevant class of sphingolipids that participate in numerous cellular processes, including apoptosis,

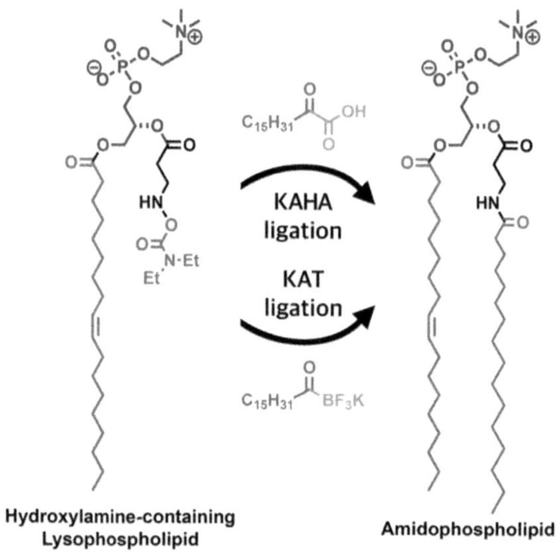

FIGURE 10.13 De novo formation of synthetic membrane-forming phospholipids using KAT and KAHA ligations.

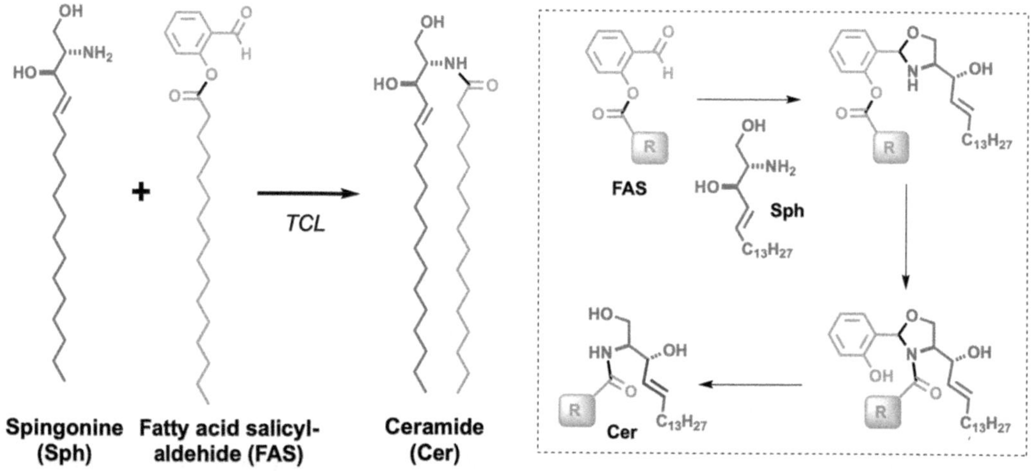

FIGURE 10.14 Formation of ceramides via traceless ceramide ligation between sphingosine and fatty acid salicylaldehyde esters. The mechanism of the reaction is highlighted.

gene regulation, and differentiation, and methods for their controlled production in living cells are in high demand. Spatiotemporal control of ceramide formation was achieved by TCL using photocaged coumarin sphingosine precursors, thus triggering apoptosis in tumor cells and revealing novel saturation-dependent apoptotic effects.

10.4.6 Chemoenzymatic Synthesis

Several research groups have recently developed innovative approaches to constructing artificial phospholipid membranes by inspiring themselves from natural pathways [12]. These strategies involve combining enzymes with chemical reactions, resulting in simpler methodologies.

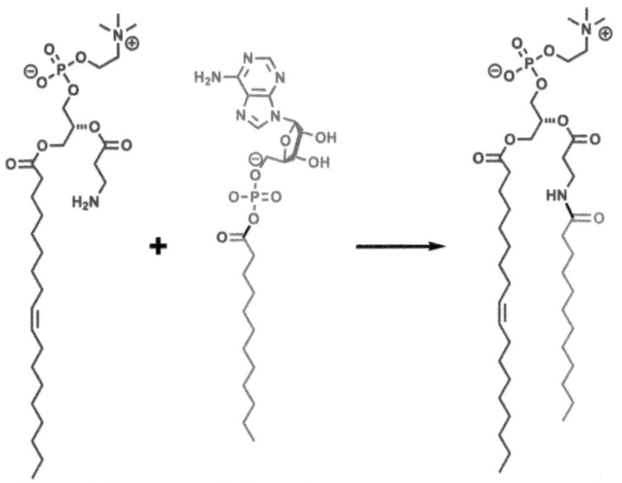

Amine-containing lysophospholipid
Fatty acyl adenylate
Phospholipid

FIGURE 10.15 Synthesis of amidophospholipids by chemoselective reaction between amine-functionalized lysophospholipids and dodecanoyl-AMP generated in situ by FadD10 using dodecanoic acid as precursor.

For instance, Devaraj et al. reported a chemoenzymatic methodology for generating synthetic membrane-forming phospholipids from water-soluble simple precursors (Figure 10.15) [46]. This breakthrough also sheds light on the origins of the molecular synthesis machinery within cells. The approach utilizes FadD10 [47]. a soluble mycobacterial ligase capable of catalyzing the formation of fatty acid adenylates (FAAs) from fatty acids, Mg2+, and ATP. FadD10 possesses an open conformation in its active site, facilitating the easy extraction and diffusion of the FAA product [47]. Once the FAA is formed, it selectively reacts with a synthetic amine-functionalized lysophospholipid, creating a new category of noncanonical amidophospholipids (Figure 10.15). These molecules spontaneously organize themselves into vesicles bound by membranes. This approach eliminates the requirement for preexisting membranes, enabling the synthesis of membranes without the need for a preexisting one.

More recently, Khanal et al. pioneered a groundbreaking chemoenzymatic approach to produce unconventional phospholipids from highly reactive building blocks, utilizing NCL as a pivotal chemoselective reaction (Figure 10.16a) [48]. In this biomimetic pathway, the ingenious deployment of a bacterial fatty acid synthase (cgFAS I) was employed to catalytically generate palmitoyl-CoA in situ from acetyl-CoA and malonyl-CoA, precursors that had hitherto remained untapped for direct reconstitution of phospholipid membranes. Subsequently, the formed palmitoyl-CoA underwent a transformative reaction with a tailor-made cysteine-functionalized lysophospholipid through the mechanism of NCL, thus affording an amidophospholipid that exhibits a remarkable propensity for spontaneous self-assembly into large-scale, micron-sized vesicles. Intriguingly, guanidine hydrochloride, cholesterol, and 1-decanol catalyzed the formation of more robust vesicles.

Taking into consideration the previous approaches, Bhattacharya et al. designed a one-pot route to prepare membrane-forming artificial phospholipids by repurposing the activity of FadD2, a mycobacterial fatty acyl-CoA ligase (Figure 10.16b) [49]. During an initial enzymatic step, fatty acids are successfully converted into coenzyme A thioesters (FA-CoAs) by FadD2. Next, the enzymatically generated FA-CoAs undergo NCL with cysteine-functionalized lysophospholipids to rapidly produce biomimetic phospholipids. Interestingly, the synthesis of noncanonical sphingolipids was also achieved by using an analogous cysteine-functionalized lysosphingomyelin as precursor. This chemoenzymatic lipid synthesis proved to be extremely effective at lipid synthesis in a cell-free transcription–translation system.

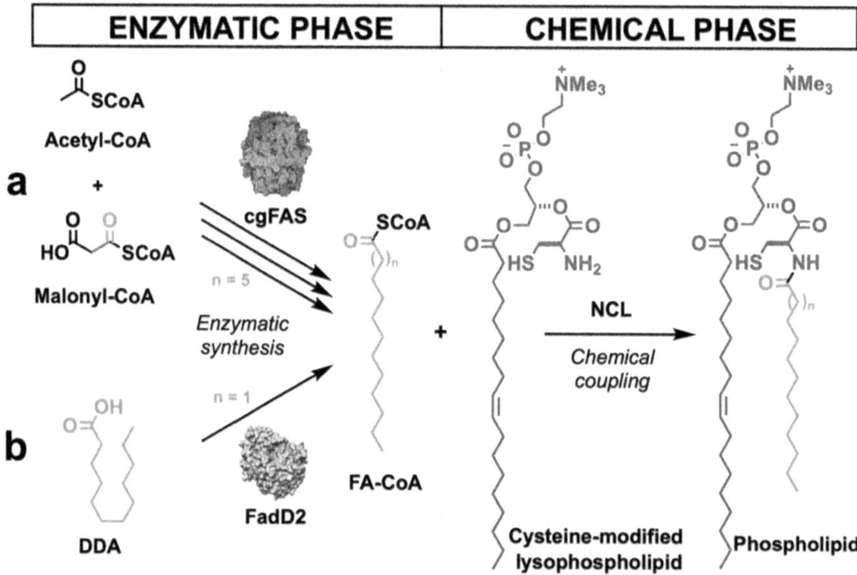

FIGURE 10.16 Chemoenzymatic synthesis of membrane-forming artificial phospholipids: (a) synthesis of amidophospholipids via NCL reaction between a cysteine-modified lysophospholipid and palmitoyl-CoA in situ generated by a bacterial type I FAS using acetyl-CoA, malonyl-CoA, and NADPH as precursors; (b) generating noncanonical phospholipids in which a cysteine-functionalized lysophospholipid undergoes NCL with fatty acyl-CoA thioesters generated enzymatically by a fatty acyl-CoA ligase (FadD2) using dodecanoic acid as precursor.

10.4.7 Other Strategies

The thiol-Michael click reaction offers enhanced biocompatibility by avoiding the need for copper (as happens with CuAAC) [50]. This reaction can be catalyzed by relatively weak bases and nucleophiles, accommodating the aqueous conditions required for in situ lipid formation. Moreover, it proceeds rapidly in mild conditions, affording the desired product without significant side reactions. Additionally, it is compatible with multiple thiol and alkene functionalities [50]. Considering this, Konetski et al. incorporated acrylate and thiol functionalities into complimentary lipid partners, facilitating the thiol-Michael formation of noncanonical tioether-containing phospholipids through the mediation of basic catalysts (Figure 10.17). Interestingly, the use of photobases for photoinitiated thiol-Michael reactions allows for spatiotemporal control in synthetic membrane formation.

Using a similar approach, Zhou et al. designed a photoinitiated thiolyne click chemistry reaction to rapidly produced ithioether analogues of phospholipids in aqueous solution [51]. These biomimetic phospholipids presented high resistance to phospholipases, making them potential candidates for drug delivery applications.

Combining aldehyde and amine amphiphiles afforded elegantly generated synthetic lipids through a chemoselective imine linkage [52]. For instance, Matsuo et al. reported an imine-based protocellular system formed by the self-assembly of a noncanonical oleyl phospholipid [22]. Interestingly, the reversibility of the imine bond added dynamic features to the lipids. Thus, Seoane et al. employed this approach to fabricating synthetic liposomes that respond to external stimuli, thus showing controlled release of relevant encapsulated biomolecules [53]. Further substitution of imines for oxime linkages allowed the construction of robust biomimetic dioxime-based liposomes in water via dual-oxime bond formation between two alkyl chains and a phosphocholine head group (Figure 10.18) [54]. Remarkably, the rapid reaction rates of both oxime ligations facilitate the dialkylation of the

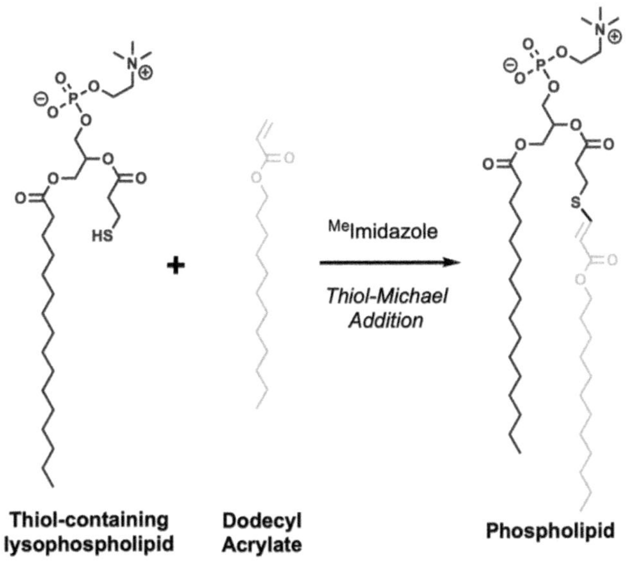

FIGURE 10.17 Synthesis of noncanonical phospholipids via thiol-Michael addition.

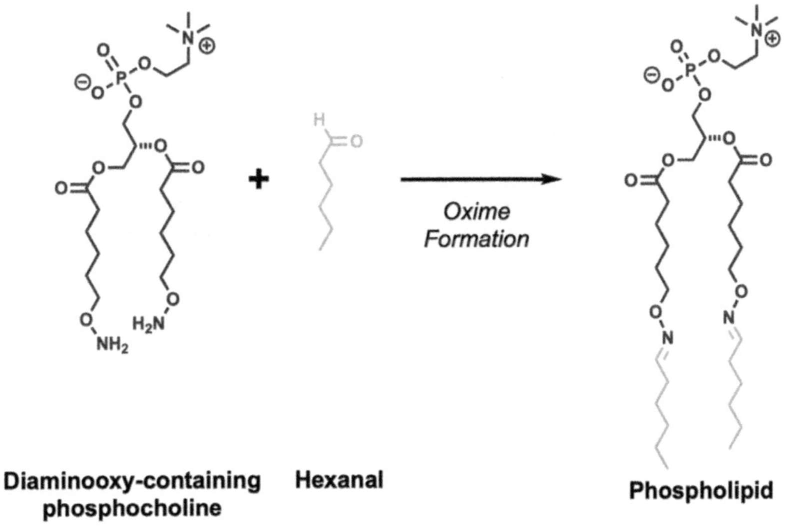

FIGURE 10.18 De novo synthesis of dioxime-based phospholipids via dual-oxime bond formation.

head group without the need for amphiphilic precursors. Moreover, this methodology was employed for direct protein reconstitution into synthetic membranes from bacterial extracts.

Histidine ligation has also been used to produce noncanonical phospholipids [55]. Interestingly, the imidazole ring of a histidine-modified lysophospholipid acted as a catalyst and promoted the coupling with long-chain acyl thioesters, thus generating novel biomimetic membrane-forming phospholipids. More recently, Xiong et al. reported on the formation of a new class of artificial phospholipids using a two-step procedure based on sulfur transfer and subsequent [2,3]-sigmatropic rearrangement [23]. The approach combines an oxyactamide moiety with a thiophthalimide group to rapidly produce an *ortho*-sulfiliminyl phenol product.

10.5 SYNTHETIC LIPIDS FOR IMAGING

The applications of artificial lipids in the field of imaging are extensive, continuously progressing alongside the advancements made in imaging technologies and our comprehension of cellular mechanisms. Interestingly, fully synthetic lipid analogues can act as efficient small-molecule probes for imaging the subcellular locations and dynamics of different classes of lipids within living cells.

10.5.1 Fluorescence Imaging

The most frequently employed technique for visualizing biomolecules of interest involves introducing an artificial fluorescent label on them. This labeling process uses of a fluorescent tag, a molecule that emits light upon excitation with a specific wavelength; adding such tags allows for the visualization of the molecule through the use of microscopy techniques. By attaching a fluorescent tag to the molecule of interest, it becomes possible to track its movement, localization, and interactions within a biological system. This method has been widely adopted in various fields of research, including biology, chemistry, and medicine, due to its versatility and applicability in studying molecular processes at a microscopic level.

Click chemistry has proven to be an incredibly effective tool to produce fluorescently labeled lipids [12]. To achieve this, lipids undergo a process wherein azide or alkyne functional groups are introduced into their structure. Following this modification, these lipids can then be reacted with fluorescent dyes that contain the complementary functional group, either an alkyne or an azide, through a CuAAC reaction. The outcome of this reaction is the formation of a covalent bond between the lipid and the fluorescent dye, ultimately producing a lipid molecule that is labelled with fluorescence. For instance, Neef et al. described a new class of synthetic phosphatidic acids modified with terminal alkynes or cyclooctyne groups that can be fluorescently tagged via CuAAC or strain-promoted azide-alkyne cycloaddition, showing lipid fluorescence distributed in several intracellular membranes (Figure 10.19) [56].

More recently, inverse electron-demand Diels–Alder (IEDDA) [57] bioorthogonal reactions between strained alkenes and tetrazines have inspired a novel collection of synthetic lipid imaging probes [58]. Taking this into consideration, Devaraj et al. described a phospholipid analogue containing a methyl cyclopropane on its head group that they used to label and subsequently visualize intracellular organelle membranes via IEDDA tagging with a fluorophore (Figure 10.20) [59].

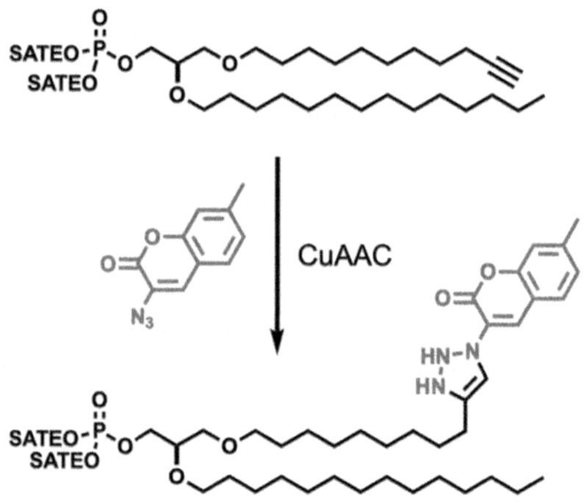

FIGURE 10.19 Synthesis of fluorescent phosphatidic acid derivatives for imaging of living cells. Azidocoumarins and synthetic phosphatidic acids are coupled via CuAAC to produce fluorescently labeled lipids.

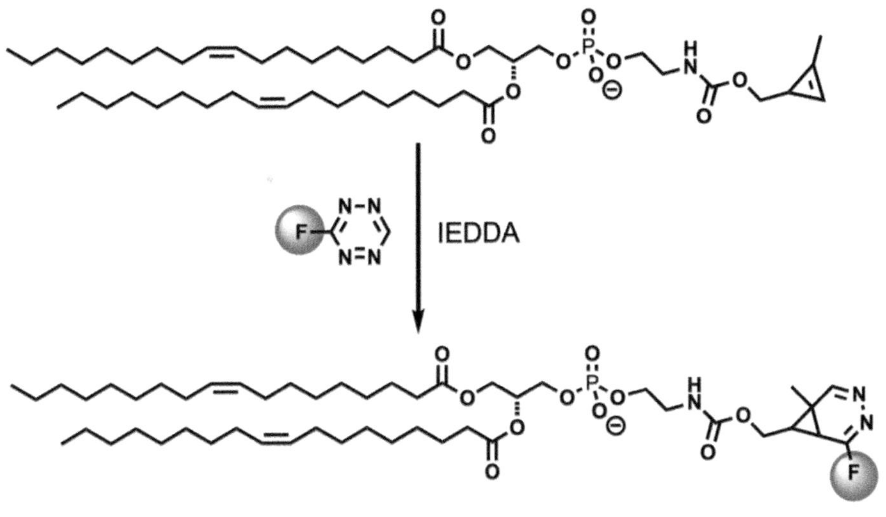

FIGURE 10.20 Synthesis of fluorescent phospholipids for imaging living cells. A phospholipid modified with a methyl cyclopropenein its head group is tagged with a tetrazine fluorophore via IEDDA, affording fluorescently labeled phospholipids.

10.5.2 Isotope-Labeled Lipids

Isotope-labelled artificial lipids are lipid species that have been altered by replacing one or more atoms with stable isotopes [60]. These labeled lipids are immensely valuable tools in biochemical and biological research, especially in studies related to lipid metabolism, interactions between lipids and proteins, as well as lipid signaling pathways [61]. Unlike radioactive isotopes, stable isotopes do not undergo decay, making them safe to use in experiments.

Isotope labeling of lipids can be achieved by employing either chemical synthesis or metabolic labeling [62, 63]. In chemical synthesis, stable isotopes are incorporated into lipid molecules during their chemical production [62]; metabolic labeling consists of feeding cells or organisms with substrates that contain isotope-labeled precursors (Figure 10.21) [63]. Subsequently, these precursors are assimilated into lipids via metabolic pathways that occur within the cellular milieu.

Common stable isotopes used for labeling lipids include deuterium (^{2}H), carbon-13 (^{13}C), nitrogen-15 (^{15}N), and oxygen-18 (^{18}O). These isotopes are frequently utilized due to their nonradioactive nature and the absence of any potential health hazards. Isotope-labeled lipids are often analyzed using techniques such as mass spectrometry, nuclear magnetic resonance (NMR) spectroscopy, stimulated Raman spectroscopy [64], and various chromatographic methods. These techniques allow researchers to detect and quantify labeled lipids in biological samples.

10.6 NONCANONICAL LIPIDS FOR THERAPEUTICS

10.6.1 Cationic Lipids

Cationic lipids are amphiphilic molecules that possess a positive charge [65]. They consist of three fundamental parts: a polar head group (typically amino groups), a hydrophobic domain (alkyl chains or cholesterol), and a linker that connects the polar head group with the nonpolar tail. These lipids form liposomes that are positively charged, usually in combination with a neutral helper lipid. However, certain drawbacks and limitations have been reported that hinder their broader successful application [66]. In contrast to liposomes, which are neutral or negatively charged, cationic liposomes

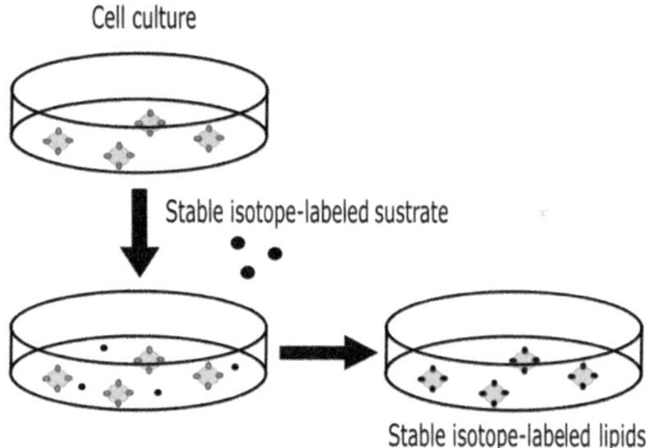

FIGURE 10.21 Production of stable isotope-labelled lipids using metabolic strategies.

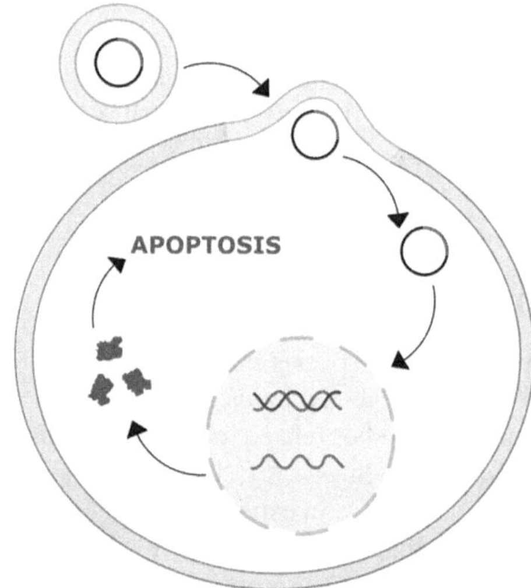

FIGURE 10.22 Apoptosis mediated by cationic lipid nanoparticles loaded with therapeutic drugs.

establish electrostatic interactions with molecules that have a negative charge, such as nucleic acids (including plasmids, messenger RNAs, and synthetic oligonucleotides) [65] as well as proteins and peptides. Remarkably, cationic lipid-based assemblies facilitate cellular uptake of these active biomolecules, making them extremely useful for the controlled apoptosis of tumor cells (Figure 10.22) [67].

Cationic liposomes have proven to be highly effective synthetic agents for delivering genes in laboratory settings [67]. Feigner and colleagues demonstrated that when DNA is mixed with cationic liposomes, it forms a cationic particle (referred to as a lipoplex) that can safeguard the DNA from enzymatic degradation and deliver it into cells by interacting with the cell membrane, which has a negative charge [68].

Cationic lipid species exhibit limited efficacy in delivery and gene silencing and heightened toxicity at elevated concentrations. Moreover, negatively charged macromolecules in serum and cell surfaces gives rise to unfavorable interactions, and the ability to penetrate tissues beyond the vasculature is hindered unless direct injection into the tissue occurs. The toxicity observed can be attributed to the large size of the complexes and the requisite high positive zeta potential for uptake. This arises from the charge ratio between cationic lipid species and nucleic acids, as well as the administration of lipoplexes at high doses. A good balance between the number of cationic residues and the chemical structure of the lipid tails enables the efficient fabrication of a new collection of cationic lipid nanostructures with optimal properties for both the encapsulation and the delivery of functional biomolecules.

10.6.2 PEG Lipids

Polyethylene glycol (PEG) lipids, which are also referred to as PEGylated lipids, belong to a category of PEG derivatives that are affixed to a lipid moiety, including 1,2-dimyristoyl-*rac*-glycerol and 1,2-distearoyl-*sn*-phosphoethanolamine(DSPE) [69]. PEG lipids have been extensively employed to enhance the duration of circulation for liposome-encapsulated medications and diminish instances of nonspecific uptake [69].

PEG lipids have the potential to present various influences on the characteristics of lipid nanoparticles [69]. The quantity of PEG lipids influences the size of the particles; furthermore, PEG lipids contribute to the stability of the particles by reducing their aggregation. In addition, the optimization of PEG can extend the duration of nanoparticles in the bloodstream by diminishing their clearance mediated by the kidneys and mononuclear phagocytes. Moreover, PEG lipids serve as a means of attaching specific ligands to particles to achieve targeted delivery. The magnitude of these effects is contingent upon the proportions and properties of the PEG lipids, including their molar mass and lipid chain length.

PEG-lipid micelles, which are primarily composed of conjugates of PEG and DSPE, or PEG-DSPE (Figure 10.23), have emerged as promising vehicles for drug delivery [70]. These micelles aim to address the limitations associated with new molecular entities that possess suboptimal biopharmaceutical attributes. The flexibility in PEG-DSPE design, combined with the simplicity of physically entrapping drugs, has distinguished PEG-lipid micelles as versatile and effective carriers for cancer therapy. They have been shown to overcome various constraints associated with poorly soluble drugs.

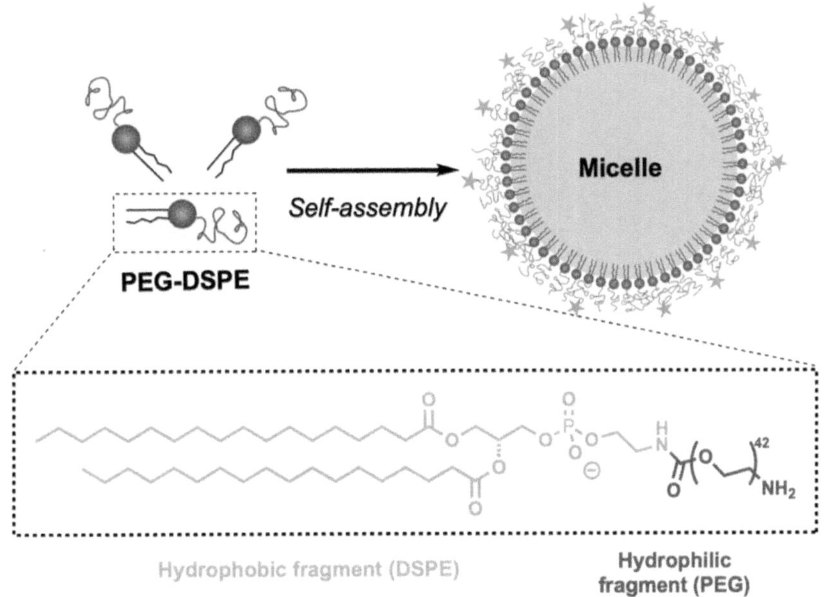

FIGURE 10.23 Construction of a PEG-DSPE micelle. The chemical structure of the block copolymer of PEG–DSPE is highlighted.

10.7 ARTIFICIAL PHOSPHOLIPIDS: LIPOSOMAL DRUG DELIVERY SYSTEMS

Phospholipids have a notable function in drug transport, specifically in liposomal drug transport [12]. Liposomes are small fluid-filled sacs that consist of one or more layers of phospholipids. These lipid vesicles can encapsulate both hydrophilic and hydrophobic drugs within their aqueous core and lipid bilayers, respectively (Figure 10.13) [71].

Liposomal drug delivery offers numerous benefits including enhanced drug solubility, regulated release, and precise targeting. The incorporation of ligands onto liposome surfaces enables the directed delivery of drugs to particular cells or tissues, thereby maximizing therapeutic efficacy. Moreover, this delivery method minimizes the risk of toxicity by limiting the exposure of healthy tissues to the administered drugs [72].

Conjugation of membrane-forming lipids to active biomolecules has recently undergone enormous interest due to potential drug delivery applications (Figure 10.24) [12]. The hydrophobic character of artificial lipids enhances both the drug loading and the stability of the corresponding lipid-drug conjugates. Lipids can self-associate into multiple assemblies, including emulsions, liposomes, micelles, and nanoparticles (Figure 10.24a). Taking these features into consideration, a wide collection of lipids (fatty acids, phospholipids, glycerides) have been conjugated to diverse drugs (e.g., doxorubicin, taxol) using prominent chemical strategies (Figure 10.24b) [12, 73]. Dynamic methodologies are generally preferred for selectively releasing the cargo upon the cleavage of the lipid assembly by an endogenous enzyme [74].

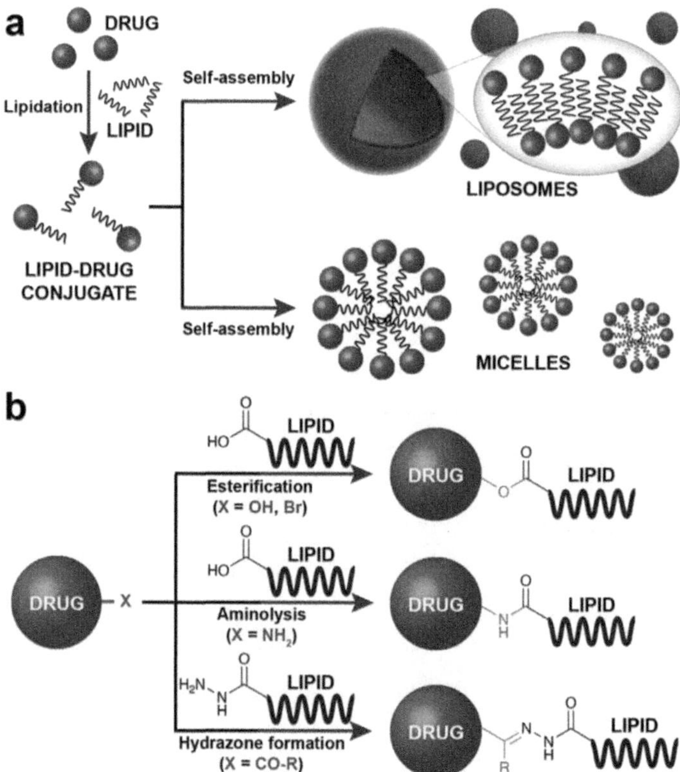

FIGURE 10.24 Lipid modification for the fabrication of drug carriers: (a) schematic representation of the construction of drug delivery systems based on the lipidation of the drug. First, the drug is lipidated, affording a lipid-drug conjugate that self-assembles into different structures such as liposomes and micelles; (b) common strategies for preparing lipid-drug conjugates.

10.8 CONCLUSIONS AND FUTURE OUTLOOK

In this chapter, we covered recent strategies for the synthesis of noncanonical lipids, which have potential applications on imaging, therapeutics, drug delivery carriers, and even origin of life studies [75–79]. Given the biological importance of mimicking natural membrane generation, we highlighted several methodologies for the in situ formation of membrane-forming artificial lipids.

Chemoselective reactions allow for artificial lipid synthesis, as well as conjugation of lipids to other lipids, fluorophores, and small-molecule drugs. Interestingly, synthetic tools that promote the formation of biomimetic cells and/or help monitor lipid biosynthesis and trafficking dynamics have become valuable for shedding light on fundamental biological signaling pathways and membrane behavior. Advances in synthetic lipid synthesis will provide more extensive research on these relatively novel families of biomimetic compounds. Moreover, use of artificial lipids will shed light on multiple aspects of lipid biology, as well as offering deeper understanding of membrane biophysics.

ACKNOWLEDGMENTS

This work was supported by the Agencia Estatal de Investigación and the Ministerio de Ciencia e Innovación [PID2021-128113NA-I00]. R.J.B. also thanks the Agencia Estatal de Investigación and the Ministerio de Ciencia e Innovación for his Ramón y Cajal contract (RYC2020-030065-I).

REFERENCES

[1] Blanco, A.; Blanco, G. 05—Lipids. In *Medical Biochemistry*; 2017; pp. 99–119. https://doi.org/10.1016/B978-0-12-803550-4/00005-7.

[2] Flores, J.; White, B. M.; Brea, R. J.; Baskin, J. M.; Devaraj, N. K. Lipids: Chemical Tools for Their Synthesis, Modification, and Analysis. *Chem Soc Rev* **2020**, *49* (14), 4602. https://doi.org/10.1039/D0CS00154F.

[3] Casares, D.; Escribá, P. V.; Rosselló, C. A. Membrane Lipid Composition: Effect on Membrane and Organelle Structure, Function and Compartmentalization and Therapeutic Avenues. *Int J Mol Sci* **2019**, *20* (9). https://doi.org/10.3390/IJMS20092167.

[4] Santos, A. L.; Preta, G. Lipids in the Cell: Organisation Regulates Function. *Cell Mol Life Sci* **2018**, *75* (11), 1909–1927. https://doi.org/10.1007/S00018-018-2765-4.

[5] Natesan, V.; Kim, S. J. Lipid Metabolism, Disorders and Therapeutic Drugs—Review. *BiomolTher (Seoul)* **2021**, *29* (6), 596–604. https://doi.org/10.4062/biomolther. 2021.122.

[6] Roh, J.; Subramanian, S.; Weinreb, N. J.; Kartha, R. V. Gaucher Disease—More than Just a Rare Lipid Storage Disease. *J Mol Med* **2022**, *100* (4), 499–518. https://doi.org/10.1007/s00109-021-02174-z.

[7] Pfrieger, F. W. The Niemann-Pick Type Diseases—A Synopsis of Inborn Errors in Sphingolipid and Cholesterol Metabolism. *Prog Lipid Res* **2023**, *90* (February), 101225. https://doi.org/10.1016/j.plipres.2023.101225.

[8] Enomoto, T.; Brea, R. J.; Bhattacharya, A.; Devaraj, N. K. In Situ Lipid Membrane Formation Triggered by Intramolecular Photoinduced Electron Transfer. *Langmuir* **2018**, *34* (3), 750–755. https://doi.org/10.1021/ACS.LANGMUIR.7B02783/ASSET/IMAGES/LARGE/LA-2017-02783J_0005.JPEG.

[9] Kraft, J. C.; Freeling, J. P.; Wang, Z.; Ho, R. J. Y. Emerging Research and Clinical Development Trends of Liposome and Lipid Nanoparticle Drug Delivery Systems. *J Pharm Sci* **2014**, *103* (1), 29–52. https://doi.org/10.1002/JPS.23773.

[10] Bhattacharya, A.; Brea, R. J.; Devaraj, N. K. De Novo Vesicle Formation and Growth: An Integrative Approach to Artificial Cells. *Chem Sci* **2017**, *8* (12), 7912–7922. https://doi.org/10.1039/C7SC02339A.

[11] Lomba-Riego, L.; Calvino-Sanles, E.; Brea, R. J. In Situ Synthesis of Artificial Lipids. *CurrOpin Chem Biol* **2022**, *71*, 102210. https://doi.org/10.1016/j.cbpa.2022.102210.

[12] Flores, J.; White, B. M.; Brea, R. J.; Baskin, J. M.; Devaraj, N. K. Lipids: Chemical Tools for Their Synthesis, Modification, and Analysis. *Chem Soc Rev* **2020**, *49* (14), 4602–4614. https://doi.org/10.1039/d0cs00154f.

[13] Brea, R. J.; Hardy, M. D.; Devaraj, N. K. Towards Self-Assembled Hybrid Artificial Cells: Novel Bottom-up Approaches to Functional Synthetic Membranes. *Chem Eur J* **2015**, *21* (36), 12564–12570. https://doi.org/10.1002/chem.201501229.

[14] Fahy, E.; Cotter, D.; Sud, M.; Subramaniam, S. Lipid Classification, Structures and Tools. *Biochim Biophys Acta Mol Cell Biol Lipids* **2011**, *1811* (11), 637–647. https://doi.org/10.1016/j.bbalip.2011.06.009.
[15] Mao, Y.; Lee, Y. Y.; Xie, X.; Wang, Y.; Zhang, Z. Preparation, Acyl Migration and Applications of the Acylglycerols and Their Isomers: A Review. *J Funct Foods* **2023**, *106* (February), 105616. https://doi.org/10.1016/j.jff.2023.105616.
[16] Jala, R. C. R.; Vudhgiri, S.; Kumar, C. G. A Comprehensive Review on Natural Occurrence, Synthesis and Biological Activities of Glycolipids. *Carbohydr Res* **2022**, *516* (April), 108556. https://doi.org/10.1016/j.carres.2022.108556.
[17] Osawa, T.; Fujikawa, K.; Shimamoto, K. Structures, Functions, and Syntheses of Glycero-Glycophospholipids. *Front Chem* **2024**, *12* (February), 1–15. https://doi.org/10.3389/fchem.2024.1353688.
[18] Lafiandra, D.; Masci, S.; Sissons, M.; Dornez, E.; Delcour, J. A.; Courtin, C. M.; Caboni, M. F. Kernel Components of Technological Value. In *Durum Wheat Chemistry and Technology: Second Edition*; 2012; pp. 85–124. https://doi.org/10.1016/B978-1-891127-65-6.50011-8.
[19] Braun, V.; Hantke, K. *Lipoproteins: Structure, Function, Biosynthesis*; Springer International Publishing. https://doi.org/10.1007/978-3-030-18768-2.
[20] Griffin, B. A. Lipid Metabolism. *Surgery (Oxford)* **2009**, *27* (1), 1–5. https://doi.org/10.1016/J.MPSUR.2008.12.003.
[21] Brea, R. J.; Cole, C. M.; Devaraj, N. K. In Situ Vesicle Formation by Native Chemical Ligation. *Angew Chem—Int Ed* **2014**, *53* (51), 14102–14105. https://doi.org/10.1002/anie.201408538.
[22] Matsuo, M.; Ohyama, S.; Sakurai, K.; Toyota, T.; Suzuki, K.; Sugawara, T. A Sustainable Self-Reproducing Liposome Consisting of a Synthetic Phospholipid. *Chem Phys Lipids* **2019**, *222*, 1–7. https://doi.org/10.1016/j.chemphyslip.2019.04.007.
[23] Xiong, F.; Lu, L.; Sun, T. Y.; Wu, Q.; Yan, D.; Chen, Y.; Zhang, X.; Wei, W.; Lu, Y.; Sun, W. Y.; Li, J. J.; Zhao, J. A Bioinspired and Biocompatible Ortho-Sulfiliminyl Phenol Synthesis. *Nat Commun* **2017**, *8* (15912). https://doi.org/10.1038/ncomms15912.
[24] Rostovtsev, V. V.; Green, L. G.; Fokin, V. V; Sharpless, K. B. A Stepwise Huisgen Cycloaddition Process: Copper()-Catalyzed Regioselective “Ligation” of Azides and Terminal Alkynes. *Angew Chem Int Ed* **2002**, *41* (14), 2596–2599.
[25] Castro, V.; Rodríguez, H.; Albericio, F. CuAAC: An Efficient Click Chemistry Reaction on Solid Phase. *ACS Comb Sci* **2016**, *18* (1), 1–14. https://doi.org/10.1021/acscombsci.5b00087.
[26] Devaraj, N. K. In Situ Synthesis of Phospholipid Membranes. *J Org Chem* **2017**, *82* (12), 5997–6005. https://doi.org/10.1021/ACS.JOC.7B00604/ASSET/IMAGES/LARGE/JO-2017-00604B_0001.JPEG.
[27] Hardy, M. D.; Yang, J.; Selimkhanov, J.; Cole, C. M.; Tsimring, L. S.; Devaraj, N. K. Self-Reproducing Catalyst Drives Repeated Phospholipid Synthesis and Membrane Growth. *Proc Natl Acad Sci USA* **2015**, *112* (27), 8187–8192. https://doi.org/10.1073/pnas.1506704112.
[28] Hardy, M. D.; Konetski, D.; Bowman, C. N.; Devaraj, N. K. Ruthenium Photoredox-Triggered Phospholipid Membrane Formation. *Org Biomol Chem* **2016**, *14* (24), 5555–5558. https://doi.org/10.1039/c6ob00290k.
[29] Konetski, D.; Gong, T.; Bowman, C. N. Photoinduced Vesicle Formation via the Copper-Catalyzed Azide-Alkyne Cycloaddition Reaction. *Langmuir* **2016**, *32* (32), 8195–8201. https://doi.org/10.1021/acs.langmuir.6b02043.
[30] Enomoto, T.; Brea, R. J.; Bhattacharya, A.; Devaraj, N. K. In Situ Lipid Membrane Formation Triggered by Intramolecular Photoinduced Electron Transfer. *Langmuir* **2018**, *34* (3), 750–755. https://doi.org/10.1021/acs.langmuir.7b02783.
[31] Vazquez, O.; Seitz, O. Templated Native Chemical Ligation: Peptide Chemistry Beyond Protein Synthesis. *J Pept Sci* **2014**, *20*, 78–86. https://doi.org/10.1002/psc.2602.
[32] Dawson, P. E.; Muir, T. W.; Clark-Lewis, I.; Kent, S. B. H. Synthesis of Proteins by Native Chemical Ligation. *Science (1979)* **1994**, *266*, 776–779.
[33] Cole, C. M.; Brea, R. J.; Kim, Y. H.; Hardy, M. D.; Yang, J.; Devaraj, N. K. Spontaneous Reconstitution of Functional Transmembrane Proteins during Bioorthogonal Phospholipid Membrane Synthesis. *Angew Chem Int Ed* **2015**, *54* (43), 12738–12742. https://doi.org/10.1002/anie.201504339.
[34] Brea, R. J.; Cole, C. M.; Lyda, B. R.; Ye, L.; Prosser, R. S.; Sunahara, R. K.; Devaraj, N. K. In Situ Reconstitution of the Adenosine A2A Receptor in Spontaneously Formed Synthetic Liposomes. *J Am Chem Soc* **2017**, *139* (10), 3607–3610. https://doi.org/10.1021/jacs.6b12830.
[35] Brea, R. J.; Rudd, A. K.; Devaraj, N. K. Nonenzymatic Biomimetic Remodeling of Phospholipids in Synthetic Liposomes. *Proc Natl Acad Sci USA* **2016**, *113* (31), 8589–8594. https://doi.org/10.1073/pnas.1605541113.

[36] Payne, R. J.; Ficht, S.; Greenberg, W. A.; Wong, C. H. Cysteine-Free Peptide and Glycopeptide Ligation by Direct Aminolysis. *Angew Chem Int Ed* **2008**, *47* (23), 4411–4415. https://doi.org/10.1002/anie.200705298.

[37] Agrigento, P.; Albericio, F.; Chamoin, S.; Dacquignies, I.; Koc, H.; Eberle, M. Facile and Mild Synthesis of Linear and Cyclic Peptides via Thioesters. *Org Lett* **2014**, *16* (15), 3922–3925. https://doi.org/10.1021/ol501669n.

[38] Burke, H. M.; McSweeney, L.; Scanlan, E. M. Exploring Chemoselective S-to-N Acyl Transfer Reactions in Synthesis and Chemical Biology. *Nat Commun* **2017**, *8* (May), 1–16. https://doi.org/10.1038/ncomms15655.

[39] Kawakami, T.; Yoshimura, S.; Aimoto, S. Synthesis of Reaper, a Cysteine-Containing Polypeptide, Using a Peptide Thioester in the Presence of Silver Chloride as an Activator. *Tetrahedron Lett* **1998**, *39* (43), 7901–7904. https://doi.org/10.1016/S0040-4039(98)01752-3.

[40] Li, Y.; Yongye, A.; Giulianotti, M.; Martinez-Mayorga, K.; Yu, Y.; Houghten, R. A. Synthesis of Cyclic Peptides through Direct Aminolysis of Peptide Thioesters Catalyzed by Imidazole in Aqueous Organic Solutions. *J Comb Chem* **2009**, *11* (6), 1066–1072. https://doi.org/10.1021/cc900100z.

[41] Souto-Trinei, F. A.; Brea, R. J.; Devaraj, N. K. Biomimetic Construction of Phospholipid Membranes by Direct Aminolysis Ligations. *Interface Focus* **2023**, *13* (5). https://doi.org/10.1098/RSFS.2023.0019.

[42] Chen, J.; Brea, R. J.; Fracassi, A.; Cho, C. J.; Wong, A. M.; Salvador-Castell, M.; Sinha, S. K.; Budin, I.; Devaraj, N. K. Rapid Formation of Non-Canonical Phospholipid Membranes by Chemoselective Amide-Forming Ligations with Hydroxylamines. *Angew Chem Int Ed* **2024**, *63* (1). https://doi.org/10.1002/anie.202311635.

[43] Bode, J. W.; Fox, R. M.; Baucom, K. D. Chemoselective Amide Ligations by Decarboxylative Condensations of N-Alkylhydroxylamines and α-Ketoacids. *Angew Chem Int Ed* **2006**, *45* (8), 1248–1252. https://doi.org/10.1002/anie.200503991.

[44] Dumas, A. M.; Molander, G. A.; Bode, J. W. Amide-Forming Ligation of Acyltrifluoroborates and Hydroxylamines in Water. *Angew Chem Int Ed* **2012**, *51* (23), 5683–5686. https://doi.org/10.1002/anie.201201077.

[45] Rudd, A. K.; Devaraj, N. K. Traceless Synthesis of Ceramides in Living Cells Reveals Saturation-Dependent Apoptotic Effects. *Proc Natl Acad Sci USA* **2018**, *115* (29), 7485–7490. https://doi.org/10.1073/pnas.1804266115.

[46] Bhattacharya, A.; Brea, R. J.; Niederholtmeyer, H.; Devaraj, N. K. A Minimal Biochemical Route towards de Novo Formation of Synthetic Phospholipid Membranes. *Nat Commun* **2019**, *10* (1), 1–8. https://doi.org/10.1038/s41467-018-08174-x.

[47] Liu, Z.; Ioerger, T. R.; Wang, F.; Sacchettini, J. C. Structures of Mycobacterium Tuberculosis FadD10 Protein Reveal a New Type of Adenylate-Forming Enzyme. *J Biol Chem* **2013**, *288* (25), 18473–18483. https://doi.org/10.1074/jbc.M113.466912.

[48] Khanal, S.; Brea, R. J.; Burkart, M. D.; Devaraj, N. K. Chemoenzymatic Generation of Phospholipid Membranes Mediated by Type I Fatty Acid Synthase. *J Am Chem Soc* **2021**, *143* (23), 8533–8537. https://doi.org/10.1021/jacs.1c02121.

[49] Bhattacharya, A.; Cho, C. J.; Brea, R. J.; Devaraj, N. K. Expression of Fatty Acyl-CoA Ligase Drives One-Pot de Novo Synthesis of Membrane-Bound Vesicles in a Cell-Free Transcription-Translation System. *J Am Chem Soc* **2021**, *143* (29), 11235–11242. https://doi.org/10.1021/jacs.1c05394.

[50] Konetski, D.; Baranek, A.; Mavila, S.; Zhang, X.; Bowman, C. N. Formation of Lipid Vesicles in Situ Utilizing the Thiol-Michael Reaction. *Soft Matter* **2018**, *14* (37), 7645–7652. https://doi.org/10.1039/C8SM01329B.

[51] Zhou, C. Y.; Wu, H.; Devaraj, N. K. Rapid Access to Phospholipid Analogs Using Thiol-Yne Chemistry. *Chem Sci* **2015**, *6* (7), 4365–4372. https://doi.org/10.1039/c5sc00653h.

[52] Takakura, K.; Yamamoto, T.; Kurihara, K.; Toyota, T.; Ohnuma, K.; Sugawara, T. Spontaneous Transformation from Micelles to Vesicles Associated with Sequential Conversions of Comprising Amphiphiles within Assemblies. *Chem Commun* **2014**, *50* (17), 2190–2192. https://doi.org/10.1039/c3cc47786j.

[53] Seoane, A.; Brea, R. J.; Fuertes, A.; Podolsky, K.; Devaraj, N. K. Biomimetic Generation and Remodeling of Phospholipid Membranes by Dynamic Imine Chemistry. *J Am Chem Soc* **2018**, *140* (27), 8388–8391. https://doi.org/10.1021/jacs.8b04557.

[54] Flores, J.; Brea, R. J.; Lamas, A.; Fracassi, A.; Salvador-Castell, M.; Xu, C.; Baiz, C. R.; Sinha, S. K.; Devaraj, N. K. Rapid and Sequential Dual Oxime Ligation Enables De Novo Formation of Functional

Synthetic Membranes from Water-Soluble Precursors. *Angew Chem—Int Ed* **2022**, *61* (29). https://doi.org/10.1002/anie.202200549.

[55] Brea, R. J.; Bhattacharya, A.; Devaraj, N. K. Spontaneous Phospholipid Membrane Formation by Histidine Ligation. *Synlett* **2017**, *28* (1), 108–112. https://doi.org/10.1055/s-0036-1588634.

[56] Neef, A. B.; Schultz, C. Selective Fluorescence Labeling of Lipids in Living Cells. *Angew Chem Int Ed* **2009**, *48* (8), 1498–1500. https://doi.org/10.1002/anie.200805507.

[57] Brea, R. J.; Devaraj, N. K. Diels-Alder and Inverse Diels-Alder Reactions. In *Chemoselective and Bioorthogonal Ligation Reactions: Concepts and Application*; Algar, R., Dawson, P., Medintz, I., Eds.; Wiley-VCH, 2017; pp. 67–95.

[58] Liang, D.; Wu, K.; Tei, R.; Bumpus, T. W.; Ye, J.; Baskin, J. M. A Real-Time Click Chemistry Imaging Approach Reveals Stimulus-Specific Subcellular Locations of Phospholipase D Activity. *Proc Natl Acad Sci USA* **2019**, *116* (31), 15453–15462. https://doi.org/10.1073/pnas.1903949116.

[59] Yang, J.; Šečkute, J.; Cole, C. M.; Devaraj, N. K. Live-Cell Imaging of Cyclopropene Tags with Fluorogenic Tetrazine Cycloadditions. *Angew Chem—Int Ed* **2012**, *51* (30), 7476–7479. https://doi.org/10.1002/anie.201202122.

[60] Argus, J. P.; Yu, A. K.; Wang, E. S.; Williams, K. J.; Bensinger, S. J. An Optimized Method for Measuring Fatty Acids and Cholesterol in Stable Isotope-Labeled Cells. *J Lipid Res* **2017**, *58* (2), 460–468. https://doi.org/10.1194/jlr.D069336.

[61] Triebl, A.; Wenk, M. R. Analytical Considerations of Stable Isotope Labelling in Lipidomics. *Biomolecules* **2018**, *8* (4). https://doi.org/10.3390/biom8040151.

[62] He, X.; Zhang, Q. Synthesis, Purification, and Mass Spectrometric Characterization of Stable Isotope-Labeled Amadori-Glycated Phospholipids. *ACS Omega* **2018**, *3* (11), 15725–15733. https://doi.org/10.1021/acsomega.8b01893.

[63] Takahashi, R.; Fujioka, S.; Oe, T.; Lee, S. H. Stable Isotope Labeling by Fatty Acids in Cell Culture (SILFAC) Coupled with Isotope Pattern Dependent Mass Spectrometry for Global Screening of Lipid Hydroperoxide-Mediated Protein Modifications. *J Proteom* **2017**, *166*, 101–114. https://doi.org/10.1016/j.jprot.2017.07.006.

[64] Shen, Y.; Zhao, Z.; Zhang, L.; Shi, L.; Shahriar, S.; Chan, R. B.; Di Paolo, G.; Min, W. Metabolic Activity Induces Membrane Phase Separation in Endoplasmic Reticulum. *Proc Natl Acad Sci USA* **2017**, *114* (51), 13394–13399. https://doi.org/10.1073/pnas.1712555114.

[65] Zhi, D.; Bai, Y.; Yang, J.; Cui, S.; Zhao, Y.; Chen, H.; Zhang, S. A Review on Cationic Lipids with Different Linkers for Gene Delivery. *Adv Colloid Interface Sci* **2018**, *253*, 117–140. https://doi.org/10.1016/j.cis.2017.12.006.

[66] Mahato, R. I.; Rolland, A.; Tomlinson, E. Cationic Lipid-Based Gene Delivery Systems: Pharmaceutical Perspectives. *Pharm Res* **1997**, *14* (7), 853–859. https://doi.org/10.1023/A:1012187414126.

[67] Simões, S.; Filipe, A.; Faneca, H.; Mano, M.; Penacho, N.; Düzgünes, N.; de Lima, M. P. Cationic Liposomes for Gene Delivery. *Expert Opin Drug Deliv* **2005**, *2* (2), 237–254. https://doi.org/10.1517/17425247.2.2.237.

[68] Miller, A. D. Cationic Liposomes for Gene Therapy. *Angew Chem Int Ed* **1998**, *37* (13–14), 1768–1965. https://doi.org/10.1002/(SICI)1521-3773(19980803)37:13/14<1768::AID-ANIE1768>3.0.CO;2-4.

[69] Tenchov, R.; Sasso, J. M.; Zhou, Q. A. PEGylated Lipid Nanoparticle Formulations: Immunological Safety and Efficiency Perspective. *Bioconjug Chem* **2023**, *34* (6), 941–960. https://doi.org/10.1021/acs.bioconjchem.3c00174.

[70] Gill, K. K.; Kaddoumi, A.; Nazzal, S. PEG-Lipid Micelles as Drug Carriers: Physiochemical Attributes, Formulation Principles and Biological Implication. *J Drug Target* **2015**, *23* (3), 222–231. https://doi.org/10.3109/1061186X.2014.997735.

[71] Mitchell, M. J.; Billingsley, M. M.; Haley, R. M.; Wechsler, M. E.; Peppas, N. A.; Langer, R. Engineering Precision Nanoparticles for Drug Delivery. https://doi.org/10.1038/s41573-020-0090-8.

[72] Singh, R. P.; Gangadharappa, H. V.; Mruthunjaya, K. Phospholipids: Unique Carriers for Drug Delivery Systems. *J Drug Deliv Sci Technol* **2017**, *39*, 166–179. https://doi.org/10.1016/J.JDDST.2017.03.027.

[73] Battistella, C.; Liang, Y.; Gianneschi, N. C. Innovations in Disease State Responsive Soft Materials for Targeting Extracellular Stimuli Associated with Cancer, Cardiovascular Disease, Diabetes, and Beyond. *Adv Mater* **2021**, *33* (46). https://doi.org/10.1002/adma.202007504.

[74] Li, F.; Snow-Davis, C.; Du, C.; Bondarev, M. L.; Saulsbury, M. D.; Heyliger, S. O. Preparation and Characterization of Lipophilic Doxorubicin Pro-Drug Micelles. *J Vis Exp* **2016**, *2016* (114), 1–8. https://doi.org/10.3791/54338.

[75] Vance, J. A.; Devaraj, N. K. Membrane Mimetic Chemistry in Artificial Cells. *J Am Chem Soc* **2021**, *143* (22), 8223–8231. https://doi.org/10.1021/jacs.1c03436.
[76] Podolsky, K. A.; Devaraj, N. K. Synthesis of Lipid Membranes for Artificial Cells. *Nat Rev Chem* **2021**, *5* (10), 676–694. https://doi.org/10.1038/s41570-021-00303-3.
[77] Brea, R. J.; Rudd, A. K.; Devaraj, N. K. Nonenzymatic Biomimetic Remodeling of Phospholipids in Synthetic Liposomes. *Proc Natl Acad Sci USA* **2016**, *113* (31), 8589–8594. https://doi.org/10.1073/pnas.1605541113.
[78] Flores, J.; White, B. M.; Brea, R. J.; Baskin, J. M.; Devaraj, N. K. Lipids: Chemical Tools for Their Synthesis, Modification, and Analysis. *Chem Soc Rev* **2020**, *49* (14), 4602–4614. https://doi.org/10.1039/d0cs00154f.
[79] Bhattacharya, A.; Brea, R. J.; Devaraj, N. K. De Novo Vesicle Formation and Growth: An Integrative Approach to Artificial Cells. *Chem Sci* **2017**, *8* (12), 7912–7922. https://doi.org/10.1039/C7SC02339A.

11 Continuous Flow Valorization of Lipids, Fatty Acids, and Glycerol

Christophe Len

11.1 INTRODUCTION

The chemical industry produces a diverse array of goods. It encompasses i) fundamental chemicals such as polymers, petrochemicals, and basic inorganic substances; ii) chemicals used for specialized purposes like safeguarding crops; formulating paints, inks, and dyes; crafting textiles and paper, and enabling engineering endeavors; and iii) consumer-oriented chemicals like detergents and soaps.

Presently, the chemical sector heavily depends on fossil fuel resources; however, the depletion of petroleum reserves is fast approaching. In a bid to replace petroleum-dependent chemical processes and enhance overall process efficiency, chemists have recently pioneered chemical reactions grounded in renewable resources, atom efficiency, less-hazardous processes, safer and less toxic solvents, auxiliary agents, and alternative technologies such as continuous flow. In stark contrast to the chemical processes observed in humans and living organisms, wherein nature champions numerous bioorganic reactions conducted within flow systems rather than in batch-type conditions, the academic and industrial chemistry communities have devised a multitude of chemical reactions within batch-type reactors ranging from conventional flasks to beakers. Continuous flow systems can be conveniently classified into four categories (Figure 11.1): i) the passage of all reagents through the reactor; ii) the immobilization of one of the reactants onto a solid substrate within the reactor, through which the substrate subsequently traverses; iii) the use of a homogeneous catalyst, where both the catalyst and the reactant traverse the reactor; and iv) the deployment of a heterogeneous catalyst within the reactor [1].

Various parameters can be effectively controlled in flow chemistry, including fluid dynamics, residence time, reaction duration, temperature, flow rate, pumped volume, and pressure. Flow reactions can encompass a range of fluid dynamics, including gas flow, liquid flow, and even the use of critical fluids, depending on the specific requirements of the process. In certain cases, a combination of gas and liquid flow is employed when working with heterogeneous solid catalysts, as is typically observed in hydrogenation reactions. Residence time, a crucial factor in flow chemistry, represents the duration for which the solution remains within the tube or reactor. It can be calculated based on the reactor's volume and the rate of flow through it (Figure 11.2). The reaction commences when the reagents are introduced at one end of the tube and subsequently undergo chemical transformation within the tube. It is essential to note that there is a distinction between reaction time and residence time. Reaction time typically denotes the period during which most materials participate in the reaction and convert into the desired target.

A reaction in a flow reactor fundamentally follows a process akin to that in a batch reactor (Figure 11.3), but there is a key distinction: In a flow reactor, the concentration at each location remains constant over time. The reactants and starting materials are continuously introduced at the inlet, typically at a consistent concentration and a constant rate. This continuous flow ensures that the concentrations of all substances involved in the reaction remain stable at each location within

DOI: 10.1201/9781003437413-11

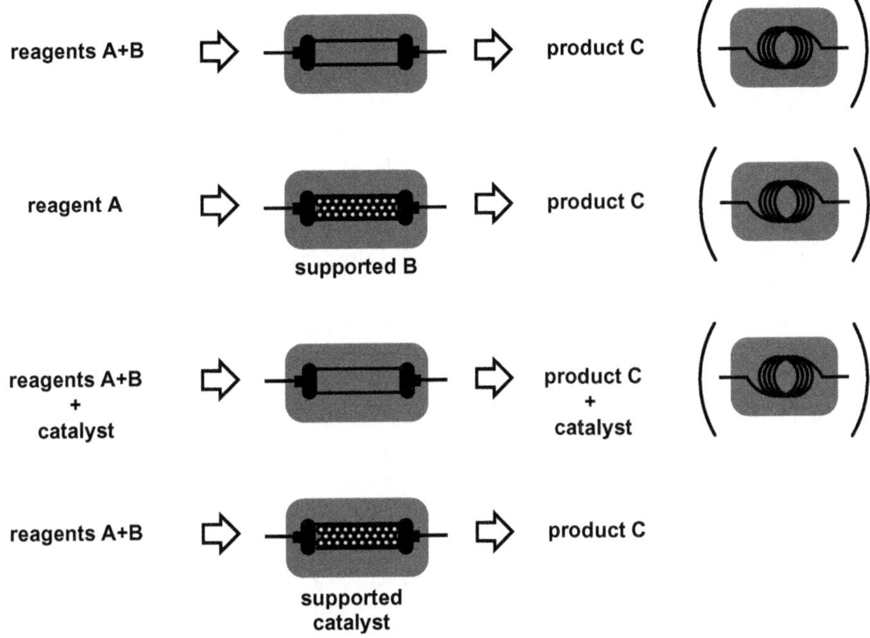

FIGURE 11.1 Types of continuous-flow reactors.

FIGURE 11.2 Residence time in a continuous-flow system.

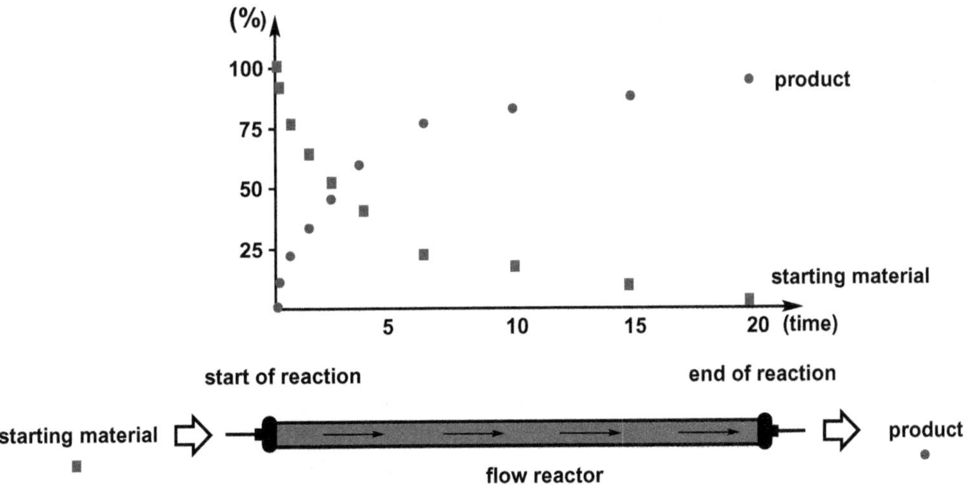

FIGURE 11.3 Progress of a reaction in a flow reactor.

the reactor. Consequently, residence time emerges as a pivotal regulatory parameter in flow chemistry. When the residence time is set too short, the solution often exits the flow reactor before the reaction reaches completion. Conversely, excessively long residence time within the flow reactor can decompose target products or initiate unwanted side reactions; the reaction has concluded when the solution is discharged from the reactor, the concentration of the starting material is nearly zero, and the concentration of the desired product has sufficiently increased. In continuous flow chemistry, synthesis can be rapidly optimized because the results of any specific set of reaction parameters can be evaluated from the initial eluent drop. Naturally, the residence time can be adjusted by altering the flow rate and the volume of the reactor. Speaking of flow rate, its adjustment affects not only the mixing rate but also the reaction kinetics.

Continuous flow chemistry, as an alternative technology, presents numerous advantages for chemical processing including improved thermal management, precise mixing control, suitability for a broad range of reaction conditions, scalability, enhanced energy efficiency, reduced waste generation, safety, and the ability to employ heterogeneous catalysis and execute multistep syntheses. Two primary types of reactors are prevalent: micro-reactors and meso-reactors; the choice depends on the channel dimensions: Micro-reactors typically range from 10 to 300 μm, while meso-reactors span from 300 μm to over 5 mm. Each has its own set of advantages and disadvantages. Micro-reactors offer advantages like low material input, minimal waste output, excellent mass transfer properties, and rapid diffusive mixing. However, they come with limitations, including low throughput, susceptibility to channel blockage, and a high pressure drop. In contrast, meso-reactors provide high throughput, low pressure drop, and the capacity to handle solids for heterogeneous catalysis, but their drawbacks include subpar mass transfer properties and slower mixing.

Looking at the global vegetable oil industry, the US Department of Agriculture projects that the world's vegetable oil production for 2023/2024 will reach 222.8 million tons, marking a 2.7% increase from 2022/2023. In 2022, the global vegetable oil market was valued at US$2.192 billion, and it is poised for substantial growth, with an estimated value of US$4.46 billion by 2030. The industry is anticipated to achieve a compound annual growth rate of 9.3% during the forecast period from 2022 to 2028. The market is categorized by type, encompassing soybean oil, rapeseed oil, sunflower oil, palm oil, olive oil, corn oil, peanut oil, coconut oil, and other variants. Moreover, it is segmented based on application, with food products accounting for 27%, followed by animal feed at 21%, cosmetics at 19%, pharmaceuticals at 13%, biodiesel at 10%, and other applications at 10%. Various distribution channels are utilized, including direct sales, supermarkets, convenience stores, specialty stores, e-commerce, and others. Key players in this market include Cargill, Wilmar International, Archer Daniels Midland Company, Avril Group, and several others. The fundamental constituents of vegetable oils are triglycerides, which consist of a glycerol moiety and three fatty acid chains that can be identical or different. An overview of the most common natural fatty acids found in vegetable oils is presented in Figure 11.4.

The majority of vegetable oils are employed either directly or with minimal processing, although some necessitate chemical alteration prior to use. This modification involves diverse chemical reactions aimed at the carboxy groups, ester-adjacent methylene sites, unsaturated bonds, or the allylic carbon atom, customized to fulfill particular industrial needs (Figure 11.5). In the case of ricinolein, an alternative option is the modification of the hydroxyl group at position 12.

In industrial processing, vegetable oils undergo transformation into glycerol, fatty acids, methyl esters, and fatty alcohols. To achieve specific chemical products, subsequent modification steps are often employed. These reactions primarily target the unsaturated bonds and encompass processes like epoxidation, hydroformylation, dimerization, thiol-ene coupling, oxidative cleavage (ozonolysis), olefin metathesis, pericyclic reactions, and radical additions, as well as transition-metal catalysis and Diels–Alder syntheses of aromatic compounds. The resulting array of chemicals includes, but is not limited to, soaps, surfactants, emollients, fuels, pesticide formulations, and lubricants. These chemical modifications play a vital role in tailoring vegetable oil derivatives for a broad spectrum of industrial applications. In this chapter, we do not exhaustively cover all continuous flow

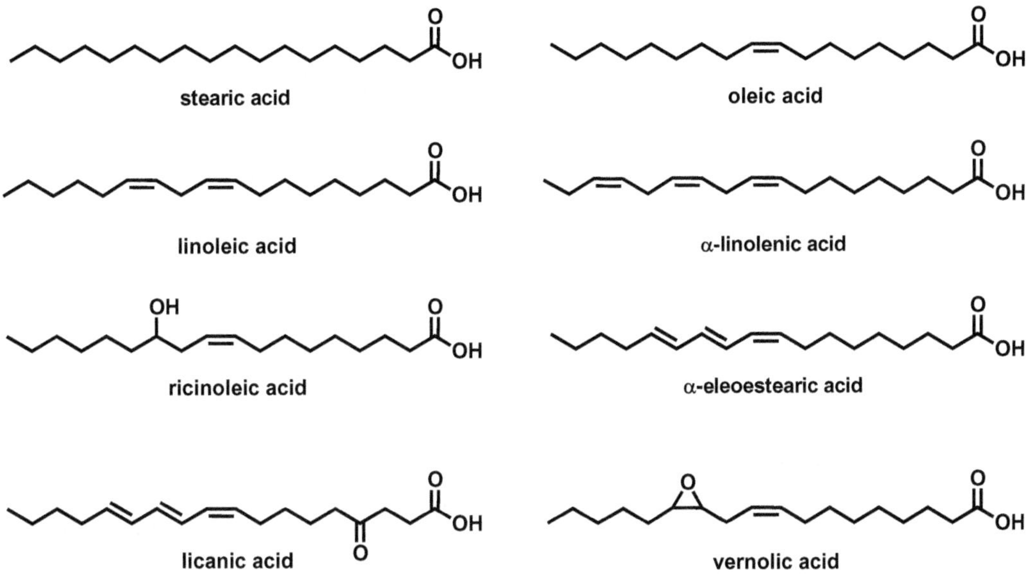

stearic acid

oleic acid

linoleic acid

α-linolenic acid

ricinoleic acid

α-eleoestearic acid

licanic acid

vernolic acid

FIGURE 11.4 The main fatty acids in vegetable oils.

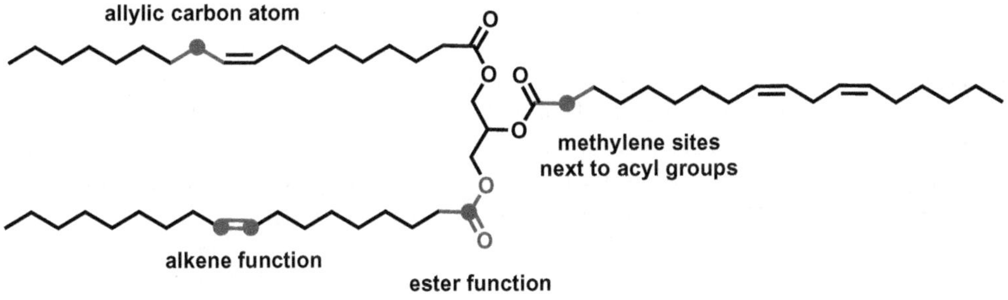

allylic carbon atom

methylene sites next to acyl groups

alkene function

ester function

FIGURE 11.5 Main chemical reactive sites of unsaturated triglycerides.

reactions that produce molecules of interest from vegetable oils, free fatty acids, and glycerol; we only detail selected reactions as examples.

11.2 CONTINUOUS TRANSFORMATIONS OF VEGETABLE OILS INTO HIGH-VALUE SMALL CHEMICALS

11.2.1 Biodiesel

Biodiesel has emerged as a compelling alternative to petrodiesel, garnering increased attention due to mounting environmental concerns and the push for renewable energy sources. This sustainable fuel is produced through the transesterification of various vegetable oils, encompassing edible, inedible, and residual sources and typically using methanol or ethanol as reactants. The resulting biodiesel consists of a blend of monoalkyl esters derived from long-chain fatty acids, typically ranging from C_8 to C_{24} carbon atoms [2]. These fatty acid methyl/ethyl esters offer several notable environmental advantages, notably reducing emissions of CO, unburned hydrocarbons, particulates,

triglyceride + 3 R^4OH (alcohol) $\rightleftharpoons$ (catalyst) glycerol + R^1COOR^4, R^2COOR^4, R^3COOR^4 (biodiesel)

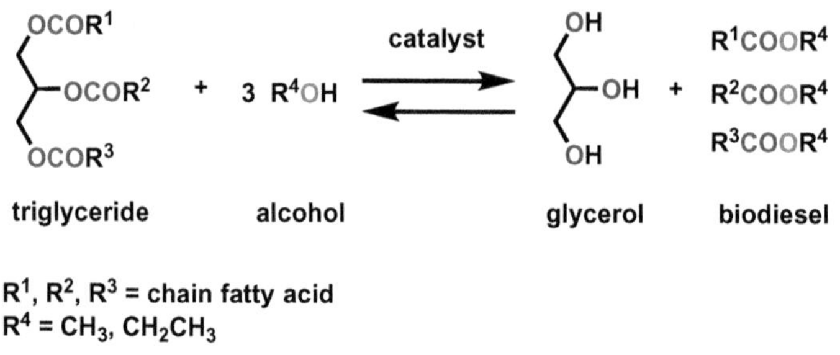

R^1, R^2, R^3 = chain fatty acid
R^4 = CH_3, CH_2CH_3

SCHEME 11.1 General chemical equation of the biodiesel formation.

and SO_2. While both homogeneous and heterogeneous catalytic methods have been proposed, homogeneous catalysis is the preferred approach for achieving high yields of methyl/ethyl esters.

Homogeneous catalysis allows for the use of base catalysts (such as NaOH, KOH, NaOMe, KOMe) and acid catalysts (such as H_2SO_4, HCl) [3, 4]. Among these, base-catalyzed transesterification stands out as a quicker and commercially favored option thanks to its operation under moderate conditions (Scheme 11.1). However, a common challenge in biodiesel production, regardless of the catalyst used, is the limited miscibility of vegetable oils and methanol, resulting in decreased reaction rates, particularly in continuous flow processes. In response to this constraint, novel approaches employing mechanochemistry in a semi-continuous flow system have been documented.

Mechanochemistry has gained substantial attention as a potent, environmentally friendly, cost-effective, and time-saving method for synthesizing novel functional materials [5]. This approach relies on chemical and physicochemical transformations induced by the application of mechanical energy through grinding and milling. Mechanochemistry enables chemical reactions through mechanical energy, which can be imparted manually or through automated milling, often without the need for solvents. These pioneering methods exhibit the potential to significantly boost the overall efficiency of biodiesel production. They offer promising solutions to the issues stemming from the limited miscibility of reactants, thereby playing a crucial role in further advancing the sustainability of this energy source.

The use of calcium diglyceroxide (CaDG) in conjunction with oil to produce fatty acid methyl ester through transesterification was reported [6, 7]. The process involved employing a near-stoichiometric molar ratio of methanol to oil (4:1) and introducing 1.5% by weight of CaDG as a basic catalyst. The researchers used a semi-continuous mechanochemical reactor with a volume of 0.5 L and containing yttrium-doped zirconia beads (ranging from 0.3 mm to 2.0 mm in diameter) occupying 55%–70% of the volume. The reaction proceeded with continuous input flow rates ranging from 4 to 45 L h–1. Subsequent stirring for 4 h at 50 °C or 24 h at room temperature facilitated the production of biodiesel with a yield exceeding 90% (Scheme 11.2). This innovative biodiesel production process exhibits scalability and can effectively convert used cooking oils with minimal yield loss. A cost analysis revealed that this approach is more cost-effective than conventional batch stirred processes. These findings indicate the potential to reduce the methanol–oil molar ratio from 12:1 to 4:1 and decrease the catalyst weight from 4% to 1.5% in standard laboratory-scale reactors, offering a more economically attractive method than conventional batch stirring procedures.

Similar approach was reported by the same team starting from both sunflower oil and used cooking oil to produce FAME [6, 7]. In this study, a slightly elevated molar ratio of methanol to oil (5:1 instead of 4:1) was employed, along with a lower flow rate (4 L h^{-1}). The results revealed consistent catalytic activity up to 120 minutes, yielding an 80% conversion rate. However, beyond the 120 minute mark, ester yields from the oils saw a decline of 17%. The authors suggested that this decrease was attributable to the neutralization of the basic sites on CaDG catalyst by the free fatty acids present in the waste oils.

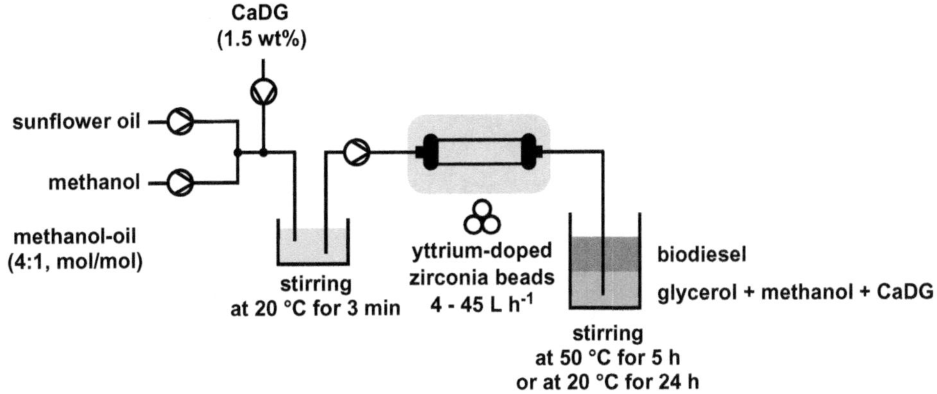

SCHEME 11.2 Semi-continuous biodiesel production catalyzed by CaDG using a high-throughput reactor filles with beads.

Triacetin, alternatively referred to as glycerol triacetate or glyceryl triacetate, is a versatile chemical compound widely acclaimed for its multifaceted utility. This compound finds applications across various industries and is an indispensable ingredient in the flavor and fragrance sector. In addition to its role in the flavor and fragrance industry, triacetin stands out as a vital plasticizer and humectant. Beyond these applications, it plays a crucial role in the pharmaceutical industry, where it serves both as an excipient and a valuable solvent for a broad spectrum of formulations, including those designed for oral, injectable, and topical use.

Triacetin's adaptability and wide-ranging properties make it a pivotal component in a diverse array of industrial processes and products. The simultaneous production of biodiesel and triacetin was a two-step multifaceted chemical process that harnessed the potential of vegetable oils and glycerol as feedstock materials. The transesterification of vegetable oils with methanol or ethanol yields biodiesel, while the glycerol obtained during this reaction is concurrently transformed into triacetin through esterification. This integrated approach optimizes resource utilization and minimizes waste byproduct in a sustainable and ecofriendly process (Scheme 11.3). The interesterification of palm oil into fatty acid ethyl esters or biofuel using supercritical ethylacetate (EtOAc) was carried out in a continuous tubular reactor [8]. Various reaction conditions, such as molar ratios of palm oil to ethyl acetate, temperature, and pressure, were explored at different mass flow rates of the mixtures and optimized. Results revealed that reaction temperatures exceeding 380 °C and prolonged residence times had an impact on the fatty acid ethyl esters and the production of triacetin, a valuable byproduct (two steps and one process). Introducing water into the mixture at a 1:30:10 molar ratio of palm oil to EtOAc to water at 380°C, 160 bar, and a mixture mass flow rate of 1.5 $gmin^{-1}$ increased the biodiesel content (89 wt%) more than the triacetin due to the hydrolysis of EtOAc and the transesterification reaction.

Another strategy involved the continuous transesterification of triglycerides into methyl esters and then the conversion of the glycerol byproduct into solketal, a value-added compound, through an *in-situ* reaction with acetone [9]. Solketal, or 1,3-di-*O*-isopropylidene-glycerol, is a chemical compound with notable properties that render it valuable in various fields. It serves as an essential intermediate in pharmaceutical chemistry, a protective group in C3-based organic chemistry, and a versatile solvent and plasticizer in polymer chemistry. Furthermore, it finds application as a fuel additive. Solketal's versatility stems from its excellent solubility in both aqueous and organic solvents, allowing it to function effectively in various reaction conditions. Solketal exhibits low toxicity, making it a safe choice for many applications. Its isopropylidene groups protect hydroxyl functionalities, allowing for selective transformations in organic synthesis. These attributes, combined with its ease

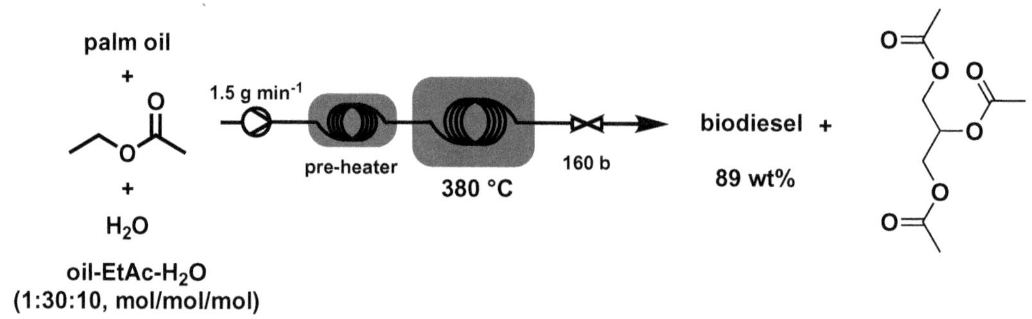

palm oil
+
+
H_2O
oil-EtAc-H_2O
(1:30:10, mol/mol/mol)
1.5 g min^{-1}
pre-heater
380 °C
160 b
biodiesel +
89 wt%

SCHEME 11.3 Simultaneous continuous biodiesel and triacetin production catalyzed in supercritical EtOAc.

of synthesis through acetalization reactions, make solketal a valuable and widely used chemical in different scientific and industrial contexts. The research was conducted using mesoscale oscillatory baffled reactors (meso-OBRs) for one-stage and two-stage catalytic transesterification of triglycerides and methanol [9]. In the two-stage process, two meso-OBRs were employed in tandem, each packed with AmberlystTM resin catalysts. The first stage incorporated a basic AmberlystTM A26-OH catalyst to facilitate the transesterification of triglycerides with methanol, while the second stage featured an acidic AmberlystTM 70-SO_3H catalyst to drive the coupling of glycerol and acetone, resulting in the formation of solketal.

11.2.2 Free Fatty Acids

Free fatty acids (FFAs) are organic compounds that consist of a carboxylic acid functional group attached to a hydrocarbon chain. They exhibit diverse properties and have multiple applications in various fields.

FFAs4 can have different chain lengths, ranging from fewer than 6 carbon atoms to more than 12, and this influences their physical and chemical properties. FFAs are generally hydrophobic, meaning they repel water and tend to dissolve in organic solvents. Depending on the chain lengths, FFAs are key raw materials for soap manufacturing and can function as surfactants due to their amphiphilic nature, with a polar carboxylic acid head and a nonpolar hydrocarbon tail. Some FFAs contribute to the aroma and taste of foods and beverages; for example, butyric acid is responsible for the smell of rancid butter [10].

FFAs are utilized in various industrial processes, such as metalworking, where they act as lubricants and cutting fluids. FFAs derived from renewable sources have been explored as precursors for biodiesel production. In summary, free fatty acids exhibit a wide range of properties and applications, making them essential components in biological systems, industrial processes, and various consumer products. Noncatalytic pyrolysis of triacylglyceride oils presents an appealing avenue for the production of renewable fuels and chemicals [10]. During this process, approximately 20 wt%–30 wt% of $C_2 - C_{10}$ free fatty acids are generated, primarily due to the presence of ester functions of oil. Of course, the prevalence of these size groups varies among the feedstocks subjected to pyrolysis, reflecting differences in the abundance of triunsaturated (linolenic), diunsaturated (linoleic), and monounsaturated (oleic) acids within the original oil.

11.2.3 Epoxides

Epoxidation of pure oil or epoxidation of oil derived from waste is a promising avenue for creating high-value products from sustainable sources. This process aligns with principles of green chemistry and sustainable resource management, making it an important area of research and industrial development. The epoxidation of oil derived from waste cocoa butter, a byproduct of food processing, is interesting starting material [11]. Ishii–Venturello catalyst in both batch and continuous flow reactors was studied, and the batch reactor yielded higher epoxide formation than flow due to the significant degradation of hydrogen peroxide in the laminar flow tubular reactor. Then, epoxides of triglycerides permitted to produce a large range of polymers *via* ring-opening and polymerization.

11.2.4 Triglycerides

The application of metathesis to oils showcases the versatility of this reaction in transforming renewable resources into a wide range of valuable chemicals and materials. This approach aligns with the principles of green chemistry and sustainable resource utilization, making it an essential area of scientific and industrial exploration. Metathesis or olefin metathesis is a chemical reaction that rearranges carbon–carbon double bonds in organic compounds breaking and forming double bonds to create new chemicals. Researchers reported on the direct chemical conversion of cocoa butter triglycerides, a readily available post-manufacture waste product from the food industry, into 1-decene through ethenolysis (olefin metathesis with ethylene) [12]. This conversion of raw waste material was achieved by employing $RuCl_2(iBu\text{-}phoban)_2(3\text{-}phenylindenyl)$ as a catalyst. The research delves into both batch and flow conditions; in the latter approach, a Teflon AF-2400 tube-in-tube gas–liquid membrane contactor was utilized to introduce ethylene into the reaction system (Scheme 11.4). These initial investigations culminated in a continuous process that demonstrated consistent output throughout the 150-minute testing period.

11.3 CONTINUOUS TRANSFORMATIONS OF FREE FATTY ACIDS INTO HIGH-VALUE SMALL CHEMICALS

11.3.1 Biodiesel

The synthesis of biodiesel from oil and the synthesis of biodiesel from FFAs are two distinct yet interconnected processes in the production of this renewable fuel source (Table 11.1). The choice between biodiesel from oil and biodiesel from fatty acids depends on factors such as feedstock availability, feedstock quality, production scale, and economic viability.

Researchers carried out two-step esterification for biodiesel production from palm fatty acid distillate (PFAD) using an ultrasound clamp reactor [13]. This setup consisted of 12 pairs of ultrasound clamps evenly distributed along the stainless-steel tube, providing maximum power of 4800 W (12 × 400 W) via an ultrasonic generator. The process yielded final methyl ester purity of 93 wt%.

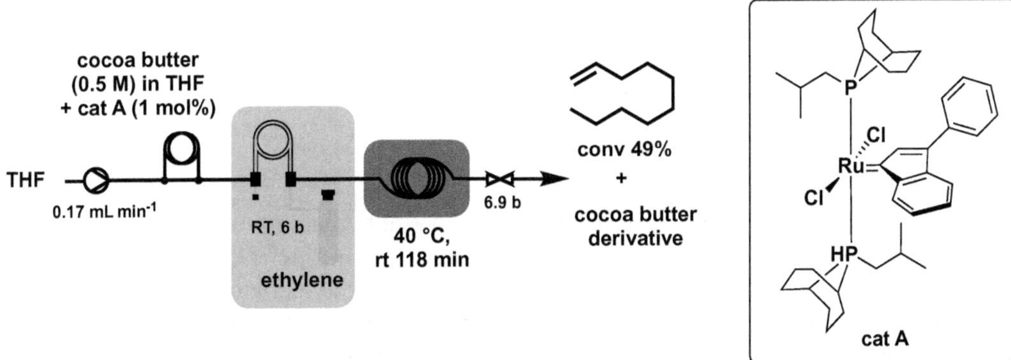

SCHEME 11.4 Continuous metathesis of cocoa butter using a tube-in-tube gas–liquid membrane contactor.

TABLE 11.1
Advantages and Drawbacks of Biodiesel from Oil and from Fatty Acids

	Biodiesel from Oil	Biodiesel from Fatty Acids
Feedstocks	Triglyceride-rich oils such as vegetable oils or animal fats, serve as the primary feedstock.	Purified fatty acids, often obtained from hydrolyzed or esterified oils, are the main starting material.
Processing complexicity	Oils are converted into biodiesel two steps: transesterification of triglycerides into fatty acid methyl esters (FAMEs) and glycerol, followed by the separation of FAMEs from glycerol.	Working with fatty acids simplifies the process as they are closer to the desired end product, reducing the need for a separate glycerol separation step.
Purity	The crude oil feedstock may contain impurities and non-triglyceride compounds that require additional purification.	Using purified fatty acids eliminates the need to deal with many of these impurities, resulting in higher-purity biodiesel.
Reaction efficiency	Transesterification reactions can be influenced by the varying compositions of different oils, affecting reaction efficiency.	Focusing on fatty acids with known compositions can lead to more consistent and efficient reactions.
Resource utilization	Utilizes the entire oil feedstock, including the glycerol byproduct, in a more comprehensive resource utilization approach.	Typically, only the fatty acids are used, and the glycerol is often discarded or used for other applications.
Flexibility	Suitable for a wide range of feedstocks but requires more rigorous purification.	Provides more control over the feedstock's composition and purity, offering greater process consistency.
Economic considerations	May be more cost-effective when using low-cost, non-food-grade feedstocks.	Reduces the costs associated with feedstock purification but may require higher-purity starting materials.

In the first step, a molar ratio of 3.75:1 of methanol to PFAD (with 46 vol% methanol), 6.6 vol% H_2SO_4, and a 400 mm length of ultrasound clamp at a 25 Lh^{-1} PFAD flow rate were recommended. This step converted PFAD into 60 wt% methyl ester.

In the second step, esterification was repeated with a molar ratio of 2.87:1 (61.6 vol% methanol) for the first esterified oil, 5.6 vol% H_2SO_4, and a 400 mm long ultrasound clamp at an flow rate of esterified oil. The ultrasound clamp reactor achieved high yields of esterified oil and crude biodiesel in a short residence time of 32 s. The maximum yields in the double-step esterification process were 104 wt% for the first esterified oil, 108 wt% for crude biodiesel, and 98 wt% for purified biodiesel, calculated based on the initial PFAD volume of 100%. Therefore, this approach demonstrates high potential for producing biodiesel with reduced energy consumption and shorter conversion times for achieving high-purity methyl ester from PFAD [13].

11.3.2 Alkanes

The synthesis of alkanes from fatty acids provides a versatile route for obtaining hydrocarbons with a range of applications, including biodiesel, lubricants, cosmetics, and pharmaceuticals. This process contributes to the development of more sustainable and environmentally friendly products. Decarboxylation is a key process to convert fatty acids into high-quality fuels, typically requiring high pressure or the addition of external hydrogen. Researchers examined continuous low-pressure (<500 psi) decarboxylation using oleic acid in a fixed-bed reactor with activated carbon [14]. Surprisingly, this process produced high-quality fuel-like hydrocarbons without the need for external hydrogen. The results demonstrated that activated carbon acted as a catalyst for both decarboxylation and in situ hydrogen production. The optimal conditions for achieving a maximum degree of decarboxylation (91%) were determined to be 400 °C, 2 hours reaction time, and a water-to-oleic acid ratio of 4:1 (mol/mol). Importantly, the liquid product exhibited density and higher heating value that were similar to those of commercial diesel and jet fuel.

11.3.3 Dimeric Fatty Acids

Oleochemical dicarboxylic acids constitute around 0.5% of the total dicarboxylic acid market for monomers, with phthalic and terephthalic acids dominating the market at 87%. These dicarboxylic acids derived from oleochemicals have a unique chemical nature that allows them to modify or alter condensation polymers, placing them in a distinct niche market. They possess several desirable properties, including elasticity, flexibility, high impact strength, hydrolytic stability, hydrophobicity, lower glass transition temperatures, and increased flexibility.

Researchers recently reported on the continuous production of dimeric fatty acids from soybean fatty acids (specifically, an oleic acid:linoleic acid weight ratio of 33:65) using an innovative approach [15]. The reaction was facilitated by employing acidic montmorillonite clay and lithium carbonate in a specially designed ball-mill reactor. This reactor featured three baffles to enhance radial turbulence and ten clapboards to divide it into 11 small reactors in series, resembling a plug-flow reactor. A solution consisting of oil, clay, and lithium carbonate in a weight ratio of 200:24:1 was pumped into the horizontal continuous ball mill reactor, filled with zirconia beads (bead-to-mass ratio = 12), and operated at 300 °C. The residence time in this system was 2.16 h, with a rotation speed of 20 rpm. The result was a 51% yield of dimeric fatty acids (Scheme 11.5). Interestingly, the authors discovered that using similar conditions in a stirred flask (500 mL) at 400 rpm produced the diacid with a higher yield of 63%; the lower yield obtained in the continuous flow process was attributed to the dissolution of Fe^{3+} and Ni^{2+} from the continuous ball-mill reactor.

2 + Li_2CO_3 montmorillonite clay

zirconia beads
300 °C, 20 rpm
rt 2.16 h

oil-clay-Li_2CO_3
(200:24:1, wt/wt/wt)

(idealized structure)
yield: 51%

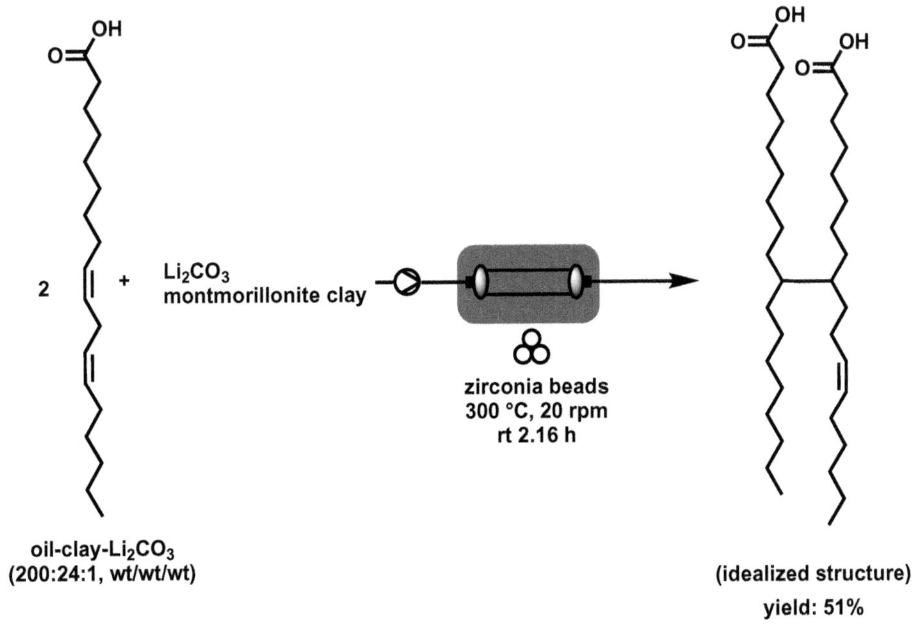

SCHEME 11.5 Continuous dimerization of soybean fatty acid using a horizontal ball-mill reactor.

11.4 CONTINUOUS TRANSFORMATIONS OF GLYCEROL INTO HIGH-VALUE SMALL CHEMICALS

11.4.1 Acrolein

The dehydration of glycerol forms various valuable chemicals including acrolein, which serves as an intermediate in the synthesis of numerous chemical compounds such as acrylic acid and quinoline. This reaction can be effectively carried out under continuous flow, with factors like temperature distribution significantly impacting the resulting product. Scientists reported on a microwave-assisted system for the catalytic dehydration of glycerol, aiming for the sustainable production of acrolein using WO_3/ZrO_2@SiC-coated microwave-absorbing catalyst [16].

This approach achieved remarkable selectivity for acrolein, exceeding 70%, while ensuring complete conversion of the starting material at 250°C under microwave heating. Moreover, a high recycling efficiency and stability were reported, as the conversion value remained at 98% and acrolein selectivity at 64% even after 24 hours of continuous operation. To achieve low coke formation in high-temperature reactions, maintaining uniform temperature distribution within the solid catalyst particles is crucial. The combination of microwave heating and a microwave-absorbing catalyst presented a novel and effective method for both investigating and enhancing the sustainable production of acrolein. This approach holds promise for advancing the development of chemicals crucial to various industrial processes [16].

11.4.2 Quinoline

Researchers designed a pilot-scale microwave (MW) apparatus for heating substantial volumes of reaction mixtures under controlled conditions and pressure while enabling continuous flow. The key advantage of this configuration was the reactor's large volume, which facilitated relatively long residence times and higher mass flow rates of up to 1 kg h^{-1}. This equipment was then employed for the synthesis of quinoline from glycerol using a modified Skraup reaction [16]. The system's significant advantage was its capacity for continuous chemical synthesis on a large pilot scale at elevated temperatures 200 °C–220 °C. This allowed for precise pressure control, with a maximum of 19 bar achieved by regulating the power absorbed by the reaction medium. Consequently, the experiment took merely 32 minutes using MW heating, while conventional heating in an oven at 325 °C required 122 minutes. MW heating yielded similar volumes of quinoline to that from thermal heating, with 41% yield using MW versus 37% with thermal heating. Remarkably, the energy consumption was significantly lower for MW heating, measuring 11,014 kJ $mole^{-1}$ as opposed to 425,200 kJ $mole^{-1}$ for thermal heating. This highlights the substantial energy-efficiency benefits of employing MW technology in this chemical synthesis process.

11.4.3 Oligomerization–Polymerization

Microwave-assisted continuous flow synthesis of glycerol oligomers and polymers showed reduced reaction times when scaling up the production of glycerol. The researchers tackled issues of process handling and investigated the impacts of temperature, flow rate, and residence time when employing a low-cost, homogeneous, and commercially available catalyst, K_2CO_3. Under the optimized conditions, the researchers achieved a 45% conversion of glycerol into diglycerol with selectivity exceeding 54%. These conditions involved using 240 grams of glycerol, 10.0 grams of K_2CO_3, and operating at 238°C for 250 minutes with a flow rate of 0.5 mL min^{-1} (Scheme 11.6) [18]. Following a purification step involving short-distance evaporation, the result was a mixture of dimers enriched to 50%, free of glycerol and cyclic diglycerol. The mechanistic investigation unveiled that the oligomerization proceeded through an SN^2-type pathway, with direct regioselective nucleophilic attack by CO_3^{2-} on the primary or secondary hydroxyl group of polyol, forming the corresponding

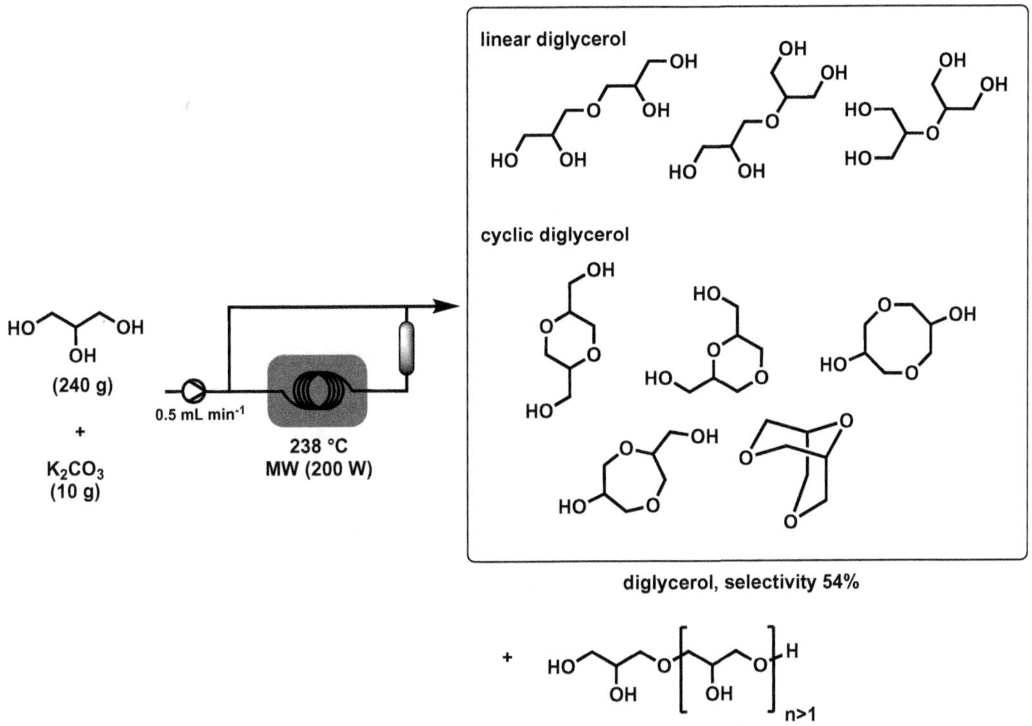

(240 g)
+
K_2CO_3
(10 g)
0.5 mL min^{-1}
238 °C
MW (200 W)
linear diglycerol
cyclic diglycerol
diglycerol, selectivity 54%
n>1

SCHEME 11.6 Microwave-assisted continuous oligomerization of glycerol in the presence of K_2CO_3.

glycerolate. The authors noted that the oligomerization exhibited first-order kinetics with consecutive conversion of diglycerols into higher oligoglycerols. They suggested that further exploration with a larger reactor and higher power would likely produce more selectively enriched oligoglycerol mixtures.

11.4.4 Solketal

Solketal (1,3-di-*O*-isopropylidene-glycerol) exhibits significant versatility, finding applications in diverse fields. It serves as a valuable compound in pharmaceutical chemistry as a synthetic intermediate, acts as an isopropylidene protective group in C_3-based organic chemistry, functions as a solvent and plasticizer in polymer chemistry, and even plays a role as a fuel additive. The synthesis of solketal involves an acetalization reaction between glycerol and acetone under acidic conditions. This reaction occurs at equilibrium, potentially producing 1,3-di-*O*-isopropylidene-glycerol as a byproduct. Typically, homogeneous acid catalysts like H_2SO_4or HCl are used, although both homogeneous Lewis acids and heterogeneous acid catalysts may also find application. In light of the industrial perspective, considerable efforts have been focused on the development of continuous-flow processes for solketal synthesis [19].

Nevertheless, persistent challenges are associated with these continuous-flow methods. These include the immiscibility of glycerol with acetone due to differences in physicochemical properties, often necessitating the use of organic co-solvents. Furthermore, homogeneous catalysis requires substantial quantities of catalysts, while heterogeneous catalysis encounters limitations related to flow rates. The presence of water, a byproduct of the reaction, can lead to catalyst deactivation. In addition to the transfer of mechanical energy and milling, bead mechanochemistry has emerged as a highly efficient method for achieving homogenization [20].

Patented studies have indeed shown that this innovative technique effectively addresses the miscibility challenges encountered in solketal synthesis under homogeneous acid catalysis. This method has been applied to continuous-flow mechanochemical solketal synthesis, using equipment and processes similar to those for the previously described continuous-flow biodiesel mechanochemistry. Various conditions and homogeneous acid catalysts were examined, with the most favorable results observed for $FeCl_3 \bullet 6H_2O$ as a Lewis acid in both loop and continuous-flow processes at 56 °C. This led to an impressive 99% solketal yield within residence times of less than 15 minutes [20]. These investigations in the presence of a solvent serve as pioneering examples of the effectiveness of continuous-flow bead mechanochemistry in facilitating organic synthesis.

11.5 CONCLUDING REMARKS AND FUTURE PERSPECTIVES

In closing, the chemical industry's role in shaping our world is undeniable, as it fuels the production of a vast array of essential goods spanning from fundamental chemicals like polymers and petrochemicals to specialized formulations for various purposes and consumer-oriented products such as detergents and soaps. However, the industry faces a critical juncture due to its heavy dependence on finite fossil fuel resources, particularly petroleum, which are fast depleting. The chemical sector is adapting to this challenge by pioneering innovative approaches, such as shifting to renewable resources, optimizing atom efficiency, reducing hazardous steps, employing safer solvents, and embracing alternative technologies like continuous-flow systems. This transformation not only enhances sustainability but also boosts overall process efficiency, making it a beacon of hope for a more sustainable future.

Flow chemistry has emerged as a transformative approach in the realm of chemical synthesis, offering unprecedented control and efficiency in a wide range of applications, including those involving oil, fatty acids, and glycerol. This innovative technique represents a paradigm shift in how we process these materials, bringing about numerous advantages and opportunities. In the context of oil, flow chemistry enables the conversion of crude or refined oil feedstocks into valuable products. The continuous flow of reactants through precisely controlled reactors allows for real-time monitoring and optimization, resulting in increased yields and selectivity.

Flow systems are especially suitable for processes such as transesterification, interesterification, saponification, epoxidation, and metathesis, which are integral in the production of specialty chemicals, surfactants, and fuels from oil derivatives. Fatty acids, abundant in natural oils, play a crucial role in the chemical industry. Flow chemistry facilitates the efficient conversion of fatty acids into a broad spectrum of chemicals, including biodiesel, alkanes, and dimeric fatty acids. Glycerol, a co-product of biodiesel production, can also be effectively processed using flow chemistry. Continuous-flow systems allow for the conversion of glycerol into high-value chemicals like acrolein, quinoline, glycerol oligomers, and solketal as valuable compounds in the pharmaceutical and polymer industries. The ability to control reaction parameters such as temperature and flow rate is particularly advantageous in glycerol transformation processes.

The future of flow chemistry and oleochemistry is marked by exciting prospects and potential advancements that could reshape the chemical industry:

- *Sustainable Production*: The integration of flow chemistry and oleochemistry will contribute significantly to sustainable and ecofriendly production processes, reducing the environmental footprint and resource consumption.
- *Advanced Reactor Design*: Expect to see novel reactor designs that enhance the efficiency of transforming oil, fatty acids, and glycerol into high-value chemicals, offering improved selectivity and control.
- *Automated Systems*: Automation and digitalization will become more prevalent, allowing for precise control, real-time monitoring, and optimization of chemical reactions, leading to improved yields and reduced waste.

- *Multistep Syntheses*: Flow systems will continue to evolve to accommodate complex, multistep syntheses, enabling the efficient production of a wide range of chemicals from oleochemical feedstocks.
- *Green Chemistry*: The principles of green chemistry will be further integrated, promoting the use of safer solvents, reducing hazardous steps, and minimizing the environmental impact of chemical processes.
- *Customization of Products*: Flow chemistry and oleochemistry will enable the tailoring of chemical products to meet specific industrial and consumer needs, facilitating the creation of more personalized and functional compounds.
- *Resource Diversification*: The shift to alternative feed stocks, such as algae-based oils and waste oils, will reduce the industry's dependence on traditional oil sources and promote resource diversification.
- *Collaboration and Innovation*: Increased collaboration between researchers, academia, and industry will drive innovation in flow chemistry and oleochemistry, accelerating the development of novel processes and products.
- *Market Expansion*: As the demand for sustainable and bio-based chemicals grows, the market for oleochemical products produced through flow chemistry will expand, presenting new business opportunities.
- *Regulatory Compliance*: Adherence to stringent environmental regulations and sustainability goals will drive the adoption of flow chemistry and oleochemistry, positioning them as critical technologies in the chemical industry's future.

REFERENCES

[1] Luis, S.V., and Garcia-Verdugo, E. (ed.). *Chemical reactions and processes under flow conditions, RSC green chemistry, No. 5*, 2010, The Royal Society of Chemistry.

[2] Ma, F., and Hanna, M.A. Biodiesel production: A review. *Bioressour. Technol.*, 1999, 70, 1–15.

[3] Rizwanul Fattah, I.M., Ong, H.C., Mahlia, T.M.I., Mofir, M., Silitonga, A.S., Ashrafur Rahman, S.M., and Ahmad, A. State of the art of catalysts for biodiesel production. *Front. Energy Res.*, 2020, 8, 101.

[4] Nag, A. (ed.). *Principles of biofuels and hydrogen gas: Production and engine performance*, 2021, New York: McGraw Hill Publishers.

[5] Rightmire, N.R., and Hanusa, T.P. Advances in organometallic synthesis with mechanochemical methods. *Dalton Trans.*, 2016, 45, 2352–2362.

[6] Malpartida, I., Maireles-Torres, P., Lair, V., Halloumi, S., Thiel, J., and Lacoste, F. New high-throughput reactor for biomass valorization. *Chem. Proc.*, 2020, 2, 31.

[7] Malpartida, I., Maireles-Torres, P., Vereda, C., Rodriguez-Maroto, J.M., Halloumi, S., Lair, V., Thiel, J., and Lacoste, F. Semi-continuous mechanochemical process for biodiesel production under heterogeneous catalysts using calcium diglyceroxide. *Renew. Energy*, 2020, 159, 117–126.

[8] Komintarachat, C., Sawangkeaw, R., and Ngamprasertsith, S. Continuous production of palm biofuel supercritical ethyl acetate. *Energy Convers. Manage.*, 2015, 93, 332–338.

[9] Eze, V.C., and Harvey, A.P. Continuous reactive coupling of glycerol and acetone—a strategy for triglyceride transesterification and in-situ valorisation of glycerol by-product. *Chem. Eng. J.*, 2018, 347, 41–51.

[10] Kubatova, A., Geetla, A., Casey, J., Linnen, M.J., Seames, W.S., Smoliakova, I.P., and Kozliak, E.I. Cleavage of carboxylic acid moieties in triacylglycerides during non-catalytic pyrolysis. *J. Am. Oil Chem. Soc.*, 92, 755–767.

[11] Plaza, D.D., Strobel, V., Keer, P.K.K.S., Sellars, A.B., Hoong, S.S., Clark, and A.J., Lapkin, A.A. Direct valorisation of waste cocoa butter triglycerides via catalytic epoxidation, ring-opening and polymerization. *J. Chem. Technol. Biotechnol.*, 2017, 92, 2254–2266.

[12] Schotten, C., Plaza, D., Manzini, S., Nolan, S.P., Ley, S.V., Browne, D.L., and Lapkin, A. Continuous flow metathesis for direct valorization of food waste: An example of Cocoa butter triglycerides. *ACS Sustain. Chem. Eng.*, 2015, 3, 1453–1459.

[13] Oo, Y.M., Prasit, T., Thawornprasert, J., and Somnuk, K. Continuous double-step esterification production of palm fatty acid distillate methyl ester using ultrasonic tubular reactor. *ACS Omega*, 2022, 7, 14666–14677.

[14] Hossain, M.Z., Chowdhury, M.B.I., Jhawar, A.K., Xu, W.Z., and Charpentier, P.A. Continuous low pressure decarboxylation of fatty acids to fuel-range hydrocarbons with in situ hydrogen production. *Fuel*, 2018, 212, 470–478.
[15] Lu, X., Wang, Z., Hu, D., Liang, X., and Ji, J. Design of a horizontal ball-mill reactor and its application in dimerization of natural fatty acids. *Ind. Eng. Chem. Res.*, 2019, 58, 10768–10775.
[16] Xie, Q., Li, S., Gong, R., Zheng, G., Wang, Y., Xu, P., Duan, Y., Yu, S., Lu, M., Ji, W., Nie, Y., and Ji, J. Microwave-assisted catalytic dehydration of glycerol for sustainable production of acrolein over a microwave absorbing catalyst. *Appl. Catal., B*, 2019, 243, 455–462.
[17] Saggadi, H., Polaert, I., Luart, D., Len, C., and Estel, L. Microwaves under pressure for the continuous production of quinoline from glycerol. *Catal. Today*, 2015, 255, 66–74.
[18] Nguyen, R., Galy, N., Alasmary, F.A., and Len, C. Microwave-assisted continuous flow for the selective oligomerization of glycerol. *Catalysts*, 2021, 11, 166.
[19] Khodadadi, M.R., Thiel, J., Varma, R.S., and Len, C. Innovative continuous synthesis of solketal. *J. Flow Chem.*, 2021, 11, 725–735.
[20] Len, C., Khodadadi, M.R., Thiel, J., and Lacoste, F. *Procédé de fabrication du (2,2-dimethyl-1,3-dioxolan-4-yl)methanol.* WO2021110688A1, 2021.

12 Glucocorticosteroids as a New Target in the Pharmacotherapy for Metabolic Disorders in Humans

Daria Kupczyk, Renata Studzińska,
Hanna Pawluk, and Renata Kołodziejska

12.1 METABOLIC DISEASES

Metabolic diseases are a very broad term covering diseases that result from disorders of biochemical changes in the body. They have so rapidly increased in frequency around the world that their incidence resembles an epidemic. Such diseases include diabetes and dyslipidemias, which are elements of metabolic syndrome [1]. These, together with obesity and elevated blood pressure, are significant health and social problems.

Metabolic syndrome (MetS) refers to the coexistence of major and modifiable cardiovascular risk factors including obesity and high blood pressure [2]. The key causes of metabolic disorders are lifestyle factors such as improper diet, lack of physical activity, use of stimulants, and poor sleep hygiene [3]. This state of affairs was caused, among other factors, by the civilizational changes of societies taking place over the years. The phenomenon has become so extensive that many fields related to human biology have been involved in effective therapy and treatment of metabolic diseases, examining their causes and courses to improve the therapeutic and healing process.

MetS, as mentioned earlier, is a complex disease that is characterized, among other things, by disorders of carbohydrate metabolism. The risk of developing type 2 diabetes is multiple times higher in people diagnosed with MetS. The common pathomechanism of type 2 diabetes and MetS is insulin resistance, along with excess adipose tissue and its dysfunction [4, 5]. The result of reduced tissue sensitivity to insulin is primarily the development of hyperglycemia, but also hypertriglyceridemia. In the beta cells of the pancreatic islets, there is an increased secretion of insulin that over time leads to their depletion; then prediabetes develops, and then diabetes. As a result of ongoing hyperglycemia, hyperinsulinemia, and lipid disorders, the endothelial function is impaired, and the vascular walls are remodeled, which leads to the development of atherosclerosis [6, 7].

Obesity is a key criterion in the diagnosis of MetS [8]. In this disease, the energy homeostasis of the body is disturbed, which results in excessive accumulation of adipose tissue. To a large extent, obesity is responsible for the development of insulin resistance, as it leads to the synthesis of pro-inflammatory cytokines in adipose tissue. Obesity and metabolic disorders result in the development of other diseases and increase cardiovascular risk [9]. Obesity also contributes to the development of a number of complications, including type 2 diabetes, lipid disorders, and hypertension, all key elements in MetS (Figure 12.1).

DOI: 10.1201/9781003437413-12

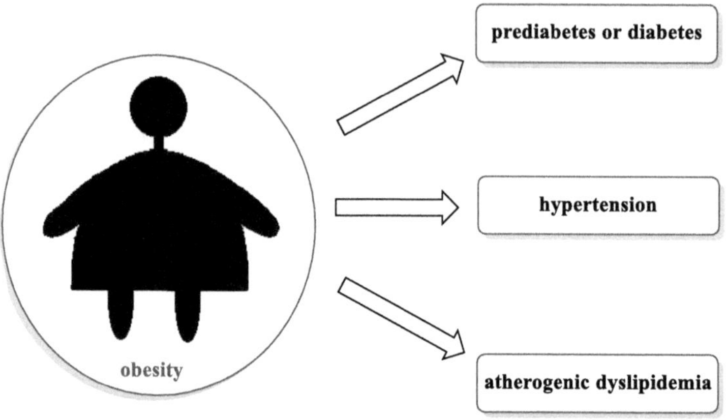

FIGURE 12.1 Major components of metabolic syndrome.

12.2 GLUCOCORTICOIDS

An important role in the regulation of carbohydrate metabolism is played by glucocorticosteroids (GCs), which belong to the group of steroid hormones [10]. Their biosynthesis takes place in the banded layer of the adrenal cortex. The chemical structure of GCs is based on the hexadecahydro-*1H*-cyclopenta[a]phenanthrene ring [11]. They are derivatives of cholesterol, and their most important representative in humans is cortisol [12] (Figure 12.2).

GC secretion is regulated by the hypothalamic–pituitary–adrenal axis in a double-negative feedback system (Figure 12.3) [13].

Cortisol secretion is controlled by adrenocorticotropic hormone (ACTH), which means that changes in plasma cortisol levels are related to changes in ACTH levels. Changes in plasma cortisol levels follow a circadian rhythm that exhibits high individual variability. Normally, its secretion increases, with the peak of secretion, which is observed in the first hour after waking up [14, 15]. GCs act via intracellular glucocorticoid receptors that are mainly located in the cytoplasm. These receptors belong to a family of proteins activated by hormone binding. They bind to identical DNA sequences of the genes they regulate [16].

Glucocorticosteroids have a wide range of activity, from regulating reproduction, growth, and inflammatory reactions to participating in the metabolism of fats and proteins and maintaining water and electrolyte homeostasis [17–20]. In addition, they also affect the functioning of the cardiovascular system, and what is important, they are responsible for the distribution of adipose tissue. The effect of glucocorticosteroids is the increased release and burning of fatty acids, which are then used for energy production instead of glucose. In addition, cortisol, by affecting the differentiation of fat cells, or adipocytes, leads to their hypertrophy and the accumulation of lipids. The result is high amounts of triglycerides in the liver, dyslipidemia, and uneven deposition of adipose tissue [21].

Glucocorticosteroids, in addition to their lipolytic effect, intensify the process of gluconeogenesis in the liver and increase its reactivity to glucagon. Gluconeogenesis is promoted by activating the transcription of genes that encode enzymes in this pathway, and signaling is mediated by the glucocorticoid receptor. By activating glucose-6-phosphatase and phosphoenolpyruvate carboxykinase, they increase the release of substrates for gluconeogenesis from peripheral tissues. As a result, peripheral glucose uptake is reduced and insulin secretion increases, which in turn leads to the development of insulin resistance [22, 23]. Secreted by pancreatic β cells, insulin exerts the opposite effect on these processes by inhibiting hepatic gluconeogenesis and promoting glucose utilization in skeletal muscle and adipose tissue; among the wide spectrum of action of glucocorticosteroids,

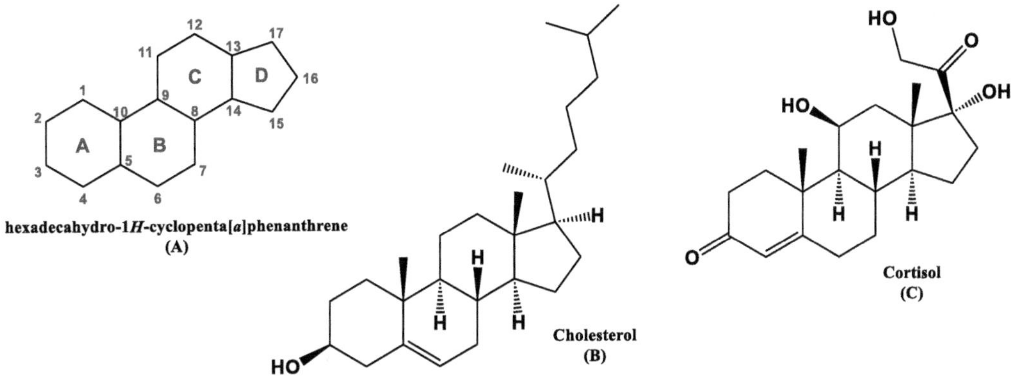

hexadecahydro-1H-cyclopenta[a]phenanthrene
(A)
Cholesterol
(B)
Cortisol
(C)

FIGURE 12.2 The basic structures of glucocorticoids: (**A**), cholesterol (**B**), cortisol (**C**).

HYPOTHALAMIC–PITUITARY–ADRENAL axis
HYPOTHALAMIC
CRH
PITUITARY
ACTH
ADRENAL
Cortisol

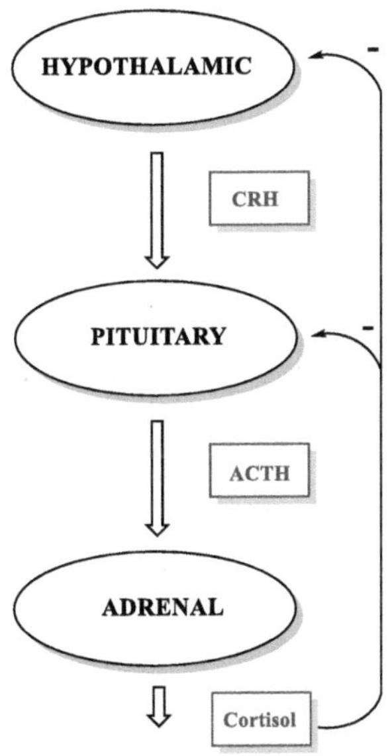

FIGURE 12.3 Regulation of glucocorticosteroid secretion.

anti-insulin activity is mentioned. However, chronic exposure to glucocorticoids leads to the development of hyperglycemia and insulin resistance [24].

GCs inhibit glucose utilization by reducing both glucose uptake and oxidation in skeletal muscle and adipose tissue, the tissues involved in glucose utilization in response to insulin [25]. Undoubtedly, this counteracts the effects of insulin, promoting glycolysis, glucose uptake, and oxidation. In addition, glucocorticosteroids, by promoting protein degradation in skeletal muscle to produce glucogenic amino acids, provide precursors for gluconeogenesis, so their gluconeogenic effect is closely related to protein metabolism [26]. As mentioned earlier, excess glucocorticosteroids disturb glucose homeostasis; therefore, it's important to know and understand their mechanisms of action in order to develop pharmacotherapy aimed at treating metabolic diseases.

12.3 PARTICIPATION OF 11β-HYDROXYSTEROID DEHYDROGENASE IN INTERCONVERSION OF GLUCOCORTICOSTEROID METABOLITES

The conjugation and conversion of glucocorticosteroids to their inactive metabolites occurs in the liver, and 90% of their excretion is renal [27]. The interconversion of the metabolites, i.e. biologically active cortisol to inactive cortisone, involves the enzyme 11β-hydroxysteroid dehydrogenase (11β-HSD), which has two isoforms (Figure 12.4) [28]. Isoform 1 (11β-HSD1) occurs mainly in the liver but also in adipose tissue, brain, vessels, and gonads, 11β-HSD2 occurs mainly in the kidneys, and also in the large intestine, placenta, and salivary glands [29, 30].

The first descriptions of 11β-HSD1 date to 1950 and are related to the discovery of cortisone, for which Hench, Kendall and Reichstein received the Nobel Prize. In turn, Cope and Black showed that the activity of glucocorticosteroids is conditioned by the presence of a hydroxyl group at the C-11 position in the steroid structure, and its oxidation inactivates a given glucocorticoid [30]. As mentioned earlier, cortisol is the active glucocorticoid, and cortisone is the inactive form. The rodent equivalent of cortisol is corticosterol, while 11-dehydrocorticosterone is the equivalent of cortisone (Figure 12.5) [31].

In 1953, Amelung and his colleagues reported that the enzyme 11β-HSD was involved in the interconversion of cortisol; in response, a number of researchers have confirmed the bidirectional action of this enzyme. Although both isoenzymes control the interconversion of biologically active glucocorticoids to their inactive forms, they differ in substrate affinity, cofactor specificity, and, of course, reaction direction. In addition, 11β-HSD1 is the product of a gene located on chromosome 1 in humans and rodents, while 11β-HSD2 is encoded by a gene located on chromosome 16 in humans

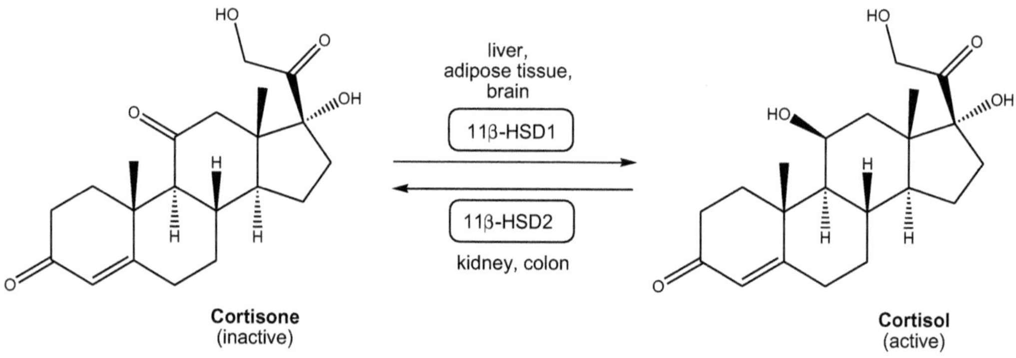

FIGURE 12.4 The physiological roles of 11β-HSD1 and 11β-HSD2.

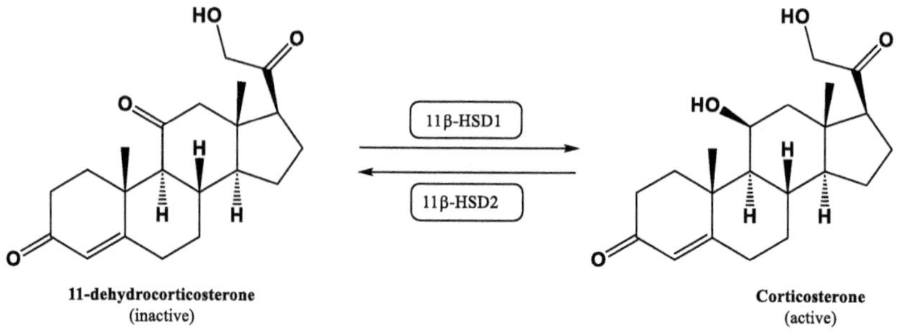

FIGURE 12.5 Structures of cortisol (corticosterone) and cortisone (11-dehydrocorticosterone) counterparts in rodents.

and on chromosome 8 in rodents [32, 33]. Isoenzymes show only about 21% amino acid sequence homology. Both isoforms are microsomal enzymes and are associated with the plasma membrane of the endoplasmic reticulum. The role of the NADPH-dependent isoform 11β-HSD1 is to activate the glucocorticoid receptor by increasing the concentration of the active form of glucocorticoids in the tissue. In turn, NAD-dependent 11β-HSD2 inactivates glucocorticoids, preventing the activation of the mineralocorticoid receptor [34].

12.4 β-HSD1

The first characterized isoenzyme was 11β-HSD1, which is present primarily in tissues rich in glucocorticoids; it is most abundant in the liver and adipose tissue, but it is also present in the pancreatic islets, lungs, brain, gonads, and bones [35]. It is the product of the HSD11B1 gene, which is located on chromosome 1 in both humans and rodents. Researchers used Edman degradation to determine its amino acid sequence, which has 291 amino acid residues. An important condition for the proper functioning of 11β-HSD1 is the appropriate concentration of NADPH, which is its cofactor. In vitro, it catalyzes a bidirectional redox reaction and exhibits NADP+-dependent 11β-dehydrogenase activity in the oxidation reaction and NADPH-dependent 11-oxoreductase activity in the reduction reaction [30]. In vivo, however, it acts as a reductase, leading to the conversion of inactive cortisone into active cortisol and increasing intracellular GC levels. Reductase activation requires hexose 6-phosphate dehydrogenase (H6PDH), which regenerates the cofactor NADPH (Figure 12.6) [36].

11β-HSD1 has a lower affinity for cortisol than for cortisone. It catalyzes the conversion of cortisone into cortisol and thus ensures a sufficiently high level of glucocorticosteroids in the tissues. The basic role of 11β-HSD1 is to increase the concentration of the active form of GCs in the tissue, which in turn activates the glucocorticoid receptor. 11β-HSD1 regulates the physiological functions of glucocorticosteroids such as hepatic gluconeogenesis and processes occurring in adipose tissue [37]. The role of 11β-HSD1 includes both endocrine regulation of circulating corticosteroids and systemic exposure to GCs, as well as conjugation of local tissue- and cell-specific exposure through autocrine activation of cortisol, independent of circulating cortisol. Hepatic 11β-HSD1, which is constitutively and strongly expressed in the liver, regulates endocrine cortisol [38]. In contrast, 11β-HSD1 in adipose, bone, muscle, and inflammation sites is dynamically regulated in a highly cell-specific manner [39, 40]. Disturbances in the function of this isoform are associated with the development of metabolic disorders. For instance, 11β-HSD1 present in adipose tissue cells promotes the development of abdominal obesity, which is associated with consequences such as type 2 diabetes, hypertension, dyslipidemia, and cardiovascular complications [41, 42].

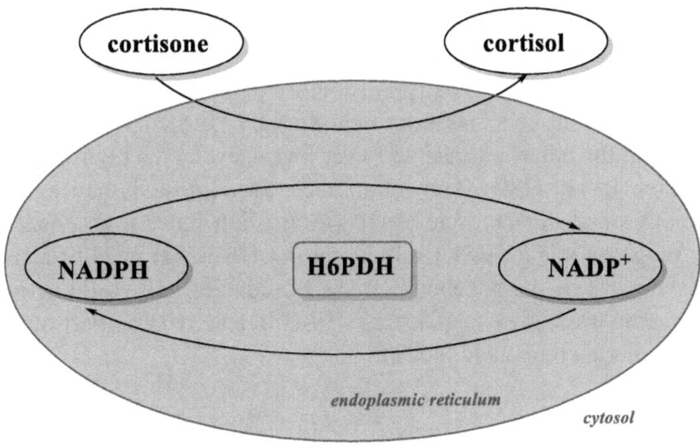

FIGURE 12.6 Action of hexose-6-phosphate dehydrogenase.

12.5 ASSOCIATION OF 11β-HSD1 WITH OBESITY, INSULIN RESISTANCE, AND METABOLIC SYNDROME

Obesity is a major risk factor for the development of type 2 diabetes and cardiovascular disease, leading to premature mortality; increased visceral fat correlates with glucose intolerance and the development of insulin resistance. All these factors, along with impaired lipid metabolism and increased blood pressure, are components of MetS [43]. The role of 11β-HSD1 in obesity-related disorders is to increase peripheral cortisol production [44]. Notably, it plays a greater role in visceral adipose tissue than in subcutaneous adipose tissue; visceral adipose tissue is characterized by greater metabolic activity, which is of great importance in glucocorticosteroid metabolism. It is visceral, not peripheral, obesity that is associated with the serious complications we described earlier [45].

To fully understand the role of 11β-HSD1 in the pathogenesis of metabolic syndrome, researchers constructed transgenic mouse models with 11β-HSD1 overexpression in visceral adipose tissue. Regenerating the active form of glucocorticosteroids exacerbated all disorders that characterize MetS in visceral obesity, including arterial hypertension and increased insulin resistance, triglyceride concentration, and pro-inflammatory cytokine concentration. The mice in the study were voracious, and their high-fat diet exacerbated insulin resistance [46]. Therefore, it can be concluded that in a mouse model, overexpression of 11β-HSD1 in adipose tissue translated into features of MetS. As in transgenic mice, overexpression of 11β-HSD1 in obese individuals is associated with the development of insulin resistance [47].

All of these factors work by regulating 11β-HSD1 transcription. Post-translational modifications within this enzyme result from differences in the availability of the NADPH cofactor. Researchers showed that a mutation in the H6PDH gene, which as previously mentioned controls the local availability of NADPH in the endoplasmic reticulum, determines the direction of the reaction catalyzed by 11β-HSD1 [48]. In addition, the presence of genetic polymorphisms at the 11β-HSD1 locus was associated with insulin resistance, body build, and waist-to-hip ratio [49].

The effects of glucocorticoids are context dependent and can be very different during acute versus chronic inflammation; 11β-HSD1 is induced early during the inflammatory response and shapes its later course. Deficiency or inhibition of 11β-HSD1 worsens acute inflammation but contributes to reduced inflammation in chronic conditions such as obesity or atherosclerosis. This is because in physiological concentrations, GCs are not harmful and have anti-inflammatory effects, while chronic stress has an immunosuppressive effect, GCs inhibit Th1 functions and increase Th2, leading to the development of a pro-inflammatory state [50].

Obesity, which should be understood as a chronic inflammation, contributes to the overexpression of 11β-HSD1, especially in metabolically active adipose tissue. In a transgenic mouse model, overexpression of 11β-HSD1 in visceral adipose tissue exacerbated all disorders characterizing MetS. In contrast, mice lacking 11β-HSD1 activity (knock out) were resistant to hyperglycemia induced by stress and a high-fat diet [51]. Phenotypically, the adipose tissue of these mice was characterized by a local decrease in corticosterone concentrations, with slightly elevated levels in the blood. These mice were insulin sensitive at the adipose tissue and liver levels; even on a high-fat diet, they gained less weight [52]. In addition, in 11β-HSD1 (knock out) mice, the adipose tissue was located peripherally, not visceral, especially since a greater role of 11β-HSD1 is attributed to the visceral location [53].

These findings suggest a role for selectively inhibiting 11β-HSD1 in adipose tissue. Indeed, there have been previous reports on the beneficial effects of reducing 11β-HSD1 activity for therapeutic purposes. There is some possibility of reducing 11β-HSD1 activity as part of treatment of type 2 diabetes, which is the main component of MetS.

12.6 β-HSD1 INHIBITION

As demonstrated earlier, 11β-HSD1 (cortisone reductase) is a microsomal enzyme, NADPH-dependent oxidoreductase, which catalyzes the transformation of inactive cortisone into physiologically active cortisol and also to a lesser extent the reverse reaction. Together with 11β-HSD2, located mainly in the kidneys and colon, the system regulates the level of cortisol in the body. Cushing's syndrome, otherwise called hypercortisolism or primary hyperactivity of the adrenal cortex, is a disease in which the adrenal cortex secretes excessive glucocorticosteroids; it is characterized by a chronic excess of cortisol in the blood that displaces deposits of fat tissue (buffalo neck, moon face, abdominal obesity) and causes thin skin, stretch marks with a characteristic pink color, acne, and insulin resistance [54]. Excess cortisol also causes the characteristic symptoms of MetS: visceral obesity, insulin resistance, diabetes, dyslipidemia, hypertension, and hyperuricemia [55]. MetS, as we have discussed, is a collection of abnormalities such as insulin resistance, obesity, dyslipidemia, hyperglycemia, and hypertension that are the major risk factors for type 2 diabetes and cardiovascular disease [56]. It is well known that intracellular concentration of cortisol is determined not only by plasma concentration but also by 11β-HSD1 activity, which catalyzes the conversion of inactive cortisone to active cortisol, especially in the liver and fat tissue [57].

Increased 11β-HSD1 expression is associated with the pathogenesis of lipid disorders as well as insulin resistance. Inhibition of 11β-HSD1 reduces cortisol levels and in turn fat tissue mass; compounds specifically inhibiting 11β-HSD1 in patients with type 2 diabetes reduce blood glucose concentrations, insulin resistance, and central obesity [58]. Studies have shown that compounds that inhibit 11β-HSD1 decrease glycogenolysis and total cholesterol levels [59]. In recent years, significant activity in the academic and pharmaceutical community has led to the discovery of many new chemical compounds as specific inhibitors of 11β-HSD1. Selective inhibitors have significant potential as treatments for type 2 diabetes, obesity, and cardiovascular disease [60, 61].

Carbenoxolone is a known inhibitor of 11β-HSD1 (Figure 12.7). It is a hemisuccinate ester and a derivative of glycyrrhizic acid, a natural product found in licorice root. Not only does carbenoxolone inhibit 11β-HSD1, it also to a lesser degree inhibits 11β-HSD2 [62]. Unlike 11β-HSD1, 11β-HSD2 is NAD^+ dependent; it allows oxidation of cortisol to inactive cortisone and thus protects tissues with mineralocorticosteroid receptors before cortisol. Inhibiting 11β-HSD2 leads to hypertension by activating the mineralocorticoid receptor in kidneys, as well as by decreasing the gene expression of endothelial nitric oxide synthase. This is particularly important to patients with type 2 diabetes and elements of MetS [63]. In addition, excess mineralocorticosteroids and supposed hyperaldosteronism caused by taking carbenoxolone can cause peripheral edema, hypokalemia, and metabolic alkalosis. Nonselective activity of carbenoxolone affects its limited range of clinical applications, prompting a search for new selective inhibitors of 11β-HSD1.

In the last two decades, many compounds have been synthesized and tested in the search for selective 11β-HSD1 inhibitors. Some of them have reached various phases of clinical trials (Figure 12.8).

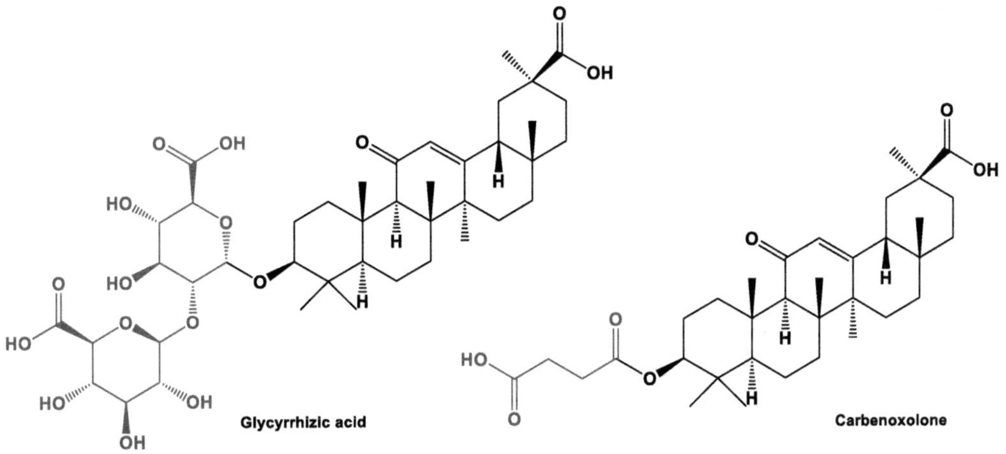

Glycyrrhizic acid

Carbenoxolone

FIGURE 12.7 Glycyrrhizic acid and its derivative carbenoxolone, a known 11β-HSD1 inhibitor.

The first selective 11β-HSD1 inhibitor that was tested in clinical trials was BVT-3498 (AMG-331). Structurally, it is a 2-aminothiazole derivative containing a morpholine ring and sulfonamide moiety. The compound was in phase II clinical development, but the work was later terminated [64]. Unfortunately, the literature lacks information on the exact values that would allow the assessment of the inhibitory activity of this compound. The Biovitrum patent only provides the information that "BVT3498 has *Ki*value for 11β-HSD1 in the nanomolar range" [65]. INCB13739 from Incyte Corporation companies was another compound that reached the clinical trial stage; this compound, like BVT-3498, contains a sulfonamide moiety. INCB13739 is an orally active, potent, selective, and tissue-specific 11β-HSD1 inhibitor with IC_{50} of 1.1 nM [66]. In the phase I clinical trials, it was well tolerated by healthy volunteers and patients with type 2 diabetes.

In subsequent years, researchers studied the effectiveness of adding INCB13739 to monotherapy with metformin. After 12 weeks, 200 mg INCB13739 with metformin significantly decreased glycated hemoglobin and glucose in fasting plasma compared with placebo. Total cholesterol, LDL cholesterol, and triglycerides were significantly reduced in patients with hyperlipidemia. Body weight was lower in patients who received INCB13739 therapy than in those who received placebo [67].

Pfizer's PF-915275, a sulfonamide, also reached phase I clinical trials. Pfizer tested it as a 11β-HSD1 inhibitor in 60 healthy adult volunteers. The compound inhibited the endogenous conversion of prednisone to prednisolone, which indicated that 11-HSD1 was inhibited. ACTH and androgen concentrations in plasma did not increase, which indicates that the hypothalamic–pituitary–adrenal axis was not activated [64, 68].

Biovitrum tested its AMG-221, a 2-aminothiazol-4(5H)-one derivative, in 44 obese patients who were given a compound in doses of 3 mg, 30 mg, or 100 mg. AMG-221 potently blocked 11β-HSD1 activity, producing sustained inhibition for the 24 hour study duration as measured in ex vivo adipose samples [69, 70]. However, further research on these inhibitors has stopped.

Despite many years of searching and clinical trials, none of the examined compounds has been introduced as a drug because the studies have been interrupted in different phases. Hence there is

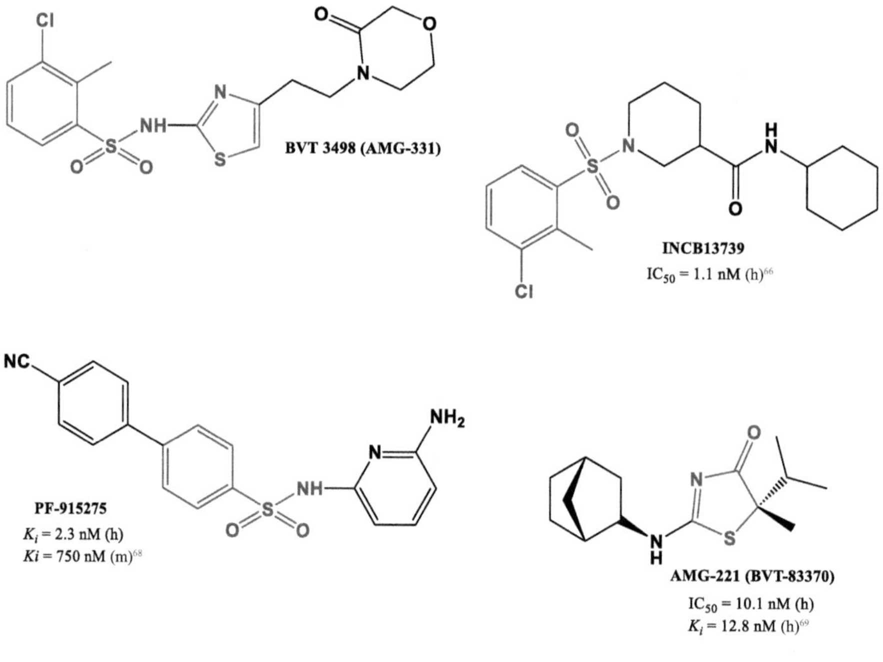

BVT 3498 (AMG-331)

INCB13739
IC_{50} = 1.1 nM (h)[66]

PF-915275
K_i = 2.3 nM (h)
Ki = 750 nM (m)[68]

AMG-221 (BVT-83370)
IC_{50} = 10.1 nM (h)
K_i = 12.8 nM (h)[69]

FIGURE 12.8 Selective 11β-HSD1 inhibitors used in clinical trials: BVT-3498 (AMG-331), INCB13739, PF-915275, and AMG-221 (BVT-83370). Data refer to tests for human (h) or murine (m) 11β-HSD1.

a need for further searches for selective inhibitors, and the subsequent studies on synthesis of new compounds and their inhibitory activity increase the chance of selecting new candidates for clinical trials. The biological activity of the compounds depends on their chemical structure, hence in the compounds tested in recent years, the structural fragments characteristic for previously examined inhibitors that have achieved the stage of clinical trials are often observed.

In vitro research was conducted for various structures, among the most numerous a group of compounds containing thiazole ring [55, 71–73]. or its partially hydrogenated form, thiazol-4-one, with an amino group in position 2 [74–81]. The most active thiazole derivatives contain sulfonamide groups, just like the inhibitors described earlier: BVT-3498, INCB13739, and PF-915275 [71] (Figure 12.9).

From among the many 2-amino-4(5*H*)-thiazol-4-one derivatives, the most active in in vitro research were compounds containing spatially large hydrophobic substituents (mainly alicyclic and aromatic rings), which interact with the active center of the enzyme, increasing the activity of the inhibitor (Figure 12.10).

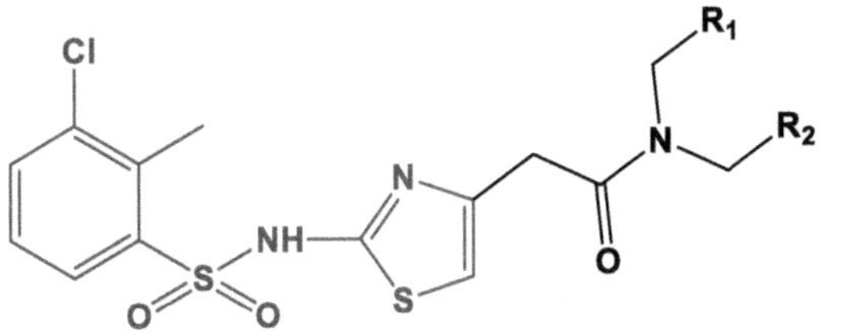

R_1, R_2 = CH_3 - BVT.14225
IC_{50} = 52 nM (h); 284 nM (m)[71]
R_1, R_2 = -CH_2-N(CH_3)-CH_2-(cycl) BVT.2733
IC_{50} = 3.34 μM (h); 96 nM (m)[71]

FIGURE 12.9 Selective 11β-HSD1 inhibitors with thiazole and sulfonamide moiety in in vitro studies. Data refer to tests for human (h) or murine (m) 11β-HSD1.

R = allyl: IC_{50} = 2.50 μM (h)[75a]
R = methyl: IC_{50} = 18.18 μM (h)[75b]
R = isopropyl: IC_{50} = 9.35 μM (h)[78]
R = t-butyl: IC_{50} = 1.60 μM (h)[79]
R = adamantyl: IC_{50} = 0.31 μM (h)[76]

K_i = 3.0 nM (h)[80]

IC_{50} = 10 nM (h)[74]

Amgen 2922
IC_{50} = 33 nM (h)[81]

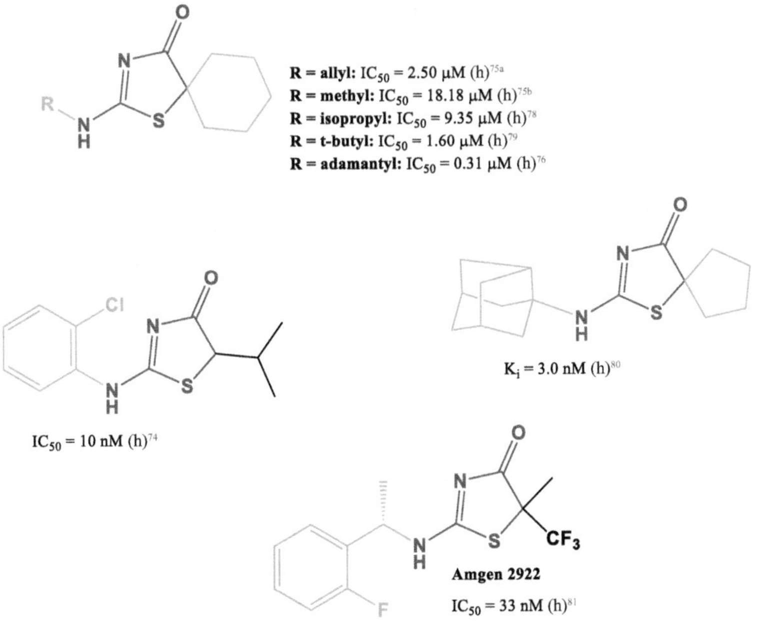

FIGURE 12.10 Selected selective 11β-HSD1 inhibitors with thiazole and thiazol-4-one moiety in in vitro studies. Data refer to tests for human (h) or murine (m) 11β-HSD1.

As potential selective 11β-HSD1 inhibitors, compounds containing condensed systems with a thiazole ring were also analyzed (Figure 12.11), but their activity was not satisfactory [82, 83]. The most active of the compounds only inhibited just over 50% of 11β-HSD1 activity at 10 μM, and all showed much lower values than those observed for inhibitors in which the thiazol ring is not condensed with other rings.

Researchers have also studied cyclic *N*-substituted sulfonamides as 11β-HSD1 inhibitors (Figure 12.12). The presence of functionalized phenyl rings and the spatially large cyclic hydrophobic substitutes in these compounds affect high inhibition activity (IC_{50} values in nanomols). These compounds are selective toward 11β-HSD1 [84–88].

Amide groups were observed in inhibitors containing two hydrophobic alicyclic systems. However, these compounds showed lower activity than compounds with the sulfonamide and disulfonamide moiety (Figure 12.13) [89].

Some compounds containing heterocyclic systems such as five-member saturated and unsaturated rings with sulfur, nitrogen, and oxygen atoms have shown high inhibition activity (Figure 12.14) [73, 90–93].

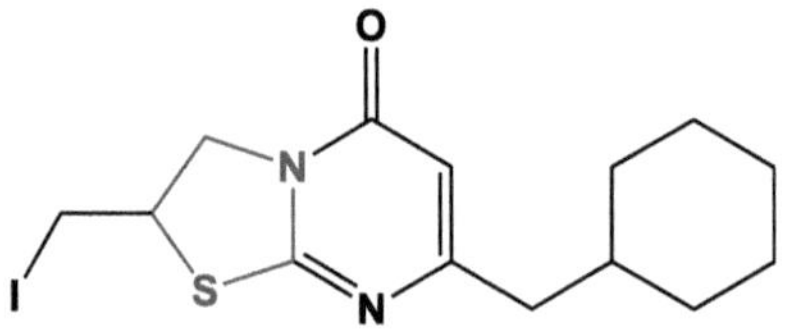

59.15% inhibition (conc. 10 μM)
IC_{50} = 6.9 μM (h)[83]

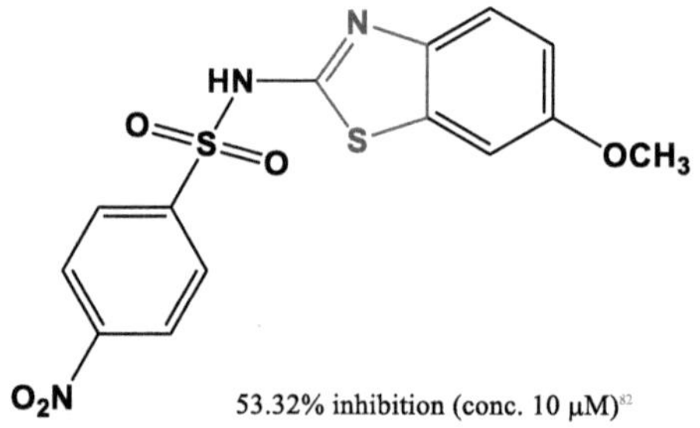

53.32% inhibition (conc. 10 μM)[82]

FIGURE 12.11 Selected selective 11β-HSD1 inhibitors with fused thiazole moiety in *in vitro* studies. Data refer to tests for human (h) or murine (m) 11β-HSD1.

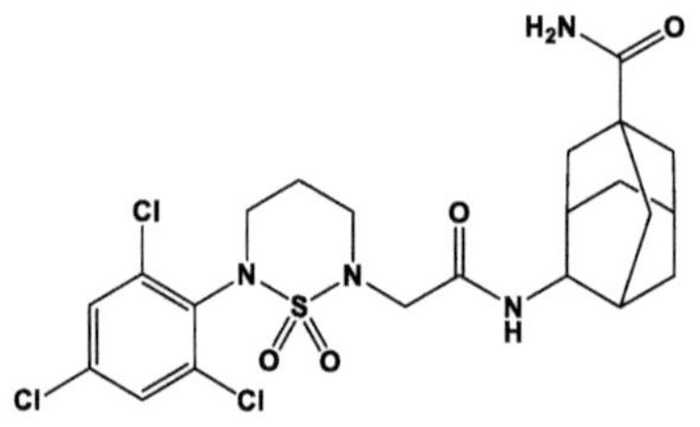

IC_{50} = 1.0 nM (h), 4.0 nM (m)[86, 87]

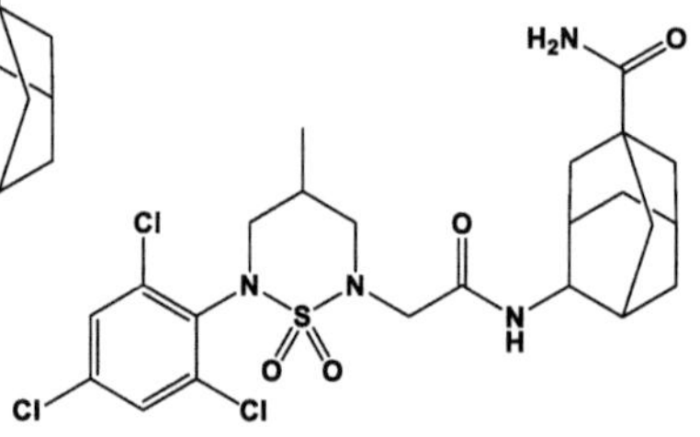

IC_{50} = 1.0 nM (h), 2.0 nM (m)[86, 88]

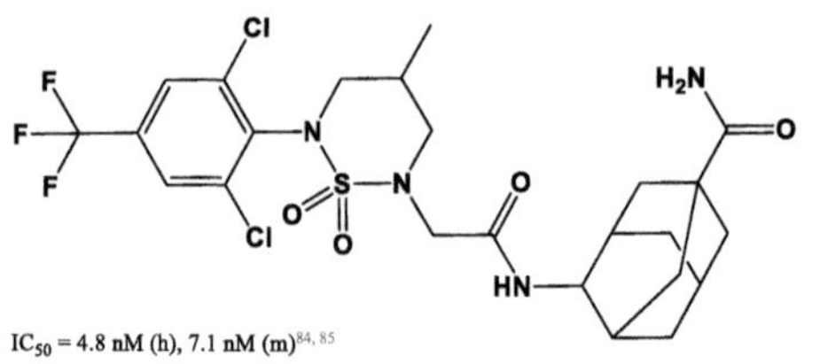

IC_{50} = 4.8 nM (h), 7.1 nM (m)[84, 85]

FIGURE 12.12 Cyclic sulfondiamides as selected selective 11β-HSD1 inhibitors in in vitro studies. Data refer to tests for human (h) or murine (m) 11β-HSD1.

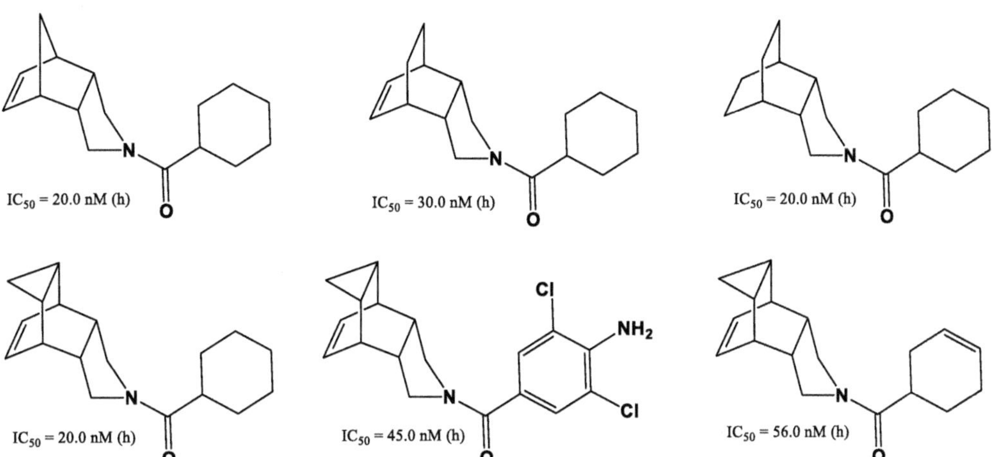

FIGURE 12.13 Selected selective 11β-HSD1 inhibitors with different heterocyclic moiety in in vitro studies. Data refer to tests for human (h) or murine (m) 11β-HSD1.

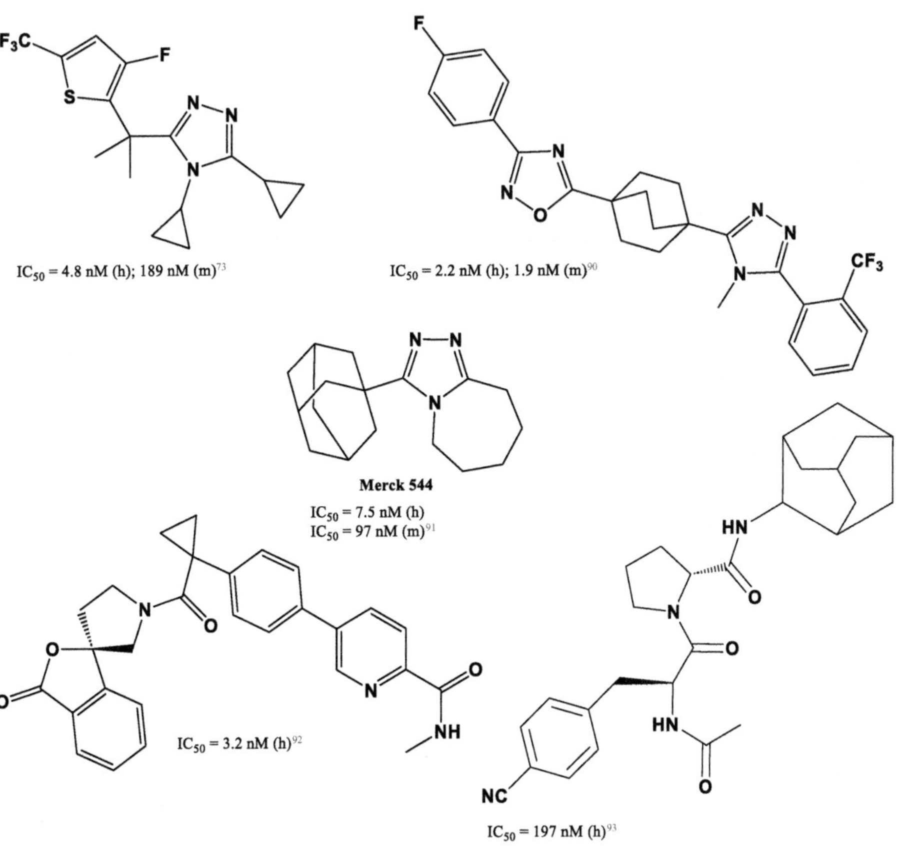

FIGURE 12.14 Selected selective 11β-HSD1 inhibitors with various five-membered heterocyclic rings in in vitro studies. Data refer to tests for human (h) or murine (m) 11β-HSD1.

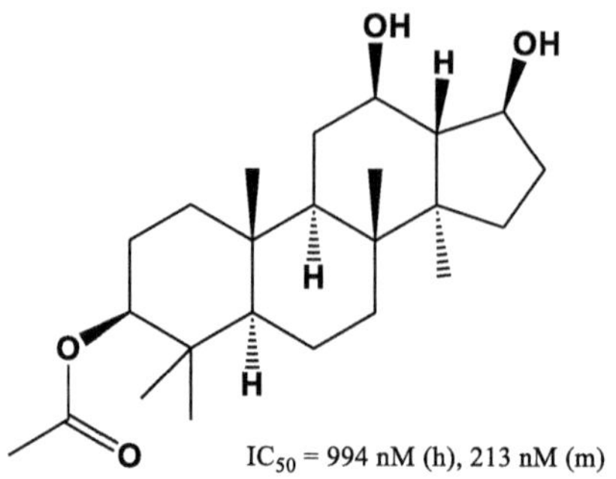

IC_{50} = 994 nM (h), 213 nM (m)

FIGURE 12.15 11β-HSD1 inhibitor containing dammarane core. Data refer to tests for human (h) or murine (m) 11β-HSD1.

The search for selective 11β-HSD1 inhibitors also included carbenoxolone in four-member condensed systems containing dammarane core. The most active compound from this group turned out to be a selective inhibitor (Figure 12.15). However, relatively poor IC_{50} values keep these compounds from being potential candidates for clinical trials [94].

The data we have presented in this chapter reflect that ongoing research in this area is highly needed. New knowledge can help us better understand the pathological processes of metabolic syndrome and design new therapeutic paths.

REFERENCES

[1] Saklayen, M.G. The Global Epidemic of the Metabolic Syndrome. *Curr. Hypertens. Rep.* **2018**,Feb 26,20(2),pp.12. Doi: 10.1007/s11906-018-0812-z.

[2] Lemieux, I.; Després, JP. Metabolic Syndrome: Past, Present and Future. *Nutrients.* **2020**,Nov 14,12(11),pp.3501. Doi: 10.3390/nu12113501.

[3] Nilsson, P.M.; Tuomilehto, J.; Rydén, L. The Metabolic Syndrome—What is It and How Should It be Managed? *Eur. J. Prev. Cardiol.* **2019**,26(2_suppl),pp.33–46. Doi: 10.1177/2047487319886404.

[4] Bovolini, A.; Garcia, J.; Andrade, MA.; Duarte, JA. Metabolic Syndrome Pathophysiology and Predisposing Factors. *Int. J. Sports Med.* **2021**,42(3),pp.199–214. Doi: 10.1055/a-1263-0898.

[5] Kahn, C.R.; Wang, G.; Lee, K.Y. Altered Adipose Tissue and Adipocyte Function in the Pathogenesis of Metabolic Syndrome. *J. Clin. Invest.* **2019**,129(10),pp.3990–4000. Doi: 10.1172/JCI129187.

[6] Iqbal, J.; Al Qarni, A.; Hawwari, A.; Alghanem, A.F.; Ahmed, G. Metabolic Syndrome, Dyslipidemia and Regulation of Lipoprotein Metabolism. *Curr. Diabetes Rev.* **2018**,14(5),pp.427–433. Doi: 10.2174/1573399813666170705161039.

[7] Aboonabi, A.; Meyer, R.R.; Singh, I. The Association between Metabolic Syndrome Components and the Development of Atherosclerosis. *J. Hum. Hypertens.* **2019**,33(12),pp.844–855. Doi: 10.1038/s41371-019-0273-0.

[8] Engin, A. The Definition and Prevalence of Obesity and Metabolic Syndrome. *Adv. Exp. Med. Biol.* **2017**,960,pp.1–17. Doi: 10.1007/978-3-319-48382-5_1.

[9] Mandviwala, T.; Khalid, U.; Deswal, A. Obesity and Cardiovascular Disease: A Risk Factor or a Risk Marker? *Curr. Atheroscler. Rep.* **2016**,18(5),pp.21. Doi: 10.1007/s11883-016-0575-4.

[10] Kuo, T.; McQueen, A.; Chen, T.C.; Wang, J.C. Regulation of Glucose Homeostasis by Glucocorticoids. *Adv. Exp. Med. Biol.* **2015**,872,pp.99–126. Doi: 10.1007/978-1-4939-2895-8_5.

[11] Samuel, S.; Nguyen, T.; Choi, A. Pharmacologic Characteristics of Corticosteroids. *J. Neurocrit. Care* **2017**,10,pp.53–59.

[12] Nicolaides, N.C.; Galata, Z.; Kino, T.; Chrousos, G.P.; Charmandari, E. The Human Glucocorticoid Receptor: Molecular Basis of Biologic Function. *Steroids* **2010**,75,pp.1–12. Doi: 10.1016/j.steroids.2009.09.002.

[13] Paragliola, R.M.; Papi, G.; Pontecorvi, A.; Coresllo, S.M. Treatment with Synthetic Glucocorticoids and the Hypothala-mus-Pituitary-Adrenal Axis. *Int. J. Mol. Sci.* **2017**,18(10),pp.2201. Doi: 10.3390/ijms18102201.

[14] Anderson Elverson, C.; Wilson, M.E. Cortisol: Circadian Rhythm and Response to a Stressor. Newborn. *Infant Nurs. Rev.* **2005**,5,pp.159–169. Doi: 10.1053/j.nainr.2005.09.002.

[15] Levine, A.; Zagoory-Sharon, O.; Feldman, R.; Lewis, J.G.; Weller, A. Measuring Cortisol in Human Psychobiological Studies. *Physiol. Behav.* **2007**,90,pp.43–53. Doi: 10.1016/j.physbeh.2006.08.025.

[16] Lewis-Tuffin, L.J.; Cidlowski, J.A. The Physiology of Human Gluco-Corticoid Receptor b (hGR b) and Glucocorticoid Resistance. *Ann. N.Y. Acad. Sci.* **2006**,1069,pp.1–9. Doi: 10.1196/annals.1351.001.

[17] Timmermans, S.; Souffriau, J.; Libert, C. A General Introduction to Glucocorticoid Biology. *Front. Immunol.* **2019**,10,pp.1545. Doi: 10.3389/fimmu.2019.01545.

[18] Cain, D.W.; Cidlowski, J.A. Immune Regulation by Glucocorticoids. *Nat. Rev. Immunol.* **2017**,17,pp.233–247. Doi: 10.1038/nri.2017.1.

[19] Nussinovitch, U.; de Carvalho, J.F.; Pereira, R.M.; Shoenfeld, Y. Glucocorticoids and the Cardiovascular System: State of the Art. *Curr. Pharm. Des.* **2010**,16,pp.3574–3585. Doi: 10.2174/138161210793797870.

[20] Reichardt, S.D.; Amouret, A.; Muzzi, C.; Vettorazzi, S.; Tuckermann, J.P.; Lühder, F.; Reichardt, H.M. The Role of Glucocorti-Coids in Inflammatory Diseases. *Cells* **2021**,10,pp.1–30. Doi: 10.3390/cells10112921.

[21] Peckett, A.J.; Wright, D.C.; Riddell, M.C. The Effects of Glucocorticoids on Adipose Tissue Lipid Metabolism. *Metabolism* **2011**,60(11),pp.1500–1510. Doi: 10.1016/j.metabol.2011.06.012.

[22] Kokkinopoulou, I.; Diakoumi, A.; Moutsatsou, P. Glucocorticoid Receptor Signaling in Diabetes. *Int. J. Mol. Sci.* **2021**,22(20),pp.11173. Doi: 10.3390/ijms222011173.

[23] Akalestou, E.; Genser, L.; Rutter, G.A. Glucocorticoid Metabolism in Obesity and Following Weight Loss. *Front. Endocrinol. (Lausanne)* **2020**,11,pp.59. Doi: 10.3389/fendo.2020.00059.

[24] Perez, A.; Jansen-Chaparro, S.; Saigi, I.; Bernal-Lopez, M.R.; Miñambres, I.; Gomez-Huelgas, R. Glucocorticoid-Induced Hyperglycemia. *J. Diabetes* **2014**,6(1),pp.9–20. Doi: 10.1111/1753-0407.12090.

[25] Magomedova, L.; Cummins, C.L. Glucocorticoids and Metabolic Control. *Handb. Exp. Pharmacol.* **2016**,233,pp.73–93. Doi: 10.1007/164_2015_1.

[26] Kuo, T.; Harris, C.A.; Wang, J.C. Metabolic Functions of Glucocorticoid Receptor in Skeletal Muscle. *Mol. Cell Endocrinol.* **2013**,380(1–2),pp.79–88. Doi: 10.1016/j.mce.2013.03.003.

[27] Sorensen, B.K.; Link, J.T.; von Geldern, T.; Emery, M.; Wang, J.; Hickman, B.; Grynfarb, M.; Goos-Nilsson, A.; Carroll, S. An Evaluation of a C-Glucuronide as a Liver Targeting Group: Conjugate of a Glucocorticoid Antagonist. *Bioorg. Med. Chem. Let.* **2004**,13(14),pp.2307–2310. Doi: 10.1016/s0960-894x(03)00431-1.

[28] Tomlinson, J.W.; Walker, E.A.; Bujalska, I.J.; Draper, N.; Lavery, G.G.; Cooper, M.S.; Hewison, M.; Stewart, P.M. 11β-Hydroxysteroid Dehydrogenase Type 1: A Tissue Specific Regulator of Glucocorticoid Response. *Endocr. Rev.* **2004**,25(5),pp.831–866. Doi: 10.1210/er.2003-0031.

[29] Davani, B.; Khan, A.; Hult, M.; Martensson, E.; Okret, S.; Efendic, S.; Jornvall, H.; Oppermann, U.C. Type 1 11beta-Hydroxysteroid Dehydrogenase Mediates Glucocorticoid Activation and Insulin Release in Pancreaticislets. *J. Biol. Chem.* **2000**,275,pp.34841–34844. Doi: 10.1074/jbc. C000600200

[30] Draper, N.; Stewart, P.M. 11β-Hydroxysteroid Dehydrogenase and the Pre-Receptor Regulation of Corticosteroid Hormone Action. *J. Endocrinol.* **2005**,186,pp.251–271. Doi: 10.1677/joe.1.06019.

[31] Hunter, R.W.; Bailey, M.A. Glucocorticoids and 11β-Hydroxysteroid Dehydrogenases: Mechanisms for Hypertension. *Curr. Opin. Pharmacol.* **2015**,21,pp.105–114. Doi: 10.1016/j.coph.2015.01.005.

[32] Paterson, J.M.; Seckl, J.R.; Mullins, J.J. Genetic Manipulation of 11beta-Hydroxysteroid Dehydrogenases in Mice. *Am. J. Physiol. Regul. Integr. Comp. Physiol.* **2005**,289,pp.642–652. Doi: 10.1152/ajpregu.00017.2005.

[33] Tannin, G.M.; Agarwal, A.K.; Monder, C.; Nowy, M.I.; Biały, P.C. The Human Gene for 11beta-Hydroxysteroid Dehydrogenase. Structure, Tissue Distribution, and Chromosomal Localization. *J. Biol. Chem.* **1991**,266,pp.16653–16658. Doi: 10.1016/S0021-9258(18)55351-5.

[34] Palermo, M.; Quinkler, M.; Stewart, P.M. Apparent Mineralocorticoid Excess Syndrome: An Overview. *Arq. Bras. Endocrinol. Metab.* **2004**,48,pp.687–696. Doi: 10.1590/s0004-27302004000500015.

[35] Chapman, K.; Holmes, M.; Seckl, J. 11β-Hydroxysteroid Dehydrogenases: Intracellular Gate-Keepers of Tissue Glucocorticoid Action. *Physiol. Rev.* **2013**,93(3),pp.1139–1206. Doi: 10.1152/physrev.00020.2012.

[36] White, P.C.; Rogoff, D.; McMillan, D.R. Physiological Roles of 11beta-Hydroxysteroid Dehydrogenase Type 1 and Hexose-6-Phosphate Dehydrogenase. *Curr. Opin. Pediatr.* **2008**,20(4),pp.453–457. Doi: 10.1097/MOP.0b013e328305e439.

[37] Krozowski, Z.; Chai, Z. The Role of 11β-Hydroxysteroid Dehydrogenases in the Cardiovascular System. *Endocr. J.* **2003**,50,pp.485–489. Doi: 10.1507/endocrj.50.485

[38] Martin, C.S.; Cooper, M.S.; Hardy, R.S. Endogenous Glucocorticoid Metabolism in Bone: Friend or Foe. *Front. Endocrinol.* **2021**,12,pp.1–13. Doi: 10.3389/fendo.2021.73361

[39] Hardy, R.S.; Fenton, C.; Croft, A.P.; Naylor, A.J.; Begum, R.; Desanti, G.; Buckley, C.D.; Lavery, G.; Cooper, M.S.; Raza, K. 11Beta-Hydroxysteroid Dehydrogenase Type 1 Regulates Synovitis, Joint Destruction, and Systemic Bone Loss in Chronic Polyarthritis. *J. Autoimmun.* **2018**,92,pp.104–113. Doi: 10.1016/j.jaut.2018.05.010

[40] Ahasan, M.M.; Hardy, R.; Jones, C.; Kaur, K; Nanus, D.; Juarez, M.; Morgan, S.A.; Hassan-Smith, Z.; Bénézech, C.; Caamaño, J.H.; Hewison, M.; Lavery, G.; Rabbitt, E.H.; Clark, A.R.; Filer, A.; Buckley, C.D.; Raza, K.; Stewart, P.M.; Cooper, M.S. Inflammatory Regulation of Glucocorticoid Metabolism in Mesenchymal Stromal Cells. *Arthritis Rheum.* **2012**,64(7),pp.2404–2413. Doi: 10.1002/art.34414

[41] Stimson, R.H.; Walker, B.R. Glucocorticoids and 11Beta-Hydroxysteroid Dehydrogenase Type 1 in Obesity and the Metabolic Syndrome. *Minerva Endocrinol.* **2007**,32(3),pp.141–159.

[42] Morton, N.M.; Seckl, J.R. 11Beta-Hydroxysteroid Dehydrogenase Type 1 and Obesity. *Front. Horm. Res.* **2008**,36,pp.146–164. Doi: 10.1159/000115363.

[43] Grundy, S.M. Metabolic Syndrome: A Multiplex Cardiovascular Risk Factor. *J. Clin. Endocrinol. Metab.* **2007**,92(2),pp.399–404. Doi: 10.1210/jc.2006-0513, Indexed in Pubmed: 17284640.

[44] Wake, D.J.; Walker, B.R. Inhibition of 11Beta-Hydroxysteroid Dehydrogenase Type 1 in Obesity. *Endocrine* **2006,**29(1),pp.101–108. Doi: 10.1385/ENDO:29:1:101.

[45] Sukhija, R.; Kakar, P.; Mehta, V.; Mehta, J.L. Enhanced 11β-Hydroxysteroid Dehydrogenase Activity, the Metabolic Syndrome and Systemic Hypertension. *Am. J. Cardiol.* **2006**,98,pp.544–548. Doi: 10.1016/j.amjcard.2006.03.028

[46] Masuzaki, H.; Paterson, J.; Shinyama, H.; Morton, N.M.; Mullins, J.J.; Seckl, J.R.; Flier, J.S. A Transgenic Model of Visceral Obesity and the Metabolic Syndrome. *Science* **2001**,294(5549),pp.2166–2170. Doi: 10.1126/science.1066285.

[47] Wake, D.J.; Rask, E.; Livingstone, D.E.; Söderberg, S.; Olsson, T.; Walker, B.R. Local and Systemic Impact of Transcriptional Up-Regulation of 11Beta-Hydroxysteroid Dehydrogenase Type 1 in Adipose Tissue in Human Obesity. *J. Clin. Endocrinol. Metab.* **2003,**88(8),pp.3983–3988. Doi: 10.1210/jc.2003-030286.

[48] Draper, N.; Walker, E.A.; Bujalska, I.J.; Tomlinson, J.W.; Chalder, S.M.; Arlt, W.; Lavery, G.G.; Bedendo, O.; Ray, D.W.; Laing, I.; Malunowicz, E.; White, P.C.; Hewison, M.; Mason, P.J.; Connell, J.M.; Shackleton, C.H.; Stewart, P.M. Mutations in the Genes Encoding 11Beta-Hydroxysteroid Dehydrogenase Type 1 and Hexose-6-Phosphate Dehydrogenase Interact to Cause Cortisone Reductase Deficiency. *Nat. Genet.* **2003**,34(4),pp.434–439. Doi: 10.1038/ng1214.

[49] Nair, S.; Lee, Y.H.; Lindsay, R.S.; Walker, B.R.; Tataranni, P.A.; Bogardus, C.; Baier, L.J.; Permana, P.A. 11Beta-Hydroxysteroid Dehydrogenase Type 1: Genetic Polymorphisms are Associated with Type 2 Diabetes in Pima Indians Independently of Obesity and Expression in Adipocyte and Muscle. *Diabetologia* **2004**,47(6),pp.1088–1095. Doi: 10.1007/s00125-004-1407-6.

[50] Chapmana, K.E.; Coutinho, A.E.; Zhang, Z.; Kipari, T.; Savill, J.S.; Seckl, J.R. Changing Glucocorticoid Action: 11-Hydroxysteroid Dehydrogenase Type 1 in Acute and Chronic Inflammation. *J. Steroid Biochem. Mol. Biol.* **2013**,137,pp.82–92. Doi: 10.1016/j.jsbmb.2013.02.002

[51] Kotelevtsev, Y.; Holmes, M.C.; Burchell, A.; Houston, P.M.; Schmoll, D.; Jamieson, P.; Best, R.; Brown, R.; Edwards, Ch.R.W.; Seckl, J.R.; Mullins, J.J. 11beta-hydroxysteroid dehydrogenase type 1 knockout mice show attenuated glucocorticoid-inducible responses and resist hyperglycemia on obesity or stress. *Proc. Natl. Acad. Sci. USA.* **1997**,Dec 23,94,pp.14924–14929. Doi: 10.1073/pnas.94.26.14924

[52] Kershaw, E.E.; Morton, N.M.; Dhillon, H.; Ramage, L.; Seckl, J.R.; Flier, J.S. Adipocyte-Specific Glucocorticoid Inactivation Protects Against Diet-Induced Obesity. *Diabetes* **2005**,54(4),pp.1023–1031. Doi: 10.2337/diabetes.54.4.1023.

[53] Sewter, C.; Vidal-Puig, A. PPARgamma and the Thiazolidinediones: Molecular Basis for a Treatment of 'Syndrome X'? *Diabetes Obes. Metab.* **2002**,4(4),pp.239–248. Doi: 10.1046/j.1463-1326.2002.00187.x.

[54] Sharma, S.T.; Nieman, L.K.; Feelders, R.A. Cushing's Syndrome: Epidemiology and Developments in Disease Management. *Clin. Epidemiol.* **2015**,17(7),pp.281–293. Doi: 10.2147/CLEP.S44336.

[55] Navarrete-Vázquez, G.; Morales-Vilchis, M.G.; Estrada-Soto, S.; Ramírez-Espinosa, J.; Hidalgo-Figueroa, S.; Nava-Zuazo, C.; Tlahuext. H.; Leon-Rivera, I.; Medina-Franco, J.L.; López-Vallejo, F.; Webster, S.P.; Binnie, M.; Ortiz-Andrade, R.; Moreno-Diaz, M. Synthesis of 2-{2-[(α/β-naphthalen-1-ylsulfonyl)amino]-1,3-thiazol-4-yl} Acetamides with 11β-Hydroxysteroid Dehydrogenase Inhibition and in Combo Antidiabetic Activities. *Eur. J. Med. Chem.* **2014**,74,pp.179–186. Doi: 10.1016/j.ejmech.2013.12.042.

[56] Scott, J.S.; Goldberg, F.W.; Turnbull, A.V. Medicinal Chemistry of Inhibitors of 11β-Hydroxysteroid Dehydrogenase Type 1 (11β-HSD1). *J. Med. Chem.* **2013**,57,pp.4466–4486. Doi: 10.1021/jm4014746.

[57] Anagnostis, P.; Katsiki, N.; Adamidou, F.; Athyros, V.G.; Karagiannis, A.; Kita, M.; Mikhailidis, D.P. 11Beta-Hydroxysteroid Dehydrogenase Type 1 Inhibitors: Novel Agents for the Treatment of Metabolic Syndrome and Obesity-Related Disorders? *Metab. Clin. Exp.* **2013**,16,pp.21–33. Doi: 10.1016/j.metabol.2012.05.002.

[58] Böhme, T.; Engel, C.K.; Farjot, G.; Güssregen, S.; Haack, T.; Tschank, G.; Ritter, K. 1,1-Dioxo-5,6-Dihydro-[4,1,2]Oxathiazines, a Novel Class of 11β-HSD1 Inhibitors for the Treatment of Diabetes. *Bioorg. Med. Chem. Lett.* **2013**,23,pp.4685–4691. Doi: 10.1016/j.bmcl.2013.05.102.

[59] Andrews, R.C.; Rooyackers, O.; Walker, B.R. Effects of the 11β-Hydroxysteroid Inhibitor Carbenoxolone on Insulin Sensitivity in Men with Type 2 Biabetes. *J. Clin. Endocrinal. Metab.* **2003**,88,pp.285–291. Doi: 10.1210/jc.2002-021194.

[60] Anderson, A.; Walker, B.R. 11β-HSD1 Inhibitors for the Treatment of Type 2 Diabetes and Cardiovascular Disease. *Drugs* **2013**,73,pp.1385–1393. Doi: 10.1007/s40265-013-0112-5.

[61] Vicker, N.; Su, X.; Ganeshapillai, D.; Smith, A.; Purohit, A.; Reed, M.J.; Potter, B.V.L. Novel Non-Steroidal Inhibitors of Human 11β-Hydroxysteroid Dehydrogenase Type 1. *J. Steroid Biochem. Mol. Biol.* **2007**,104,pp.123–129. Doi: 10.1016/j.jsbmb.2007.03.023.

[62] Diederich, S.; Hanke, B.; Quinkler, M.; Herrmann, M.; Bahr, V., Oelkers, W. In the Search for Specific Inhibitors of Human 11Beta-Hydroxysteroid-Dehydrogenases (11Beta-HSDs): Chenodeoxycholic Acid Selectively Inhibits 11Beta-HSD-I. *Eur. J. Endocrinol.* **2000**,142,pp.200–207. Doi: 10.1530/eje.0.1420200.

[63] Kotelevtsev, Y.; Brown, R.W.; Fleming, S.; Kenyon, C.; Edwards, C.R.W.; Seckl, J.R.; Mullins, J.J. Hypertension in Mice Lacking 11β-Hydroxysteroid Dehydrogenase Type 2. *J. Clin. Invest.* **1999**,103,pp.683–689. Doi: 10.1172/JCI4445.

[64] Fotsch, C.; Wang, M.; Blockade of Glucocorticoid Excess at the Tissue Level: Inhibitors of 11β-Hydroxysteroid Dehydrogenase Type 1 as a Therapy for Type 2 Diabetes. *J. Med. Chem.* **2008**,51,pp.4852–4857. Doi: 10.1021/jm800369f.

[65] Abrahmsen, L.; Nilsson, J.; Opperman, U.; Svensson, S. Methods for the Protein Production and Purification of Soluble Recombinant Polypeptides. Patent WO2005068646A1, 28 July 2005.

[66] Chemicals. https://www.dcchemicals.com/product_show-DC11207.html accessed on 20 July 2022.

[67] Rosenstock, J.; Banarer, S.; Fonseca, V.; Inzucchi, S.; Sun, W.; Yao, W.; Hollis, G.; Flores, R.; Levy, R.; Williams, W.V.; Seckl, J.R.; Huber, R. The 11-β-Hydroxysteroid Dehydrogenase Type 1 Inhibitor INCB13739 Improves Hyperglycemia in Patients with Type 2 Diabetes Inadequately Controlled by Metformin Monotherapy. *Diabetes Care* **2010**,33,pp.1516–1522. Doi: 10.2337/dc09-2315.

[68] Courtney, R.; Stewart, P.M.; Toh, M.; Ndongo, M.N.; Calle, R.A.; Hirshberg, B. Modulation of 11*β*-Hydroxysteroid Dehydrogenase (11*β*HSD) Activity Biomarkers and Pharmacokinetics of PF-00915275, a Selective 11_HSD1 Inhibitor. *J. Clin. Endocrinol. Metab.* **2008**,93,pp.550–556. Doi: 10.1210/jc.2007-1912

[69] Gibbs, J.P.; Emery, M.G.; McCaffery, I.; Smith, B.; Gibbs, M.A.; Akrami, A.; Rossi, J.; Paweletz, K.; Gastonguay, M.R.; Bautista, E.; Wang, M.; Perfetti, R.; Daniels, O. Population Pharmokinetic/Pharmacodynamic Model of Subcutaneous Adipose 11β-Hydroxysteroid Dehydrogenase Type 1 (11β-HSD1) Activity After Oral Administration of AMG 221, a Selective 11β-HSD1 Inhibitor. *J. Clin. Pharmacol.* **2011**,51(6),pp.830–841. Doi: 10.1177/0091270010374470.

[70] Joharapurkar, A.; Dhanesha, N.; Shah, G.; Kharul, R.; Jain, M. 11β-Hydroxysteroid Dehydrogenase Type 1: Potential Therapeutic Target for Metabolic Syndrome. *Pharmacol. Rep.* **2012**,64,pp.1055–1085. Doi: 10.1016/s1734-1140(12)70903-9.

[71] Barf, T.; Vallgårda, J.; Emond, R.; Häggström, C.; Kurz, G.; Nygren, A.; Larwood, V.; Mosialou, E.; Axelsson, K.; Olsson, R.; Engblom, L.; Edling, N.; Rönquist-Nii, Y.; Öhman, B.; Alberts, P.; Abrahmsén, L. Arylsulfonamidothiazoles as a New Class of Potential Antidiabetic Drugs. Discovery of Potent and Selective Inhibitors of the 11β-Hydroxysteroid Dehydrogenase Type 1. *J. Med. Chem.* **2002**,45,pp.3813–3815. Doi: 10.1021/jm025530f.

[72] Goldberg, F.W.; Dossetter, A.G.; Scott, J.S.; Robb, G.R.; Boyd, S.; Groombridge, S.D.; Kemmitt, P.D.; Sjögren, T.; Gutierrez, P.M.; deSchoolmeester, J.; Swales, J.G.; Turnbull, A.V.; Wild, M.J. Optimization of Brain Penetrant 11β-Hydroxysteroid Dehydrogenase Type I Inhibitors and In Vivo Testing in Diet-Induced Obese Mice. *J. Med. Chem.* **2014**,57,pp.970–986. Doi: 10.1021/jm4016729.

[73] Koike, T.; Sasuga, D.; Hosaka, M.; Kawano, T.; Fukodome, H.; Kurosawa, K.; Mortitomo, A.; Mimasu, S.; Ishii, H.; Yoshimura, S. Discovery and Biological Evaluation of Potent Orally Active Human 11β-Hydroxysteroid Dehydrogenase Type 1 Inhibitors for the Treatment of Type 2 Diabetes Mellitus. *Chem. Pharm. Bull.* **2019**,67,pp.824–838. Doi: 10.1248/cpb.c19-00211.

[74] Yuan, C.; St Jean, D.J., Jr; Liu, Q.; Cai, L.; Li, A.; Han, N.; Moniz, G.; Askew, B.; Hungate, R.W.; Johansson, L.; et al. The Discovery of 2-Anilinothiazolones as 11β-HSD1 Inhibitors. *Bioorg. Med. Chem. Lett.* **2007**,17,pp.6056–6061. Doi: 10.1016/j.bmcl.2007.09.070.

[75] a) Studzińska, R.; Kołodziejska, R.; Kupczyk, D.; Plaziński, W.; Kosmalski, T. A Novel Derivatives of Thiazol-4(5H)-One and Their Activity in the Inhibition of 11β-Hydroxysteroid Dehydrogenase Type 1. *Bioorg. Chem.* **2018**,79,pp.115–121. Doi: 10.1016/j.bioorg.2018.04.014. b) Studzińska, R.; Kołodziejska, R.; Płaziński, W.; Kupczyk, D.; Kosmalski, T.; Jasieniecka, K.; Modzelewska-Banachiewicz, B. Synthesis of the *N*-Methyl Derivatives of 2-Aminothiazol-4(5*H*)-One and Their Interactions with 11*β*HSD1-Molecular Modeling and *In Vitro* Studies. *Chem. Biodiversity* **2019**,16,pp.e1900065. Doi: 10.1002/cbdv.201900065

[76] Studzińska, R.; Kupczyk, D.; Płaziński, W.; Baumgart, S.; Bilski, R.; Paprocka, R.; Kołodziejska, R. Novel 2-(adamantan-1-yloamino)thiazol-4(5H)-One Derivatives and Their Inhibitory Activity Towards 11β-HSD1—Synthesis Molecular Docking and In Vitro Studies. *Int. J. Mol. Sci.* **2021**,22,pp.8609. Doi: 10.3390/ijms22168609.

[77] Baumgart, S.; Kupczyk, D.; Archała, A.; Koszła, O.; Sołek, P.; Płaziński, W.; Płazińska, A.; Studzińska, R. Synthesis of Novel 2-(Cyclopentylamino)thiazol-4(5*H*)-One Derivatives with Potential Anticancer, Antioxidant, and 11β-HSD Inhibitory Activities. *Int. J. Mol. Sci.* **2023**,24,pp.7252. Doi: 10.3390/ijms24087252

[78] Kupczyk, D.; Studzińska, R.; Bilski, R.; Baumgart, S.; Kołodziejska, R.; Woźniak, A. Synthesis of Novel 2-(Isopropylamino)thia-zol-4(5H)-One Derivatives and Their Inhibitory Activity of 11β-HSD1 and 11β-HSD2 in Aspect of Carcinogenesis Prevention. *Molecules* **2020**,25,pp.4233. Doi:10.3390/molecules25184233.

[79] Kupczyk, D.; Studzińska, R.; Baumgart, S.; Bilski, R.; Kosmalski, T.; Kołodziejska, R.; Woźniak, A. A Novel N-tert-butyl Derivatives of Pseudothiohydantoin as Potential Target in Anti-Cancer Therapy. *Molecules* **2021**,26,pp.2612. Doi:10.3390/molecules26092612.

[80] Johansson, L.; Fotsch, C.; Bartberger, D.M.; Castro, V.M.; Chen, M.; Emery, M.; Gustafsson, S.; Hale, C.; Hickman, D.; Homan, E.; et al. 2-Amino-1,3-thiazol-4(5H)-Ones as Potent and Selective 11β-Hydroxysteroid Dehydrogenase Type 1 Inhibitors: Enzyme-Ligand Co-Crystal Structure and Demonstration of Pharmacodynamic Effects in C57Bl/6 Mice. *J. Med. Chem.* **2008**,51,pp.2933–2943. Doi:10.1021/jm701551j.

[81] St Jean, D.J., Jr; Yuan, C.; Bercot, E.A.; Cupples, R.; Chen, M.; Fretland, J.; Hale, C.; Hungate, R.W.; Komorowski, R.; Veniant, M.; et al. 2-(S)-Phenethylaminothiazolones as Potent, Orally Efficacious Inhibitors of 11β-Hydroxysteriod Dehydrogenase Type 1. *J. Med. Chem.* **2007**,50,pp.429–432. Doi: 10.1021/jm061214f.

[82] Moreno-Díaz, H.; Villalobos-Molina, R.; Ortiz-Andrade, R.; Díaz-Coutiño, D.; Medina-Franco, J.L.; Webster, S.P.; Binnie, M.; Estrada-Soto, S.; Ibarra-Barajas, M.; León-Rivera, I.; Navarrete-Vázquez, G. Antidiabetic Activity of N-(6-Substituted-1,3-Benzothiazol-2-yl)Benzenosulfonamides. *Bioorg. Med. Chem. Lett.* 18,pp.2871–2877. Doi: 10.1016/j.bmcl.2008.03.086.

[83] Studzińska, R.; Kupczyk, D.; Płazińska, A.; Kołodziejska, R.; Kosmalski, T.; Modzelewska-Banachiewicz, B. Thiazolo[3,2-α]Pyrimidin-5-One Derivatives as a Novel Class of 11β-Hydroxysteroid Dehydrogenase Inhibitors. *Bioorg. Chem.* **2018**,81,pp.21–26. Doi: 10.1016/j.bioorg.2018.07.033.

[84] Choi, K.J.; Na, Y.; Jung, W.H.; Park, S.B.; Kang, S.; Nam, H.J.; Ahn, J.H.; Kim, K.Y. Protective Effect of a Novel Selective 11β-HSD1 Inhibitor on Eye Ischemia-Reperfusion Induced Glaucoma. *Biochem. Pharmacol.* **2019**,169,pp.113632. Doi: 10.1016/j.bcp.2019.113632.

[85] Lee, J.H.; Bok, J.H.; Park, S.B.; Pagire, H.S.; Na, Y.; Rim, E.; Jung, W.H.; Song, J.S.; Kang, N.S.; Seo, H.W.; Jung, K; Lee, B.H.; Kim, K.Y.; Ahn, J.H. Optimization of Cyclic Sulfamide Derivatives as 11β-Hydroxysteroid Dehydrogenase 1 Inhibitors for the Potential Treatment of Ischemic Brain Injury. *Bioorg. Med. Chem. Lett.* **2020**,30,pp.126787. Doi: 10.1016/j.bmcl.2019.126787.

[86] Kim, S.H.; Bok, J.H.; Lee, J.H.; Kim, I.H.; Kwon, S.W.; Lee, G.B.; Kang, S.K.; Park, J.S.; Jung, W.H.; Kim, H.Y., Rhee, S.D.; Ahn, S.H.; Bae, M.A.; Ha, D.C.; Kim, K.Y.; Ahn, J.H. Synthesis and Biological Evaluation of Cyclic Sulfamide Derivatives as 11β-Hydroxysteroid Dehydrogenase 1 Inhibitors. *Med. Chem. Lett.* **2012**,3,pp.88–93. Doi: 10.1021/ml200226x.

[87] Park, J.S.; Rhee, S.D.; Jung, W.H.; Kang, N.S.; Kim, H.Y.; Kang, S.K.; Ahn, J.H.; Kim, K.Y. Anti-Diabetic and Anti-Adipogenic Effects of a Novel Selective 11β-Hydroxysteroid Dehydrogenase Type 1 Inhibitor in the Diet-Induced Obese Mice. *Eur. J. Pharmacol.* **2012**,691,pp.19–27. Doi: 10.1016/j.ejphar.2012.06.024.

[88] Park, J.S.; Bae, S.J.; Choi, S.W.; Son, Y.H.; Park, S.B.; Rhee, S.D.; Kim, H.Y.; Jung, W.H.; Kang, S.K.; Ahn, J.H.; Kim, S.H.; Kim, K.Y. A Novel 11β-HSD1 Inhibitor Improves Diabesity and Osteoblast Differentiation. *J. Mol. Endocrinol.* **2014**,52(2),pp.191–202. Doi: 10.1530/JME-13-0177.

[89] Leiva, R.; Grinan-Ferre, C.; Seira, C.; Valverde, E.; McBride, A.; Binnie, A.; Perez, B.; Luque, F.J.; Pallas, M.; Bidon-Chanal, A.; Webster, S.P.; Vazquez, S. Design, Synthesis and In Vivo Study of Novel Pyrrolidine-Based 11β-HSD1 Inhibitors for Age-Related Cognitive Dysfunction. *Eur. J. Med. Chem.* **2017**,139,pp.412–428. Doi: 10.1016/j.ejmech.2017.08.003

[90] Gu, X.; Dragovic, J.; Koo, G.C.; Koprak, S.L.; LeGrand, C.; Mundt, S.S.; Shah, K.; Springer, M.S.; Tan, E.Y.; Thieringer, R.; Hermanowski-Vosatka, A.; Zokian, H.J.; Balkovec, J.M.; Waddell, S.T. Discovery of 4-Heteroarylbicyclo[2.2.2]Octyltriazoles as Potent and Selective Inhibitors of 11β-HSD1: Novel Therapeutic Agents for the Treatment of Metabolic Syndrome. *Bioorg. Med. Chem. Lett.* **2005**,15,pp.5266–5269. Doi: 10.1016/j.bmcl.2005.08.052.

[91] Hermanowski-Vosatka, A.; Balkovec, J.; Cheng, K.; Chen, H.Y.; Hernandez, M.; Koo, G.C.; LeGrand, C.B.; Li, Z.; Metzger, J.M.; Mundt, S.S.; Noonan, H.; Nunes, C.N.; Olson, S.H.; Pikounis, B.; Ren, N.; Robertson, N.; Schaeffer, J.M.; Shah, K.; Springer, M.S.; Strack, A.M.; Strowski, M.; Wu, K.; Wu, T.; Xiao, J.; Zhang, B.B.; Wright, S.D.; Thieringer, R. 11β-HSD1 Inhibition Ameliorates Metabolic Syndrome and Prevents Progression of Atherosclerosis in Mice. *J. Exp. Med.* **2005**,202,pp.517–527. Doi: 10.1084/jem.20050119.

[92] Zhang, C.; Xu, M.; He, C.; Zhuo, J.; Burns, D.M.; Qian, D-Q.; Lin, Q.; Li, Y-L.; Chen, L.; Shi, E.; Agrios, C.; Weng, L.; Sharief, V.; Jalluri, R.; Li, Y.; Scherle, P.; Diamond, S.; Hunter, D.; Covington, M.; Marando, C.; Wynn, R.; Katiyar, K.; Contel, N.; Vaddi, K.; Yeleswaram, S.; Hollis, G.; Huber, R.; Friedman, S.; Metcalf, B.; Yao, W. Discovery of 1′-(1-Phenylcyclopropane-Carbonyl)-3H-Spiro[Isobenzofuran-1,3′-Pyrrolidin]-3-One as a Novel Steroid Mimetic Scaffold for the Potent and Tissue-Specific Inhibition of 11β-HSD1 Using a Scaffold-Hopping Approach. *Bioorg. Med. Chem. Lett.* **2022**,69,pp.128782. Doi: 10.1016/j.bmcl.2022.128782.

[93] Boudon, S.; Heidl, M.; Vuorinen, A.; Wandeler, E.; Campiche, R.; Odermatt, A.; Jackson, E. Design, Synthesis, and Biological Evaluation of Novel Selective Peptide Inhibitors of 11β-Hydroxysteroid Dehydrogenase 1. *Bioorg. Med. Chem.* **2018**,26,pp.5128–5139. Doi: 10.1016/j.bmc.2018.09.009.

[94] Shao, L.-D.; Bao, Y.; Shen, Y.; Su, J.; Leng, Y.; Zhao, Q.-S. Synthesis of Selective 11β-HSD1 Inhibitors Based on Dammarane Scaffold. *Eur. J. Med. Chem.* **2017**,135,pp.324–338. Doi: 10.1016/j.ejmech.2017.04.059.

13 Metabolic Disorders in Strokes

Renata Kołodziejska, Hanna Pawlu, Daria Kupczyk, and Renata Studzińska

13.1 STROKE

Stroke, according to WHO, is a clinical syndrome characterized by a sudden onset of focal or generalized disturbances of brain function that if not fatal persist for more than 24 hours and have no cause other than vascular [1]. It is one of the most common causes of death after cancer and myocardial infarction, and the leading cause of disability worldwide.

According to its etiology, stroke can be divided into ischemic, hemorrhagic, or subarachnoid. Ischemic stroke accounts for 80%–85% of all cases; it is an acute condition of cerebral circulation failure caused by a thrombus in the artery wall, cerebral artery embolism, or hemodynamic disorders. Hemorrhagic stroke accounts for about 15% of all strokes; it breaks the continuity of the cerebral vessel causing extravasation of blood within the brain. Finally, subarachnoid hemorrhage accounts for about 6% of all cases and is the result of spontaneous blood flow into the fluid spaces of the brain located between the arachnoid and pia mater, less often into the brain tissue and ventricles [2].

13.2 METABOLIC DISORDERS AS A CAUSE OF STROKE

In addition to nonmodifiable stroke risk factors such as age, sex, race and genetic predisposition, the most important regulated factors include hypertension, heart disease, diabetes, dyslipidemia, and blood coagulation disorders. Most of these factors are part of the so-called metabolic syndrome (MetS), which most generally refers to metabolic abnormalities including i) central obesity, ii) atherogenic dyslipidemia, iii) high blood pressure, and iv) hyperglycemia (Figure 13.1) [3].

There is no conclusive information on causes of MetS. One proposal is that insulin resistance (IR), a common physiological abnormality, predisposes individuals to the metabolic abnormalities that contribute to MetS [4]. MetS is also identified in people with visceral obesity, who have a higher long-term risk of atherosclerotic vascular disease [5]. IR and central obesity together or separately contribute to the formation of MetS, and obesity can lead to IR, hypoadiponectinemia, and disorders of the secretion of pro-inflammatory factors, which together contribute to the formation of MetS [6, 7]. MetS is an important risk factor for major cardiovascular events including stroke [8–10], and ischemic stroke risk increases with increasing number of MetS components [11, 12]. We discuss in detail both insulin resistance and obesity, the two most important components of MetS, which both predispose to cerebrovascular disease.

13.2.1 Insulin Resistance as a Metabolic Disorder in the Context of Stroke

Thrombosis and atherosclerosis are the main mechanisms of ischemic stroke. During IR, there is not only energy disturbance but also excessive stimulation of the miotic pathway, which results in inflammation, oxidative, stress and neuronal damage. IR accelerates the development of thrombosis and atherosclerosis, which suggests that it is an independent risk factor for ischemic stroke [3]. Here,

DOI: 10.1201/9781003437413-13

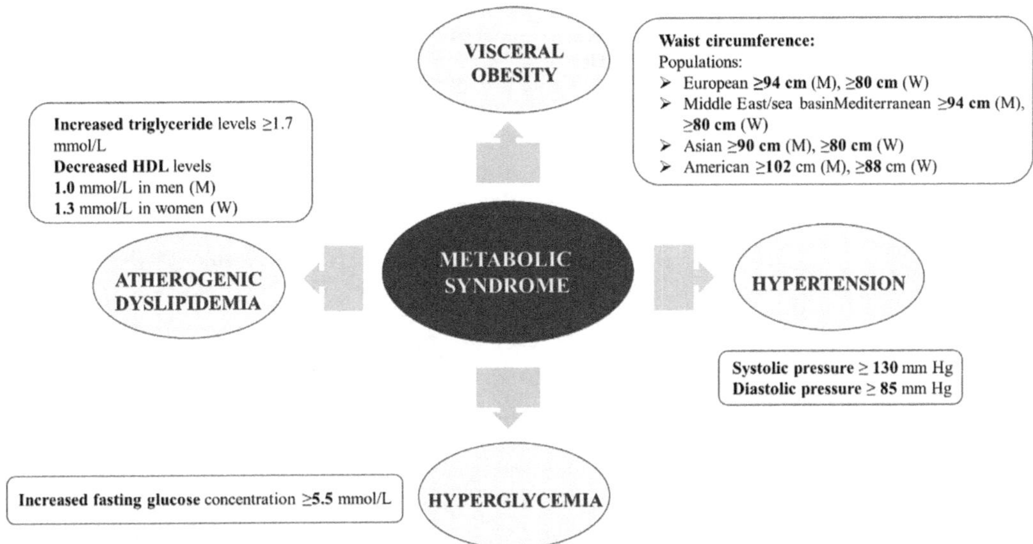

FIGURE 13.1 Metabolic syndrome.

we discuss the physiological function of insulin, its dysfunctions, and their consequences underlying the mechanisms of stroke.

13.2.1.1 Insulin's Mechanisms of Action

Under physiological conditions, insulin regulates glucose metabolism, stimulating glucose transfer to peripheral tissues. It stimulates glucose oxidation, glycogenogenesis, and inhibits gluconeogenesis and glycogenolysis. Insulin causes the deposition of glycogen in the liver and skeletal muscles, increases the uptake of amino acids by tissues, and enhances protein synthesis. It also stimulates lipogenesis and inhibits lipolysis, which leads to the storage of free fatty acids (FFA) in the form of triacylglycerols (TAG) in adipose tissue. In addition to its metabolic effect, it is also a mitogenic factor. It promotes cell growth, proliferation, migration and inhibits apoptosis. It also has vasodilating properties; it stimulates the production of nitric oxide (NO) and has anti-inflammatory effects [13].

Insulin acts through a specific insulin receptor (IRTK/INSR), belonging to the family of receptors with tyrosine kinase activity. It consists of two α and two β subunits connected by disulfide bridges [14]. Hepatocytes, adipocytes, and muscles have the richest receptor distribution on the surface of the membrane. In brain tissue, insulin receptors are distributed mainly in the hypothalamus, olfactory bulbs, hippocampus, striatum, cerebral cortex and cerebellum [15, 16].

Insulin binding to the extracellular α domain of the IRTK receptor induces a conformational change, which leads to autophosphorylation of tyrosine residues of intracellular β subunits. Active tyrosine kinase mediated by adenosine triphosphate (ATP) phosphorylates tyrosine residues of substrate proteins such as insulin receptor substrate (IRS), growth factor receptor-associated protein 2 (GRB-2), growth factor receptor-associated protein 10 (GRB-10), SHC transforming protein, and SH2B adapter protein (SH2B-2). Upon phosphorylation, these substrates interact with a series of effector or adaptive molecules containing Src homology 2 (SH2) domains that specifically recognize various phosphotyrosine motifs. Among these substrates, the IRS protein family is the best characterized [17]. Phosphorylation of IRS plays a major role in insulin-mediated cellular signaling. The variety of interactions in which the IRS proteins enter determine the pleiotropic effect of insulin on the body. The phosphorylation of tyrosine residues in IRS-1 and IRS-2 stimulates two insulin signaling pathways (Figure 13.2).

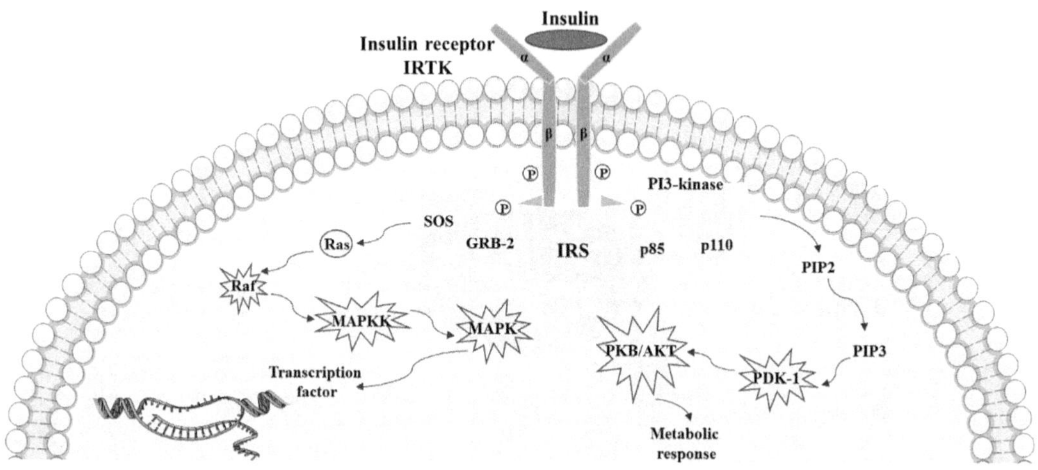

FIGURE 13.2 Insulin signaling pathways: GRAB-2—Growth factor receptor-associated protein 2; IRS—Insulin receptor substrate; IRTK—Insulin receptor; MAPK—Mitogen-activated protein kinase; MAPKK—Mitogen-activated protein kinase kinase; PI3K—Phosphatidylinositol 3-kinase; PDK-1—protein 3-phosphoinositide-dependent protein kinase-1; PKB/AKT—Protein kinase B; Ras—GTPases; PIP2—Phosphatidylinositol-(4,5)-diphosphate; PIP3—Phosphatidylinositol-(3,4,5)-triphosphate; Raf—Serine/threonine-specific protein kinases; SOS—Son of sevenless. This figure was created using Servier Medical Art (available at https://smart.servier.com/).

Activating the phosphatidylinositol 3-kinase (PI3K) pathway generates a metabolic response to the hormone [18]. The second pathway, due to the involvement of mitogen-activated protein kinase (MAPK), is responsible for the mitogenic effect of insulin. In the Ras/MAPK-dependent pathway, adapter proteins GRB-2 and SOS (Son of sevenless) first attach to the phosphorylated IRS. This activates the Ras protein (GTPases, GTP-binding proteins), which in turn activates Raf protein kinase. Raf then activates the MAPKK and MAPK phosphorylation cascade. This leads to the phosphorylation of transcription factors in the cell nucleus that promote cell growth, proliferation, and differentiation.

In the PI3K pathway, IRS proteins bind to the p85 subunit of phosphatidylinositol 3-kinase (p85 subunit plays a regulatory role), activating the p110 subunit (p110 subunit has a catalytic role), which not only increases activity but also displaces the kinase to the cell membrane near its substrate phosphatidylinositol-(4,5)-diphosphate (PIP2). Activated PI3K catalyzes the phosphorylation of PIP2 to phosphatidylinositol-(3,4,5)-triphosphate (PIP3), which activates protein kinase B (PKB), also known as AKT. AKT regulates the translocation of insulin-dependent glucose transporter type 4 (GLUT4) from the cytoplasm to the plasma membrane. In addition, AKT inhibits glycogen synthesis by phosphorylating glycogen synthase kinase 3 (GSK3), while dephosphorylation of GSK3 inhibits its activity and increases glycogen synthesis. The rate of glycogen synthesis is influenced by the supply of glucose-6-phosphate (G6P, a product of glucokinase (GCK)). In contrast, the activity of GCK, the enzyme that catalyzes glucose phosphorylation, can be increased or suppressed in response to changes in glucose supply, usually occurring during times of hunger and satiety [19]. Generally, activated AKT phosphorylates many downstream substrates in a variety of functional pathways, making it a key node in the branching of insulin signaling (Figure 13.3).

AKT phosphorylates downstream substrates in a variety of metabolic tissues, including skeletal muscle, liver, and adipose tissue, inducing an insulin-induced nutrient reserve in these tissues (Figure 13.4).

In skeletal muscle, AKT promotes glucose uptake through highly coordinated translocation and fusion of the GLUT4 glucose transporter, packaged in GSV vesicles, to the plasma membrane. AKT

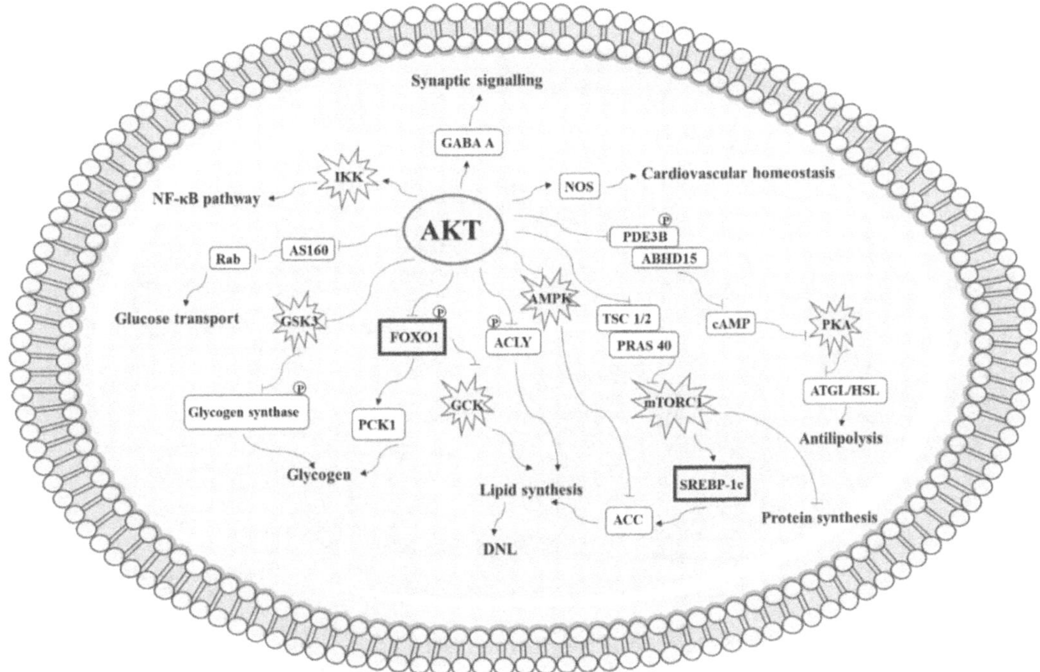

FIGURE 13.3 AKT pathways: ABHD15—Abhydrolase Domain Containing 15; ACC—Acetyl-CoA carboxylase; ACLY—ATP citrate lyase; AKT—Protein kinase B; AMPK—5'AMP-activated protein kinase; AS160—AKT substrate of 160 kDa; ATGL—Adipose triglyceride lipase; cAMP—Cyclic adenosine-3′,5′-monophosphate; DNL—De novo lipogenesis; FOXO1—Forkhead box O1; GABA A—γ-Aminobutyric acid receptor; GCK—Germinal centre kinase; GSK3—Serine/threonine protein kinase; PCK1–Phosphoenolpyruvate carboxykinase; HSL—Hormone-sensitive lipase; IKK—IκB kinase; NFκB—Nuclear factor kappa-light-chain-enhancer of activated B cells; NOS—Nitric oxide synthases; mTORC1—Mammalian target of rapamycin complex 1; PKA—Protein kinase A; PDE3B—Phosphodiesterase 3B; PRAS 40—Proline-rich Akt substrate of 40 kDa; Rab– Proteins possess a GTPase fold; SREBP-1c—Sterol regulatory element-binding protein 1; TSC 1/2—Tuberous sclerosis proteins 1 and 2. This figure was created using Servier Medical Art (available at https://smart.servier.com/).

phosphorylates the AS160 protein (the 160 kDa AKT substrate of the GTPase activating protein (GAP), also known as TBC1D4, and the related GAP substrate TBC1D1. Phosphorylation by AKT blocks TBC1D4/TBC1D1 inactivation of small Rab GTPase protein switches that control vesicle trafficking, which promotes GSV translocation [19, 20].

Rho GTPase RAC1 (a family of small G-signaling proteins, a subfamily of the Ras superfamily) coordinates a second PI3K-dependent glucose uptake signaling mechanism in skeletal muscle. RAC1 signaling promotes GLUT4 translocation by inducing reorganization of the actin cytoskeleton. Glucose that enters the myocyte after insulin stimulation can undergo glycolysis or be a substrate for glycogen synthesis. The main role of insulin as a hormone is to store energy in the form of glycogen, and glycogen synthesis is regulated by covalent modification and allosteric modification by G6P. Covalent modification takes place as a result of insulin-promoted dephosphorylation of both glycogen synthase (GS) and glycogen phosphorylase (GP). Insulin stimulates glycogen synthesis by inhibiting GSK3, which prevents GS inactivation because dephosphorylation activates GS. It also inactivates GP by dephosphorylation with phosphorylase kinase. Dephosphorylated GS is more sensitive to the allosteric effects of G6P, facilitating the activation of insulin-stimulated glycogen synthesis. In contrast, allosteric control of phosphorylase by G6P is the primary mechanism for controlling glycogenolysis [19].

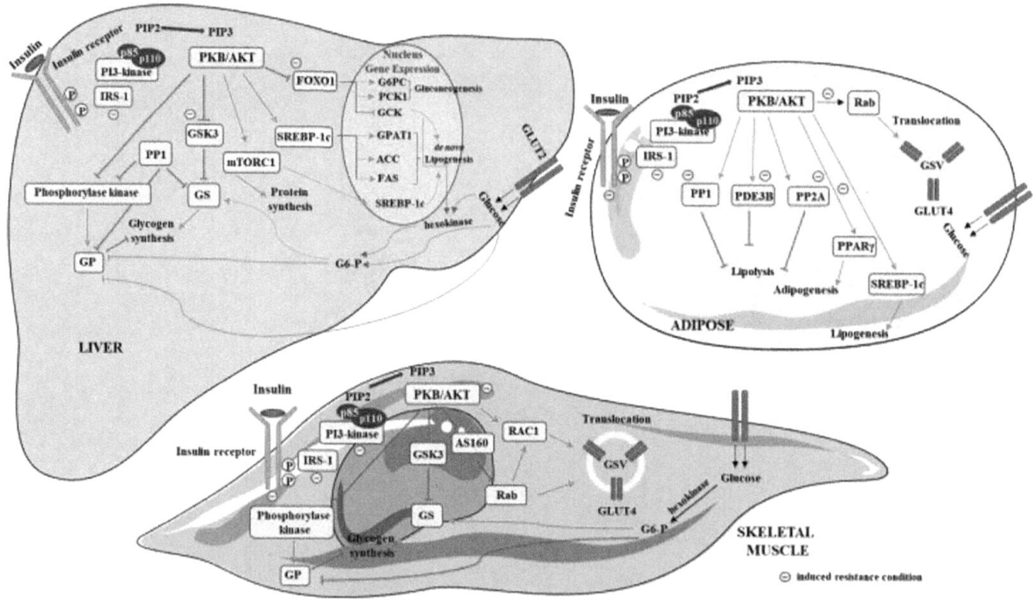

FIGURE 13.4 Insulin signaling in skeletal muscle, liver, and adipose tissue: ACC—Acetyl-CoA carboxylase; AS160—AKT substrate of 160 kDa; FAS—Fatty acid synthase; FOXO1—Forkhead box O1; GCK—glucokinase; GP—Glycogen phosphorylase; G6PC—Glucose 6-phosphatase; GS—Glycogen synthase; GPAT1—Glycerol-3-phosphate acyltransferase; 1GSK3—Glycogen synthase kinase 3; IRS-1—Insulin receptor substrate; mTORC1—Mammalian target of rapamycin complex 1; PCK1– Phosphoenolpyruvate carboxykinase; PDE3B—Phosphodiesterase 3B; PI3-kinase—Phosphatidylinositol 3-kinase; PIP2—Phosphatidylinositol-(4,5)-diphosphate; PIP3—Phosphatidylinositol-(3,4,5)-triphosphate; PKB/AKT—Protein kinase B; PPARγ– Peroxisome proliferator-activated transcription factor γ; PP1—Protein phosphatase 1; PP2A—Protein phosphatase-2A; Rab—Proteins possess a GTPase fold; RAC1—Ras-related C3 botulinum toxin substrate 1; SREBP-1c—Sterol regulatory element-binding protein 1. This figure was created using Servier Medical Art (available at https://smart.servier.com/).

In the liver, insulin promotes the synthesis of all major classes of metabolic macromolecules: glycogen, lipids, and proteins. AKT substrates include GSK3, which regulates glycogen synthesis; Forkhead box O1 (FOXO1) transcription factor, which regulates gluconeogenic gene transcription, and multiple regulators of mTORC1 activity (Mammalian target of rapamycin complex 1), which in turn control lipogenic gene expression and protein synthesis.

Insulin increases hepatic glycogen synthesis by regulating GS and GP by GSK3 and protein phosphatase 1 (PP1). Moreover, hyperglycemia is sufficient to inactivate hepatic glycogen phosphorylase by glucose allostery and thereby promote hepatic glycogen synthesis. It also translocates glucokinase from the nucleus to the cytoplasm, allowing G6P to flow [21]. Still, hepatic insulin signaling via IRTK is required for proper glycogen synthesis; insulin regulates glycogen synthase allostery through G6P. It follows that both insulin and glucose act together in the regulation of hepatic glycogen metabolism [22].

In the liver, AKT reduces gluconeogenesis by suppressing FOXO1, i.e. it prevents the transcriptional activation of the expression of gluconeogenic genes. Active nuclear FOXO1 binds the transcriptional coactivator peroxisome proliferative activated receptor-γ coactivator 1-α (PGC1α) to coordinate a gluconeogenic transcription program involving increased expression of G6P and cytosolic phosphoenolpyruvate carboxykinase (PCK1) [23, 24]. Active FOXO1 also binds the corepressor SIN3A (Paired amphipathic helix protein) to reduce glucokinase expression, further promoting glucose export [25].

Furthermore, insulin increases lipogenesis by up-regulating sterol regulatory element binding protein 1c (SREBP-1c). SREBP-1c increases glycolysis by activating glucokinase.

SREBP-1c promotes *de novo* lipogenesis (DNL) by increasing the transcription of several lipogenic enzymes, in particular acetyl-CoA carboxylase 1 (ACC), fatty acid synthase (FAS) and glycerol-3-phosphate acyltransferase 1 (GPAT1) [26, 27].

Insulin also activates the DNL flux by regulating the phosphorylation of lipogenic enzymes, although the specific signal transduction pathways are not fully understood. For example, ACC is rapidly activated in response to insulin [28, 29]. Insulin also regulates the phosphorylation of ATP citrate lyase (ACLY). ACLY converts the citrate intermediate of the tricarboxylic acid cycle (TCA) to the lipogenic precursor of acetyl CoA, thus linking glucose metabolism to DNL [30, 31]. ACLY phosphorylation activates the enzyme, preventing its allosteric inhibition by citrate. However, it is not clear whether insulin is used to increase ACLY activity [32].

The regulation of protein synthesis by insulin is largely dependent on signalling to the mammalian target of rapamycin (mTOR) network. mTOR is a protein kinase that can form two functional complexes mTORC1 and mTORC2. Together, mTORC1 and mTORC2 regulate processes that control cell growth and proliferation, including protein synthesis, autophagy, and metabolism. Insulin-mediated mTOR-mediated protein synthesis takes place in many types of insulin-responsive cells, including hepatocytes, adipocytes, and myocytes [33].

mTOR activation is strongly integrated with the PI3K-AKT pathway. Activation of AKT mTORC1 may involve phosphorylation of AKT and inactivation of the mTORC1 activation inhibitors tuberous sclerosis complex 2 (TSC2) and/or the proline-rich 40 kDa AKT substrate (PRAS40) [34, 35].

Activated mTORC1 phosphorylates protein kinases involved in signal transduction, including ribosomal protein S6 kinase (S6K) and the eukaryotic translation initiation factor binding proteins 1 and 2 (4EBP1/2). Thus, it initiates the translation of proteins crucial for the progression of the cell cycle. mTORC1 signalling also exerts negative feedback on proximal insulin signalling by promoting S6K phosphorylation and IRS1 destabilization. Excessive anabolic activity of mTOR limits (*via* S6K1) the anabolic activity of insulin. Activation of mTORC1-S6K promotes PI3K negative feedback inhibition not only by phosphorylation of the IRS but also by mTORC1-mediated phosphorylation of the negative regulator of PI3K signalling, the Grb10 adapter protein. mTORC1 also regulates the synthesis of non-protein macromolecules, including phosphatidylcholine needed for the secretion of VLDL (very low-density lipoprotein) triglycerides. Insulin-activated mTORC1 also increases the production of SREBP-1c, which facilitates the storage of fatty acids as triglycerides (TGA) [36, 37]. mTORC2 phosphorylation of AKT Ser473 upstream activates AKT kinase and may alter its substrate specificity [38–40].

The most important physiological functions of insulin action in white adipose tissue include inhibition of lipolysis and stimulation of glucose uptake. Lipolysis suppression requires phosphodiesterase 3B (PDE3B) and a protein-containing abhydrolase domain 15 (ABHD15) and is activated by cAMP, leading to the phosphorylation of perillipin (PLIN), triglyceride lipase (ATGL) and hormone-sensitive lipase (HSL). PDE3B degrades cAMP to attenuate prolipolytic PKA (protein kinase A) signalling towards HSL and PLIN [41, 42].

Furthermore, protein phosphatase 1 and protein phosphatase-2A mediate the suppression of PI3K-dependent insulin-induced lipolysis by dephosphorylation of lipolytic regulatory proteins [43, 44]. Insulin also promotes lipogenesis in white adipose tissue by activating SREBP-1c, signaling translocation of glucose or fatty acid transport proteins, promoting fatty acid esterification, and stimulating adipogenesis via peroxisome proliferator-activated transcription factor γ (PPARγ) [45, 46]. Insulin-stimulated glucose uptake is mediated through phosphoinositide-3-kinase (PI3K)-dependent and PI3K-independent pathways using multiple effectors to promote translocation [19].

13.2.1.2 Insulin Resistance

Under normal physiological conditions, elevated plasma glucose levels increase insulin secretion and circulating insulin levels, thereby stimulating glucose transfer to peripheral tissues and inhibiting hepatic gluconeogenesis. This homeostasis is disturbed in individuals with IR. Insulin-resistant patients with normal plasma insulin levels are unable to achieve a coordinated target tissue glucose-lowering response. As compensation, insulin levels increase, leading to hyperinsulinemia [13].

What is the cause of insulin resistance? One cause of IR may be a reduction in glucose utilization due to increased fatty acid oxidation, which is referred to as the glucose–fat cycle or the Randle cycle (Figure 13.5). According to Randle et al., FFA oxidation increases mitochondrial acetyl-CoA levels via β-oxidation and subsequently inactivate pyruvate dehydrogenase. High levels of citrate inhibit phosphofructokinase, leading to the accumulation of intracellular G6P. Increased concentrations of G6P, in turn, allosterically inhibit hexokinase, leading to the accumulation of intracellular glucose and a decrease in its uptake [47].

Another explanation for the development of IR is the hexosamine biosynthetic pathway. Fructose-6-phosphate (F6P) is synthesized from G6P, which is mainly metabolized to fructose-1,6-bisphosphate during glycolysis; only 5% of F6P is converted to glucosamine-6-phosphate (glucosamine-6P) by a reaction catalyzed by glutamine:fructose-6-phosphate amidotransferase. Then, glucosamine-6P is converted into *N*-acetylglucosamine uridine 5'-diphosphate (UDP-GlcNAc), which serves as a donor for glycosylation and *O*-GlcNAcylation of lipids and proteins [48]. These modifications, especially O-GlcNAcylation, can affect target proteins by regulating gene expression or enzyme activity [49]. *O*-GlcNAc also modifies components of the insulin signaling pathway, such as IRS-1/2, PI3K, PDK1, AKT, and Munc18-c (mammalian uncoordinated-18), a protein necessary for insulin-stimulated translocation of GLUT4 [50–52].

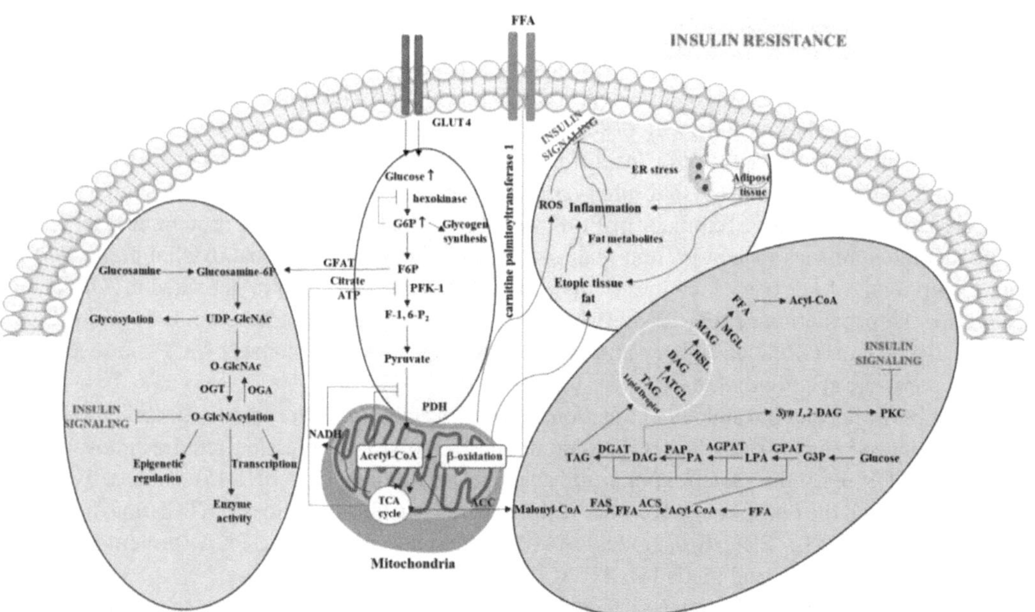

FIGURE 13.5 Potential mechanisms for insulin resistance: ACC—Acetyl-CoA carboxylase; ACS—Acyl-CoA synthetases; AGPAT—Acylglycerolphosphate acyltransferase; ATGL—Adipose triglyceride lipase; ATP—Adenosine triphosphate; DAG—Diacylglycerol; DGAT—Aiacylglycerol acyltransferase; FAS—Fatty acid synthase; FFA—Free fatty acid; F-1,6-P2—Fructose-1,6-bisphosphate; GLUT4—Glucose transporter type 4; GPAT—Glycerol-3-phosphate acyltransferase; G3P—Glyceraldehyde-3-phosphate; G6P—Glucose-6-phosphate; HSL—Hormone-sensitive lipase; MAG—Monoacylglycerol; MGL—Monoacylglycerol lipase; NADH—Reduced nicotinamide adenine dinucleotide; OGA—O-GlcNAcase; OGT—O-GlcNAc transferase; PA—Phosphatidic acid; PAP—Phosphatidic acid phosphatases; PDH—Pyruvate dehydrogenase; PFK-1—Phosphofructokinase 1; PKC—Protein kinase C; TAG—Triacylglycerols; TCA—Tricarboxylic acid cycle. This figure was created using Servier Medical Art (available at https://smart.servier.com/).

High ectopic lipid accumulation in the liver and skeletal muscle, which impairs insulin signaling, could also be responsible for IR. In the diacylglycerol (DAG)-protein kinase C (PKC) hypothesis, acetyl-CoA obtained in the FFA esterification reaction is transferred to the glycerol backbone to form lysophosphatidic acid (LPA), DAG, and TAG by lipogenesis. Increased DAG in the liver induces the translocation of nPKC (PKCε and PKCθ in liver and skeletal muscle, respectively) to the plasma membrane and inhibits IRTK [53, 54]. The most likely hypothesis of the mechanism by which ectopic lipid accumulation induces IR is that several lipid metabolites, including DAG, LPA, ceramides and acylcarnitines, are involved in the pathogenesis of IR. Impaired mitochondrial function is another suggested mechanism by which IR develops. The most common causes of mitochondrial dysfunction include i) increased production of reactive oxidative species (ROS), ii) impaired glucose and fatty acid oxidation due to metabolic inflexibility, and iii) endoplasmic reticulum stress [55].

13.2.1.3 How IR Leads to Stroke

Insulin signaling from the central nervous system (CNS) regulates energy balance through complex mechanisms [56]. The main function of insulin in the brain is appetite suppression [57].

Increasing central insulin action in the brain modulates peripheral metabolism, increasing whole body insulin sensitivity and suppressing endogenous glucose production. Mechanisms linking the brain and peripheral insulin action are elusive but likely involve the sympathetic and parasympathetic systems and the hypothalamic–pituitary–adrenal axis. Based on experimental animal studies, insulin in some hypothalamic nuclei strongly inhibits hepatic glucose production, promote muscle glucose uptake, inhibit adipose tissue lipolysis, and inhibit glucagon secretion [58–62].

The concentration of insulin in the cerebrospinal fluid is correlated with the concentration in the circulating plasma. While it is still uncertain whether insulin can be produced in the CNS, circulating insulin can enter brain tissue via the blood–brain barrier (BBB) [63]. Glucose metabolism in brain tissue is particularly important, as much as 20% of the total body energy is used by the brain. Insulin maintains the balance of energy metabolism in brain tissue and promotes glucose by translocating GLUT4 from the cytoplasm to the plasma membrane in neurons in the hippocampus and cerebral cortex [64].

Insulin stimulates neurite growth; modulates the release and uptake of catecholamines; and regulates the expression and localization of *N*-methyl-D-aspartate, α-amino-3-hydroxy-5-methyl-4-isoxazolepropionic acid, and γ-aminobutyric acid. It modulates synaptic plasticity to increase neuronal survival by inhibiting apoptosis. It protects brain tissue by preventing ischemia and oxidative stress and by regulating cholesterol metabolism in neurons and astrocytes [65]. The lack of protective action of insulin can significantly increase the risk of cardiovascular disease (CVD), including stroke (Figure 13.6). IR promotes inflammation, atherosclerosis, and embolism and leads to and/or aggravates other components of MetS (Figure 13.7) [66, 67].

Atherosclerosis is a disease of the arteries. The wall of the artery is made of three layers. The inner layer is made up of vascular endothelial cells resting on the basement membrane made of collagen and proteoglycans. The middle layer of the arterial wall is densely filled with smooth muscle cells; in large arteries, this layer is reinforced by elastic fibers. The outer membrane is formed by loose connective tissue. The endothelium is the largest para- and autocrine organ; vasoactive substances produced within it including NO play an important role in the local regulation of the tension of the blood vessel wall. Insulin receptors in the membrane of endothelial cells participate in vascular homeostasis. Insulin, in a pathway dependent on PI3K and AKT, stimulates the release of NO. By activating AKT, it is anti-apoptotic and increases endothelial cell survival [68]. Additionally, by reducing the activity of nuclear factor-κB (NF-κB), insulin reduces the expression of the pro-inflammatory monocyte chemoattractant protein-1 (MCP-1) in aortic endothelial cells [69]. It also reduces the expression of mRNA and protein of intercellular adhesion molecule 1 (ICAM-1), a factor that promotes inflammation within the endothelium, by increasing the expression of NOS and NO production [70].

As insulin resistance develops, there is an imbalance between two pathways: PI3K and MAPK. PI3K activity decreases, which reduces endothelial NO release, while excessive MAPK stimulation increases production of vasoconstrictor endothelin 1 and activates intracellular pathways involved in cell proliferation and inflammation, including IKKβ-NF-κB. As a consequence, there is inflammation in the vascular wall, endothelial dysfunction, proliferation of vascular smooth muscle cells, and atherosclerosis. Atherosclerotic lesions cause arterial hypertension, which intensifies the progression of atherosclerotic plaque [71].

Insulin can raise blood pressure by promoting antidiuretic effects and stimulating the activation of the sympathetic nervous system [72]. Increased sympathetic activity is mediated by leptin, which can cause short- and long-term changes in blood pressure through central and peripheral effects [73]. In addition, increased FFAs in the blood increase the activity of the IKKβ protein in endothelial cells and reduce NO production; long-term high concentrations of lipotoxic FFAs damage β cells [63].

Endothelial dysfunction is also caused by pro-inflammatory cytokines released from adipose tissue, which activate the NF-κB pathway and increase the expression of adhesins including ICAM-1. This increased adhesin expression triggers the adhesion of mononuclear leukocytes (monocytes, T lymphocytes) to endothelial cells, forming atherosclerotic plaque. Toxic forms of lipids such as

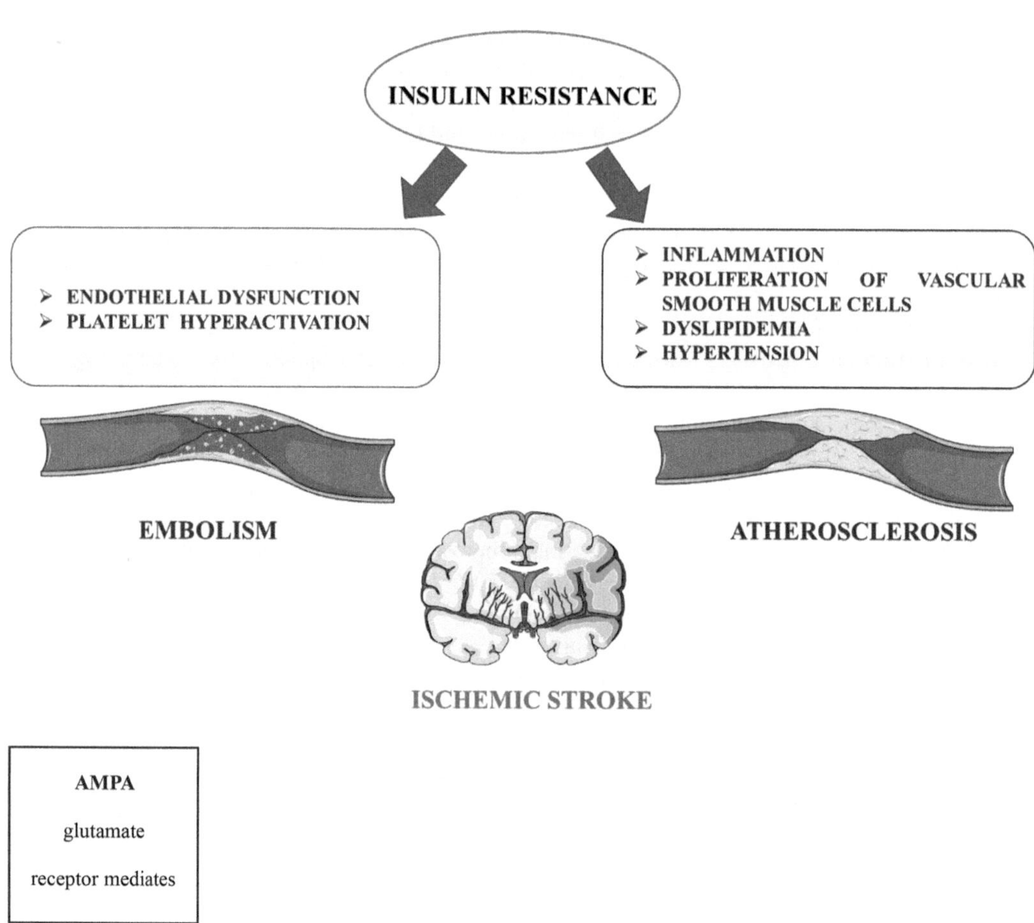

FIGURE 13.6 Influence of insulin resistance on arterial embolism and atherosclerosis. This figure was created using Servier Medical Art (available at https://smart.servier.com/).

long-chain fatty acyl CoA, diacylglycerol, and ceramides play an important role in the development of inflammation and atherosclerotic plaque within the arterial wall; these lipids activate pro-inflammatory pathways and induce oxidative stress. Atherogenic dyslipidemia associated with IR also plays an important role in the development of atherosclerosis. The most atherogenic particles are the small, dense LDL particles; they easily penetrate the basement membrane and undergo oxidation and digestion by proteolytic enzymes and glycolization [63].

Modified lipoproteins cause an inflammatory reaction in the intima and stop leukocytes from flowing on the inner surface of the vessel. The monocytes pass into the arteries' intima under the influence of cytokines with chemotactic effect [74]. The greatest importance in this process is attributed to MCP-1 and its receptor CCR2. Oxidative stress, oxidized LDL particles, and transcription factors such as NF-κB and activator protein 1 are responsible for the expression of MCP-1 in the vascular wall. Monocytes transform into macrophages in the intima, which is associated with the appearance of scavenger receptors on their cell membrane. This allows macrophages to absorb them, which is beneficial as it reduces the harmful effect of LDL on endothelial and smooth muscle cells [75].

The unlimited absorption of cholesterol by macrophages forms cholesterol clusters within them. It then takes the form of a foam cell that produces ROS; metalloproteinases (MMPs); and pro-inflammatory cytokines such as tumor necrosis factor alpha (TNF-α), interleukin 6 (IL-6), and interleukin 1 (IL-1), as well as coagulation initiating factors. The accumulation of cholesterol in the foam cells breaks them down to form extracellular cholesterol deposits. Macrophages with the M1 phenotype stimulate the atherosclerotic process; they are a constant source of numerous chemokines, cytokines, growth factors, and proteolytic enzymes. The activity of MCP-1 and macrophage colony activation factor ensures a continuous influx of new monocytes to the atherosclerotic lesion. Macrophages also activate T lymphocytes, mediated mainly by interleukin 2 (IL-2) [75].

Overall, IR is promoted at the molecular level when macrophage polarization shifts from the M2 activation state (anti-inflammatory) maintained by signal transducer and transcription activator 6 (STAT6) and PPAR to the classic M1 activation state (pro-inflammatory) induced by NF-κB, AP1. Thus, inflammation favors the development of IR. Hyperinsulinemia, via MAPK, increases the secretion of plasminogen activator inhibitor 1 (PAI-1) [76]. Inhibiting the formation of plasmin from plasminogen impairs fibrinolytic activity and forms a thrombus [77]. Pro-inflammatory cytokines and excessive release of FFA caused by IR can cause lipotoxicity, leading to increased expression of PAI-1. Long-term elevated concentration of PAI-1 in the blood promotes a chronic inflammatory reaction in the vascular wall and atherosclerosis, as well as increased prothrombotic readiness [78].

IR also promotes vascular occlusion and development of CVD by affecting platelet adhesion and aggregation; platelet adhesion, aggregation, and release are the three steps in the process of platelet activation. Adhesion to the subendothelial extracellular matrix occurs through the matrix's initial interaction with specific receptors on platelets, including the GP1b/V/IX complex bound to von Willebrand factor and GPVI and αIIβ1 receptors bound to the collagen component of the extracellular matrix on the platelet surface. Strong adhesion leads to thrombus formation, and the activated platelets bound within the thrombus will take up new platelets from the circulation via platelet–platelet interactions mediated by the αIIbβ3 integrin receptor. Platelets express two purinergic receptors, P2Y1 and P2Y12 [79]. Hyperglycemia can activate platelets to downregulate IRS1 and increase P2Y12 expression in platelets via the IRS1/PI3K/AKT pathways. Platelet hyperactivation promotes macrovascular and microvascular events [80].

In addition, hyperinsulinemia increases the expression of angiotensin II receptor type 1 (AT1 receptor) mRNA in cultured vascular smooth muscle cells, which enhances the vasoconstrictive effect of Ang II [81]. The source of Ang II may also be adipose tissue itself [82]. Ang II activates the NF-κB pathway, increases the expression of adhesion molecules and chemokines in the vascular wall, stimulates ROS production, and promotes multiple inflammatory pathways involved in atherosclerosis. It also causes serine phosphorylation of IRS-1, which inhibits insulin signaling. This

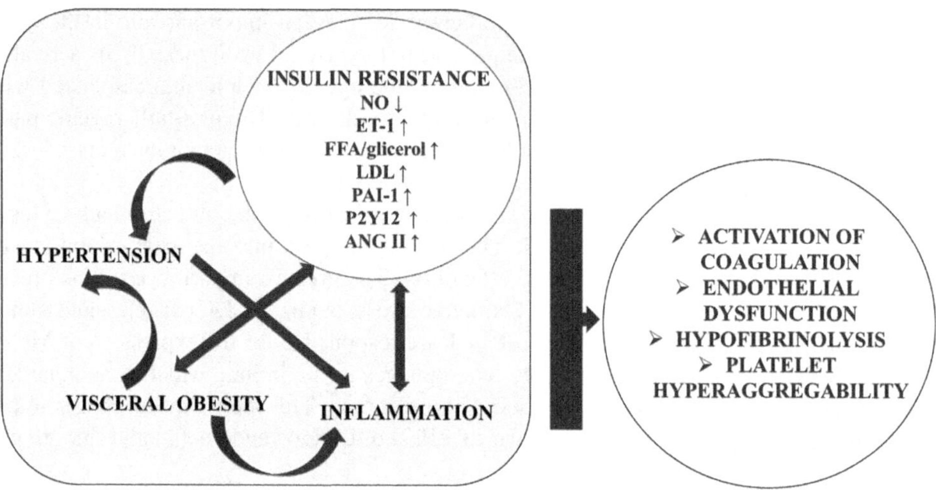

FIGURE 13.7 Effect of IR on CVD.

confirms the pathophysiological relationship between IR, atherosclerosis, and essential hypertension (Figure 13.7) [83].

13.2.1.4 How IR Influences the Progression of Ischemic Stroke

Ischemic stroke is a complex cycle of interrelated molecular and cellular mechanisms (Figure 13.8) [84–86]. Inflammation as well as oxidative stress play an important role at all stages of the ischemic cascade.

Three distinct phases of ROS production have been identified in ischemic stroke: i) oxygen and glucose deprivation, ii) xanthine oxidase XO) activation, and iii) reperfusion [87, 88]. In the first stage, blockage of cerebral blood flow deprives the brain tissue of oxygen and glucose; the energy deficit becomes the main mechanism of cell death in the infarcted area of the brain. The activity of the Na^+/K^+ ATPase decreases and there is an increase in the extracellular concentration of K^+ ions and an uncontrolled influx of Na^+, Ca^{2+}, and Cl^- ions. Dysfunction of energy-dependent ion pumps and channels leads to the release and impaired reuptake of glutamate. Excessive activation of glutamate receptors results in excitotoxicity and accumulation of Ca^{2+} ions, which activates catabolic enzymes and increases ROS. Calcium overload causes the release of apoptotic factors from the mitochondria, which induces apoptosis [89, 90].

In the second step, adenine nucleotides are converted to hypoxanthine and xanthine, substrates for xanthine oxidase (XO). XO generates ROS after the first burst of ROS; this step correlates with the timing of ATP depletion [88]. The production of ROS and/or RNS is also associated with reperfusion, or restoration of blood flow; at this stage, the influx of oxygen, calcium ions, and leukocytes can cause excessive production of ROS and generate an inflammatory response. In addition, activation of NADPH oxidase in immune cells and elevated ROS concentrations in mitochondria can induce prolonged opening of the mitochondrial permeability transition pore (mPTP), leading to a new burst of ROS, known as ROS-induced ROS release. The short openings of the mPTP play an important physiological role by regulating intramitochondrial Ca^{2+}, maintaining healthy mitochondrial homeostasis. Disruption of mitochondrial bioenergetics, which develops during reoxygenation, when resting cytosol Ca^{2+} concentration is high and ATP level is low, leads to excessive mitochondrial Ca^{2+} uptake in mitochondria and, consequently, to mPTP induction. In contrast, mitochondrial de-energization, if caused by mPTP opening, was associated with higher ROS production [91–93].

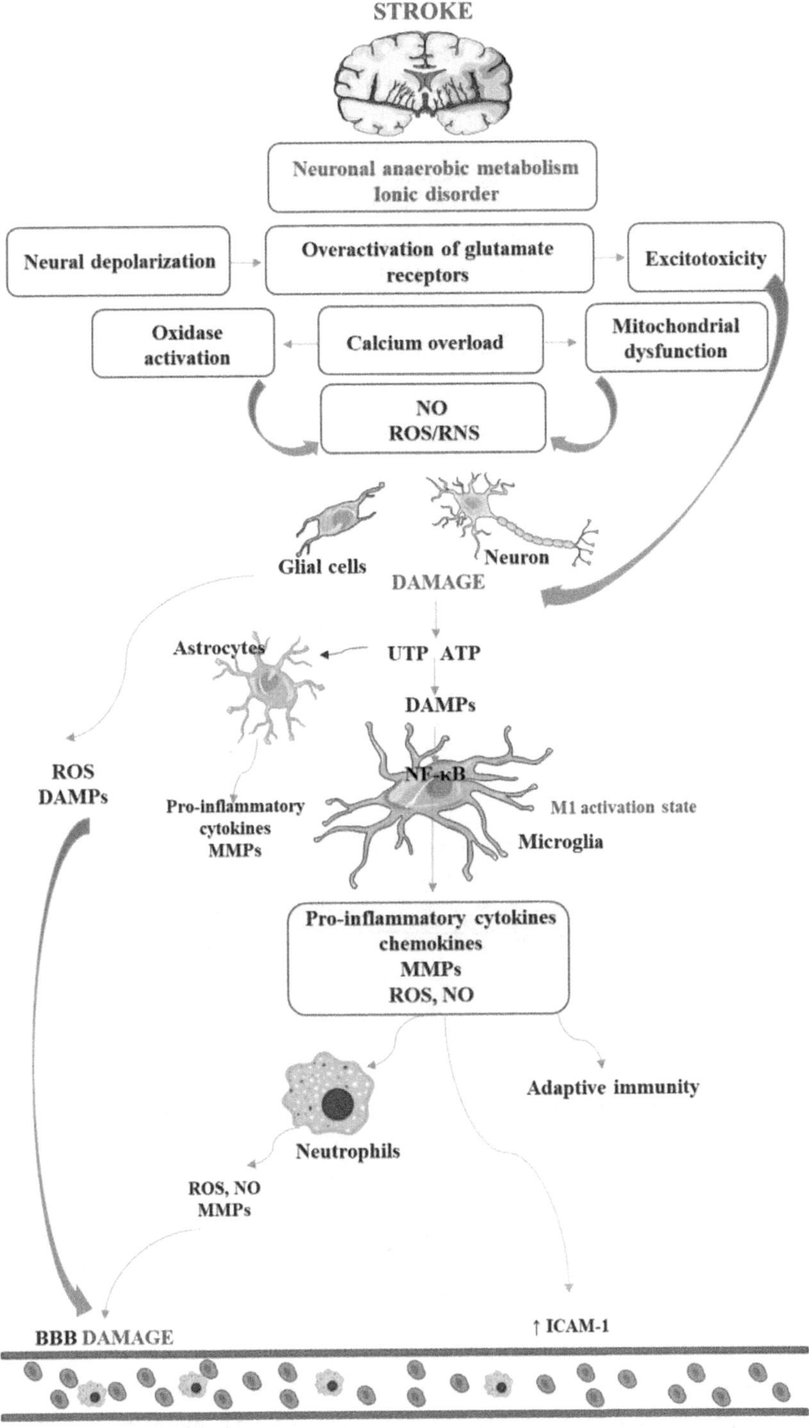

FIGURE 13.8 Stroke pathology mechanism. ATP—Adenosine 5'-triphosphate; BBB—Blood-brain barrier; DAMPs—Molecular patterns; ICAM-1—Intercellular adhesion molecule 1; MMPs—Matrix metalloproteinases; NF-κB—Nuclear factor-κB; NO—Nitrogen oxide; UTP—Uridine-5'-triphosphate; RNS—Reactive nitrogen species; ROS—Reactive oxygen species. This figure was created using Servier Medical Art (available at https://smart.servier.com/).

Inflammatory cells including microglia and infiltrating neutrophils/macrophages are also a source of ROS that activates inflammatory cells and exacerbates inflammatory response. Activated microglia and astrocytes secrete inflammatory mediators, cytokines, chemokines, and MMPs, which induce the expression of cell adhesion molecules on the endothelial surface and ICAM-1. This enables the influx of neutrophils to the ischemic areas of the brain and the transport of leukocytes to sites of damage. Infiltrating immune cells, regardless of their beneficial role, can also damage the ischemic brain by producing various destructive cytotoxic mediators (including NO, ROS, and prostanoids) that prolong the inflammatory response, increasing the brain damage [94]. This can lead to secondary complications such as edema and hemorrhagic transformation. Cytotoxic edema is related to the function of ion transporters, and these membrane complexes, such as Na^+/K^+-ATPase, Ca^{2+}-ATPase, and Na^+/Ca^{2+} exchanger, can be inhibited by ROS through the peroxidation of membrane phospholipids and protein modification [95].

Lymphocytopenia, in turn, contributes to significant immunosuppression, which increases the risk of infection after a stroke. In addition, the disproportionate concentration of pro-inflammatory mediators activates the autonomic nervous system, inhibits pro-inflammatory pathways, and stimulates anti-inflammatory mechanisms through the release of interleukins and growth factors. In addition, necrotic neurons release "hazard signals" that activate the immune system and secrete molecular patterns (DAMPs). DAMPs induce Toll-like receptors on microglia and in turn stimulates NF-κB to synthesize most of the pro-inflammatory cytokines (IL-1β,IL-6, IL-18, TNF-α) and chemotactic chemokines. Pro-inflammatory cytokines disrupt the BBB, while MMPs mediate the destruction of the basal plate by increasing the permeability of the BBB and facilitating the entry of additional peripheral immune cells into the stroke-affected area of the brain. This is because acute cerebral ischemia causes not only a local inflammatory response but also a systemic immune response [95].

After ischemic stroke, the transport of oxygen and glucose is limited, and the main source of neuronal ATP is anaerobic oxidative metabolism [96]. Oxygen–glucose deprivation inhibits the AKT pathway, reducing in mTORC1 activity and blocking autophagy. Autophagy protects neuronal cells from ischemic damage and removes damaged mitochondria during reperfusion, which attenuates mitochondria-induced neuronal apoptosis and ischemic damage. However, overactivated autophagy promotes neural cell atrophy and worsens ischemic brain damage [97].

Excess accumulation of ROS due to mitochondrial dysfunction activates FOXO3, increasing the abundance of LC3 (protein 1A/1B-light chain 3) to generate autophagosome, excessive autophagy that increases neuronal apoptosis [98]; dysfunction of the autophagy response results in adipocyte dysfunction. IR induces glucotoxicity that exacerbates oxidative stress, inflammation, and endoplasmic reticulum stress due to lipotoxicity, further attenuating the autophagy response [99]. Moreover, by inhibiting GLUT4 membrane translocation, IR leads to neuronal apoptosis due to insufficient glucose uptake. IR can also affect neuronal function. Neural damage is responsible for poor clinical outcomes such as worsening neurological function and poor functional outcome at three months in patients with ischemic stroke due to IR [100].

Inflammation is known to play a significant role in promoting IR; inflammation and IR feed each other in a vicious circle [101]. The inflammatory response plays a key role in brain damage caused by ischemic stroke, and IR exacerbates it. IR causes local accumulation of pro-inflammatory macrophages, produces MCP-1, which recruits monocytes, and activates pro-inflammatory macrophages [102].

Blocking the oxidative metabolism of glucose prevents polarization of macrophages to the M2 phenotype and shifts them to the activated M1 state. Obesity coexisting with IR induces IKKβ activation, leading to NF-kB nuclear translocation and, as a result, to the production of various inflammatory markers and potential mediators. Obesity also co-promotes the phosphorylation of IRS1 at serine sites by activating c-Jun N-terminal kinases (JNK) [103].

Metabolic disorders caused by peripheral IR are a source of oxidative stress in brain tissue. IR increases the production of ROS by increasing FFA concentration, which promotes glucotoxicity, lipotoxicity, and nuclear translocation of NF-kB [104]. The effects of oxidative stress and IR appear to be mutual; oxidative stress reduce insulin sensitivity in skeletal muscle cells and adipocytes. It also phosphorylates IRS proteins by activating the IKKβ/NF-kB and JNK pathways, causing IRS degradation [105].

Insulin is essential for mitochondrial function, and low mitochondrial function is a major cause of IR; mitochondria are also known to be the main source of ROS. Insulin deprivation not only reduces the efficiency of ATP production by isolated muscle mitochondria but also increases the degradation of mitochondrial proteins [106]. Insulin therefore appears to be essential for brain function. IR, which leads to insulin dysregulation, can cause neurological damage through a variety of mechanisms and is associated with poor prognosis in patients with ischemic stroke, exacerbating inflammatory response, oxidative stress, and neuronal damage [16]. Lifestyle changes including a healthy diet, weight loss, smoking cessation, and increasing physical activity can improve peripheral insulin sensitivity and potentially prevent ischemic stroke.

13.2.2 Obesity as a Metabolic Disorder in the Context of Stroke

The key abnormality that characterizes MetS is being overweight or obese, especially abdominal obesity. MetS develops over many years, most often starting with overweight and then obesity followed by individual components of MetS including elevated blood pressure, atherogenic dyslipidemia, and abnormal glycemia.

The most commonly used indicator to detect overweight and obesity is BMI (Body Mass Index). It should be emphasized, however, that obesity does not result from excess body weight (overweight), but from excess body fat. The etiology of overweight and eventually obesity is heterogeneous and includes the strong development of muscle tissue, excessive accumulation of body water, and induced development of adipose tissue. For this reason, BMI is often not a good indicator of fat content [107].

Adipose tissue is described as physiological ("due") and excess ("ballast"—supra-physiological). It is important when assessing the risk of CVD due to obesity to examine whether the patient has abnormal body fat content and how this tissue is distributed in the body. Intra-abdominal or visceral fat is associated with IR and CVD [108–110]. Adipose tissue is the body's largest energy reservoir because it has high energy density and is poorly hydrated, unlike glycogen or proteins. It performs three basic functions: It is insulating and shock-absorbing (it protects organs from mechanical injuries), it is energetic, and it is also a very good source of metabolic water. The main mass of adipose tissue is cells called adipocytes, although there are also pre-adipocytes, fibroblasts, leukocytes, monocytes, macrophages, endothelial cells, and subpopulations of stem cells known as stromal vascular fraction cells. All cells play a role in the physiology and pathophysiology of this tissue [111].

Adipose tissue is a very active endocrine organ that secretes many different hormone-like substances, biologically active peptides (adipokines) that show autocrine and paracrine action within the tissue as well as endocrine action in distant tissues, organs, and cytokines (TNF-α, IL-6, IL-1β, MCP-1). Adipose tissue can grow in two ways: hypertrophy (increase in cell size) and hyperplasia (increase in cell number), and there are three types of adipocytes: white (also referred to as yellow), brown, and pink (in pregnant and lactating women). The main role of white adipocytes is to store triglycerides during increased energy supply and use them during periods of increased energy expenditure; brown adipocytes produce heat, and pink adipocytes produce milk [112].

Adipocytes communicate with other organs through secreted adipokines such as adiponectin, leptin, chemerin, resistin, visfatin, and waspin. They also receive signals from other tissues. The reception of information is possible due to the expression of numerous receptors on their surface. The reception of signals and the secretion of adipokines make the white adipose tissue a place of integration and communication with the endocrine, nervous, muscular, immune, circulatory, and reproductive systems [113].

The metabolism of adipose tissue is influenced by hormonal factors (mainly insulin and catecholamines), autocrine and paracrine factors, nutritional status (food intake, fasting), stress factors, and physical exercise. Adipose tissue is responsible for maintaining a constant concentration of fatty acids in the blood. In adipose tissue, lipogenesis and accumulation of triglycerides and lipolysis, i.e. degradation of TAG to FFA and glycerol, occur [107]. Adipokine dysregulation is associated with the onset of obesity, IR, type 2 diabetes, CVD, hypertension, and MetS [113].

Hypertrophic adipocytes become resistant to the antilipolytic effects of insulin and have a reduced ability to store lipids. When their storage capacity is exceeded, fats accumulate, among other places in muscle and liver cells, causing lipotoxicity. Of the numerous classes of lipids involved in eliciting lipotoxicity, ceramides are among the most deleterious; they modulate the signaling pathways that regulate glucose metabolism, triglyceride synthesis, and fibrosis leading to cellular dysfunction and apoptosis in all cell types, including neurons and neurovascular cells [114, 115].

Excessive accumulation of lipids in adipocytes activates NADPH oxidase, forms ROS, and triggers inflammatory reactions with low activity and insulin resistance [116]. Adipose hyperplasia increases the demand for oxygen, but due to the inability to increase perfusion, hypoxia occurs. Hypoxic adipocytes undergo necrosis more quickly, which is associated with the infiltration of adipose tissue by macrophages, which leads to overproduction of biologically active metabolites known as adipocytokines. Adiopocytokines mainly include free FFA, pro-inflammatory mediators (TNF-α, IL-1, IL-6), PAI-1, leptin, angiotensinogen/angiotensin II. In adipose tissue, inflammatory mediators activate pro-inflammatory pathways related to NF-κB and JNK. Pro-inflammatory cytokines bind to surface receptor adipocytes, and macrophages activate the NF-κB and JNK pro-inflammatory pathways, contributing to further increased production of inflammatory cytokines by these cells. Hypertrophic adipocytes release increased amounts of FFA that activate inflammatory pathways through TLR-4 receptors located on the surface of adipocytes and M1 macrophages. A destructive loop is triggered between macrophages and adipocytes that increases the inflammatory response in adipose tissue [117, 118].

TNF-α causes oxidative stress and activates other cytokines involved in the acute-phase response. Moreover, TNF-α as well as other pro-inflammatory cytokines (IL-1, IL-6) are effective regulators of tissue factor (TF), the main activator of the blood coagulation cascade. TF overexpression by endothelial cells, monocytes, and macrophages promotes thromboembolic complications [119]. In general, adipocytokines integrate endocrine, autocrine, and paracrine signals to mediate multiple processes including insulin sensitivity, oxidative stress, energy metabolism, blood coagulation, and inflammatory responses that accelerate the atherosclerotic process [120]. T lymphocytes also promote inflammation with low activity in adipose tissue. The toll-like receptor 4 (TLR4) signaling pathway is recognized as one of the main drivers of the obesity-induced systemic inflammatory response associated with MetS [117]. As adipocyte cells increase in size, they secrete more active substances that interfere with insulin signals and thus induce IR [121].

Larger adipocytes cause intracellular modifications that increase the cholesterol content in the envelope surrounding the fat droplet and decrease its content in the cell membrane. SREBP-2 is expressed and ABCA1 (ATP-binding cassette transporter; regulator of cellular cholesterol and phospholipid homeostasis) is repressed. Lower cholesterol content in the cell membrane reduces the expression of GLUT4, reducing glucose uptake and triggering IR. Lower cholesterol content in the cell membrane also increases TNF-α, IL-6 and angiotensinogen [122].

Activation of pro-inflammatory factors such as TNF-α, IL-6, and ROS reduces the plasma levels of adiponectin, a protein with a strong anti-inflammatory effect. Adiponectin is a polypeptide hormone that is produced by mature fat cells as a result of the activation of the PPAR-γ nuclear receptor. It regulates lipid and glucose metabolism, and thus body weight, and increases insulin sensitivity. It has anti-atherosclerotic effects that include impeding endothelial activation, reducing macrophage conversion to foam cells, and inhibiting smooth muscle proliferation and arterial remodeling. Adiponectin activates the AMPK and PPAR-α signaling pathways through adiponectin receptor 1 (AdipoR1) and AdipoR2, respectively, increasing fatty acid oxidation and glucose uptake in muscles and suppressing gluconeogenesis in liver tissues. Targeted disruption of AdipoR1 inhibited AMPK activation, increased endogenous glucose production, and increased IR [123, 124].

Leptin, as another adipokine secreted by adipocytes, is involved in the regulation of satiety and energy intake. Most overweight and obese people have elevated leptin levels, which results in increased appetite. It is important in glucose-insulin homeostasis. Leptin binds to the leptin receptor (LepRb) and activates the JAK2/STAT3 pathway to reduce body weight and normalize blood glucose levels. JAK2 stimulates the phosphorylation of the insulin receptor substrate (IRS1/2), then activates the PI3K/AKT pathway and directly influences insulin action [125].

Leptin reduces insulin synthesis and secretion from pancreatic β cells, increasing insulin sensitivity and decreasing hepatic glucose production and glucagon levels. In contrast, stimulates leptin synthesis [126, 127]. High serum leptin levels associated with IR promote the release of pro-inflammatory compounds including IL-6, TNF-α, and IL-12 by monocytes and macrophages. Excess leptin activates the synthesis of transforming growth factor-β1 (TGF-β1) and the proliferation of vascular endothelial cells [128].

Generally, with adipose tissue dysfunction, the adipokine secretion profile is significantly altered toward pro-inflammatory, atherogenic, and diabetogenic. In addition, patients with MetS have abnormal lipid profiles that include more lipoproteins containing apolipoprotein B (a component of 80% of LDL proteins and 40% of VLDL proteins and an indicator of atherosclerotic heart disease risk), elevated TAG, elevated LDL (the main cholesterol transporter), and low HDL-C (high-density lipoprotein cholesterol). The basis of this abnormality is the activation of lipolysis, which increases FFA. It is worth noting that HDL particles, in addition to directing the reverse transport of free cholesterol from peripheral tissues to the liver, also have antioxidant and anti-inflammatory properties. Using an enzymatic mechanism, they reduce the oxidative modification of LDL particles, the main stage of atherogenesis, thus inhibiting the initiation and progression of atherosclerosis [129, 130].

The antioxidant properties of HDL are associated with enzymes such as paraoxonase 1 (PON-1), lecithin-cholesterol acyltransferase and platelet activating factor acetylhydrolase. The anti-inflammatory and antioxidant capacity of HDL may be impaired as a result of chronic hyperglycemia by reducing the activity of PON-1, an independent risk factor for the development of ischemic heart disease. This may be due to the increased concentration of FFA, which, by activating several pathways, increase the degree of phosphorylation of IRS1 serine, thereby inhibiting the transmission of signals from the insulin receptor [129, 130].

Dyslipidemia can be mediated by the cholesteryl ester transfer protein, which enables the exchange of cholesterol esters between VLDLs, LDLs, and HDLs, forming VLDL particles rich in cholesterol esters and HDL and LDL in triglycerides. Hepatic lipase-hydrolyzing TAG forms small, dense HDL and LDL particles. The changed HDLs are removed from the bloodstream, disrupting the reverse transport of cholesterol and reducing their protective function against the vascular endothelium. In addition, atherogenic LDL particles form that are associated with increased CVD risk [131].

Obesity is therefore one of the risk factors for stroke. The main pathogenetic factor of these abnormalities in obese people is the excessive content of visceral adipose tissue, which shows pro-inflammatory activity. This initiates a cascade of metabolic reactions culminating in IR and hyperinsulinism as well as abnormalities in plasma lipid concentrations. Thus, stroke risk is related to the metabolic consequences of obesity [132].

Despite obesity being a risk factor for stroke, several studies have shown a protective effect of obesity on stroke outcome in patients. Increased BMI was associated with better post-stroke outcomes, reduced risk of re-stroke, reduced morbidity, better functional outcomes, and reduced mortality [133–138]. However, not all studies have confirmed the existence of the obesity paradox in stroke [139–143]. In contrast, in animal models of obesity, obese rodents suffer increased ischemic brain damage and have poorer behavioral outcomes compared to control animals. In addition to increased ischemic damage, obese rodents often show increased permeability of the BBB and an increased rate of hemorrhagic transformations [133]. In short, abdominal obesity is an important risk factor in stroke (Figure 13.9), and it can be said with certainty that reducing abdominal obesity will reduce stroke risk.

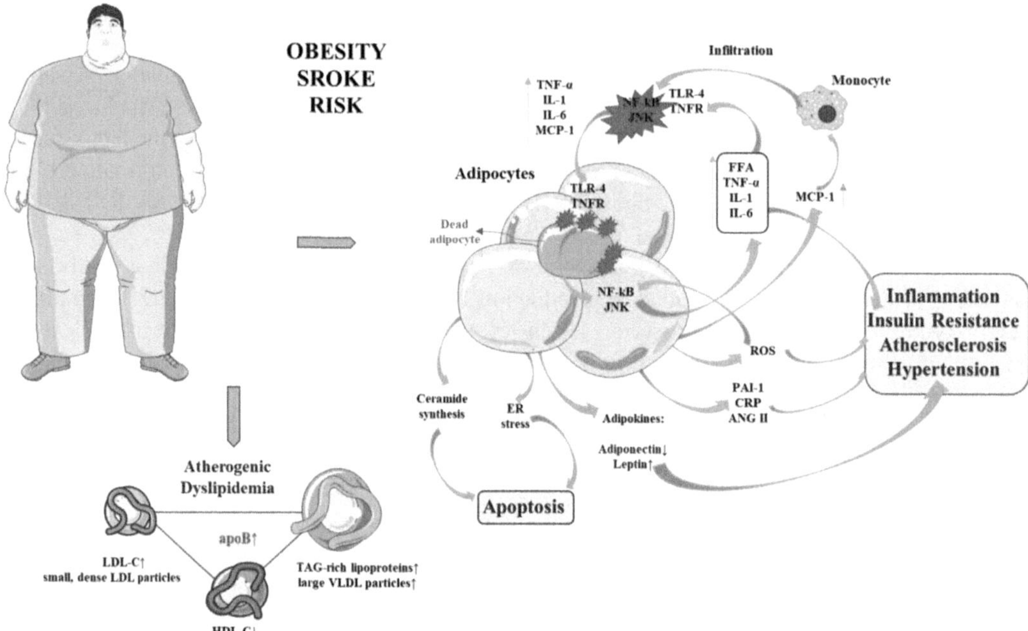

FIGURE 13.9 Obesity stroke risk. ANG II—Angiotensin II; apoB– Apolipoprotein B; CRP—C-reactive protein; ER—Endoplasmic reticulum; FFA—Free fatty acids; HDL-C—High-density lipoprotein cholesterol; IL-1—Interleukin 1; IL-6—Interleukin 6; JNK—c-Jun N-terminal kinases; LDL-C—Low-density lipoprotein cholesterol; LDL—Low-density lipoprotein; MCP-1—Monocyte chemoattractant protein-1; NF-κB—Nuclear factor-κB; ROS—Reactive oxygen species; TAG—Triacylglycerols; TNF-α–Tumour necrosis factor alpha; TNFR—TNF receptor; TLR-4—Toll-like receptor 4; VLDL—Very low-density lipoprotein. This figure was created using Servier Medical Art (available at https://smart.servier.com/).

13.3 SUMMARY

Stroke is the second leading cause of death and the third leading cause of disability in the adult population. The coexistence of obesity with elevated blood pressure and disorders of carbohydrate and lipid metabolism, referred to as metabolic syndrome (MetS), can contribute to stroke risk, and IR and central adiposity can influence the development of MetS. Obesity can be responsible for IR, hypoadiponectinemia, and disorders of the secretion of pro-inflammatory factors. Individual components of MetS, including obesity and IR, affect the formation of blood clots and atherosclerosis, promote inflammation, and can cause oxidative stress in brain tissue. Changing eating habits and increasing physical activity can bring benefits in the form of weight loss, improving lipid parameters, reducing hypertension, and improving insulin sensitivity, which ultimately translates into a reduced risk of cardiovascular diseases including strokes.

REFERENCES

[1] S. Hatano, Experience from multicentre stroke register. A preliminary report, *Bull. WHO*, **1976**, *54*, 541.

[2] R. Mazur, M. Świerkocka-Miastkowska, Udar mózgu—pierwsze objawy, *Choroby Serca i Naczyń*, **2005**, 2, 84.

[3] J. F. Arenillas, M. A. Moro, A. Dávalos, The metabolic syndrome and stroke, *Stroke*, **2007**, *38*, 2196.

[4] K. G. M. M. Alberti, P. Zimmet, J. Shaw, Metabolic syndrome—a new world-wide definition. A consensus statement from the international diabetes federation, *Diabet. Med.*, **2006**, *23(5)*, 469.

[5] S. M. Grundy, J. I. Cleeman, S. R. Daniels, K. A. Donato, R. H. Eckel, B. A. Franklin, D. J. Gordon, R. M. Krauss, P. J. Savage, S. C. Jr. Smith, J. A. Spertus, F. Costa; American Heart Association; National Heart, Lung, and Blood Institute. Diagnosis and management of the metabolic syndrome: An American heart association/national heart, lung, and blood institute scientific statement, *Circulation*, **2005**, *112*, 2735.
[6] T. Fulop, D. Tessier, A. Carpentier, The metabolic syndrome, *Pathol. Biol.*, **2006**, *54(7)*, 375.
[7] I. Kowalska, M. Strączkowski, A. Nikolajuk, A. Adamska, M. Karczewska-Kupczewska, E. Otziomek, I. Kinalska, M. Górska, Insulin resistance, serum adiponectin, and proinflammatory markers in young subjects with the metabolic syndrome, *Metabolism*, **2008**, *57(11)*, 1539.
[8] J. K. Ninomiya, G. L'Italien, M. H. Criqui, J. L. Whyte, A. Gamst, R. S. Chen, Association of the metabolic syndrome with history of myocardial infarction and stroke in the third national health and nutrition examination survey, *Circulation*, **2004**, *109*, 42.
[9] H. J. Milionis, E. Rizos, J. Goudevenos, K. Seferiadis, D. P. Mikhailidis, M. S. Elisaf, Components of the metabolic syndrome and risk for first-ever acute ischemic nonembolic stroke in elderly subjects, *Stroke*, **2005**, *36*, 1372.
[10] S.-H. Suk, R. L. Sacco, B. Boden-Albala, J. F. Cheun, J. G. Pittman, M. S. Elkind, M. C. Paik, Abdominal obesity and risk of ischemic stroke. The Northern Manhattan stroke study, *Stroke*, **2003**, *34*, 1586.
[11] H. J. Chen, C. H. Bai, W. T. Yeh, H. C. Chiu, W. H. Pan, Influence of metabolic syndrome and general obesity on the risk of ischemic stroke, *Stroke*, **2006**, *37*, 1060.
[12] S. Kurl, J. A. Laukkanen, L. Niskanen, D. Laaksonen, J. Sivenius, K. Nyyssonen, J. T. Salonen, Metabolic syndrome and the risk of stroke in middle-aged men, *Stroke*, **2006**, *37*, 806.
[13] M. C. Petersen, G. I. Shulman, Mechanisms of insulin action and insulin resistance, *Physiol. Rev.*, **2018**, *98*, 2133.
[14] P. De Meyts, Insulin and its receptor: Structure, function and evolution, *Bioessays*, **2004**, *26*, 1351.
[15] T. Scherer, K. Sakamoto, C. Buettner, Brain insulin signalling in metabolic homeostasis and disease, *Nat. Rev. Endocrinol.*, **2021**, *17(8)*, 468.
[16] P.-F. Ding, H.-S. Zhang, J. Wang, Y.-Y. Gao, J.-N. Mao, Ch.-H. Hang, W. Li, Insulin resistance in ischemic stroke: Mechanisms and therapeutic approaches, *Front. Endocrinol.*, **2022**, *13*, 1092431.
[17] J. F. Youngren, Regulation of insulin receptor function, *Cell Mol. Life Sci.*, **2007**, *64*, 873.
[18] L. C. Cantley, The phosphoinositide 3-kinase pathway, *Science*, **2002**, *296*, 1655.
[19] M. C. Petersen, G. I. Shulman, Mechanisms of insulin action and insulin resistance, *Physiol. Rev.*, **2018**, *98(4)*, 2133.
[20] A. Klip, Y. Sun, T. T. Chiu, K. P. Foley, Signal transduction meets vesicle traffic: The software and hardware of GLUT4 translocation, *Am. J. Physiol. Cell. Physiol.*, **2014**, *306*, 879.
[21] P. B. Iynedjian, Molecular physiology of mammalian glucokinase, *Cell. Mol. Life Sci.*, **2009**, *66*, 27.
[22] M. Bollen, S. Keppens, W. Stalmans, Specific features of glycogen metabolism in the liver, *Biochem. J.*, **1998**, *336*, 19.
[23] J. Nakae, T. Kitamura, D. L. Silver, D. Accili, The forkhead transcription factor Foxo1 (Fkhr) confers insulin sensitivity onto glucose-6-phosphatase expression, *J. Clin. Invest.*, **2001**, *108*, 1359.
[24] P. Puigserver, J. Rhee, J. Donovan, C. J. Walkey, J. C. Yoon, F. Oriente, Y. Kitamura, J. Altomonte, H. Dong, D. Accili, B. M. Spiegelman, Insulin-regulated hepatic gluconeogenesis through FOXO1-PGC-1α interaction, *Nature*, **2003**, *423*, 550.
[25] F. Langlet, R. A. Haeusler, D. Linděn, E. Ericson, T. Norris, A. Johansson, J. R. Cook, K. Aizawa, L. Wang, C. Buettner, D. Accili, Selective inhibition of FOXO1 activator/repressor balance modulates hepatic glucose handling, *Cell*, **2017**, *171(4)*, 824.
[26] J. Ericsson, S. M. Jackson, J. B. Kim, B. M. Spiegelman, P. A. Edwards, Identification of glycerol-3-phosphate acyltransferase as an adipocyte determination and differentiation factor 1- and sterol regulatory element-binding protein-responsive gene, *J. Biol. Chem.*, **1997**, *272*, 7298.
[27] J. R. Krycer, L. J. Sharpe, W. Luu, A. J. Brown, The Akt-SREBP nexus: Cell signaling meets lipid metabolism, *Trends Endocrinol. Metab.*, **2010**, *21*, 268.
[28] R. W. Brownsey, A. N. Boone, J. E. Elliott, J. E. Kulpa, W. M. Lee, Regulation of acetyl-CoA carboxylase, *Biochem. Soc. Trans.*, **2006**, *34*, 223.
[29] L. A. Witters, T. D. Watts, D. L. Daniels, J. L. Evans, Insulin stimulates the dephosphorylation and activation of acetyl-CoA carboxylas, *Proc. Natl. Acad. Sci. USA*, **1998**, *85*, 5473.
[30] R. N. Bergman, Non-esterified fatty acids and the liver: Why is insulin secreted into the portal vein? *Diabetologia*, **2000**, *43*, 946.

[31] K. Hughes, S. Ramakrishna, W. B. Benjamin, J. R. Woodgett, Identification of multifunctional ATP-citrate lyase kinase as the alpha-isoform of glycogen synthase kinase-3, *Biochem. J.*, **1992**, *288*, 309.
[32] I. A. Potapova, M. R. El-Maghrabi, S. V. Doronin, W. B. Benjamin, Phosphorylation of recombinant human ATP: Citrate lyase by cAMP-dependent protein kinase abolishes homotropic allosteric regulation of the enzyme by citrate and increases the enzyme activity. Allosteric activation of ATP: Citrate lyase by phosphorylated sugars, *Biochemistry*, **2000**, *39*, 1169.
[33] S. Sengupta, T. R. Peterson, D. M. Sabatini, Regulation of the mTOR complex 1 pathway by nutrients, growth factors, and stress, *Mol. Cell.*, **2010**, *40*, 310.
[34] K. Inoki, Y. Li, T. Zhu, J. Wu, K.-L. Guan, TSC2 is phosphorylated and inhibited by Akt and suppresses mTOR signalling, *Nat. Cell. Biol.*, **2002**, *4*, 648.
[35] E. Vander Haar, S. I. Lee, S. Bandhakavi, T. J. Griffin, D.-H. Kim, Insulin signalling to mTOR mediated by the Akt/PKB substrate PRAS40, *Nat. Cell. Biol.*, **2007**, 9, 316.
[36] P. P. Hsu, S. A. Kang, J. Rameseder, Y. Zhang, K. A. Ottina, D. Lim, T. R. Peterson, Y. Choi, N. S. Gray, M. B. Yaffe, J. A. Marto, D. M. Sabatini, The mTOR-regulated phosphoproteome reveals a mechanism of mTORC1-mediated inhibition of growth factor signaling, *Science*, **2011**, *332*, 1317.
[37] Y. Yu, S.-O. Yoon, G. Poulogiannis, Q. Yang, X. M. Ma, J. Villén, N. Kubica, G. R. Hoffman, L. C. Cantley, S. P. Gygi, J. Blenis, Phosphoproteomic analysis identifies Grb10 as an mTORC1 substrate that negatively regulates insulin signaling, *Science*, **2011**, *332*, 1322.
[38] D. R. Alessi, M. Andjelkovic, B. Caudwell, P. Cron, N. Morrice, P. Cohen, B. A. Hemmings, Mechanism of activation of protein kinase B by insulin and IGF-1, *EMBO J.*, **1996**, *15*, 6541.
[39] D. A. Guertin, D. M. Stevens, C. C. Thoreen, A. A. Burds, N. Y. Kalaany, J. Moffat, M. Brown, K. J. Fitzgerald, D. M. Sabatini, Ablation in mice of the mTORC components raptor, rictor, or mLST8 reveals that mTORC2 is required for signaling to Akt-FOXO and PKCalpha, but not S6K1, *Dev. Cell*, **2006**, *11*, 859.
[40] E. Jacinto, V. Facchinetti, D. Liu, N. Soto, S. Wei, S. Y. Jung, Q. Huang, J. Qin, B. Su, SIN1/MIP1 maintains rictor-mTOR complex integrity and regulates Akt phosphorylation and substrate specificity, *Cell*, **2006**, *127*, 125.
[41] Y. H. Choi, S. Park, S. Hockman, E. Zmuda-Trzebiatowska, F. Svennelid, M. Haluzik, O. Gavrilova, F. Ahmad, L. Pepin, M. Napolitano, M. Taira, F. Sundler, L. Stenson Holst, E. Degerman, V. C. Manganiello, Alterations in regulation of energy homeostasis in cyclic nucleotide phosphodiesterase 3B-null mice, *J. Clin. Invest.*, **2006**, *116*, 3240.
[42] K. Jaworski, E. Sarkadi-Nagy, R. E. Duncan, M. Ahmadian, H. S. Sul, Regulation of triglyceride metabolism. IV. Hormonal regulation of lipolysis in adipose tissue, *Am. J. Physiol. Gastrointest. Liver. Physiol.*, **2007**, *293*, 1.
[43] N. Begum, Stimulation of protein phosphatase-1 activity by insulin in rat adipocytes. Evaluation of the role of mitogen-activated protein kinase pathway, *J. Biol. Chem.*, **1995**, *270*, 709.
[44] S. Resjo, O. Goransson, L. Harndahl, S. Zolnierowicz, V. Manganiello, E. Degerman, Protein phosphatase 2A is the main phosphatase involved in the regulation of protein kinase B in rat adipocytes, *Cell Signal.*, **2002**, *14*, 231.
[45] S. Kersten, Mechanisms of nutritional and hormonal regulation of lipogenesis, *EMBO Rep.*, **2001**, *2*, 282.
[46] J. Rieusset, F. Andreelli, D. Auboeuf, M. Roques, P. Vallier, J. P. Riou, J. Auwerx, M. Laville; H. Vidal, Insulin acutely regulates the expression of the peroxisome proliferator-activated receptor-gamma in human adipocytes, *Diabetes*, **1999**, *48*, 699.
[47] P. J. Randle, P. B. Garland, C N. Hales, E. A. Newsholme, The glucose fatty-acid cycle. Its role in insulin sensitivity and the metabolic disturbances of diabetes mellitus, *Lancet*, **1963**, *1*, 785.
[48] S. Marshall, V. Bacote, R. R. Traxinger, Discovery of a metabolic pathway mediating glucose-induced desensitization of the glucose transport system. Role of hexosamine biosynthesis in the induction of insulin resistance, *J. Biol. Chem.*, **1991**, *266*, 4706.
[49] M. Hawkins, N. Barzilai, R. Liu, M. Hu, W. Chen, L. Rossetti, Role of the glucosamine pathway in fat-induced insulin resistance, *J. Clin. Invest.*, **1997**, *99*, 2173.
[50] R. J. Copeland, J. W. Bullen, G. W. Hart, Cross-talk between GlcNAcylation and phosphorylation: Roles in insulin resistance and glucose toxicity, *Am. J. Physiol. Endocrinol. Metab.*, **2008**, *295*, 17.
[51] L. E. Ball, M. N. Berkaw, M. G. Buse, Identification of the major site of O-linked beta-N-acetylglucosamine modification in the C terminus of insulin receptor substrate-1, *Mol. Cell Proteomics*, **2006**, *5*, 313.
[52] E. S. Kang, D. Han, J. Park, T. K. Kwak, M. A. Oh, S. A. Lee, S. Choi, Z. Y. Park, Y. Kim, J. W. Lee, O-GlcNAc modulation at Akt1 Ser473 correlates with apoptosis of murine pancreatic beta cells, *Exp. Cell Res.*, **2008**, *314*, 2238.

[53] E. Fabbrini, F. Magkos, B. S. Mohammed, T. Pietka, N. A. Abumrad, B. W. Patterson, A. Okunade, S. Kleina, Intrahepatic fat, not visceral fat is linked with metabolic complications of obesity, *Proc. Natl. Acad. Sci. USA*, **2009**, *106*, 15430.
[54] S.-H. Lee, S.-Y. Park, C. S. Choi, Insulin resistance: From mechanisms to therapeutic strategies, *Diabetes Metab. J.*, **2022**, *46*, 15.
[55] S. C. da Silva Rosa, N. Nayak, A. M. Caymo, J. W. Gordon, Mechanisms of muscle insulin resistance and the cross-talk with liver and adipose tissue, *Physiol. Rep.*, **2020**, *8*, e14607.
[56] S. Kullmann, A. Kleinridders, D. M. Small, A. Fritsche, H.-U. Häring, H. Preissl, M. Heni, Central nervous pathways of insulin action in the control of metabolism and food intake, *Lancet Diabetes Endocrinol.*, **2020**, *8(6)*, 524.
[57] S. C. Woods, E. C. Lotter, L. D. McKay, D. Jr. Porte, Chronic intracerebroventricular infusion of insulin reduces food intake and body weight of baboons, *Nature*, **1979**, *282*, 503.
[58] L. Koch, F. T. Wunderlich, J. Seibler, A. C. Könner, B. Hampel, S. Irlenbusch, G. Brabant, C. R. Kahn, F. Schwenk, J. C. Brüning, Central insulin action regulates peripheral glucose and fat metabolism in mice, *J. Clin. Invest.*, **2008**, *118*, 2132.
[59] S. Obici, B. B. Zhang, G. Karkanias, L. Rossetti, Hypothalamic insulin signaling is required for inhibition of glucose production, *Nat. Med.*, **2002**, *8*, 1376.
[60] S. A. Paranjape, O. Chan, W. Zhu, A. M. Horblitt, E. C. McNay, J. A. Cresswell, J. S. Bogan, R. J. McCrimmon, R. S. Sherwin, Influence of insulin in the ventromedial hypothalamus on pancreatic glucagon secretion in vivo, *Diabetes*, **2010**, *59*, 1521.
[61] A. Pocai, T. K. T. Lam, R. Gutierrez-Juarez, S. Obici, G. J. Schwartz, J. Bryan, L. Aguilar-Bryan, L. Rossetti, Hypothalamic K(ATP) channels control hepatic glucose production, *Nature*, **2005**, *434*, 1026.
[62] T. Scherer, J. O'Hare, K. Diggs-Andrews, M. Schweiger, B. Cheng, C. Lindtner, E. Zielinski, P. Vempati, K. Su, S. Dighe, T. Milsom, M. Puchowicz, L. Scheja, R. Zechner, S. J. Fisher, S. F. Previs, C. Buettner, Brain insulin controls adipose tissue lipolysis and lipogenesis, *Cell Metab.*, **2011**, *13*, 183.
[63] M. Li, X. Chi, Y. Wang, S. Setrerrahmane, W. Xie, H. Xu, Trends in insulin resistance: Insights into mechanisms and therapeutic strategy, *Signal. Transduct. Target. Ther.*, **2022**, *7(1)*, 216.
[64] G. Ashrafi, Z. Wu, R. J. Farrell, T. A. Ryan, Glut4 mobilization supports energetic demands of active synapses, *Neuron*, **2017**, *93(3)*, 606.
[65] T. Scherer, K. Sakamoto, C. Buettner, Brain insulin signalling in metabolic homeostasis and disease, *Nat. Rev. Endocrinol.*, **2021**, *17(8)*, 468.
[66] S. J. Hierons, J. S. Marsh, D. Wu, C. A. Blindauer, A. J. Stewart, The interplay between non-esterified fatty acids and plasma zinc and its influence on thrombotic risk in obesity and type 2 diabetes, *Int. J. Mol. Sci.*, **2021**, *22(18)*.
[67] M.-J. van Rooy, E. Pretorius, Metabolic syndrome, platelet activation and the development of transient ischemic attack or thromboembolic stroke, *Thromb. Res.*, **2015**, *135*, 434.
[68] G. Zeng, F. H. Nystrom, L. V. Ravichandran, L. N. Cong, M. Kirby, H. Mostowski, M. J. Quon, Roles for insulin receptor, PI3-kinase, and Akt in insulin-signaling pathways related to production of nitric oxide in human vascular endothelial cells, *Circulation*, **2000**, *101*, 1539.
[69] A. Aljada, H. Ghanim, R. Saadeh, P. Dandona, Insulin inhibits NFκB and MCP-1 expression in human aortic endothelial cells, *J. Clin. Endocrinol. Metab.*, **2001**, *8*, 450.
[70] A. Aljada, R. Saadeh, E. Assian, H. Ghanim, P. Dandona, Insulin inhibits the expression of intercellular adhesion molecule-1 by human aortic endothelial cells through stimulation of nitric oxide, *J. Clin. Endocrinol. Metab.*, **2000**, *85*, 2572.
[71] R. A. DeFronzo, Insulin resistance, lipotoxicity, type 2 diabetes and atherosclerosis: The missing links. The Claude Bernard lecture 2009, *Diabetologia*, **2010**, *53*, 1270.
[72] S. Söderberg, B. Ahrén, B. Stegmayr, O. Johnson, P.-G. Wiklund, L. Weinehall, G. Hallmans, T. Olsson, Leptin is a risk marker for first-ever hemorrhagic stroke in a population-based cohort, *Stroke*, **1999**, *30(2)*, 328.
[73] A. A. da Silva, J. M. do Carmo, X. Li, Z. Wang, A. J. Mouton, J. E. Hall, Role of hyperinsulinemia and insulin resistance in hypertension: Metabolic syndrome revisited, *Can. J. Cardiol.*, **2020**, *36(5)*, 671.
[74] P. Libby, P. Theroux, Pathophysiology of coronary artery disease, *Circulation*, **2005**, *111*, 3481.
[75] E. Maury, S. M. Brichard, Adipokine dysregulation, adipose tissue inflammation and metabolic syndrome, *Mol. Cell Endocrinol.*, **2010**, *314*, 1.
[76] C. Banfi, P. Eriksson, G. Giandomenico, L. Mussoni, L. Sironi, A. Hamsten, E. Tremoli, Transcriptional regulation of plasminogen activator inhibitor type 1 gene by insulin: Insights into the signaling pathway, *Diabetes*, **2001**, *50*, 1522.

[77] M. L. Correia, W. G. Haynes, A role for plasminogen activator inhibitor-1 in obesity: From pie to PAI? *Arterioscler. Thromb. Vasc. Biol.*, **2006**, *26*, 2183.
[78] A. M. Potenza, S. Gagliardi, C. Nacci, R. M. Carratu, M. Montagnani, Endothelial dysfunction in diabetes: From mechanisms to therapeutic targets, *Curr. Medicinal. Chem.*, **2009**, *16(1)*, 94.
[79] R. Kaur, M. Kaur, J. Singh, Endothelial dysfunction and platelet hyperactivity in type 2 diabetes mellitus: Molecular insights and therapeutic strategies, *Cardiovasc. Diabetol.*, **2018**, *17(1)*,121.
[80] S. Yang, J. Zhao, Y. Chen, M. Lei, Biomarkers associated with ischemic stroke in diabetes mellitus patients, *Cardiovasc. Toxicol.*, **2016**, *16(3)*, 213.
[81] G. Nickenig, J. Röling, K. Strehlow, P. Schnabel, M. Böhm, Insulin induces upregulation of vascular AT1 receptor gene expression by posttranscriptional mechanisms, *Circulation*, **1998**, *98*, 2453.
[82] S. Engeli, R. Negrel, A. M. Sharma, Physiology and pathophysiology of the adipose tissue renin-angiotensin system, *Hypertension*, **2000**, *35*, 1270.
[83] M. S. Zhou, I. H. Schulman, L. Raij, Vascular inflammation, insulin resistance, and endothelial dysfunction in salt-sensitive hypertension: Role of nuclear factor kappa B activation, *J. Hypertens.*, **2010**, *28*, 527.
[84] H. Pawluk, A. Woźniak, G. Grześk, R. Kołodziejska, M. Kozakiewicz, E. Kopkowska, E. Grzechowiak, G. Kozera, The role of selected pro-inflammatory cytokines in pathogenesis of ischemic stroke, *Clin. Interv. Aging*, **2020**, *15*, 469.
[85] H. Pawluk, R. Kołodziejska, G. Grześk, M. Kozakiewicz, A. Woźniak, M. Pawluk, A. Kosinska, M. Grześk, J. Wojtasik, G. Kozera, Selected mediators of inflammation in patients with acute ischemic stroke, *Int. J. Mol. Sci.*, **2022**, *23*, 10614.
[86] H. Pawluk, R. Kołodziejska, G. Grześk, A. Woźniak, M. Kozakiewicz, A. Kosinska, M. Pawluk, E. Grzechowiak, J. Wojtasik, G. Kozera, Increased oxidative stress markers in acute ischemic stroke patients treated with thrombolytics, *Int. J. Mol. Sci.*, **2022**, *23*, 15625.
[87] J. Wang, S. Gao, C. Lenahan, Y. Gu, X. Wang, Y. Fang, W. Xu, H. Wu, Y. Pan, A. Shao, J. Zhang, Melatonin as an antioxidant agent in stroke: An updated review, *Aging Dis.*, **2022**, *13(6)*, 1823.
[88] A. Y. Abramov, A. Scorziello, M. R. Duchen, Three distinct mechanisms generate oxygen free radicals in neurons and contribute to cell death during anoxia and reoxygenation, *J. Neurosci.*, **2007**, *27*, 1129.
[89] P. Lipton, Ischemic cell death in brain neurons, *Physiol. Rev.*, **1999**, *79*, 1431.
[90] S. Kahlert, G. Zündorf, G. Reiser, Glutamate-mediated influx of extracellular Ca^{2+} is coupled with reactive oxygen species generation in cultured hippocampal neurons but not in astrocytes, *J. Neurosci. Res.*, **2005**, *79*, 262.
[91] D. B. Zorov, M. Juhaszova, S. J. Sollott, Mitochondrial reactive oxygen species (ROS) and ROS-induced ROS release, *Physiol. Rev.*, **2014**, *94*, 909.
[92] M. Crompton, L. Andreeva, On the involvement of a mitochondrial pore in reperfusion injury, *Basic Res. Cardiol.*, **1993**, *88*, 513.
[93] M. Crompton, A. Costi, A heart mitochondrial Ca^{2+}-dependent pore of possible relevance to re-perfusion-induced injury. Evidence that ADP facilitates pore interconversion between the closed and open states, *Biochem. J.*, **1990**, *266*, 33.
[94] S. M. Lucas, N. J. Rothwell, R. M. Gibson, The role of inflammation in CNS injury and disease, *Br. J. Pharmacol.*, **2006**, *147*, 232.
[95] A. Bustamante, A. Simats, A. Vilar-bergua, T. García-Berrocoso, J. Montaner, Blood/brain biomarkers of inflammation after stroke and their association with outcome: From C-reactive protein to damage-associated molecular patterns, *Neurotherapeutics*, **2016**, *13*, 671.
[96] P. Scheinberg, Survival of the ischemic brain: A progress report, Circulation, **1979**, *60(7)*, 1600.
[97] C. Gu, J. Yang, Y. Luo, D. Ran, X. Tan, P. Xiang, H. Fei, Y. Lu, W. Guo, Y. Tu, X. Liu, H. Wang, Znrf2 attenuates focal cerebral Ischemia/reperfusion injury in rats by inhibiting Mtorc1-mediated autophagy, *Exp. Neurol.*, **2021**, 342, 113759.
[98] A. Ahsan, M. Liu, Y. Zheng, W. Yan, L. Pan, Y. Li, S. Ma, X. Zhang, M. Cao, Z. Wu, W. Hu, Z. Chen, X. Zhang, Natural compounds modulate the autophagy with potential implication of stroke, *Acta Pharm. Sin. B*, **2021**, *11(7)*, 1708.
[99] M. Kitada, D. Koya, Autophagy in metabolic disease and ageing, *Nat. Rev. Endocrinol.*, **2021**, *17(11)*, 647.
[100] L. P. van der Heide, A. Kamal, A. Artola, W. H. Gispen, G. M. Ramakers, Insulin modulates hippocampal activity-dependent synaptic plasticity in a n-methyl-D-Aspartate receptor and phosphatidyl-inositol-3-kinase-dependent manner, *J. Neurochem.*, **2005**, *94(4)*, 1158.
[101] J. M. Olefsky, C. K. Glass, Macrophages, inflammation, and insulin resistance, *Annu. Rev. Physiol.*, **2010**, *72(1)*, 219.

[102] M. Shimobayashi, V. Albert, B. Woelnerhanssen, I. C. Frei, D. Weissenberger, A. C. Meyer-Gerspach, N. Clement, S. Moes, M. Colombi, J. A. Meier, M. M. Swierczynska, P. Jenö, C. Beglinger, R. Peterli, M. N. Hall, Insulin resistance causes inflammation in adipose tissue, *J. Clin. Invest.*, **2018**, *128(4)*, 1538.
[103] S. E. Shoelson, J. Lee, A. B. Goldfine Inflammation and insulin resistance, *J. Clin. Invest.*, **2006**, *116(7)*, 1793.
[104] M. Maciejczyk, E. Żebrowska, A. Chabowski, Insulin resistance and oxidative stress in the brain: What's new?, *Int. J. Mol. Sci.*, **2019**, *20(4)*.
[105] H. Yaribeygi, F. R. Farrokhi, A. E. Butler, A. Sahebkar, Insulin resistance: Review of the underlying molecular mechanisms, *J. Cell Physiol.*, **2019**, *234(6)*, 8152.
[106] M. M. Robinson, S. Dasari, H. Karakelides, H. R. Bergen, K. S. Nair, Release of skeletal muscle peptide fragments identifies individual proteins degraded during insulin deprivation in type 1 diabetic humans and mice, *Am. J. Physiol. Endocrinol. Metab.*, **2016**, *311(3)*, 628.
[107] E. Murawska-Ciałowicz, Adipose tissue—morphological and biochemical characteristic of different depots, *Postepy Hig. Med. Dosw.*, **2017**, *71*, 466.
[108] H. Bays, L. Mandarino, R. A. DeFronzo, Role of the adipocytes, FFA, and ectopic fat in the pathogenesis of type 2 diabetes mellitus: PPAR agonists provide a rational therapeutic approach, *J. Clin. Endocrinol. Metab.*, **2004**, *89*, 463.
[109] L. Lapidus, C. Bengtsson, B. Larsson, K. Pennert, E. Rybo, L. Sjöström, Distribution of adipose tissue and risk of cardiovascular disease and death: A 12 year follow up of participants in the population study of women in Gothenburg, Sweden, *Br. Med. J.*, **1984**, *289*, 1257.
[110] J. P. Despre, S. Moorjani, P. J. Lupien, A. Tremblay, A. Nadeau, C. Bouchard, Regional distribution of body fat, plasma lipoproteins, and cardiovascular disease, *Arterioscler. Thromb. Vas. Biol.*, **1990**, *10*, 497.
[111] L. Siemińska, Tkanka tłuszczowa. Patofizjologia, rozmieszczenie, różnice płciowe oraz znaczenie w procesach zapalnych i nowotworowych, *Pol. J. Endocrinol.*, **2007**, 58, 330.
[112] W. P. Cawthorn, E. L. Scheller, O. A. MacDougald, Adipose tissue stem cells meet preadipocyte commitment: Going back to the future, *J. Lipid Res.*, **2012**, *53*, 227.
[113] K. Rabe, M. Lehrke, K. G. Parhofer, U. C. Broedl, Adipokines and insulin resistance, *Mol. Med.*, **2008**, *14*, 741.
[114] J. K. Sethi, A. J. Vidal-Puig, Thematic review series: Adipocyte biology. Adipose tissue function and plasticity orchestrate nutritional adaptation, *J. Lipid Res.*, **2007**, *48*, 1253.
[115] A. Kennedy, K. Martinez, C.-C. Chuang, K. LaPoint, M. McIntosh, Saturated fatty acid-mediated inflammation and insulin resistance in adipose tissue: Mechanisms of action and implications, *J. Nutr.*, **2009**, *139(1)*, 1.
[115] Y. Li, Ch. L. Talbot, B. Chaurasia, Ceramides in Adipose Tissue, *Frontiers in Endocrinology*, **2020**, *11*, 407.
[116] S. Furukawa, T. Fujita, M. Shimabukuro, M. Iwaki, Y. Yamada, Y. Nakajima, O. Nakayama, M. Makishima, M. Matsuda, I. Shimomura, Increased oxidative stress in obesity and its impact on metabolic syndrome, *J. Clin. Invest.*, **2004**, *114*, 1752.
[117] M.-J. van Rooy, E. Pretorius, Metabolic syndrome, platelet activation and the development of transient ischemic attack or thromboembolic stroke, *Thrombosis Research*, **2015**, *135*, 434.
[118] D. C. W. Lau, B. Dhillon, H. Yan, P. E. Szmitko, S. Verma, Adipokines: Molecular links between obesity and atherosclerosis, *Am. J. Physiol. Heart Circ. Physiol.*, **2005**, *288(5)*, 2031.
[119] G. Grignani, A. Maiolo, Cytokines and hemostasis, *Haematologica*, **2000**, *85*, 967.
[120] H. S. Schipper, B. Prakken, E. Kalkhoven, M. Boes, Adipose tissue-resident immune cells: Key players in immunometabolism, *Trends Endocrinol. Metab.*, **2012**, *23*, 407.
[121] H. Hauner, The new concept of adipose tissue function, *Physiol. Behav.*, **2004**, *83*, 653.
[122] S. Le Lay, S. Krief, C. Farnier, I. Lefrere, X. Le Liepvre, R. Bazin, P. Ferré, I. Dugail, Cholesterol, a cell size-dependent signal that regulates glucose metabolism and gene expression in adipocytes, *J. Biol. Chem.*, **2001**, *276*, 16904.
[123] S. Li, H. J. Shin, E. L. Ding, R. M. van Dam, Adiponectin levels and risk of type 2 diabetes: A systematic review and meta-analysis, *JAMA*,**2009**, *302*, 179.
[124] K. Hotta, T. Funahashi, N. L. Bodkin, H. K. Ortmeyer, Y. Arita, B. C. Hansen, Y. Matsuzawa, Circulating concentrations of the adipocyte protein adiponectin are decreased in parallel with reduced insulin sensitivity during the progression to type 2 diabetes in rhesus monkeys, *Diabetes*, **2001**, *50*, 1126.
[125] M. Amitani, A. Asakawa, H. Amitani, A. Inui, The role of leptin in the control of insulin-glucose axis, *Front. Neurosci.*, **2013**, *7*, 51.

[126] R. B. Ceddia, H. A. Koistinen, J. R. Zierath, G. Sweeney, Analysis of paradoxical observations on the association between leptin and insulin resistance, *FASEB J.*, **2002**, *16*, 1163.
[127] S. D. Covey, R. D. Wideman, C. McDonald, S. Unniappan, F. Huynh, A. Asadi, M. Speck, T. Webber, S. C. Chua, T. J. Kieffer, The pancreatic beta cell is a key site for mediating the effects of leptin on glucose homeostasis, *Cell Metab.*, **2006**, *4*, 291.
[128] T. Gainsford, et al. Leptin can induce proliferation, differentiation, and functional activation of hemopoietic cells, *Proc. Natl. Acad. Sci. USA*, **1996**, *93*, 14564.
[129] G. F. Lewis, G. Steiner, Acute effects of insulin in the control of VLDL production in humans: Implications for the insulin-resistant state, *Diabetes Care*, **1996**, *19(4)*, 390.
[130] H. N. Ginsberg, Y.-L. Zhang, A. Hernandez-Ono, Regulation of plasma triglycerides in insulin resistance and diabetes, *Arch. Med. Res.*, **2005**, *36(3)*, 232.
[131] C. N. Manjunath, J. R. Rawal, P. M. Irani, K. Madhu, Atherogenic dyslipidemia, *Indian J. Endocrinol. Metab.*, **2013**, *17(6)*, 969.
[132] J. W. Horn, T. Feng, B. Mørkedal, L. B. Strand, J. Horn, K. Mukamal, I. Janszky, Obesity and risk for first ischemic stroke depends on metabolic syndrome: The HUNT study, *Stroke*, **2021**, *52(11)*, 3555.
[132] G. A. Quiñones-Ossa, C. Lobo, E. Garcia-Ballestas, W. A. Florez, L. R. Moscote-Salazar, A. Agrawal, Obesity and Stroke: Does the Paradox Apply for Stroke?, *Neurointervention*, **2021**, *16*, 9.
[133] M. J. Haley, C. B. Lawrence, Obesity and stroke: Can we translate from rodents to patients?, *J. Cereb. Blood. Flow. Metab.*, **2016**, *36(12)*, 2007.
[134] K. Vemmos, G. Ntaios, K. Spengos, P. Savvari, A. Vemmou, T. Pappa, E. Manios, G. Georgiopoulos, M. Alevizaki, Association between obesity and mortality after acute first-ever stroke: The obesity-stroke paradox, *Stroke*, **2011**, *42*, 30.
[135] W. Doehner, J. Schenkel, S. D. Anker, J. Springer, H. J. Audebert, Overweight and obesity are associated with improved survival, functional outcome, and stroke recurrence after acute stroke or transient ischaemic attack: Observations from the TEMPiS trial, *Eur. Heart J.*, **2013**, *34*, 268.
[136] K. K. Andersen, T. S. Olsen, The obesity paradox in stroke: Lower mortality and lower risk of readmission for recurrent stroke in obese stroke patients, *Int. J. Stroke*, **2015**, *10*, 99.
[137] R. Barba, J. Marco, J. Ruiz, J. Canora, J. Hinojosa, S. Plaza, A. Zapatero-Gaviria, The obesity paradox in stroke: Impact on mortality and short-term readmission, *J. Stroke Cerebrovasc. Dis.*, **2015**, *24*, 766.
[138] A. E. Hassan, S. A. Chaudhry, V. Jani, M. Grigoryan, A. A. Khan, M. M. Adil, A. I. Qureshi, Is there a decreased risk of intracerebral hemorrhage and mortality in obese patients treated with intravenous thrombolysis in acute ischemic stroke?, *J. Stroke Cerebrovasc. Dis.*, **2013**, *22*, 545.
[139] Y. Kim, C. K. Kim, S. Jung, et al. Obesity-stroke paradox and initial neurological severity, *J. Neurol. Neurosurg. Psychiatry*, **2015**, *86*, 743.
[140] C. Dehlendorff, K. K. Andersen, T. S. Olsen, Body mass index and death by stroke: No obesity paradox, *JAMA Neurol.*, **2014**, *71*, 978.
[141] M. Katsnelson, T. Rundek, Obesity paradox and stroke: Noticing the (fat) man behind the curtain, *Stroke*, **2011**, *42*, 3331.
[142] W. Doehner, H. J. Audebert, The impact of body weight on mortality after stroke: The controversy continues, *JAMA Neurol.*, **2015**, *72*, 126.
[143] A. Brzecka, M. Ejma, Obesity paradox in the course of cerebrovascular diseases, *Adv. Clin. Exp. Med.*, **2015**, *24*, 379.

14 Carbohydrates
Classification and Biosynthesis

Simran Sandhu, Samina Naz,
Ikechukwu P. Ejidike, and Athar Ata

14.1 INTRODUCTION

Carbohydrates are an essential part of a healthy and balanced diet. They are one of the three main macronutrients, along with protein and fat, and are composed of carbon, hydrogen, and oxygen atoms [1]. Carbohydrates play a crucial role in the human body by providing energy, helping to regulate blood glucose and insulin metabolism, and participating in cholesterol and triglyceride metabolism [2–4]. During digestion, carbohydrates are hydrolyzed into glucose or individual sugar molecules that are used by the body as a primary source of energy. Any excess glucose is stored in the liver and muscle tissue for later use [5–8]. Carbohydrates encompass a variety of foods, including sugar, fruits, vegetables, fibers, and legumes. It is important to choose healthy sources like whole grains, fruits, and vegetables to get the most benefit from carbohydrates [9].

14.2 CLASSIFICATION OF CARBOHYDRATES

Carbohydrates are mainly classified into two groups: simple and complex. Simple carbohydrates are further classified as monosaccharides or disaccharides [10, 11]. Figure 14.1 shows the classification of carbohydrates. Monosaccharides, also known as simple sugars, are the building blocks of complex carbohydrates; these sugars are identified by a prefix indicating the number of carbon atoms and the suffix "-ose," for instance, triose (three carbon atoms), tetrose (four), pentose (five), and hexose (six). Representative examples of these naming are shown in Figure 14.2. The most abundant monosaccharide in nature is D-glucose, a hexose. Other common hexose monosaccharides include galactose and fructose. Galactose is used to produce lactose, the disaccharide sugar found in milk, while fructose is found in many fruits and is frequently used as a sweetener in processed foods [12].

Upon examining the structure of glucose, galactose, and fructose, it becomes apparent that the molecular formula of each compound is $C_6H_{12}O_6$. Glucose and galactose are classified as aldoses due to their inclusion of C-1 aldehyde functionality, while fructose is classified as a ketose due to its ketone moiety at C-2. These three compounds possess six carbon atoms and are classified as hexoses. Carbohydrates containing five or six carbons exist in cyclic form; they are cyclized using hemiacetal formation chemistry [13]. Figure 14.3 shows the cyclic forms of a few carbohydrates.

A disaccharide is a sugar that contains two molecules of monosaccharide. When two monosaccharide molecules combine, they undergo a dehydration reaction, forming a glycosidic bond. Some examples of disaccharides include maltose, which is found in grains and consists of two glucose molecules; lactose, which is found in milk and formed by a dehydration reaction between galactose and glucose; and sucrose, which is table sugar and contains both glucose and fructose molecules [14]. Structures of these examples are shown in Figure 14.4.

Polysaccharides are complex carbohydrates composed of multiple monosaccharide units bonded by glycosidic bonds. They are large, linear, or branched polymers with hundreds of repeating

DOI: 10.1201/9781003437413-14

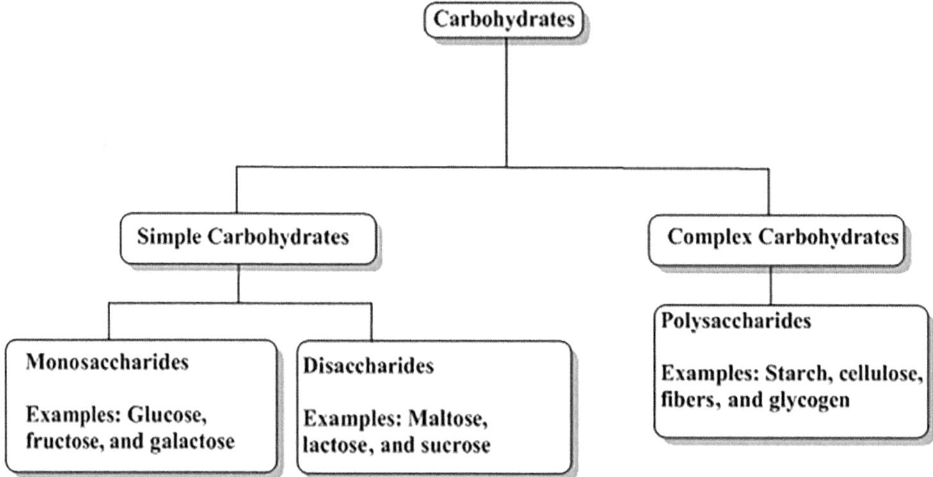

FIGURE 14.1 Classification of carbohydrates.

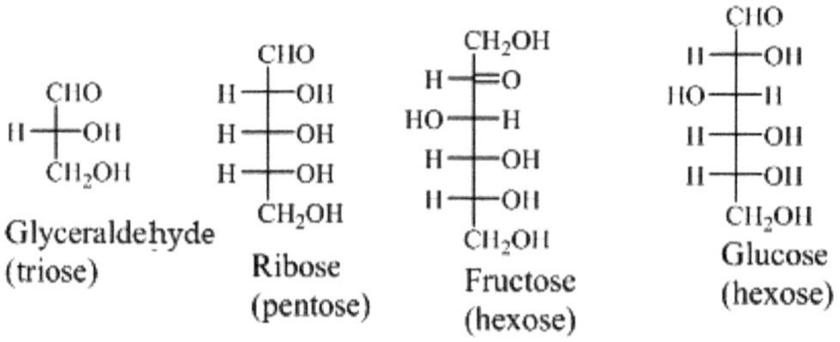

FIGURE 14.2 Linear structures of glyceraldehyde, ribose, fructose, and glucose.

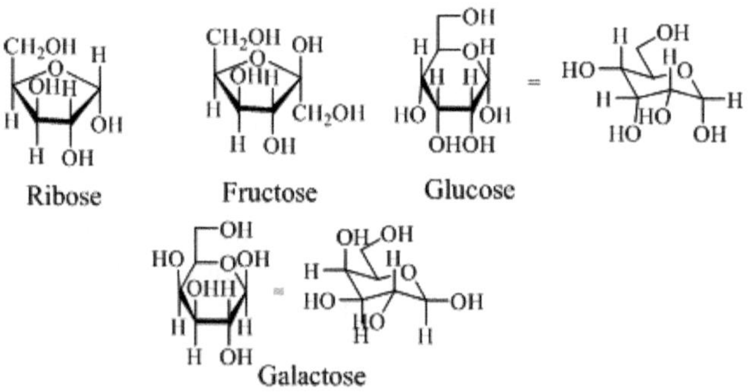

FIGURE 14.3 Cyclic structures of ribose, fructose, glucose, and galactose.

monosaccharide units. Unlike mono- and disaccharides, polysaccharides are not sweet and are generally insoluble in water. However, they play a crucial role in the structure and function of living organisms [15]. Polysaccharides are highly diverse in their structure and function; some of the most biologically important include starch, glycogen, and cellulose, all composed of repetitive glucose units. Despite having the same composition, however, these polysaccharides differ in their structure and orientation of glycosidic linkages [16].

Cellulose, for example, is a linear polymer of glucose molecules and is a major structural component of plant cell walls; it provides rigidity and strength to the cell walls, allowing plants to maintain shape and withstand external pressures. Glycogen and starch are both branched polymers of glucose; glycogen is the primary energy-storage molecule in animals and bacteria, while plants store energy in starch. The orientation of the glycosidic linkages in these two polysaccharides is different, resulting in distinct properties of the linear and branched macromolecules. Figure 14.5 demonstrates the structure of two polysaccharide carbohydrates: cellulose and starch. Cellulose and starch have similar structures and only differ in the type of 1,4 linkage present. Specifically, in cellulose, glucose units are connected by a beta acetal linkage, while starch contains an alpha linkage for this acetal moiety. These linkages are stereoisomers, resulting in distinct structural differences between the two polysaccharides [16].

14.3 BIOSYNTHESIS OF CARBOHYDRATES

Plants use a complicated process called photosynthesis to produce carbohydrates, which serve as their primary energy source. This process involves two distinct pathways that combine to convert light into chemical energy. The first pathway is called the light reaction, which absorbs

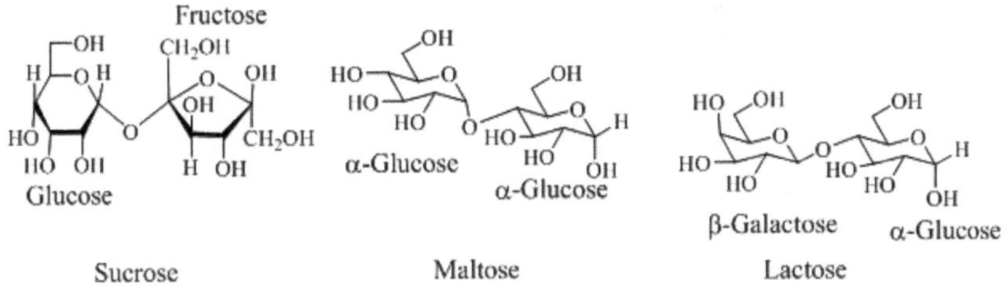

FIGURE 14.4 Structures of sucrose, maltose, and lactose.

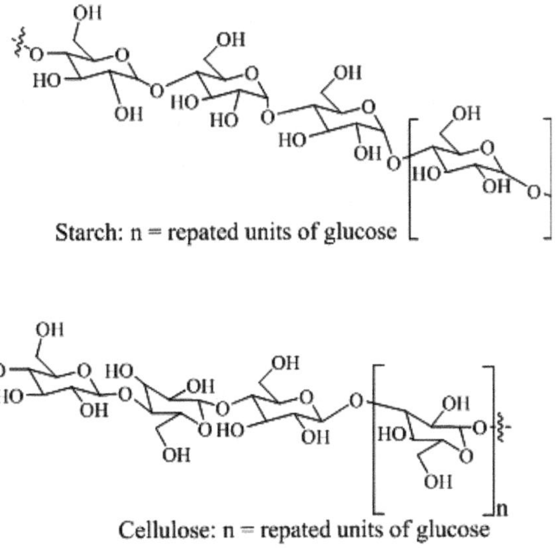

FIGURE 14.5 Structures of starch and cellulose.

electromagnetic energy from the sun and converts it into chemical energy in the form of ATP and NADPH. The second pathway is called the enzymatic reaction, which uses the energy produced by the light reaction to convert carbon dioxide into sugar through complex biochemical reactions. Essentially, photosynthesis is the process by which plants utilize the power of the sun to sustain life and produce the necessary nutrients that support our ecosystem. This process is summarized in the following equation [17]:

$$xCO_2 + xH_2O \xrightarrow[\text{Green Plant}]{\text{Sun Light}} \text{Carbohydrates (e.g glucose;} C_6H_{12}O_6) + xO_2$$

While this equation offers a broad understanding of the connections between reactants and products, delving into the chemical intermediates that drive the process is crucial. Further investigation is necessary to gain a complete understanding of how carbon dioxide is assimilated into an organic compound and converted into sugars while also replenishing the carbon dioxide acceptor. The elucidation of these reactions is a remarkable accomplishment in biosynthetic research.

14.3.1 Biosynthesis of Fructose and Glucose

Glucose is an essential energy source for animals, and it is created through a metabolic process called gluconeogenesis. This process involves producing sugars, such as glucose, from non-carbohydrate sources like lactate, glycerol, and amino acids to fuel reactions that break down molecules (catabolic reactions [18].

It's important to note that glucose is the primary energy source for several vital organs, including the brain, testes, erythrocytes, and kidney medulla, except during fasting, when ketone bodies are used. In mammals, the liver and kidneys are the primary organs responsible for gluconeogenesis. The liver generates glucose from lactate, glycerol, and amino acids, while the kidneys produce glucose from amino acids and lactate. Organisms have developed sophisticated mechanisms to regulate their metabolic pathways, balancing the energy required for survival with the need to conserve resources [18].

Seven out of ten steps occur at or near equilibrium in the glycolysis and gluconeogenesis pathways. However, the conversions of pyruvate to phosphoenolpyruvate (PEP), fructose-1,6-bP to fructose-6-P, and glucose-6-P to glucose are highly regulated due to their spontaneity; these reactions are carefully managed to ensure that the organism conserves energy wherever possible. When excess power is available, gluconeogenesis is suppressed, but when energy is needed, gluconeogenesis is activated. The conversion of pyruvate to PEP is controlled by acetyl-CoA. When the tricarboxylic acid cycle produces an abundance of acetyl-CoA, pyruvate carboxylase is activated, redirecting pyruvate away from the cycle. This diversion of metabolites helps the organism store energy for later use [18].

The conversion of fructose-1,6-bP to fructose-6-P is negatively regulated and inhibited by AMP and fructose-2,6-bP. These molecules act as reciprocal regulators to glycolysis' phosphofructokinase, which is positively regulated by AMP and fructose-2,6-bP. When energy levels exceed the organism's needs, producing a high ATP-to-AMP ratio, gluconeogenesis increases and glycolysis decreases; conversely, when energy levels are insufficient (i.e., low ATP-to-AMP ratio), glycolysis increases and gluconeogenesis decreases. The conversion of glucose-6-P to glucose is controlled by substrate-level regulation; glucose-6-P is the critical metabolite responsible for this type of regulation. As glucose-6-P levels rise, so does glucose-6-phosphatase activity, producing more glucose. This process halts glycolysis, conserving energy [18]. Gluconeogenesis is visually outlined in Scheme 14.1.

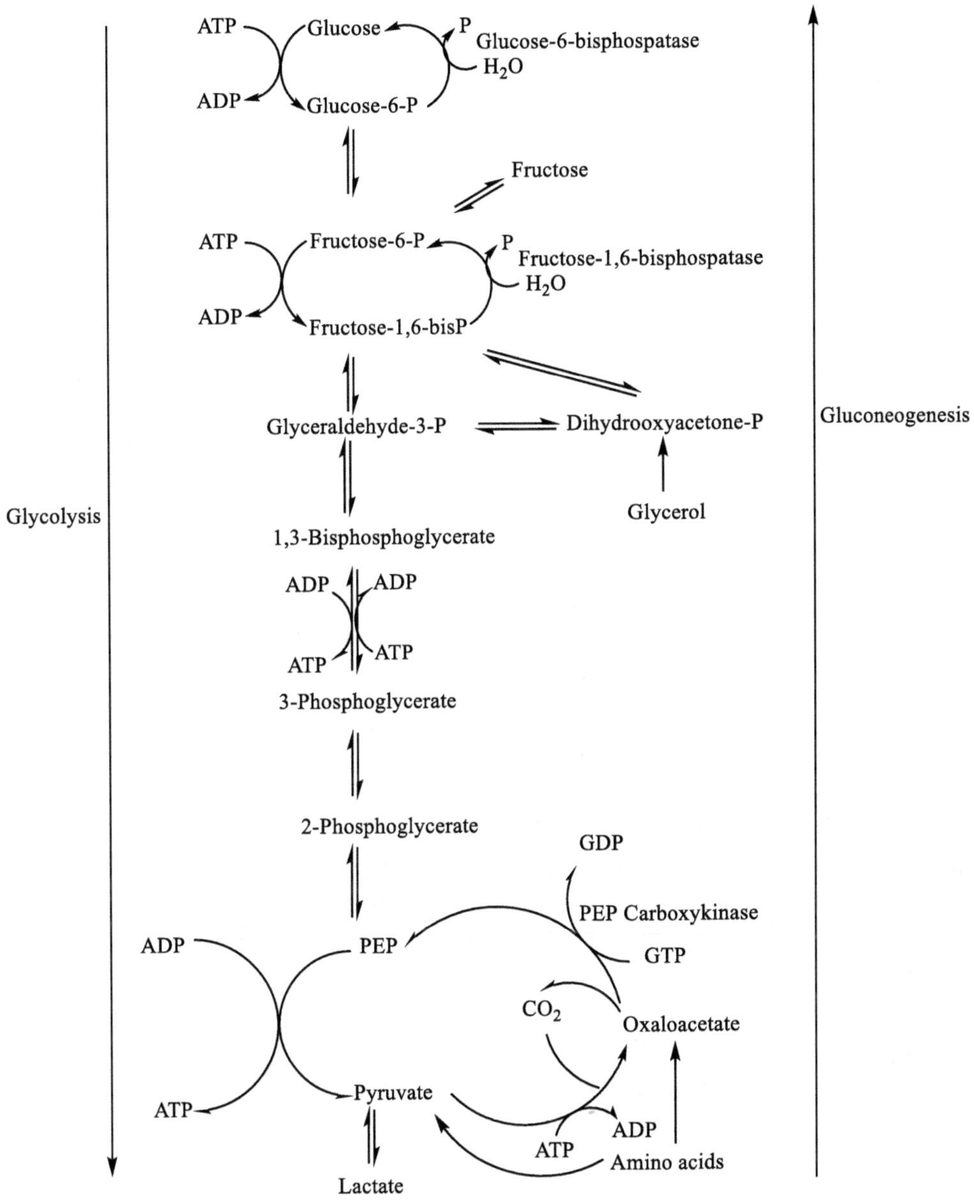

ATP Glucose P Glucose-6-bisphospatase
H_2O
ADP Glucose-6-P
Fructose
ATP Fructose-6-P P Fructose-1,6-bisphospatase
H_2O
ADP Fructose-1,6-bisP
Glyceraldehyde-3-P Dihydrooxyacetone-P Gluconeogenesis
Glycolysis Glycerol
1,3-Bisphosphoglycerate
ADP ADP
ATP ATP
3-Phosphoglycerate
2-Phosphoglycerate
GDP
PEP Carboxykinase
ADP PEP GTP
CO_2 Oxaloacetate
ATP Pyruvate
ADP
ATP Amino acids
Lactate

SCHEME 14.1

14.3.2 Biosynthesis of Sucrose

Sucrose is a crucial component in the metabolic processes of higher plants. It is a sugar produced during photosynthesis and is the primary transportation material of energy in plants. It is also a precursor for the synthesis of polysaccharides. During the process, fructose 6-phosphate is converted to glucose 1-phosphate. Glucose 1-phosphate then undergoes a reaction with uridine-5′-triphosphate to produce uridine diphosphate glucose (UDP-glucose). This UDP-glucose can react with either fructose 6-phosphate to form sucrose phosphate that is then converted to sucrose [19] or react directly with fructose to create sucrose (Scheme 14.2).

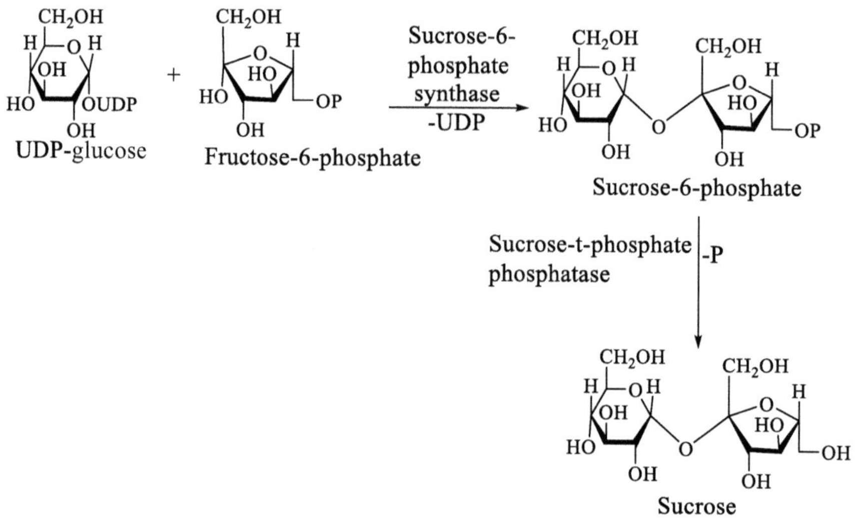

SCHEME 14.2

Once formed, the free sucrose can remain in place or be transported through sieve tubes to different plant parts. Additionally, multiple reactions such as hydrolysis by invertase and reversal of the synthetic sequence can convert sucrose to monosaccharides. These monosaccharides can then be used to make other oligosaccharides or polysaccharides. Overall, the production of sucrose in higher plants plays an important role in their growth and development, making it an essential metabolite for their survival [19].

14.3.3 Biosynthesis of Xylose and Arabinose

Xylose and arabinose are formed through the conversion of UDP-glucose into UDP-glucuronic acid. The decarboxylation of UDP-glucuronic acid produces UDP-xylose, which upon hydrolysis forms xylose. In nature, UDP-xylose is epimerized to produce arabinose [20] (Scheme 14.3).

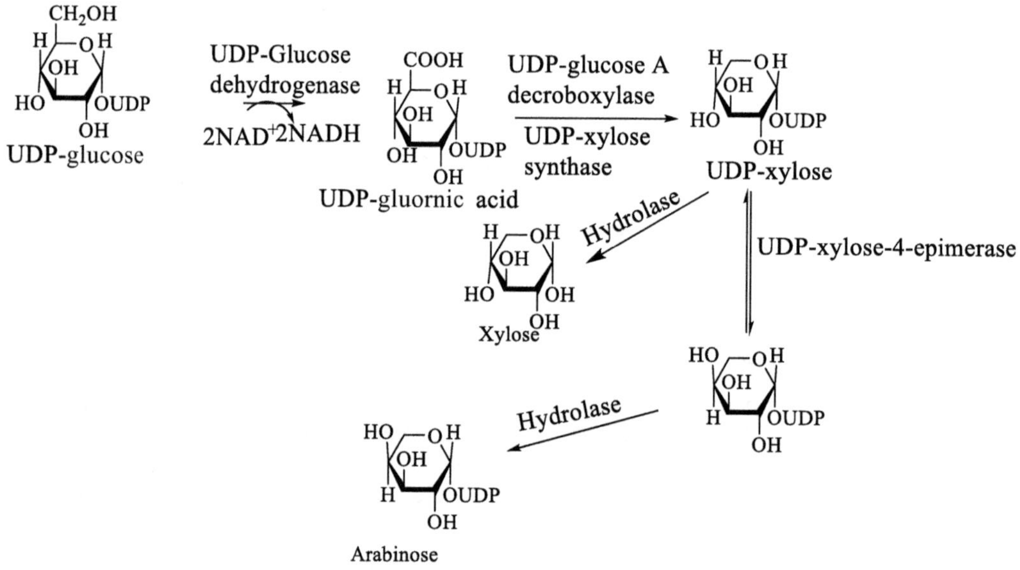

SCHEME 14.3

14.3.4 Biosynthesis of Glycosylated Natural Products

Bioactive natural products often contain basic molecular structures such as terpenes, steroids, flavonoids, and alkaloids, typically attached to various sugars; these sugars are responsible for the specific bioactivity of the compounds. Interestingly, the sugar moiety can increase the hydrophilicity of compounds, making them more soluble and enabling them to pass through cell membranes more easily, thus increasing their effectiveness and usefulness in various biomedical applications. For instance, pseudopterosins such as pseudopterosin A are diterpenes with a sugar moiety derived from the soft coral *Pseudopterogorgiaelisabethae.* These compounds exhibit antimicrobial and anti-inflammatory activities [21, 22].

Similarly, N-β-D-glucopyranosyl-*p*-hydroxyphenylacetamide, isolated from the medicinally important plant *Drypetesgossweileri*, exhibited anti-enzymatic activities beneficial to human health [23]. Camelliasaponin B is a triterpenoid saponin, and camelliasides A and B are flavonoid glycosides; all are found in the plant *Camelliaoleifera* and demonstrated antifungal activity against *Rhizoctoniasolani* [24, 25]. The structures of compounds 1–5 are displayed in Figure 14.6.

Diterpenes are a diverse group of organic compounds biosynthesized from geranylgeranyl pyrophosphate (GGPP), a molecule derived from the mevalonate pathway. This process involves a series of enzymatic reactions that convert GGPP into a diterpene skeleton, such as compound 6. The biosynthesis of diterpenes is highly complex and consists of various enzymes, including terpene synthases, cytochrome P450s, and prenyltransferases. Once the diterpene skeleton has been synthesized, it undergoes further modifications such as enzymatic glycosylation [26], which produces glycosylated diterpenes. One example of this process is the biosynthesis of pseudopterosin A (**1**), which is illustrated in Scheme 14.4.

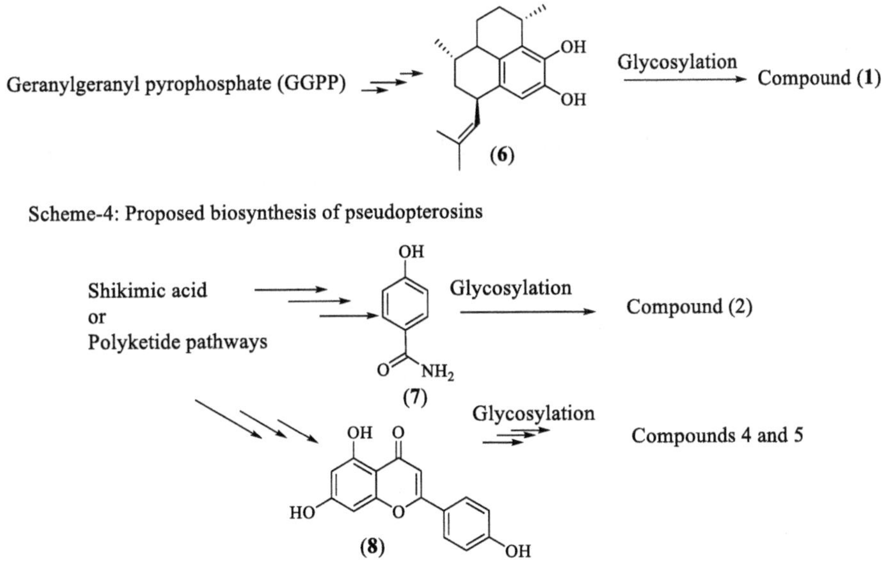

Geranylgeranyl pyrophosphate (GGPP) → (6) —Glycosylation→ Compound (**1**)

Scheme-4: Proposed biosynthesis of pseudopterosins

Shikimic acid or Polyketide pathways → (7) —Glycosylation→ Compound (2)

→ (8) —Glycosylation→ Compounds 4 and 5

SCHEME 14.4

Aromatic compounds, a class of organic molecules with a unique ring-shaped structure, have various applications across various fields. These compounds are highly valued for their distinct odors and flavors and are commonly used in foods and fragrances. Additionally, they are used as dyes in the textile industry and as neurotransmitters in the medical field. Furthermore, aromatic compounds are widely used as building blocks in the chemical industry. These compounds can be

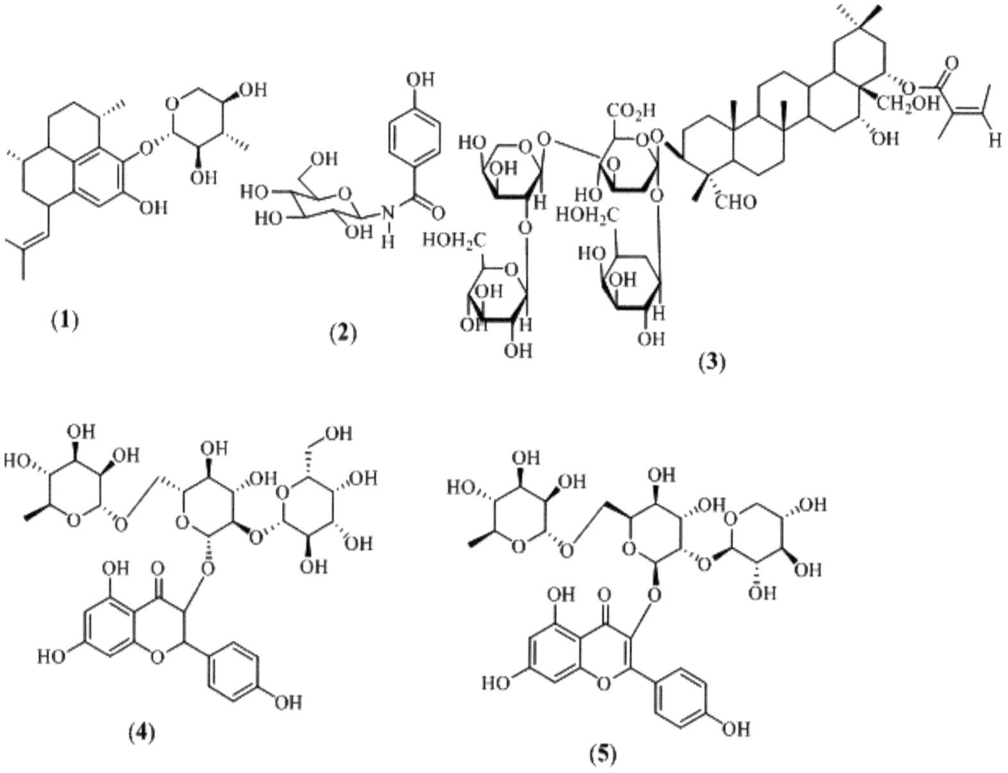

FIGURE 14.6 Structures of compounds 1–5.

precursors to synthesize various other chemicals, including pharmaceuticals, plastics, and synthetic fibers. In fact, it is estimated that over 40% of all bulk chemicals comprise an aromatic moiety [27].

The production of aromatic compounds involves the shikimic acid or polyketide pathways. The latter pathway forms a core aromatic skeleton through head-to-tail condensation of polyketide precursors [28, 29]. Once an aromatic skeleton such as acetamide (**7**), flavonoids (**8**), or any other aromatic compound is biosynthesized, it undergoes reactions to produce glycosylated bioactive natural products. In Scheme 14.5, we graphically display the biosynthesis of compounds (**2**), (**4**), and (**5**). This process involves intricate biochemical pathways and enzymatic reactions that ultimately form complex molecular structures. These pathways are followed by all glycosylated aromatic natural products.

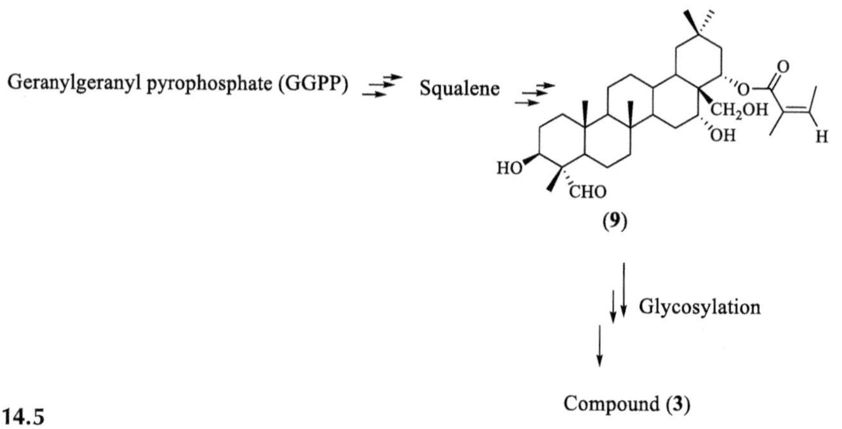

Geranylgeranyl pyrophosphate (GGPP) → Squalene → (**9**) → Glycosylation → Compound (**3**)

SCHEME 14.5

The production of saponins, biologically active compounds found in many plants, is a complex biosynthetic process that begins with converting acetyl-CoA, a small molecule building block for many organic compounds, into squalene via the mevalonate pathway. Squalene, in turn, serves as a precursor for synthesizing steroids and terpenoids, which are important classes of compounds in many biological systems. However, squalene is further transformed into cycloartane aglycone in saponin biosynthesis, which serves as the precursor for synthesizing steroidal and triterpenoid saponins. The transformation of squalene into cycloartane, an important precursor to steroids and triterpenoid, involves a series of enzymatic reactions that modify the molecule's structure [30]. These reactions are mediated by various enzymes including oxidosqualene cyclase and lanosterol synthase that catalyze the formation of specific chemical bonds.

Once aglycone (**8**) has been synthesized, it undergoes a final glycosylation, a process in which sugar molecules are attached to the aglycone backbone. This step is catalyzed by glycosyltransferases, which add the sugar molecules in a specific sequence and pattern to produce the final saponins [31]. Scheme 14.5 illustrated this process.

In this chapter, we explored the classification and biosynthesis of different sugars and glycosylated natural products. By understanding their biosynthesis comprehensively, we can identify the enzymes involved in their production in the natural world. This enzymology knowledge can be further leveraged to aid in purifying these critical enzymes, which can be utilized in organic synthesis to create essential bioactive saponins and other carbohydrates of immense industrial value. Through this process, we can unlock the full potential of these natural products, paving the way for new discoveries and innovations in organic synthesis.

ACKNOWLEDGEMENTS

Athar Ata's (AA) research at the University of Winnipeg was supported by the Natural Sciences and Engineering Research Council of Canada. This funding helped AA to host Simran Sandhu (SS) as a graduate student and Ikechukwu P. Ejidike (IPE) as a postdoctoral fellow in his research group at the University of Winnipeg, Winnipeg, Manitoba, Canada. AA moved to Michigan Technological University (MTU), Houghton, MI, in August 2023. AA is also grateful to MTU for funding. All authors (SS, SN, and IPE) contributed equally to this chapter.

REFERENCES

[1] Chandel, N. S. 2021. Carbohydrate Metabolism. *Cold Spring Harb Perspect Biol* 13(1): a040568.
[2] Shah, R., Sabir, S., Alhawaj, A. F. Sep. 19, 2022. *Physiology, Breast Milk*. StatPearls Publishing; Treasure Island (FL). [PubMed]
[3] Bolla, A. M., Caretto, A., Laurenzi, A., Scavini, M., Piemonti, L. 2019. Low-Carb and Ketogenic Diets in Type 1 and Type 2 Diabetes. *Nutrients* 11(5): 962.
[4] Mills, S., Stanton, C., Lane, J. A., Smith, G. J., Ross, R. P. April 24, 2019. Precision Nutrition and the Microbiome, Part I: Current State of the Science. *Nutrients* 11(4): 923.
[5] Nawale, R. B., Mourya, V. K., Bhise, S. B. 2006. Non-enzymatic Glycation of Proteins: A Cause for Complications in Diabetes. *Indian J Biochem Biophys* 43(6): 337–344.
[6] Gurung, P., Zubair, M., Jialal, I. Jan. 18, 2023. *Plasma Glucose – StatPearls* [Internet]. StatPearls Publishing; Treasure Island (FL).
[7] Daghlas, S. A., Mohiuddin, S. S. May 1, 2023. *Biochemistry, Glycogen – StatPearls* [Internet]. StatPearls Publishing; Treasure Island (FL). [PubMed]
[8] Holesh, J. E., Aslam, S., Martin, A. May 12, 2023. *Physiology, Carbohydrates – StatPearls* [Internet]. StatPearls Publishing; Treasure Island (FL). [PubMed]
[9] Mozaffarian, R. S., Lee, R. M., Kennedy, M. A., Ludwig, D. S., Mozaffarian, D., Gortmaker, S. L. 2013. Identifying Whole Grain Foods: A Comparison of Different Approaches for Selecting More Healthful Whole Grain Products. *Public Health Nutr* 16: 2255–264.

[10] Asp, N.-G. 1995. Classification and Methodology of Food Carbohydrates as Related to Nutritional Effects. *Am J Clin Nutr* 61(Suppl 4): S980–S987.
[11] Champ, M., Langkilde, A. M., Brouns, F., Kettlitz, B., Le Bail-Collet, Y. 2003. Advances in Dietary Fiber Characterization. 2. Consumption, Chemistry, Physiology and Measurement of Resistant Starch; Implications for Health and Food Labelling. *Nutr Res Rev* 16: 143–161.
[12] Malik, V. S., Hu, F. B. 2015. Fructose and Cardiometabolic Health: What the Evidence From Sugar-Sweetened Beverages Tells Us. *J Am Coll Cardiol* 66(14): 1615–1624.
[13] Chenghua, S., Zukang, F., John, D. W., Ezra, Peisach, J. B., Yasuyo, I., Genji Kurisu, S. V., Stephen, K. B., Jasmine, Y. Y. 2021. Modernized Uniform Representation of Carbohydrate Molecules in the Protein Data Bank. *Glycobiology* 31(9): 1204–1218.
[14] Haynes, W. M., Lide, D. R., Bruno, T. J., eds. 2014–2015. *CRC Handbook of Chemistry and Physics* (95th ed.). CRC Press, pp. 5–40.
[15] May, F., Weinland, H. 1953. Glycogen Formation in the Galactogen-Containing Eggs of Helix pomatia During Embryonal Period. *Zeitschrift für Biologie* 105(5): 339–347.
[16] Mischnick, P., Momcilovic, D. 2010. Chemical Structure Analysis of Starch and Cellulose Derivatives. *Adv Carbohydr Chem Biochem* 64: 117–210.
[17] Calvin, M. 1989. Forty Years of Photosynthesis and Related Activities. *Photosynth Res* 21(1): 3–16.
[18] Atkinson, D. E. 1968. The Energy Charge of the Adenylate Pool as a Regulatory Parameter. Interaction with Feedback Modifiers. *Biochemistry* 7: 4030–4034.
[19] Stein, O., Granot, D. 2019. An Overview of Sucrose Synthases in Plants. *Front Plant Sci* 10: 95.
[20] Postma, P., Lengeler, J., Jacobson, G. 1993. Phosphoenol Pyruvate: Carbohydrate Phosphotransferase Systems of Bacteria. *Microbiol Rev* 57: 543–594.
[21] Look, S. A., Fenical, W., Jacobs, R. S., Clardy, J. 1986. The Pseudopterosins: Anti-Inflammatory and Analgesic Natural Products from the Sea Whip Pseudopterogorgia Elisabethae. *Proc Natl Acad Sci USA* 83(17): 6238–6240.
[22] Ata, A., Win, H. Y., Holt, D., Holloway, P., Segstro, E. P, Jayatilake, G. S. 2004. Antibacterial Diterpenes from Pseudopterogorgia Elisabethae. *Helv Chim Acta* 87: 1090–1098.
[23] Ata, A., Tan, D. S., Matochko, W. L., Adesanwo, J. K. 2011. Chemical Constituents of Drypetes Gossweileri and Their Enzyme Inhibitory and Antifungal Activities. *Phytochem Lett* 4: 34–37.
[24] Luan, F., Zeng, J., Yang, Y., He, X., Wang, B., Gao, Y., Zeng, N. 2020. Recent Advances in Camellia oleifera Abel: A Review of Nutritional Constituents, Biofunctional Properties, and Potential Industrial Applications. *J Funct Foods* 104242.
[25] Sekine, T., Arita, J., Yamaguchi, A., Saito, K., Okonogi, S., Morisaki, N., Iwasaki, S., Murakoshi, I. 1991. Two Flavonol Glycosides from Seeds of Camellia Sinensis. *Phytochemistry* 30(3): 991–995.
[26] Kohl, A. C., Kerr, R. G. 2003. Pseudopterosin Biosynthesis: Aromatization of the Diterpene Cyclase Product, Elisabethatriene. *Mar Drugs* 1(4): 54–65.
[27] El Hadi, M., Zhang, F.-J., Wu, F.-F., Zhou, C.-H., Tao, J. 2013. Advances in Fruit Aroma Volatile Research. *Molecules* 18(7): 8200–8229.
[28] Herrmann, K. M., Weaver, L. M. 1999. The Shikimate Pathway. *Physiol Plant Mol Biol* 50: 473–503.
[29] Hong, F., Hopwood, D. A., Khosla, C. 1993. Engineered Biosynthesis of Novel Polyketides. *Science* 262: 1546–1550.
[30] Noushahi, H. A., Khan, A. H., Noushahi, U. F., Hussain, M., Javed, T., Zafar, M., Batool, M., Ahmed, U., Liu, K., Harrison, M. T., Saud, S., Fahad, S., Shu, S. 2022. Biosynthetic Pathways of Triterpenoids and Strategies to Improve their Biosynthetic Efficiency. *Plant Growth Regul* 97(3): 439–454.
[31] Moses, T., Papadopoulou, K. K., Osbourn, A. 2014. Metabolic and Functional Diversity of Saponins, Biosynthetic Intermediates and Semi-Synthetic Derivatives. *Crit Rev Biochem Mol Biol* 49(6): 439–462.

15 Biosynthesis of Fats, Fatty Acids, and Lipids

Ahindra Nag

15.1 INTRODUCTION

Animals can synthesize and store large quantities of triacylglycerols that are synthesized within various organs such as the liver, kidney, lactating mammary gland, aorta, and intestinal mucosa. They are used later as fuel tǒ supply the body's energy for 12 hours. Plants also manufacture triacylglycerols as an energy-rich fuel, stored especially in fruits, nuts, and seeds.

15.2 BIOSYNTHESIS OF FATS

Fat is an essential part of the body that is an essential source of fatty acids and helps the body to absorb vitamins A, D, and E. The body uses fats to build cell membranes, nerve tissue (including the brain), and hormones and also uses fat as fuel. Biosynthesis of fat initially takes place in the formation of glycerol-3-phosphate in one of two processes. In one process, free-glycerol-containing lipids are phosphorylated by ATP and Mg^{2+} in the presence of enzyme glycerol kinase to form L-glycerol-3-phosphate (Figure 15.1). In the other process (Figure 15.2), glycerol-3-phosphate is obtained by reducing glyceraldehyde-3-phosphate or dihydroxyacetone phosphate produced in the pentose phosphate pathway or by glycolysis. Fatty acids, meanwhile, are activated by the respective thiokinase in the presence of ATP and coenzyme A.

Glycerol-3-phosphate is then esterified with 2 moles of fatty acyl-CoA to form 1, 2-diacyl glycerol phosphate (phosphatidate) (Figure 15.4) via 1-acyl glycerol-3-phosphate (lysophosphatidate). The enzyme (glycerophosphate acyl transferase) for this reaction is not very specific for definite chain length of the activated acids. However, it reacts more swiftly with C_{16}–, C_{17}–, and C_{18} – fatty acids. This fact provides one possible explanation for the predominance of C_{16} – and C_{18} – fatty acids in neutral fats: In that reaction, the common intermediate phosphatidate forms in the synthesis of triacylglycerols (fats) and phosphoglycerides.

In the next step, the phosphatidate is dephosphorylated with a phosphatase and then esterified with another mole of acyl-CoA to form the neutral fat triacylglycerol (Figure 15.5).

$$\begin{array}{c} CH_2OH \\ | \\ CHOH \\ | \\ CH_2OH \end{array} \xrightarrow[\text{ATP, } Mg^{2+}]{\text{Glycerol kinase}} \begin{array}{c} CH_2OH \\ | \\ CHOH \\ | \\ CH_2OP \end{array}$$

Glycerol → L-Glycerol-3-Phosphate

FIGURE 15.1 Biosynthesis of L-glycerol-3-phosphate.

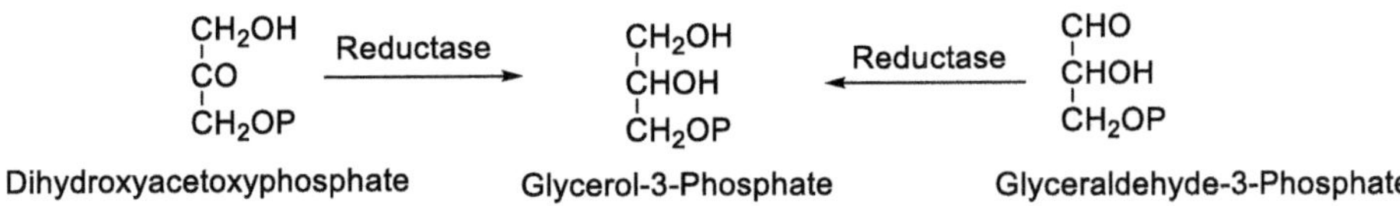

FIGURE 15.2 Biosynthesis of glycerol-3-phosphate.

DOI: 10.1201/9781003437413-15

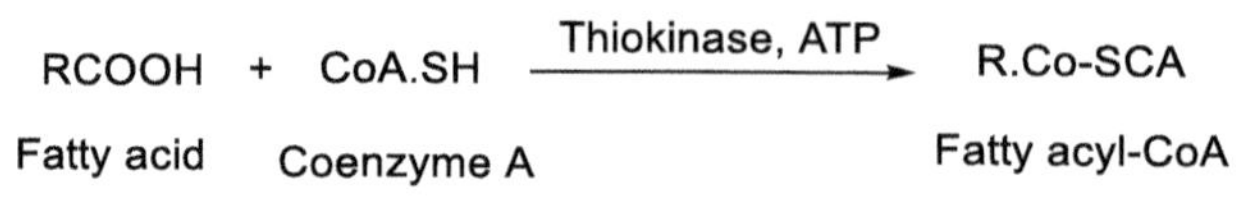

$$\underset{\text{Fatty acid}}{RCOOH} + \underset{\text{Coenzyme A}}{CoA.SH} \xrightarrow{\text{Thiokinase, ATP}} \underset{\text{Fatty acyl-CoA}}{R.Co\text{-}SCA}$$

FIGURE 15.3 Activation of fatty acyl-CoA.

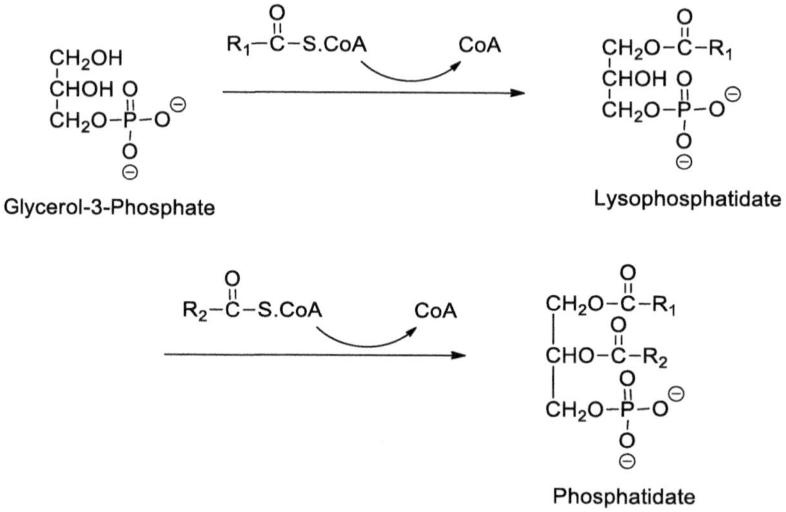

Glycerol-3-Phosphate + $R_1-C(=O)-S.CoA$ → Lysophosphatidate + CoA

Lysophosphatidate + $R_2-C(=O)-S.CoA$ → Phosphatidate + CoA

FIGURE 15.4 Biosynthesis of phosphatidate.

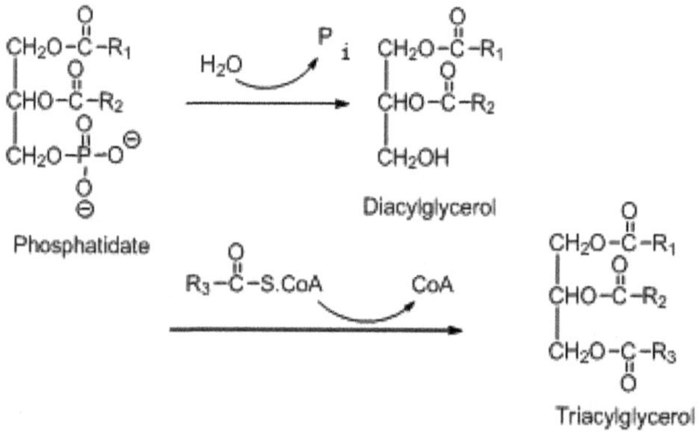

Phosphatidate + H_2O → Diacylglycerol + P_i

Diacylglycerol + $R_3-C(=O)-S.CoA$ → Triacylglycerol + CoA

FIGURE 15.5 Biosynthesis of triglycerol.

15.3 BIOSYNTHESIS OF MALONYL CO-A

The major source of fatty acids in mammals is carbohydrates predominately in the liver, adipose tissue as well as mammary glands. The biosynthesis of fatty acids takes place in two steps First, carbohydrates degrade into pyruvic acid by glycolysis as an important intermediary in the conversion of fatty acid into acetyl coenzyme A (acetyl-CoA) by oxidative decarboxylation under the influence of thiamine pyrophosphate (lipolate, NAD^{+}·coenzyme A, and Mg^{++}ions). Then, the acetyl-CoA is carboxylated in the presence of NADPH, ATP, Mn^{++}, and HCO_3^- (source of CO_2) to form malonyl-CoA, in which the reaction is catalyzed by the biotin-containing enzyme as acetyl-CoA carboxylase (Figure 15.6).

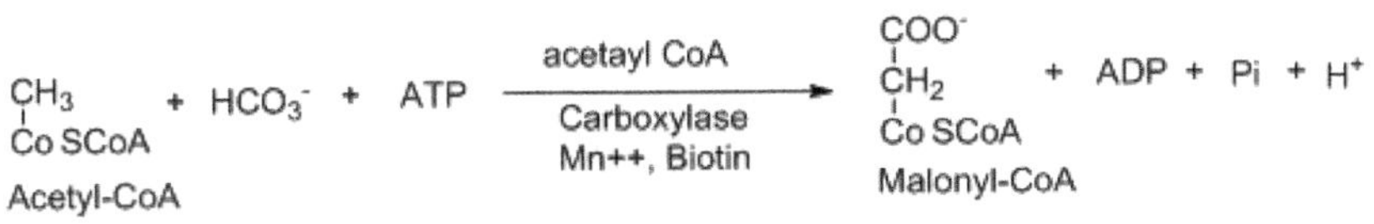

$$CH_3\text{-}CoSCoA\ (\text{Acetyl-CoA}) + HCO_3^- + ATP \xrightarrow[\text{Mn++, Biotin}]{\text{acetayl CoA Carboxylase}} COO^-\text{-}CH_2\text{-}CoSCoA\ (\text{Malonyl-CoA}) + ADP + Pi + H^+$$

FIGURE 15.6 Biosynthesis of malonyl Co-A.

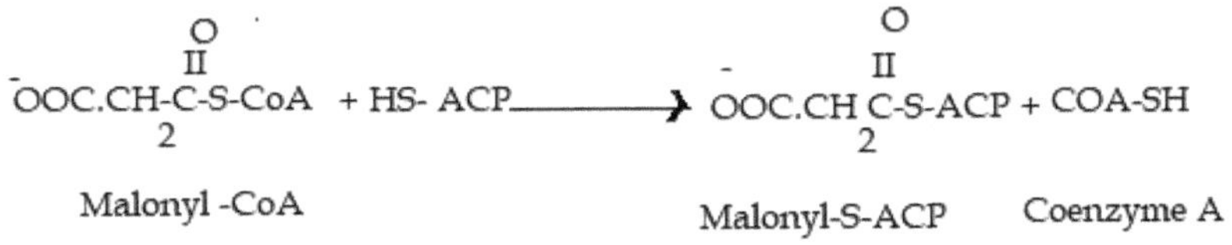

$$^-OOC.CH_2\text{-}\overset{O}{\overset{\|}{C}}\text{-S-CoA}\ (\text{Malonyl -CoA}) + \text{HS- ACP} \rightarrow {}^-OOC.CH_2\ \overset{O}{\overset{\|}{C}}\text{-S-ACP}\ (\text{Malonyl-S-ACP}) + \text{COA-SH}\ (\text{Coenzyme A})$$

FIGURE 15.7 Biosynthesis of malonyl S-ACP.

$$CH_3\text{-C-S-synthase}\ (\text{Acetyl synthase}) + {}^-OOC.CH_2\ \overset{O}{\overset{\|}{C}}\text{-S-ACP -}\ (\text{Malonyl-S-ACP})$$

$$\downarrow \text{Condensation} \rightarrow \text{Synthase-SH} + CO_2$$

$$H_3C\text{-}\overset{O}{\overset{\|}{C}}\text{-CH - }\overset{O}{\overset{\|}{C}}\text{- S-ACP}\ (\text{Acetoacyl-ACP})$$

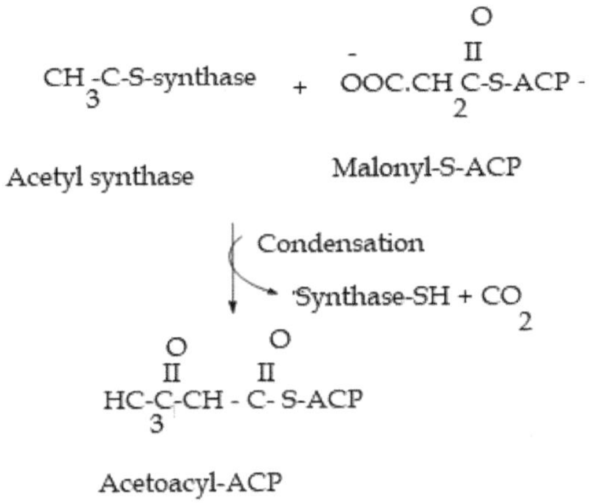

FIGURE 15.8 Biosynthesis of acetoacyl-ACP.

15.4 BIOSYNTHESIS OF MALONYL S-ACP

The reaction sequence for fatty acid synthesis begins with the combination of acetyl-CoA molecule with the –SH group of cysteinyl residue of ketoacyl synthase of monomer 1. Coenzyme A is removed by *transacylase*, which catalyzes the reaction. This step is called acyl transfer. Malonyl-CoA combines with the adjacent –SH group of ACP to form malonyl-ACP-enzyme (monomer 2). Here, also, coenzyme A of malonyl-CoA is removed, and the reaction is catalyzed by the same transacylase. This step is called malonyl transfer (Figure 15.6).

In the condensation step, the acetyl-S-synthase condenses from monomer 1 with the malonyl-S-ACP of monomer 2 to form acetoacyl-S-ACP (Figure 15.8). CO_2 is released, and the reaction is catalyzed by ketoacyl synthase.

In this step, the acetoacetyl-ACP is reduced to D-hydroxy butyryl-ACP (Figure 15.9). Here, NADPH, a reducing agent generated in energy-producing reactions, is consumed in biosynthetic reactions.

The next step is dehydration; D-3-hydroxybutyryl-ACP is dehydrated to form crotonyl-S-ACP, which is trans-Δ^2-enoyl-ACP (Figure 15.10).

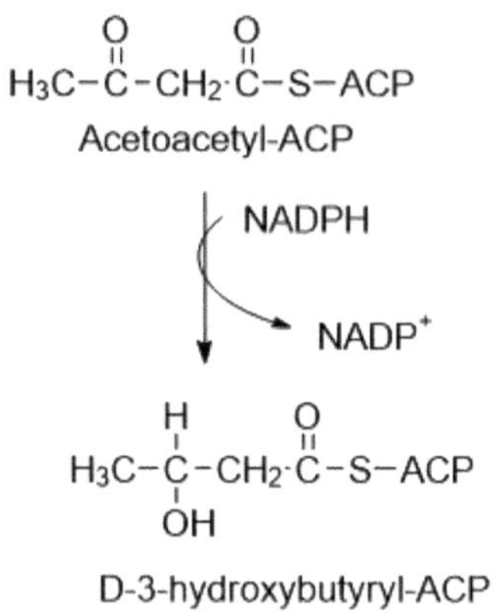

$H_3C-C(=O)-CH_2-C(=O)-S-ACP$

Acetoacetyl-ACP

NADPH

$NADP^+$

$H_3C-CH(OH)-CH_2-C(=O)-S-ACP$

D-3-hydroxybutyryl-ACP

FIGURE 15.9 Biosynthesis of D-3-hydroxylbutyryl-ACP.

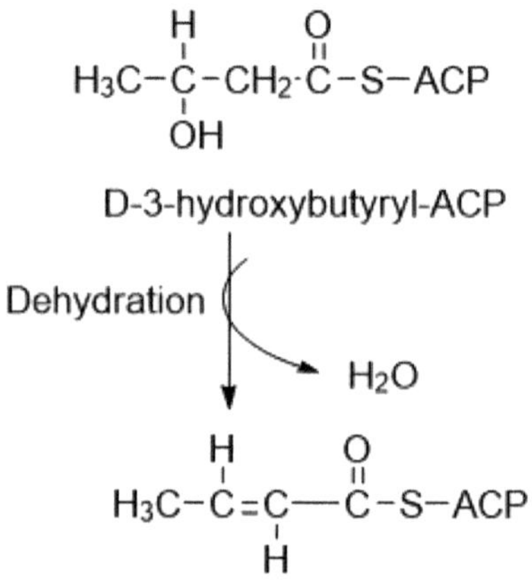

$H_3C-CH(OH)-CH_2-C(=O)-S-ACP$

D-3-hydroxybutyryl-ACP

Dehydration

H_2O

$H_3C-CH=CH-C(=O)-S-ACP$

FIGURE 15.10 Biosynthesis of crotonyl-S-ACP.

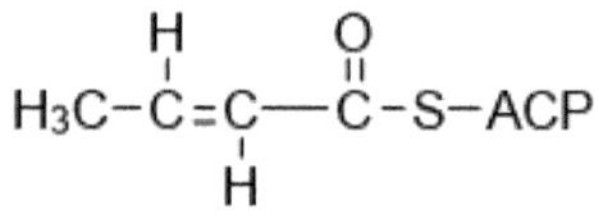

$H_3C-CH=CH-C(=O)-S-ACP$

FIGURE 15.11 Biosynthesis of butyryl-S-ACP.

In the reduction step, crotonyl-S-ACP is reduced to form butyryl-ACP; NADPH is used as the reducing agent (Figure 15.11).

Butyryl-S-ACP converts to palmitoyl-S-ACP in the following reaction routes: Butyryl-S-ACP condenses with malonyl-ACP to form C_6-β-ketoacyl-ACP, similar to the condensation of acetyl-ACP with malonyl-ACP to form C_4-β-ketoacyl-ACP. Reduction, dehydration, and a second reduction convert C_6-β-ketoacyl-ACP into C_6-β-acyl-ACP for the third round of elongation, and elongation continues until palmityl-ACP (C_{16}-acyl=ACP) is formed. In the next step, the palmitoyl group is completely removed by *deacylase* to form palmitic acid (enzymatic complex). This step continues to form unsaturated fatty acid (Figure 15.12).

$$CH_3(CH_2)_{14}..CO\text{-}SCOA \xrightarrow{\text{Deacylase}} CH_3(CH_2)_{14}COOH + \text{Enzyme}$$

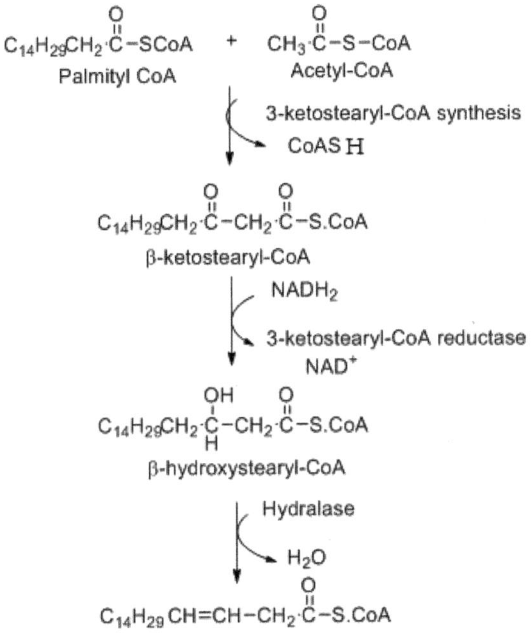

$$C_{14}H_{29}CH_2{\cdot}\overset{O}{\overset{\|}{C}}-SCoA + CH_3{\cdot}\overset{O}{\overset{\|}{C}}-S-CoA$$

Palmityl CoA Acetyl-CoA

3-ketostearyl-CoA synthesis

CoAS H

$$C_{14}H_{29}CH_2{\cdot}\overset{O}{\overset{\|}{C}}-CH_2{\cdot}\overset{O}{\overset{\|}{C}}-S.CoA$$

β-ketostearyl-CoA

$NADH_2$

3-ketostearyl-CoA reductase

NAD^+

$$C_{14}H_{29}CH_2{\cdot}\underset{H}{\overset{OH}{C}}-CH_2{\cdot}\overset{O}{\overset{\|}{C}}-S.CoA$$

β-hydroxystearyl-CoA

Hydralase

H_2O

$$C_{14}H_{29}CH{=}CH-CH_2{\cdot}\overset{O}{\overset{\|}{C}}-S.CoA$$

FIGURE 15.12 Biosynthesis of Δ^2-unsaturated stearyl-S-COA.

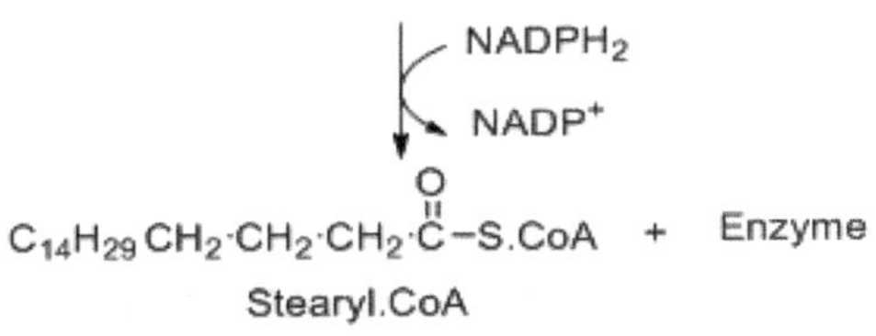

$NADPH_2$

$NADP^+$

$$C_{14}H_{29}CH_2{\cdot}CH_2{\cdot}CH_2{\cdot}\overset{O}{\overset{\|}{C}}-S.CoA + \text{Enzyme}$$

Stearyl.CoA

FIGURE 15.13 Biosynthesis of steryl CoA.

The lengthening of fatty acids in the mitochondria of fractions of the cells forms steryl CoA (Figure 15.13).

15.5 BIOSYNTHESIS OF LIPIDS

Lipids are essential structural constituents of cellular membranes, and they play roles in cell signaling and in the generation of bioactive metabolites. Lipids regulate many physiological functions of cells, and alterations in membrane lipids act as chemical messenger and hormone regulation. Lipids transport fat-soluble nutrients and also help the body absorb fat-soluble vitamins like A, D, and E. Lipids help regulate body temperature. Lipids help form prostaglandin, which plays a role in inflammation. Lipid metabolism are associated with major diseases including cancer, type 2 diabetes, cardiovascular disease, and immune disorders. The endoplasmic reticulum is the main site for lipid synthesis and for their hydrophobic nature, most lipids can be effectively transformed by free diffusion from one compartment to another and rely on an active mechanism to facilitate intercompartmental transport.

15.5.1 Lecithins

Lecithins are composed of phosphoric acid, glycerol, cholines, ester of glycerol, and two fatty acids. Chain length, position, and degree of FFA unsaturation vary to produce different lecithins that show different biological functions. Lecithin biosynthesis starts from serine (amino acid) derived from pathways of carbohydrate and protein metabolism. Serine, through stepwise decarboxylation and methylation, forms choline (Figure 15.14).

Next, serine undergoes stepwise reactions in the presence of the enzyme choline phosphokinase. Subsequent action by enzyme cytidyl transferase then forms CDP-choline (CDP-ethanolamine). Then CDP-choline reacts with diglyceride and transfers phosphocholine onto the free alcoholic group of the diglyceride. Cytidine monophosphate is released and rephosphorylated to cytidine triphosphate by ATP to lecithin (Figure 15.15).

15.5.2 Cephalin

Biosynthesis of cephalin takes place in a similar way to lecithin, but the starting material is ethanolamine (Figure 15.16).

Cephalin converts into lecithin with the removal of three methyl groups from the cephalin (Figure 15.17).

15.5.3 Plasmalogenes

Plasmalogenes are considered phosphatidal rather than phosphatidyl derivatives in lecithins and cephalins. Kiyseu and Kennady demonstrated plasmalogenic diglyceride biosynthesis in rat liver in 1960. The synthesis is similar to that for lecithins and cephalins: CDP-choline or CDP-ethanolamne condenses with an alkenyl glycerol ester. In a plasmogenic diglyceride, the α or β position has an aldehydogenic group (α, β unsaturated ether) in place of fatty acid. The aldohydogenic group ($CH_2COCH{=}CHR''$) derived from the reduction of the fatty acid-CoA (Figure 15.18).

15.5.4 Spingnomyelins

This lipid contains a fatty acid, phosphoric acid, Choline, and a complex amino alcohol (sphingosine) instead of glycerol. Spingnomyelin biosynthesis takes place as follows: The amino acid is activated by combining with pyridoxal phosphate in the presence of Mn^{+2}, and then it condenses with palmityoyl-CoA to 3-ketodihydrophinogosine that converts to sphingosine (Figure 15.19).

Next, sphinogosine is acylated at the amino acid group by a fatty acyl-CoA to form N-acylsphnogosine (a ceramide), which is a common intermediate of sphingomylelin. Next, in forming sphingomylin, the terminal hydroxyl group of sphingosine is substituted by phosohrocholine derived from CDP-choline (Figure 15.20).

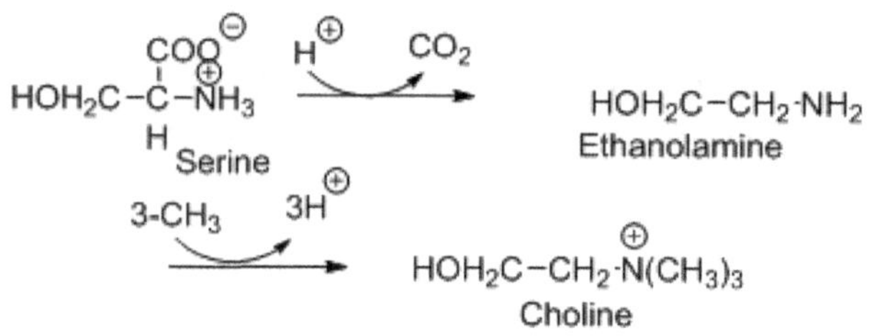

$HOH_2C{-}CH(COO^{\ominus}){-}NH_3^{\oplus}$ (Serine) $\xrightarrow{H^{\oplus},\ -CO_2}$ $HOH_2C{-}CH_2{\cdot}NH_2$ (Ethanolamine)

$\xrightarrow{3\text{-}CH_3,\ -3H^{\oplus}}$ $HOH_2C{-}CH_2{\cdot}N^{\oplus}(CH_3)_3$ (Choline)

FIGURE 15.14 Biosynthesis of choline.

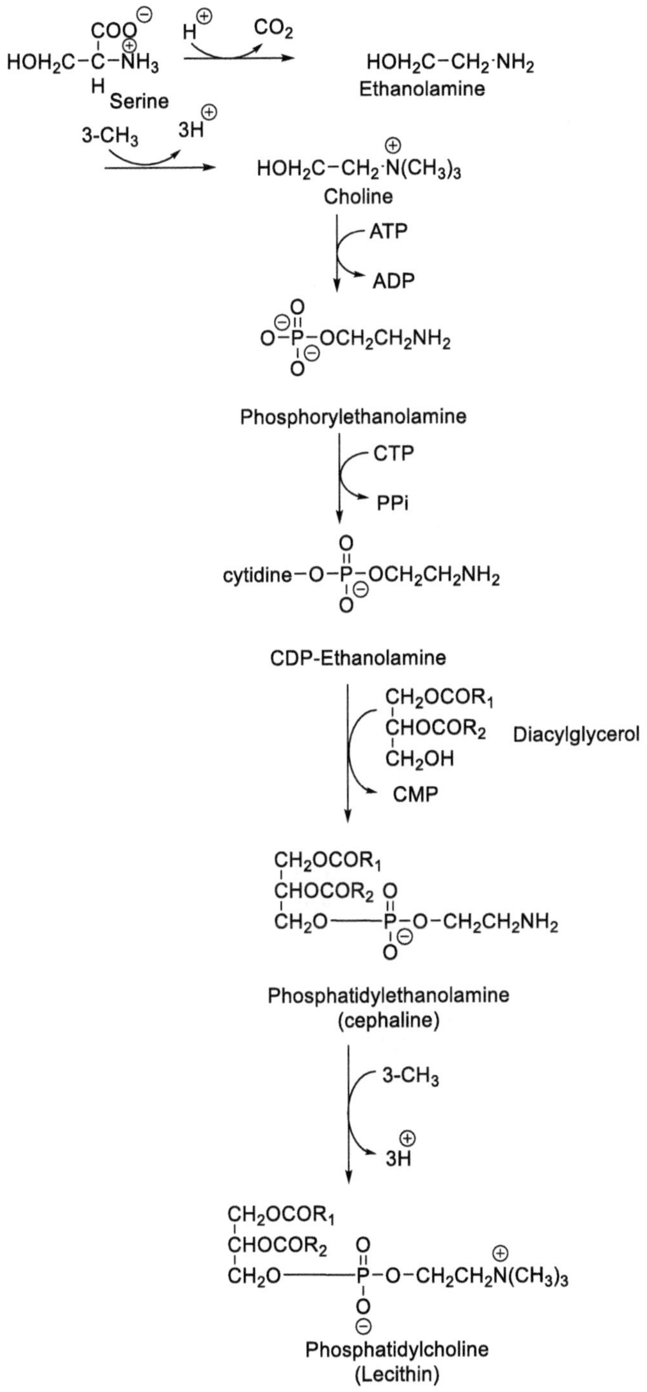

$HOH_2C-CH(COO^{\ominus})-NH_3^{\oplus}$ (Serine) $\xrightarrow{H^{\oplus}\;\;CO_2}$ $HOH_2C-CH_2\cdot NH_2$ (Ethanolamine)

$\xrightarrow{3\text{-}CH_3\;\;3H^{\oplus}}$ $HOH_2C-CH_2\cdot N^{\oplus}(CH_3)_3$ (Choline)

$\downarrow$ ATP → ADP

$^{\ominus}O-P(=O)(O^{\ominus})-OCH_2CH_2NH_2$

Phosphorylethanolamine

$\downarrow$ CTP → PPi

$\text{cytidine}-O-P(=O)(O^{\ominus})-OCH_2CH_2NH_2$

CDP-Ethanolamine

$\downarrow$ $CH_2OCOR_1-CHOCOR_2-CH_2OH$ Diacylglycerol → CMP

$CH_2OCOR_1-CHOCOR_2-CH_2O-P(=O)(O^{\ominus})-O-CH_2CH_2NH_2$

Phosphatidylethanolamine (cephaline)

$\downarrow$ 3-CH_3 → $3H^{\oplus}$

$CH_2OCOR_1-CHOCOR_2-CH_2O-P(=O)(O^{\ominus})-O-CH_2CH_2N^{\oplus}(CH_3)_3$

Phosphatidylcholine (Lecithin)

FIGURE 15.15 Biosynthesis of lecithin.

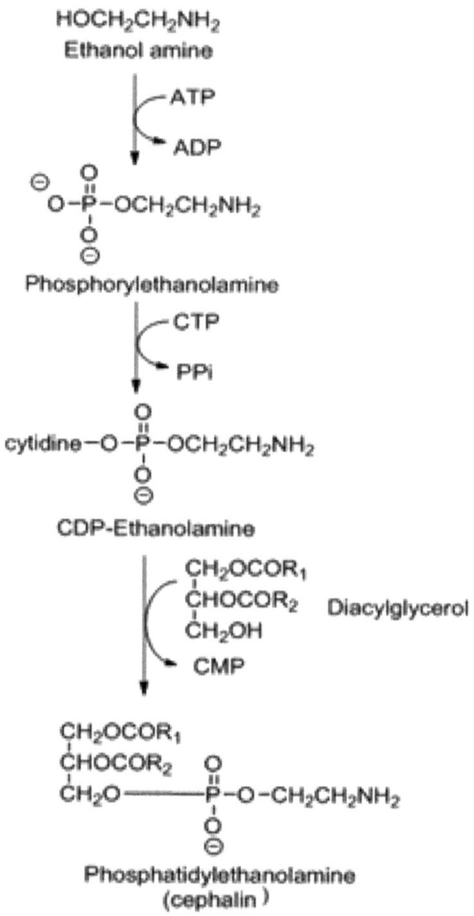

$HOCH_2CH_2NH_2$
Ethanol amine

ATP → ADP

$^{\ominus}O-P(=O)(O^{\ominus})-OCH_2CH_2NH_2$
Phosphorylethanolamine

CTP → PPi

$cytidine-O-P(=O)(O^{\ominus})-OCH_2CH_2NH_2$
CDP-Ethanolamine

CH_2OCOR_1 / $CHOCOR_2$ / CH_2OH Diacylglycerol
→ CMP

CH_2OCOR_1 / $CHOCOR_2$ / $CH_2O-P(=O)(O^{\ominus})-O-CH_2CH_2NH_2$
Phosphatidylethanolamine
(cephalin)

FIGURE 15.16 Biosynthesis of cephalin.

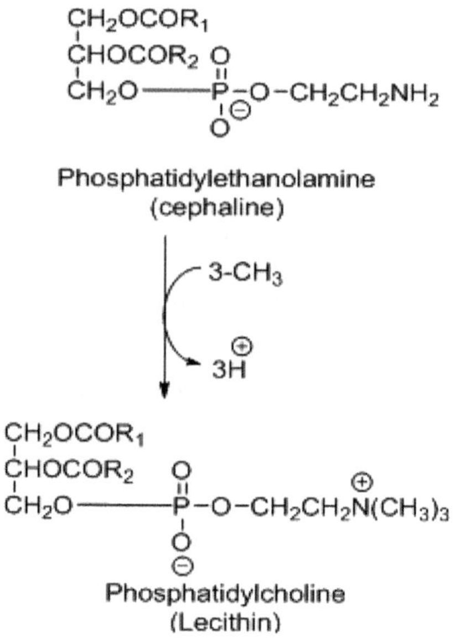

CH_2OCOR_1 / $CHOCOR_2$ / $CH_2O-P(=O)(O^{\ominus})-O-CH_2CH_2NH_2$
Phosphatidylethanolamine
(cephaline)

$3\text{-}CH_3$ → $3H^{\oplus}$

CH_2OCOR_1 / $CHOCOR_2$ / $CH_2O-P(=O)(O^{\ominus})-O-CH_2CH_2\overset{\oplus}{N}(CH_3)_3$
Phosphatidylcholine
(Lecithin)

FIGURE 15.17 Conversion of cephalin into lecithin.

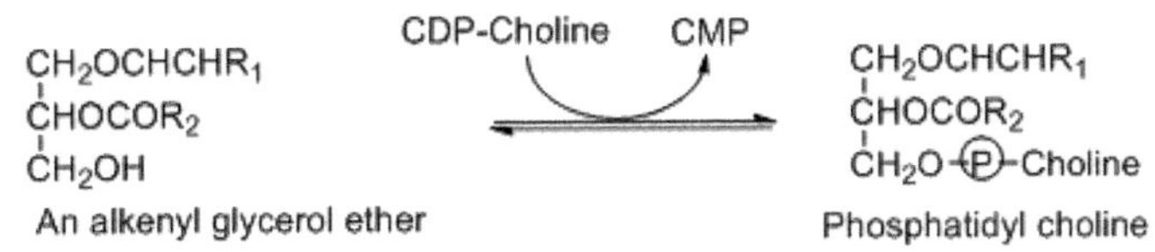

CDP-Choline CMP

$CH_2OCHCHR_1$ / $CHOCOR_2$ / CH_2OH → $CH_2OCHCHR_1$ / $CHOCOR_2$ / CH_2O-Ⓟ-Choline

An alkenyl glycerol ether → Phosphatidyl choline

FIGURE 15.18 Biosynthesis of plasmalogenes.

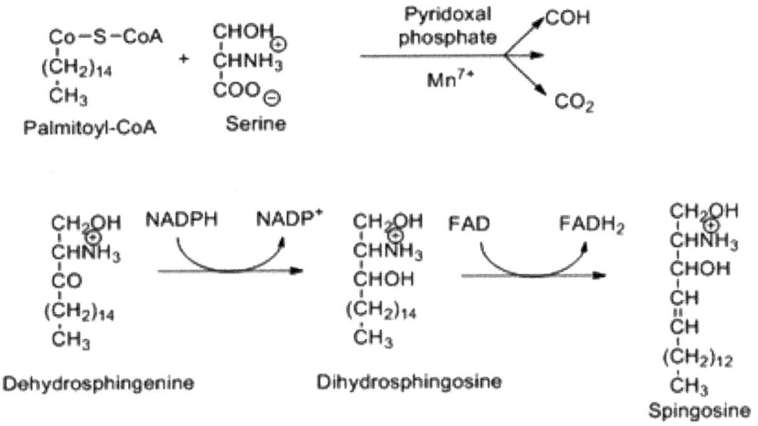

Co–S–CoA / $(CH_2)_{14}$ / CH_3 (Palmitoyl-CoA) + CHOH / $CHNH_3^{\oplus}$ / $COO^{\ominus}$ (Serine) —Pyridoxal phosphate, Mn^{7+}→ COH, CO_2

CH_2OH / $CHNH_3^{\oplus}$ / CO / $(CH_2)_{14}$ / CH_3 (Dehydrosphingenine) —NADPH → NADP⁺→ CH_2OH / $CHNH_3^{\oplus}$ / CHOH / $(CH_2)_{14}$ / CH_3 (Dihydrosphingosine) —FAD → $FADH_2$→ CH_2OH / $CHNH_3^{\oplus}$ / CHOH / CH / CH / $(CH_2)_{12}$ / CH_3 (Spingosine)

FIGURE 15.19 Biosynthesis of spingosine.

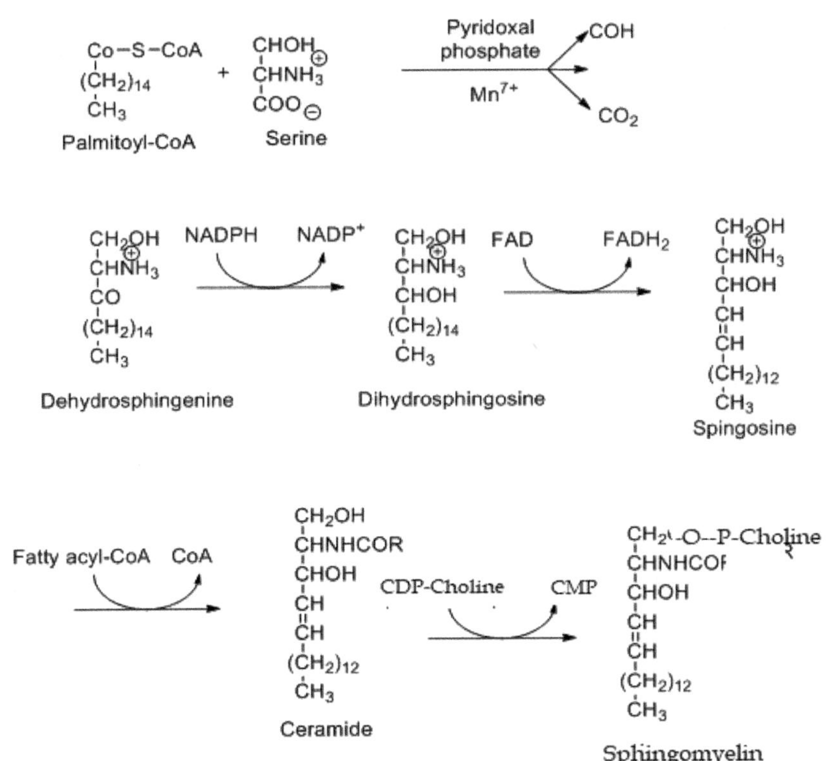

FIGURE 15.20 Biosynthesis of sphingmycelin.

REFERENCES

[1] A. Nag (ed.), *Greener Synthesis of Organic Compounds, Drugs and Natural Products*, Taylor and Francis, UK, 2022.
[2] I.L. Finer, *Organic Chemistry*, Vol. 2, Addison Wesley Longman Limited, UK, 1973.
[3] O.P. Agrawal, *Organic Chemistry Natural Products*, Vol. II, Goel Publishing House, New Delhi, 1980.
[4] F.D. Gunstone, *Fatty Acid and Lipid Chemistry*, Springer, 2020.
[5] C.C. Akoh, *Food Lipids*, Taylor and Francis, UK, 2017.
[6] F.D. Gunstone, *Fatty Acid and Lipid Chemistry*, Springer, 2020.
[7] D.L. Nelson and M.M. Cox, *Lehninger Principles of Biochemistry*, 8th edition, W.H. Freeman and Co Ltd., USA, 2021.

16 Biosynthesis of Steroids and Steroid-Related Compounds

Athar Ata

16.1 INTRODUCTION

The word steroid was initially used to describe cholesterol discovered in gallstones [1]. Cholesterol, an amorphous solid, got its name from the ancient Greek words *chole* ("bile") and *stereos* ("solid"). Structurally, cholesterol (**1**) has a tetracyclic structure consisting of three fused six-membered and one five-membered ring, as shown in Figure 16.1. Steroids contain a basic skeleton of 17 carbon atoms. The functional groups attached to C-17 vary considerably, with the most common locations being C-3, C-4, C-7, C-11, C-12, and C-17. The presence of cholesterol in arterial plaques was detected for the first time in 1843 [2]. Later, it was discovered in animals and plants [3].

These compounds perform various physiological activities, including growth, development, sexual differentiation, and reproduction. Steroids are a class of physiologically active compounds classified into four types: androgens, estrogens, progestogens, and glucocorticoids; these compounds have varying effects on the brain. Androgens and estrogens, for example, play a role in sexual differentiation of the brain and impact cognition; testosterone (**2**) and estrone (**3**) are representative examples of androgens and estrogens, respectively. Progestogens are considered neuroprotective, while glucocorticoids play an important role in the body's glucose, protein, and fat metabolism. Lynestrenol (**4**) and prednisone (**5**) are representative examples of progestogens and glucocorticoids, respectively. Structures of representative examples (**2–5**) are shown in Figure 16.1. Dysfunctions in steroid function have been reported in various disease pathogeneses. However, regulating steroid signaling can be a potential treatment for several neurodevelopmental, psychiatric, and neurodegenerative disorders [4].

Understanding the role of steroids in physiology has been one of the biggest challenges in scientific research due to their limited availability in nature. These molecules are incredibly small in animals and plants, making it difficult to synthesize them on a large scale. Moreover, the presence of chiral centers in their structures further complicates the synthesis process. However, overcoming these hurdles is necessary to unlock the physiological and pharmaceutical potential of steroids. Developing biotechnological processes for their production is crucial, and the first step towards achieving this goal is to elucidate the biosynthetic origin of these compounds. Biosynthetic studies provide valuable information on the enzymology of steroid synthesis in plants and animals. By purifying and cloning these enzymes into a suitable vector, they can be overexpressed to produce steroids on a large scale. This approach can help us better understand the detailed physiological and pharmaceutical activities of these compounds, paving the way for new discoveries in medicine and beyond [5].

Both steroids and triterpenoids have a common biosynthesis origin. The major difference between these compounds is the presence of two methyl groups at C-4. Lanosterol (**6**) has a unique steroid–terpenoid structure and is a biosynthetic precursor to sterols found in animals and plants. Steroids and triterpenoids can be easily identified with the help of ^{1}H-NMR spectroscopy. The ^{1}H-NMR spectrum of steroids exhibits two singlets at δ 0.5 to 1.15 due to the protons of C-18 and C-19 methyl groups bonded to quaternary carbons C-13 and C-10, all respectively. The ^{1}H-NMR spectrum of triterpenoids shows more than two singlets in this range due to methyl groups bonded to quaternary carbons [6]. The basic building block of steroids and triterpenoids is isopentenyl pyrophosphate (IPP) [7]. In this chapter, I describe the biosynthetic origin of steroids and steroid-related compounds.

DOI: 10.1201/9781003437413-16

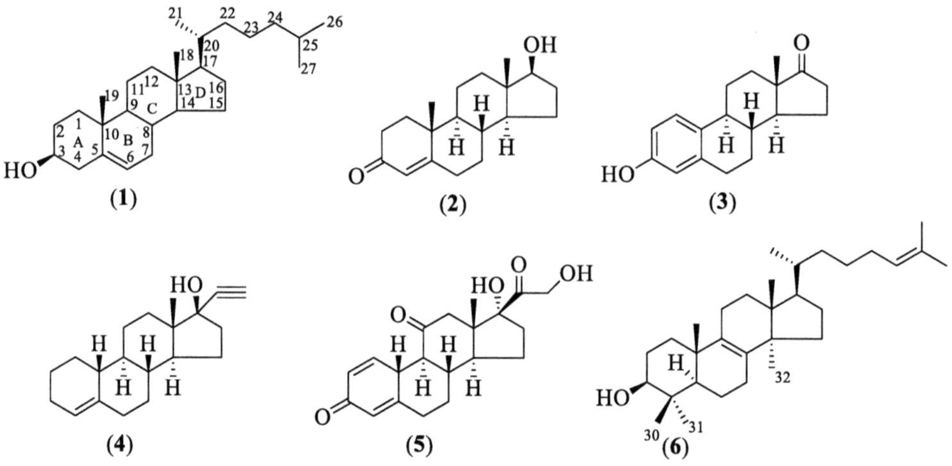

FIGURE 16.1 Structures of compounds (**1–6**).

16.2 BIOSYNTHESIS OF LANOSTEROL (6)

Lanosterol (**6**) is a biosynthetic precursor to all sterols and triterpenoid compounds. Its biosynthesis starts from the head-to-tail coupling of two IPP (**7**) units; the IPP is also a building block of all terpenoids. Isoprenoids are vital to all organismal classes, supporting core cellular functions such as aerobic respiration and membrane stability by making cholesterol. Isoprenoids also form the largest group of so-called secondary metabolites, such as the extremely diverse classes of plant-defensive terpenoids that are widely exploited as perfumes, food additives, and pharmaceutical agents (for example, the antimalarial compound artemisinin). The number of IPP units involved in the biosynthesis of terpenoids determines their class, one of four: Two IPP units produce monoterpenoid (C10), three molecules of IPP make sesquiterpenoids, four make diterpenoid (C20), and six units make triterpenoid (C30). Similarly, tetraterpenoid (C40) contains eight units of IPP [7].

IPP biosynthesis has two routes in nature: i) the mevalonate pathway and ii) the methylerythritol phosphate (MEP) pathway. The mevalonate pathway starts by reacting two molecules of acetyl-CoA (**8**) to afford acetoacetyl-CoA (**9**). Compound (**9**) reacts with a third molecule of (**8**) to give 3-hydroxy-3-methyl-glutaryl-CoA (HMG-CoA) (**10**). This series of reactions is catalyzed by thiolase and hydroxymethylglutaryl-CoA synthase. The reduction of (**10**) gives mevalonic acid (**11**). Compound (**11**) is converted to mevalonate-5-diphosphate (**12**) after two consecutive phosphorylation steps. These steps are catalyzed by a mevalonic acid kinase and phosphomevalonate kinase, respectively. Compound (**11**), upon decarboxylation and dehydration, produces IPP (**7**), which is isomerized to dimethylallyl pyrophosphate (DMAPP (**13**)) [7–10]. These reactions are summarized in Scheme 16.1.

The MEP pathway is a complex series of reactions that involves the conversion of D-glyceraldehyde 3-phosphate (**14**) and pyruvate (**15**) into 1-deoxy-D-xylulose 5-phosphate (DXP) (**16**) through a thiamin diphosphate-dependent reaction. This DXP (**16**) is then reductively isomerized to MEP (**17**) by DXP reducto-isomerase. The next step involves the conversion of (**17**) into methylerythritol cytidyl diphosphate (CDP-ME, **18**) by reacting it with cytidine 5'-triphosphate, CTP, which is catalyzed by CDP-ME synthetase. This is followed by the phosphorylation of the C-2 hydroxyl group of (**17**) by an ATP-dependent enzyme (IspE), forming 4-diphosphocytidyl-2-C-methyl-D-erythritol-2-phosphate (**18**). The latter is then cyclized to 2-C-methyl-D-erythritol-2,4-cyclodiphosphate (MEcPP (**19**)). The cyclic pyrophosphate of MEcPP (**19**) is then opened by IspG-catalyzed ring-opening, followed by the C-3-reductive dehydration of MEcPP (**19**) producing 4-hydroxy-3-methyl-butenyl-1-diphosphate (HMBPP, **20**). Compound **20** is finally converted to IPP (**7**) and DMAPP (**8**) [11–14]. This entire series of reactions is illustrated in Scheme 16.2.

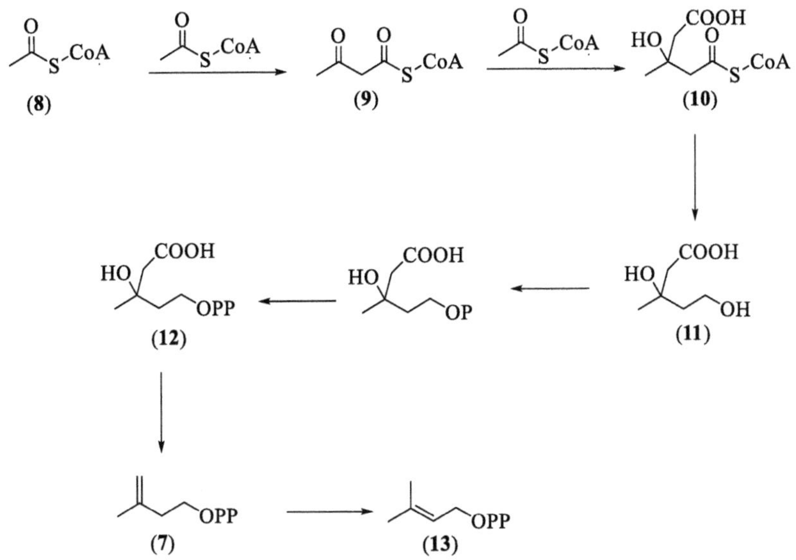

SCHEME 16.1 Biosynthesis of IPP (7) via the mevalonate pathway.

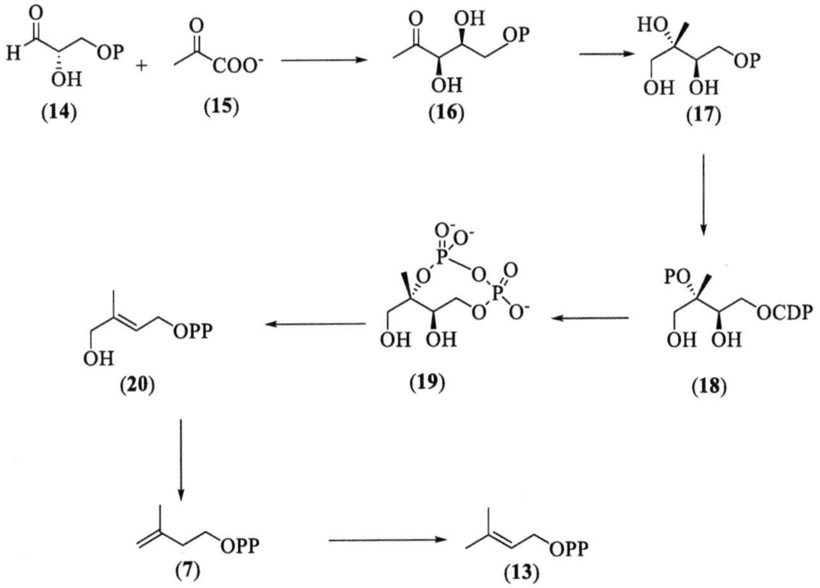

SCHEME 16.2 Biosynthesis of IPP (7) via the MEP pathway.

Reaction of DMAPP (**13**) with IPP (**7**) is catalyzed head to tail by prenyltransferase produces geranyl pyrophosphate (**21**). Compound **21** reacts with another molecule of IPP (**7**) to biosynthesize farnesyl pyrophosphate (FPP (**22**)). These reactions are summarized in Scheme 16.3 [15–17].

The coupling of two FPP (**21**) molecules produces squalene (**22**), which, upon oxidation, yields 2, 3-epoxysqualene (**24**). Cyclization of (**24**) produces lanosterol (**6**). These reactions are shown in Scheme 16.4 [18–21].

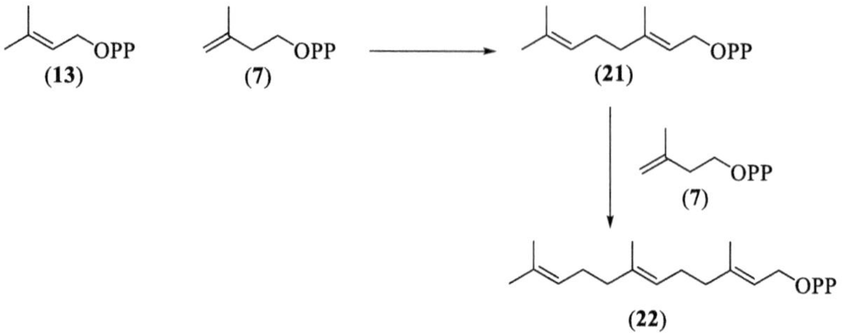

OPP (13) OPP (7) → OPP (21) ↓ OPP (7) OPP (22)

SCHEME 16.3 Biosynthesis of FPP.

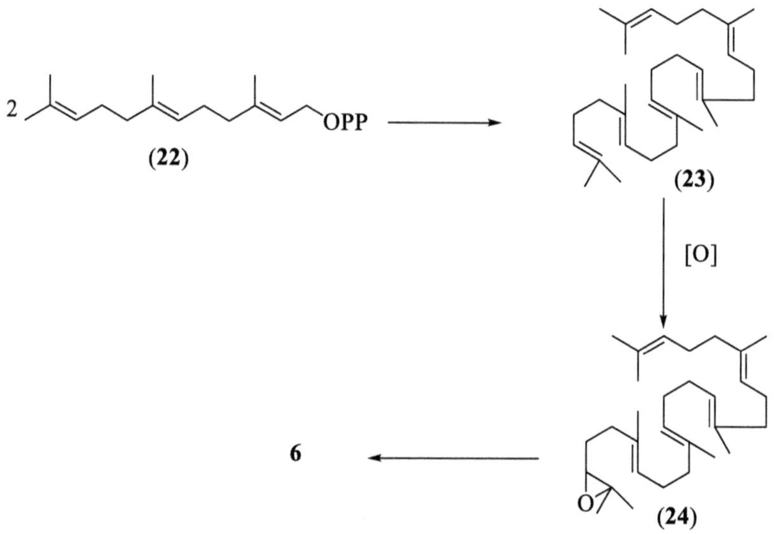

2 OPP (22) → (23) ↓ [O] (24) → 6

SCHEME 16.4 Biosynthesis of lanosterol.

16.3 BIOSYNTHESIS OF CHOLESTEROL (1) FROM LANOSTEROL (6)

The biosynthesis of compound (**1**) from lanosterol (**6**) is a complex process that involves several enzymatic oxidation and reduction steps. The first step begins with the demethylation of the C-32 methyl group, which gives compound (**29**). The formation of compound (**28**) involves the oxidation of C-32 methyl to produce a 32-hydroxymethylene group (**25**). Further oxidation of the C-32-hydroxymethylene functionality gives a 32-aldehyde analogue of compound (**6**), compound (**26**). The enzymatic Baeyer–Villiger oxidative rearrangement of the 32-aldehyde group in (**26**) converts it to the formyloxy group at C-32, as shown in compound (**27**). The removal of the formyloxy group produces the Δ [14–15] analogue of lanosterol (**28**). Reducing Δ [14–15] double bond produces 14-demthyllanosterol (**29**) [22–27]. In this class of compounds, if there are α and β methyl groups present, then the α-methyl group is more likely to be oxidized initially compared to the β-methyl group. This biogenetic information helps to elucidate the structures and stereochemistry of naturally occurring compounds [28]. The biosynthesis of intermediate (**29**) is illustrated in Scheme 16.5.

In the next step of the biosynthetic pathway, (**29**) undergoes a demethylation reaction at C-4. This is a crucial step in the process and requires careful attention to detail. As mentioned, the

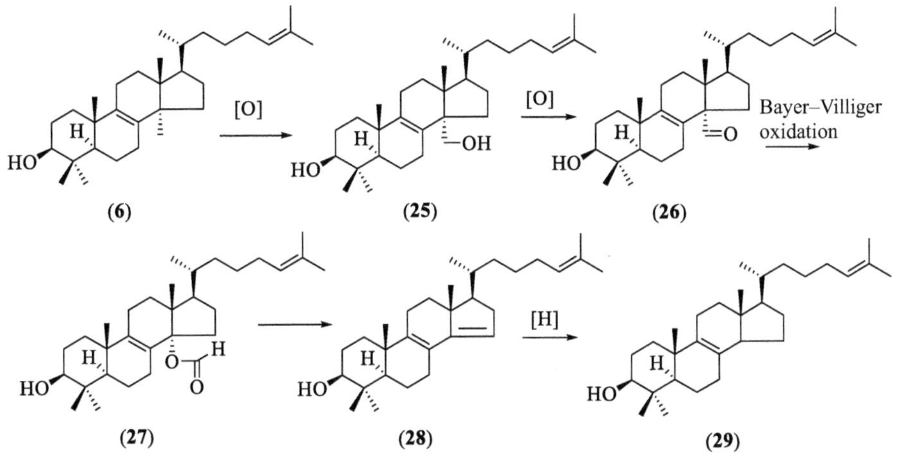

SCHEME 16.5 Biosynthesis of compound (29).

C-4α-methyl group will be oxidized first, followed by the oxidation of the C-4β-methyl group. It is important to note that the demethylation reaction will follow the same oxidation reaction sequence as those of (**29**), except for the Bayer-Villger rearrangement of aldehyde to the formyloxy group. During the demethylation process, the C-4α aldehyde group in compound (**31**) will be oxidized to carboxylic acid (**33**), which will then undergo decarboxylation to afford compound (**32**). Likewise, the demethylation of C-4 methyl in compound **33** follows the same sequence of reactions, ultimately leading to (**34**) [29–33]. These biosynthetic reactions are outlined in detail in Scheme 16.6.

The double bond Δ [8–9] in **34** is isomerized to Δ [7–8] as shown in compound **35**. A dehydrogenation reaction on **35** affords compound **36** containing a conjugated homocyclic diene system (Δ [5–8]) in ring B. A selective reduction of a Δ [7–8] double bond biosynthesizes cholesterol (**1**) [34–36]. These reactions are shown in Scheme 16.7.

16.4 BIOSYNTHESIS OF STEROIDAL ALKALOIDS

Steroidal alkaloids are a diverse group of compounds found in various families of plants such as *Solanaceae*, *Liliaceae*, *Apocynaceae*, *Buxaceae*, and even in certain amphibians and marine organisms. These alkaloids contain nitrogen either in the ring or side chain and have been found to exhibit a wide range of biological activities. For instance, they possess anticancer, anticholinergic, antimicrobial, anti-inflammatory, analgesic, antimyocardial ischemia, anti-angiogenesis, and many other beneficial health-related effects [37–45]. Their versatility and potential therapeutic applications make them an exciting study area in natural product chemistry. These compounds are produced in nature either from steroids or lanosterol.

16.4.1 BIOSYNTHESIS OF *SOLANACEAE* ALKALOIDS

Steroidal alkaloids of *Solanaceae* are divided into two groups based on the structure of their side chain: i) steroidal base side chains have an oxa-azasipro structure, and ii) steroidal alkaloids contain cyclic amine as a side chain.

Tomatine (**37**) is a representative member of the oxa-aza spiro class of steroidal bases, and solanine (**38**) is an example of a cyclic amine-containingsteroidal base. Compound (**37**) is biosynthesized from cholesterol (**1**). These compounds can easily be identified by the ^{1}H-NMR spectrum that shows two singlets for the C-18 and C-19 methyl groups and two doublets for the C-21 and C-27 methyl groups in the range of δ 0.5 to 1.40. C-26 methyl is oxidized to the hydroxymethylene group to

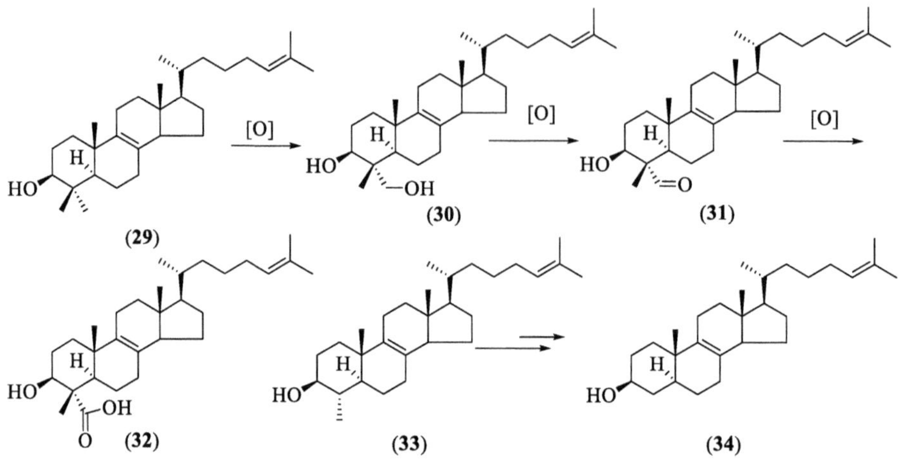

SCHEME 16.6 Biosynthesis of compound (**34**).

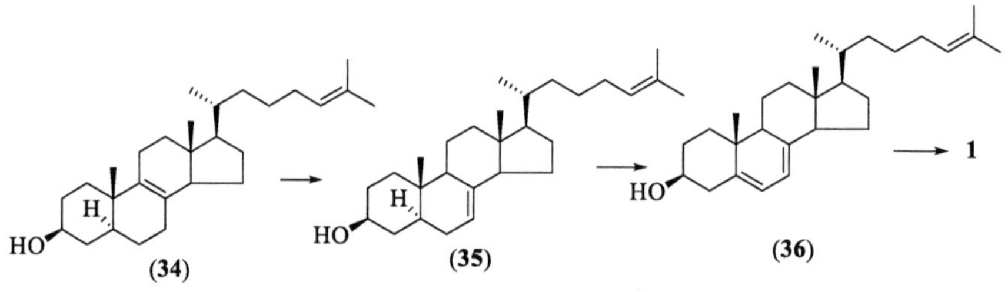

SCHEME 16.7 Biosynthesis of compound (**1**).

afford (**39**). Further oxidation of the C-26 hydroxyl group yields C-26 aldehyde (**40**). Transamination of aldehyde moiety in **40** and hydroxylation reaction at C-22 yields intermediate (**41**). The latter compound is cyclized to produce a compound (**42**). Hydroxylation at C-16 of **42** gives compound **43**. Its cyclization biosynthesizes compound (**37**) [46–52]. The substitution reaction at C-16 produces compound (**38**). Biosynthesis of (**37**) and (**38**) are shown in Scheme 16.8.

16.4.2 Biosynthesis of *Apocynaceae* Alkaloids

Apocynaceae contains steroidal alkaloids that exhibit various biological activities such as antihypertensive and antiarrhythmic effects. These alkaloids can be classified into four types based on their main structure: pregnane (**44**), conanine (**45**), paravallarine (**46**), and amino-glycosteroid (**47**) (Figure 16.2) [53–57].

The pregnane class (**44**) alkaloids can be identified in the ^{1}H-NMR spectrum by identifying protons of two tertiary methyl groups and protons of ethyl side on ring D in the methyl range of the spectrum. Similarly, (**45**) series alkaloids exhibit one C-19 methyl group and one C-18 methylene group in their ^{1}H-NMR spectrum. On the other hand, steroidal bases of the (**46**) class only show the resonance of the C-19 methyl group in their ^{1}H-NMR spectrum. Lastly, the ^{1}H-NMR spectrum of compounds in group (**47**) is similar to those in class (**45**) but with additional signals due to the presence of a sugar moiety. These compounds are believed to be biosynthesized from cholesterol (**1**) [58–60].

Hydroxylation at C-20 and C-22 affords compound (**48**), which undergoes oxidative cleavage to give intermediate (**49**). Replacement of hydroxyl group with amine produces compound (**50**). Reduction of C-20 carbonyl group gives (**44**). Transamination reaction at C-20 carbonyl gives (**51**),

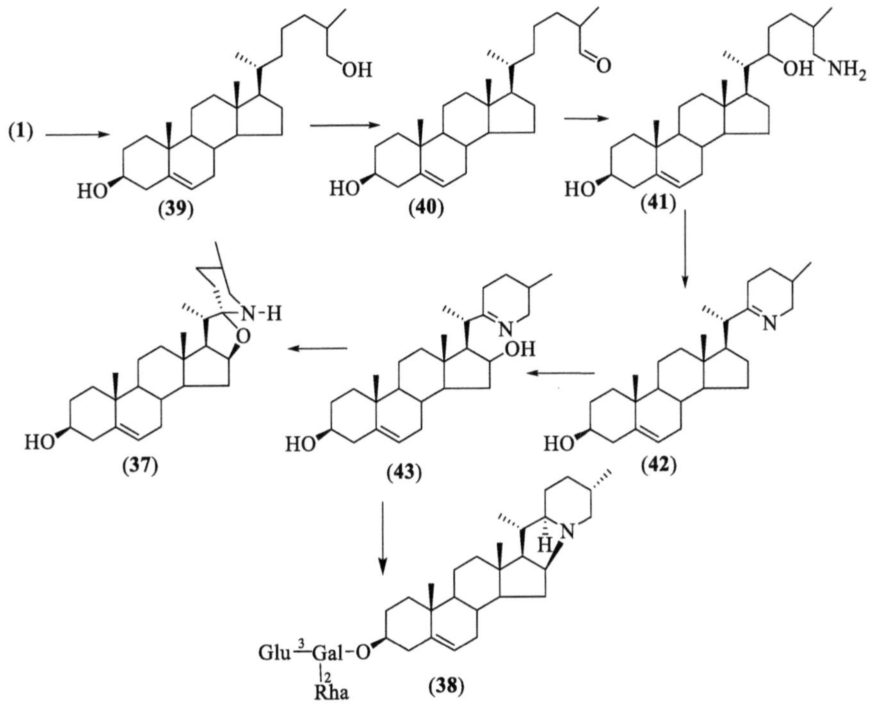

SCHEME 16.8 Biosynthesis of steroidal alkaloids (**37**) and (**38**).

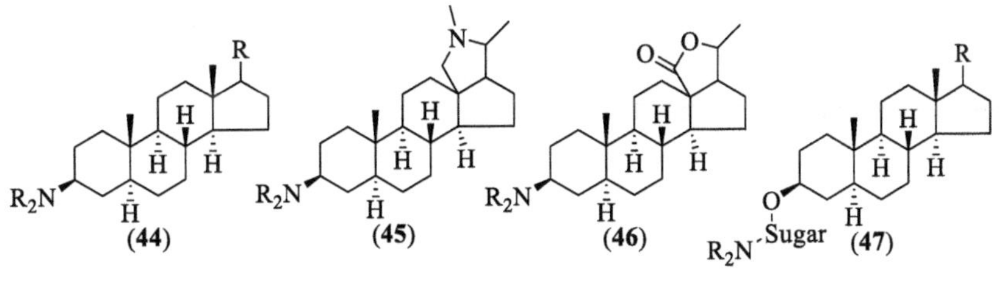

FIGURE 16.2 Structures of four classes of *Apocynaceae* alkaloids: (**44**)–(**47**).

which upon oxidation of C-18 methyl group yields intermediate (**52**). Cyclization reaction of C-20 amino group with C-18 hydroxymethylene groups affords (**45**). Alkaloids of class (**46**) are biosynthesized from (**50**). Amino-glycosteroid (**47**) are produced from (**49**) [58–60]. The biosynthesis of compounds (**44**)–(**47**) is shown in Scheme 16.9.

16.4.3 Biosynthesis of *Buxaceae* Alkaloids

The genus *Sarcococca* belongs to the family *Buxaceae* and comprises around 20 species that are primarily found in the southwestern region of China and South Asian countries. Traditional folk medicine has long used these plants to treat ailments such as stomach pain, rheumatism, swollen sore throat, and traumatic injury [61–66]. These medicinal properties are attributed to steroidal alkaloids in the plants. This class of alkaloids contains a basic pregnane-type structure with amine functionality at C-3 and C-20 (**55**) [67–70]. These compounds are produced in nature from compound (**49**) as outlined in Scheme 16.10 [71, 72].

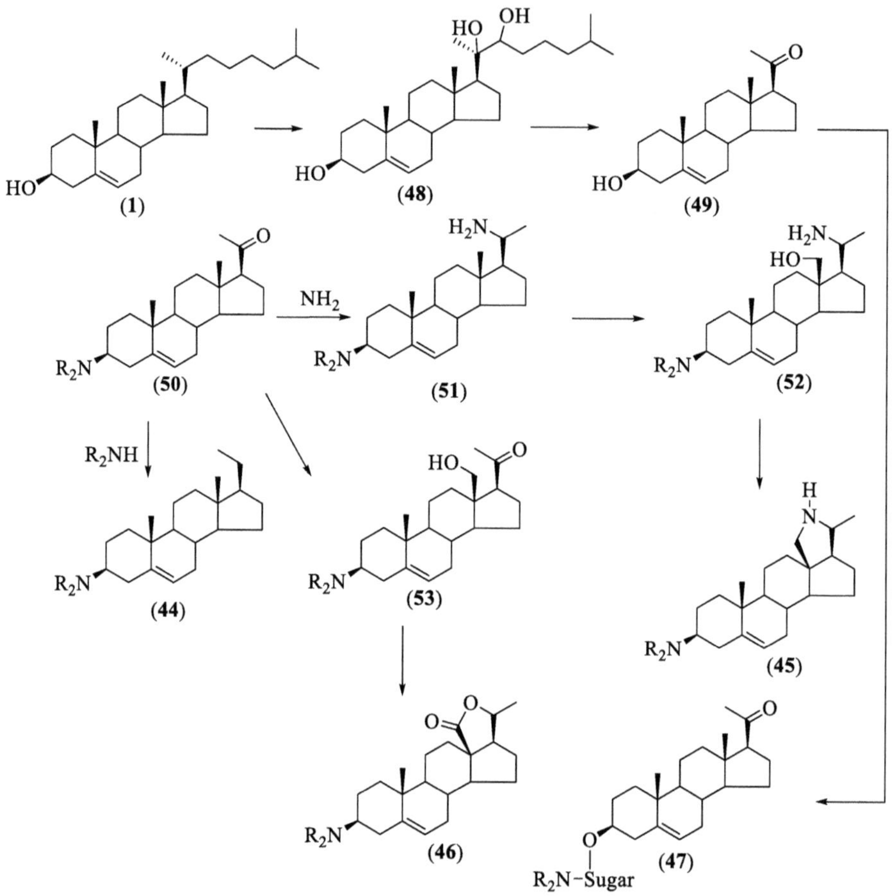

SCHEME 16.9 Biosynthesis of compounds (**44**)–(**47**).

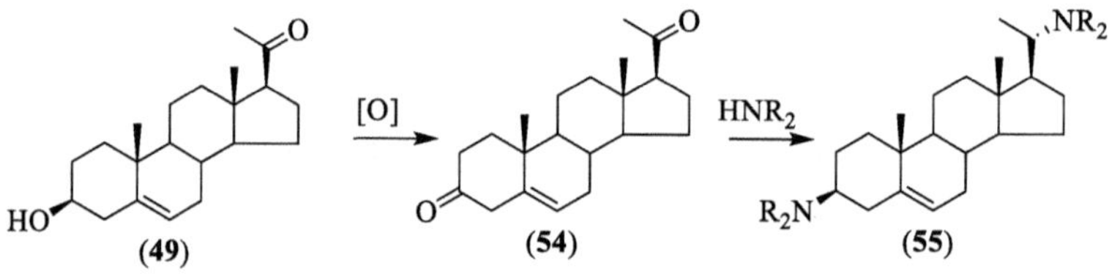

SCHEME 16.10 Biosynthesis of compound (**55**).

The steroidal alkaloids of genus *Buxus* are known for their wide range of health-related bioactivities, including anti-fatigue, antirheumatism, antimalaria, antidepression, anti-HIV, antiTB, and anti-acetylcholinesterase (AChE) properties [73–80]. Alkaloids with AChE inhibitory activities have shown potential in treating early symptoms of Alzheimer's disease and as lead candidate molecules against neurodegenerative diseases [81–85]. Their unique triterpenoid–steroidal pregnane basic structure contains C-4 methyl groups, a 9β,10β-cycloartenol system, and a degraded C-20 side chain. *Buxus* alkaloids are divided into two classes: derivatives of 9β,10β-cyclo-4,4,14α-trimethyl-5α-pregnane system (**56**) and derivatives of 9(10→19) *abeo* 4,4,14α-trimethyl-5α-pregnane system (**57**).

Spectral studies such as ^{1}H NMR and UV spectra can aid in distinguishing between the two classes.[86] The ^{1}H-NMR spectrum of members of steroidal bases of structure (**56**) shows an up-field (δ 0.0 0.2) pair of AB doublets for cyclopropyl protons, whereas compounds of basic structure (**57**) exhibit signals for the C-11 and C-19 vinylic protons in the olefinic range of the spectrum instead. The presence of a 9(10→19) abeo-diene system can be determined from the UV spectrum as these compounds exhibit absorption maxima at 238 nm and 245 nm with shoulders at 228 nm and 252 nm. Structural modifications are also observed in each case, including the absence of one or both methyls, the presence of a methylene group at C-4, and different oxygen functions and points of unsaturation. Figure 16.3 displays the structures of compounds (**56**) and (**57**) and modifications made in *Buxus* alkaloids (**58**)–(**62**).

The biosynthesizing of these compounds involves the conversion of lanosterol (**6**) to cycloartenol (**65**) through enzymatic reactions. This compound naturally occurs in various plants [87, 88]. The bioconversion of compound (**63**) forms C-20 keto compound (**64**) by the oxidative cleaving of the C-20 side chain. Further oxidation of the C-3 hydroxyl group forms 3,20 diketo compound (**65**). These important mono- and diketo compounds are biosynthetic precursors of *Buxus* alkaloids. Scheme 16.11 shows the bioconversion process from compound (**6**) to (**56a-c**).

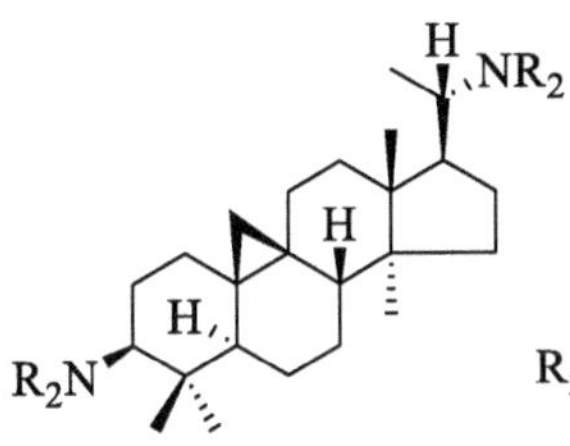

(**56**) 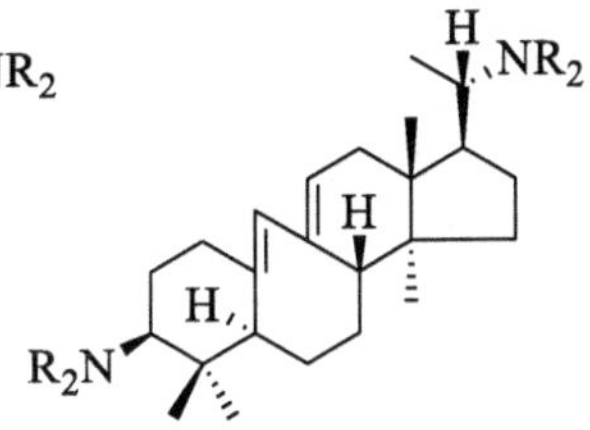 (**57**) 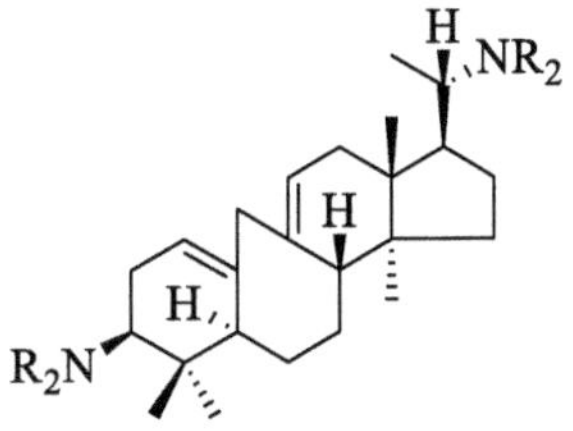 (**58**)

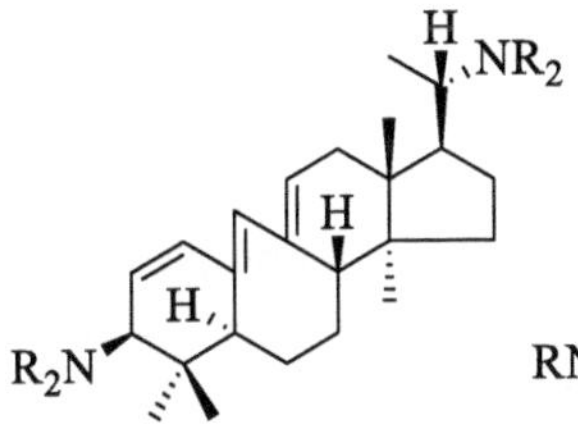

(**59**) 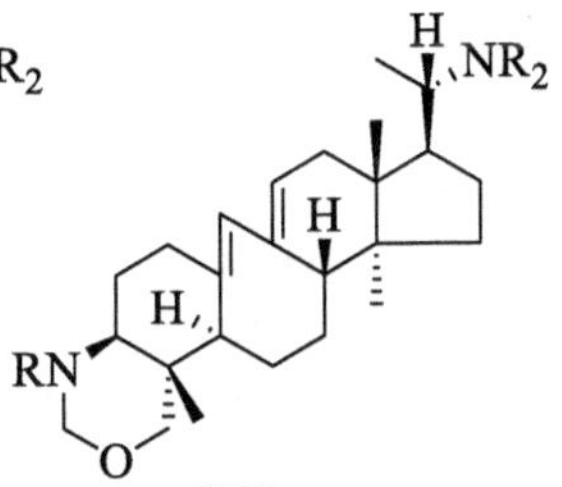 (**60**) 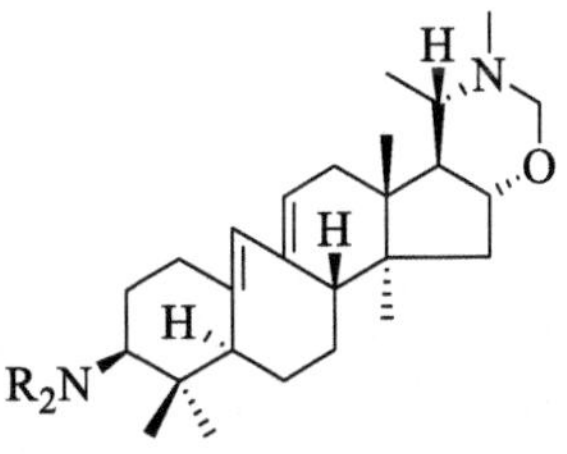 (**61**)

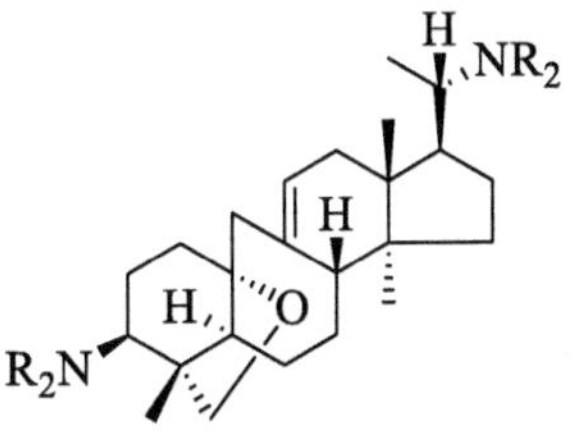

(**62**)

R groups represent amino or keto groups

FIGURE 16.3 The structures of *Buxus* alkaloids (**56**)–(**62**).

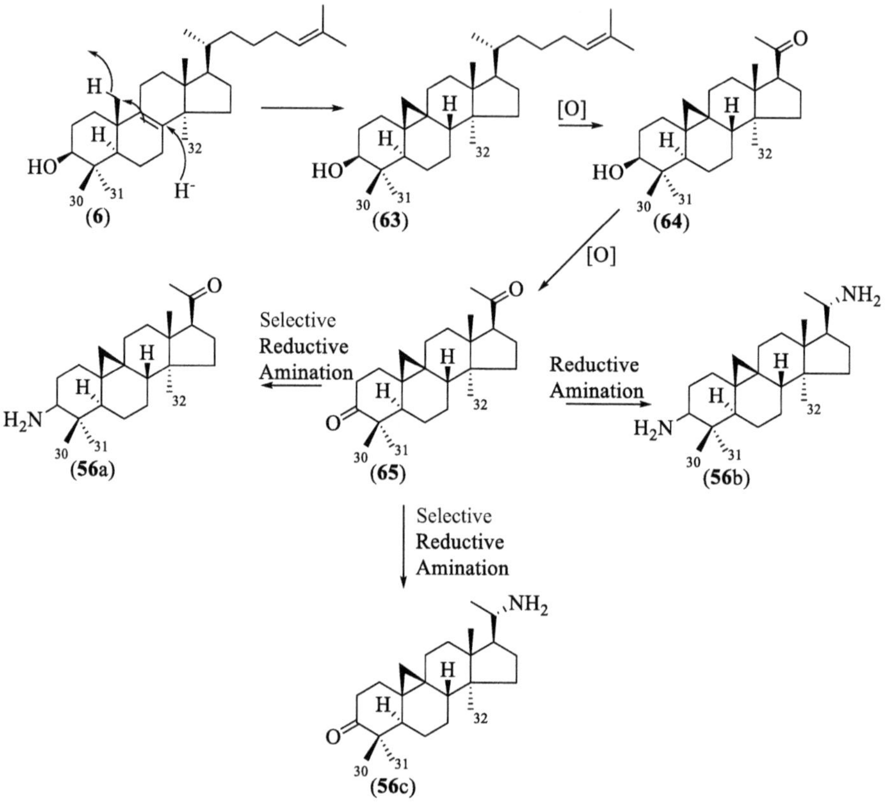

SCHEME 16.11 Biosynthesis of *Buxus* alkaloids of class (**56**).

The basic skeleton (**57**) comprises 9(10→19) *abeo* conjugated diene alkaloids and 9(10→19) *abeo* nonconjugated diene alkaloids. Two representative examples of these compounds are (**57**) and (**58**); these compounds are naturally produced when the 9β,10β-cyclopropane ring in compounds (**56a-c**) undergoes de-cyclization. Compounds belonging to basic structure (**57**) are derived from (**56a-c**), which undergo oxidation reaction at C-11 to produce the C-11 hydroxy derivative (**66**) of steroidal alkaloid (**56**). The expansion of ring B to a seven-membered ring is initiated by the loss of an α-hydrogen from the cyclopropane ring, followed by the loss of a C-11 hydroxyl group, resulting in compound (**57**). Similarly, *Buxus* alkaloids of structure (**58**) are produced by following a similar path as that of compound **57** through the loss of hydrogen from C-1 of ring A. Compounds containing 9(10→19) *abeo*triene system (**59**) arise from compounds of class (**57**) [89]. A compound of (**57**) series produces C-2 hydroxy analogue (**67**) upon oxidation at C-2. This compound undergoes dehydration to yield compound (**59**). These biosynthetic pathways are shown in Scheme 16.12.

Several *Buxus* alkaloids contain tetrahydrooxazine ring (**60**) [90–93]. This ring is produced from (**68**), which on condensation of an aldehyde with the C-3 amino group gives ketamine (**69**). An attack of the C-31 hydroxyl group on ketamine forms a tetrahydrooxazine ring in compound (**60**). Similarly, cyclization can occur in compounds with C-20 amino and C-16 hydroxy groups to produce steroidal alkaloids containing tetrahydrooxazine rings at C-16/C-20 in compound (**71**). The formation of the tetrahydrooxazine ring in these compounds is shown in Scheme 16.13.

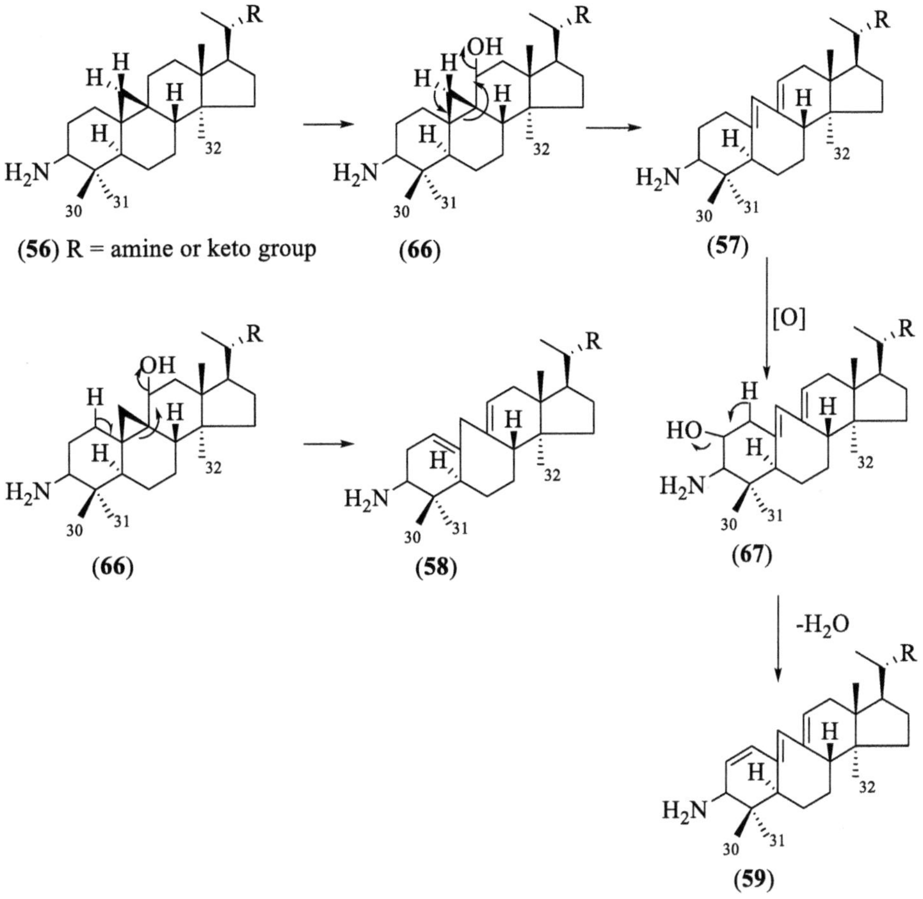

(56) R = amine or keto group (66) (57) [O] (66) (58) (67) $-H_2O$ (59)

SCHEME 16.12 Biosynthesis of compounds **57–59**.

Buxus steroidal alkaloids also contain a tetrahydrofuran ring incorporated in their structures. This ether linkage is present at C-2 (**73**), C-6 (**75**), and C-10 (**77**). These compounds are biosynthesized by oxygen attacking the C-31 hydroxyl group on a double bond at Δ [1–2] or Δ [1–10] to give (**72**) and (**74**), respectively, whereas (**73**) arises in nature by the nucleophilic substitution reaction of the C-31 hydroxyl group at C-6. The formation of a tetrahydrofuran ring is proposed as several biosynthetic intermediates are reported in the literature [93, 94]. The presence of these biosynthetic precursors in *Buxus* plants strongly support the proposed biosynthesis. These biosynthesis reactions are summarized in Scheme 16.14.

In closing, in this chapter, I have summarized the biosynthesis of steroids and steroid-related compounds. The biosynthesis experiments I have discussed can be used to gain knowledge about enzymes involved in the biosynthesis of steroids and their derivatives. This knowledge can help to design bioassays to purify the key enzyme to use them for in vitro biochemical reactions to produce these compounds in the laboratory. This green chemistry will not only be safe for our environment but also help to preserve plants for our future generation. This biosynthetic knowledge can also help in designing biomimetic type synthesis of bioactive natural products to provide on large scale for their detailed pharmacological studies.

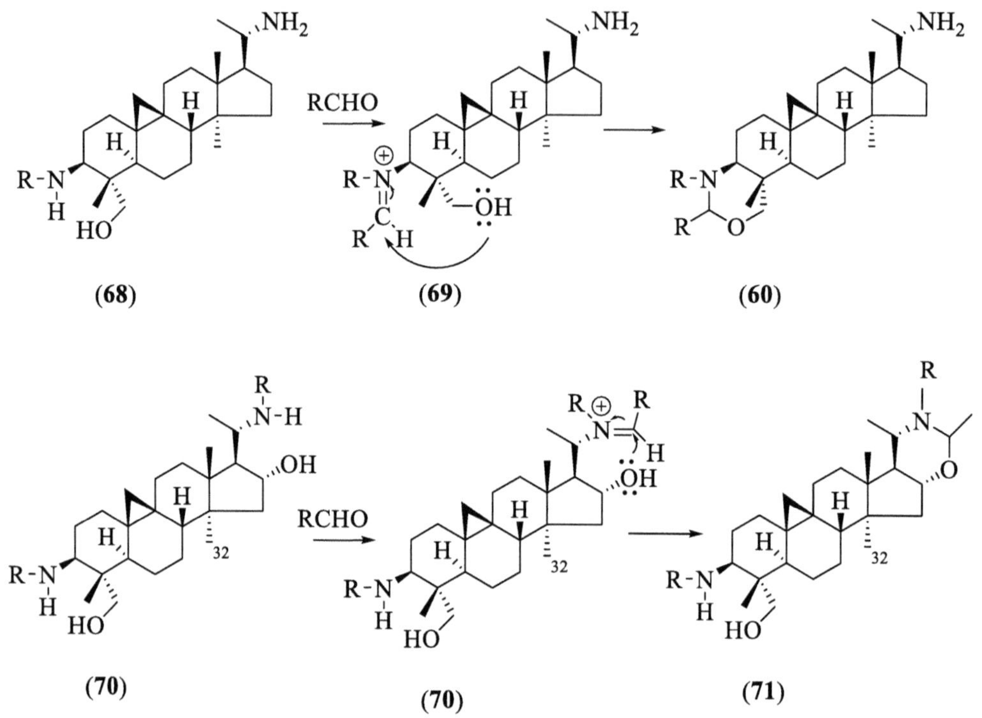

(68) (69) (60)

(70) (70) (71)

SCHEME 16.13 Biosynthesis of compounds (**68**)–(**71**).

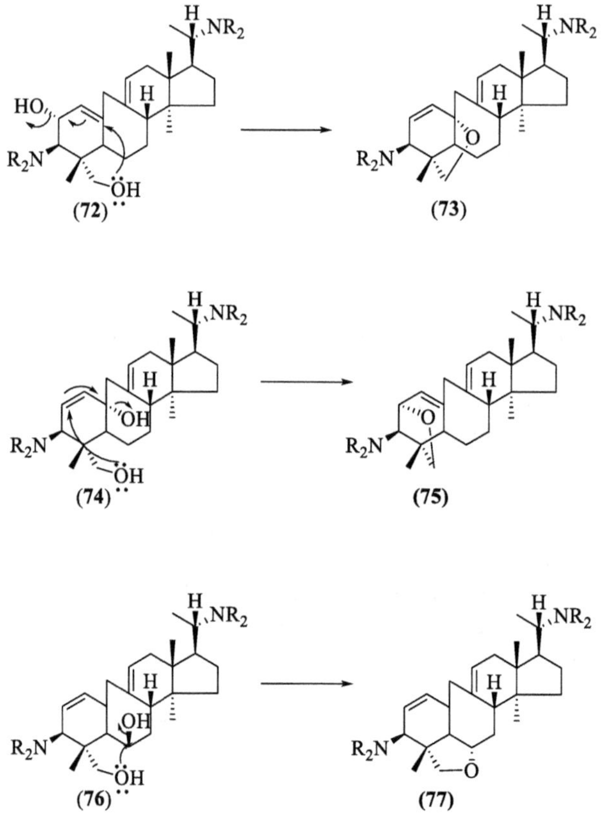

(72) (73)

(74) (75)

(76) (77)

SCHEME 16.14 Biosynthesis of compounds (**72**)–(**77**).

REFERENCES

[1] Mathew, B., Daniel, R. 2008. *Cholesterol:* A Century of Research and Debate. *Libyan Journal of Medicine* 3 (2): 63.

[2] Nes, D. W. 2011. Biosynthesis of Cholesterol and Other Sterols. *Chemical Reviews* 111: 6423–645.

[3] Brown, M. S., Goldstein, J. L. 1986. A Receptor-Mediated Pathway for Cholesterol Homeostasis. *Science* 232: 34–47.

[4] Alemany, M. 2022. The Roles of Androgens in Humans: Biology, Metabolic Regulation and Health. *International Journal of Molecular Sciences* 23: 11952.

[5] Park, D., Swayambhu, G., Lyga, T., Pfeifer, B. A. 2021. Complex Natural Product Production Methods and Options. *Synthetic and Systems Biotechnology* 6: 1–11.

[6] Jaeger, M., Asper, R. L. E. G. 2012. Steroids and NMR. *Annual Reports on NMR Spectroscopy* 77: 115–258.

[7] Ninkuu, V., Zhang, L., Yan, J., Fu, Z., Yang, T., Zeng, H. 2021. Biochemistry of Terpenes and Recent Advances in Plant Protection. *International Journal of Molecular Sciences* 22: 5710.

[8] Diao, H., Chen, N., Wang, K., Zhang, F., Wang, Y.-H., Wu, R. 2020. Biosynthetic Mechanism of Lanosterol: A Completed Story. *ACS Catalysis* 10: 2157–2168.

[9] Kolesnikova, M. D., Xiong, Q., Lodeiro, S., Hua, L., Matsuda, S. P. 2006. Lanosterol Biosynthesis in Plants. *Archives of Biochemistry and Biophysics* 447: 87–95.

[10] Bae, S.-H., Lee, J. N., Fitzky, B. U., Seong, J., Paik, Y.-K. 1999. Cholesterol Biosynthesis from Lanosterol. *Journal of Biological Chemistry* 274: 14624–14631.

[11] Testa, C. A., Brown, M. J. 2003. The Methylerythritol Phosphate Pathway and Its Significance as a Novel Drug Target. *Current Pharmaceutical Biotechnology* 4: 248–259.

[12] Singh, N., Chevé, G., Avery, M. A., McCurdy, C. R. 2007. Targeting the Methyl Erythritol Phosphate (MEP) Pathway for Novel Antimalarial, Antibacterial and Herbicidal Drug Discovery: Inhibition of 1-Deoxy-D-Xylulose-5-Phosphate Reductoisomerase (DXR) Enzyme. *Current Pharmaceutical Design* 13: 1161–1177.

[13] Lange, B. M., Croteau, R. 1999. Isoprenoid Biosynthesis via a Mevalonate-Independent Pathway in Plants: Cloning and Heterologous Expression of 1-Deoxy-d-xylulose-5-phosphate Reductoisomerase from Peppermint. *Archives of Biochemistry and Biophysics* 365: 170–174.

[14] Lichtenthaler, H. K. 2000. Non-Mevalonate Isoprenoid Biosynthesis: Enzymes, Genes and Inhibitors. *Biochemical Society Transactions* 28: 785–789.

[15] Dhar, M. K., Koul, A., Kaul, S. 2013. Farnesyl Pyrophosphate Synthase: A Key Enzyme in Isoprenoid Biosynthetic Pathway and Potential Molecular Target for Drug Development. *New Biotechnology* 30: 114–123.

[16] Davis, E. M., Croteau, R. 2000. Cyclization Enzymes in the Biosynthesis of Monoterpenes, Sesquiterpenes, and Diterpenes. *Topics in Current Chemistry* 209: 53–95.

[17] Russell, R. G. 2006. Bisphosphonates. *Annals of the New York Academy of Sciences* 1068: 367–401.

[18] Abe, I. 2007. Enzymatic Synthesis of Cyclic Triterpenoids. *Natural Product Reports* 24: 1311–1331.

[19] Dean, P. D., Ortiz de Montellano, P. R., Bloch, K., Corey, E. J. 1967. *A Soluble 2,3-Oxidosqualene Sterol Cyclase. The Journal of Biological Chemistry* 242: 3014–3015.

[20] Wendt, K. U., Poralla, K., Schulz, G. E. 1997. Structure and Function of a Squalene Cyclase. *Science* 277: 1811–1815.

[21] Corey, E. J., Gross, S. K. 1967. Formation of Sterols by the Action of 2,3-Oxidosqualene-Sterol Cyclase on the Factitious Substrates, 2,3: 22,23-Dioxidosqualene and 2,3-Oxido-22,23-Dihydrosqualene. *Journal of the American Chemical Society* 89: 4561–4562.

[22] Tuck, S. F., Patel, H., Safi, E., Robinson, C. H. 1991. Lanosterol 14 Alpha-Demethylase (P45014DM): Effects of P45014DM Inhibitors on Sterol Biosynthesis Downstream of Lanosterol. *Journal of Lipid Research* 32: 893–902.

[23] Trzaskos, J. M., Bowen, W. D., Shafiee, A., Fischer, R. T., Gaylor, J. L. 1984. Cytochrome P-450-Dependent Oxidation of Lanosterol in Cholesterol Biosynthesis. Microsomal Electron Transport and C-32 Demethylation. *Journal of Biological Chemistry* 259: 13402–13412.

[24] Fukushima, H., Grinstead, G. F., Gaylor, J. L. 1981. Total Enzymic Synthesis of Cholesterol from Lanosterol. Cytochrome b5-Dependence of 4-Methyl Sterol Oxidase. *Journal of Biological Chemistry* 256: 4822–4826.

[25] Sekigawa, Y., Fukuhara, M., Sonoda, Y., Sato, Y. 1995. Purification and Characterization of a Cytochrome P450 Isozyme Catalyzing Lanosterol 14α-Demethylation (P45014DM) in Hamster Liver. *Lipids* 30: 1067–1073.

[26] Lepesheva, G. I., Waterman, M. R. 2007. Sterol 14Alpha-Demethylase Cytochrome P450 (CYP51), a P450 in all Biological Kingdoms. *Biochimica et Biophysica Acta (BBA)—General Subjects* 1770: 467–477.
[27] Lepesheva, G. I., Waterman, M. R. 2004. CYP51–the Omnipotent P450. *Endocrinology* 215: 165–170.
[28] Lin, X., Cane, D. E. 2009. Biosynthesis of the Sesquiterpene Antibiotic Albaflavenone in Streptomyces Coelicolor. Mechanism and Stereochemistry of the Enzymatic Formation of Epi-Isozizaene. *Journal of American Chemical Society* 131: 6332–6333.
[29] Clayden, J., Greeves, N., Warren, S., Wothers, P. 2001. *Organic Chemistry*. Oxford University Press, Oxford, Oxfordshire.
[30] Roland, S. L., Sigel, A., Sigel, H. 2007. *The Ubiquitous Roles of Cytochrome P450 Proteins: Metal Ions in Life Sciences*. Wiley, New York.
[31] Ziogas, B. N., Malandrakis, A. A. 2015. Sterol Biosynthesis Inhibitors: C14 Demethylation (DMIs). In: Ishii, H., Hollomon, D. (eds) *Fungicide Resistance in Plant Pathogens*. Springer, Tokyo.
[32] Hu, D., Gao, Y.-H., Yao, X.-Y., Gao, H. 2020. Recent Advances in Dissecting the Demethylation Reactions in Natural Product Biosynthesis. *Current Opinion in Chemical Biology* 59: 47–53.
[33] Darnet, S., Schaller, H. 2019. Metabolism and Biological Activities of 4-Methyl-Sterols. *Molecules* 24: 451.
[34] Moebius, F. F., Fitzky, B. U., Lee, J. N., Paik, Y.-K., Glossmann, H. 1998. Molecular Cloning and Expression of the Human Δ7-Sterol Reductase. *Proceedings of National Academy of Sciences, USA* 95: 1899.
[35] Bae, S.-H., Lee, J. N., Fitzky, B. U., Seong, J., Paik, Y.-K. 1999. Cholesterol Biosynthesis from Lanosterol. *Journal of Biological Chemistry* 274: 14624.
[36] Wasif, C. A., Maslen. C., Kacchilel-Linjewile, S., Lin, D., Linck, L. M., Connor, W. E., Steiner, R. D., Porter, F. D. 1998. Mutations in the Human Sterol Delta7-Reductase Gene at 11q12-13 Cause Smith-Lemli-Opitz Syndrome. *American Journal of Human Genetics* 63: 55.
[37] Hollman, A. 1991. Plants in Cardiology. *British Heart Journal* 65: 286.
[38] Dey, P., Kundu, A., Chakraborty, H. J., Kar, B., Choi, W. S., Lee, B. M., Bhakta, T., Atanasov, A. G., Kim, H. S. 2019. Therapeutic Value of Steroidal Alkaloids in Cancer: Current Trends and Future Perspectives. *International Journal of Cancer* 145: 1731–1744.
[39] Jiang, Q. W., Chen, M. W., Cheng, K. J., Yu, P. Z., Wei, X., Shi, Z. January 2016. Therapeutic Potential of Steroidal Alkaloids in Cancer and Other Diseases. *Medicinal Research Reviews* 36 (1): 119–143.
[40] Furbee, B. 2009. *Neurotoxic Plants. Clinical Neurotoxicology: Syndromes, Substances, Environments*. Elsevier Inc., Amsterdam.
[41] Heilpern, K. L. 1995. Zigadenus Poisoning. *Annals of Emergency Medicine* 25: 259–262.
[42] Hegnauer, R. 1986. *Chemotaxonomie der Pflanzen: Eine Übersichtüber die Verbreitung und die systematischeBedeutung der Pflanzenstoffe*, vol. 7. Springer Basel AG, Basel, p. 711.
[43] Xiang, M. L., Hu, B. Y., Qi, Z. H., Wang, X. N., Xie, T. Z., Wang, Z. J., Ma, D. Y., Zeng, Q., Luo, X. D. 2022. Chemistry and Bioactivities of Natural Steroidal Alkaloids. *Natural Products and Bioprospecting* 12: 23.
[44] Tang, J., Li, H. L., Shen, Y. H., Jin, H. Z., Yan, S. K., Liu, R. H., Zhang, W. D. 2008. Antitumor Activity of Extracts and Compounds from the Rhizomes of *Veratrum dahuricum*. *Phytotherapy Research* 22: 1093–1096.
[45] Bai, S. T., Zhu, G. L., Peng, X. R., Dong, J. R., Yu, M. Y., Chen, J. C., Wan, L. S., Qiu, M. H. 2016. Cytotoxicity of Triterpenoid Alkaloids from *Buxus microphylla* against Human Tumor Cell Lines. *Molecules* 21: 1125–1130.
[46] Ohyama, K., Okawa, A., Moriuchi, Y., Fujimoto, Y. 2013. Biosynthesis of Steroidal Alkaloids in Solanaceae Plants: Involvement of an Aldehyde Intermediate During C-26 Amination. *Phytochemistry* 89: 26–31.
[47] Ohyama, K., Okawa, A., Fujimoto, Y. 2014. Biosynthesis of Steroidal Alkaloids in Solanaceae Plants: Incorporation of 3β-Hydroxycholest-5-en-26-al into Tomatine with Tomato Seedlings. *Bioorganic Medicinal Chemistry Letters* 24: 3556–3558.
[48] Nakayasu, M., Umemoto, N., Ohyama, K., Fujimoto, Y., Lee, H. J., Watanabe, B., Muranaka, T., Saito, K., Sugimoto, Y., Mizutani, M. 2017. A Dioxygenase Catalyzes Steroid 16α-Hydroxylation in Steroidal Glycoalkaloid Biosynthesis. *Plant Physiology* 175: 120–133.
[49] Akiyama, R., Watanabe, B., Nakayasu, M., Lee, H. J., Kato, J., Umemoto, N., Muranaka, T., Saito, K., Sugimoto, Y., Mizutani, M. 2021. The Biosynthetic Pathway of Potato Solanidanes Diverged from that of Spirosolanes due to Evolution of a Dioxygenase. *Nature Communications* 12: 1300.

[50] Akiyama, R., Umemoto, N., Mizutani, M. 2023. Recent Advances in Steroidal Glycoalkaloid Biosynthesis in the Genus Solanum. *Plant Biotechnology* 40: 185–191.

[51] Liu, Y., Hu, H., Yang, R., Zhu, Z., Cheng, K. 2023. Current Advances in the Biosynthesis, Metabolism, and Transcriptional Regulation of α-Tomatine in Tomato. *Plants (Basel)* 12: 3289.

[52] Itkin, M., Heinig, U., Tzfadia, O. Bhide, A. J., Shinde, B., Cardenas, P. D., Bocobza, S. E., Unger, T., Malitsky, S., Finkers, R., Tikunov, Y., Bovy, A., Chikate, Y., Singh, P., Rogachev, I., Beekwilder, J., Giri, A. P., Aharoni, A. 2013. Biosynthesis of Antinutritional Alkaloids in Solanaceous Crops is Mediated by Clustered Genes. *Science* 341, 175–179.

[53] Abd Karim, H. A., Ismail, N. H., Osman, C. P. 2022. Steroidal Alkaloids from the Apocynaceae Family: Their Isolation and Biological Activity. *Natural Product Communications* 17: 1–14.

[54] Liu, L., Cao, J. X., Yao, Y. C., Xu, S. P. 2013. Progress of Pharmacological Studies on Alkaloids from Apocynaceae. *Journal of Asian Natural Product Research* 15: 166–184.

[55] Martin-Smith, M. 1965. Steroidal Alkaloids in Apocynaceae. *Nature* 206: 752–753.

[56] Dickel, D., Lucas, R., Macphillamy, H. B. 1959. A New Hypotensive Steroid Alkaloid from *Conopharyngia pachysiphon. Journal of American Chemical Society* 81: 3154–3155.

[57] Raffauf, R. F., Flagler, M. B. 1960. Alkaloids of the Apocynaceae. *Economic Botany* 14: 37–55.

[58] Pinder, A. R. 1964. Steroidal alkaloids. In: Coffey, S. (ed) *Rodd's Chemistry of Carbon Compounds*. Elsevier Inc., Amsterdam, pp. 381–465.

[59] Rahman, A., Muzaffar, A. 1988. Steroidal Alkaloids of Apocynaceae and Buxaceae. In: Brossi, A. (ed) *The Alkaloids: Chemistry and Pharmacology*. Elsevier Inc., pp. 79–239.

[60] Cheenpracha, S., Jitonnom, J., Komek, M., Ritthiwigrom, T., Laphookhieo, S. 2016. Acetylcholinesterase Inhibitory Activity and Molecular Docking Study of Steroidal Alkaloids from *Holarrhena pubescens* Barks. *Steroids* 108: 92–98.

[61] Devkota. K. P., Lenta, B. N., Fokou, P. A., Sewald, N. 2008. Terpenoid Alkaloids of the Buxaceae Family with Potential Biological Importance. *Natural Products Reports* 25: 612–630.

[62] Devkota, K. P., Choudhary, M. I., Ranjit, R., Samreen, S. N. 2007. Structure Activity Relationship Studies on Antileishmanial Steroidal Alkaloids from *Sarcococca hookeriana. Natural Product Research* 21: 292–297.

[63] Huo, S., Wu, J., He, X., Pan, L., Du, J. 2018. Two New Cytotoxic Steroidal Alkaloids from *Sarcococca hookeriana. Molecules* 24: 11.

[64] Rahman, A., Feroz, F., Naeem, I., Haq, Z., Nawaz, S. A., Khan, N., Khan, M. R., Choudhary, M. I. 2004. New Pregnane-Type Steroidal Alkaloids from *Sarcocca saligna* and their Cholinesterase Inhibitory Activity. *Steroids* 69: 735–741.

[65] Rahman, A., Anjum, S., Farooq, A., Khan, M. R., Choudhary, M. I. 1997. Two New Pregnane-Type Steroidal Alkaloids from *Sarcococca saligna. Phytochemistry* 46: 771–775.

[66] Devkota, K. P., Lenta, B. N., Choudhary, M. I., Naz, Q., Fekam, F. B., Rosenthal, P. J. 2007. Cholinesterase Inhibiting and Antiplasmodial Steroidal Alkaloids from *Sarcococca hookeriana. Chemical Pharmaceutical Bulletin* 55: 1397–1401.

[67] Zhang, P., Shao, L., Shi, Z., Zhang, Y., Du, J., Cheng, K. J. 2015. Pregnane Alkaloids from *Sarcococca ruscifolia* and Their Cytotoxic Activity. *Phytochemistry Letters* 14: 31–34.

[68] Devkota, K. P., Lenta, B. N., Wansi, J. D., Choudhary, M. I., Kisangau, D. P., Naz, Q., Sewald, N. 2008. Bioactive 5 Alpha-Pregnane-Type Steroidal Alkaloids from *Sarcococca hookeriana. Journal of Natural Products* 71: 1481–1484.

[69] Kalauni, S. K., Choudhary, M. I., Shaheen, F., Manandhar, M. D., Rahman, A., Gewali, M. B., Khalid, A. 2001. *Steroidal Alkaloids from the Leaves of Sarcococca coriacea of Nepalese Origin. Journal Natural Products* 64: 842–844.

[70] He, K., Du, J. 2010. Two New Steroidal Alkaloids from the Roots of *Sarcococca ruscifolia. Journal of Asian Natural Products Research* 12: 233–238.

[71] Adhikari, A. 2009. *Bioprospecting Studies on Sarcococca Coriacea* of Nepalese Origin, Ph.D. Thesis, University of Karachi, Karachi, Pakistan.

[72] Habermehl, G. 1962. Konstitution und Konfiguration des Samandaridins. *Angewandte Chemie* 74: 154.

[73] Lam. C. W., Wakeman, A., James, A., Ata, A. Gengan, R. M., Ross, S. A. 2015. Bioactive *Steroidal* Alkaloids from *Buxus macowanii* Oliv. *Steroids* 95: 73–79.

[74] Ata, A., Andersh, B. J. 2008. *Buxus Steroidal Alkaloids: Chemistry and Biology; The Alkaloids: Chemistry and Biology*. Academic Press, London, UK, pp. 191–213.

[75] Meshkatalsadat, M. H., Mollataghi, A., Ata, A. 2006. New Triterpenoidal Alkaloids from *Buxus hyrcana. Zeitschrift für Naturforschung B* 61: 201–206.

[76] Althaus, J. B., Jerz, G., Winterhalter, P., Kaiser, M., Brun, R., Schmidt, T. J. 2014. Antiprotozoal Activity of *Buxus sempervirens* and Activity-Guided Isolation of O-tigloylcyclovirobuxeine-B as the Main Constituent Active against *Plasmodium falciparum. Molecules* 19: 6184–6201.
[77] Rahman, A. U., Choudhary, M. I. 1999. Diterpenoid and Steroidal Alkaloids. *Natural Product Reports* 16: 619–635.
[78] Matochko, W. L. James, A., Lam, C. W., Kozera, D. J., Ata, A., Gengan, R. M. 2010. *Triterpenoidal Alkaloids from Buxus natalensis* and Their Acetylcholinesterase Inhibitory Activity. *Journal of Natural Products* 73: 1858–1862.
[79] Szabó, L. U., Schmidt, T. J. 2021. Investigation of the Variability of Alkaloids in *Buxus sempervirens* L. Using Multivariate Data Analysis of LC/MS Profiles. *Molecules* 27: 82.
[80] Khodzhaev, B. U., Sharafutdinova, S. M., Dzhabbarov, A., Yunusov, S. Y. 1983. Alkaloids of *Buxus sempervirens. Chemistry of Natural Compounds* 19: 512–513.
[81] Moss, D. E., Perez, R. G. 2021. Anti-Neurodegenerative Benefits of Acetylcholinesterase Inhibitors in Alzheimer's Disease: Nexus of Cholinergic and Nerve Growth Factor Dysfunction. *Current Alzheimer Research* 18: 1010–1022.
[82] Bachurin, S. O., Bovina, E. V., Ustyugov, A. A. 2017. Drugs in Clinical Trials for Alzheimer's Disease: The Major Trends. *Medicinal Research Reviews* 37: 1186–225.
[83] Mehta, D., Jackson, R., Paul, G., Shi, J., Sabbagh, M., Network, V. H., et al. 2017. *Why* Do Trials for Alzheimer's Disease Drugs Keep Failing? A Discontinued Drug Perspective for 2010–2015. *Expert Opinion on Investigational Drugs* 26: 735–739.
[84] Ferreira, D., Nordberg, A., Westman, E. 2020. Biological Subtypes of Alzheimer Disease: A Systematic Review and Meta-Analysis. *Neurology* 94: 436–448.
[85] Naz, S., Holloway, P., Ata, A., Sener, B. 2022. Exploring Medicinal Plants for the Development of Natural Enzyme Inhibitors. In: Mukherjee, P. K. (ed) *Evidence-Based Validation of Herbal Medicines.* Elsevier Inc., pp. 670–691.
[86] Ata, A. 1995. Phytochemical and Structural Studies on *Buxus hildebrandtii* and *B. smepervires*, Ph.D. Thesis, University of Karachi, Karachi, Pakistan.
[87] Schaller, H, 2003. The Role of Sterols in Plant Growth and Development. *Progress in Lipid Research* 42: 163–175.
[88] Zhang, Z. L., Luo, Z. L., Shi, H. W., Zhang, L. X., Ma, X. J. 2017. Research Advance of Functional Plant Pharmaceutical Cycloartenol About Pharmacological and Physiological Activity. *Zhongguo Zhong Yao Za* 42: 433–437.
[89] Asif, E., Ali, S. S., Nasir, H., Jamal, S. A., Ata, A., Farooq, A., Choudhary, M. I., Sener, B. 1992. New Steroidal Alkaloids from the Roots of *Buxus papillosa. Journal of Natural Products* 55: 1063–1066.
[90] Calegario, G., Pollier, J., Arendt, P., de Oliveira, L. S., Thompson, C., Soares, A. R., Pereira, R. C., Goossens, A., Thompson, F. L. 2016. Cloning and Functional Characterization of Cycloartenol Synthase from the Red Seaweed *Laurencia dendroidea. PLoS One* 11: e0165954.
[91] Bai, S. T., Zhu, G. L., Peng, X. R., Dong, J. R., Yu, M. Y., Chen, J. C., Wan, L. S., Qiu, M. H. 2016. Cytotoxicity of Triterpenoid Alkaloids from *Buxus microphylla* against Human Tumor Cell Lines. *Molecules* 21: 1125.
[92] Ata, A., Iverson, C. D., Kalhari, K. S., Akhter, S., Betteridge, J., Meshkatalsadat, M. H. 2010. Triterpenoidal Alkaloids from *Buxus hyrcana* and Their Enzyme Inhibitory, Anti-fungal and Anti-leishmanial Activities. *Phytochemistry* 71: 1780–1786.
[93] Yan, Y. X., Chen, J. C., Sun, Y., Wang, Y. Y., Su, J., Li, Y., Qiu, M. H. 2010. Triterpenoid Alkaloid Derivatives from *Buxus rugulosa. Chemistry & Biodiversity* 7: 1822–1827.
[94] Rahman, A.-U., Choudhary, M. I., Ata, A. 1992. New Tetrahydrofuranoid Steroidal Alkaloids from the Leaves of *Buxus hildebrandtii. Heterocycles* 34: 157–171.

Problems with Answers

1. What is the difference between α-D-glucose and β-D-glucose?

   Answer: α-D-glucose is prepared by recrystallizaion of glucose from methanol. Its melting point is 146 °C, and its specific rotation is $[\alpha]_D = +\ 112°$, whereas β-D-glucose is prepared by recrystallization of glucose from water. Its melting point is 150 °C, and its specific rotation is $[\alpha]_D = +\ 19°$.

2. What is amino sugar?

   Answer: Amino sugars are widely distributed in living organisms from microorganisms to humans. They are obtained by replacing the hydroxyl group of a monosaccharide with an amino group, with being D-glucosamine being the most common example. More than 60 amino sugars are known, with one of the most abundant being N-acetyl-D-glucosamine, the main component of chitin.

3. What is sorbitol?

   Answer: Sorbitol is a polyhydroxy compound that is nearly 60% sucrose. It is stable at high temperatures and can be effectively used in cooking and baking. It can be used as a sweeting agent by diabetic patients as its metabolism is insulin independent.

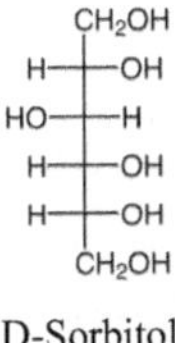

D-Sorbitol

4. What are grape sugar, fruit sugar, malt sugar, milk sugar, and table sugar?

   Answer: Glucose is known as grape sugar as it is naturally occurring in the fruits as well as other parts of the plants in free form.

   Fructose is known as fruit sugar and is digested and absorbed into the blood stream more slowly than added sugars.

   Maltose is known as malt sugar and is important in brewing beer, whisky, and malt vinegar.

   Lactose is known as milk sugar and is the only common sugar of milk origin.

   Sucrose is known as table sugar.

5. What is glycosylation?

   Answer: Glycosylation is a reaction in which a carbohydrate, a glycosyl donor, couples with a glycosyl acceptor to construct a glycoconjugate. Glycosylation needs a suitable activating condition to form glycoside. In the process, a glycosyl donor generally has a leaving group at the anomeric position, and the glycosyl acceptor has a nucleophilic acceptor that is coupled under optimum conditions to undergo glycosylation.

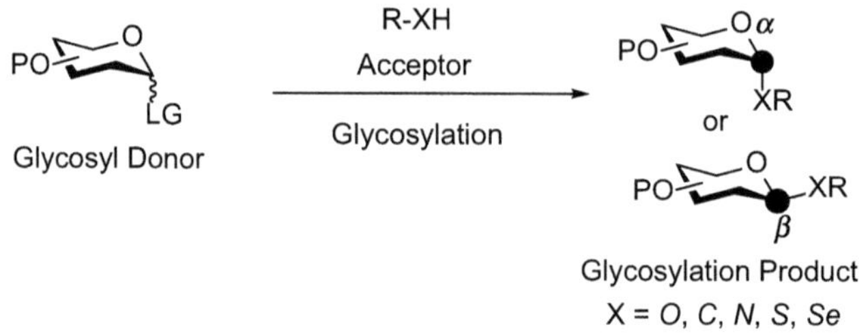

PO LG
Glycosyl Donor
R-XH
Acceptor
Glycosylation
PO O α XR
or
PO O XR β
Glycosylation Product
X = O, C, N, S, Se

6. What is the anomeric effect?

Answer: The anomeric effect is a stereoelectronic effect wherein the anomeric substituents prefer the axial orientation instead of the less-hindered equatorial orientation under steric considerations. For example, the anomeric effect in glucose is that α-methyl glucoside is more stable than ß-methyl glucoside due to hyperconjugation; in β-methyl glucoside, the methoxy group is at the equatorial position and cannot be involved in hyperconjugation. The anomeric effect was originally called the Edward–Lemieux effect, and Lemieux and Chu renamed it in 1958.

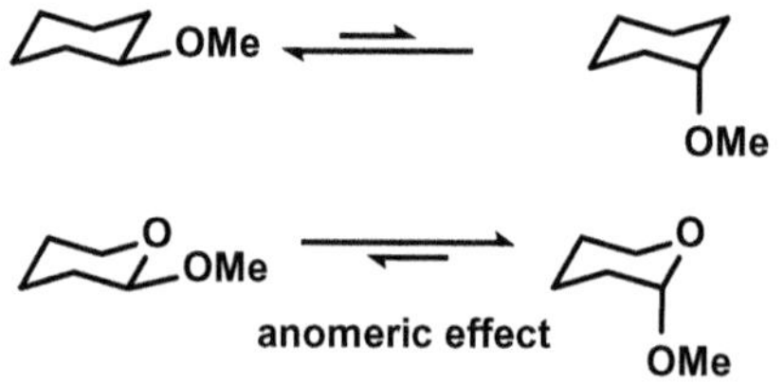

OMe
OMe
O OMe
O
anomeric effect
OMe

7. What is mutarotation?

Answer: The gradual change in the specific rotation of α and β forms of sugars until the constant specific rotation is attained in a solution. In 1844, mutarotation was discovered by Dubrunfaut, a French chemist. The α and β anomers are diastereomers of each other that show different specific rotations; the process involves a solution or liquid sample of a pure α anomer that will rotate the plane polarized light by a different amount and/or in the opposite direction than the pure β anomer of that compound. The optical rotation of the solution depends on the optical rotation of each anomer and its ratio in the solution. For example, if a solution of β-D- glucopyranose is dissolved in water, its specific optical rotation will be +18.7°. Over time, some of the β-D-glucopyranose will undergo mutarotation to become α-D-glucopyranose, which has an optical rotation of +112.2°. The rotation of the solution will increase from +18.7° to an equilibrium value of +52.7°.

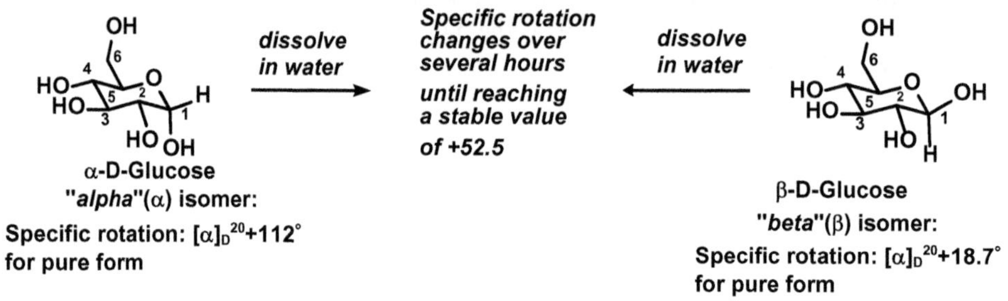

dissolve in water

Specific rotation changes over several hours until reaching a stable value of +52.5

dissolve in water

α-D-Glucose
"*alpha*"(α) isomer:
Specific rotation: $[\alpha]_D^{20}$+112° for pure form

β-D-Glucose
"*beta*"(β) isomer:
Specific rotation: $[\alpha]_D^{20}$+18.7° for pure form

8. What is the fate of mutarotation in sucrose?

   Answer: Sucrose, ordinary table sugar, is an example of a disaccharide, having α-D-glucopyranose and β-D-fructofuranose linked by a glycosidic bond at their anomeric carbons. However, sucrose does not undergo mutarotation because it is not a reducing sugar, and because of the absence of the hemiacetals, it does not undergo mutarotation.

9. Name and discuss the process when mutarotation has occurred at a position other than an anomeric position?

   Answer: Epimerization is the mutarotation process that occurs at positions other than anomeric positions and involves the interconversion of one epimer to another. For example, β-D-glucopyranose and β-D-mannopyranose are epimers because they differ only in the stereochemistry at the C-2 position: The C-2 hydroxy group is axial (up from the plane of the ring), while in β-D-glucopyranose is equatorial (in the plane of the ring). These two molecules are epimers, but because they are not mirror images of each other, they are not enantiomers.

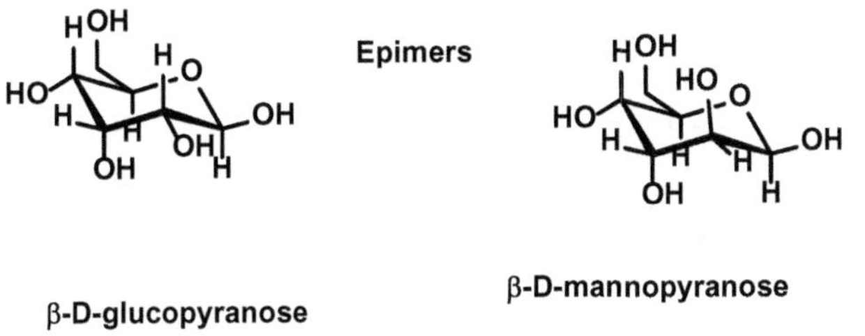

Epimers

β-D-glucopyranose

β-D-mannopyranose

10. Explain the difference between D- and L- sugars.

    Answer: In Fischer's projection, the position of the hydroxyl group on the chiral carbon farthest from the carbonyl group determines a D or an L sugar. In D- sugars, the –OH is on the right side, while in L- sugars, it is positioned on the left. D- sugars predominate in nature, although L-forms of some sugars, such as fucose, do exist.

D-Glucose

L-Glucose

11. What is the difference between lactose, maltose, and cellobiose? What kind of linkage is present?

    Answer: All three sugars are disaccharides. Lactose comprises condensed galactose and glucose and forms a β-1→4 glycosidic linkage. Maltose and cellobiose comprise two glucose units that are condensed and form an α-1→4 and β-1→4 glycosidic linkages, respectively.

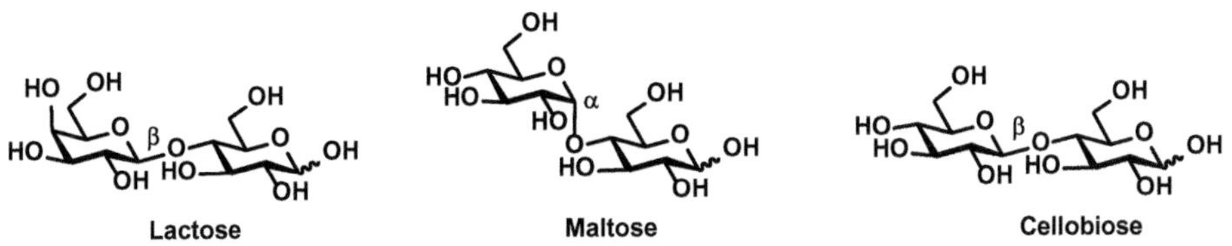

Lactose

Maltose

Cellobiose

12. What is the difference between glycone and aglycone?

Answer: The sugar part is known as glycone, and the nonsugar part is the aglycone. There are two basic classes of glycosides: C- glycosides in which the sugar is attached to the aglycone through C-C bond and O- glycosides in which the sugar is connected to the aglycone through the oxygen–carbon bond.

13. How can you distinguish α- and β- glycosides by NMR?

Answer: After separation, determining which anomer is the α- glucoside and which is the β-is possible in several ways. The most reliable method is J_{CH} analysis of the C-1 and H-1 atoms. An equatorial arrangement typically has a higher J_{CH} (~170 Hz) than an axial arrangement (~160 Hz).

10. How can you distinguish α-D- and β-D-glucpyranose by NMR?

Answer: One-dimensional proton nuclear magnetic resonance spectroscopy (^{1}H-NMR) has been a source of valuable structural information of polysaccharides, such as the number and the configuration of anomeric protons. The signals arising due to anomeric protons usually appear in the range of 4.3 ppm–5.9 ppm, and the protons of α-glycosides typically resonate 0.3 ppm–0.5 ppm downfield from those of the corresponding β-glycosides. In addition, the α-anomeric proton resonates farther downfield (5.1 ppm) from the β-anomeric proton (4.5 ppm), making these two anomer distinguishable by ^{1}H-NMR even at low field. The proton NMR spectrum also gives an idea about the constituent monosaccharides based on chemical shifts and spin–spin coupling constants ($J_{H\text{-}H}$).

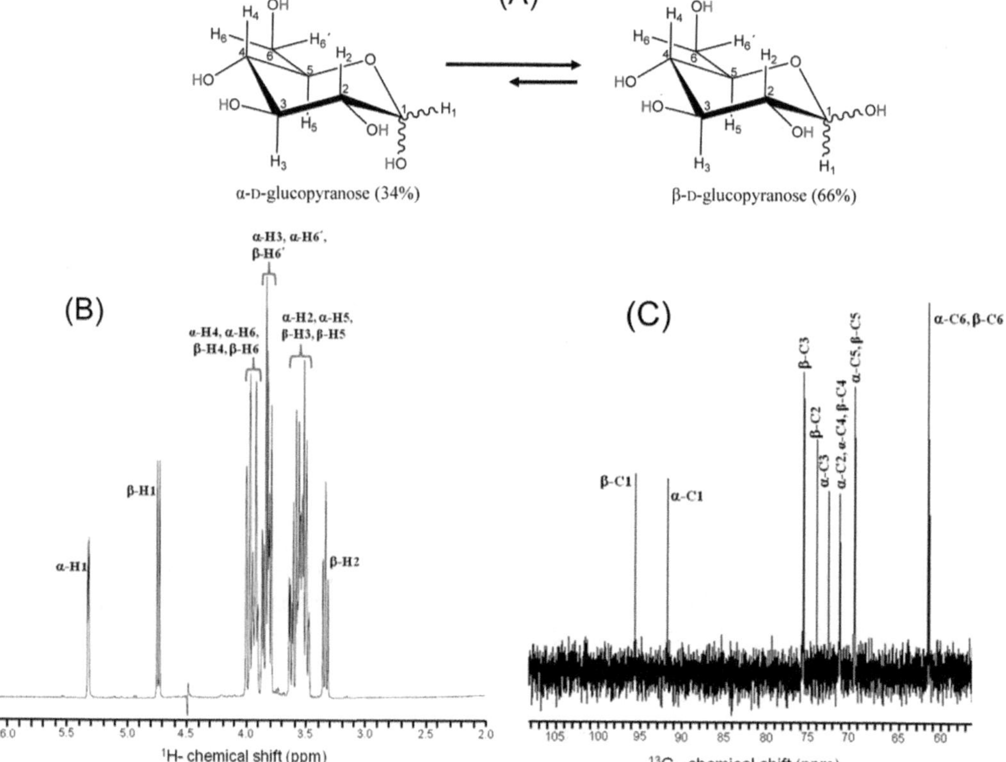

14. What are the anomers of glucose NMR?
    Answer: In glucose, the α-anomer has a dihedral angle of 60° and a coupling constant of 2.7 Hz, while the β-anomer has an angle of 180° and a coupling constant of about 7.2 Hz.

15. What is the difference between hemiacetal and acetal?
    Answer: Hemiacetal is an intermediate with one –OH and one –OR group at the central carbon atom bonded with four groups; the general formula for hemiactal is $R_1R_2C(OH)OR$. Actal hgas two –OR groups at the central carbon atom, and its general formula is $R_1R_2C(OR)_2$. Additionally, acetals are more stable than hemiacetals.

R_1, OH, R_2, OR Hemiacetal	R_1, OR, R_2, OR Acetal

16. Identify 1-thioglucopyranoside based on ^{1}H and ^{13}C-NMR spectra?
    Answer: The upper figure displays ^{1}H-NMR, and the lower figure displays ^{13}C-NMR.

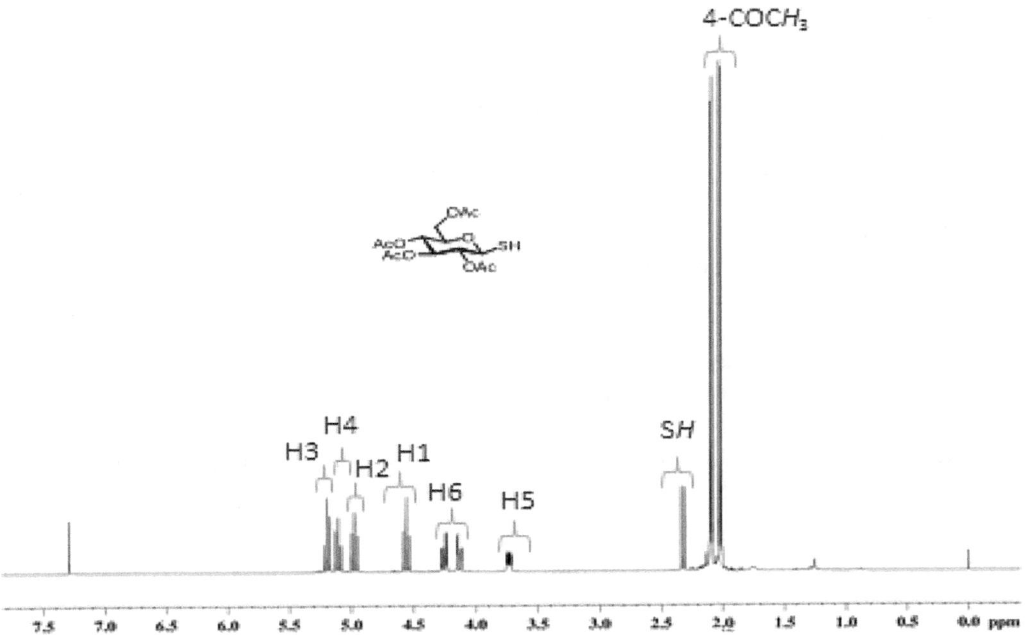

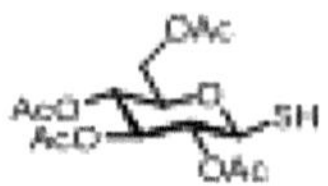

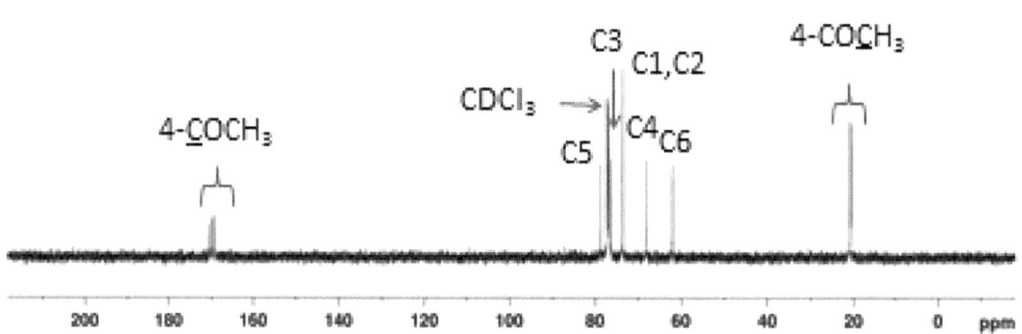

17. What is a sweet vaccine?
    Answer: A carbohydrate-based vaccine

18. Name a few members of *Enterobacteriaceae.*
    Answer: *Salmonella*, *Escherichia*, *Shigella*, *Klebsiella*, etc.

19. What is enterobacterial common antigen or ECA?
    Answer: A polysaccharide composed of the repeating unit [α-D-Fucp4NAc-(1→4)-β-D-ManpNAcA-(1→4)-α-D-GlcpNAc], present in all members of *Enterobacteriaceae.*

20. What is K-antigen?
    Answer: The capsule coating on the surfaces of bacteria surrounding the cell wall and constituted by polysaccharide.

21. Name a few *Shigella* strains.
    Answer: *Shigella sonnei, S. dysentriae, S. flexneri*

22. Why has the development of new synthetic strategies to convert carbohydrates into HMF attracted attention in the scientific community?
    Answer: Because of fossil resource depletion and because HMF is considered the most important platform molecule in the utilization of biomass.

23. Name three transformations of carbohydrates via catalysis into valuable chemical derivatives.
    Answer: i) One-step conversion of cellulose into alkyl-α,β-glycoside surfactants; ii) complete C–O-hydrogenolysis of sugars and sugar alcohols to alkanes; iii) conversion of inulin to HMF.

24. What is the NMR technique that makes it possible to distinguish between signals due to CH_3, CH_2, CH, and quaternary carbons in the steroidal frameworks?
    Answer: Distortionless enhancement by polarization transfer, a fundamental technique in combination with other 1D and 2D NMR experiments for the structural assignment of steroids.

25. Deuterium-labeled *myo*-inositol probes such as PI4P and PI5P (phosphatidylinositol-5-phosphate) allow the quantification of corresponding endogenous phospholipids and can additionally function as tracers for studying PIP metabolism. S.J. Conway et al. optimized the large-scale synthesis of deuterated *myo*-inositol, which was utilized in the synthesis of the PI4P and PI5P probes.[1]
Propose a synthesis of deuterated *myo*-inositol **2** starting from hydroquinone **1**.

1 → (6 steps) → **2**   D = Deuterium

Answer: The synthesis of deuterated *myo*-inositol **2** starting from hydroquinone **1** was described by Zimmermann and Trost.[2–4]

1 → (D_2SO_4, D_2O, reflux, 95%, ref.[2]) → **7** 93% D_4 → (I_2, 35% H_2O_2, *i*-PrOH, 45 °C, 89%, ref.[3]) → **8** 93% D_4 → (Br_2, $CHCl_3$, 0 °C, ref.[4]) → **9**

→ ($NaBD_4$, D_2O, Et_2O, 0 °C) → **10** → (Ac_2O, K_2CO_3, AcOH, reflux, 29% over 3 steps) → **rac 11** 90% D_6 → ($NaIO_3$, $RuCl_3$, MeCN, H_2O, 0 °C) → **rac 12**: R^1 = H, R^2 = Ac; **rac 13**: R^1 = Ac, R^2 = H

→ (Et_3N, MeOH/H_2O, r.t., 55% over 2 steps) → **2** 84–90% D_6

1. Joffrin, A. M.; Saunders, A. M.; Barneda, D.; Flemington, V.; Thompson, A. L.; Sanganee, H. J.; Conway, S. J. Development of Isotope-Enriched Phosphatidylinositol-4- and 5-Phosphate Cellular Mass Spectrometry Probes. *Chem. Sci.* **2021**, *12* (7), 2549–2557. doi.org/10.1039/d0sc06219g.
2. Zimmermann, H. Specifically Deuteriated Intermediates for the Synthesis of Liquid Crystals and Liquid-Crystalline Polymers. *Liq. Cryst.* **1989**, *4* (6), 591–618. doi.org/10.1080/02678298908033195.
3. Saito, T., Ikemoto, K., Tsunomachi, H.; Sakaguchi, K. *US Pat.*, 4973720, **1990**.
4. Trost, B. M.; Patterson, D. E.; Hembre, E. J. Dynamic Kinetic Asymmetric Transformations of Conduritol B Tetracarboxylates: An Asymmetric Synthesis of d-*mYo*-Inositol 1,4,5-Trisphosphate. *J. Am. Chem. Soc.* **1999**, *121* (46), 10834–10835. doi.org/10.1021/ja992960v.

26. Explain why *C2*-symmetric phosphoramidites are used as desymmetrizing agents.
Answer: Phosphorus, like nitrogen, bears a lone electron pair. However, the energy barrier for electron pair inversion (pyramidal inversion) is much higher in phosphorus;

room temperature is often not enough to cross this energetic barrier. Thus, a phosphorus atom bearing three different substituents (and one electron pair) can act as a center of chirality in the same way as carbon with four different substituents. C_2 symmetry at phosphoramidite ligands avoids chirality at phosphorus because the superimposability of the molecule is independent of electron pair orientation.

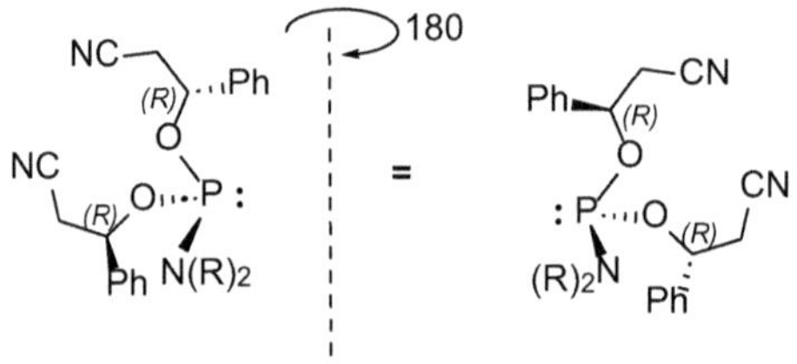

27. The global supply 1*H*-tetrazole is becoming scarce. Explain why (draw intermediate) dicyanoimidazole (DCI) can be used as 1*H*-tetrazole analogue in phosphoramidite transfer reactions.
    Answer: DCI **14** shares chemical similarities with phosphoramidite activating agent 1*H*-tetrazole **15** (pKa = 4.8). DCI is slightly less acidic (pKa = 5.2) but significantly more soluble in acetonitrile and more nucleophilic than 1*H*-tetrazole. Its ability to be a good leaving group along with having acidic hydrogen present on the nitrogen atom makes it possible to participate in the catalytic cycle like 1*H*-tetrazole.

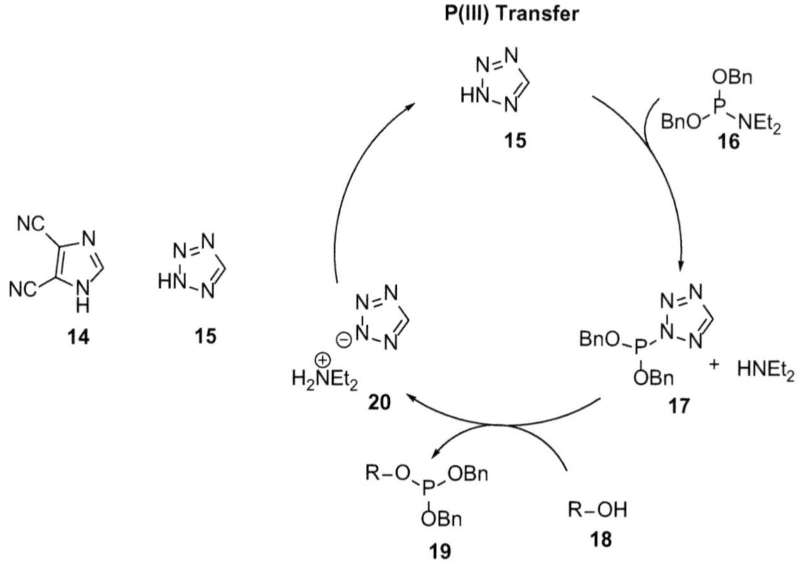

P(III) Transfer

28. In the total synthesis of tetrodotoxin, the inositol-skeleton bearing intermediate **4** was synthesized from protected D-glucose derivative **3**. Give the reaction sequence.

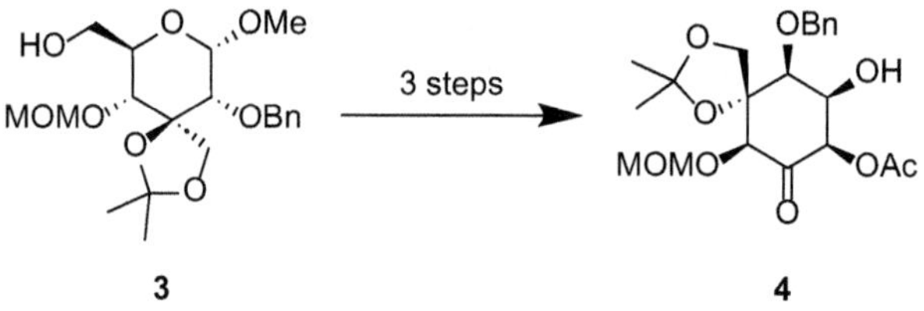

Answer: The proposed reaction sequence is based on Ferrier carbocyclization as a key step, developed by Sato et al.[5]

3 → TFAA, DMSO, Et_3N, CH_2Cl_2 → 21 → K_2CO_3, Ac_2O, MeCN, 72% over 2 steps → 22 → 1) $Hg(OAc)_2$, AcOH; 2) NaCl, H_2O-Acetone, 58% → 4

5. Akai, S.; Seki, H.; Sugita, N.; Kogure, T.; Nishizawa, N.; Suzuki, K.; Nakamura, Y.; Kajihara, Y.; Yoshimura, J.; Sato, K.-I. Total Synthesis of (–)-Tetrodotoxin from d-Glucose: A New Route to Multi-Functionalized Cyclitol Employing the Ferrier(II) Reaction toward (–)-Tetrodotoxin. *Bull. Chem. Soc. Jpn.* **2010**, *83* (3), 279–287. doi.org/10.1246/bcsj.20090194.

29. Racemic intermediate 6, useful for the synthesis of several natural products, has been synthesized from *myo*-inositol orthoformate 5. Propose the reaction sequence.

5 → 4 steps → 6

Answer: Intermediate **6** was synthesized from *myo*-inositol orthoformate **5** using the protocol developed by Sureshan and Potter.[6, 7]

5 → NaH, PMBCl, DMF, r.t., 40% → 23 → DMSO, $(COCl)_2$, CH_2Cl_2, Et_3N, -60 °C, 92% → 24 → PPh_3CH_2OMeCl, *t*-BuOK, THF, 0 °C, 95% → 25 → aq. HCl, THF, quant. → 6

6. Mondal, S.; Prathap, A.; Sureshan, K. M. Vinylogy in Orthoester Hydrolysis: Total Syntheses of Cyclophellitol, Valienamine, Gabosine K, Valienone, Gabosine G, 1-*Epi*-Streptol, Streptol, and Uvamalol A. *J. Org. Chem.* **2013**, *78* (15), 7690–7700. doi.org/10.1021/jo401272j.
7. Riley, A. M.; Potter, B. V. L. Synthesis of a Conformationally Restricted Cyclic Phosphate Analog of Inositol Triphosphate. *J. Org. Chem.* **1995**, *60* (16), 4970–4971. doi.org/10.1021/jo00121a006

30. What are nanocatalysts?

Answer: Nanocatalysts are homogeneous or heterogeneous catalysts synthesized from nanoparticles. Nanoparticles have larger surface-to-volume ratios than bulk materials, which makes them attractive candidates for use as catalysts.

31. What is starch? Suggest a reaction mechanism for the synthesis of *N*-substituted pyrroles using starch.

Answer: Starch is a mixture of two polysaccharides, amylase, and amylopectin. Its reaction mechanism for the synthesis of *N*-substituted pyrroles is as follows:

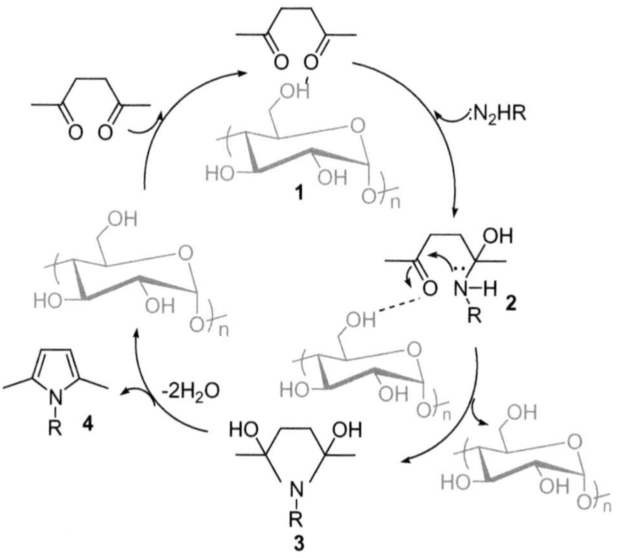

32. What is the difference between starch and cellulose?

Answer: Cellulose and starch have similar structures, but they differ in the type of 1,4 linkage present. Specifically, in cellulose, glucose units are connected by a beta acetal linkage, while starch contains an alpha linkage for this acetal moiety. These linkages are stereoisomers, resulting in distinct structural differences between the two polysaccharides.

33. What is the best solvent in the synthesis of *N*-substituted pyrroles using starch? Explain the rule of substituted effects on the rate reaction in the synthesis of *N*-substituted pyrroles using starch. What are advantages of this method?

Answer: The best solvent is none. The electron-donating groups on phenyl ring increase the yield and reduce the reaction time relative to electron-withdrawing groups. The major advantages this procedure are high atom economy because of the solvent-free reaction, the reusability of the catalyst, the short reaction time, the absence of a heat requirement, and its waste-free nature given that H_2O is the only byproduct.

34. There is wide and continuous availability of carbohydrates in nature as renewable raw materials. They are essential for human survival [1].
Discuss some applications of carbohydrates in foods and food packaging.
Answer: Carbohydrates have a wide range of properties that make them applicable in the food-related industries including solubility, sweetening, hygroscopicity, crystallization inhibition, flavor, and coating ability. They serve as energy sources and fuel and metabolic intermediates (fructans, starch, and glycogen) [2, 3]. Carbohydrates in the forms of gums, starches, and pectin are used in cakes, jams, noodles, cookies and other baked goods, and canned food as thickening agents. Sucrose in its pure form is simple table sugar, and it is used as a sweetener in baked goods and ice cream; corn syrup and sucrose combine to impart sweetness, a smooth texture, and a glossy appearance [4]. Sucralose, a trichloro-trideoxygalactosucrose, is 650 times sweeter than sucrose and has been marketed as a noncaloric sweetener under the brand name Splenda® [5]. Isomaltulose, a 6-*O*-α-D-glucosyl-D-fructose, is isomeric with sucrose, from which it is produced at an approximate 60 000-t/yr scale by Südzucker [6]. The industrial process involves a glucosyl shift from the 2^f-*O* of sucrose to the 6^f-OH, effected by action of an immobilized *Protaminobacter rubrum*-derived α(1→6)-glucosyltransferase [7], a transformation that most likely proceeds through a closed-shell intermediate, i.e. without full separation of the glucose and fructose portions. The isomaltulose produced is subsequently hydrogenated to isomalt, an approximate 1:1 mixture of the terminally α-glucosylated glucitol and mannitol [7] which is on the market as a low-caloric sweetener [8].

Sucrose [9, 10] is a valuable raw material used in the preparation of various fats and oils both by itself and as an extender. In this process, the sucrose is hydrolyzed by acid to generate glucose and fructose, which are mixed into other fatty acids or their salts to obtain triglycerides. The mixed triglycerides are then esterified with fatty acids or their salts to obtain oils such as cod liver oil, oleic acid, and glycerol esters. Alginate is a slow-releasing carbohydrate that gives foods texture and shape.

Polysaccharides are used for thickening, stabilization, and preservation of food items, and in processing frozen foods, modified starch maintains solubility and homogenizes texture to prevent dressings from weeping and other products bound and intact. Carboxy methyl-cellulose[3,4] (CMC) is used as a thickening and binding agent that extends shelf life in cakes, gelatinous desserts, cottage cheese, dressings and toppings, doughs, etc. Sodium CMC forms a water-soluble coating that is used in making greaseproof paper, commonly known as butter paper, for packaging oily foodstuffs. Chitosan is used as a coating that gives paper of higher wet strength and antibacterial and grease barrier properties.

Industrially, carbohydrates are used in food packaging to keep food from spoiling. Starch- and cellulose-based biopolymers are promising alternatives to nonrenewable resins for sustainable, environmentally friendly single-use food packaging products. Xanthan is a pentasaccharide that controls viscosity in food packaging. Agar shows excellent film-forming ability, thermoplasticity, environmental adaptability, biodegradability, and water resistance [11]. Chitosan, pectin, and galactomannans are also useful in food packaging films and coatings [12].

REFERENCES

[1] D. de Wit, L. Maat, A.P.G. Kieboomc, *Carbohydrates as Industrial Raw Materials: Industrial Crops and Products*, Vol 2, 1993, pp. 1–12.
[2] F.W. Lichtenthaler (Ed.), *Carbohydrates as Organic Raw Materials*, VCH Publ., Weinheim/New York, Vol. I, 1991, pp. 365–374.
[3] H.-D. Belitz, W. Grosch, P. Schieberle, *Text Book of Food Chemistry*, Springer, Berlin, Heidelberg, 2009.
[4] L.H. Meyer, *Food Chemistry*, CBS Publication, 2006.
[5] F.W. Lichtenthaler, Carbohydrates as Green Raw Materials for the Chemical Industry. *Carbohydr. Res.*, 313 (1998), p. 69; M.R. Jenner, T.H. Grenby (Ed.), *Sweeteners*, Elsevier Applied Science, London/New York, 1989, p. 121.
[6] R.S. Dwivedi, Saccharide Sweet (SS) Principles, Classification and Structural and Functional Details of SS Sweeteners and Plants. *Alternative Sweet and Supersweet Principles* (2022), pp. 113–122. Springer Nature Singapore.
[7] M. Kunz, *Ullmann's Encyclopedia of Industrial Chemistry*, 5th Ed., A25, Wiley VCH, 1994, p. 426.
[8] F.W. Lichtenthaler, S. Peters, Carbohydrates as Green Raw Materials for Chemical Industry. *Comptes. Rendus. Chimie.*, 7, no. 2 (2004), pp. 65–90.
[9] F.W. Lichtenthaler, P. Pokinskyj, S. Immel, *Sucrose as a Renewable Organic Raw Material*, Zucker Industrie, Vol. 121, 1996, pp. 174–190.
[10] J. Plaza-Diaz, A. Gil, Sucrose: Dietary Importance, in *Encyclopedia of Food and Health*, Academia Press, 2016, pp. 2205–2211.
[11] R. Sothornvit, Nanostructured Materials for Food Packaging Systems: New Functional Properties, *Curr. Opin. Food Sci.,* 34 (2019), p. 312.
[12] M. Mesqari, A.H. Alami, T. Sathyapalan, A. Sahebker, A Comprehensive Review of the Development of Carbohydrate Macromolecules and Copper Oxide Nanocomposite Films in Food Nanopackaging, *Bioinorg. Chem. Appl.*, 2022 (2022), pp. 28–32. | Article ID 7557825. https://doi.org/10.1155/2022/7557825.

35. Discuss some applications of carbohydrates in textiles.

Answer: Starch products are used in textile industries to strengthen [1] warp yarns and improve their resistance to abrasion during weaving. They are also employed in finishing fabrics and glazing sewing threads as well as in printing. Cellulose is used in rayon and bamboo fabrics, and nanocellulose is increasingly being used as a filler textile nanocomposites. Some biodegradable fibers include bacterial cellulose and other microbial polysaccharides. Chitosan is used in spinning, dyeing, printing [2], and finishing.

Several derivatives of β-cyclodextrin have been synthesized to enable it to attach to textile fibers permanently. Also, different methods have been developed for grafting of native and modified β-cyclodextrin on textile fibers.

REFERENCES

[1] F.W. Lichtenthaler, S. Peters, Carbohydrates as green raw materials for chemical industry. *Compts Rendus Chimie*, 7, no. 2, 2004, 65–90.
[2] A. Banerjee, *Use of Novel Polysaccharides in Textile Printing*. PhD thesis, Colorado State University, Colorado, 1913.

36. Discuss some applications of carbohydrates in bioplastics.

Answer: Recently degradable bioplastics are made of non-food carbohydrates such as straw, corn stover, bagasse or protein materials. The first known bioplastic (PHB) was discovered by a French scientist Maurice Lemoigne from bacterium Bacillus megaterium. Bioplastics are three types as (1) Starch-based Bioplastics: Simple bioplastic derived from corn starch (2) Cellulose-based Bioplastics: Produced using cellulose esters and cellulose derivatives (3) Protein-based Bioplastics: Produced using protein sources such as wheat gluten, casein, and milk.

Bioplastics vary in their solubility in water depending on their composition and structure. Some bioplastics are designed to be water – soluble for specific applications, while others are intended be more resistance. The advantages of carbohydrates made bioplastics as (1) Bioplastics will preserve existing fuels (2) It gives first decomposition in weeks or months (3) It emits less carbon compared to traditional plastics (4) Better for health (PLA and PHB) as they are suitable for food packaging.

Starch-, cellulose-, and protein-based bioplastics are widely used in food packing [1, 2]. Approximately 50% of the bioplastics used commercially are prepared from starch because starches have high tensile strength, biodegradability, and ease of manufacturing.

The disadvantages of carbohydrates made bioplastics as (1) The price of bioplastic is more than traditional plastics (2) The land required for growing crops competes with land for food production (3) Some bioplastics require a specific disposal procedure and industrial compositing with advanced machine and facilities (4) the main resources of bioplastics are crops, the fertilizers and pesticides used in growing the crops and the chemical processing needed to turn organic material into plastic will emit pollutants to the environment.

Approximately 50% of the bioplastics used commercially are prepared from starch because the starch gives several advantages such as better tensile strength, biodegradability, and ease of manufacturing. The production of starch-based bioplastics is simple, and they are widely used for packaging applications. For trade applications, the starch-based plastics are regularly mixed with eco-friendly polyesters. Cellulose based bioplastics are made of cellulose esters and cellulose derivatives. Protein based bioplastics are made of wheat gluten

Starch-based bioplastics are non-toxic, biodegradable, and have good mechanical properties. They are also semi-permeable to carbon dioxide and insoluble in cold water or alcohol. Starch-based materials are used in loose fill foams for transport packaging and in service ware like cups, plates, and cutlery.

REFERENCES

[1] Abidin, M.Z.A.Z.; Julkapli, N.M.; Juahir, H.; Azaman, F.; Sulaiman, N.H.; Abidin, I.Z. Fabrication and properties of chitosan with starch for packaging application. *Malays. J. Anal. Sci.* **2015**, *19*, 1032–1042.

[2] Edhirej, A.; Sapuan, S.M.; Jawaid, M.; Zahari, N.I. Cassava: Its polymer, fiber, composite, and application. *Polym. Compos.* **2017**, *38*, 555.

37. What are designer lipids?

    Answer: Novel, health-friendly lipids with potential application in foods, nutraceuticals, pharmaceuticals including for treating obesity, cancer, heart disease, and inflammation.

38. What are important parameters and components for the quality of foods in the commercial food industry?

    Answer: Moisture, proteins, carbohydrates, glucose, starch, cellulose, and fat.

DETERMINATION OF MOISTURE CONTENT

The importance of moisture content determination in food is required for the texture and taste, microbial stability, economic, and food processing operations. For the determination of the moisture content of a food, it is important to prevent any loss or gain of water. (a) Exposure of a sample to the atmosphere and excessive temperature fluctuations should be minimized. (b) When samples are stored in containers, it is common practice to fill the container to the top to prevent a large headspace, because this reduces changes in the sample due to equilibration with its environment.

The moisture content of a food material is determined by the following equation:

$$\text{Moisture}\ (100\%) = \frac{m_{int} - m_{final}}{m_{int}} \times 100$$

Here, $m_{initial}$ and m_{final} are the mass of the sample before and after drying, respectively.

Total Solid Content

The total solids content is a measure of the amount of material remaining after all the water has been evaporated.

Thus, total solids (100%) = 100–100% Moisture

$$\%\text{Total Solids} = \frac{M_{DRIED}}{M_{INITIAL}} \times 100$$

Total Solids (%) = m_{final} × 100

$m_{initial}$

Thus, %Total solids = (100–%Moisture)

DETERMINATION OF PROTEIN

Protein content is also important in the food industry, and the Kjeldahl method is internationally recognized for estimating the protein content in foods. The method consists of heating a substance with sulfuric acid and catalyst to decompose the organic substance by oxidation and liberate the reduced nitrogen as ammonium sulfate, here, potassium sulfate and copper sulfate. Potassium sulfate reduces the boiling point and thus ensures complete reaction, and copper sulfate acts as catalyst. An excess of sodium hydroxide is added that converts ammonium salt into ammonia. The liberated ammonia is absorbed by an excess of standard acid, and the unused acid is titrated back with sodium hydroxide solution.

$$\text{Sample digestion} + H_2SO_4 \rightarrow (NH_4)_2SO_4(aq) + CO_2(g) + SO_2(g) + H_2O(g)$$

$$\text{Evolution of ammonia: } (NH_4)_2SO_4(aq) + 2NaOH \rightarrow Na_2SO_4(aq) + 2H_2O(l) + 2NH_3(g)$$

Materials required: Purified protein containing sample, Kjeldahl flask, (1v) concentrated sulfuric acid catalyst (potassium sulfate and copper sulfate [1:1], Standard sodium hydroxide solution.

Methodology

Accurately weighed (0.5 gm to 1 gm) of substance is digested with 10 ml of concentrated sulfuric acid in a Pyrex Kjeldahl flask with a small amount of potassium sulfate and copper sulfate. The heating is continued under a fume board till the brown color of the initial liquid disappears and the contents are clear as before. Generally, this takes place within 60 to 90 minutes, and all the nitrogen in the substance is converted to ammonium sulfate.

The flask is cooled, and 50 ml of ammonia-free distilled water is added. The solution is then carefully transferred into a 1 L flask. An excess of chilled sodium hydroxide solution is poured slowly down the side of the flask and refluxed with a Kjeldahl trap and a water condenser. The flask is heated, and the liberated ammonia is absorbed into sulfuric acid. The Kjeldahl trap prevents sputtering of alkali into acid under vigorous boiling. After one hour, when no more ammonia evolves (tested with red litmus), the receiver is removed. The excess acid is determined by titration with standard alkali, using phenolphthalein as indicator.

V ml of N H_2S0_4 is required for complete neutralization of ammonia evolved

= V ml of N H_2S0_4 = V ml of N NH_3

1000 ml of NH_3 contains 1 gm of ammonia or 14 gm of nitrogen

Amount of nitrogen present in V ml of N NH_3 = 14 × V × W gm

$$\text{Percentage of nitrogen} = \frac{\text{Weight of nitrogen}}{\text{Amount of substance taken}} \times 100$$

$$= \frac{14 \times V \times 1N \times 100}{1000 \times W}$$

Where N stands for normality of acid used

V stands for volume of acid used

W stands for organic sample Kjeldahlised

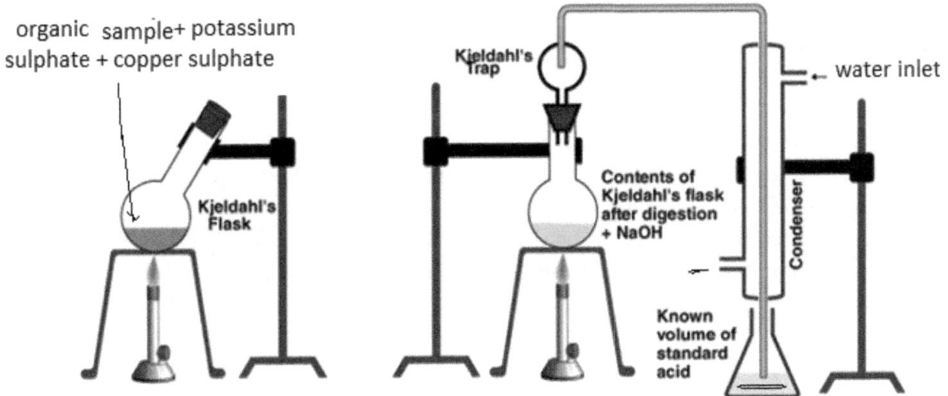

Nitrogen estimation by Kjeldahl's method

where N stands for normality of acid used, V for volume of acid used, and W for amount of substance taken in Kjeldahl flask.

It is also possible to calculate the amount of crude protein in the sample. Although there are differences between samples, crude protein (CP) can be measured by multiplying the percent nitrogen by a factor (PF, usually 6.25), Correct factor for a sample can be found by dividing 100 by the % nitrogen content of protein. CP = % N × 6.25 (PF).

DETERMINATION OF CARBOHYDRATE

THEORY

Carbohydrates are carbon compounds that contain hydrogen and oxygen in the ratio of 2:1. These compounds are of fundamental importance in living organisms as a source of metabolic energy. Their content can be estimated by reacting anthrone with carbohydrate to produce a blue-green solution. This reaction is not suitable for carbohydrate–protein mixtures; the output is red instead.

MATERIALS REQUIRED

Anthrone reagent (2 g/liter in conc H_2SO_4), standard glucose solution (0.1 g/1 liter), unknown glucose, colorimeter

PROCEDURE

Add 5 ml of the anthrone to 1 ml protein-free glucose solution in a test tube. Place the test tube in water bath for 15 minutes with a marble on top to prevent the loss of water by evaporation. Cool the solution and read the extinction at 620 nm against reagent blank. Repeat the experiment with a different concentration of glucose and prepare the standard curve for the glucose solution. Find the unknown concentration of glucose solution.

Prepare the DNS reagent by mixing 500 ml of sodium potassium tartrate (600 g/liter) and 200 ml of DNS solution. Add 1 ml of the reagent to 4 ml sugar solution. Prepare a blank by adding 1 ml of reagent to 4 ml of distilled water. Cover the test tube with a marble and place it in boiling water bath for 10 minutes, cool it to room temperature, and read the extinction at 540 nm against the blank. Repeat the experiment with different concentrations of glucose solution and prepare a standard curve of sugar. From the curve, estimate the concentration of unknown sugar solution.

ESTIMATION OF GLUCOSE

THEORY

The method is highly specific for glucose. In this method, glucose oxidase catalyzes the oxidation of β-D-glucopyranose to D-glucono-1, 5-lactone. With the formation of hydrogen peroxide, the lactone is then slowly hydrolyzed to D-gluconic acid. Peroxidase is incorporated into the reaction and catalyzes the hydrogen peroxide reaction with the chromogen ABTS (2,2,-azino-di-(3-ethylbenze-thiozoline)-6-sulphonate) to give color; its extinction is measured at 437 nm. Peroxidase is used to break down hydrogen peroxide to half molecules of oxygen and water. The oxygen combines with ABTS to produce color.

MATERIALS REQUIRED

- Glucose oxidase reagent: This reagent is prepared fresh. Place 50 mg of glucose oxidase in 280 mg of ABTS in sodium phosphate buffer (0.5 mol/liter). Add 4 mg of peroxidase and make up the volume to 250 ml.
- Sodium sulfate and zinc sulfate reagent: Dilute 28 ml of zinc sulfate solution. (10 g/100 ml) in 1 litre of sodium sulfate (93 ml mol/liter) solution.
- Standard glucose solution (0.5 ml mol/liter)

PROCEDURE

Take 0.2 ml of glucose solution in a sterilized test tube and add 4 ml of sodium sulfate and zinc sulfate reagent. Mix the solution thoroughly and then add 0.2 ml of 0.5 mol/liter sodium hydroxide, centrifuge at 1000 rpm for 5 minutes, and take 1 ml of the supernatant. Add 10 ml of glucose oxidase reagent and incubate it for 1 hour at 30 °C; read the extinction at 437 nm against distilled water blank. Prepare a standard curve with different concentration of glucose solution and repeat the method. From the curve, find the unknown concentration of glucose solution.

ESTIMATION OF STARCH

THEORY

Starch is hydrolyzed to glucose and dehydrated to hydroxymethyl furfural in hot acid medium. The compound forms a green product in anthrone.

MATERIALS REQUIRED

Anthrone reagent prepared as follows: 200 mg anthrone in 100 ml of ice cold 95% sulfuric acid, 80% ethanol, 52% perchloric acid; standard glucose stock prepared by 100 mg in 100 ml water. Working standard is 10 ml prepared stock diluted in 100 ml of water.

PROCEDURE

Homogenize 0.1 g to 0.5 g of the sample in hot 80% ethanol to remove sugars and then centrifuge. Wash the residue repeatedly with hot 80% ethanol till washing with anthrone produces a colorless solution. Dry the residue over a water bath. Next add 5 ml of water and 6.5 ml of 52 % perchloric acid. Extract at 0 °C for 20 minute. Centrifuge and save the supernatant. Repeat the process with fresh percholric acid, centrifuge the supernatants, and make up the volume to 100 ml. Pipette out 0.1 or 0.2 ml of the supernatant and make up to 1 ml with water. Prepare the standards by taking 0.2 ml, 0.4 ml, 0.6 ml, 0.8 ml, and 1 ml of the working standard and make up the volume to 1 ml in each tube with water. Add 4 ml of anthrone reagent to each tube. Heat for eight minutes in a boiling water bath, cool rapidly, and check the intensity of the green color at 650 nm. Find the glucose in the sample using the standard graph and multiply the result by 0.9 to arrive at the starch content.

ESTIMATION OF CELLULOSE

THEORY

Cellulose is a polysaccharide composed of glucose units joined in the form of repeating units of the disaccharide cellobiose with numerous crosslinkage.

MATERIALS

Mix 15 ml of 18% acetic acid and 15 ml of concentrated nitric acid. Prepare anthrone reagent is by dissolving 200 mg anthrone in 100 ml concentrated sulfuric acid; prepare the reagent fresh and chill for 2 hours. Prepare 67% sulfuric acid with water.

PROCEDURE

Add 3 ml acetic/nitric reagent to a known amount (0.5 g or 1 mg) of the sample in a test tube and mix in a vortex mixer. Place the tube in a water bath at 100 °C for 30 minutes. Cool and centrifuge the contents for 15–20 minutes. Discard the supernatant and wash the residue with distilled water. Add 10 ml of 67% sulfuric acid and allow it to stand for 1 hour. Dilute 1 ml of the solution to 100 ml. To 1 ml of the diluted solution, add 10 ml of anthrone reagent and mix well. Heat the tubes in a boiling water bath for 10 minutes. Cool and measure the color at 630 nm. Set a blank with anthrone reagent and distilled water. Take 100 mg cellulose in a test tube and add 10 ml of 67% sulfuric acid for standard. Take a series of volumes (say 0.4 to 2 ml corresponding to 40 mg–200 mg cellulose) and develop color. Draw the standard graph and calculate the amount of cellulose in the sample.

DETERMINATION OF HEMICELLULOSE

THEORY

Hemicelluloses are noncellulosic, nonpectic cell wall polysaccharides including xylans, mannans, glucomannans, galactans, and arabinogalactans. Here the method is applied gravimetric method for determination of cellulose [1, 2].

MATERIALS REQUIRED

Weigh 18.61 g disodium ethylenediamine tetraacetate and 6.81 g sodium borate decahydrate and transfer the weighed compounds to a beaker. Dissolve the product in about 200 ml of distilled water by heating in a water bath. Add to this about 100 ml solution containing 30 g of sodium lauryl, 4.5 g of disodium hydrogen phosphate, and 10 ml of 2 –ethoxy ethanol. Make up the volume to 1 L and adjust the pH to 7.0 (this is neutral detergent solution). Other required chemicals are decahydronapthelene, sodium sulfate, and acetone.

PROCEDURE

Take 1 g of the powdered sample in a refluxing flask and add 10 ml of cold neutral detergent solution. Add 2 ml of decahydronapthelene, and 0.5 g sodium sulfate. Heat to boiling and reflux for 60 minute. Filter the content through sintered glass crucible by suction and wash with hot water. Finally give two washings with acetone. Transfer the residue to a crucible and dry at 100° C for 8 hours. Cool the desiccator and weigh.

REFERENCES

[1] Nag, A., *Analytical and Instrumental Techniques in Agriculture, Environmental and Engineering*, Prienticel, 2016.

[2] Goering, H.D. and Vanscest, P.J., *Forage Fiber Analysis*, US Dept of Agriculture Research Service, Wahington, 1975.

DETERMINATION OF FAT

Fat content of foods is important to know, and the Soxhlet method is recognized by the Association of Official Analytical Chemists as the standard method for crude fat analysis. Crude fat content is determined by extracting the fat from the sample using a solvent. Solvent is heated, volatilized, and condensed above the sample; solvent drips onto the sample and soaks it to extract the fat. The solvent in flask is removed by rotary evaporator. Fat content is measured by weight loss of sample or weight of fat removed.

MATERIALS REQUIRED

Soxhlet apparatus, analytical balance, thimbles, vacuum oven, water bath, rotary evaporator, glass beads, 500 ml graduated cylinder, and mortar and pestle

METHODS

First the sample is ground and dried, and its moisture content is determined. Approximately 2–3 gm of weighed sample is placed in the thimble. The thimble is kept in a Soxhlet extractor. Place

200 ml hexane or petroleum ether in 500 ml round bottom flak and add several glass beads; heat the solvent in the flask until it boils. Adjust the heat source so that solvent drips from the condenser into the sample chamber at about 6 to 8 drops per second; continue the extraction for 6 to 8 hours. At the end of the extraction period, remove the thimble containing the sample using tongs and distill the ether from the collection tube. Dry the thimble in a vacuum oven at 70 °C for 24 hours. Cool the dried sample in the thimble in a desiccator then reweigh. The difference in the weights gives the fat-soluble material present in the sample.

$$\% \text{ Fat} = \frac{\text{Weight of fat soluble materials} \times 100}{\text{Weight of Sample}}$$

FAT IODINE NUMBER

Iodine reacts with the double bond of unsaturated fatty acid, and the amount of iodine that reacts with fat indicates the amount of the fat. The iodine number is defined as the weight of iodine absorbed by 100 g of the fat.

Materials Required

Fat, Wij's solution, potassium iodide (10%), sodium thiosulfate (0.1 N), starch indicator

Procedure

Prepare Wij's by dissolving 7.9 g of pure iodine trichloride in 100 ml of glacial acetic acid in a beaker and warm in a water bath. In another flask, dissolve 8.7 g of resublimed iodine in a second portion of warm 100 cc glacial acetic acid. Mix the two solutions in a 1000 ml measuring flask and fill the volume to the mark with glacial acetic acid. Store the Wij's solution in a well stopper amber bottle.

Determine the iodine number. First weigh 0.1 g–0.2 g of oil or fat in a 250 ml iodine flask and add 15 cc of chloroform or carbon tetrachloride to dissolve it; along with this carry out a blank experiment, omitting the fat. Add 25 cc of Wij's solution to the flasks containing the fat and the blank. Close the mouth of the flasks tightly and keep the flasks in the dark for about 30 minutes after adding 25 ml of 10% solution of potassium iodide at the neck of the flasks. Then dilute the mixture with 50 ml–60 ml water. Titrate it immediately against 0.1 N sodium thiosulfate solution. Shake vigorously and again titrate until the yellow color almost disappears. Add starch solution and titrate again until the blue color disappears. Let Vb and Vw be the number of cc of 0.1 N sodium thiosulfate required for 'W' g of fat. The amount of thiosulfate is equivalent to the amount of iodine that reacts.

$$\text{Iodine number} = \frac{(V_b - V_w) \times 127}{10 \times W} \times 100$$

where 127 is the equivalent weight of iodine.

DETERMINATION OF CRUDE FIBER

Crude fiber refers to indigestible cellulose, pentosans, lignin, and other components of this type in foods; it is a useful parameter in food and feed analyses. Crude fiber is commonly used as an index of the feeding value of poultry and stock feeds to evaluate the efficiency of milling and separating bran from starchy endosperm and in the chemical determination of succulence of fresh vegetables and fruits. It is insoluble in dilute acid and dilute alkali under specific conditions.

Materials

Round bottom flask, Soxhlet apparatus, heating mantle, sulfuric acid (5 gm of 98% concentrated to 1000 ml with distilled water), dilute sodium hydroxide solution, petroleum ether, ethyl alcohol (95% volume), grouch crucible

Method

Accurately weigh about 2.5 gm of the ground material into a thimble and extract for one hour with petroleum ether using Soxhlet apparatus. Transfer the material from thimble into 1 L flask. Take 200 ml of dil. H_2SO_4 in a beaker and bring to boil. Transfer all of the boiling acid to the flask containing fat-free material and immediately with a reflux condenser and boil the solution in a heating mantel for 20 minutes. Cool the flask and filter the solution through coarse, acid-washed, hardened filter paper held in a funnel. Wash the residue on the filter paper with water until the washings no longer acidic to blue litmus paper. After that, wash the residue on filter paper with 200 ml diluted Na0H solution; immediately connect the flask with a reflux condenser and boil for 30 minutes. Cool the flask and immediately filter through the filter paper. Wash the residue with hot water and about 15 ml of ethyl alcohol and a minimum three successive washings with petroleum ether. Transfer the contents in a dried grouch crucible and heat it in air oven at 105 °C ± 1 °C for three hours and weigh the total amount of crucible. Repeat the process of heating for 30 minutes until constant weight is obtained. The contents of the crucible are heated in a muffle furnace at 550 °C until all the carbonaceous material is burnt. Cool the crucible containing the ash in a desiccator and weigh. Calculate crude fiber content as follows:

$$\text{Iodine number} = \frac{(V_b - V_w) \times 127}{10 \times W} \times 100$$

where M_1 = mass of grouch crucible and contents before ashing
M_2 = mass of grouch crucible and contents after ashing
M = mass in gm of the material taken for the test

DETERMINATION OF ASCORBIC ACID

Ascorbic acid is a naturally organic compound with antioxidant properties. Ascorbic acid and its sodium, potassium, and calcium salts are commonly used as antioxidant food additives. These compounds are water soluble and therefore cannot protect fats from oxidation.

The fat-soluble esters of ascorbic acid with long-chain fatty acids (ascorbyl palmitate or ascorbyl stearate) can be used as food antioxidants. Fruits and vegetables are important sources of ascorbic acid. ascorbic acid oxidizes to dehydro form by air at alkaline pH, but it is stable in acid form. Concentration of ascorbic acid is determined in 2,6-dichlorophenol indophenol dye by titration. Only ascorbic acid reduces indophenol dye; other constitutions of citrus fruit have no effect on dye.

Materials

Prepare fresh standard ascorbic acid solution by dissolving 10 mg L ascorbic acid in 100 ml distilled water. Prepare dye solution by dissolving 50 mg sodium salt of 2,6– Dichloro phenol indophenol in 150 ml hot distilled water containing 42 mg sodium bicarbonate; Cool and dilute the solution to 500 ml with water and store solution in a refrigerator. For the lemon juice, juice and filter fresh lemons to produce a clean solution; dilute this with water so that approximately 1 ml of diluted juice contains at least 0.1 mg of ascorbic acid.

PROCEDURE

Pipette 10 ml lemon juice into a 100 ml conical flask and add 1 ml chloroform and 1 ml glacial acetic acid. Titrate the solution against the dye solution in a burette until the solution is pink for 15 seconds. If the solution is colored, observe the color in chloroform layer. Similarly, titrate standard ascorbic acid solution with dye solution and record the volume of dye used. Perform blank titration also using 10 ml of water. Calculate the acid content as follows:

Weight of ascorbic acid dissolved in 100 ml of solution = W gm
Volume of dye solution consumed in 10 ml of lemon juice – V ml
Volume of dye solution consumed in standard asocorbic acid (10 ml) = V_1
Volume of dye solution consumed in blank (water) = V_2

$$\text{Concentration (mg/100 ml) of ascorbic acid in the solution} = \frac{V - V_1}{V_1 - V_2} \times W \times \text{Dilution of Juice}$$

Sulfur dioxide in a sample interacts with 2,6-dichlorophenol indophenol dye, hampering the analysis of ascorbic acid. It can be eliminated by formaldehyde condensation as follows: Add 1 ml 40% formaldehyde and 0.1 ml HCl to 10 ml filtrate in a test tube, hold it for 10 minutes, and titrate as before.

39. What are pericyclic reactions? Discuss different types of pericyclic reactions. Discuss the molecular rearrangements of the following compounds.

i)

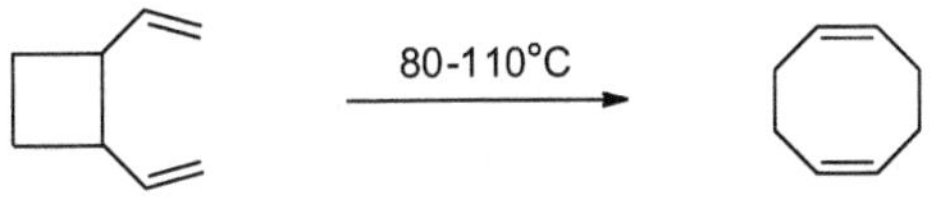

80-110°C

Cis-1,2-divinylcyclobutane → Cis,cis-1,5-cyclooctadiene

ii)

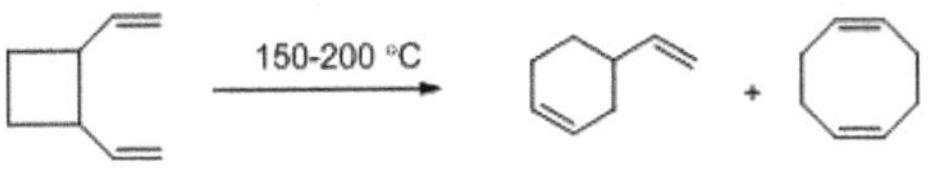

150-200 °C

Trans-1,2-divinylcyclobutane

iii)

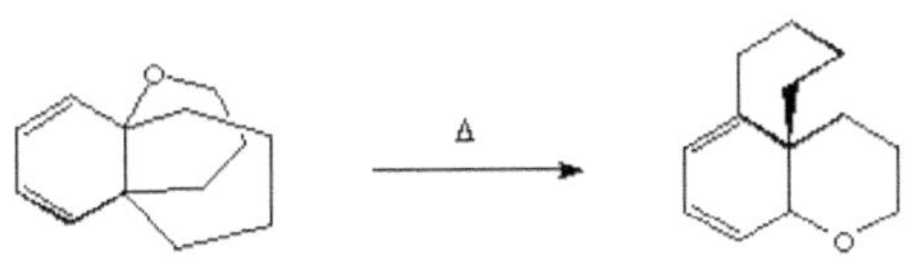

Δ

iv)

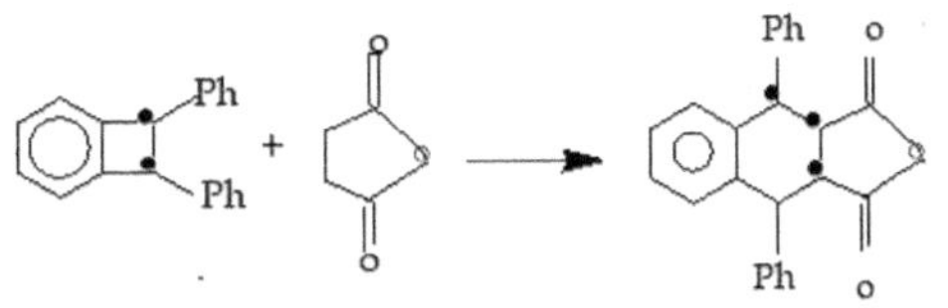

Ph Ph + O O → Ph O Ph O

Answer: A pericyclic reaction is a concerted reaction that proceeds through a cyclic transition state that is insensitive to polar solvents and catalysts but requires light or heat. It has the following properties: concentrated, reversible, single transition structure with no intermediates, mechanisms that can be explained through analysis of frontier molecular orbitals (FMOs), and reactions that provide the same analysis whether forward or reverse. The reaction is completely stereospecific: A single stereoisomer of the reactant forms a single stereoisomer of the product. In a pericyclic reaction, half of the molecular orbitals have lower energies than the isolated p orbitals and are called bonding molecular orbitals, and the other half have higher energies than the isolated p orbitals and are called antibonding molecular orbitals. The two molecular orbitals with the highest energy among the occupied molecular orbitals are called the highest occupied molecular orbital (HOMO), and the unoccupied molecular orbital with the lowest energy is the lowest unoccupied molecular orbital (LUMO). FMOs allow for analysis of a molecular interaction to be restricted to the interactions between the HOMO and LUMO of the reacting partner. There are three types of pericyclic reactions: electrolcyclic, cycloaddition, and sigmatropic reactions.

In electrolcyclic reactions, ð and π bonds are interconnected. For instance, 1,3 butadiene becomes a ð in cyclobutene, and the ð bond of cyclobutene becomes a π bond in 1,3 butadiene. Two features determine the course of the reactions: the number of π bonds involved and whether the reaction occurs in the presence of heat (thermal conditions) or light.

1,3-Butadiene Cyclobutene Cyclobutene 1,3-Butadiene

A cycloaddition is a reaction between two compounds with π bonds to form a cyclic product with two new σ bonds. The reactions are concentrated and stereospecific. The reactions can be initiated by heat or light and are characterized by the numbers of π electrons in the two reactants. Cycloaddition can take place either across the same face (called superfacial) or across opposite faces (called antarafacial) of planes in each reacting system.

(2+2) Cycloaddition

Ethylene Ethylene Cyclobutane

(4+2) Cycloaddition

1,3-Butadiene Ethylene Cyclobutene

Sigmatropic reactions are uncatalyzed reactions in which a ð bond migrates along with the atom or group attached to it a new position within a π framework.

Here, one end of ð moves over five atoms.

H 1 5 2 4 3 H

Sigmatropic shift (1,5)

Here, both ends of a ð bond migrate over three atoms.

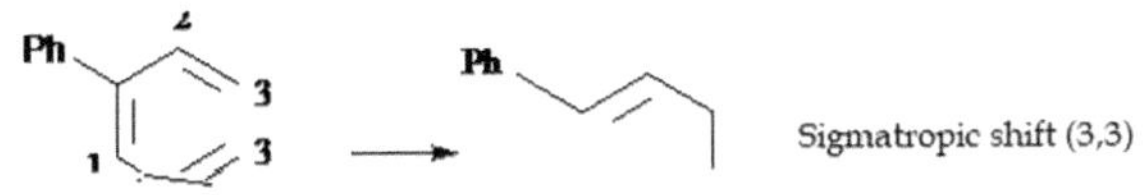

Sigmatropic shift (3,3)

Here, cis 1,2 –divenyl cycloclobutane rearranges not via the normal pathway (i.e., chairlike transition state) as steric reason but via boatlike six-membered transition state to give *cis, cis* 1, 5 cyclooctadiene.

80-110°C

Cis,cis-1,5-cyclooctadiene

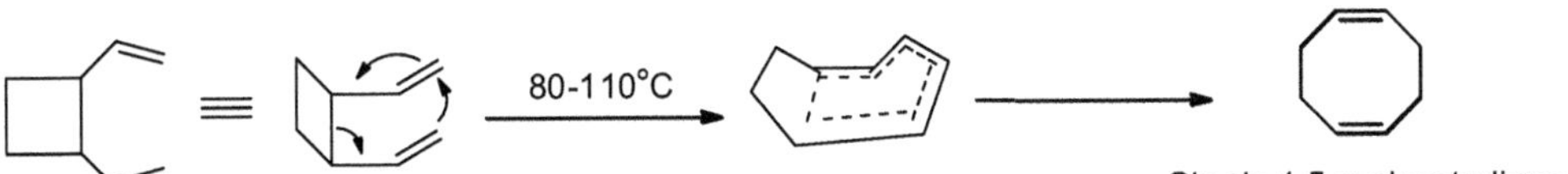

Here is a molecular arrangement reaction of the *trans* isomer of 1,2-divnyl cyclobutane in which the compound does not proceed through a cyclic transition state for steric reasons; instead, the reaction proceeds by a nonconcentrated pathway involving dissociating into allylic radicals to give products 4-vinylcyclohexen and *cis, cis* -1,5-cyclooctadiene.

4-vinyl cyclohexane

Cis,cis-1,5-cyclooctadiene

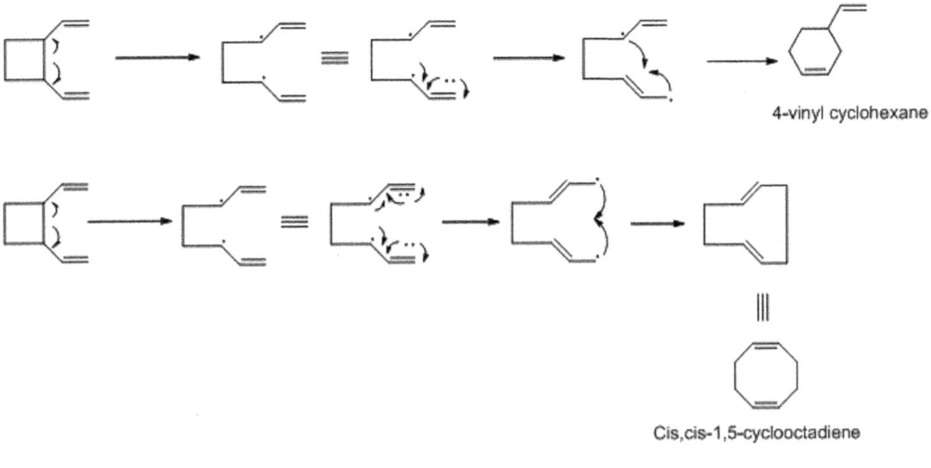

This is a 1,5 sigmatropic carbon shift reaction.

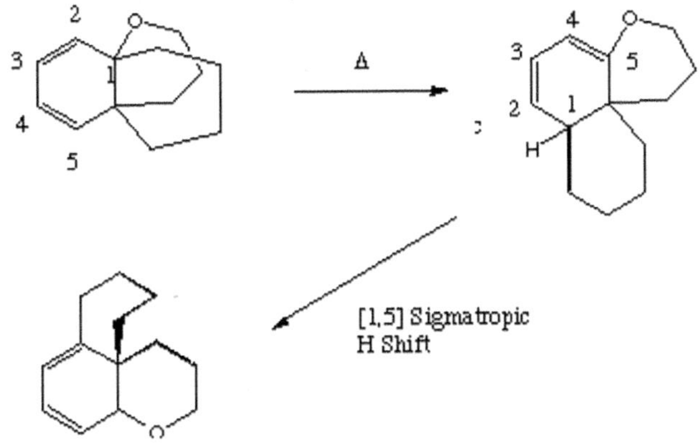

[1,5] Sigmatropic
H Shift

Here, the conrotatory ring opening of the cyclobutane ring to a 4π electron system occurs to give a nonaromatic intermediate. This intermediate undergoes a Diels–Alder reaction with maleic anhydride in the endo sense, ultimately forming the aromatic product.

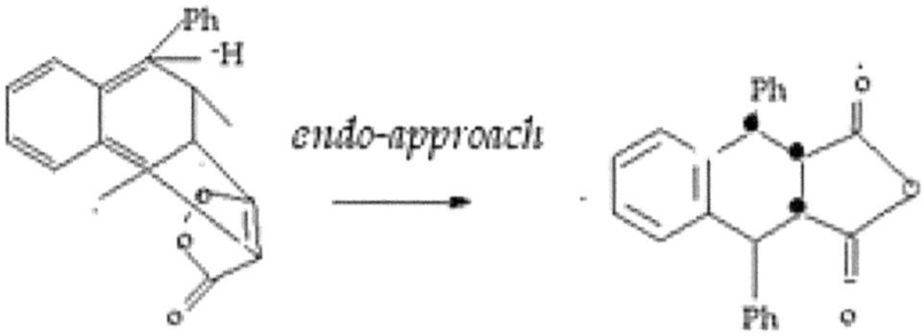

40. Discuss the properties and synthesis of the sex hormones estrogen, progesterone, androsterone, and testosterone. What are oral contraceptives?

Answer: There are about 60 known hormones so far, about 30 of which are steroidal; the remainder are not [1–5]. Sex hormones are produced in the gonads, in testes in males and in ovaries in females. Their production is stimulated by another group of hormones secreted by the anterior lobe of the pituitary gland, and they are carried to the testes and ovaries through the blood stream. Sex hormones are involved in reproduction; puberty and secondary sex characteristics; sexual development and desire; bone, muscle, and hair growth; menstruation, body fat distribution and cholesterol levels, and inflammatory responses.

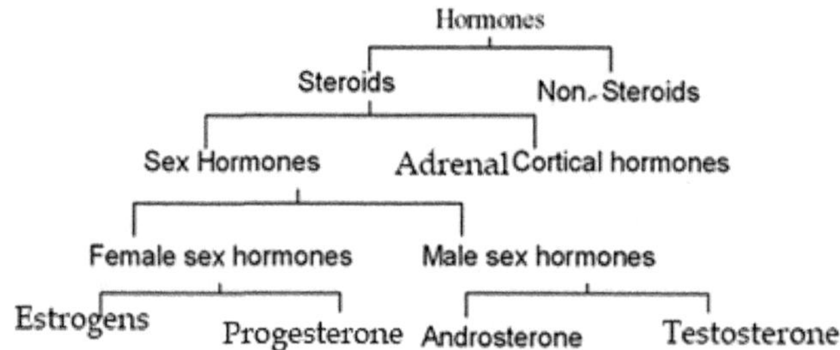

ESTROGEN

Estrogen (mol. formula $C_{18}H_{22}O_{2}$)) is responsible for the development of the reproductive system and secondary sex characteristics in females. In males, estrogen plays a role in modulating libido, erectile function, and spermatogenesis. This hormone is produced in the adrenal glands, ovaries, and fatty tissue.

Estrogen is soluble in alkali; it has one keto group and forms mono-oxime by treatment with hydroxyl amine and semicarbozone by treatment with semicarbazide. There is one phenolic group, and the compound forms monoacetate and mono-oxime. X-ray analysis shows keto and phenolic groups on opposite sides of the molecule. On distillation with zinc dust, estrogen gives chrysene, which indicates the presence of steroid nucleus. There are four major endogenous estrogen structures: estrone, estradiol, estriol, and estetrol. Estrone was synthesized by S.N. Ancchenko and I.V. Torgov [6] starting with 6-methoxy-1-tetralone and vinyl-magnesium bromide to produce acid–labile allyl alcohol following the reaction shown here.

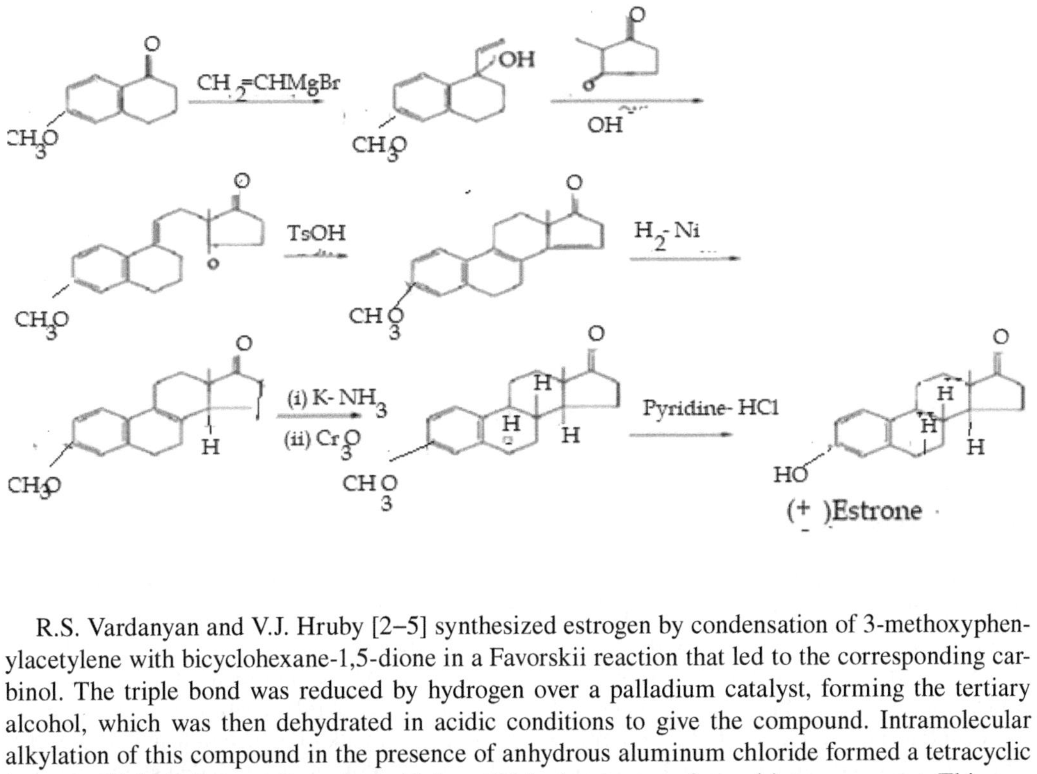

R.S. Vardanyan and V.J. Hruby [2–5] synthesized estrogen by condensation of 3-methoxyphenylacetylene with bicyclohexane-1,5-dione in a Favorskii reaction that led to the corresponding carbinol. The triple bond was reduced by hydrogen over a palladium catalyst, forming the tertiary alcohol, which was then dehydrated in acidic conditions to give the compound. Intramolecular alkylation of this compound in the presence of anhydrous aluminum chloride formed a tetracyclic ketone, which during condensation with benzaldehyde was transformed into an eneone. This was methylated at the β-position relative to the keto group by methyl iodide in the presence of potassium tert-butylate, and the resulting compound underwent ozonolysis, forming dicarboxylic acid. Cyclization of this compound to a cyclopentanone derivative formed methyl ester of the desired estrone and demethylation of the phenolic hydroxyl group by hydrobromic acid formed the desired estrone, as shown.

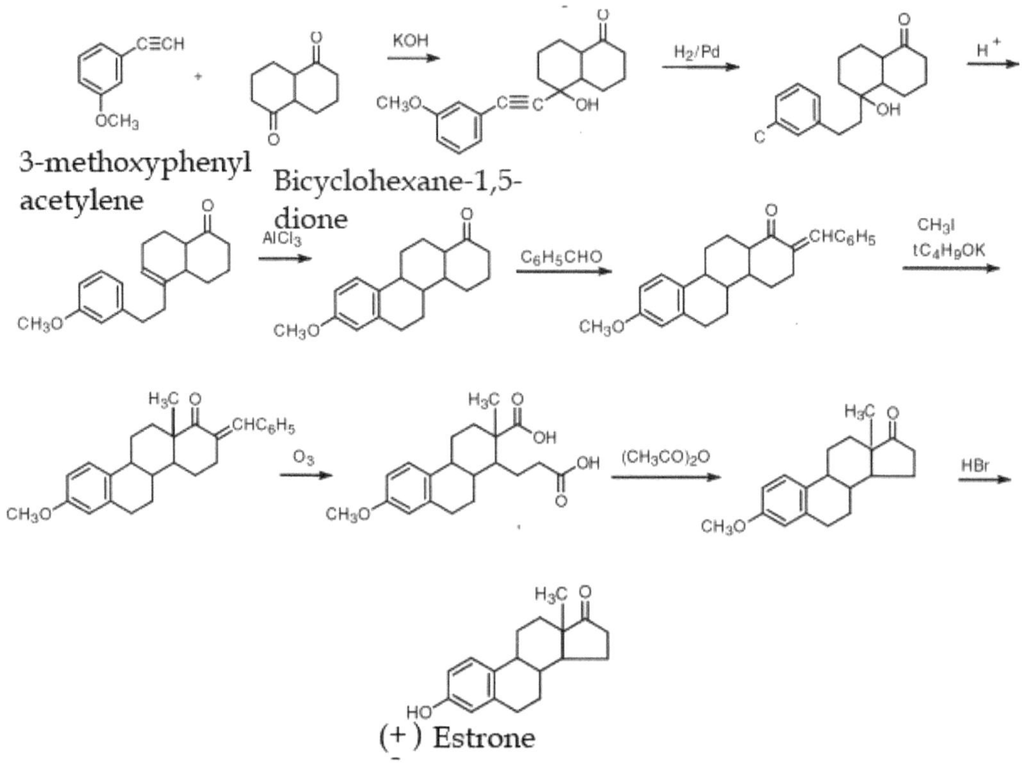

3-methoxyphenyl acetylene + Bicyclohexane-1,5-dione $\xrightarrow{KOH}$ $\xrightarrow{H_2/Pd}$ $\xrightarrow{H^+}$ $\xrightarrow{AlCl_3}$ $\xrightarrow{C_6H_5CHO}$ $\xrightarrow{CH_3I,\ tC_4H_9OK}$ $\xrightarrow{O_3}$ $\xrightarrow{(CH_3CO)_2O}$ $\xrightarrow{HBr}$

(+) Estrone

PROGESTERONE

Progesterone plays a major role in regulating female reproduction [1–3]. It is mainly secreted by the corpus luteum of the ovary. The principal function of placental progesterone is to maintain a quiescent, noncontractile uterus; it acts directly on the myometrium to prevent forceful and coordinated uterine contractions. It also functions as a ligand of progesterone receptor membrane component 1, which impacts tumor progression, metabolic regulation, and viability control of nerve cells. Its molecular formula is $C_{21}H_{30}O_2$. Progesterone is very sensitive to alkali, which indicate the presence of α, β unsaturated ketonic group. Progesterone undergoes a haloform reaction that shows the presence of a ketonic group.

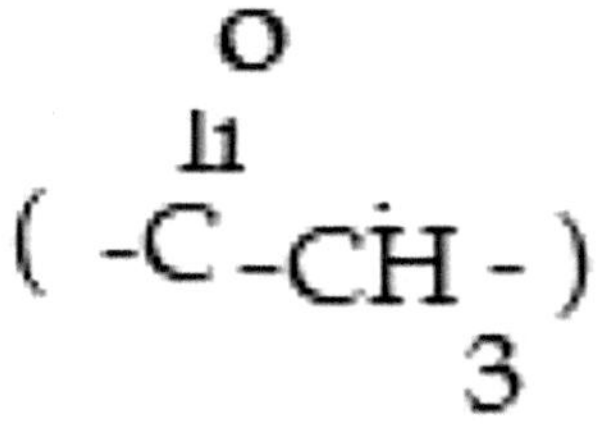

$$(-\overset{O}{\overset{\|}{C}}-CH_3 -)$$

Butenadt et al. [2] synthesized progesterone from cholesterol in 1939. The process involves treating cholesterol with acetic anhydride and bromine, yielding a dibromo derivative that on oxidation followed by debromination yieldws dehydro-epiandrosterone. One carbon chain was introduced by treatment with HCN. Grignard reaction and selective hydrogenation of the conjugated double bond affords pregnenolone, which on Oppenauer oxidation gives progesterone.

Cholesterol

(i) AC_2O

(ii) Br_2

CrO_3 - CH_3COOH

Zn-CH_3COOH

HCN

Dehydroepiandrosterone acetate

$POCl_3$

($-H_2O$)

(i) CH_3MgBr

(ii) Hydrolysis ($-NH_3$)

Pregenenolone

Oppenauer Oxidatin

Progesterone

Shepherd et al. [2] synthesized progesterone from egosterol in 1955. This method yielded 77% ergosterone by the Oppenauer oxidation of ergosterol with cyclohexane and aluminum isopropoxide in boiling toluene.

Ergosterol

Oppenauer oxidation

Cyclohexane in tolune

HCl in CH_3OH

HCL

Heat

Isoergosterone

H_2

Pd/C

(Piperidine)

$NaCr_2O_7$

CH_3COOH

Progesterone

Progesterone is commercially produced by semisynthesis. Two main routes are used: One from yam diosgenin first pioneered by Marker [6] in 1940 (known as Marker degradation), and one based on soy phytosterols scaled up in the 1970s. Additional (not necessarily economical) semisyntheses of progesterone have also been reported starting from a variety of steroids. For the example, cortisone [7] can be simultaneously deoxygenated at the C-17 and C-21 position by treatment with iodotrimethylsilane in chloroform cto produce 11-keto-progesterone (ketogestin), which in turn can be reduced at position-11 to yield progesterone.

ANDROSTERONE

Androsterone is an endogenous steroid hormone, and its main source is the urine of male and female animals. Androsterone and its 5β-isomer, etocholanolone, are produced in the body as metabolites of testosterone. It is a weak neurosteroid that crosses into the brain and affects brain function; its molecular formula is $C_{19}H_{30}O_2$. On oxidation, androsterone gives diketone, which indicates the presence of secondary alcoholic nature. Furthermore, it does not add bromine, and it forms mono-oxime and monoacetate, showing that it is a saturated hydroxyl ketone. This hormone was synthesised by L. Ruicka in 1934 [7] from cholesterol.

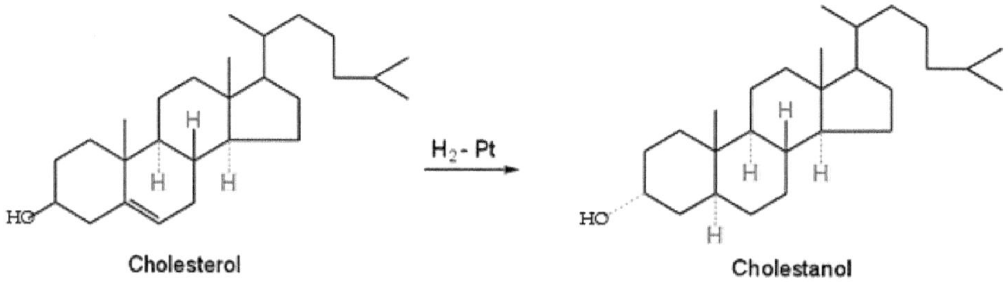

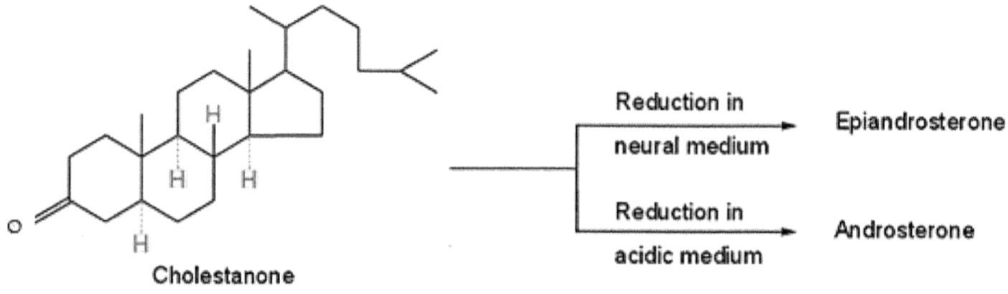

Cholesterol $\xrightarrow{H_2-Pt}$ Cholestanol

Cholestanone
Reduction in neural medium → Epiandrosterone
Reduction in acidic medium → Androsterone

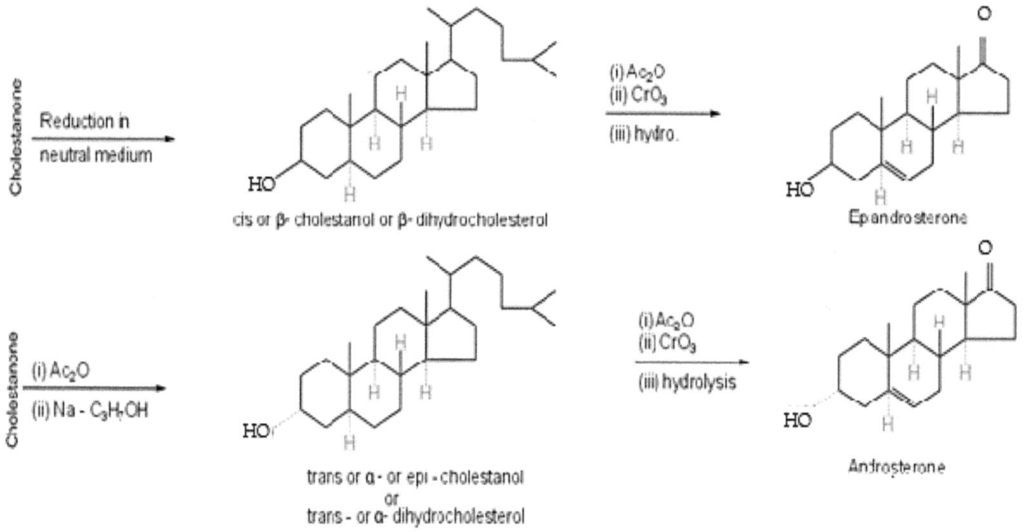

Cholestanone → Reduction in neutral medium → cis or β- cholestanol or β- dihydrocholesterol → (i) Ac_2O (ii) CrO_3 (iii) hydro. → Epandrosterone

Cholestanone → (i) Ac_2O (ii) Na - C_3H_7OH → trans or α- or epi -cholestanol or trans - or α- dihydrocholesterol → (i) Ac_2O (ii) CrO_3 (iii) hydrolysis → Androsterone

TESTOSTERONE

Testosterone is a primary male sex hormone that is an anabolic–androgenic steroid. Anabolic refers to muscle building, and androgenic refers to increasing male sex characteristics. The hormone plays a key role in the development of testicles and prostates and male secondary sexual characteristics; it also signals the body to make new blood cells and ensures that bones and muscles strong. Its molecular formula is $C_{19}H_{28}O_2$, and it shows one double bond, one ketone, and one secondary alcoholic group. It is very sensitive to alkalis, showing an α, β-unsaturated group on UV spectrum. Acetylation, bromination, and oxidation of cholesterol yield dehydroepiandrosterone. Acetylation followed by selective hydrogenation reduces the C-17 carbonyl group. Benzoylation followed by selective hydrolysis and Oppenauer oxidation yields testosterone [7].

Cholesterol

(i) Ac_2O (ii) Br_2

CrO_3, CH_3COOH

Zn- CH_3COOH (ii) Hydrolysis

Dehydrosterone

(i) Ac_2O (ii) Na-C_3H_7OH

(i) C_6H_5COCl (ii) Partial hydrolysis (CH_3OH-NaOH

(i) Oppenauer Oxidation (ii) Hydrolysis(KOH)

Testosterone

ORAL CONTRACEPTIVES

Oral contraceptives block the release of eggs from ovaries to prevent pregnancy. Thy are also sometimes used to treat heavy or irregular menstruation and endometriosis, a condition in which the tissue that lines the uterus grows in other areas of the body and causes pain, heavy or irregular menstruation, and other symptoms). The structures of some contraceptives are shown here.

Norethynodrel

Lynoestrenol

Ethynodiol

Megestrol

Ethinyloestradiol

Mestranol

REFERENCES

[1] L. Fieser and M.P. Fieser, *Steroids Chemistry*, Reinhold Publishing Company, New York, USA, 1967.
[2] W. Klyne, *The Chemistry of Steroids*, Methuen and Company, London, 1960.
[3] R. Robison (ed.), *Introduction to Steroid Chemistry*, Elsevier, 1968.
[4] I.L. Finer, *Organic Chemistry*, Vol. 2, Addison Wesley Longman limited, UK, 1973.
[5] O.P. Agrawal, *Organic Chemistry Natural Products*, Vols. 1 and II, Goel Publishing House, New Delhi, 1980.
[6] S.N. Ancchenko and I.V. Torgov, New syntheses of estrone, d,1-8-iso-oestrone and d,1-19-nortestosterone. *Tetrahedron Letters* 4 (23), 1553–1558, 1953..
[7] L. Ruzicka and A. Wettstein, *Helvetica Chimica Acta* 18 1264–1275, 1935.

41. List metabolic abnormalities in MetS.

    Answer: MetS encompasses the following metabolic abnormalities: central obesity, atherogenic dyslipidemia, hypertriglyceridemia, decreased high-density lipoprotein cholesterol and the presence of small, dense lipoprotein particles, high blood pressure, and hyperglycemia.

42. Why does insulin resistance (IR) increase the risk of stroke?

    Answer: IR accelerates the formation of thrombosis and promotes the development of atherosclerosis.

43. Which enzymes stimulate de novo lipogenesis?

    Answer: Acetyl-CoA carboxylase 1, fatty acid synthase, and glycerol-3-phosphate acyltransfer 1.

44. How does IR affect the development of atherosclerosis?
    Answer: IR reduces NO release by the endothelium and increases the production of endothelin 1. IR overstimulates MAPK, leading to inflammation in the vascular wall, endothelial dysfunction, and proliferation of vascular smooth muscle cells. IR increases blood pressure and affects atherogenic dyslipidemia. It also promotes the polarization of macrophages from the alternative M2 activation state (anti-inflammatory) to the classic M1 activation state (pro-inflammatory).
    IR also increases the expression of angiotensin-type receptor mRNA, which potentiates the vasoconstrictive effect of Ang II. Ang II activates the NF-κB pathway, increases the expression of adhesion molecules and chemokines in the vascular wall, stimulates ROS production, and promotes multiple inflammatory pathways involved in atherosclerosis.

45. How does IR promote ischemic stroke? Select all that apply.
    A. IR induces glucotoxicity, which exacerbates oxidative stress, inflammation, and endoplasmic reticulum stress due to lipotoxicity.
    B. IR attenuates the autophagy response.
    C. IR has an anti-apoptotic effect.
    D. IR produces MCP1, which recruits monocytes and activates pro-inflammatory macrophages.
    E. IR reduces the production of ROS by increasing the concentration of free fatty acids FFA, which promote glucotoxicity, lipotoxicity and nuclear translocation of NF-kB.
    Answer: A, B, D

46. How does obesity affect stroke risk? Select all that apply.
    A. Obesity affects the excessive accumulation of lipids in adipocytes, activating NADPH oxidase, forming reactive oxygen species, and triggering an inflammatory reaction with low activity and insulin resistance.
    B. Obesity reduces the production of adipocytokines that promote inflammation.
    C. Obesity is associated with elevated apolipoprotein-containing lipoproteins.
    D. Obesity increases the level of leptin, which stimulates the release of pro-inflammatory compounds.
    E. Obesity activates pro-inflammatory factors such as TNF-α and IL-6, which increase the level of plasma adiponectin, a protein with a strong anti-inflammatory effect.
    Answer: A, C, D

47. Under physiological conditions, insulin (select all that apply)
    A. increases the uptake of amino acids by tissues and enhances protein synthesis
    B. stimulates lipogenesis and lipolysis
    C. promotes cell growth, proliferation, migration and inhibits apoptosis
    D. stimulates glucose oxidation and gluconeogenesis
    Answer: A, C

48. AKT regulates (select all that apply)
    A. translocation of the insulin-dependent glucose transporter 4
    B. glycogen synthesis by phosphorylation of glycogen synthase kinase 3
    C. lipogenic gene expression and protein synthesis
    D. nitric oxide synthesis
    Answer: all

49. What are glucocorticoids (GCs)?
    Answer: Steroid hormones that play an important role in regulating carbohydrate metabolism. Their biosynthesis takes place in the banded layer of the adrenal cortex.

The chemical structure of GCs is based on the hexadecahydro-1H-cyclopenta[a] phenanthrene ring. They are cholesterol derivatives, and their most important representative in humans is cortisol. GCs have a wide range of activity, regulating reproduction, growth, inflammatory reactions, fat and protein metabolism, and water and electrolyte homeostasis. They also affect the functioning of the cardiovascular system and are responsible for the distribution of adipose tissue.

50. What are the difference in the structures of
    A. cortisol and cortisone
    B. cortisol and corticosterone?

    Answer: A. Cortisone has an oxygen atom bonded with a double bond to carbon C-11 of the hexadecahydro-1H-cyclopenta[a]phenanthrene system (it is a ketone). Cortisol has a hydroxyl group at C-11 (it is an alcohol).

    B. Cortisol at C-17 has an OH group, and corticosterone does not; in place of the OH group, there is a hydrogen atom.

51. What is the role of 11β-hydroxysteroid dehydrogenase(11β-HSD) in the human body?

    Answer: 11β-HSD exists in two isoforms: 11β-HSD1 and 11β-HSD2. 11β-HSD1 (cortisone reductase) is a microsomal, NADPH-dependent oxidoreductase enzyme that catalyzes the conversion of inactive cortisone to physiologically active cortisol and, to a lesser extent, the reverse reaction. Together with 11β-HSD2 dependent on NAD, located mainly in the kidneys and colon, it regulates the level of cortisol in the body.

52. What are the consequences of excessive levels of cortisol in the body?

    Answer: Chronic excess cortisol in the blood leads to the characteristic symptoms of Cushing's syndrome displacement of adipose tissue deposits (causing buffalo hump, full moon face, abdominal obesity), thinning of the skin, form stretch marks with a characteristic pink color, acne, and IR. Excess cortisol also causes characteristic symptoms of metabolic syndrome: visceral obesity, insulin resistance, diabetes, dyslipidemia, hypertension, and hyperuricemia.

53. What is metabolic syndrome?

    Answer: MetS is a complex disorder defined by a group of combined factors including insulin resistance, visceral obesity, dyslipidemia, hyperglycemia, and hypertension that increase the risk of atherosclerotic cardiovascular disease and type 2 diabetes.

54. List the chemical groups that occur in compounds that inhibit the activity of 11β-HSD.

    Answer:

    - hexadecahydro-*1H*-cyclopenta[a]phenanthrene system
    - thiazole i thiazolone systems
    - sulfonamide and cyclicsulfondiamidesmoiety
    - other heterocyclic systems, among them 5-membered saturated and unsaturated rings with sulfur, nitrogen and oxygen atoms (eg. 2,5-dihydrofuran, pyridine, pyrrolidine, triazole, oxadiazole, thiophene)

55. What is insulin resistance

    Answer: Decreased sensitivity of peripheral tissues to insulin secretion

56. What is the role of adrenocorticotropic hormone?

    Answer: Adrenocorticotropic hormone stimulates the secretion of hormones produced by the adrenal glands, mainly cortisol, a glucocorticosteroid.

57. What is metabolism? Discuss the synthesis of carbohydrate metabolism.

Answer: Metabolism is the chemical processes occurring within a living cell that take place in two stages, anabolism and catabolism.

Anabolism is the synthesis of complex molecules from simple compound molecules, and it requires energy, making it an endergonic reaction. Catabolism breaks down large molecules such as polysaccharides and proteins. This releases energy and is called an exergonic reaction.

Carbohydrate metabolism [1] is a fundamental biochemical process responsible for the metabolic formation, breakdown, and interconversion of carbohydrates in living organisms, and it takes place in three steps: glycolysis, Kreb's cycle, and electron transport system.

Glycolysis [2–5], the degradation of glycose takes place in three stages: energy investment, splitting, energy generation. This is called the EMP pathway because it was discovered by three German scientists Embden, Meyerhof, and Parnas. The free energy released in this process is used to form the high-energy molecules ATP and $NADH_2$. In glycolysis, one molecule of glucose is oxidized to form two molecules of pyruvate; four molecules of ATP form and expend two molecules of ATP, giving a net gain of two molecules. One step of the process forms two molecules of $NADH_2$ per molecule of glucose oxidation. Glycolysis occurs in the cytosol (cytoplasm matrix) of the cell.

Energy investment: First, glucose is phosphorylated (activated) by ATP in presence of the enzymes hexokinase or glucokinase and Mg^{++} to form gluco-6-phosphate and ADP. Gluco–6-phosphate is converted to fructose -6-phosphate in the presence of phosphohexose isomerase and Mg^{+2}. Fructose 6-phosphate is converted to fructose 1,6-bishphoshphate by phosphofructokinase and ATP.

$$\text{Glucose} + \text{ATP} \xrightarrow[Mg^{+2}]{\text{Glucokinase}} \text{Gluco 6-phosphate} + \text{ADP}$$

Splitting: Fructose 1,6 –bishphosphate is converted to glyceraldehyde 3-phosphate and dihydroxy acetone phosphate by aldolase.

Energy generation: First, glyceraldehyde -3-phosphate transforms into 1,3-bisphosphoglyceral dihydrate by glyceraldehyde 3-phosphate hydrogenase enzyme. Then, 1,3-bisphophoglyceral dehydrate is converted to 2-phophoglycerate by phosphoglycerate mutase. Then, 2-phosphoglycerate is converted to phosphoenol pyruvate by enolase, Mg^{+2}, and Mn^{+2}. Phosphoenol pyruvate is then converted to pyruvic acid by pyruvate kinase enzyme.

Glucose + ATP

↓ Mg^{+2} + Glucokinase

Gluco-6-phosphate + ADP

↓ *phosphohexose isomerase* + Mg^{+2}

Fructose-6-phosphate

↓ phosphofructokinase + ATP

Fructose 1,6-bishphoshphate

↓ *Aldolase*

Glyceralelehyde-3-phosphate + Dihydroxy acetone

↓ Glyceraldehyde-3-phosphate hydrogenase

1.3-bisphosplonglyceral dilydrate

↓ *Phosphoglycerate mutase*

2-Phosphoglycerate

↓ *Enolase* + Mg^{+2} + Mn+2

Phosphoenol pyruvate + pyruvate kinase

↓ *Pyruvate kinase*

Pyruvic acid

Krebs cycle: Named after scientist Hans Kreb, who described the reactions in 1937. This cycle is also called the citric acid cycle because the first product of the cycle is citric acid. Reactions of the cycle take place within mitochondrial matrix. The cycle utilizes two thirds of total oxygen consumed by the body. This cycle is amphibolic as this cycle plays roles in both oxidative and synthetic processes. The reaction of this cycle takes place in four stages.

In stage one, process pyruvic acid is decarboxylated in presence of oxygen, uniting with coenzyme A to form acetyl coenzyme. This acetyl coenzyme reacts with oxaloacetic acid and water to form citric acid by the condensing enzyme citrate synthetase; the acid is then converted to isocitrate by aconitate hydratase for dehydration and rehydration. This isomerization leads to an interchange of an H, and OH is inhibited by fluroacetate.

CH_3CO-S-CoA
Acetyl coenzyme A
+
COCOOH
|
CH_2COOH
Oxaloacetic acid

Citrate -Synthatase ($+H_2O$) →

CH_2COOH
|
HO — C — COOH
|
COOH + COA-SH

Citric acid Coenzyme A

Coenzyme-A liberated in the last reaction reacts with more pyruvate compound formed from glycolysis, and isocitrate forms.

COO^-
|
H—C—H
|
^-OOC—C—OH
|
CH_2
|
COO^-
Citrate

Dehydeation →

COO^-
|
H—C
||
^-OOC—C
|
CH_2
|
COO^-
Cis-Aconitrate(enzyme bound)

Rehradtion →

COO^-
|
H—C—OH
|
^-OOC—C—H
|
CH_2
|
COO^-
Isocitrate

In the second stage, isocitrate is converted to α-ketoglutarate and then to succinyl-CoA. First, the reactant undergoes oxidation to give oxalosuccinic acid; then it undergoes decarboxylation to give α-ketoglutaric acid. Both steps are catalyzed by isocitrate dehydrogenase, which is present in two types: One is specific for NAD^+ and found only in mitochondria; the other is specific for $NADP^+$ and found both in mitochondria and cytoplasm.

$$\begin{array}{l} H \\ HO-C-COOH \\ H-C-COOH \\ H-C-COOH \\ H \end{array} \xrightleftharpoons[]{NADP^+ \quad NADPH + H^+} \begin{array}{c} COCOOH \\ | \\ H-C-COOH \\ | \\ CH_2COOH \end{array} \xrightleftharpoons[]{CO_2} \begin{array}{c} COCOOH \\ | \\ CH_2 \\ | \\ CH_2COOH \end{array}$$

Next, α-ketoglutaric acid oxidative decarboxylation gives succinyl-CoA by α-ketoglutarate dehydrogenase.

$$\underset{\text{α-Ketoglutaric acid}}{\begin{array}{c} COCOOH \\ | \\ CH_2 \\ | \\ CH_2COOH \end{array}} + HS\text{-}CoA \xrightarrow[(-NADH,\ H^+)]{\text{α-ketoglutarate dehydrogenase}\ NAD^+} \underset{\text{Succinyl coenzyme A}}{\begin{array}{c} CO\text{-}S\text{-}CoA \\ | \\ CH_2 \\ | \\ CH_2COO^- \end{array}} + CO_2$$

In the third stage of the Krebs cycle, succinyl-CoA is hydrolyzed to succinate in the presence of succinate thiokinase and energy-rich GDP that transfers its high energy phosphate bond to ADP to form ATP (substrate-level phosphorylation).

$$GTP + ADP \Rightarrow ATP + GDP$$

$$\underset{\text{Succinyl-CoA}}{\begin{array}{c} CO\text{-}S\text{-}CoA \\ | \\ CH_2 \\ | \\ COOH \end{array}} + H_3PO_4 + GDP + H_2O \longrightarrow \underset{\text{Succinate}}{\begin{array}{c} COO^- \\ | \\ CH_2 \\ | \\ CH_2 \\ | \\ COOH \end{array}} + CoA\text{-}SH + GTP$$

In the fourth stage, succinate is converted to fumarate by succinate dehydrogenase, and fumarate is then reversibly dehydrated to form L-malate in the presence of fumarate hydratase. The malate is then dehydrogenated by malate dehydrogenase and NAD^+ to give oxaloacetate. In this way, oxaloacetate is regenerated on completion of the cycle and with another molecule of acetyl coenzyme A in a repeating cycle.

COO⁻ | CH2 | CH2 | COOH
Succinate
FAD FAD2
Succinate dehydrogenase
COOH | CH ‖ CH | COOH
Fumaric acid
H O 2
Fumare
COOH | CO | CH2 | COOH
Oxaloacetic acid
NADPH +H⁺ NAD⁺
Malate dehydrogenase
COOH | HO—CH | CH2 | COOH
Maleic acid

The citric acid cycle is entirely aerobic and 12 ATP molecules are produced per molecule of acetyl-CoA, 3 NADH, and 1 $FADH_2$ in the following reaction:

$$\text{Acetyl-CoA} + 3\ NAD^+ + FAD + GDP + P_i$$

$$\downarrow$$

$$CoA + 2CO_2 + 3NADH + FADH_2 + GTP + 2H^+$$

Glycolysis → Pyruvic acid (3C)
CO 2
NAD NADH +H⁺
Oxaloacetate+Acetyl-CoA(2C)
H O 2
Malic acid(4C)
Citric acid (6
H O 2
Fumaric acid(4C)
FADH 2
Citric acid Cycle
(Kreb s cycle)
FAD
H O 2
Cis -aconit
Succinic acid (4C)
H O 2
H O 2
GTP
Isocitric acid (6C)
GDP+P i
NAD⁺
Succinyl-CoA(4C)
NADH +H⁺ +CO 2
α- Ketoglutaric acid(5C)

Electron transfer: The last step in the cycle is the oxidation of NADH + H^+ and $FADH_2$ produced from the citric acid cycle by molecular oxygen through electron transfer, which takes place on the inner mitochondrial membrane. This is a sequence of oxidation reactions in which an electron is accepted from reduced compounds and is finally transferred to oxygen with the formation of ATP and water. The synthesis of oxidative pathway is called oxidative phosphorylation. Each molecule of NADH + H^+ yields 3 molecules of ATP, and each molecule of $FADH_2$ yields 2 molecules of ATP. The electron carriers include NAD, FAD, and cytochromes. Because many cytochromes are involved, this process is called the cytochrome pathway.

REFERENCES

[1] R.D. Guthrie and J. Honeyman, *An Introduction to Carbohydrate Chemistry*, Clarendon Press Partner Oxford University Press, UK, 1974.
[2] I.L. Finer, *Organic Chemistry*, Vol. 2, Addison Wesley Longman Limited, UK, 1973.
[3] O.P. Agrawal, *Organic Chemistry of Natural Products*, Vol. 2, Goel Publishing House, New Delhi, 1980.
[4] D.L. Nelson and M.M. Cox, *Lehninger, Principles of Biochemistry*, 8th edition, Palgrave Macmillan, UK, 2021.
[5] *Harper's, Illustrated Biochemistry*, 31st edition, McGraw Hill Medical, 2018.

58. Discuss lipid metabolism.

Answer: Lipid metabolism refers to either breaking down or storing fats for energy. Lipids are nonpolar and soluble in organic solvents; they have significant structural varieties, but the major dietary lipids are triacyl glycerol, cholesterol, and phospholipids. Gaucher disease and Tay–Sachs disease are disorders of lipid metabolism; high lipid levels can also cause diabetes, alcoholism, kidney disease, hypothyroidism, liver disease, and stress. High lipid counts adhere easily to the blood's circulating nerve walls, and the growing fatty scale causes a variety of atherosclerosis disorders such as stroke and heart attack. The liver and pancreas are important sites for lipid metabolism and play important roles in lipid digestion, absorption, synthesis, decomposition, and transport.

Lipid digestion takes place in three stages: digestion of triglyceride, digestion of phospholipids, and digestion of cholesterol ester.

Digestion of triglycerides: Lipid degustation starts from the stomach where lingual lipase (active pH 2.5 to 5) from the mouth (secreted by the dorsal surface of the tongue; Von Ebner's glands) enters into the stomach with the food; it is active in butter, ghee, and milk. Gastric lipase, an acid-stable lipase is secreted by chief cells of stomach stimulated by gastrin; 30% of digestion of triglyceride is completed in the stomach; the other lipids are emulsified in the intestine to make them smaller droplets, reducing surface tension and increasing surface of droplets. Bile salts (sodium glycocholate and sodium taurocholate) are responsible for that process. The process of emulsification is also facilitated by peristalsis (mechanical mixing) and phospholipids. Bile (pH 7.7) neutralizes the acid of stomach environmental food. It provides favorable pH and stimulates the action of pancreatic juice. In intestine, pancreatic lipase hydrolyzes triglyceride to form 2-monoacylglycerol and two fatty acids. In short, binding to the triacylglycerol molecules at the oil/water interface is obligatory for the action of zymogen secreted by the pancreas activated by trypsin.

$$\text{Triacyl glycerol} \xrightarrow[H_2O]{\text{Lipase}} \text{2,3 Diacyl glycerol} + \text{Fatty acid} \xrightarrow[H_2O]{\text{Lipase}} \text{2-monoacyl glycerol} + \text{Fatty acid}$$

Digestion of phospholipids: Intestinal lipase acts within intestinal mucosal cells that hydrolyze the absorbed primary (α) monoglycerides that form glycerol and FFA. Here, phospholipase A2 acts on phospholipids activated by trypsin and requiring bile salts for its activity.

$$\text{Phospholipids} \xrightarrow{\text{Phospholipase}} \text{Lysophospholids}$$

Digestion of cholesterol esters: This process is imitated by cholesterol esterase enzyme, which breaks down cholesterol esters into free fatty acid and cholesterol.

$$\text{Cholesterol esters} \xrightarrow{\text{Cholestrol esterase}} \text{Free fatty acids+cholesterol}$$

The major end product is 2-monoacyl glyceride (76%) after intestinal digestion by the enzymatic pancreatic lipase. Other end products are 1-monoacyl glyceride (6%) and glycerol and fatty acids (14%).

Lipid absorption takes place via micelle formation by 2-mono acyl glycerol, long-chain fatty acids, cholesterol, phospholipids, and lysophospholipids, which occurs due to the detergent action of bile salts. Micelle is essential for the absorption of fat-soluble vitamins A, D, and K, whereas fatty acids, 2-monoacyl glycerol, and other digested products passively diffuse into the mucosal cells. About 98% dietary lipids are absorbed. Bile salts are reabsorbed from the ileum and returned to the liver to be re-excreted for micelle formation.

Short-chain fatty acids are not re-esterified, and their absorption is faster than long-chain fatty acids. Short-chain fatty acids directly enter into blood vessels, then to portal vein, finally to liver, where they are immediately utilized for energy. But long-chain fatty acids are re-esterified to form triglyceride. Fatty acids synthesis starts in cytoplasm with the help of acetyl-CoA and NADPH produced from mitochondria and the pentose phosphate pathway, respectively, using fatty acid synthases. First acetyl-CoA transported from mitochondria through citrate–malatepyruvate shuttle in form of citrate, which further breaks into acetyl-CoA and oxaloacetate. Two activated fatty acids react with monoacylglycerol (MAG) to form the triglyceride, but free glycerol is absorbed from intestinal lumen directly and not available for re-esterification.

Immediately after lipid absorption, there is turbidity in plasma due to circulating chylomicrons that appear in plasma up to 2 hours after meals. Chylomirons are ultra-low-density lipoproteins that consist of triglycerides (85%–92%), phospholipids (6%–12%), cholesterol (1%–3%), and proteins (1%–2%). The chyle (milky fluid) from the intestinal mucosal cells loaded with chylomicrons is transported through the lacteals into thoracic duct and then emptied into lymph circulation. This turbidity is cleared within a few hours by lipoprotein lipase enzyme (clearing factor). These chylomicrons then reach either the adipose tissue or the muscle via blood depending upon their requirement. In the capillaries of the adipose tissue and muscles, apolipoprotein C-II activates the enzyme lipoprotein lipase, which breaks down the chylomicrons and hydrolyzes the triacylglycerols to fatty acids and glycerol. These end products are utilized by the muscles for energy, whereas in the adipose tissue, they are re-esterified to triacylglycerols for storage. The remaining chylomicrons that still have the cholesterol and apolipoproteins travel to the liver via blood and via the specific receptors are taken up by endocytosis. Some triacylglycerols that might have entered via the remnants of chylomicron in the liver are used up to synthesize other cellular components.

REFERENCES

[1] Jeremy M. Berg, John L. Tymockzo and Luber Stryer, *Biochemistry*, 7th Edition W.H. Freeman & Co Ltd, USA, 2011.

[2] David L. Nelson and Michael M. Cox, *Lehninger Principles of Biochemistry*, 8th Edition, McMillan, UK, 2021.

[3] Reginald H. Garrett and Charles M. Grisham, *Biochemistry*, 5th Edition, Brooks & Cole, UK, 2013.

[4] F. D. Gunstone, *Fatty acid and Lipid Chemistry*, Springer, 2020.

59. Discuss terpenoids.

Terpenoids, an important class of natural products, have diverse structures. These compounds are derived from 5-carbon-containing compound, known as isoprene unit (**1** and **2**).

These compounds are classified as monoterpenes (containing two isoprene units), sesquiterpenes (three units), diterpenes (having four units), and triterpenoids based on the number of carbon atoms derived from isoprene units.

Monoterpenes are volatile compounds that play an important role in the perfume industry [1]. Geranyl pyrophosphate (**3**) is a biosynthetic precursor of monoterpenes that undergoes cyclization to afford limonene (**4**), which on hydroxylation affords *trans*-isopiperitenol (**5**). The latter on oxidation of hydroxyl group gives isopiperitenone (**6**), which after reduction of endocyclic double bond yields *cis*-isopulegone (**7**). Isomerization of a double bond in (**7**) affords pulegone (**8**). The reduction of double a double in (**8**) produces both enantiomers of menthone (**9**) and (**10**)) [2]. The biosynthesis of monoterpenes is shown in Figure 1.

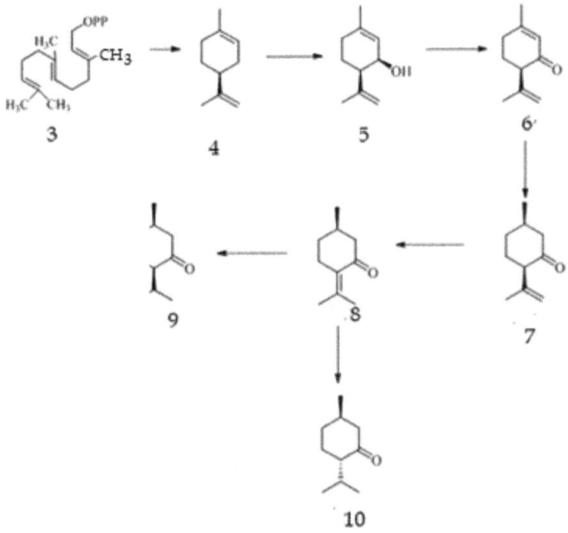

Sesquiterpenes are known to have various biological activities. For instance, artemisinin (**16**) exhibits antimalarial activity, and its ether analogue is used in clinics [3]. These compounds are produced in nature from FPP (**3**), which is cyclized by cyclase to afford amorphan-4, 11-diene (**12**); on hydroxylation, this yields its hydroxy analogues (**13**). Further oxidation of primary alcohol gives aldehyde analogue (**14**). Oxidation of aldehyde affords carboxylic acid analogue (**15**), on photooxidation produces a of artemisinin (**16**) [4]. This whole sequence of biosynthesis is shown Figure 2.

Farnesyl pyrophosph
11

Amorpha-4,11-diene synthase

12

Cytochrome P450 monooxygenase

13

Alcohol dehydrogenase

14

Aldehyde dehydrogenase

15

16

Diterpenes are members of terpenoidal class of natural products. These compounds are made up of four isoprene units. They are biosynthesized in nature using geranylpyrophosphate (GGP) (**17**) as an intermediate [5]. Compound (**18**) is produced in nature by coupling FPP (**17**) with another unit of isoprene (**1**). A number of diterpene cyclase enzymes are present in the nature to fold/cyclize GGP (**17**) to give various classes of diterpenes [5]. For instance, momilactone (**21**) is produced from *syn*-primara-7,15-diene (**20**). GGP cyclized to afford intermediate (**19**) that on further cyclization gave (**20**). Similarly, cyclization of GGP (**17**) by ent-kaurene diterpene cyclase affords intermediate (**22**) which on further cyclization yields ent-kaurene type diterpenoid structure (**23**). All of these biosynthetic steps involved in the biosynthesis of diterpenes are shown in Figure 3 [6].

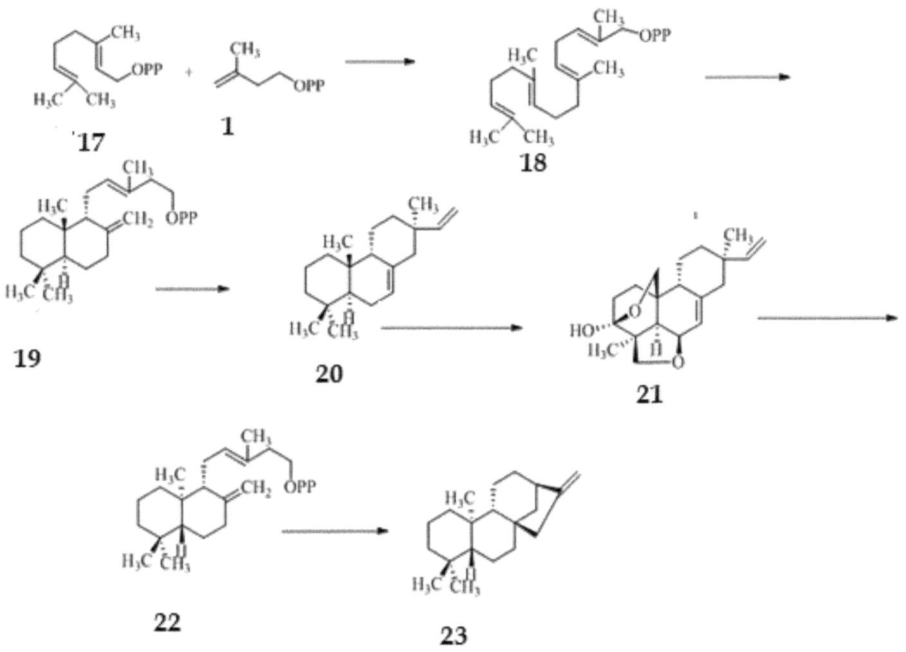

REFERENCES

[1] Banthorpe, D. V., Charlwood, B. V., Francis, M. J. O. 1972. The biosynthesis of monoterpene. *Chem. Rev.* 72, 115–155.

[2] Turner, G., Gershenzon, J., Croteau, R. B. 2000. Developmental regulation of monoterpene biosynthesis in the glandular trichomes of peppermint. *Plant Physiol.* 124, 655–664.

[3] Ata, A., Nazand, S., Friesen, K. 2022. Biosynthesis of natural products. In *Greener Synthesis of Organic Compounds*, edited by Ahindra Nag, Taylor and Francis.

[4] Wen, W., Yu, R. 2011. Artemisinin biosynthesis and its regulatory enzymes: Progress and perspective. *Pharmacogn. Rev.* 5, 189–194.

[5] Heskes, A. M., Sundram, T. C. M., Boughton, B. A., Jensen, N. B., Hansen, N. L., Crocoll, C., Cozzi, F., Rasmussen, S., Hamberger, B., Hamberger, B., Staerk, D., Møller, B. L., Pateraki, I. 2018. Biosynthesis of bioactive diterpenoids in the medicinal plant Vitex agnus-castus. *Plant J.* 93, 943–958.

[6] Okada, K., Kawaide, H., Miyazaki, S. 2016. HpDTC1, a stress-inducible bifunctional diterpene cyclase involved in momilactone biosynthesis, functions in chemical defence in the Moss Hypnum plumaeforme. *Sci. Rep.* 6, 25316.

60. Discuss biosynthesis of bile acid.

Answer: Bile acids are steroid acids found predominantly in the bile of mammals and other vertebrates. Biosynthesis of bile acids such as monohydroxy cholanoic acid or lithocholic acid ($C_{23}H_{38}(OH)COOH$) and dihydroxycholanic acid or deoxycholic acid($C_{23}H_{37}(OH)_2COOH$ is formed from cholesterol in liver. Cholesterol 7 alpha-hydroxylase is the rate-limiting enzyme in the synthesis of bile acid from cholesterol catalyzing the formation of 7-hydroxy cholesterol by 7α-hydrolase (CYP7A1). These images describe the process [1–3].

HO

Cholesterol

$(CH_3Co)_2O$

CH_3COOH

O

H_3C O

Cholesteryl acetate

Hydroxylatim

CYPTAI

HO OH

7 - α - Hydroxy Cholesterol

(i)Inversion of C_3

(ii) H_2

HO OH H

3α, 7α - dihydroxy coprostane

hydroxylation

HO

HO OH H

3α, 7α, 12α – trihydroxy coprostane

Oxydation

HO

COOH

HO OH H

3α, 7α, 12α – trihydroxy coprostanic acid

Oxydation

COOH

HO OH H

3α, 7α - dihydroxy coprostanic acid

Oxydation

COOH

HO OH H

chenodeoxycholic acid

Bacterial reduction

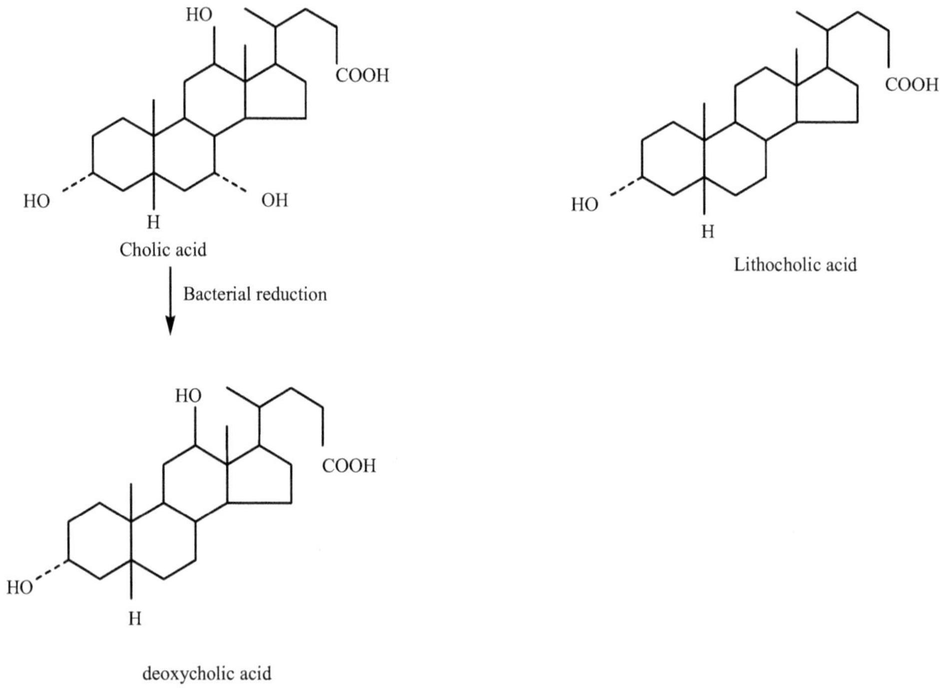
HO
COOH
HO
OH
H
Cholic acid
Bacterial reduction
HO
COOH
HO
H
deoxycholic acid
COOH
HO
H
Lithocholic acid

Synthesis and interconversions of bile acid

REFERENCES

[1] Nag, A. (ed.), *Greener Synthesis of Organic Compounds, Drugs and Natural Products*, Taylor and Francis, UK, 2022.

[2] Sekigawa, Y., Fukuhara, M., Sonoda, Y., Sato, Y., Purification and characterization of a cytochrome P450 isozyme catalyzing lanosterol 14α-demethylation (P450*14DM*) in hamster liver. *Lipids* 30: 1067–1073, 1995.

[3] Nelson, D. L., Cox, M. M., *Lehninger, Principles of Biochemistry*, 8th edition, Palgrave Macmillan, UK, 2021.

[4] Lepesheva, G. I., Waterman, M. R., Sterol 14alpha-demethylase cytochrome P450 (CYP51), a P450 in all biological kingdoms. *Biochimica et Biophysica Acta (BBA)—General Subjects* 1770: 467–477, 2007.

61. What is carbohydrate xylose? Comparison physical properties of xylose with glucose, fructose, maltose, lactose, and sucrose

Answer: Xylose is a monosaccharide which means that it contains five carbon atoms and includes an aldehydic functional groups derived from hemicelluloses. With its free aldehyde group, it is a reducing sugar. Xylose gives Fehling's, Tollen's and Benedict tests D-xylose is a common sweetener used in food and drinks for people with diabetes.

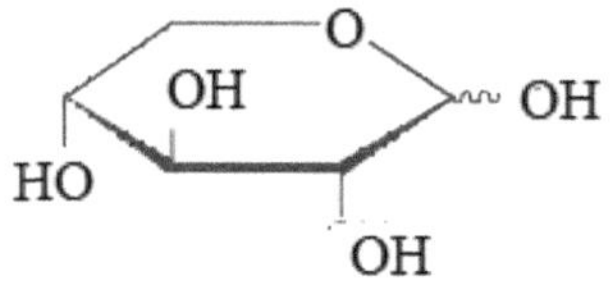
O
OH
OH
HO
OH

Structure of D-xylose

Comparison physical properties of xylose with glucose, fructose, maltose, lactose, and sucrose.

Sugar substrate	Specific rotation in water at 20°C	Test for Fehling's Tollen's tests	Derivatives and time for formation
D-Glucose	+53	Positive	Osazone (M.pt 205 °C) (4–5 minutes)
D-Fructose	–92	Positive	Osazone (M.pt 203 °C) (3 minutes)
Maltose	+129	Positive	Soluble (No formation)
Lactose	+52	Positive	No formation
Sucrose	+66	Negative	Osazone (M.pt 205 °C) (30 minutes)
D-Xylose	+19	Positive	Osazone formation (7 minutes)

62. Discuss Seliwanoff's test, Barfoed's reagent test, Benedict's reagent test and Molisch's test?

Answer: Seliwanoff's test: Add 1 gm of sample in 2c,c conc. HCl and a pinch of resorcinol in a dry test tube. Place the test tube in boiling water for 3 minutes deep red color usually followed by brownish precipitate. Dissolve precipitate in alcohol yielding a deep red solution. Fructose responds to this test but glucose does not respond this test.

Sucrose $\xrightarrow[\text{Heat}]{\text{HCl}}$ Glucose + Fructose $\xrightarrow[-3\,H_2O]{[H^+]}$ 5-(hydroxymthyl furfural) + 2 resorcinol $\xrightarrow[3\,H_2]{[H^+],\ 0.5\,O_2}$ Wine-red colour

Barfoed's reagent test: This reagent is prepared by dissolving 13.3 gm of crystallized neutral copper (II) acetate in 200 ml of 1 % acetic acid solution. It should be freshly prepared.

Heat a test tube containing 1 ml of the reagent and 1 ml of a dilute solution of the carbohydrate in a beaker of boiling water. Red copper (I) oxide is formed within two minutes. Lactose gives a negative test of this reagent.

Benedict's reagent: This test is given by reducing carbohydrate. When a reducing sugar is subjected to heat in the presence of an alkali, it gets converted into an enediol (which is a relatively powerful reducing agent). Therefore, when reducing sugars are present in the carbohydrate solution, the cupric ions (Cu^{2+}) in Benedict's reagent are reduced to cuprous ions (Cu^{+}). These cuprous ions form copper(I) oxide with the reaction mixture and precipitate out as a brick-red colored compound.

$$R{-}CHO + Cu(citrate)_2^{-2} \longrightarrow R{-}COO^- + Cu_2O$$

Aldose; Benedict's reagent; Carboxylate anion; Cpper oxide (brick red)

Take 1–2 ml of aqueous solution of carbohydrate in a test tube and add 1–2 ml of Benedict's reagent. Keep the test tube for 1–2 minutes on boiling water of a water bath. Formation of a reddish precipitate indicates the presence of reducing sugar.

Benedict's solution is prepared as follows: Take a solution of 4.3 gm sodium citrate add 0.25 gm sodium carbonate dissolved in 50 ml distilled water in a 250 measuring cylinder. Again add in 0.5 gm of copper sulphate in 40 ml of water in the measuring cylinder and make the volume of the solution 125 ml by adding distilled water.

Molisch's test: This test is used to detect the presence of carbohydrate in a given sample. It is based on when a carbohydrate-containing sample is treated with sulphuric acid, the hydroxyl group is removed from the sugar molecule. After the hydroxyl group is removed from a pentose sugar molecule, furfural is formed. The resulting furfural reacts with Molisch's reagent (sulphonated – naphthol) to produce a purplish red product. If the carbohydrate is poly- or disaccharide, a glycoprotein or glycolipid, the acid first hydrolyses it into component monosaccharides, which get dehydrated to form furfural or its derivatives. A green ring might be observed if any impurities are present in the reagent as they might interact with the α-naphthol and the acid. A rind ring is seen if a concentrated sugar solution is used. This might be due to the charring of the sugar due to the acid.

Molisch's reagent is prepared by adding 3.75 gm of α-naphthol in 25 ml of absolute ethanol 99%. In a test tube a solution of carbohydrate in water (suspension in the case of starch) add a few drops of Molisch's reagent. Next add concentrated sulphuric acid (1 ml) slowly along the side of the test tube. A violet ring is formed at the interface of two liquids.

conc. H_2SO_4

- 3 H_2O

Hydroxyl groups of sugar are dehydrated in prence of acid.

D-Glucose

or

Hydroxymethylfurfural

Hydroxymethylfurfural

conc. H_2SO_4

- H_2O

Hydroxymethylfurfural

α-Naphthol

[O], H_2SO_4

- H^+, 2e^-

Purple/violet colored complex

Index

Note: Page numbers in *italics* indicate a figure and page numbers in **bold** indicate a table on the corresponding page.

D

G

H

O

P

Q

R

U

V

W

X

Y

Z

For Product Safety Concerns and Information please contact our EU representative GPSR@taylorandfrancis.com Taylor & Francis Verlag GmbH, Kaufingerstraße 24, 80331 München, Germany

Batch number: 10392095

Printed by Printforce, the Netherlands